D1573850

TENSOR NORMS AND OPERATOR IDEALS

NORTH-HOLLAND MATHEMATICS STUDIES 176
(Continuation of the Notas de Matemática)

Editor: Leopoldo NACHBIN

Centro Brasileiro de Pesquisas Físicas
Rio de Janeiro, Brazil
and
University of Rochester
New York, U.S.A.

NORTH-HOLLAND – AMSTERDAM • LONDON • NEW YORK • TOKYO

TENSOR NORMS AND OPERATOR IDEALS

Andreas DEFANT
Fachbereich Mathematik
Universität Oldenburg
Oldenburg, Germany

Klaus FLORET
Fachbereich Mathematik
Universität Oldenburg
Oldenburg, Germany
and
IMECC/Unicamp
Campinas/S.P., Brasil

1993

NORTH-HOLLAND – AMSTERDAM • LONDON • NEW YORK • TOKYO

ELSEVIER SCIENCE PUBLISHERS B.V.
Sara Burgerhartstraat 25
P.O. Box 211, 1000 AE Amsterdam, The Netherlands

ISBN: 0 444 89091 2

© 1993 ELSEVIER SCIENCE PUBLISHERS B.V. All rights reserved.

No part of this publication may be reproduced, stored in a retrieval system or transmitted in any form or by any means, electronic, mechanical, photocopying, recording or otherwise, without the prior written permission of the publisher, Elsevier Science Publishers B.V., Copyright & Permissions Department, P.O. Box 521, 1000 AM Amsterdam, The Netherlands.

Special regulations for readers in the U.S.A. – This publication has been registered with the Copyright Clearance Center Inc. (CCC), Salem, Massachusetts. Information can be obtained from the CCC about conditions under which photocopies of parts of this publication may be made in the U.S.A. All other copyright questions, including photocopying outside of the U.S.A., should be referred to the publisher.

No responsibility is assumed by the publisher for any injury and/or damage to persons or property as a matter of products liability, negligence or otherwise, or from any use or operation of any methods, products, instructions or ideas contained in the material herein.

This book is printed on acid-free paper.

Printed in The Netherlands

Contents

Introduction 1

Chapter I: Basic Concepts 7

1. Bilinear Mappings 7
1.2. Continuous bilinear mappings, 1.5. non–validity of Hahn–Banach theorem, 1.7. non–validity of open–mapping theorem, 1.8. canonical extension to the bidual.

2. The Algebraic Theory of Tensor Products 15
2.2. Universal property and construction of tensor products, 2.4. examples, 2.5. trace, 2.6. trace duality, 2.7. tensor product of operators.

3. The Projective Norm 26
3.1. Minkowski gauge functional, 3.2. basic properties, 3.3. Bochner integrable functions, theorem of Dunford–Pettis, 3.4. compact sets, 3.6. nuclear operators, 3.7. trace, 3.9. π does not respect subspaces, 3.10. extension property, 3.11. Grothendieck's characterization of L_1, 3.12. lifting problems, 3.13. $\mathcal{L}_p^g$-spaces, Ex 3.24. Radon–Nikodým theorem for operator valued measures.

4. The Injective Norm 46
4.1. Basic properties, 4.2. examples, 4.3. ε does not respect quotients, 4.5. lifting of vector–valued continuous functions, compact extension property, 4.6. integral bilinear forms, Ex 4.3. Fourier matrices.

5. The Approximation Property 58
5.2. Survey about counterexamples, 5.3. compact operators, 5.4. characterization with nuclear operators and the trace, 5.5. injectivity of completions of tensor products of operators, 5.6. operators with nuclear dual, Ex 5.17. compactly approximable operators.

6. Duality of the Projective and Injective Norm 70
6.3. Dense sequences of finite dimensional Banach spaces, Johnson spaces C_p, 6.4. embedding theorem, 6.5. weak principle of local reflexivity, 6.6. principle of local reflexivity, 6.7. extension lemma for integral bilinear forms, Ex 6.4. Lindenstrauss' compactness argument.

7. **The Natural Norm on the p–Integrable Functions**   77
7.1. Bochner p–integrable functions, Δ_p, 7.2. continuous triangle inequality, 7.3. positive operators, density lemma, 7.4. Δ_p respects subspaces and quotients, quotient lemma, 7.5. Fourier transform, 7.6. Hilbert transform, 7.7. type and cotype, 7.9. a Beckner–like result, Ex 7.1. averaging operator in L_p.

8. **Absolutely and Weakly p–Summable Series and Averaging Techniques**   90
8.1. Absolutely p–summable and weakly p–summable sequences, 8.2. representations of operators on or into ℓ_p, 8.3. unconditional summability, 8.4. general scheme of averaging, 8.5. Rademacher functions, Khintchine inequality, 8.6. type and cotype of L_p, 8.7. Gauss functions, 8.9. Orlicz property, Ex 8.9. Rademacher versus Gauss averaging, Ex 8.12. absolutely (r,s)–summing operators.

9. **Operator Ideals**   108
9.2. Quasinorms, 9.4. criterion, 9.6. examples, 9.7. injective ideals and the injective hull, 9.8. surjective ideals and the surjective hull, 9.9. dual ideals, 9.10. composition ideals, Ex 9.8. space ideals, Ex 9.13. quasinuclear operators, Ex 9.16. K–convex operators.

10. **Integral Operators**   118
10.3. Characterization with the trace, 10.4. examples, 10.5. factorization, 10.7. characterization with $T \otimes id_G$, 10.8. and 10.9. summability of the diagonal of infinite matrices, Ex 10.4. extension and lifting properties of integral operators.

11. **Absolutely p–Summing Operators**   127
11.1. Basic characterizations, 11.2. positive operators, 11.3. Grothendieck–Pietsch domination theorem and factorization, 11.4. Dvoretzky–Rogers theorem, 11.5. composition, 11.6. Hilbert–Schmidt operators, 11.7. Pietsch lemma, 11.8. Kwapień's test, 11.9. absolutely 2–summing norm of id_E, 11.10. absolutely p–summing norm of the identity of finite dimensional Hilbert spaces, 11.11. little Grothendieck theorem, 11.12. operators with absolutely 2–summing duals and a characterization of Hilbert spaces, Ex 11.10. extension property of absolutely 2–summing operators, Ex 11.13. the ideal of Hilbert–Schmidt operators, Ex 11.16. Banach–Mazur distances between finite dimensional Banach spaces and the Kadec–Snobar result about projections, Ex 11.18. factorization of Hilbert–Schmidt operators.

Contents vii

Chapter II: Tensor Norms 146

12. Definition and Examples 146
12.1. Reasonable norms and the metric mapping property, 12.2. criterion, 12.4. finite and cofinite hull, 12.5. Lapresté's tensor norms $\alpha_{p,q}$, 12.6. completion with respect to $\alpha_{p,q}$, 12.8. the diagram of Lapresté's tensor norms, 12.9. tensor product representation of weakly p-summable sequences, Ex 12.7. tensors of finite rank.

13. The Five Basic Lemmas 159
13.1. Approximation lemma, 13.2. extension lemma, 13.3. embedding lemma, 13.4. density lemma, 13.5. $\mathfrak{L}_p$-local technique lemma.

14. Grothendieck's Inequality 166
14.1. Idea of proof, 14.4. proof of Grothendieck's inequality in tensor form, 14.5. matrix form, 14.7. estimates for the Grothendieck constant K_G, Ex 14.1. the original proof – more or less.

15. Dual Tensor Norms 177
15.1. Trace duality, 15.2. dual norms, 15.5. duality theorem, 15.6. accessibility of tensor norms, 15.7. conditions for the good behaviour of duality, 15.9. tensor norms and their duals on "symmetric" finite dimensional spaces, 15.10. duality of Δ_p, the Chevet–Persson–Saphar inequalities, 15.11. tensor norms closest to Δ_p, 15.12. another proof of the Beckner result, Ex 15.10. weakly conditionally compact subsets of tensor products.

16. The Bounded Approximation Property 190
16.1. Topologies on $\mathfrak{L}(E, F)$, 16.2. characterization with the cofinite hull $\overleftarrow{\pi}$ of the projective norm, 16.4. results involving the Radon–Nikodým property, 16.5. and 16.6. duality of ε and π, 16.7. duality of the operator ideals $\mathfrak{K}, \mathfrak{N}$, and $\mathfrak{L}$, 16.8. non–nuclear operators with nuclear dual, Ex 16.4. – Ex 16.8. reflexivity of $\mathfrak{L}(E, F)$ for special spaces.

17. The Representation Theorem for Maximal Operator Ideals . . . 200
17.2. Maximal operator ideals, 17.3. tensor norms and operator ideals which are associated with each other, 17.4. right–tensor norms and a general way of constructing maximal operator ideals, 17.5. the representation theorem, 17.6. the embedding theorem, 17.7. the transfer argument, 17.8. the dual ideal, 17.9. the adjoint ideal, 17.10. – 17.13. examples, 17.14. Grothendieck's theorem, 17.15. and 17.16. characterization with tensor product operators $T \otimes id_G$, 17.19. ideal norms of identity operators in symmetric finite dimensional spaces, 17.20. injective embedding of $E \tilde{\otimes}_\alpha F$ into the space of operators, 17.21. unit ball of $\mathfrak{A}(E, F)$, Ex 17.16. continuity of $S \otimes T''$, Ex 17.17. density lemma for maximal normed operator ideals.

Contents

18. (p,q)–Factorable Operators 223
18.2. The norm of the integrating functional, 18.4. ultraproducts, 18.5. factorization through positive functionals, 18.6. p–factorable operators, 18.7. p–integral operators, 18.9. Maurey's factorization theorem, 18.11. the factorization theorem, Ex 18.4. – Ex 18.11. properties of ultraproducts.

19. (p,q)–Dominated Operators 241
19.2. The basic estimates, 19.3. Kwapień's factorization theorem, 19.4. composition of factorable and dominated operators, 19.6. Grothendieck inequality for C^*–algebras.

20. Projective and Injective Tensor Norms 250
20.3. Duality relations, 20.4. examples, 20.6. the projective associate, 20.7. the injective associate, 20.9. finite dimensional characterization, 20.10. general rules for associates, 20.11. – 20.13. relations with operator ideals, 20.14. results about g_p, 20.15. table for w_2, d_2, g_2 and their adjoints, 20.17. Grothendieck's inequality in its original formulation (operator version), finite dimensional Grothendieck constants, 20.18. Banach spaces satisfying Grothendieck's theorem, 20.19. a result of Saphar and the best constants in the little Grothendieck theorem.

21. Accessible Tensor Norms and Operator Ideals 275
21.2. Accessible operator ideals, 21.4. total accessibility of certain composition ideals, 21.5. accessibility of Lapresté's tensor norms, 21.6. a result about the bounded approximation property, 21.7. the α–approximation properties, Reinov's results in the case $\alpha = g_p$, 21.11. some results of Kisljakov and Bourgain–Reinov, H^∞, Ex 21.3. principle of local reflexivity for operator ideals.

22. Minimal Operator Ideals 287
22.1. The minimal kernel of an operator ideal, 22.2. the representation theorem for minimal operator ideals, 22.3. examples, 22.6. the dual of $\mathfrak{A}^{min}(E,F)$, 22.7. weak–*–continuous operators in $\mathfrak{A}^{min}(F',E)$, 22.8. the dual ideal of the minimal kernel, 22.9. a counterexample, Ex 22.7. extension and lifting properties of minimal operator ideals.

23. $\mathcal{L}_p^g$–Spaces 300
23.1. Local techniques, 23.2. various characterizations, 23.3. relations with the $\mathcal{L}_p$–spaces of Lindenstrauss–Pełczyński, 23.4. dual characterization, 23.5. $\mathcal{L}_\infty^g$– and $\mathcal{L}_1^g$–spaces, 23.6. the projection constant, 23.7. quasinuclear operators, 23.8. characterization of space ideals with integral operators, 23.9. coincidence of absolutely 1–summing and nuclear operators, Hardy spaces, 23.10. Grothendieck's theorem for $\mathcal{L}_p^g$–spaces, a characterization of Hilbert spaces, Ex 23.8. the extension norm of an operator.

24. Stable Measures 314

24.1. The linear dimension of ℓ_p and L_q, 24.2. positive definite functions and Bochner's theorem, 24.3. moments of stable measures, 24.4. Lévy's theorem about the embedding of ℓ_p into L_q (Lévy embeddings), 24.5. embeddings into L_q, 24.6. – 24.7. results due to Saphar, Kwapień and Maurey about absolutely q–summing operators with values in ℓ_p or spaces with cotype, 24.8. stable type and Rademacher type, Ex 24.1. Schur product, Ex 24.5. stable type.

25. Composition of Accessible Operator Ideals 327

25.1. Representation of the minimal kernel of accessible operator ideals, 25.4. cyclic composition theorem, 25.5. Persson–Pietsch multiplication table, 25.6. quotient ideals, 25.7. quotient formula, 25.8. the adjoint of composition ideals, 25.9. the regular hull and characterizations of the associates of $\mathfrak{L}_{p,q}$ and $\mathfrak{L}_p$, in particular, of the associated space ideals — results of Kwapień, 25.10. isomorphic characterization of subspaces, quotients, etc. of L_p, 25.11. the minimal kernel of the injective resp. surjective hull of an operator ideal.

26. More About L_p and Hilbert Spaces 344

26.1. Inequalities about $\alpha_{p,q}$ coming from the Khintchine and Grothendieck inequalities, 26.2. factorization through Hilbert spaces of the identity mapping $\ell_s^n \to \ell_r^n$, 26.3. continuity of operators between spaces of Bochner p–integrable functions, complexification of operators, 26.5. Kwapień's result about the factorization of operators $L_r \to L_s$ through L_p, 26.6. tensor norms and ideals on Hilbert spaces, 26.7. the Hilbert–Schmidt tensor norm σ, 26.8. Schatten's result about self-adjoint, symmetric extensions of σ to Banach spaces, 26.10. limit orders of tensor norms, Puhl's result, 26.11. unconditional bases in $\ell_2 \tilde\otimes_\alpha \ell_2$, Ex 26.6. unconditionally summable sequences in $L_p(\mu \otimes \nu)$.

27. Grothendieck's Fourteen Natural Norms 361

27.2. Grothendieck's diagram, 27.3. the original notations.

Chapter III: Special Topics 365

28. More Tensor Norms 365

28.1. Three new classes of tensor norms, 28.3. description of the projective associate of $\alpha_{p,q}^*$, 28.4. characterization of operators in $\mathfrak{L}_{p,q}^{inj\ sur}$, 28.5. a characterization of operators factoring through a Hilbert space, 28.6. – 28.8. description of the composition ideals $\mathfrak{L}_p \circ \mathfrak{L}_q$ and its adjoints, 28.9. table of results, Ex 28.14. complexification of operators.

29. **The Calculus of Traced Tensor Norms** 378
   29.1. The tensor contraction, 29.4. the associated operator ideal of a traced tensor norm, 29.5. characterization of p–dominated operators, T_p–spaces, 29.6. the calculus of traced tensor norms, 29.7. the tensor product of two tensor norms and of maximal normed operator ideals, 29.8. properties of $\mathfrak{A} \otimes \mathfrak{B}$, 29.9. – 29.11. tensor products of special tensor norms, Ex 29.8. ultrastable ideals.

30. **The Vector Valued Fourier Transform** 394
   30.1. Fourier operators, 30.3. their characterization, 30.5. Rademacher and Gauss type and cotype, Kwapień's type/cotype theorem, characterization of Hilbert spaces, 30.6. the main theorem, 30.8. type and cotype with respect to orthonormal bases.

31. **Pisier's Factorization Theorem** 407
   31.1. K–convex operators, 31.2. duality of type and cotype, 31.4. Pisier's factorization theorem, 31.5. factorization of compactly approximable operators through Hilbert spaces, 31.6. non–accessible tensor norms/operator ideals, 31.7. "abstract" proof of Grothendieck's inequality.

32. **Mixing Operators** 415
   32.2. Reformulation of former results, 32.3. tensor product characterization, continuity of tensor product operators between spaces of Bochner p–integrable functions, 32.4. a domination theorem, 32.5. Maurey–Pisier extrapolation theorem, 32.6. a characterization of $\mathfrak{M}_{q,p}^{dual}$, 32.7. Maurey's splitting theorem, 32.10. and 32.11. relation with absolutely (r, s)–summing operators, 32.12. a finite dimensional result.

33. **The Radon–Nikodým Property for Tensor Norms and Reflexivity** 430
   33.1. Duality of ε and π revisited, 33.3. Lewis' theorem, 33.4. permanence properties, 33.5. Lapresté's tensor norms, 33.6. coincidence of p–nuclear and p–integral operators, 33.7. p–strong operators, 33.8. – 33.11. reflexivity of tensor products and components of operator ideals.

34. **Tensorstable Operator Ideals** 445
   34.1. β–tensorstable and metrically β–tensorstable operator ideals, 34.2. strongly β–tensorstable operator ideals, 34.3. examples, 34.4. permanence properties, 34.5. factorization arguments, 34.6. projection constant of the injective tensor product, 34.7. stability of space ideals, 34.8. double Khintchine inequality, stability of tensor products of Hilbert spaces, 34.9. ε– and π–stability of (p, q)–factorable operators and their relatives, 34.10. distribution of eigenvalues, Pietsch's tensor product trick, 34.11. a result of Kwapień unifying Orlicz's, Littlewood's and Grothendieck's inequalities, 34.12. improving weak inequalities with tensor products, mixing operators $\ell_1 \to \ell_p$.

35. Tensor Norm Techniques for Locally Convex Spaces 469
35.2. Tensor norm topologies, 35.3. traced tensor norms, 35.4. locally convex space ideals, 35.6. injective and projective tensor norms on locally convex spaces, 35.7. tensor product of direct sums, 35.8. lifting of bounded sets, property (BB), 35.9. problème des topologies, Taskinen's counterexample, 35.10. injective tensor product of (DF)-spaces.

Appendices:

A. Some Structural Properties of Banach Spaces 489
A1. Subspaces and quotients, A2. dual systems, A3. lemma of Ky Fan, A4. bases, A5. Banach algebras, A6. lattices, A7. abstract L_p-spaces and their representation.

B. Integration Theory 493
Extension procedure and the basic theorems: B1. Daniell functionals, B2. the convergence theorems, B3. measurable functions, the fundamental theorem of the Daniell–Stone integration theory, B4. product measures; **The L_p–spaces:** B5. Hölder and continuous triangle inequality, B6. duality, Radon–Nikodým theorem, Segal's localization theorem, Lebesgue decomposition, B7. strictly localizable measures; **Borel–Radon measures and Riesz representation theorem:** B8. τ–continuity and Bourbaki's extension procedure, Kölzow's theorem, B9. representation of Borel–Radon measures, B10. L_1 is complemented in its bidual; **Bochner integration:** B11. measurability of Banach space valued functions, B12. $\mathfrak{L}_p(\mu, E)$, B14. Pettis integrability, B15. variation lemma.

C. Representable Operators 508
Grothendieck's characterization: C1. Riesz–densities, C3. nuclear operators, C4. factorization through ℓ_1; **The Dunford–Pettis theorem:** C6. a general result about representability, C7. strong version of the Dunford–Pettis theorem.

D. The Radon–Nikodým Property 517
Basic properties and examples: D1. reduction to the Lebesgue measure, D3. examples, D4. dual of $L_p(\mu, E)$; **Pietsch integral operators:** D5. and D6. relation with other operator ideals; **The Radon–Nikodým property and operator ideals:** D7. and D8. characterizations in terms of (Pietsch) integral=nuclear, D9. integral operators which are not Pietsch integral.

Bibliography 527

List of Symbols 545

Index 555

Introduction

1. Grothendieck's "Résumé de la théorie métrique des produits tensoriels topologiques", submitted in 1954 and published in 1956 in the Bulletin of the Mathematical Society of São Paulo, is one of very few papers which have deeply influenced the course of Functional Analysis: It not only demonstrated the enormous possibilities for the using of tensor products in Banach space theory, but it anticipated the study of Banach spaces in terms of finite dimensional subspaces — the so-called "local" theory, which has so much enriched our understanding of Banach spaces.

The "Résumé" (as this paper is nowadays called) contained no proofs (with the exception of the main theorem) and apparently few people at the time understood it — or even noted its importance. This, in spite of the fact that Grothendieck's theory of locally convex tensor products and nuclear spaces (also poorly understood when it first appeared) had already won great appreciation. There was simply some reluctance in the Functional Analysis community to accepting the idea of thinking in terms of tensor products.

It was only in 1968 that the "Résumé" received more attention through the famous Studia Mathematica paper of Lindenstrauss and Pełczyński [173] "Absolutely summing operators in $\mathcal{L}_p$-spaces and applications", which presented Grothendieck's deep "théorème fondamental de la théorie métrique des produits tensoriels" (the main result of the "Résumé") as an inequality concerning $n \times n$ matrices and Hilbert spaces. Without using tensor products, various fascinating applications were given, mainly dealing with the class of absolutely p-summing operators. Banach space theory (which in the mid-sixties had been considered to be nearly complete by some people) was reactivated in an incredible way — and many of its important results nowadays trace their roots back to the "Résumé". It is astonishing to see that many (certainly not all) of the ideas of Banach space theory in the last 20 years are already contained in Grothendieck's paper, although sometimes in a hidden way. The phrase "this result is implicitly contained in the Résumé" became almost a cliché for a certain time, but was nevertheless quite often true.

2. Tensor products apparently appeared in Functional Analysis for the first time during the late thirties in the work of Murray and John von Neumann on Hilbert spaces. The first systematic study of classes of norms on tensor products of Banach spaces is due to Schatten who in 1943 continued his work in a series of papers (some together with von Neumann). Schatten's influential monograph "A Theory of Cross-Spaces" contains what was known in 1950: the most beautiful applications of the theory dealt with operator ideals on Hilbert spaces [249], the Hilbert–Schmidt operators, the trace-class or more generally the Schatten–von Neumann-classes $\mathfrak{S}_p$. Many of the more elementary aspects of Grothendieck's theory were known to Schatten, but he was not aware of the important role of the finite dimensional behaviour of tensor norms. Therefore, he did not succeed in developing — for general Banach spaces — a useful theory of duality; the reader will get a more precise idea of why this happened in section 15. On the other hand, the idea of ideals of operators was always apparent in the study of tensor products — in Schatten's as well as in Grothendieck's work. When writing the "Résumé", Grothendieck was not aware of Schatten's work.

In 1968 Pietsch and his school began a systematic investigation of operator ideals on the class of Banach spaces and, ignoring tensor products, developed a method of thinking in a "categorical" manner which is just as powerful as thinking in terms of tensor products — but has the advantage that it is certainly much easier to learn the basics of operator ideal theory than the basics of the theory of tensor products. This development culminated in 1978 in the publication of Pietsch's book "Operator Ideals" which contains in a nearly encyclopaedic way everything that was known about operator ideals at the time. Although many of the ideas and results clearly came from the "Résumé", tensor products were never used in his book.

3. There was a bias against tensor products in Banach space theory, but the projective tensor norm π and the injective norm ε proved their usefulness rather quickly through simple examples involving vector valued function spaces. New results about general tensor norms were only obtained around 1970; may be the most important of these "early" papers are those of Saphar and Kwapien's Bordeaux–paper on operators factoring through L_p–spaces. This latter paper, however, was written in the language of operator ideals. Later on (in 1973) Lotz based his investigation of maximal normed operator ideals (he had called them Grothendieck ideals) on tensor products. The rich paper of Gordon–Lewis–Retherford on operator ideals was stimulated by Schatten's and Grothendieck's work and uses tensor product methods explicitly.

It was Pisier's work in Banach space theory (starting around 1975) which finally drew new attention to the tensor product approach. A highlight is certainly his solution of the last problem stated in the "Résumé": There exist infinite dimensional Banach spaces P such that $P \otimes_\varepsilon P = P \otimes_\pi P$ holds isomorphically.

Pisier's book "Factorization of Linear Operators and Geometry of Banach Spaces" is centered around the problem of determining conditions under which an operator between Banach spaces factors through a Hilbert space. These investigations finally led to solutions of all six of the problems stated at the end of the "Résumé" except that the exact constant in the Grothendieck inequality (as the "théorème fondamental" is nowadays called) is not yet known (the approximation problem was solved in the negative by Enflo in 1972). Reading Pisier's book, it becomes strikingly clear that thinking in terms of operator ideals *and* in terms of tensor products is quite useful. Further support for this way of dual thinking is given by a trick due to Pietsch from 1983 where he uses tensor products of operators in order to give a simple proof of the famous result concerning the distribution of eigenvalues of absolutely p–summing operators due to Johnson, König, Maurey, and Retherford (see 34.10. for some details).

4. Unfortunately, the beauty and power of "tensorial" thinking, becomes clear only after one is really accustomed to using it. The "Résumé" is very difficult to read and for this reason various attempts have been made to present the theory of tensor norms in a clearer fashion (Amemiya–Shiga [3], Lotz [180], Losert–Michor [179], Michor [191], Gilbert–Leih [81] are known to us), but there seems to exist none which is easily accessible and, at the same time, incorporates the wonderful theory of operator ideals as it exists nowadays. We have already expressed our point of view in a sort of pre-book in [43]. Our central aim with the present book is to convince the reader that the theory of tensor norms is much less difficult than it sometimes seems and that it gives one a deep and beautiful insight into various questions in Banach space theory — in particular into the theory of operator ideals, which is our main concern apart from the forementioned objective. We hope that we can convince the reader (and the historical development gives clear evidence for this) that both theories, the theory of tensor norms and of normed operator ideals (if we consider them for a moment to be really different), are more easily understood and also richer if one works with both simultaneously. It should become obvious that certain phenomena have their natural framework in tensor products and others in operator ideals.

5. We assume that the reader of this book knows the basics of Banach space theory as they are usually presented in an introductory course on Functional Analysis. Some additional information is collected in the four appendices — in particular, our point of view for the theory of measure and integration. Appendices C and D about the Radon–Nikodým property will not be needed before section 16.

The book is divided into three chapters: Basic Concepts, Tensor Norms, and Special Topics. The first chapter may serve as part of an introductory course in Functional Analysis since it already shows the powerful use of the projective and injective tensor norms π and ε, as well as the basics of the theory of operator ideals. The second chapter is the main part of the book: it presents the theory of tensor norms as designed by Grothendieck in the "Résumé" and deals with the relation between tensor norms and operator ideals. This relation is dominated by the following definition (17.3.) which constitutes a one-to-one correspondence between finitely generated tensor norms α and maximal normed operator ideals $(\mathfrak{A}, \mathbf{A})$: They are called associated (in symbols $\alpha \sim (\mathfrak{A}, \mathbf{A})$) if
$$\mathfrak{A}(M, N) = M' \tilde{\otimes}_\alpha N$$
holds isometrically for all finite dimensional Banach spaces M and N. We believe that the following theorems are fundamental for the understanding of the interplay between operator ideals $\mathfrak{A}$ and their associated tensor norms α :

The *representation theorem for maximal operator ideals* (17.5.)
$$\mathfrak{A}(E, F') \overset{1}{=} (E \otimes_{\alpha'} F)' \qquad\qquad (\textit{isometry})$$
and the *representation theorem for minimal operator ideals* (22.2.)
$$E' \tilde{\otimes}_\alpha F \overset{1}{\twoheadrightarrow} \mathfrak{A}^{min}(E, F) \qquad\qquad (\textit{metric surjection}),$$

where E and F are arbitrary Banach spaces and $\mathfrak{A}^{min}$ is the minimal kernel of $\mathfrak{A}$.

In view of applications it is natural to first study tensor norms on finite dimensional normed spaces and later extend them to arbitrary normed or Banach spaces. There are two ways to do this — an inductive procedure

$$\overrightarrow{\alpha}(z; E, F) := \inf\{\alpha(z; M, N) \mid z \in M \otimes N; M, N \text{ finite dim.}\}$$

and a projective procedure

$$\overleftarrow{\alpha}(z; E, F) := \sup\{\alpha(Q_L^E \otimes Q_K^F(z); E/L, F/K) \mid E/L, F/K \text{ finite dim.}\} \ .$$

Both coincide if (and in some sense only if, see 16.2.) both spaces have the metric approximation property. Grothendieck chose the first approach and this is justified when one looks at examples. However, we have found it very useful for our investigations to also have the "cofinite hull" $\overleftarrow{\alpha}$ at hand and we hope that we can convince the reader that it is helpful for structuring one's thoughts and is often quite useful for finding and working out proper statements and proofs. For operator ideals the cofinite hull gains importance through the *embedding theorem*, which states that

$$E' \tilde{\otimes}_{\overleftarrow{\alpha}} F \overset{1}{\hookrightarrow} \mathfrak{A}(E, F) \qquad\qquad (metric\ injection)$$

if α and $\mathfrak{A}$ are associated (see 17.6.). As has already been mentioned, these three theorems describe the relationship between tensor norms and operator ideals.

Chapter III deals with special questions; for this, as well as a detailed account of the contents of the other two chapters, the reader may consult the table of contents.

6. Each section is accompanied by a series of *exercises* and the reader is strongly advised to solve them or at least give them a close look. They are of varying degrees of difficulty: if the result is not immediate from what was said in the same section, then there will be hints for the solution, which — we hope — make the solution accessible. Clearly, after a while (more or less after section 23) the reader should be continually mindful of the three fundamental theorems just mentioned as well as the local techniques.

We do not claim that a result presented as an exercise is easy ("just an exercise") — we only say that at this stage of the development of the theory this result can be obtained (together with the hint) without too much difficulty.

7. The book treats only tensor products of two "variables" — and not norms on tensor products $E_1 \otimes \cdots \otimes E_n$ of more than two Banach spaces. Though there is certainly a good part of the theory which easily extends to the case of several variables, there exist new interesting phenomena which we do not study. Moreover, we deal mostly with normed, not quasinormed operator ideals (with a few exceptions: for example the cyclic composition theorem in section 25): this is a consequence of the fact that tensor norms and their duality theory cannot be used to study general quasinormed ideals. One may consider this to be a disadvantage, but one should not forget that most of the interesting operator ideals are actually normed and that the theory of operator ideals (as it is presented in Pietsch's book for example) also deals mainly with normed (maximal or minimal) ideals.

The following list contains some topics which we did not include in the book — partly because of lack of space (we never thought that the book would grow so large), partly because we did not feel competent:

— Pisier's example of a space P with $P \otimes_\varepsilon P = P \otimes_\pi P$
— Geometry of tensor products, stability properties
— Tensor products of Banach lattices
— Tensor products of C^*-algebras
— Applications to Harmonic Analysis.

8. A short word about citations: We are very concerned that we state the origin of those results and methods which we consider to be important for the ideas which are presented. Nevertheless, we find it boring to constantly repeat "due to Grothendieck" during most of the first two chapters. Primarily in the first chapter we label many results as now being folklore — a judgement which may be rather subjective; we apologize for any omissions which may occur in this way — although we hope that there are none.

During the preparation of this book we received great support from many colleagues and friends — and we learned very much from them: In particular, for our understanding of the basic ideas of tensor norms and the "Résumé" we owe much to the work of Harksen, Lotz and Saphar. Our knowledge about operator ideals stems for the most part from Pietsch and his school — his book on operator ideals serves as a permanent reference for us. We profited much from the study of the papers of Gilbert–Leih, Gordon–Lewis–Retherford, Kwapień, Lewis, Lindenstrauss–Pełczyński, Maurey and Pełczyński. And our vision of the topic was widened when we read in Pisier's work.

We are also grateful to those colleagues who helped us in solving special problems during the preparation of the book — in particular, R. Alencar, B. Carl, M. Defant, P. Harmand, H. Jarchow, A. Pełczyński, N. Tomczak–Jaegermann and J. Voigt; we are happy that G. Pisier provided us (shortly before finishing the book) with an example of a non–accessible tensor norm.

We thank the universities of Ann Arbor, Campinas, Lecce, Niteroi, Oldenburg, Pretoria, Rio de Janeiro and Zürich for having given us the opportunity to present a series of lectures or seminars on our ideas; we acknowledge the critical assistance of colleagues and students at these universities. Various students from the University of Oldenburg wrote diploma theses related to our work: those of L. Behrens (section 27) and M. Meile (section 33) directly influenced our presentation.

We gratefully acknowledge the support of the Land Niedersachsen, the Volkswagen foundation, the Gesellschaft für Mathematik und Datenverarbeitung GMD, the Brazilian CNPq and the DAAD during the preparation of this book.

There were many colleagues and friends who spent a great deal of time and energy helping us in the difficult task of proof–reading: Klaus Bierstedt, Bernd Carl, Jesús Castillo, Martin Defant, Ute Defant, Peter Harmand, Hans Jarchow, Mario Matos, Jorge Mujica, Peter Stollmann, João Prolla and Jürgen Voigt. Bruce Hanson had a close look at our English. We thank all of them wholeheartedly — not only for the errors which they found, but also for their many observations which helped improve the readability of the text; in this respect we are particularly grateful to Hans Jarchow.

Nowadays when the publisher's task is often reduced to copying and selling (we do not say that the latter were easy) the secretarial work of producing a manuscript has gained special importance. We feel that Mrs. I. Matziwitzki from the University of Oldenburg has done an excellent job — with dedication and a good sense of aesthetics. We are deeply grateful for this — it is always a pleasure to work with her.

Our feelings of gratitude go to the first author's wife Ute for her understanding and patience.

It is a particular pleasure for us to have this book published in the series Notas de Matemática edited by our friend and colleague Leopoldo Nachbin; according to the introduction of the Résumé Grothendieck and he intended at the time to write a monograph about locally convex spaces including many aspects of the metric theory of tensor products. Unfortunately, this never happened. We thank Leopoldo Nachbin for always having encouraged us to write our book and the North Holland/Elsevier Science Publishers for accepting it for publication.

Campinas and Oldenburg, September 1991

Andreas Defant
Klaus Floret

Chapter I

Basic Concepts

After some remarks on bilinear mappings and the introduction of algebraic tensor products, the projective and injective tensor norms and some of their most important properties are studied in sections 3 – 6. Vector valued p-integrable functions, summability of sequences and averaging methods are treated in section 7 and section 8 – always from the perspective of tensor products. The remaining sections are devoted to the basics of the theory of operator ideals and integral and absolutely p-summing operators between Banach spaces, culminating in the so-called little Grothendieck theorem.

The reader should be familiar with some fundamental tools from Banach space theory, such as the Hahn–Banach theorem, the Mackey theorem / uniform boundedness principle, the weak– and weak–$*$–topologies, continuous linear operators and the open mapping and closed graph theorems. This clearly includes some simple knowledge about the classical Banach spaces $C(K)$ and L_p, as well as the sequence spaces c_o and ℓ_p. In Appendix A some additional information about the structure theory of Banach spaces is collected.

1. Bilinear Mappings

This first section treats bilinear mappings and puts special emphasis on the fact that in many respects they behave differently than linear mappings, although they are intimately related with them.

1.1. Let E, F, G be vector spaces over the same scalar field $\mathbb{K} = \mathbb{R}$ or $\mathbb{C}$ of the real or complex numbers. A mapping

$$\Phi : E \times F \longrightarrow G$$

is called *bilinear* if the mappings

$$_x\Phi : F \longrightarrow G \qquad \text{and} \qquad \Phi_y : E \longrightarrow G$$
$$ y \rightsquigarrow \Phi(x,y) \qquad\qquad\qquad\quad x \rightsquigarrow \Phi(x,y)$$

are linear for each $x \in E$ and $y \in F$, in symbols: $\Phi \in Bil(E, F; G)$ if

$$_x\Phi \in L(F, G) \quad \text{and} \quad \Phi_y \in L(E, G)$$

for each $x \in E$ and $y \in F$. For simplicity: $Bil(E, F) := Bil(E, F; \mathbb{K})$. If E, F, G are normed spaces (or more generally: topological vector spaces), the set of *continuous bilinear* mappings $E \times F \to G$ will be denoted by $\mathfrak{Bil}(E, F; G)$ and $\mathfrak{Bil}(E, F)$ if $G = \mathbb{K}$.
From

$$\Phi(x, y) - \Phi(x_o, y_o) = \Phi(x - x_o, y - y_o) + \Phi(x - x_o, y_o) + \Phi(x_o, y - y_o)$$

the following result is easily deduced:

PROPOSITION: *For $\Phi \in Bil(E, F; G)$ the following assertions are equivalent:*
(a) *Φ is continuous.*
(b) *Φ is continuous at $(0, 0)$.*
(c) *There is a constant $c \geq 0$ with $\|\Phi(x, y)\|_G \leq c \|x\|_E \|y\|_F$ for all $(x, y) \in E \times F$.*

It is easy to see that

$$\|\Phi\| := \min\{c \geq 0 \mid c \text{ as in (c)}\} = \sup\{\|\Phi(x, y)\|_G \mid x \in B_E, y \in B_F\}$$

defines a norm on $\mathfrak{Bil}(E, F; G)$ which is even a complete norm if $\|\cdot\|_G$ is. Note that continuous bilinear mappings are *not uniformly continuous* since, e.g., the restriction of $\mathbb{R}^2 \longrightarrow \mathbb{R}$, $(x, y) \rightsquigarrow xy$ to the diagonal is the function $\mathbb{R} \ni x \rightsquigarrow x^2 \in \mathbb{R}$.

1.2. A bilinear mapping $\Phi \in Bil(E, F; G)$ is *separately continuous* if all $_x\Phi : F \to G$ and $\Phi_y : E \to G$ are continuous.

THEOREM: *Let E, F, G be normed spaces and E complete. Every separately continuous bilinear mapping $\Phi \in Bil(E, F; G)$ is continuous.*

PROOF: The set
$$D := \{z' \circ \Phi_y \mid z' \in B_{G'}, y \in B_F\} \subset E'$$
is $\sigma(E', E)$–bounded since for each $x \in E$
$$|\langle z' \circ \Phi_y, x \rangle| = |\langle z', \Phi(x, y)\rangle| \leq \|z'\| \, \|\Phi(x, y)\| \leq \|_x\Phi\|.$$

Mackey's theorem / the uniform boundedness principle shows that D is uniformly bounded, i.e. there is a $c \geq 0$ such that for all $z' \in B_{G'}$ and $y \in B_F$

$$|\langle z', \Phi(x, y)\rangle| = |\langle z' \circ \Phi_y, x\rangle| \leq c\|x\|_E \qquad \text{for all } x \in E.$$

It follows that $\|\Phi\| \le c$. $\square$

1.3. Some examples of bilinear mappings:
(1) For $x' \in E'$ and $y' \in F'$
$$[x' \otimes y'](x,y) := \langle x', x \rangle \langle y', y \rangle$$
defines a continuous bilinear form and $\|x' \otimes y'\| = \|x'\| \, \|y'\|$. If x'_n and y'_n are in the unit balls and $(\lambda_n) \in \ell_1$, then
$$\varphi(x,y) := \sum_{n=1}^{\infty} \lambda_n [x'_n \otimes y'_n](x,y)$$
is well-defined and $\|\varphi\| \le \sum_{n=1}^{\infty} |\lambda_n|$; bilinear forms of this kind are called *nuclear* bilinear forms.
(2) The evaluation map on the space $\mathcal{L}(E,F)$ of continuous linear operators
$$\mathcal{L}(E,F) \times E \longrightarrow F$$
$$(T,x) \rightsquigarrow Tx$$
has norm 1 (if E and F are not trivial).
(3) If E and F are finite dimensional, then all bilinear mappings $E \times F \to G$ are continuous (use bases).
(4) The convolution mapping
$$L_1(\mathbb{R}) \times L_1(\mathbb{R}) \longrightarrow L_1(\mathbb{R})$$
$$(f,g) \rightsquigarrow f * g$$
is bilinear.
(5) Take the continuous functions on a compact space K and E a normed space. Then
$$C(K) \times E \longrightarrow C(K,E)$$
$$(f,x) \rightsquigarrow f(\cdot)x$$
is bilinear.

1.4. The mappings

$$\begin{array}{cccccc} Bil(E,F) & \longrightarrow & L(E,F^*) & \quad & L(E,F^*) & \longrightarrow & Bil(E,F) \\ \varphi & \rightsquigarrow & L_\varphi & & T & \rightsquigarrow & \beta_T \\ \langle L_\varphi x, y \rangle & := & \varphi(x,y) & & \beta_T(x,y) & := & \langle Tx, y \rangle \end{array}$$

give isomorphisms of vector spaces and are inverse to each other. Since

$$\|\varphi\| = \sup\{|\varphi(x,y)| \mid x \in B_E, y \in B_F\} = \sup\{\|L_\varphi x\| \mid x \in B_E\} = \|L_\varphi\| \in [0,\infty],$$

this isomorphism reduces for continuous bilinear forms to an isometry of normed spaces

$$\mathfrak{Bil}(E, F) \stackrel{1}{=} \mathcal{L}(E, F')$$

$$\|L_\varphi\| = \|\varphi\| \quad \text{and} \quad \|\beta_T\| = \|T\|.$$

This relationship is basic for the understanding of the ideas which will be presented in this book: *the continuous bilinear forms on $E \times F$ are exactly the continuous operators $E \to F'$.*

1.5. Since there is *no Hahn-Banach theorem* for operators, there is none for bilinear continuous forms in the following sense: Let $G \subset E$ be a subspace and $\varphi \in \mathfrak{Bil}(G, F)$; does there exist an extension $\tilde{\varphi} \in \mathfrak{Bil}(E, F)$ of φ? This would mean, by the identification of bilinear forms and operators, that every $T \in \mathcal{L}(G, F')$ would have an extension $\tilde{T} \in \mathcal{L}(E, F')$:

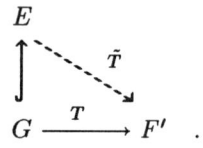

$$\begin{array}{c} E \\ \uparrow \quad \searrow^{\tilde{T}} \\ G \xrightarrow{T} F' \end{array}.$$

To see some examples of operators which are not extendable take the special case in which $G = F'$ and $T = id_G$ the identity-mapping: the extension $\tilde{T}$ would be a projection of E onto G.

(1) Due to a famous result of Lindenstrauss-Tzafriri [175] every infinite dimensional Banach space which is not isomorphic to a Hilbert space has a non–complemented closed subspace.

(2) To see a more concrete example take the Rademacher functions defined on $[\,0,1]$

$$r_n(t) := (-1)^k \quad \text{if} \quad t \in \left[\frac{k}{2^n}, \frac{k+1}{2^n}\right[$$

(they form an orthonormal system in $L_2[0,1]$, Lebesgue measure) and consider the injection

$$\ell_2 \longrightarrow L_1[0,1]$$

$$(\xi_n) \rightsquigarrow \sum_{n=1}^\infty \xi_n r_n\ .$$

The Khintchine inequality (8.5. and 8.6.) will show that L_1 induces an equivalent norm on ℓ_2. But ℓ_2 cannot be complemented in L_1: a projection $P : L_1 \to \ell_2 \hookrightarrow L_1$ would be weakly compact so $P = P^2$ would be compact by the Dunford–Pettis property of L_1 (see Appendix C7.), which is not possible since ℓ_2 is infinite dimensional. (Another argument for the non–complementation follows from the fact — which will be proved in section 11 — that every operator $L_1 \to \ell_2$ is absolutely 2–summing.)

1.6. Extension to the completions, fortunately, is no problem. Recall that there is no uniform continuity!

PROPOSITION: *Let E, F, G be normed spaces and G complete. Every $\Phi \in \mathfrak{Bil}(E, F; G)$ has a unique extension $\tilde{\Phi} \in \mathfrak{Bil}(\tilde{E}, \tilde{F}; G)$. Moreover, $\|\Phi\| = \|\tilde{\Phi}\|$.*

This follows easily from the isometric relation

$$\mathfrak{Bil}(E, F; G) = \mathfrak{L}(E, \mathfrak{L}(F, G))$$

and the extension of linear continuous operators.

1.7. There is also *no open mapping theorem* for bilinear continuous surjective mappings.

REMARK: *If E is a normed space of dimension at least 2, then the scalar multiplication $\mathbb{K} \times E \to E$ is bilinear, continuous, surjective but not open.*

PROOF: Take an open non-void set $V \subset E$ and a functional $y' \in E'$ with

$$\inf |\langle y', V \rangle| > 0.$$

If U is the open unit ball in $\mathbb{K}$, then $0 \in U \cdot V$, but 0 is not an interior point of $U \cdot V$ since

$$U \cdot V \cap \ker y' = \{0\}. \quad \Box$$

It is also possible to give examples $\Phi \in \mathfrak{Bil}(E, F; G)$ which are surjective and not open in zero, i.e. zero is not an interior point of $\Phi(B_E, B_F)$: In 1973 Cohen [33] constructed a relatively complicated bilinear mapping $\ell_1 \times \ell_1 \to \ell_1$ of this kind, but two years later Horowitz [118] found the following simple example $\Phi : \mathbb{K}^3 \times \mathbb{K}^3 \to \mathbb{K}^4$

$$\Phi(x, y) := (x_1 y_1, x_1 y_2, x_1 y_3 + x_3 y_1, x_2 y_2, x_3 y_2 + x_2 y_1)$$

(see also Ex 1.8.). There is, however, a closed graph theorem (Ex 1.11.).

1.8. Another negative property of continuous bilinear mappings is that they do not remain continuous for the weak topologies: the unit vectors e_n in ℓ_2 (real) converge weakly to zero but $(e_n | e_n)_{\ell_2} = 1$ does not (see also Ex 1.2.).

1.9. For normed spaces E and F the following isometries hold:

$$\Phi : \mathfrak{Bil}(E, F) \stackrel{1}{=} \mathfrak{L}(E, F') \stackrel{1}{\hookrightarrow} \mathfrak{L}(F'', E') \stackrel{1}{=} \mathfrak{Bil}(F'', E) \stackrel{1}{=} \mathfrak{Bil}(E, F''),$$
$$\phantom{\Phi : \mathfrak{Bil}(E, F) \stackrel{1}{=}} T \quad\leadsto\quad T'$$

where the last equality is the obvious "transposition" $U^t(x,y) := U(y,x)$. For each $\varphi \in \mathfrak{Bil}(E,F)$ define $\varphi^\wedge := \Phi(\varphi) \in \mathfrak{Bil}(E,F'')$; it satisfies $\|\varphi^\wedge\| = \|\varphi\| = \|L_\varphi\|$ and

$$\varphi^\wedge(x,y'') = \langle L'_\varphi(y''), x \rangle_{E',E} = \langle y'', L_\varphi(x) \rangle_{F'',F'} = \langle y'', \varphi(x,\cdot) \rangle_{F'',F'}$$

for all $(x,y'') \in E \times F''$. Since, by definition $\varphi(x,y) = \langle y, L_\varphi(x) \rangle_{F,F'}$ for all $x \in E$ and $y \in F$, the map $\varphi^\wedge$ extends φ from $E \times F$ to $E \times F''$ with equal norm. $\varphi^\wedge$ is called the *canonical (right-) extension* of φ.

PROPOSITION: *Let E and F be normed spaces and $\varphi \in \mathfrak{Bil}(E,F)$. Then $\varphi^\wedge$ is the unique bilinear separately $\sigma(E,E')$-$\sigma(F'',F')$-continuous mapping $\psi: E \times F'' \longrightarrow \mathbb{K}$ which extends φ.*

PROOF: That $\varphi^\wedge$ is such an extension is clear from the equation following the definition; uniqueness follows from the $\sigma(F'',F')$–density of F in F''. □

Clearly, there is also a *canonical left-extension* $^\wedge\varphi$ on $E'' \times F$ defined by $^\wedge\varphi := ((\varphi^t)^\wedge)^t$; it follows that

$$^\wedge\varphi(x'',y) = \langle x'', (L'_\varphi \circ \kappa_F)y \rangle_{E'',E'} = \langle x'', \varphi(\cdot,y) \rangle_{E'',E'}.$$

How are the functionals $(^\wedge\varphi)^\wedge$ and $^\wedge(\varphi^\wedge)$ on $E'' \times F''$ related? Surprisingly enough, the following is true:

COROLLARY: *For $\varphi \in \mathfrak{Bil}(E,F)$ the following three statements are equivalent:*
(a) *The two "canonical" extensions $(^\wedge\varphi)^\wedge$ and $^\wedge(\varphi^\wedge)$ of φ to $E'' \times F''$ coincide.*
(b) *There is a $\psi \in Bil(E'',F'')$ which is separately $\sigma(E'',E')$-$\sigma(F'',F')$-continuous and extends φ.*
(c) $L_\varphi : E \to F'$ *is weakly compact.*
In this case the ψ in (b) is equal to $(^\wedge\varphi)^\wedge = {}^\wedge(\varphi^\wedge)$.

PROOF: The proposition implies easily that (a) is equivalent to (b). To see the equivalence of (a) and (c) first note that, by Ex 1.12. and Ex 1.14.

$$L_{(^\wedge\varphi)^\wedge} = \kappa_{F'} \circ P_{F'} \circ L''_\varphi : E''' \longrightarrow F'''$$
$$L_{^\wedge(\varphi^\wedge)} = P_{F'''} \circ L''_{\varphi^\wedge} = P_{F'''} \circ (\kappa_{F'} \circ L_\varphi)'' =$$
$$= P_{F'''} \circ \kappa''_{F'} \circ L''_\varphi = L''_\varphi.$$

Now recall that L_φ is weakly compact if and only if $L''_\varphi(E'') \subset F'$. □

Exercises:

Ex 1.1. If E and F are finite dimensional, then every $\varphi \in Bil(E, F)$ has the form

$$\varphi = \sum_{k=1}^{n} x'_k \otimes y'_k \qquad \text{for some } x'_k \in E^* \text{ and } y'_k \in F^* .$$

Ex 1.2. (a) If E and F are normed spaces, then every $\varphi \in \mathfrak{Bil}(E, F)$ which is continuous for the weak topologies has the form as in Ex 1.1. with continuous $x'_k \in E'$ and $y'_k \in F'$.
(b) If E and F are vector spaces, then every $\sigma(E^*, E)$-$\sigma(F^*, F)$-continuous bilinear form φ on $E^* \times F^*$ has the form

$$\varphi = \sum_{k=1}^{n} x_k \otimes y_k \qquad \text{for some } x_k \in E \text{ and } y_k \in F .$$

Ex 1.3. Let E be a Banach space and F be normed. If $S \subset \mathfrak{Bil}(E, F)$ is a set of separately continuous forms such that $\{{}_x\varphi \mid \varphi \in S\} \subset F'$ is equicontinuous for each $x \in E$, then $\sup\{\|\varphi\| \mid \varphi \in D\}$ is finite.

Ex 1.4. Using the space c_f of "finite" sequences (ξ_n) (i.e. $\xi_n \in \mathbb{K}$ and $\xi_n = 0$ except for finitely many n), show that theorem 1.2. implies the uniform boundedness principle: If E is a Banach space, then every pointwise bounded subset $D \subset \mathcal{L}(E, G)$ is uniformly bounded. Hint: $\sum_n \xi_n T_n x$.

Ex 1.5. Let $G \subset F$ be a subspace. Show that every $\varphi \in \mathfrak{Bil}(E, G)$ is extendable to a $\tilde{\varphi} \in \mathfrak{Bil}(E, F)$ if and only if every operator $T \in \mathcal{L}(E, G')$ is liftable to an operator $\tilde{T} \in \mathcal{L}(E, F')$:

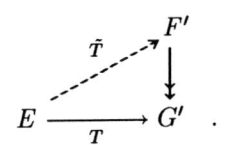

$$\begin{array}{ccc} & & F' \\ & \tilde{T} \nearrow & \downarrow \\ E & \xrightarrow{T} & G' \end{array}$$

Ex 1.6. Let $G \subset E$ be a subspace and $I : G \xrightarrow{1} E$ the isometric injection. Every $\varphi \in \mathfrak{Bil}(G, G')$ is extendable to a $\tilde{\varphi} \in \mathfrak{Bil}(E, G')$ if and only if there is a $U \in \mathcal{L}(E, G'')$ with $U \circ I = \kappa_G$.

Ex 1.7. A Banach space F is called *injective* if for every Banach space E, every subspace $G \subset E$ and every $T \in \mathcal{L}(G, F)$ there is an extension $\tilde{T} \in \mathcal{L}(E, F)$ of T. If there is a constant $\lambda \geq 1$ such that $\|\tilde{T}\| \leq \lambda \|T\|$ can be achieved, then F is said to have the λ-*extension property*; F has the *metric extension property* if $\lambda = 1$.
(a) $\ell_\infty(\Gamma)$ has the metric extension property for every set Γ.
(b) F is injective if and only if F is complemented in some $\ell_\infty(\Gamma)$. If $P : \ell_\infty(\Gamma) \to F$ is a projection, then F has the $\|P\|$-extension property. Hint: $F \xrightarrow{1} \ell_\infty(B_{F'})$.

Ex 1.8. If $\Phi : \mathbb{K}^3 \times \mathbb{K}^3 \to \mathbb{K}^4$ is Horowitz' surjective bilinear mapping which is not open in zero (see 1.7.) and E and F are normed, then

$$\hat{\Phi} : (\mathbb{K}^3 \times \mathbb{K} \times E) \times (\mathbb{K}^3 \times F) \longrightarrow \mathbb{K}^4 \times F$$
$$(x_1, x_2, x_3; y_1, y_2) \rightsquigarrow (\Phi(x_1, y_1), x_2 y_2)$$

is also continuous, surjective, and not open in zero. In particular, there are such mappings $\ell_p \times \ell_p \to \ell_p$ for each $p \in [1, \infty]$.

Ex 1.9. Let $\mathcal{F}(X)$ be the space of all scalar-valued functions $f : X \to \mathbb{K}$ with finite support $[f \neq 0] := \{x \in X \mid f(x) \neq 0\}$. Then

$$\mathcal{F}(X) \times \mathcal{F}(Y) \longrightarrow \mathcal{F}(X \times Y)$$
$$(f, g) \rightsquigarrow [f \otimes g : (x, y) \rightsquigarrow f(x) g(y)]$$

is bilinear. Use $[f \otimes g \neq 0] = [f \neq 0] \times [g \neq 0]$ in order to construct a bilinear mapping $\mathbb{R}^2 \times \mathbb{R}^2 \to \mathbb{R}^4$ the range of which is not a linear subspace; other examples of this type are easily deduced from 1.3.(5).

Ex 1.10. Under what circumstances is a bilinear mapping injective?

Ex 1.11. Let E, F and G be Banach spaces and $\Phi_n, \Phi \in Bil(E, F; G)$.

(a) *Closed graph theorem:* If Φ has a closed graph, then Φ is continuous.

(b) *Banach–Steinhaus theorem:* If $\Phi_n(x, y) \to \Phi(x, y)$ for all $(x, y) \in E \times F$ and all Φ_n are continuous, then Φ is continuous.

Hint: In both cases show the separate continuity of Φ.

Ex 1.12. Show that for the canonical embedding $\kappa_E : E \hookrightarrow E''$ of a Banach space and the canonical projection $P_{E'} : E''' \to E'$ the following relations hold:

(a) $\kappa_E' = P_{E'}$.
(b) $P_{E''} \circ \kappa_E'' = id_{E''}$.
(c) $P_{E'}' = \kappa_{E''}$ if and only if E is reflexive.
(d) $\kappa_E'' = \kappa_{E''}$ if and only if E is reflexive.

Ex 1.13. If $\varphi \in Bil(E, E')$ is defined by $\varphi(x, x') := \langle x', x \rangle$, then $L_\varphi = \kappa_E$.

Ex 1.14. Let $\varphi \in Bil(E, F)$. Show that

$$L_{(\varphi^\wedge)} = \kappa_{F'} \circ L_\varphi : E \longrightarrow F'''$$

$$L_{(^\wedge \varphi)} = P_{F'} \circ L_\varphi'' : E'' \longrightarrow F'.$$

Hint: Use proposition 1.9. for $^\wedge \varphi$.

Ex 1.15. Let $\Phi \in Bil(E_1, E_2; F)$ and let $K_i \subset E_i$ be compact convex sets; then for the set of extreme points the following holds:

$$\text{ext}(\text{conv}(\Phi(K_1, K_2))) \subset \Phi(\text{ext } K_1, \text{ext } K_2).$$

Hint: Use the fact that the extreme points of conv(L) with L compact are in L and the concept of an extremal set D of a convex set C (i.e. $\alpha x + (1-\alpha)y \in D$ with $\alpha \in]0,1[$ and $x, y \in C$ implies $x, y \in D$; see e.g. [70]).

2. The Algebraic Theory of Tensor Products

With the aid of a certain universal object the study of bilinear mappings can be reduced to the study of linear mappings. The construction of this new vector space $E \otimes F$ is, at least for an analyst, very simple.

2.1. For an arbitrary set A define $\mathcal{F}(A)$ to be the set of all functions $f : A \to \mathbb{K}$ with finite support, i.e. $f(\alpha) = 0$ except on a finite subset of A. For fixed $\alpha \in A$ the "α–unit vector" is the function $e_\alpha \in \mathcal{F}(A)$ defined by

$$e_\alpha(\beta) := \delta_{\alpha\beta} \quad ,$$

the Kronecker delta. It is clear that each $f \in \mathcal{F}(A)$ has the unique representation

$$f = \sum_{\alpha \in A} f(\alpha) e_\alpha \ ,$$

in other words: $(e_\alpha)_{\alpha \in A}$ is an algebraic basis of $\mathcal{F}(A)$. Now take two sets A and B and consider, as in Ex 1.9., the bilinear map

$$\Psi_o : \mathcal{F}(A) \times \mathcal{F}(B) \longrightarrow \mathcal{F}(A \times B)$$
$$(f, g) \rightsquigarrow f(\cdot)g(\cdot\cdot) \quad .$$

Since $e_{(\alpha,\beta)} = \Psi_o(e_\alpha, e_\beta)$, it is clear that

$$\operatorname{span} \operatorname{im} \Psi_o = \mathcal{F}(A \times B) \quad .$$

(Note that e_α has a double meaning if $A \cap B \neq \emptyset$ — but this clearly does not cause any problems.) For $\Phi \in \operatorname{Bil}(\mathcal{F}(A), \mathcal{F}(B); G)$ define $T \in L(\mathcal{F}(A \times B), G)$ by

$$T(e_{(\alpha,\beta)}) := \Phi(e_\alpha, e_\beta)$$

(and linear extension). It is obvious that T is the unique linear map $\mathcal{F}(A \times B) \to G$ with

$$T \circ \Psi_o = \Phi \quad .$$

It follows that
$$L(\mathcal{F}(A \times B), G) = Bil(\mathcal{F}(A), \mathcal{F}(B); G)$$
$$T \rightsquigarrow T \circ \Psi_o$$
is a linear isomorphism of vector spaces. It is worthwhile to look at $G := \mathbb{K}$:
$$\mathcal{F}(A \times B)^* = Bil(\mathcal{F}(A), \mathcal{F}(B)) = L(\mathcal{F}(A), \mathcal{F}(B)^*)$$
– the linear functionals on $\mathcal{F}(A \times B)$, the bilinear forms on $\mathcal{F}(A) \times \mathcal{F}(B)$ and the linear mappings $\mathcal{F}(A) \to \mathcal{F}(B)^*$ coincide ! This fundamental algebraic relation allows one to consider certain operators T as linear functionals on a new space depending on the domain and range of T – and this idea is essential to the theory presented in this book.

2.2. The universal property of the pair $(\mathcal{F}(A \times B), \Psi_o)$ is expressed in the

DEFINITION: *Let E and F be $\mathbb{K}$-vector spaces. A pair (H, Ψ_o) of a $\mathbb{K}$-vector space H and a bilinear map $\Psi_o \in Bil(E, F; H)$ is called a tensor product of the pair (E, F) if for each $\mathbb{K}$-vector space G and each $\Phi \in Bil(E, F; G)$ there is a unique $T \in L(H, G)$ with $\Phi = T \circ \Psi_o$:*

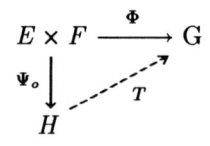

$$\begin{array}{ccc} E \times F & \xrightarrow{\Phi} & G \\ {\scriptstyle \Psi_o} \downarrow & \nearrow {\scriptstyle T} & \\ H & & \end{array}$$

In other words,
$$L(H, G) \longrightarrow Bil(E, F; G)$$
$$T \rightsquigarrow T \circ \Psi_o$$
is injective, surjective (and linear) for all G. This easily implies the following properties:

PROPOSITION: (1) *If (H, Ψ_o) is a tensor product of (E, F), then* span im $\Psi_o = H$.
(2) *If (H_1, Ψ_o^1) and (H_2, Ψ_o^2) are two tensor products of (E, F), then there is a unique linear isomorphism $S : H_1 \to H_2$ with $S \circ \Psi_o^1 = \Psi_o^2$.*
(3) *If $T_i : E_i \to F_i$ are linear isomorphisms (onto) and (H, Ψ_o) is a tensor product of (F_1, F_2), then $(H, \Psi_o \circ (T_1 \times T_2))$ is a tensor product of (E_1, E_2).*

If E is a vector space and $(x_\alpha)_{\alpha \in A}$ an algebraic basis of E, then
$$\mathcal{F}(A) \longrightarrow E$$
$$f \rightsquigarrow \sum_\alpha f(\alpha) x_\alpha$$

is obviously an isomorphism (onto). The construction in 2.1. and the proposition give the

THEOREM: *Every pair (E, F) of vector spaces has (up to a linear isomorphism as in (2)) a unique tensor product (H, Ψ_o).*

It will be shown in 2.4.(4) that the space

$$E \otimes F := \mathfrak{Bil}((E^*, \sigma(E^*, E)), (F^*, \sigma(F^*, F)))$$

with the bilinear mapping

$$\begin{aligned} E \times F &\longrightarrow \mathfrak{Bil}((E^*, \sigma(E^*, E)), (F^*, \sigma(F^*, F))) \\ (x, y) &\rightsquigarrow [(x^*, y^*) \rightsquigarrow \langle x^*, x \rangle \langle y^*, y \rangle] \end{aligned}$$

is a tensor product of (E, F). This space has the advantage that it is constructed "naturally", not with "coordinates" (i.e. with the help of a basis) — let us call it *the* tensor product of (E, F). Clearly, much of the power of the theory of tensor products stems from the fact that in specific situations other spaces H can serve as a tensor product – it is customary to speak of other "realizations" of the tensor product and to write

$$H = E \otimes F$$

— the bilinear mapping $\Psi_o : E \times F \to H$ being either obvious from the specific type of spaces or being given explicitly. In this sense

$$\mathcal{F}(A \times B) = \mathcal{F}(A) \otimes \mathcal{F}(B) \quad .$$

Recall other instances in mathematics where we are used to working with a universal object – without always thinking of a specific realization: the completion of a metric or normed space, the inductive or projective limits of spaces – or even the real numbers! The same will be done with $E \otimes F$.

Before going on to various examples and methods of checking whether or not a space is *a / the* tensor product, note again the fundamental properties of tensor products:

$$Bil(E, F; G) = L(E \otimes F, G)$$

$$(E \otimes F)^* = Bil(E, F) = L(E, F^*) .$$

The tensor product mapping

$$E \times F \longrightarrow E \otimes F$$

will be denoted by $(x, y) \rightsquigarrow x \otimes y$. By (1) of the proposition every element $z \in E \otimes F$ has the form

$$z = \sum_{k=1}^{n} x_k \otimes y_k \quad ;$$

elements of the form $x \otimes y$ are called *elementary tensors*. If $\Phi \in Bil(E, F; G)$ and $T \in L(E \otimes F, G)$ with
$$\Phi(x, y) = T(x \otimes y) ,$$
then T is called *the linearization* of Φ (sometimes T is denoted by Φ^L) and Φ the *associated bilinear mapping* of T. From this one–to–one correspondence it is clear that defining a linear $T : E \otimes F \to G$ is the same as defining a bilinear $\Phi : E \times F \to G$. So definitions of T very often will look like

$$T : E \otimes F \longrightarrow G$$
$$x \otimes y \rightsquigarrow ...$$

and one must check whether $(x, y) \rightsquigarrow ...$ is bilinear.

2.3. It is clear from the construction in 2.1. and (3) of the last proposition that for bases $(x_\alpha)_{\alpha \in A}$ of E and $(y_\beta)_{\beta \in B}$ of F the family $(x_\alpha \otimes y_\beta)_{(\alpha,\beta) \in A \times B}$ is a basis of $E \otimes F$. This observation leads to the following test:

CRITERION: *Let E, F and H be vector spaces and $\Psi_o : E \times F \to H$ bilinear. Then the space* $(\mathrm{span}\Psi_o(E \times F), \Psi_o)$ *is a tensor product of* (E, F)

$$\mathrm{span\ im}(\Psi_o) = E \otimes F \qquad (via\ \Psi_o)$$

if and only if $(\Psi_o(x_i, y_j))_{i=1,...,n; j=1,...,m}$ *is linearly independent whenever* $(x_1, ..., x_n)$ *and* $(y_1, ..., y_m)$ *are linearly independent in E and F, respectively.*

PROOF: The condition is clearly necessary. Conversely, let $(x_\alpha)_{\alpha \in A}$ and $(y_\beta)_{\beta \in B}$ be bases of E and F. Then, by the condition, $(\Psi_o(x_\alpha, y_\beta))_{(\alpha,\beta) \in A \times B}$ is a basis of the linear hull of $\Psi_o(E, F)$, and hence this space is isomorphic to

$$\mathcal{F}(A \times B) = E \otimes F$$

and Ψ_o serves as the tensor product map. $\square$

An easy calculation with the criterion gives

REMARK: *If $(e_\alpha)_{\alpha \in A}$ is a basis of F with coefficient functionals $e_\alpha^* : F \to \mathbb{K}$, then $E^{(A)} := \oplus_{\alpha \in A} E$ is a tensor product of (E, F)*

$$E \otimes F = E^{(A)}$$

via $E \times F \ni (x, y) \rightsquigarrow (\langle e_\alpha^*, y \rangle x)_{\alpha \in A} \in E^{(A)}$. *In particular, for every $z \in E \otimes F$ there exists a unique family* $(x_\alpha) \in E^{(A)}$ *with*

$$z = \sum_{\alpha \in A} x_\alpha \otimes e_\alpha .$$

Now assume $z = \sum_{j=1}^n x_j \otimes e_{\alpha_j}$ and order $(x_1, ..., x_n)$ such that $(x_1, ..., x_k)$ is a basis of $\text{span}\{x_1, ..., x_n\}$. Then

$$z = \sum_{j=1}^k x_j \otimes e_{\alpha_j} + \sum_{j=k+1}^n \sum_{i=1}^k \beta_{ij} x_i \otimes e_{\alpha_j} = \sum_{j=1}^k x_j \otimes \left[e_{\alpha_j} + \sum_{i=k+1}^n \beta_{ji} e_{\alpha_i} \right],$$

hence the

COROLLARY: *For every $0 \neq z \in E \otimes F$ there are an $n \in \mathbb{N}$ and linearly independent $(x_1, ..., x_n)$ and $(y_1, ..., y_n)$ with*

$$z = \sum_{j=1}^n x_j \otimes y_j.$$

2.4. Some examples of tensor products (use the criterion or remark 2.3. — readers who are not familiar with tensor products should check the details; in order to become acquainted with a mathematical object one also has to get used to it ...):

(1) The tensor product with finite dimensional spaces involved is as follows:

$$E \otimes \mathbb{K}^n = E^n \qquad x \otimes (\xi_k)_{k=1}^n := (\xi_k x)_{k=1}^n$$
$$\mathbb{K}^m \otimes \mathbb{K}^n = \mathbb{K}^{m \cdot n} \qquad (\eta_\ell)_{\ell=1}^m \otimes (\xi_k)_{k=1}^n := (\eta_\ell \xi_k)_{\ell,k=1}^{m,n}$$
$$\mathbb{K} \otimes E = E \qquad \lambda \otimes x := \lambda x.$$

(2) If K and L are compact, then

$$C(K) \otimes C(L) =$$
$$= \left\{ f \in C(K \times L) \mid \exists n \in \mathbb{N}, g_i \in C(K), h_i \in C(L) : f(\cdot, \cdot\cdot) = \sum_{i=1}^n g_i(\cdot) h_i(\cdot\cdot) \right\}.$$

This is why the notation

$$g \otimes h := g(\cdot) h(\cdot\cdot)$$

is used for the product of two functions in different variables. The Stone–Weierstrass theorem shows that

$$C(K) \otimes C(L) \subset C(K \times L)$$

is dense with respect to uniform convergence; is there a natural norm on $E \otimes F$ which, in this case, is the sup–norm? For the polynomials

$$\mathcal{P}(\mathbb{C}^n) \otimes \mathcal{P}(\mathbb{C}^m) = \mathcal{P}(\mathbb{C}^{n+m})$$

holds and for the entire holomorphic functions

$$\mathcal{H}(\mathbb{C}^n) \otimes \mathcal{H}(\mathbb{C}^m) \subset \mathcal{H}(\mathbb{C}^{n+m}),$$

where, by Taylor–expansions, it is easily seen that $\mathcal{H}(\mathbb{C}^n)\otimes\mathcal{H}(\mathbb{C}^m)$ is dense in $\mathcal{H}(\mathbb{C}^{n+m})$ with respect to the topology of uniform convergence on compact sets.

(3) If $\mathcal{S}(X)$ is any vector space of functions on X of a special type (continuous, differentiable, integrable — or sequences with special properties) and E a vector space, then

$$\mathcal{S}(X)\otimes E = \left\{f : X \to E \mid \exists n \in \mathbb{N}, (g_i, x_i) \in \mathcal{S}(X) \times E : f(\cdot) = \sum_{i=1}^n g_i(\cdot)x_i\right\}.$$

Clearly, the notation $g \otimes x$ for $g(\cdot)x$ will be used.

(4) Let E and F be vector spaces. Then, using the criterion and Ex 1.2.(b) it is clear that
$$\mathfrak{Bil}((E^*, \sigma(E^*, E)), (F^*, \sigma(F^*, F)))$$
is a tensor product of (E, F) via the map

$$(x,y) \rightsquigarrow \left[\begin{array}{rl} x\otimes y : & E^* \times F^* \longrightarrow \mathbb{K} \\ & (x^*, y^*) \rightsquigarrow \langle x^*, x\rangle\langle y^*, y\rangle\end{array}\right].$$

This was formally "the" tensor product of (E, F), see 2.2.. This means that $x\underline{\otimes}y$ is $x \otimes y$ and the notation $\underline{\otimes}$ is not necessary. Nevertheless we shall from time to time use the notation $x\underline{\otimes}y$ for the bilinear form on $E^* \times F^*$ since it has some advantages in the future; for example, if the mapping

$$\text{tensor product} \longrightarrow \text{bilinear forms/operators/ functions}$$

were extended to larger spaces by density, the injectivity might be lost! This could happen easily in the following example as will be seen in section 5.

(5) $\qquad E^* \otimes F = \{T \in L(E, F) \mid \operatorname{rank}(T) < \infty\} =: \mathcal{F}(E, F)$

via $(x^*, y) \rightsquigarrow [x^*\underline{\otimes}y : x \rightsquigarrow \langle x^*, x\rangle y]$ (still easy with the criterion). If E and F are normed spaces, then

$$E' \otimes F = \{T \in \mathcal{L}(E, F) \mid \operatorname{rank}(T) < \infty\} =: \mathfrak{F}(E, F),$$

the space of continuous finite rank operators. This identification will be often written as

$$E' \otimes F \ni z = \sum_{k=1}^n x'_k \otimes y_k \rightsquigarrow T_z := \sum_{k=1}^n x'_k\underline{\otimes}y_k \in \mathfrak{F}(E, F),$$

hence $T_z(x) = \sum_{k=1}^n \langle x'_k, x\rangle y_k$; note, that by the very definition of the tensor product, this is independent from the special representation of z. If E or F is finite dimensional,

$$E' \otimes F = \mathcal{L}(E, F),$$

in particular,
$$\mathbb{K}^n \otimes \mathbb{K}^m = \mathcal{L}(\mathbb{K}^n, \mathbb{K}^m) = \mathbb{K}^{nm},$$
the space of matrices as was already seen in example 1.

In summary, the basic examples of tensor products, also from the point of view of applications, are
— vector valued functions
— functions of two variables
— finite rank operators .

2.5. If E is a vector space, then the evaluation map
$$E^* \times E \longrightarrow \mathbb{K} \ ; \qquad (x^*, x) \rightsquigarrow \langle x^*, x \rangle$$
has a unique linearization in $(E^* \otimes E)^*$: it is called the *trace* tr_E on E

$$\begin{array}{ccc} E^* \times E & \xrightarrow{\text{evaluation}} & \mathbb{K} \\ \downarrow & \nearrow & \\ \mathcal{F}(E, E) = E^* \otimes E & \text{trace } tr_E & \end{array}.$$

To check that this is the usual trace of matrices for dim $E = n$, take a basis $(e_1, ..., e_n)$ of E with coefficient functionals e_k^*. For $T \in L(E, E)$ with
$$T = \sum_{\ell=1}^{m} x_\ell^* \otimes x_\ell$$
and representing matrix $(\langle e_\ell^*, T e_k \rangle)$
$$tr_E(T) = tr_E \left(\sum_{\ell=1}^{m} x_\ell^* \otimes x_\ell \right) = \sum_{\ell=1}^{m} \langle x_\ell^*, x_\ell \rangle = \sum_{\ell=1}^{m} \left\langle \sum_{k=1}^{n} \langle x_\ell^*, e_k \rangle e_k^*, x_\ell \right\rangle =$$
$$= \sum_{k=1}^{n} \left\langle e_k^*, \sum_{\ell=1}^{m} \langle x_\ell^*, e_k \rangle x_\ell \right\rangle = \sum_{k=1}^{n} \langle e_k^*, T e_k \rangle ,$$
hence the trace of T is the sum of the diagonal elements of the representing matrix of the mapping T.

2.6. Take E and F vector spaces and $\varphi \in (E \otimes F)^* = Bil(E, F) = L(E, F^*)$ with associated operator $T = L_\varphi$. Using the transposition (see Ex 2.1.), the identification
$$E \otimes F \longrightarrow F \otimes E = \mathfrak{Bil}((F^*, \sigma(F^*, F)), (E^*, \sigma(E^*, E))) \hookrightarrow \mathcal{F}(F^*, E)$$
$$z \ \rightsquigarrow \ z^t$$

is as follows: For $z = \sum_{k=1}^{n} x_k \otimes y_k$ the associated linear map is

$$S : F^* \longrightarrow E \quad , \quad S(y^*) = \sum_{k=1}^{n} \langle y^*, y_k \rangle x_k$$

or $S = \sum_{k=1}^{n} y_k \underline{\otimes} x_k$. The duality bracket between $E \otimes F \hookrightarrow \mathcal{F}(F^*, E)$ and its algebraic dual $(E \otimes F)^* = L(E, F^*)$ turns out to be the so-called trace duality between operators:

TRACE DUALITY: Let $\varphi \in (E \otimes F)^*$ with associated map $T \in L(E, F^*)$ and $z \in E \otimes F$ with associated $S \in \mathcal{F}(F^*, E)$. Then

$$\langle \varphi, z \rangle = tr_E(S \circ T) = tr_{F^*}(T \circ S) \quad .$$

PROOF: By linearity, it is enough to check this for elementary tensors $z = x_o \otimes y_o$: but this is easy since

$$\langle \varphi, x_o \otimes y_o \rangle = \langle Tx_o, y_o \rangle$$
$$S \circ T = T^* y_o \underline{\otimes} x_o : E \to E$$
$$T \circ S = y_o \underline{\otimes} Tx_o : F^* \to F^* \quad . \quad \square$$

Note that transposition $z \rightsquigarrow z^t$ means going to the dual of the associated operator (see Ex 2.2.) – more or less: minor difficulties come from the domain and range spaces (see Ex 2.3.).

2.7. The *tensor product of operators* is defined by the

PROPOSITION: Let $T_i \in L(E_i, F_i)$ for $i = 1, 2$. Then there is a unique operator $S \in L(E_1 \otimes E_2, F_1 \otimes F_2)$ with the property $S(x_1 \otimes x_2) = T_1 x_1 \otimes T_2 x_2$ for all $(x_1, x_2) \in E_1 \times E_2$. The operator S is denoted by $T_1 \otimes T_2$, hence

$$(T_1 \otimes T_2)(x_1 \otimes x_2) = T_1 x_1 \otimes T_2 x_2 \quad .$$

PROOF: This follows from the diagram

$$\begin{array}{ccc} E_1 \times E_2 & \xrightarrow{(T_1, T_2)} & F_1 \times F_2 \\ \downarrow & \searrow & \downarrow \\ E_1 \otimes E_2 & \dashrightarrow & F_1 \otimes F_2 \end{array}$$

and the universal property of tensor products. $\square$

PROPERTIES: (1) *If T_1 and T_2 are surjective, then $T_1 \otimes T_2$ is surjective.*

(2) *If T_1 and T_2 are injective, then $T_1 \otimes T_2$ is injective.*
(3) *In particular, if $G \subset E$, then $G \otimes F$ is a subspace of $E \otimes F$.*
(4) *The following formula holds:*
$$\ker(T_1 \otimes T_2) = (\ker T_1) \otimes E_2 + E_1 \otimes (\ker T_2) \subset E_1 \otimes E_2 \ .$$

PROOF: (1) is simple. (2) follows from the criterion 2.3. using a basis of the form $x_\alpha \otimes y_\beta$ on $E_1 \otimes E_2$.

To see (4) decompose
$$E_1 = (\ker T_1) \oplus G_1 \quad , \quad E_2 = (\ker T_2) \oplus G_2$$
and take $z = \sum_n x_n^1 \otimes x_n^2 \in \ker(T_1 \otimes T_2)$. Then $x_n^i = u_n^i + g_n^i$ and, by (2) and (3), $\sum_n g_n^1 \otimes g_n^2 = 0$; therefore
$$z = \sum_n (u_n^1 + g_n^1) \otimes (u_n^2 + g_n^2) = \sum_n u_n^1 \otimes (u_n^2 + g_n^2) + \sum_n g_n^1 \otimes u_n^2$$
$$\in (\ker T_1) \otimes E_2 + E_1 \otimes (\ker T_2) \ .$$
The other inclusion is obvious. □

An interesting special case of $T_1 \otimes T_2$ is given by $x_o^* \in E^*$ and $T \in L(F, G)$:
$$x_o^* \otimes T : E \otimes F \to \mathbb{K} \otimes F = F$$
with $(x_o^* \otimes T)(x \otimes y) = \langle x_o^*, x \rangle Ty$.

Exercises:

Ex 2.1. Show that the transposition map $z \rightsquigarrow z^t$ defined by
$$E \otimes F \longrightarrow F \otimes E \ , \ x \otimes y \rightsquigarrow y \otimes x$$
is an isomorphism.

Ex 2.2. Show that
$$\begin{array}{ccc} z \in E' \otimes F & \longrightarrow & \mathcal{L}(E, F) \ni T \\ \updownarrow & & \downarrow \updownarrow \\ z^t \in F \otimes E' & \longrightarrow & \mathcal{L}(F', E') \ni T' \end{array}$$

is commutative.

Ex 2.3. Show that

$$\begin{array}{ccc} z \in E' \otimes E'' & = & \mathfrak{F}(E, E'') \ni T \\ \updownarrow \downarrow & & \downarrow \updownarrow \\ z^t \in E'' \otimes E' & = & \mathfrak{F}(E', E') \ni T' \circ \kappa_{E'} \end{array}$$

commutes. This means in particular, that $tr_{E'}$ can be considered as being defined on the space $\mathfrak{F}(E, E'')$.

Ex 2.4. (a) Take two separating dual systems $\langle E_i, F_i \rangle$. Then the mapping

$$(E_1 \otimes E_2) \times (F_1 \otimes F_2) \longrightarrow \mathbb{K}$$

$$\left(\sum_k x_k^1 \otimes x_k^2, \sum_\ell y_\ell^1 \otimes y_\ell^2\right) \rightsquigarrow \sum_{k,\ell} \langle x_k^1, y_\ell^1 \rangle \langle x_k^2, y_\ell^2 \rangle$$

is bilinear and defines a separating dual system

$$\langle E_1 \otimes E_2, F_1 \otimes F_2 \rangle \quad .$$

(b) In the case $F_i = E_i^*$ use $E_1^* \otimes E_2^* \subset (E_1 \otimes E_2)^*$ to show that the duality bracket is just the restriction of the trace duality. This is why the pairing in (a) is also called trace duality.

(c) For normed spaces E and F the dual system $\langle E \otimes F, E' \otimes F' \rangle$ is separating.

Ex 2.5. Let E and F be finite dimensional normed spaces, $z \in E \otimes F = \mathfrak{F}(E', F)$ with associated operator S and $\varphi \in (E \otimes F)^* = \mathfrak{L}(E, F')$ with associated operator T. Then

$$\langle \varphi, z \rangle = tr_E(S' \circ T) = tr_{E'}(T' \circ S) = tr_{F'}(T \circ S') = tr_F(S \circ T') \ .$$

Ex 2.6. Let E and F be normed. Show that

$$E \otimes F = \{T \in \mathfrak{L}(E', F) \mid T \text{ weak} - * - \text{continuous, rank}(T) < \infty\} \ .$$

Ex 2.7. Let $\mathfrak{L}_1(\mu)$ be the space of μ–integrable functions (on a measure space (Ω, μ)) and $L_1(\mu)$ the Banach space of equivalence classes of functions. Show that for every normed space E

$$\mathfrak{L}_1(\mu) \otimes E = \{f : \Omega \to E \mid f \text{ Bochner integrable, dim span } f(\Omega) < \infty\}$$
$$L_1(\mu) \otimes E = \mathfrak{L}_1(\mu) \otimes E/\sim$$

where $\sim$ stands for the "almost everywhere equal" equivalence relation.

Ex 2.8. Show that for $1 \leq p \leq \infty$ and E normed

$$\ell_p \otimes E = \{(x_n) \in \ell_p(E) \mid \text{dim span}\{x_n\} < \infty\}.$$

Describe $c_o \otimes E$.

Ex 2.9. If $E = E_1 \oplus E_2$, then $E \otimes F = (E_1 \otimes F) \oplus (E_2 \otimes F)$; more generally,

$$\left(\bigoplus_{\alpha \in A} E_\alpha\right) \otimes F = \bigoplus_{\alpha \in A}(E_\alpha \otimes F) .$$

Is something like this true for infinite products?

Ex 2.10. (a) If G is a $\mathbb{C}$-vector space, define $\overline{G}$ to be the $\mathbb{C}$-vector space with the same addition and the new scalar multiplication

$$(\lambda, x) \rightsquigarrow \overline{\lambda} x .$$

Use this construction to show: If E, F, G are $\mathbb{C}$-vector spaces and $\Phi : E \times F \to G$ bi-antilinear (i.e. biadditive and $\Phi(\lambda x, y) = \Phi(x, \lambda y) = \overline{\lambda}\Phi(x, y)$), then there is a unique antilinear $T : E_1 \otimes E_2 \to G$ such that

$$\Phi(x, y) = T(x \otimes y) .$$

(b) Recall that the Riesz–identification $R_H : H \to H'$ of a complex Hilbert space with its dual is antilinear. Then there is a unique antilinear map

$$R_H \otimes R_K : H \otimes K \to H' \otimes K'$$

satisfying $(R_H \otimes R_K)(x \otimes y) = R_H x \otimes R_K y$ for all $x \in H$ and $y \in K$.

Ex 2.11. The formation of tensor products depends upon the underlying field. More precisely: Let $E_\mathbb{C}$ and $F_\mathbb{C}$ be complex vector spaces and $E_\mathbb{R}, F_\mathbb{R}$ the same spaces considered as real ones. Show that there is a natural map $E_\mathbb{R} \otimes^\mathbb{R} F_\mathbb{R} \to E_\mathbb{C} \otimes^\mathbb{C} F_\mathbb{C}$ which is $\mathbb{R}$-linear, and surjective but, in general, not injective.

Ex 2.12. The mapping $E \to E \otimes E$ defined by $Q(x) := x \otimes x$ is in some sense a typical quadratic form.

(a) Show that Q is nearly injective in the following sense : if $x+y \neq 0$ and $Q(x) = Q(y)$, then $x = y$. Hint: Use *symmetric* bilinear forms on $E \otimes E$, i.e. those for which $\langle \varphi, u \otimes v \rangle = \langle \varphi, v \otimes u \rangle$, or think in terms of operators.

(b) If $\varphi \in \mathfrak{Bil}(E, E)$ is symmetric, then the two canonical extensions to $E'' \times E''$ (see 1.9.) coincide on the diagonal: $(^\wedge\varphi)^\wedge(x'', x'') = {}^\wedge(\varphi^\wedge)(x'', x'')$ for all $x'' \in E''$. Hint: $(^\wedge\varphi)^t = \varphi^\wedge$ in this case.

Ex 2.13. Define a tensor product for n-linear mappings $E_1 \times \cdots \times E_n \to G$ and denote it by $E_1 \otimes \cdots \otimes E_n$. Show that

$$E_1 \otimes \cdots \otimes E_n = (E_1 \otimes \cdots \otimes E_k) \otimes (E_{k+1} \otimes \cdots \otimes E_n)$$

for each $1 \leq k \leq n-1$.

3. The Projective Norm

The projective norm π on $E \otimes F$ has the following highly desirable universal property: For normed spaces E, F, G each bilinear mapping $E \times F \to G$ is continuous if and only if its linearization $E \otimes_\pi F \to G$ is continuous.

3.1. Before going through the basic calculations, recall that if $C \subset E$ is an absolutely convex subset of a vector space, its *Minkowski gauge functional* m_C is defined by

$$m_C(x) := \inf\{\lambda \geq 0 \mid x \in \lambda C\} \qquad \in [0, \infty] \ .$$

It is easy to check that m_C is a seminorm on the linear span of C which coincides with $\{x \in E \mid m_C(x) < \infty\}$; notation:

$$[\![C]\!] := (\operatorname{span} C, m_C) \ .$$

m_C is a norm if C is linearly bounded (i.e. bounded on every 1–dimensional subspace). It follows that

$$(*) \qquad (1-\varepsilon)C \subset \overset{\circ}{B}_{[\![C]\!]} \subset C \subset B_{[\![C]\!]} \subset (1+\varepsilon)C$$

for the open and closed unit balls and each $\varepsilon > 0$. If $[\![C]\!]$ is a Banach space, C is called a *Banach disc*.

If $A_1 \subset E_1$ and $A_2 \subset E_2$ are subsets, then

$$A_1 \otimes A_2 := \{x_1 \otimes x_2 \in E_1 \otimes E_2 \mid x_i \in A_i\} \ .$$

This seems to be confusing since in the case where the A_i are subspaces this is *not* the tensor product of these two spaces – it is in general not linear! But in practice it will always be clear what is meant.

3.2. Take $\Phi \in Bil(E, F; G)$ for normed spaces E, F, G and denote its linearization for the moment by $\Phi^L \in L(E \otimes F, G)$. Then

$$\|\Phi\|_{Bil(E,F;G)} = \sup\{\|\Phi(x,y)\|_G \mid x \in \overset{\circ}{B}_E, y \in \overset{\circ}{B}_F\} =$$
$$(**) \qquad\qquad = \sup\{\|\Phi^L(x \otimes y)\|_G \mid x \in \overset{\circ}{B}_E, y \in \overset{\circ}{B}_F\} =$$
$$= \sup\{\|\Phi^L(z)\|_G \mid z \in \Gamma(\overset{\circ}{B}_E \otimes \overset{\circ}{B}_F)\} \qquad \in [0, \infty]$$

where Γ stands for the absolutely convex hull.

DEFINITION: *The projective norm π on the tensor product $E \otimes F$ of two normed spaces E and F is the Minkowski gauge functional of $\Gamma(\overset{\circ}{B}_E \otimes \overset{\circ}{B}_F)$.*

The name "projective" will become clear in 3.8.. Notations:

$$\pi(z; E, F) := m_{\Gamma(\overset{\circ}{B}_E \otimes \overset{\circ}{B}_F)}(z) \qquad \text{for } z \in E \otimes F$$

$$E \otimes_\pi F := (E \otimes F, \pi)$$

and $E \tilde{\otimes}_\pi F$ for the completion. Actually π is a norm and has an alternative and more convenient description (which the reader should always keep in mind):

PROPOSITION: *π is a norm on $E \otimes F$ and can be calculated as follows:*

$$\pi(z; E, F) = \inf\left\{ \sum_{n=1}^N \|x_n\| \|y_n\| \mid N \in \mathbb{N}, z = \sum_{n=1}^N x_n \otimes y_n \right\} =$$

$$= \inf\left\{ \sum_{n=1}^\infty \|x_n\| \|y_n\| \mid z = \sum_{n=1}^\infty x_n \otimes y_n \text{ in } E \otimes_\pi F \right\}.$$

Moreover,

(1) $$\mathfrak{Bil}(E, F; G) = \mathfrak{L}(E \otimes_\pi F, G)$$

holds isometrically — and π is the unique seminorm on $E \otimes F$ which has this property for $G = \mathbb{K}$. In particular,

$$\pi(z; E, F) = \max\{|\langle S, z \rangle| \mid S \in \mathfrak{L}(E, F'), \|S\| \leq 1\},$$

the duality bracket given by

$$(E \otimes_\pi F)' = \mathfrak{Bil}(E, F) = \mathfrak{L}(E, F'),$$

which (by 2.6.) can be calculated by trace duality.

(2) *The open unit ball in $E \otimes_\pi F$ is $\Gamma(\overset{\circ}{B}_E \otimes \overset{\circ}{B}_F)$.*
(3) *$\pi(x \otimes y; E, F) = \|x\| \|y\|$ for all $(x, y) \in E \times F$.*
(4) *π satisfies the metric mapping property: If $T_i \in \mathfrak{L}(E_i, F_i)$, then*

$$\|T_1 \otimes T_2 : E_1 \otimes_\pi E_2 \longrightarrow F_1 \otimes_\pi F_2\| \leq \|T_1\| \|T_2\|.$$

(5) *π is finitely generated, i.e. for all $z \in E \otimes F$*

$$\pi(z; E, F) = \inf\{\pi(z; M, N) \mid z \in M \otimes N; M \subset E \text{ and } N \subset F \text{ finite dimensional}\}.$$

PROOF: By definition π is a seminorm on $E \otimes F = \operatorname{span} \Gamma(\mathring{B}_E \otimes \mathring{B}_F)$. The formula (**) for the norm in $\mathfrak{Bil}(E, F; G)$ and property (*) of 3.1. for the unit balls of a Minkowski gauge functional imply that for each $\Phi \in Bil(E, F; G)$

$$\|\Phi\|_{\mathfrak{Bil}(E,F;G)} = \sup \left\{ \|\Phi^L(z)\| \mid \pi(z; E, F) \leq 1 \right\} \in [0, \infty] \; ;$$

in particular: $\mathfrak{Bil}(E, F) = (E \otimes_\pi F)'$ isometrically. To see that π is a norm note that by 2.4.(4) and the Hahn-Banach theorem the following holds: for every $0 \neq z \in E \otimes F$ there exists an $x' \otimes y' \in E' \otimes F' \subset \mathfrak{Bil}(E, F)$ with $\langle x' \otimes y', z \rangle = 1$ (this was Ex 2.4.). Hence $\langle E \otimes F, \mathfrak{Bil}(E, F) \rangle$ is a separating dual system, which implies that π is a norm. Clearly, every seminorm η on $E \otimes F$ for which $(E \otimes F, \eta)' = \mathfrak{Bil}(E, F)$ holds isometrically, has to be equal to π. So (1) has been proven.

To see (2), observe first that, again by (*) of 3.1.,

$$\mathring{B}_{E \otimes_\pi F} \subset \Gamma(\mathring{B}_E \otimes \mathring{B}_F) \; .$$

Conversely, if $z = \sum_{k=1}^n \alpha_k x_k \otimes y_k \in \Gamma(\mathring{B}_E \otimes \mathring{B}_F)$, then there is an $\varepsilon > o$ with

$$(1+\varepsilon)x_k \in \mathring{B}_E \text{ and } (1+\varepsilon)y_k \in \mathring{B}_F$$

for $k = 1, ..., n$. Hence $(1+\varepsilon)^2 z \in \Gamma(\mathring{B}_E \otimes \mathring{B}_F)$, and $\pi(z; E, F) \leq (1+\varepsilon)^{-2} < 1$.

Now π has to be calculated more explicitly: For $x \in \mathring{B}_E$ and $y \in \mathring{B}_F$ property (2) implies that $\pi(x \otimes y; E, F) \leq 1$, hence for all $(x, y) \in E \times F$

$$\pi(x \otimes y; E, F) \leq \|x\| \, \|y\| \; .$$

But there are $x' \in B_{E'}$ and $y' \in B_{F'}$ with $\langle x', x \rangle = \|x\|$ and $\langle y', y \rangle = \|y\|$, hence

$$\|x\| \, \|y\| = |\langle x' \otimes y', x \otimes y \rangle| \leq \|x' \otimes y'\| \pi(x \otimes y; E, F) \leq \pi(x \otimes y; E, F) \, .$$

This is (3) — and everything is ready for proving the initial formula for π: By the triangle inequality (and an obvious notation)

$$\pi(z; E, F) \leq \inf{}_\infty(z) \leq \inf{}_f(z) \qquad \text{for } z \in E \otimes F.$$

On the other hand, take $\pi(z; E, F) < \lambda$, i.e.

$$z = \lambda \sum_{k=1}^n \alpha_k x_k \otimes y_k \in \lambda \Gamma(\mathring{B}_E \otimes \mathring{B}_F) \, ,$$

which implies

$$\inf{}_f(z) \leq \sum_{k=1}^n \|\lambda \alpha_k x_k\| \, \|y_k\| < \lambda$$

and therefore $\inf_f(z) \leq \pi(z; E, F)$.

Clearly, the inf–formula implies (5), and it will also be used for the proof of (4) since it is handier than the original definition: Take $T_i \in \mathcal{L}(E_i, F_i)$ and

$$z = \sum_{k=1}^{n} x_k^1 \otimes x_k^2 \in E_1 \otimes E_2.$$

Then

$$\pi(T_1 \otimes T_2(z); F_1, F_2) = \pi\left(\sum_{k=1}^{n} T_1 x_k^1 \otimes T_2 x_k^2; F_1, F_2\right) \leq$$

$$\leq \sum_{k=1}^{n} \|T_1 x_k^1\| \, \|T_2 x_k^2\| \leq \|T_1\| \, \|T_2\| \sum_{k=1}^{n} \|x_k^1\| \, \|x_k^2\|,$$

which implies $\|T_1 \otimes T_2 : E_1 \otimes_\pi E_2 \to F_1 \otimes_\pi F_2\| \leq \|T_1\| \, \|T_2\|$. □

For brevity, the notation $T_1 \otimes_\pi T_2$, and $T_1 \tilde{\otimes}_\pi T_2$ for its unique extension to the completions $E_1 \tilde{\otimes}_\pi E_2 \to F_1 \tilde{\otimes}_\pi F_2$, will be used. By considering elementary tensors, it even follows that $\|T_1 \otimes_\pi T_2\| = \|T_1\| \, \|T_2\|$ holds.

3.3. Let (Ω, μ) be a measure space and E a Banach space. Then

$$\mathcal{L}_1(\mu) \otimes E \subset \mathcal{L}_1(\mu, E)$$

— the space of Bochner integrable functions and

$$L_1(\mu) \otimes E \subset L_1(\mu, E)$$

for the equivalence classes. It is clear from the definition of Bochner integrable functions (see Appendix B12.) that the vector space $\mathfrak{S}(\mu) \otimes E$ of E–valued step functions

$$\sum_{k=1}^{n} \chi_{A_k} \otimes x_k \qquad A_k \ \mu\text{–integrable}, \quad x_k \in E$$

is dense in $\mathcal{L}_1(\mu, E)$ and hence the space $S(\mu) \otimes E$ of classes is dense in $L_1(\mu, E)$. It follows that

$$\mathfrak{S}(\mu) \otimes E \subset \mathcal{L}_1(\mu) \otimes E \subset \mathcal{L}_1(\mu, E) \text{ and } S(\mu) \otimes E \subset L_1(\mu) \otimes E \subset L_1(\mu, E)$$

are dense inclusions.

PROPOSITION: *Let μ be an arbitrary measure and E a Banach space. Then*

$$L_1(\mu) \otimes_\pi E \xrightarrow{\ 1\ } L_1(\mu, E)$$

is an isometric dense embedding; in other words,

$$\pi(\tilde{f}; L_1, E) = \int_\Omega \|f(\omega)\|_E \mu(d\omega)$$

for each $\tilde{f} \in L_1(\mu) \otimes E$, and $L_1(\mu) \tilde{\otimes}_\pi E = L_1(\mu, E)$ holds isometrically.

PROOF: Obviously $L_1 \times E \to L_1(E)$ has norm ≤ 1, hence

$$\|L_1 \otimes_\pi E \hookrightarrow L_1(E)\| \leq 1 \ .$$

Since it is easy to see that $S(\mu) \otimes E$ is π–dense in $L_1(\mu) \otimes E$ (see 3.9. or Ex 3.3.), it is enough to check, by Ex 3.4., that

$$\pi(\tilde{f}; L_1, E) \leq \int_\Omega \|f(\omega)\|_E \mu(d\omega)$$

holds for step functions

$$f = \sum_{n=1}^{N} \chi_{A_n} \otimes x_n \ .$$

Clearly, the A_n can be assumed to be pairwise disjoint, therefore

$$\pi(\tilde{f}; L_1, E) \leq \sum_{n=1}^{N} \|\tilde{\chi}_{A_n}\|_{L_1} \|x_n\|_E = \int_\Omega \sum_{n=1}^{N} \chi_{A_n} \|x_n\|_E d\mu = \int_\Omega \|f(\omega)\|_E \mu(d\omega) \ . \ \Box$$

This example is quite important; for $1 < p \leq \infty$ it is not true that $L_p \tilde{\otimes}_\pi E = L_p(E)$, see sections 7 and 12 for a discussion if $p < \infty$ and 7.1. and Ex 6.8. for $p = \infty$.

What is the dual of $L_1 \otimes_\pi F$? In other words, is there a good description of $(L_1 \otimes_\pi F)' = (L_1 \tilde{\otimes}_\pi F)' = L_1(\mu, F)'$?

THEOREM OF DUNFORD-PETTIS (weak version): *If μ is a strictly localizable measure on Ω and $T : L_1(\Omega, \mu) \to F'$ a continuous linear operator into the dual of a Banach space F, then there is a function $g : \Omega \to F'$ such that*

(1) *g is weak-$*$-measurable (see Appendix B11.) and $g(\Omega) \subset \overline{TB_{L_1}}^{\sigma(F', F)}$. In particular, $\|g(w)\|_{F'} \leq \|T\|$ for all $w \in \Omega$.*

(2) *g is a weak-$*$-density of T, i.e.: for each $x \in F$ and $\tilde{f} \in L_1$*

$$\langle T\tilde{f}, x \rangle = \int_\Omega f(w) \langle g(w), x \rangle \mu(dw) \ .$$

In other words, $T\tilde{f} = \int fg d\mu$ as a $\sigma(F', F)$–Pettis integral.

PROOF: Call $\varphi \in \mathfrak{Bil}(L_1, F) = (L_1 \otimes_\pi F)' = \mathcal{L}(L_1, F')$ the bilinear form associated with T and define $\langle \varphi_x, \tilde{f} \rangle := \varphi(\tilde{f}, x) = \langle T\tilde{f}, x \rangle$ for each $x \in F$ and $\tilde{f} \in L_1$. Then $\varphi_x \in L_1' = L_\infty$. Since μ is strictly localizable, there is a "lifting" $\rho : L_\infty \to \mathcal{L}_\infty$ (see Appendix B7.), i.e. ρ is linear, positive, $\rho(\tilde{1}) = 1$ and $\rho(\tilde{f}) \in \tilde{f}$. It follows that

$$|\rho(\tilde{f})(w)| \leq \|\tilde{f}\|_{L_\infty} .$$

For each $w \in \Omega$ the functional $g(w) \in F'$ is defined by

$$\langle g(w), x \rangle := \rho(\varphi_x)(w) \qquad \text{for all } x \in F,$$

hence $\langle g(\cdot), x \rangle = \rho(\varphi_x)$ is measurable for all $x \in F$ and

$$|\langle g(w), x \rangle| = |\rho(\varphi_x)(w)| \leq \|\varphi_x\|_{L_\infty} = \sup\{|\langle T\tilde{f}, x \rangle| \mid \tilde{f} \in B_{L_1}\}$$

which means that $g(w) \in (TB_{L_1})^{00} = \overline{TB_{L_1}}^{\sigma(F', F)}$ by the bipolar theorem. Clearly, the following representation holds:

$$\langle T\tilde{f}, x \rangle = \langle \varphi_x, \tilde{f} \rangle = \int_\Omega \rho(\varphi_x)(w) f(w) \mu(dw) = \int_\Omega f(w) \langle g(w), x \rangle \mu(dw) . \square$$

If $T \in \mathcal{L}(L_1, F)$ is weakly compact, then the set $\overline{TB_{L_1}} \subset F$ is weak-$*$-closed in F'' and therefore:

COROLLARY: *If μ is strictly localizable, F a Banach space and $T : L_1(\Omega, \mu) \to F$ weakly compact, then there is a weakly measurable function $g : \Omega \to F$ with values in $\overline{TB_{L_1}}$ such that*

$$\langle x', T\tilde{f} \rangle = \int_\Omega f(w) \langle x', g(w) \rangle \mu(dw)$$

for all $(x', \tilde{f}) \in F' \times L_1$. In other words, $T\tilde{f} = \int fg d\mu$ as a Pettis integral.

In Appendix C it will be shown that the function g in this corollary can be chosen to be Bochner integrable if μ is a finite measure. Appendix D4. demonstrates conditions under which the dual of $L_1(E)$ can be described by bounded measurable E'-valued functions.

3.4. The following theorem – as most of them in this part of the theory: due to Grothendieck – describes the compact subsets of $E \tilde{\otimes}_\pi F$.

THEOREM: *Let E and F be normed spaces and $K \subset E \tilde{\otimes}_\pi F$ relatively compact. Then there are zero sequences (x_n) and (y_n) in E and F, respectively, and a compact subset $H \subset \ell_1$ such that for each $z \in K$ there is a $(\lambda_n) \in H$ with*

$$z = \sum_{n=1}^\infty \lambda_n x_n \otimes y_n .$$

Moreover, for every $\varepsilon > 0$, the set H and the sequences (x_n), (y_n) can be chosen in such a way that $x_n \in B_E$, $y_n \in B_F$ and $\sup_{\lambda \in H} \|\lambda\|_{\ell_1} \leq (1+\varepsilon) \sup_{z \in K} \|z\|_{E \tilde{\otimes}_\pi F}$.

PROOF: The idea (following Pietsch) is to expand each $z \in E \tilde{\otimes}_\pi F$ in an absolutely convergent series $z = \sum z_n$ and to choose a representation

$$z_n = \sum \alpha_k^n x_k^n \otimes y_k^n \quad \text{with} \quad \sum |\alpha_j^n| = \|x_k^n\| = \|y_k^n\| \sim \pi(z_n)^{1/3}$$

— and all of this somehow simultaneously in K, with the aid of its precompactness. For this, take $\varepsilon_o := 1$ and $\varepsilon_n > 0$ with $\sum_{n=0}^\infty \varepsilon_n = 1 + \varepsilon$ and define

$$\tilde{B}_n := \varepsilon_n^3 \, \overset{\circ}{B}_{E \tilde{\otimes}_\pi F} \quad \text{and} \quad B_n := \tilde{B}_n \cap (E \otimes_\pi F) \quad .$$

Assume $K \subset \tilde{B}_o$; then there is a finite $D_o \subset B_o$ with

$$K \subset D_o + \tilde{B}_1$$

and, by induction, finite sets $D_n \subset B_n$ with

$$F_1 := (K - D_o) \cap \tilde{B}_1 \subset D_1 + \tilde{B}_2$$
$$F_n := (F_{n-1} - D_{n-1}) \cap \tilde{B}_n \subset D_n + \tilde{B}_{n+1} \; .$$

There are (pairwise disjoint) finite indexing sets A_n and B_j^n with $D_n := \{z_j^n | j \in A_n\}$ and representations

$$z_j^n = \sum_{k \in B_j^n} \alpha_{j,k}^n x_{j,k}^n \otimes y_{j,k}^n$$

with

$$\|x_{j,k}^n\| = \|y_{j,k}^n\| = \varepsilon_n \quad \text{and} \quad \sum_{k \in B_j^n} |\alpha_{j,k}^n| \leq \varepsilon_n \; .$$

The set $\Gamma := \{(n,j,k) \mid n \in \mathbb{N} \cup \{0\}, j \in A_n, k \in B_j^n\}$ is countable and $\|x_{j,k}^n\| = \|y_{j,k}^n\| \to 0$ under some enumeration of Γ. Take as the desired compact set in $\ell_1(\Gamma)$

$$H := \left\{ (\beta_{n,j,k})_{(n,j,k) \in \Gamma} \;\Big|\; \sum_{j \in A_n} \sum_{k \in B_j^n} |\beta_{n,j,k}| \leq \varepsilon_n \text{ for } n = 0, 1, 2, ... \right\} \; .$$

For $z \in K$ there are $j_n \in A_n$ with

$$z - z_{j_o}^o \in (K - D_o) \cap \tilde{B}_1 = F_1$$
$$z - z_{j_o}^o - z_{j_1}^1 \in (F_1 - D_1) \cap \tilde{B}_2$$
$$z - z_{j_o}^o - ... - z_{j_n}^n \in (F_n - D_n) \cap \tilde{B}_{n+1} \quad ...$$

It follows that
$$z = \sum_{n=0}^{\infty} \sum_{k \in B_{j_n}^n} \alpha_{j_n,k}^n x_{j_n,k}^n \otimes y_{j_n,k}^n$$
which is the desired representation. $\square$

3.5. Bounded subsets $C \subset E \tilde{\otimes}_\pi F$ are "liftable" in the sense that there are bounded $A \subset E$ and $B \subset F$ such that
$$C \subset \overline{\Gamma(A \otimes B)}^{E \tilde{\otimes}_\pi F}$$
(why?) — the theorem implies that the same holds for relatively compact sets:

COROLLARY 1: *Let E and F be normed and $C \subset E \tilde{\otimes}_\pi F$ relatively compact. Then there are compact sets $A \subset E$ and $B \subset F$ such that*
$$C \subset \overline{\Gamma(A \otimes B)}^{E \tilde{\otimes}_\pi F} .$$

The fact that $\tilde{E} = E \tilde{\otimes}_\pi \mathbb{K}$ gives another important special case:

COROLLARY 2: *Every relatively compact subset K in the completion of a normed space E is contained in the closed absolutely convex hull of a zero sequence in E. Moreover, for every $\varepsilon > 0$ the x_n can be chosen to be in $[(1+\varepsilon) \sup_{x \in K} \|x\|] B_E$.*

Since for every $x_n \to 0$ there is $\lambda_n \uparrow \infty$ such that $\lambda_n x_n \to 0$, it follows that for every compact subset K of a Banach space ($K \subset \overline{\Gamma\{x_n\}}$) there is a compact Banach disc L (take $L := \overline{\Gamma\{\lambda_n x_n\}}$) such that K is compact even in $[L]$.

Another consequence of the theorem is that every element $z \in E \tilde{\otimes}_\pi F$ has an absolutely convergent series expansion of the form $z = \sum_{n=1}^{\infty} x_n \otimes y_n$ with $x_n \in E$ and $y_n \in F$.

COROLLARY 3: *For $z \in E \tilde{\otimes}_\pi F$*
$$\pi(z; E, F) = \inf\left\{ \sum_{n=1}^{\infty} \|x_n\| \|y_n\| \ \Big| \ z = \sum_{n=1}^{\infty} x_n \otimes y_n \right\} .$$

PROOF: By the triangle inequality for π the right hand side defines a norm $\pi_\infty \geq \pi$ on $E \tilde{\otimes}_\pi F$ which, by 3.2., coincides with π on the π_∞-dense subspace $E \otimes F$ of $E \tilde{\otimes}_\pi F$; this implies the result — e.g. by Ex 3.4.. $\square$

An application: Every $\tilde{f} \in L_1(\mu, E)$ has an expansion
$$\tilde{f} = \sum_{n=1}^{\infty} \tilde{f}_n \otimes x_n \qquad \tilde{f}_n \in L_1(\mu), \quad x_n \in E$$

such that
$$\sum_{n=1}^{\infty} \int_{\Omega} |f_n(\omega)| \mu(d\omega) \|x_n\| \leq (1+\varepsilon) \int_{\Omega} \|f(\omega)\|_E \mu(d\omega) .$$

3.6. An operator $T : E \to F$ between Banach spaces is called *nuclear* if there are $x'_n \in E'$ and $y_n \in F$ such that $\sum_{n=1}^{\infty} \|x'_n\| \|y_n\| < \infty$ and for each $x \in E$

$$Tx = \sum_{n=1}^{\infty} \langle x'_n, x \rangle y_n .$$

Clearly, T is continuous in this case and

$$\mathbf{N}(T) := \inf \left\{ \sum_{n=1}^{\infty} \|x'_n\| \|y_n\| \mid \text{as above} \right\} \geq \|T\|$$

defines a norm on the vector space $\mathfrak{N}(E, F)$ of all nuclear operators; it is easy to see that $\mathfrak{N}(E, F)$ becomes a Banach space with this norm. Recall from 2.4.(5) the injective map

$$\begin{aligned} E' \otimes F &\longrightarrow \mathfrak{F}(E, F) \subset \mathfrak{N}(E, F) \subset \mathfrak{L}(E, F) \\ x' \otimes y &\rightsquigarrow x' \underline{\otimes} y \end{aligned}$$

with $(x' \underline{\otimes} y)(x) := \langle x', x \rangle y$. With this notation nuclear operators have an expansion

$$T = \sum_{n=1}^{\infty} x'_n \underline{\otimes} y_n$$

being absolutely convergent in $\mathfrak{N}(E, F)$ and hence in $\mathfrak{L}(E, F)$, which implies that they are compact. The definition of $\mathfrak{N}(E, F)$, its norm and corollary 3 give the

PROPOSITION: *The map*

$$\begin{aligned} J : E' \widetilde{\otimes}_\pi F &\longrightarrow \mathfrak{N}(E, F) \\ x' \otimes y &\rightsquigarrow x' \underline{\otimes} y \end{aligned}$$

is a metric surjection.

Note that the shorthand definition for J was used: first on elementary tensors, then finite sums, and finally extension to the completion by continuity; it is therefore clear that for $x \in E$

$$[J(z)](x) = \left[\sum_{n=1}^{\infty} x'_n \underline{\otimes} y_n \right](x) = \sum_{n=1}^{\infty} \langle x'_n, x \rangle y_n$$

for each representation $z = \sum_{n=1}^{\infty} x'_n \otimes y_n$. Unfortunately, it will turn out that J is not in general injective – and hence is not an isometry on $E' \otimes_\pi F$. This has to do with the approximation property and will be investigated carefully in section 5 (for **N** on $\mathfrak{F}(E,F) = E' \otimes F$ see Ex 5.16.). Clearly, if E or F is finite dimensional, then

$$E' \tilde{\otimes}_\pi F = E' \otimes_\pi F = \mathfrak{N}(E, F) \quad \text{(via } J)$$

and J is an isometry.

3.7. Since $|\langle x', x \rangle| \leq \|x'\| \|x\|$, the *trace* defined in 2.5.

$$tr_E : E' \otimes_\pi E \longrightarrow \mathbb{K}$$

has norm 1 (if $E \neq \{0\}$) and hence extends to the completion

$$tr_E \in (E' \tilde{\otimes}_\pi E)' \quad , \quad \|tr_E\| = 1$$

with (by continuity)

$$\langle tr_E, z \rangle = \sum_{n=1}^\infty \langle x'_n, x_n \rangle$$

for any convergent representation $z = \sum_{n=1}^\infty x'_n \otimes x_n$. If the mapping

$$J : E' \tilde{\otimes}_\pi E \longrightarrow \mathfrak{N}(E, E)$$

is injective, then the trace is defined for nuclear operators – but it will be seen in section 5 that this is not always possible. The trace is a very useful tool for many calculations. To see an example:

REMARK: $\mathbf{N}(id_E) = \dim E$ *for all finite dimensional Banach spaces E* .

PROOF: Take an Auerbach basis $(e_1, \cdots, e_n)$ of E with coefficient functionals e'_k , i.e.

$$\|e_k\| = \|e'_k\| = 1 \quad \text{and} \quad \langle e'_k, e_\ell \rangle = \delta_{k,\ell}$$

(see Appendix A4.). Then $id_E = \sum_{k=1}^n e'_k \underline{\otimes} e_k = J(\sum_{k=1}^n e'_k \otimes e_k)$, hence $\mathbf{N}(id_E) \leq \sum_{k=1}^n \|e'_k\| \|e_k\| = n$ and

$$n = \left\langle tr_E, \sum_{k=1}^n e'_k \otimes e_k \right\rangle \leq \pi\left(\sum_{k=1}^n e'_k \otimes e_k\right) = \mathbf{N}(id_E) ,$$

since J is an isometry. □

For many calculations the following result will be helpful:

PROPOSITION: *For $z \in E \tilde{\otimes}_\pi F$ and $\varphi \in (E \tilde{\otimes}_\pi F)' = \mathfrak{L}(E, F')$ the following formula holds:*

$$\langle \varphi, z \rangle = \langle tr_F, L_\varphi \tilde{\otimes}_\pi id_F(z) \rangle = \langle tr_E, id_E \tilde{\otimes}_\pi (L'_\varphi \circ \kappa_F)(z) \rangle .$$

PROOF: By linearity and continuity it is enough to check this for elementary tensors $z = x \otimes y$:

$$\langle \varphi, x \otimes y \rangle = \langle L_\varphi x, y \rangle = \langle tr_F, L_\varphi x \otimes y \rangle =$$
$$= \langle x, (L'_\varphi \circ \kappa_F)(y) \rangle = \langle tr_E, x \otimes (L'_\varphi \circ \kappa_F)(y) \rangle. \quad \square$$

In particular,

$$\pi(z; E, F) = \sup\{|\langle tr_F, S \tilde{\otimes}_\pi id_F(z) \rangle| \mid \|S\|_{\mathfrak{L}(E,F')} \leq 1\}$$

for all $z \in E \tilde{\otimes}_\pi F$.

3.8. For $T_i \in \mathfrak{L}(E_i, F_i)$ the operator $T_1 \tilde{\otimes}_\pi T_2 : E_1 \tilde{\otimes}_\pi E_2 \longrightarrow F_1 \tilde{\otimes}_\pi F_2$ is well–defined; is it a homomorphism if the T_i are?

PROPOSITION: *If $Q_i : E_i \stackrel{1}{\twoheadrightarrow} F_i$ are two metric surjections, then*

$$Q_1 \otimes Q_2 : E_1 \otimes_\pi E_2 \longrightarrow F_1 \otimes_\pi F_2$$

is a metric surjection. In particular, $Q_1 \tilde{\otimes}_\pi Q_2$ is a metric surjection as well.

PROOF: $Q_i(\overset{\circ}{B}_{E_i}) = \overset{\circ}{B}_{F_i}$ and the description 3.2.(2) of the open unit ball in the projective tensor product imply

$$Q_1 \otimes Q_2(\overset{\circ}{B}_{E_1 \otimes_\pi E_2}) = \overset{\circ}{B}_{F_1 \otimes_\pi F_2}$$

which is the claim. $\square$

This property of π is the reason for calling it *projective*.

3.9. Let $G \subset E$ be a subspace. If $G \otimes_\pi F$ were isomorphically a subspace of $E \otimes_\pi F$, the Hahn–Banach theorem would imply that every linear operator in $\mathfrak{L}(G, F') = (G \otimes_\pi F)'$ could be extended to $E \to F'$. There have been given various examples in 1.5. showing that this is *not* true; therefore: π *does not respect subspaces, not even isomorphically* ! Later on (5.8.) it will even be shown that $T_1 \tilde{\otimes}_\pi T_2$ may fail to be injective if the T_i are. But π respects complemented subspaces, dense subspaces and the embedding into the bidual:

PROPOSITION: *Let E and F be normed.*
(1) *If $G \subset E$ is complemented with a projection $P : E \to G$, then*

$$\pi(z; E, F) \leq \pi(z; G, F) \leq \|P\| \pi(z; E, F)$$

for all $z \in G \otimes F$ and $P \otimes id_F$ is a projection of $E \otimes_\pi F$ onto $G \otimes_\pi F$.

(2) If $G \subset E$ is dense, then $G \otimes_\pi F$ is an isometric dense subspace of $E \otimes_\pi F$ and, consequently, of $\tilde{E} \otimes_\pi \tilde{F}$.

(3) π respects the natural embedding into the bidual isometrically:

$$id_E \otimes \kappa_F : E \otimes_\pi F \xhookrightarrow{1} E \otimes_\pi F''.$$

PROOF: (1) follows from the mapping property and (2) from the unique extension of a continuous bilinear form to its completion (1.6.). To see (3), observe first that $\|id_E \otimes \kappa_F\| \leq 1$, hence
$$\pi(z; E, F'') \leq \pi(z; E, F)$$
for each $z \in E \otimes F$. Take a $\varphi \in (E \otimes_\pi F)'$ of norm ≤ 1 with
$$\langle \varphi, z \rangle = \pi(z; E, F)$$
and its canonical right-extension $\varphi^\wedge \in (E \otimes_\pi F'')'$ (see 1.9.). Then
$$\pi(z; E, F) = \langle \varphi, z \rangle = \langle \varphi^\wedge, z \rangle \leq \|\varphi^\wedge\| \pi(z; E, F'') \leq \pi(z; E, F''). \quad \square$$

The result (3) also follows from Ex 3.19. on weak retractions. The following example (taken from Jameson's book [119]) shows in a simple way that π does not respect subspaces isometrically: Take for $n \geq 2$

$$G := \left\{ (\xi_i) \in \ell_1^n \mid \sum_{i=1}^n \xi_i = 0 \right\},$$

and consider $G' \otimes_\pi G = \mathfrak{N}(G, G) \hookrightarrow \mathfrak{N}(G, \ell_1^n) = G' \otimes_\pi \ell_1^n$ and z_o corresponding to id_G. Then, by remark 3.7.,
$$\pi(z_o; G', G) = n - 1 \ .$$

$z_o \in G' \otimes_\pi \ell_1^n$ corresponds to the embedding $G \hookrightarrow \ell_1^n$, hence $z_o = \sum_{k=1}^n e_k' \otimes e_k$ for the canonical basis. It is easily checked that $\|e_k'\|_{G'} = 1/2$ and therefore
$$\pi(z_o; G', \ell_1^n) = n/2 \ ,$$
since $\ell_1^n \otimes_\pi G' = \ell_1^n(G')$ holds isometrically by 3.3..

3.10. The following characterization holds:

PROPOSITION: *For every Banach space E the following assertions are equivalent:*

(a) $E \otimes_\pi \cdot$ *respects subspaces isomorphically*.

(b) *There is a $\lambda \geq 1$ such that: Whenever $G \subset F$ is a subspace, then*
$$\pi(z; E, G) \leq \lambda \pi(z; E, F) \quad \text{for all } z \in E \otimes G \ .$$
(c) *E' has the λ-extension property for some $\lambda \geq 1$.*
(d) *E' is an injective Banach space.*
Moreover, the constants λ in (b) and (c) can be taken to be the same.

PROOF: (a) $\leftrightarrow$ (d) and (b) $\leftrightarrow$ (c) follow from the Hahn–Banach theorem applied to
$$\mathcal{L}(F, E') = (F \otimes_\pi E)' \xrightarrow{\text{restriction}} (G \otimes_\pi E)' = \mathcal{L}(G, E') \ .$$
(d) $\leftrightarrow$ (c) was shown by the reader in Ex 1.7.. □

If $E = L_1(\mu)$, then for every $\tilde{f} \in L_1(\mu) \otimes G$
$$\pi(\tilde{f}; L_1, G) = \int_\Omega \|f(\omega)\| \mu(d\omega) = \pi(\tilde{f}; L_1, F) \ ,$$
hence $L_1(\mu) \otimes_\pi \cdot$ respects subspaces isometrically. For a localizable measure (i.e. $L_\infty(\mu)$ is order complete, see Appendix B6.) $L_\infty(\mu) = (L_1(\mu))'$ holds, hence:

COROLLARY: *If μ is localizable, then $L_\infty(\mu)$ has the metric extension property.*

A theorem of Nachbin–Goodner–Hasumi–Kelley states that the spaces with the metric extension property are exactly the $C(K)$ where K is compact and extremally disconnected (=Stonean), see Day [38], p.123 or Lacey [164], p.92 . This fits with the Gelfand representation of the C^*-algebra L_∞ .

Moreover, due to a theorem of Kakutani (see Appendix A7.), abstract L-spaces L are isometric to some $L_1(\mu)$ (where μ is a Borel–Radon measure on a locally compact space and, in particular, is strictly localizable). This implies that:

(1) $L \otimes_\pi \cdot$ *respects subspaces isometrically.*
(2) L' *has the metric extension property.*

The duality between abstract M- and L-spaces has various consequences; first observe that every $C(K)$ is an abstract M-space, hence $C(K)''$ is the dual of an abstract L :

(3) $C(K)''$ *has the metric extension property for each compact space K; in particular,*
(4) *For $G \subset F$ every operator $T \in \mathcal{L}(G, C(K))$ can be extended to some operator $\tilde{T} \in \mathcal{L}(F, C(K)'')$ with the same norm $\|\tilde{T}\| = \|T\|$ — but there is in general no extension $\tilde{T} : F \to C(K)$.*

The last statement follows from the fact that $c_o = C(\mathbb{N} \cup \{\infty\})$ is not complemented in ℓ_∞. So, in general, $C(K)$ is not an injective Banach space — but it will be seen in 4.5. that compact operators with values in $C(K)$ can be extended.

3.11. The following characterization is due to Grothendieck [91].

THEOREM: *Let E be a Banach space. Then $E \otimes_\pi \cdot$ respects subspaces isometrically if and only if E is isometric to some $L_1(\mu)$.*

PROOF: If $E \otimes_\pi \cdot$ respects subspaces isometrically (the other direction was shown before), then, by the last proposition, E' has the metric extension property and therefore $E' = C(K)$ by the theorem of Nachbin et al.. Grothendieck showed (see Appendix A7.) that this forces E to be isometric to some $L_1(\mu)$. □

3.12. Dual to the extension problem is the *lifting problem:* Given a surjective mapping $Q : F \to G$ between Banach spaces and $T \in \mathcal{L}(E, G)$. Is there a $\tilde{T} \in \mathcal{L}(E, F)$ with $T = Q \circ \tilde{T}$?

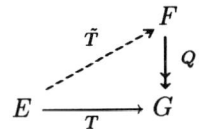

$$\begin{array}{ccc} & & F \\ & \tilde{T} \nearrow & \downarrow Q \\ E & \xrightarrow{T} & G \end{array}$$

It is easy to see (lift the Te_γ) that for $E = \ell_1(\Gamma)$ and $\varepsilon > 0$ there is always a lifting $\tilde{T}$ with $\|\tilde{T}\| \leq (1 + \varepsilon)\|T\|$ — but, in general, the same norm cannot be achieved (Ex 3.21.). Conversely, a result of Köthe [151] (with the aid of Ex 3.22.) implies that each Banach space E such that all T's are liftable (with arbitrary norm) is isomorphic to some $\ell_1(\Gamma)$.

PROPOSITION: *Let E be a Banach space such that $E \otimes_\pi \cdot$ respects subspaces with a constant λ (as in proposition 3.10.(b)). If $Q : F \to G$ is a metric surjection between normed spaces, then for every $T \in \mathcal{L}(E, G)$ there is an operator $\tilde{T} \in \mathcal{L}(E, F'')$ such that $\kappa_G \circ T = Q'' \circ \tilde{T}$ and $\|\tilde{T}\| \leq \lambda \|T\|$:*

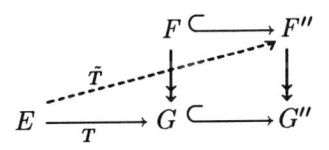

$$\begin{array}{ccc} & F \hookrightarrow & F'' \\ \tilde{T} \nearrow & \downarrow & \downarrow \\ E \xrightarrow{T} & G \hookrightarrow & G'' \end{array}$$

PROOF: Since

$$T \in \mathcal{L}(E, G) \subset (E \otimes_\pi G')' \xleftarrow{(id_E \otimes Q')'} (E \otimes_\pi F')' = \mathcal{L}(E, F''),$$

the Hahn–Banach theorem shows that there is a $\tilde{T} \in \mathfrak{L}(E, F'')$ with $\|\tilde{T}\| \leq \lambda \|T\|$ and $T = (id_E \otimes Q')'(\tilde{T})$. It is easy to check that the latter means that $\kappa_G \circ T = Q'' \circ \tilde{T}$. □

COROLLARY 1: *Let E be a Banach space such that $E \otimes_\pi \cdot$ respects subspaces with a constant λ and F, G normed spaces. If for $Q \in \mathfrak{L}(F, G)$ and $T \in \mathfrak{L}(E, G)$ there is a bounded and absolutely convex subset $C \subset F$ with $Q(C) \supset T(B_E)$, then there is a "lifting" $\tilde{T} \in \mathfrak{L}(E, F'')$ with $\kappa_G \circ T = Q'' \circ \tilde{T}$ and $\tilde{T}(B_E) \subset \lambda C^{00}$.*

PROOF: Obviously, the mapping $[C] \to [Q(C)]$ is a metric surjection; moreover, with $I : [C] \hookrightarrow F$, it is clear that $I''(B_{[C]}) \subset C^{00} \subset F''$. So it is enough to apply the proposition. □

(See also Ex 3.23..) Clearly, any space $L_1(\mu)$ satisfies the assumption with $\lambda = 1$:

COROLLARY 2: *Let $Q : F \twoheadrightarrow G$ be a metric surjection between Banach spaces.*
(1) *For every measure μ and every $T \in \mathfrak{L}(L_1(\mu), G)$ there is a $\tilde{T} \in \mathfrak{L}(L_1(\mu), F'')$ with $\|\tilde{T}\| = \|T\|$ and $\kappa_G \circ T = Q'' \circ \tilde{T}$.*
(2) *For every measure μ, every compact operator $T \in \mathfrak{L}(L_1(\mu), G)$ and every $\varepsilon > 0$ there is a compact lifting $\tilde{T} \in \mathfrak{L}(L_1(\mu), F)$ with $\|\tilde{T}\| \leq (1+\varepsilon)\|T\|$.*

PROOF: (1) is clear. For (2) observe that $T(B_{L_1}) \subset \overline{\Gamma\{y_n\}}$ with (y_n) a zero sequence in $(1+\varepsilon)\|T\|B_G$ (see corollary 2 in 3.5.). For $x_n \in F$ with $\|x_n\| \leq (1+\varepsilon)\|y_n\|$ and $Qx_n = y_n$ take the compact set $C := \overline{\Gamma\{x_n\}}$, which clearly satisfies $Q(C) \supset T(B_{L_1})$. Since $C^{00} = C \subset (1+\varepsilon)^2\|T\|B_F$ holds, corollary 1 implies that $\operatorname{im}(\tilde{T}) \subset F$ and $\|\tilde{T}\| \leq (1+\varepsilon)^2\|T\|$. □

See also Ex 3.31..

3.13. The metric results on extensions and liftings connected with the spaces $C(K)$, L_∞ and L_1 have natural isomorphic counterparts for spaces which "look locally like" ℓ_∞^n or ℓ_1^n.

DEFINITION: *For $1 \leq p \leq \infty$ and $1 \leq \lambda < \infty$ a normed space E is called an $\mathfrak{L}_{p,\lambda}^g$-space, if for every finite dimensional subspace $M \subset E$ and $\varepsilon > 0$ there are $R \in \mathfrak{L}(M, \ell_p^m)$ and $S \in \mathfrak{L}(\ell_p^m, E)$ for some $m \in \mathbb{N}$ such that $SRx = x$ for all $x \in M$ and $\|S\|\|R\| \leq \lambda + \varepsilon$:*

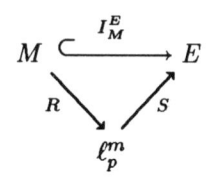

$$\begin{array}{ccc} M & \xrightarrow{I_M^E} & E \\ & \searrow{\scriptstyle R} \quad \nearrow{\scriptstyle S} & \\ & \ell_p^m & \end{array}$$

E is called an $\mathcal{L}_p^g$-space if it is an $\mathcal{L}_{p,\lambda}^g$ for some λ. In section 4 (Ex 4.7.) it will be shown that each $L_p(\mu)$ is an $\mathcal{L}_{p,1}^g$-space. The $\mathcal{L}_p^g$-spaces are very closely related to the $\mathcal{L}_p$-spaces of Lindenstrauss–Pełczyński [173] ; they will be studied to some extent in Chapter II, section 23 (where the exact relationship between $\mathcal{L}_p$– and $\mathcal{L}_p^g$-spaces will also be clarified). A starter for the "local techniques" (i.e. the method of investigating a Banach space in terms of its finite dimensional subspaces) is the following

PROPOSITION: *If E is $\mathcal{L}_{1,\lambda}^g$-space and $G \subset F$ is a subspace, then*

$$\pi(z; E, F) \leq \pi(z; E, G) \leq \lambda \pi(z; E, F)$$

for all $z \in E \otimes G$; in other words, $E \otimes_\pi \cdot$ respects subspaces up to the constant λ.

PROOF: Take $z \in E \otimes G$ and $\varepsilon > 0$. Since π is finitely generated (proposition 3.2.(5) and Ex 3.30.), there is a finite dimensional subspace $M \subset E$ with $z \in M \otimes G$ and

$$\pi(z; M, F) \leq (1 + \varepsilon)\pi(z; E, F) \quad .$$

For a factorization $I_M^E = SR$ of the embedding $I_M^E : M \hookrightarrow E$ through some ℓ_1^m with $\|S\| \|R\| \leq \lambda + \varepsilon$ the mapping property gives

$$\pi(z; E, G) = \pi(SR \otimes id_G(z); E, G) \leq \|S\|\pi(R \otimes id_G(z); \ell_1^m, G) \stackrel{1}{=}$$

$$= \|S\|\pi(R \otimes id_F(z); \ell_1^m, F) \leq \|S\| \|R\|\pi(z; M, F) \leq$$

$$\leq (\lambda + \varepsilon)\pi(z; M, F) \leq (\lambda + \varepsilon)(1 + \varepsilon)\pi(z; E, F)$$

since $\ell_1^m \otimes_\pi \cdot$ respects subspaces isometrically. □

It follows that if E is a $\mathcal{L}_{1,\lambda}^g$-space its dual E' has the λ–extension property (3.10.) and E has the lifting property expressed in 3.12. . Grothendieck's result 3.11. (and Ex 4.7.) says that the $\mathcal{L}_{1,1}^g$-spaces are exactly the $L_1(\mu)$-spaces. It will be shown in section 23 that the $\mathcal{L}_1^g$-spaces E are those such that $E \otimes_\pi \cdot$ respects subspaces isomorphically.

Exercises:

Ex 3.1. Show that $\pi(\cdot ; E, F)$ is the Minkowski gauge functional of $\Gamma(B_E \otimes B_F)$.
Ex 3.2. Show that $\|T_1 \otimes_\pi T_2\| = \|T_1\| \|T_2\|$.
Ex 3.3. Prove directly, by using the infimum–description of the π–norm, that

$$E_o \otimes_\pi F_o \stackrel{1}{\hookrightarrow} E \otimes_\pi F$$

is a dense isometric injection if $E_o \subset E$ and $F_o \subset F$ are dense.

Ex 3.4. Let G be a vector space with two norms $\|\cdot\|_1$ and $\|\cdot\|_2$ which coincide on a $\|\cdot\|_1$-dense subspace $G_o \subset G$.

(a) The norms are equal on G under each of the following three conditions:
   (1) $\|\cdot\|_2 \leq \lambda \|\cdot\|_1$ for some λ.
   (2) There is a norm $\|\cdot\|_3$ on G with $\|\cdot\|_i \leq \lambda_i \|\cdot\|_3$ for some λ_i and G_o is $\|\cdot\|_3$-dense.
   (3) $(G, \|\cdot\|_2)$ is a Banach space and there is a separated topological space X and an injective $\Phi : G \to X$ which is continuous for both norms.

(b) Give an example of two non-equivalent Banach space norms which coincide on a subspace which is dense for both norms. Hint: $\|x\|_2 := \|Sx\|_1$ defines a norm for each injective $S \in L(G,G)$ — not necessarily continuous.

Ex 3.5. A subset $K \subset E \tilde{\otimes}_\pi F$ is compact if and only if there are a compact $H \subset \ell_1$, zero sequences (x_n) in E and (y_n) in F such that

$$K = \left\{ \sum_{n=1}^{\infty} \lambda_n x_n \otimes y_n \mid (\lambda_n) \in H \right\} .$$

If $K \subset E \tilde{\otimes}_\pi F$ is absolutely convex or convex, then the set $H \subset \ell_1$ can be chosen to be of the same type.

Ex 3.6. Let $T \in \mathfrak{L}(L_1(\mu), L_1(\nu))$ and $S \in \mathfrak{L}(E, F)$. Then there is a unique continuous operator $U \in \mathfrak{L}(L_1(\mu, E), L_1(\nu, F))$ with

$$U(\tilde{f} \otimes x) = T\tilde{f} \otimes Sx .$$

Ex 3.7. Let $T \in \mathfrak{L}(L_1(\mu), L_1(\nu))$ and $(\tilde{f}_n)$ a sequence in $L_1(\mu)$. Then

$$\int \left(\sum_{n=1}^{\infty} |(T\tilde{f}_n)(s)|^p \right)^{1/p} \nu(ds) \leq \|T\| \int \left(\sum_{n=1}^{\infty} |f_n(t)|^p \right)^{1/p} \mu(dt)$$

for all $1 \leq p < \infty$. Hint: Consider $T \otimes id_{\ell_p}$.

Ex 3.8. If $\tilde{g} \in L_1(\mu, F)$ and $\tilde{f} \in L_\infty(\mu)$, then $\tilde{f}\tilde{g}$ is Bochner integrable. Show that

$$T : L_\infty(\mu) \longrightarrow F$$
$$\tilde{f} \rightsquigarrow \int fg d\mu$$

is nuclear and $\mathbf{N}(T) \leq \int \|g\| d\mu$. Actually, there is equality of norms; see Appendix C3..

Ex 3.9. Show that there is a metric surjection $\ell_2 \tilde{\otimes}_\pi \ell_2 \to \ell_1$; in particular, $\ell_2 \tilde{\otimes}_\pi \ell_2$ is not reflexive.

Ex 3.10. *Closed graph theorem:* Let E, F and G be Banach spaces. Show that every linear operator $T : E \otimes_\pi F \to G$ with closed graph is continuous. Hint: Ex 1.11..

Ex 3.11. Does $\ell_2^2 \otimes_\pi \ell_2^2 = \ell_2^4$ hold isometrically?

Ex 3.12. If $G \subset E$ has codimension n, then

$$\pi(z; G, F) \leq (n+1)\pi(z; E, F)$$

for all $z \in G \otimes F$. Can this inequality be improved in general?

Ex 3.13. (a) Show that $\ell_1 \tilde\otimes_\pi F \to \mathcal{L}(c_o, F)$ is injective (F a Banach space).
(b) If $I : G \hookrightarrow F$ is a metric injection, then: $T \in \mathfrak{N}(c_o, G)$ if and only if $I \circ T \in \mathfrak{N}(c_o, F)$; in this case, $\mathbf{N}(T) = \mathbf{N}(I \circ T) = \sum_{n=1}^\infty \|Te_n\|$.

Ex 3.14. $\mathbf{N}(T) \leq \min\{\dim E, \dim F\} \|T\|$ for all $T \in \mathcal{L}(E, F)$.

Ex 3.15. Let $E_{\mathbb{C}}$ and $F_{\mathbb{C}}$ be complex Banach spaces and $E_{\mathbb{R}}$ and $F_{\mathbb{R}}$ the same spaces considered as real spaces.
(a) For every $\mathbb{C}$-linear $T \in \mathcal{L}(E, F)$

$$\mathbf{N}_{\mathbb{C}}(T : E_{\mathbb{C}} \to F_{\mathbb{C}}) \leq \mathbf{N}_{\mathbb{R}}(T : E_{\mathbb{R}} \to F_{\mathbb{R}}) \leq 2\mathbf{N}_{\mathbb{C}}(T : E_{\mathbb{C}} \to F_{\mathbb{C}})$$

holds. Is it true that $\mathbf{N}_{\mathbb{C}}(T) = \mathbf{N}_{\mathbb{R}}(T)$? Hint: 3.7..
(b) Denote by $\otimes^{\mathbb{K}}$ the tensor product over $\mathbb{K}$. Show that the natural map

$$E_{\mathbb{R}} \otimes_\pi^{\mathbb{R}} F_{\mathbb{R}} \longrightarrow (E_{\mathbb{C}} \otimes_\pi^{\mathbb{C}} F_{\mathbb{C}})_{\mathbb{R}}$$

from Ex 2.11. is a metric surjection.

Ex 3.16. For $U \in \mathcal{L}(E_o, E)$ and $S \in \mathcal{L}(F, F_o)$ (all spaces are complete) the diagram

$$\begin{array}{ccc} E' \tilde\otimes_\pi F & \xrightarrow{1} & \mathfrak{N}(E, F) \ni T \\ U' \tilde\otimes_\pi S \downarrow & & \downarrow \quad \} \\ E'_o \tilde\otimes_\pi F_o & \xrightarrow{1} & \mathfrak{N}(E_o, F_o) \ni STU \end{array}$$

commutes. Deduce $\mathbf{N}(STU) \leq \|S\|\mathbf{N}(T)\|U\|$. Clearly, this also follows directly from the definition of nuclear operators.

Ex 3.17. If G is a closed subspace of a Banach space E with $\mathfrak{N}(F, G) = F' \tilde\otimes_\pi G$ for some F and $F' \otimes_\pi G \hookrightarrow F' \otimes_\pi E$ is not an isomorphic embedding, then there is a non-nuclear $T \in \mathcal{L}(\ell_p(F), \ell_p(G))$ which is nuclear as an operator $\ell_p(F) \to \ell_p(E)$. Hint: The operator $F \to G$ associated with $z \in F' \otimes G$ has a norm $\leq \pi(z; F', E)$.

Ex 3.18. Let E and F be Banach spaces such that $F' \subset E$ isometrically.
(a) Then there is a projection $P : E \twoheadrightarrow F'$ with norm $\leq \lambda$ if and only if $\pi(z; F', F) \leq \lambda \pi(z; E, F)$ for all $z \in F' \otimes F$.

(b) If $\rho(F', E) := \inf\{ \|P\| \mid P : E \to F' \text{ projection} \} < \infty$, then there is a projection with norm $\rho(F', E)$.

Ex 3.19. For a metric injection $I : G \hookrightarrow E$ of normed spaces the following statements are equivalent:

(a) $I \otimes id_F : G \otimes_\pi F \to E \otimes_\pi F$ is an isomorphic [resp. metric] injection for every normed space F.

(b) $I \otimes id_{G'} : G \otimes_\pi G' \to E \otimes_\pi G'$ is an isomorphic [resp. metric] injection.

(c) There is $U \in \mathfrak{L}(E, G''')$ with $U \circ I = \kappa_G$ [and $\|U\| \leq 1$].

(d) There is $S \in \mathfrak{L}(G', E')$ with $I' \circ S = id_{G'}$ [and $\|S\| \leq 1$].

Hint: Ex 1.6.. In this case, properties (c) and (d) justify calling I a *weak retraction*.

Ex 3.20. Let M be an abstract M-space and $G \subset F$. Then every operator $T : G \to M$ can be extended to an operator $\tilde{T} : E \to M''$ with $\|\tilde{T}\| = \|T\|$.

Ex 3.21. Let H be a closed subspace of a Banach space F and $Q : E \to F/H$ the metric surjection.

(a) If $T \in \mathfrak{L}(\ell_1(\Gamma), F/H)$ and $\varepsilon > 0$, then there is a "lifting" $\tilde{T} \in \mathfrak{L}(\ell_1(\Gamma), F)$ of T (this means: $T = Q \circ \tilde{T}$) with $\|\tilde{T}\| \leq (1 + \varepsilon)\|T\|$.

(b) Every $T \in \mathfrak{L}(\ell_1, F/H)$ is liftable with the same norm if and only if H is proximinal in F (i.e. every element in F has a best-approximation in H).

(c) If F is a dual space and H weak-$*$-closed, then every $T \in \mathfrak{L}(\ell_1(\Gamma), F/H)$ is liftable to F with the same norm.

Ex 3.22. Assume that the canonical surjection $\ell_1(B_E) \twoheadrightarrow E$ has a right inverse $S \in \mathfrak{L}(E, \ell_1(B_E))$. Then for every metric surjection $Q : F \twoheadrightarrow G$ and $T \in \mathfrak{L}(E, G)$ there is a lifting $\tilde{T} \in \mathfrak{L}(E, F)$ with $\|\tilde{T}\| \leq (1 + \varepsilon)\|S\| \|T\|$.

Ex 3.23. Show that in 3.12., corollary 1, the assumption $Q(C) \supset T(B_E)$ may be replaced by $\overline{Q(C)} \supset T(B_E)$. Hint: Use $E \otimes_\pi (F'/\ker m_{C^\circ})$.

Ex 3.24. *Radon–Nikodým theorem for operator valued measures:* (a) Let μ be a strictly localizable measure, E and F normed spaces and $T : L_1(\Omega, \mu) \to \mathfrak{L}(E, F)$ continuous and linear. Use the weak version of the Dunford–Pettis-theorem 3.3. to show that there is a $\sigma(\mathfrak{L}(E, F''), E \tilde{\otimes}_\pi F')$-density $D : \Omega \to \mathfrak{L}(E, F'')$ such that

$$T\tilde{f} = \int_\Omega f(\omega) D(\omega) \mu(d\omega) \qquad \text{for all } \tilde{f} \in L_1(\mu, \Omega).$$

(as a weak-$*$-Pettis integral in $(E \tilde{\otimes}_\pi F')' = \mathfrak{L}(E, F'')$).

(b) Let $\mathcal{A}$ be a σ-algebra of subsets of Ω and μ a (non–negative) measure on $\mathcal{A}$. If E and F are Banach spaces and $M : \mathcal{A} \to \mathfrak{L}(E, F)$ is σ-additive, then there is an operator $T \in \mathfrak{L}(L_1(\mu), \mathfrak{L}(E, F))$ with $T\check{\chi}_A = M(A)$ if and only if for each $x \in E$ and $y' \in F'$ the signed measure $\langle M(\cdot)x, y' \rangle$ is μ-absolutely continuous with density in $\mathcal{L}_\infty(\mu)$.

Ex 3.25. A special case of the foregoing Radon–Nikodým theorem is $E = \mathbb{K}$. Show for the Fourier transform

$$\mathfrak{F} : L_1[0, 1] \longrightarrow c_o ; \quad \mathfrak{F}\tilde{f} := \left(\int_0^1 f(t) \exp(int) \, dt \right)_{n \in \mathbb{N}}$$

that the values of the density cannot be chosen in $F = c_o$ (see also Appendix C).

Ex 3.26. Every space $L_\infty(\mu)$ has the *compact extension property*: If $F \subset E$ and $T \in \mathfrak{K}(F, L_\infty(\mu))$ (the space of compact operators), then there is, for every $\varepsilon > 0$, an extension $\tilde{T} \in \mathfrak{K}(E, L_\infty(\mu))$ with $\|\tilde{T}\| \leq (1+\varepsilon)\|T\|$. Hint: Dualize 3.12., corollary 2.

Ex 3.27. Show that for arbitrary measure spaces (Ω_i, μ_i)
$$L_1(\mu_1 \otimes \mu_2) \stackrel{1}{=} L_1(\mu_1, L_1(\mu_2)) \stackrel{1}{=} L_1(\mu_1) \tilde{\otimes}_\pi L_1(\mu_2) \ .$$
Hint: $L_1(\mu_1) \otimes_\pi L_1(\mu_2) \stackrel{1}{\hookrightarrow} L_1(\mu_1 \otimes \mu_2)$ by Fubini-Tonelli.

Ex 3.28. (a) Let (e_n) be an orthonormal system in ℓ_2 and $(\lambda_n) \in \ell_1$ with non-negative λ_n. Using the trace, show that $\pi(\sum_{n=1}^\infty \lambda_n e_n \otimes e_n; \ell_2, \ell_2) = \sum_{n=1}^\infty \lambda_n$.
(b) If (e_n) and (f_n) are two orthonormal systems and $(\lambda_n) \in \ell_1$, then
$$\pi\left(\sum_{n=1}^\infty \lambda_n e_n \otimes f_n; \ell_2, \ell_2\right) = \sum_{n=1}^\infty |\lambda_n| \ .$$
Hint: Use $Ue_n := f_n$. Recall the *spectral decomposition theorem*: Every compact operator $\ell_2 \to \ell_2$ has the form $\sum \lambda_n e_n \otimes f_n$ with $(\lambda_n) \in c_o$ and (e_n) and (f_n) orthonormal systems.
(c) Take $E = \ell_p$ (for $1 \leq p \leq \infty$) or c_o and $T := \sum_{n=1}^\infty \lambda_n e_n \otimes e_n \in \mathcal{L}(E)$ for $(\lambda_n) \in c_o$. Then T is nuclear if and only if $(\lambda_n) \in \ell_1$. Hint: Use $|tr(\ell_p^n \hookrightarrow \ell_p \xrightarrow{T} \ell_p \twoheadrightarrow \ell_p^n)| \leq \mathbf{N}(T)$ and adjust the signs.

Ex 3.29. If $G \subset F$ is a subspace and $E \otimes_\pi F$ does not induce an equivalent norm on $E \otimes_\pi G$, then there are $M_n \in \mathrm{FIN}(E), N_n \in \mathrm{FIN}(F)$ and $z_n \in M_n \otimes (N_n \cap G)$ with
$$\pi(z_n; M_n, N_n \cap G) > n\pi(z_n; M_n, N_n) \ .$$
Hint: Use the fact that π is finitely generated.

Ex 3.30. $\pi(z; E, F) = \inf\{\pi(z; M, F) \mid z \in M \otimes F, M \subset E \text{ finite dimensional}\}$.

Ex 3.31. If E is an $\mathcal{L}_{1,\lambda}^g$-space and $Q : F \twoheadrightarrow G$ a metric surjection, then there is for every compact operator $T \in \mathcal{L}(E, G)$ a compact operator $\tilde{T} \in \mathcal{L}(E, F)$ with $Q\tilde{T} = T$ and $\|\tilde{T}\| \leq (\lambda + \varepsilon)\|T\|$. Hint: As in corollary 2 in 3.12. use a lifting of compact sets.

Ex 3.32. If $G \subset E$ is a subspace and $T \in \mathfrak{N}(G, F)$, then for every $\varepsilon > 0$ there is an extension $\tilde{T} \in \mathfrak{N}(E, F)$ of T with $\mathbf{N}(\tilde{T}) \leq (1+\varepsilon)\mathbf{N}(T)$.

Ex 3.33. If $F \subset E$ is a complemented subspace (with projection P) and E an $\mathcal{L}_{p,\lambda}^g$-space, then F is an $\mathcal{L}_{p,\lambda\|P\|}^g$-space.

Ex 3.34. Let E and F be Banach spaces, E complemented in its bidual E'' with projection P. If $T \in \mathfrak{N}(E', F)$ is $\sigma(E', E)$-$\sigma(F, F')$-continuous, then for every $\varepsilon > 0$ there are $x_n \in E$ and $y_n \in F$ such that
$$T = \sum_{n=1}^\infty x_n \otimes y_n \quad \text{and} \quad \sum_{n=1}^\infty \|x_n\| \|y_n\| \leq \|P\|\mathbf{N}(T) + \varepsilon \ .$$
Hint: Look at $T' : F' \to E \hookrightarrow E'' \xrightarrow{P} E$.

4. The Injective Norm

The tensor product $E \otimes F$ of two normed spaces is canonically embedded into the normed space $\mathfrak{Bil}(E', F') = (E' \otimes_\pi F')'$ of continuous bilinear forms on $E' \times F'$. This space induces another natural norm on $E \otimes F$, the injective norm ε, which is the norm of uniform convergence on the cartesian product $B_{E'} \times B_{F'}$ of dual unit balls.

4.1. Let E and F be normed spaces. Then for $x' \in B_{E'}$ and $y' \in B_{F'}$ the linear form $x' \otimes y' : E \otimes F \to \mathbb{K} \otimes \mathbb{K} = \mathbb{K}$ is defined. For $z \in E \otimes F$

$$\varepsilon(z; E, F) := \sup\{|\langle x' \otimes y', z\rangle| \mid x' \in B_{E'}, y' \in B_{F'}\}$$

gives exactly the norm which is induced on $E \otimes F$ by the embedding into $\mathfrak{Bil}(E', F') = (E' \otimes_\pi F')'$ since $\overset{\circ}{B}_{E' \otimes_\pi F'} = \Gamma(\overset{\circ}{B}_{E'} \otimes \overset{\circ}{B}_{F'})$. Notation: $E \otimes_\varepsilon F$ and for the completion $E \tilde{\otimes}_\varepsilon F$.

PROPOSITION: *Let E and F be normed spaces. Then*

(1) $$E \otimes_\varepsilon F \overset{1}{\hookrightarrow} (E' \otimes_\pi F')'$$

— *and even* $E \otimes_\varepsilon F \overset{1}{=} (E' \otimes_\pi F')'$ *if both spaces are finite dimensional.*

(2) *For norming subsets $A \subset B_{E'}$ and $B \subset B_{F'}$*

$$\varepsilon(z; E, F) = \sup\{|\langle x' \otimes y', z\rangle| \mid x' \in A, y' \in B\}.$$

(3) $\varepsilon(x \otimes y; E, F) = \|x\| \|y\|$ *for all $x \in E$ and $y \in F$.*
(4) $\varepsilon \leq \pi$ *on $E \otimes F$.*
(5) ε *satisfies the metric mapping property: If $T_i \in \mathcal{L}(E_i, F_i)$, then*

$$\|T_1 \otimes T_2 : E_1 \otimes_\varepsilon E_2 \longrightarrow F_1 \otimes_\varepsilon F_2\| \leq \|T_1\| \|T_2\|$$

— *and even* $= \|T_1\| \|T_2\|$.

Recall that a subset $A \subset B_{E'}$ is called *norming* if $\|x\|_E = \sup\{|\langle x', x\rangle| \mid x' \in A\}$ for all $x \in E$. The set $\text{ext}B_{E'}$ of the extreme points is norming since every linear weak-$*$-continuous functional attains its supremum on the weak-$*$-compact unit ball $B_{E'}$ at an extreme point.

4. The Injective Norm

PROOF: (1) was the definition; for (2) take any representation of z and calculate:

$$\varepsilon(z; E, F) = \sup\{|\langle x' \otimes y', \sum_{n=1}^{m} x_n \otimes y_n\rangle| \mid x' \in B_{E'}, y' \in B_{F'}\} =$$

$$= \sup_{x' \in B_{E'}} \sup_{y' \in B_{F'}} |\langle y', \sum_{n=1}^{m} \langle x', x_n\rangle y_n\rangle| =$$

$$= \sup_{x' \in B_{E'}} \left\|\sum_{n=1}^{m} \langle x', x_n\rangle y_n\right\|_F = \sup_{x' \in B_{E'}} \sup_{y' \in B} |\langle x' \otimes y', \sum_{n=1}^{m} x_n \otimes y_n\rangle| =$$

$$= \sup_{y' \in B} \left\|\sum_{n=1}^{m} \langle y', y_n\rangle x_n\right\|_E = \sup_{y' \in B} \sup_{x' \in A} |\langle x' \otimes y', z\rangle| \quad .$$

(3) is clear from the definition, (3) implies (4) and (5) follows from

$$\langle x'_1 \otimes x'_2, (T_1 \otimes T_2)(x_1 \otimes x_2)\rangle = \langle T'_1 x'_1 \otimes T'_2 x'_2, x_1 \otimes x_2\rangle \quad . \quad \square$$

Since $B_E \subset B_{E''}$ is norming, it follows that

$$\varepsilon(z; E', F') = \sup\{|\langle x \otimes y, z\rangle| \mid x \in B_E, y \in B_F\} =$$
$$= \sup\{|\langle w, z\rangle| \mid w \in \Gamma(B_E \otimes B_F)\} =$$
$$= \sup\{|\langle w, z\rangle| \mid \pi(w, E, F) \leq 1\} \quad ,$$

in other words,

$$E' \otimes_\varepsilon F' \xhookrightarrow{1} (E \otimes_\pi F)' \stackrel{1}{=} \mathfrak{L}(E, F') \quad .$$

This shows that even though π and ε coincide for elementary tensors they are far from being equal : ε is connected with the operator norm, π with the nuclear norm of operators! For explicit examples see Ex 4.2. and Ex 4.3..

4.2. Examples:

(1) Since $E \otimes F \hookrightarrow \mathfrak{F}(E', F) \xhookrightarrow{1} \mathfrak{L}(E', F) \xhookrightarrow{1} (E' \otimes_\pi F')'$, it follows that

$$E \otimes_\varepsilon F \xhookrightarrow{1} \mathfrak{L}(E', F)$$

— and similarly

$$E' \otimes_\varepsilon F = \mathfrak{F}(E, F) \xhookrightarrow{1} \mathfrak{L}(E, F) \quad ,$$

and therefore (for Banach spaces E and F)

$$E' \tilde{\otimes}_\varepsilon F \stackrel{1}{=} \overline{\mathfrak{F}}(E, F) \quad ,$$

the space of all *approximable operators* $T \in \mathfrak{L}(E, F)$, i.e. those for which there exists a sequence (T_n) of finite rank operators with $\|T - T_n\| \to 0$; the space $\overline{\mathfrak{F}}(E, F)$ is equipped with the operator norm.

An element $z \in E \otimes F$ gives a $\sigma(E', E)$-$\sigma(F, F')$-continuous operator in $\mathfrak{L}(E', \tilde{F})$. It follows that
$$E \tilde{\otimes}_\varepsilon F \xrightarrow{\;1\;} \{T \in \mathfrak{L}(E', \tilde{F}) \mid \sigma(E', \tilde{E})\text{-}\sigma(\tilde{F}, F')\text{-continuous}\}$$
since the latter space is closed in the Banach space $\mathfrak{L}(E', \tilde{F})$; see also Ex 22.2.. Note that
$$E' \tilde{\otimes}_\varepsilon F \longrightarrow \text{operators}$$
is injective — a fact which was not true for π!

(2) $C(K) \otimes E$ is the space of all continuous functions $K \to E$ the image of which is contained in a finite dimensional subspace. If K is compact, then the Dirac functionals δ_t (for $t \in K$) form a norming subset for $C(K)$, hence
$$\varepsilon(f; C(K), E) = \sup\{|\langle \delta_t \otimes y', \sum f_n \otimes x_n \rangle| \mid t \in K, y' \in B_{E'}\} =$$
$$= \sup\{|\langle y', \sum f_n(t) x_n \rangle| \mid t \in K, y' \in B_{E'}\}$$
$$= \sup\{\|f(t)\|_E \mid t \in K\} = \|f\|_{C(K,E)},$$
where $C(K, E)$ is the space of continuous E-valued functions. Hence
$$C(K) \otimes_\varepsilon E \xrightarrow{\;1\;} C(K, E) \ .$$
A simple argument using uniform continuity and a partition of unity shows that this space is even dense, hence
$$C(K) \tilde{\otimes}_\varepsilon E \stackrel{1}{=} C(K, E)$$
for each Banach space E.

(3) If K and L are compact topological spaces, the same reasoning with norming subsets shows that $C(K) \otimes_\varepsilon C(L) \subset C(K \times L)$ holds isometrically. The Stone–Weierstrass theorem gives density, hence
$$C(K) \tilde{\otimes}_\varepsilon C(L) \stackrel{1}{=} C(K \times L) \stackrel{1}{=} C(K, C(L))$$
— where the latter follows from example (2) (check that the identifications are the "natural" ones!) or by direct verification of $C(K \times L) = C(K, C(L))$.

(4) It is impressive to see how the natural mappings usually fit together and give new information. This is one reason for the power of tensor product methods. Consider the following mappings for the space $\mathfrak{K}(E, C(K))$ of compact operators:
$$\mathfrak{F}(E, C(K)) \subset \mathfrak{K}(E, C(K)) \xrightarrow{\Phi} C(K, E') = C(K) \tilde{\otimes}_\varepsilon E'$$
$$T \rightsquigarrow [t \rightsquigarrow T'\delta_t]$$

— note that Φ is defined on $\mathfrak{K}$. Since

$$\sup_{t \in K} \|T'\delta_t\| = \sup_{t \in K, y \in B_E} |\langle T'\delta_t, y \rangle| = \sup_{y \in B_E} \|Ty\|_{C(K)} = \|T\|,$$

Φ is isometric. For $T = x' \underline{\otimes} f \in E' \otimes C(K) \subset \mathfrak{K}(E, C(K))$ it follows that

$$\Phi(x' \underline{\otimes} f) = f \otimes x'$$

(there was a dualization /transposition involved!) and hence, by example (1), Φ is surjective on $\overline{\mathfrak{F}}(E, C(K))$ and therefore

$$\overline{\mathfrak{F}}(E, C(K)) \stackrel{1}{=} \mathfrak{K}(E, C(K)) \stackrel{1}{=} E' \tilde{\otimes}_\varepsilon C(K) \; :$$

Every compact operator with values in $C(K)$ is approximable.

4.3. Concerning subspaces and quotients, ε behaves "dually" to π. This is in some sense clear from the definition, but will be made much more precise in the sequel.

PROPOSITION:
(1) *If $G_i \subset E_i$ are subspaces, then*

$$G_1 \otimes_\varepsilon G_2 \xhookrightarrow{1} E_1 \otimes_\varepsilon E_2$$

i.e. ε respects subspaces isometrically. In particular, ε is finitely generated, i.e.

$$\varepsilon(z; E_1, E_2) = \inf\{\varepsilon(z; M_1, M_2) \mid z \in M_1 \otimes M_2, M_i \subset E \text{ finite dimensional}\} \; .$$

(2) *If $T_i \in \mathfrak{L}(E_i, F_i)$ are injective, E_1 and E_2 Banach spaces, then*

$$T_1 \tilde{\otimes}_\varepsilon T_2 : E_1 \tilde{\otimes}_\varepsilon E_2 \longrightarrow F_1 \tilde{\otimes}_\varepsilon F_2$$

is injective.

(3) *ε does not respect quotients — not even isomorphically: If $Q : F \stackrel{1}{\twoheadrightarrow} G$ is a metric surjection, then, in general, $\mathrm{id}_E \otimes_\varepsilon Q$ need not be open and its completion $\mathrm{id}_E \tilde{\otimes}_\varepsilon Q$ need not be surjective.*

PROOF: (1) The mapping property implies that for $z \in G_1 \otimes G_2$

$$\varepsilon(z; E_1, E_2) \leq \varepsilon(z; G_1, G_2) = \sup\{|\langle x'_1 \otimes x'_2, z\rangle| \mid x'_i \in B_{G'_i}\}$$

and the Hahn–Banach theorem gives equality.
For (2) consider the commutative diagram

$$\begin{array}{ccc} E_1 \tilde{\otimes}_\varepsilon E_2 & \hookrightarrow & \mathcal{L}(E_1', E_2) \ni S \\ {\scriptstyle T_1 \tilde{\otimes}_\varepsilon T_2} \downarrow & & \downarrow \hspace{0.3cm} \} \\ F_1 \tilde{\otimes}_\varepsilon F_2 & \hookrightarrow & \mathcal{L}(F_1', \tilde{F}_2) \ni T_2 S T_1' \end{array}.$$

Since all operators from $E_1 \tilde{\otimes}_\varepsilon E_2$ are $\sigma(E_1', E_1)$-$\sigma(E_2, E_2')$-continuous (see Example 4.2.(1) — the completeness of E_1 is used here) and T_1' has a $\sigma(E_1', E_1)$-dense range, it is clear that $T_1 \tilde{\otimes}_\varepsilon T_2$ is injective.

(3) By the open mapping theorem it is enough to find a metric surjection $Q : F \twoheadrightarrow G$ and an E such that

$$id_{E'} \tilde{\otimes}_\varepsilon Q : E' \tilde{\otimes}_\varepsilon F = \overline{\mathfrak{F}}(E,F) \longrightarrow \overline{\mathfrak{F}}(E,G) = E' \tilde{\otimes}_\varepsilon G$$
$$\hspace{3.5cm} S \hspace{0.8cm} \rightsquigarrow \hspace{0.8cm} QS$$

is not surjective — in other words: an approximable operator $T : E \to G$ which is not of the form QS.

An "advanced" example runs as follows: Take $E = G = \ell_2$ and $Q : \ell_1 \twoheadrightarrow \ell_2$ (every separable Banach space is a quotient of ℓ_1). It will be seen later on (Ex 11.18.) that every operator $\ell_2 \to \ell_2$ which factors through ℓ_1 is Hilbert–Schmidt. But there are compact=approximable operators $\ell_2 \to \ell_2$ which are not Hilbert–Schmidt.

A more "elementary" example uses "local techniques" and the fact that π does not respect subspaces: By Ex 3.29. there are finite dimensional spaces $E_n, G_n \subset F_n$ and $z_n \in E_n \otimes G_n$ such that

$$\pi(z_n; E_n, F_n) < \frac{1}{n} \pi(z_n; E_n, G_n) \ .$$

For the natural quotient map $Q_n : F_n' \to F_n'/G_n^0 = G_n'$ this implies that the quotient norm of

$$id_{E_n} \otimes Q_n : E_n' \otimes_\varepsilon F_n' = (E_n \otimes_\pi F_n)' \twoheadrightarrow (E_n \otimes_\pi G_n)' = E_n' \otimes_\varepsilon G_n'$$

is not smaller than $n \cdot \varepsilon(\, \cdot \, ; E_n', G_n')$ on $E_n' \otimes G_n'$. Hence

$$id \tilde{\otimes}_\varepsilon Q : \ell_2(E_n') \tilde{\otimes}_\varepsilon \ell_2(F_n') \longrightarrow \ell_2(E_n') \tilde{\otimes}_\varepsilon \ell_2(G_n')$$

cannot be surjective, because of the open mapping theorem and the fact that all spaces $E_m' \otimes_\varepsilon F_m'$ are 1-complemented in $\ell_2(E_n') \tilde{\otimes}_\varepsilon \ell_2(F_n')$. $\square$

4.4. However,

PROPOSITION: $C(K) \otimes_\varepsilon \cdot$ respects metric surjections, i.e.: If $Q : F \stackrel{1}{\twoheadrightarrow} G$ is a metric surjection between normed spaces, then

$$id \otimes_\varepsilon Q : C(K) \otimes_\varepsilon F \longrightarrow C(K) \otimes_\varepsilon G$$

is a metric surjection and therefore also

$$id\tilde{\otimes}_\varepsilon Q : C(K)\tilde{\otimes}_\varepsilon F \longrightarrow C(K)\tilde{\otimes}_\varepsilon G \ .$$

"Philosophically" this is a consequence of the facts that $L_1 \otimes_\pi \cdot$ respects subspaces, ε is dual to π and $C(K)$ is in some sense dual to L_1 by the Kakutani theorem (at least $C(K)'' = L_1(\mu)$ for some μ). Using the "local techniques" it is possible to make this precise and prove the statement of the proposition: First assume K to be finite, then clearly

$$\ell_\infty^n \otimes_\varepsilon E = \ell_\infty^n(E) \ ,$$

i.e. E^n with the maximum norm; individual lifting of $(x_1, ..., x_n) \in G^n$ gives that

$$id \otimes Q : \ell_\infty^n \otimes_\varepsilon F \longrightarrow \ell_\infty^n \otimes_\varepsilon G$$

is a metric surjection. [If F and G are Banach spaces, this follows also from

$$id \otimes Q' : (\ell_\infty^n(G))' = \ell_1^n(G') \hookrightarrow \ell_1^n(F') = (\ell_\infty^n(F))'$$

being a metric injection!]. The strategy of the proof is to next show that $C(K)$ looks locally like ℓ_∞^m .

LEMMA: *For every $\varepsilon > 0$ and every finite dimensional subspace $M \subset C(K)$ there are $S \in \mathfrak{L}(M, \ell_\infty^m)$ and $R \in \mathfrak{L}(\ell_\infty^m, C(K))$ for some $m \in \mathbb{N}$ with $\|R\| \|S\| \leq 1 + \varepsilon$ and*

$$RSf = f$$

for all $f \in M$; in other words, $C(K)$ is a $\mathfrak{L}_{\infty,1}^g$-space (see 3.13. for the definition).

PROOF: (a) Take $(f_1, ..., f_n; x_1', ..., x_n')$ an Auerbach basis of M (i.e. $\|f_j\| = \|x_j'\| = 1$ and $\langle x_j', f_i \rangle = \delta_{ij}$) and $\varepsilon < 1/2$. Then there is a partition of unity $(\varphi_\ell)_{\ell=1}^m$ in $C(K)$ such that there are $t_\ell \in K$ with $\varphi_\ell(t_\ell) = 1$ and $g_k \in M_o := \text{span } \{\varphi_\ell\}$ with

$$\|f_k - g_k\|_\infty \leq \varepsilon/n \qquad k = 1, ..., n \ .$$

For $S_1 := \sum x_k' \otimes g_k : M \to M_o$ it follows that for each $f \in M$

$$\|f - S_1 f\|_\infty \leq \sum_{k=1}^n |\langle x_k', f \rangle| \|f_k - g_k\|_\infty \leq \varepsilon \|f\|_\infty \ ,$$

hence $\|S_1\| \leq 1 + \varepsilon$. Since M_o is isometric to ℓ_∞^m (with the mapping $g \rightsquigarrow (g(t_\ell))_\ell$), the desired factorization is nearly obtained: Only the embedding $M_o \hookrightarrow C(K)$ needs a small pertubation.
(b) To do this, first observe that

$$\|f\|_\infty \leq \|f - S_1 f\|_\infty + \|S_1 f\|_\infty \leq \varepsilon \|f\|_\infty + \|S_1 f\|_\infty \ ,$$

hence $\|S_1^{-1} : S_1 M \to M\| \leq (1-\varepsilon)^{-1}$. Now extend the functionals $(S_1^{-1})' x_k' \in (S_1 M)'$ to $y_k' \in C(K)'$ with the same norm, note

$$\langle y_k', S_1 f_\ell \rangle = \delta_{k\ell}$$

and define

$$R_1 := \sum_{k=1}^{n} y_k' \otimes (S_1 f_k - f_k) : C(K) \to C(K)$$

which has norm $\leq \|S_1^{-1}\| \sum_k \|x_k'\| \|g_k - f_k\|_\infty \leq \varepsilon(1-\varepsilon)^{-1} < 1$. Therefore the operator

$$R_2 := id_{C(K)} - R_1 : C(K) \to C(K)$$

satisfies $\|R_2\| \leq 1 + \varepsilon(1-\varepsilon)^{-1} = (1-\varepsilon)^{-1}$ and $R_2 S_1 f_k = f_k$, hence $R_2 S_1$ is the embedding $M \hookrightarrow C(K)$. Since $S_1(M) \subset M_o$ and M_o is isometric to ℓ_∞^m, the desired factorization is found. □

Note that the operator $R_2 : C(K) \to C(K)$ which was constructed in the proof is even an isomorphism and $\|R_2^{-1}\| \leq (1-\|R_1\|)^{-1} \leq (1-2\varepsilon)^{-1}$. This means that the subspace M is contained in a subspace (namely $R_2 M_o$) which is nearly isometric to ℓ_∞^m.

PROOF of the proposition: Take a metric surjection $Q : F \xrightarrow{1} G$ and $x_o \in C(K) \otimes G$. Then there is a finite dimensional subspace $M \subset C(K)$ with $x_o \in M \otimes G$. Since ε respects subspaces,

$$\varepsilon(x_o; M, G) = \varepsilon(x_o; C(K), G) \quad .$$

Now take $\delta > 0$ and a factorization $R \circ S$ of $M \hookrightarrow C(K)$ through some ℓ_∞^m with $\|R\| \|S\| \leq 1 + \delta$:

$$\begin{array}{ccccc}
 & & \ell_\infty^m \otimes_\varepsilon F & \xrightarrow{R \otimes id_F} & C(K) \otimes_\varepsilon F \\
 & & \downarrow & & \downarrow \\
M \otimes_\varepsilon G & \xrightarrow{S \otimes id_G} & \ell_\infty^m \otimes_\varepsilon G & \xrightarrow{R \otimes id_G} & C(K) \otimes_\varepsilon G \quad .
\end{array}$$

Choose $y_o \in \ell_\infty^m \otimes_\varepsilon F$ which lifts $S \otimes id_G(x_o)$ and satisfies

$$\varepsilon(y_o; \ell_\infty^m, F) \leq (1+\delta)\varepsilon(S \otimes id_G(x_o); \ell_\infty^m, G) \quad .$$

Then it is clear that $y_1 := R \otimes id_F(y_o)$ lifts $x_o = RS \otimes id_G(x_o)$ and

$$\varepsilon(y_1; C(K), F) \leq \|R\|\varepsilon(y_o; \ell_\infty^m, F) \leq \|R\|(1+\delta)\varepsilon(S \otimes id_G(x_o); \ell_\infty^m, G) \leq$$
$$\leq \|R\|(1+\delta)\|S\|\varepsilon(x_o; M, G) \leq (1+\delta)^2 \varepsilon(x_o; C(K), G) ,$$

which was to be shown. □

It is clear, from the proof, that $E \tilde{\otimes}_\varepsilon \cdot$ respects metric surjections for each $\mathcal{L}^g_{\infty,1}$-space E, and also: that $\mathcal{L}^g_{\infty,\lambda}$-spaces respect metric surjections up to the constant λ. It will be shown in 23.5. that this property charactizes the $\mathcal{L}^g_\infty$-spaces.

4.5. The proposition has interesting analytic consequences.

COROLLARY 1: *For every metric surjection $Q : F \twoheadrightarrow G$ between Banach spaces, every continuous function $f : K \to G$ on a compact space K and every $\varepsilon > 0$ there is a continuous function $g : K \to F$ with $Q \circ g = f$ and $\|g\|_\infty \leq (1+\varepsilon)\|f\|_\infty$.*

PROOF: This follows from

$$id \tilde{\otimes}_\varepsilon Q : C(K, F) = C(K) \tilde{\otimes}_\varepsilon F \xrightarrow{1} C(K) \tilde{\otimes}_\varepsilon G = C(K, G) \,. \qquad \square$$

The argument can be applied to function spaces other than $C(K)$, see the work of Kaballo [134], [135] and 23.5..

COROLLARY 2: *If $G \subset E$ is a subspace, $T \in \mathfrak{K}(G, C(K))$ and $\varepsilon > 0$, then there is an extension $\tilde{T} \in \mathfrak{K}(E, C(K))$ of T with $\|\tilde{T}\| \leq (1+\varepsilon)\|T\|$.*

PROOF: In 4.2.(4) the relation $\mathfrak{K}(F, C(K)) = F' \tilde{\otimes}_\varepsilon C(K)$ was shown, hence

$$\mathfrak{K}(E, C(K)) = E' \tilde{\otimes}_\varepsilon C(K) \xrightarrow{1} G' \tilde{\otimes}_\varepsilon C(K) = \mathfrak{K}(G, C(K)). \qquad \square$$

For $L_\infty(\mu)$ this *compact extension property* was already shown in Ex 3.26.. Lindenstrauss ([171], p.92) has given examples showing that corollary 2 does not hold for $\varepsilon = 0$ — not even for finite dimensional E. Consequently (look at the proofs above) also corollary 1 is false for $\varepsilon = 0$ and even for finite dimensional F; see also Ex 3.21.(b).

4.6. The dual of $(E \otimes_\pi F)'$ is the space of continuous bilinear forms. Which bilinear forms φ are continuous even with respect to ε, i.e. $\varphi \in (E \otimes_\varepsilon F)'$?

THEOREM: *Let $\varphi \in (E \otimes F)^*$. Then $\varphi \in (E \otimes_\varepsilon F)'$ if and only if there exists a (positive) Borel-Radon measure μ on $B_{E'} \times B_{F'}$ (equipped with the weak-$*$-topologies) such that for all $z \in E \otimes F$*

$$\langle \varphi, z \rangle = \int_{B_{E'} \times B_{F'}} \langle x' \otimes y', z \rangle \mu(d(x', y')) \,.$$

The measure μ can be chosen such that $\|\mu\| = \mu(B_{E'} \times B_{F'}) = \|\varphi\|_{(E \otimes_\varepsilon F)'}$.

This is why bilinear forms in $(E \otimes_\varepsilon F)' \subset (E \otimes_\pi F)' = \mathfrak{Bil}(E, F) = \mathfrak{L}(E, F')$ are called *integral*. The operators $E \to F'$ associated with integral forms are called "integral" as well; they will be studied in detail in section 10.

PROOF: (a) By the very definition of the injective norm

$$I: E \otimes_\varepsilon F \xhookrightarrow{1} C(B_{E'} \times B_{F'})$$
$$z \rightsquigarrow [(x', y') \rightsquigarrow \langle x' \otimes y', z \rangle]$$

is an isometric injection. Then if φ has such a representation, it follows that

$$|\langle \varphi, z \rangle| \leq \mu(B_{E'} \times B_{F'}) \varepsilon(z; E, F)$$

and, vice-versa, the Hahn–Banach theorem gives for every $\varphi \in (E \otimes_\varepsilon F)'$ a signed measure $\mu \in C(B_{E'} \times B_{F'})' =: M(B_{E'} \times B_{F'})$ which represents φ as indicated.

(b) But μ may not be positive. To achieve this, the metric surjection

$$I' : M(B_{E'} \times B_{F'}) \twoheadrightarrow (E \otimes_\varepsilon F)'$$

has to be examined; denote by M_1^+ the convex, weak-$*$-compact set of probability measures on $B_{E'} \times B_{F'}$ and $D \subset M_1^+$ the set of Dirac measures. By the definition of the injective norm

$$I'(D)^0 = B_{E \otimes_\varepsilon F}$$

holds; moreover $\lambda I'(D) \subset I'(D)$ for all $|\lambda| \leq 1$ and hence the convex hull and the absolutely convex hull of $I'(D)$ coincide. The bipolar theorem gives (closure with respect to the weak-$*$-topology in $(E \otimes_\varepsilon F)'$):

$$B_{(E \otimes_\varepsilon F)'} = I'(D)^{00} = \overline{\Gamma(I'(D))} = \overline{\operatorname{conv} I'(D)} \subset I'(M_1^+) \ .$$

It follows that for every $\varphi \in (E \otimes_\varepsilon F)'$ there is a $\mu \in M_1^+$ with $\varphi = \|\varphi\| I'(\mu)$ — this was to be shown. $\square$

4.7. If μ is a positive Borel–Radon measure on a compact set K, then

$$\beta_I : C(K) \tilde{\otimes}_\varepsilon C(K) = C(K \times K) \longrightarrow \mathbb{K}$$
$$h \rightsquigarrow \int h(\omega, \omega) \mu(d\omega)$$

is clearly continuous, i.e. integral on $C(K) \times C(K)$, and of norm $\mu(K)$. The theorem says that this is somehow the typical integral bilinear form (the details will be given in a moment).

REMARK: *If μ is a finite measure, then the "integrating functional"*

$$\beta_I : L_\infty(\Omega, \mu) \otimes_\varepsilon L_\infty(\Omega, \mu) \longrightarrow \mathbb{K}$$
$$\tilde{f} \otimes \tilde{g} \rightsquigarrow \int f(\omega) g(\omega) \mu(d\omega)$$

is continuous and $\|\beta_I\|_{(L_\infty \otimes_\varepsilon L_\infty)'} = \mu(\Omega)$.

PROOF: Choose a lifting $\rho : L_\infty(\Omega, \mu) \to \mathcal{L}_\infty(\Omega, \mu)$ (Maharam's theorem, see Appendix B7.). Hence,
$$\delta_\omega(\tilde{f}) := \rho(\tilde{f})(\omega)$$
defines an element in L'_∞ of norm 1 for every $\omega \in \Omega$. The claim follows from
$$\int \sum f_n(\omega) g_n(\omega) \mu(d\omega) = \int \sum \delta_\omega(\tilde{f}_n) \cdot \delta_\omega(\tilde{g}_n) \mu(d\omega) \le$$
$$\le \mu(\Omega) \varepsilon \left(\sum \tilde{f}_n \otimes \tilde{g}_n; L_\infty, L_\infty \right). \quad \square$$

Note that for these two bilinear forms the norm in $(E \otimes_\varepsilon F)'$ and $(E \otimes_\pi F)'$ coincide.

COROLLARY: *For* $\varphi \in (E_1 \otimes E_2)^*$ *the following statements are equivalent:*

(a) φ *is integral.*

(b) *There exists a compact set* Ω *and there exist a positive Borel–Radon measure* μ *and operators* $R_i \in \mathcal{L}(E_i, C(\Omega))$ *such that*
$$\langle \varphi, x_1 \otimes x_2 \rangle = \int_\Omega (R_1 x_1)(\omega) \cdot (R_2 x_2)(\omega) \mu(d\omega) \quad \text{for all } x_i \in E_i \ .$$

(c) *There is a finite measure space* (Ω, μ) *and operators* $R_i \in \mathcal{L}(E_i, L_\infty(\Omega, \mu))$ *with*
$$\langle \varphi, x_1 \otimes x_2 \rangle = \int_\Omega (R_1 x_1)(\omega) \cdot (R_2 x_2)(\omega) \mu(d\omega) \quad \text{for all } x_i \in E_i \ .$$

The measures μ *can be chosen such that*
$$\|\varphi\|_{(E \otimes_\varepsilon F)'} = \mu(\Omega)$$
and $\|R_1\| = \|R_2\| = 1$ *(if* $E_i \ne \{0\}$*).*

PROOF: Clearly, if β_I is the "integrating" form on $C(\Omega) \otimes C(\Omega)$ or $L_\infty(\mu) \otimes L_\infty(\mu)$, then, as noted before,
$$\varphi = \beta_I \circ (R_1 \otimes_\varepsilon R_2)$$
is integral. It remains to show (a) $\curvearrowright$ (b) : The representation theorem gives a μ on the compact set $B_{E'} \times B_{F'}$ — and the operators
$$[R_1 x_1](x'_1, x'_2) := \langle x'_1, x_1 \rangle$$
$$[R_2 x_2](x'_1, x'_2) := \langle x'_2, x_2 \rangle$$
give the factorization. $\square$

4.8. Since
$$E'\tilde{\otimes}_\pi F \twoheadrightarrow \mathfrak{N}(E,F) \subset \overline{\mathfrak{F}}(E,F) = E'\tilde{\otimes}_\varepsilon F ,$$
the norms π and ε are not equivalent on $E' \otimes F$ whenever there is an approximable operator which is not nuclear. For a long time it was an open problem whether or not there are infinite dimensional Banach spaces E such that
$$E \otimes_\pi E = E \otimes_\varepsilon E$$
holds isomorphically! Pisier [224] succeeded in 1981 to construct such Banach spaces P, thus solving one of the toughest problems Grothendieck left behind for the functional analysts; P can be chosen to be separable and even to satisfy
$$P'\tilde{\otimes}_\pi P \twoheadrightarrow \mathfrak{N}(P,P) \stackrel{1}{=} \overline{\mathfrak{F}}(P,P) = P'\tilde{\otimes}_\varepsilon P$$
and this mapping is not injective, hence π and ε are not equivalent on $P' \otimes P$. (For the very first step of Pisier's construction see Ex 4.15..) John [128] showed that every compact operator $T : P \to P'$ ist nuclear.

Exercises:

Ex 4.1. Does $\ell_2^2 \otimes_\varepsilon \ell_2^2 = \ell_2^4$ hold isometrically? Check that $\pi \neq \varepsilon$ on $\ell_2^2 \otimes \ell_2^2$.

Ex 4.2. Show that ε and π are not equivalent on $\ell_2 \otimes \ell_2$. Hint: Ex 3.28..

Ex 4.3. Consider the "Fourier matrix"
$$w_n := n^{-1/2} \sum_{k,\ell=1}^{n} \exp(2\pi i \frac{k \cdot \ell}{n}) e_k \otimes e_\ell \in \mathbb{C}^n \otimes \mathbb{C}^n .$$
Show that
(a) $\pi(w_n; \ell_1^n, \ell_1^n) = n^{3/2}$.
(b) $1 \leq \varepsilon(w_n; \ell_1^n, \ell_1^n) \leq n$.
(c) $\|id : \ell_1^n \otimes_\varepsilon \ell_1^n \to \ell_1^n \otimes_\pi \ell_1^n\| \geq n^{1/2}$.
(d) ε and π are not equivalent on $\ell_1 \otimes \ell_1$.

Hint: For (b) use the fact that w_n represents a unitary operator $\ell_2^n \to \ell_2^n$. A result essentially due to Hardy and Littlewood [100], [178] (see also Dineen–Timoney [57]) says that there is a constant $c > 0$ such that for all $n \in \mathbb{N}$
$$cn^{\gamma(p)} \leq \varepsilon(w_n; \ell_p^n, \ell_p^n) \leq n^{\gamma(p)}$$
where $\gamma(p) = 1/p - 1/2$ if $2 \leq p \leq \infty$ and $\gamma(p) = 2/p - 1$ if $1 \leq p \leq 2$ (for $p \geq 2$ even $c = 1$ works) . For (c) and (d) see also proposition 11.1..

Ex 4.4. For $\alpha = \varepsilon$ or $\alpha = \pi$ show that
$$(E \otimes_\alpha F) \otimes_\alpha G = E \otimes_\alpha (F \otimes_\alpha G).$$

Ex 4.5. Let E be a real normed space and define a complex multiplication on the real tensor product $E_{\mathbb{C}} := E \otimes_\varepsilon \mathbb{C}$ by $\alpha \circ x := (id_E \otimes \overline{\alpha} id_{\mathbb{C}})(x)$. Show that for a continuous $\mathbb{R}$-linear $E \to F$ the operator
$$T_{\mathbb{C}} := T \otimes id_{\mathbb{C}} : E_{\mathbb{C}} \to F_{\mathbb{C}}$$
is $\mathbb{C}$-linear and $\|T_{\mathbb{C}}\| = \|T\|$.

Ex 4.6. Show that every $z \in \ell_2 \tilde{\otimes}_\varepsilon \ell_2$ has a representation
$$z = \sum_{n=1}^{\infty} \lambda_n e_n \otimes f_n$$
with orthonormal systems (e_n) and (f_n) and $(\lambda_n) \in c_o$. In this case the norm of z can be calculated as follows: $\varepsilon(z; \ell_2, \ell_2) = \max |\lambda_n|$. Hint: Expansion of compact operators.

Ex 4.7. (a) Use the arguments from the proof of lemma 4.4. to show: If E is a normed $\mathcal{L}^g_{p,\lambda}$-space, then the completion $\tilde{E}$ is an $\mathcal{L}^g_{p,\lambda}$-space. For the converse see Ex 6.6..
(b) For $1 \le p \le \infty$ every $L_p(\mu)$ is an $\mathcal{L}^g_{p,1}$-space. Hint: Step functions, and (a).
(c) For $1 < p < \infty$ the Hardy spaces H^p are $\mathcal{L}^g_p$-spaces. Hint: By the M. Riesz theorem they are complemented in $L_p[0, 2\pi]$, see Duren [60], p.54 and 67.
(d) If E is a Banach space which has an increasing family (E_α) of subspaces which are $\mathcal{L}^g_{p,\lambda}$-spaces and the union of which is dense in E, then E is a $\mathcal{L}^g_{p,\lambda}$-space.
(e) Just for fun: If a (real or complex) vector space E is the union of a (not necessarily increasing) sequence of subspaces E_n, then every finite dimensional $M \subset E$ is contained in some E_n.

Ex 4.8. (a) Show that $\ell_\infty \otimes_\varepsilon E$ is isometrically contained in
$$\{(x_n) \in \ell_\infty(E) \mid \{x_n\} \subset E \text{ relatively compact}\}$$
and deduce from this that $\ell_\infty \tilde{\otimes}_\varepsilon \ell_\infty \subset \ell_\infty(\mathbb{N} \times \mathbb{N})$ is a proper subspace.
(b) If $\beta \mathbb{N}$ denotes the Stone–Čech compactification of $\mathbb{N}$ equipped with the discrete topology, then $C(\beta \mathbb{N}) = \ell_\infty$ (note that $\beta \mathbb{N}$ is the Gelfand space of the C^*-algebra ℓ_∞). Show that the natural mapping $\beta(\mathbb{N} \times \mathbb{N}) \to \beta \mathbb{N} \times \beta \mathbb{N}$ is not bijective.

Ex 4.9. Show that for every measure μ and Banach space E
$$L_\infty(\mu) \tilde{\otimes}_\varepsilon E \hookrightarrow L_\infty(\mu, E)$$
holds isometrically, but the spaces are different in general. Hint: B_{L_1} is norming for L_∞; both norms are dominated by π, hence it is enough to take step functions.

Ex 4.10. Show that $A \subset E \tilde{\otimes}_\varepsilon F$ is bounded if and only if $\sup |\langle x' \otimes y', A\rangle| < \infty$ for all $x' \in E'$ and $y' \in F'$.

Ex 4.11. A sequence (z_n) in $E \tilde{\otimes}_\varepsilon F$ is a weak Cauchy sequence if and only if for all $x' \in E'$ and $y' \in F'$ the sequence $(\langle x' \otimes y', z_n\rangle)$ converges. Hint: Lebesgue's dominated convergence theorem.

Ex 4.12. $E \otimes_\varepsilon F = E \otimes_\pi F$ holds isomorphically if and only if every continuous bilinear form on $E \times F$ is integral.

Ex 4.13. Let $G \subset E$ a subspace and φ an integral bilinear form on $G \times F$. Then there is an integral bilinear form on $E \times F$ which extends φ.

Ex 4.14. Every nuclear bilinear form (see 1.3.) is integral; the natural mappings

$$E' \tilde{\otimes}_\pi F' \longrightarrow (E \otimes_\varepsilon F)'$$
$$E' \tilde{\otimes}_\pi F \longrightarrow (E \otimes_\varepsilon F')'$$
$$E \tilde{\otimes}_\pi F \longrightarrow (E' \otimes_\varepsilon F')'$$

all have norm 1 whenever E and F are different from $\{0\}$.

Ex 4.15. Let $I_n : E_n \hookrightarrow E_{n+1}$ be a sequence of isometric injections of normed spaces E_n such that $\sup \|I_n \otimes I_n : E_n \otimes_\varepsilon E_n \to E_{n+1} \otimes_\pi E_{n+1}\| =: C < \infty$. Then

$$\varepsilon \leq \pi \leq C\varepsilon \quad \text{on} \quad E \otimes E$$

for the space $E := \bigcup_{n=1}^\infty E_n$ with its natural norm.

5. The Approximation Property

Is every compact operator approximable? Is $E' \tilde{\otimes}_\pi F = \mathfrak{N}(E, F)$? Is the trace defined for nuclear operators? Is an operator nuclear if its dual is? Is $T_1 \tilde{\otimes}_\pi T_2$ injective if T_1 and T_2 are? These questions find an answer with the help of Grothendieck's approximation property.

5.1. Let E and F be normed spaces. $\mathfrak{L}_{co}(E, F)$ denotes $\mathfrak{L}(E, F)$ equipped with the topology τ_{co} of uniform convergence on all absolutely convex compact subsets of E; it is a separated locally convex space with seminorms

$$p_K(T) := \sup \|TK\| \qquad \text{for all } T \in \mathfrak{L}(E, F),$$

where $K \subset E$ is absolutely convex and compact; $E'_{co} := \mathfrak{L}_{co}(E, \mathbb{K})$.

DEFINITION: *A normed space E has the*

(1) *approximation property if $id_E \in \overline{\mathfrak{F}(E,E)}^{\tau_{co}}$; in other words, for every absolutely convex compact K and $\varepsilon > 0$ there is a $T \in \mathfrak{F}(E,E)$ with $\|Tx - x\| \leq \varepsilon$ for all $x \in K$.*

(2) *λ-bounded approximation property for $\lambda \in [1, \infty[$ if $id_E \in \lambda \overline{B_{\mathfrak{F}(E,E)}}^{\tau_{co}}$; in other words, there is a net (T_η) of finite rank operators with $\|T_\eta\| \leq \lambda$ and $T_\eta \to id_E$ in $\mathfrak{L}_{co}(E,E)$.*

(3) *metric approximation property := 1-bounded approximation property.*

(4) *bounded approximation property if it has the λ-bounded approximation property for some $\lambda \geq 1$.*

Some consequences are immediately clear:

(1) E has the approximation property if and only if for all normed spaces F (or only $F = E$) the space $\mathfrak{F}(E,F)$ is τ_{co}-dense in $\mathfrak{L}(E,F)$.

(2) E has the λ-bounded approximation property if and only if there is a net (T_η) in $\mathfrak{F}(E,E)$ with $\|T_\eta\| \leq \lambda$ and $T_\eta x \to x$ for all $x \in E$.

This follows from the fact that on equicontinuous subsets of $\mathfrak{L}(E,F)$ the pointwise topology and τ_{co} coincide.

(3) If a Banach space E has a basis, then it has the bounded approximation property and an equivalent norm $\|\cdot\|_o$ such that $(E, \|\cdot\|_o)$ has the metric approximation property. This norm is given by $\|x\|_o := \sup \|P_n x\|$, where the operators P_n are the expansion operators of the basis.

5.2. For many years it was an open problem whether or not every separable Banach space has a basis — or at least the approximation property. This was solved in the negative by Enflo [63] in 1972; the present "state of the art" is the following:

(1) All ℓ_p (for $1 \leq p \leq \infty, p \neq 2$) and c_o have a closed subspace without the approximation property ($2 < p \leq \infty$ and c_o : Enflo; $1 \leq p < 2$: Szankowski [262]); Pisier's spaces P mentioned in 4.8. cannot have the approximation property (by theorem 5.6. below).

(2) Szankowski [261] constructed a separable Banach lattice without the approximation property.

(3) The space $\mathfrak{L}(\ell_2, \ell_2)$ does not have the approximation property (Szankowski [263], see also Pisier [221]).

That this space does not have the approximation property is scandalous! All the "usual" spaces (which means: not artificially constructed) have the metric approximation property. For $C(K)$ this follows from arguments with partition of unity, for $L_p(\mu)$ with averaging operators. Exception:

(4) It is not known whether the (also non-separable) space $H^\infty(D)$ of all bounded holomorphic functions on the open unit ball of $\mathbb{C}$ has the approximation property (but it has the g_p-metric approximation property for all $1 < p < \infty$, see Bourgain-Reinov

[17] and section 21), while the disc algebra $A(D)$ (the subspace of $H^\infty(D)$ of functions with continuous boundary values) has the metric approximation property.

All the implications between the various approximation properties and the property of having a basis are either false or trivially true:

(5) There is a Banach space with separable dual and with the approximation property but without the bounded approximation property (Figiel–Johnson [66]).

(6) There is a Banach space with separable dual and the bounded approximation property but without the metric one (this follows from theorem 1 of Figiel–Johnson [66] with the aid of a construction due to Lindenstrauss, mentioned in this paper; see also Lindenstrauss–Tzafriri [177], I, 1.e.20.). It is not known whether a Banach space with the bounded approximation property can be renormed in order to have the metric one.

(7) There is a reflexive separable Banach space with the metric approximation property and without a basis; Szarek [264]. Szarek's method was recently refined by Mankiewicz and Nielsen [181].

A positive result will be proven in section 16 with the Radon–Nikodým property:

(8) A reflexive Banach space (or a separable dual space) with the approximation property even has the metric approximation property.

5.3. The first question to be attacked is the approximation of compact operators:
$$E' \tilde{\otimes}_\epsilon F = \overline{\mathfrak{F}}(E,F) \subset \mathfrak{K}(E,F) \ .$$

When does equality hold?

PROPOSITION: (1) *A Banach space E has the approximation property if and only if $F' \tilde{\otimes}_\epsilon E = \mathfrak{K}(F,E)$ for all Banach spaces F.*

(2) *The dual E' of a Banach space E has the approximation property if and only if $E' \tilde{\otimes}_\epsilon F = \mathfrak{K}(E,F)$ for all Banach spaces F.*

It is enough to check the conditions for reflexive F (see Ex 5.8.).

PROOF: (1) Since TB_F is relatively compact for $T \in \mathfrak{K}(F,E)$, the condition is easily seen to be necessary. Conversely, take $K \subset E$ compact. Then there is a compact Banach disc L such that K is a compact subset of $[\![L]\!]$ (see 3.5., corollary 2). The embedding $J : [\![L]\!] \hookrightarrow E$ is compact = approximable by assumption, hence there is an operator $T = \sum_{k=1}^n x'_k \otimes y_k \in [\![L]\!]' \otimes E$ with
$$\left\| \sum_{k=1}^n \langle x'_k, x \rangle y_k - x \right\| \leq \varepsilon \quad \text{for all } x \in K \ .$$

The dual $J' : E' \to [\![L]\!]'$ has a weak-$*$-dense $= [\![L]\!]'_{co}$-dense range; therefore the x'_k's can be approximated on K by $y'_k \in E'$ and $\hat{T} := \sum_{k=1}^n y'_k \otimes y_k$ gives the result.

(2) If $T \in \mathfrak{K}(E, F)$, then by the approximation property of E' there is an $S \in E'' \otimes E'$ such that
$$\|T' - ST'\| \leq \varepsilon$$
and hence also $\|T'' - T''S'\| \leq \varepsilon$. If $S = \sum_{k=1}^{n} x_k'' \otimes x_k'$, then

$$T''S' = \sum_{k=1}^{n} x_k' \otimes T''x_k'' \quad \in E' \otimes F$$

since $T''(F'') \subset F$; it follows that $\|T - \sum x_k' \otimes T''x_k''\| \leq \varepsilon$. To see the converse it must be shown, by (1), that every $T \in \mathfrak{K}(F, E')$ is approximable. Since, by assumption $T' \circ \kappa_E \in \mathfrak{K}(E, F')$ is approximable, it follows that

$$T = (T' \circ \kappa_E)' \circ \kappa_F$$

is also approximable. $\square$

COROLLARY: *If E' or F has the approximation property, then*

$$E' \tilde{\otimes}_\varepsilon F = \mathfrak{K}(E, F)$$

(E, F *Banach spaces*): *each compact operator $E \to F$ is approximable.*

5.4. Is $E' \tilde{\otimes}_\pi F = \mathfrak{N}(E, F)$? Is the trace defined for nuclear operators (see 3.7.) ? To solve these types of questions, observe first that there are natural mappings

$$E' \tilde{\otimes}_\pi F \twoheadrightarrow \mathfrak{N}(E, F) \hookrightarrow \mathfrak{K}(E, F) \hookrightarrow \mathfrak{L}(E, F) \hookrightarrow \mathfrak{Bil}(E, F')$$
$$E' \tilde{\otimes}_\pi F \to E' \tilde{\otimes}_\varepsilon F \hookrightarrow \mathfrak{K}(E, F) \hookrightarrow \mathfrak{L}(E, F) \hookrightarrow \mathfrak{Bil}(E, F')$$
$$E' \tilde{\otimes}_\pi F \to (E \otimes_\varepsilon F')' \hookrightarrow (E \otimes_\pi F')' = \mathfrak{Bil}(E, F') \quad .$$

The last mapping comes from the fact that nuclear bilinear forms are integral, Ex 4.14.; all these mappings have norm 1 (provided E and F are $\neq \{0\}$, of course) and they are all the same in the following sense : just $E' \tilde{\otimes}_\pi F \to \mathfrak{Bil}(E, F')$ naturally. So, *if one is injective, all three are injective*. The same holds for the three mappings

$$E \tilde{\otimes}_\pi F \to \mathfrak{N}(E', F) \hookrightarrow \mathfrak{K}(E', F) \hookrightarrow \mathfrak{L}(E', F) \hookrightarrow \mathfrak{Bil}(E', F')$$
$$E \tilde{\otimes}_\pi F \to E \tilde{\otimes}_\varepsilon F \hookrightarrow \mathfrak{K}(E', F) \hookrightarrow \mathfrak{Bil}(E', F')$$
$$E \tilde{\otimes}_\pi F \to (E' \otimes_\varepsilon F')' \hookrightarrow (E' \otimes_\pi F')' = \mathfrak{Bil}(E', F')$$

(all are injective if one is); also the same is true for the natural mappings

$$E' \tilde{\otimes}_\pi F' \to \ldots \hookrightarrow \mathfrak{Bil}(E, F) \quad .$$

The question is under which circumstances

$$E \tilde{\otimes}_\pi F \to \text{operators / bilinear forms}$$

is injective. The main result of this section, due to Grothendieck, answers this question in a very satisfactory manner — a result which is basic for the study of tensor products and operators.

5.5. For this, a description of the dual of $\mathcal{L}_{co}(E,F)$ will be needed; the Banach space E will be fixed. Denote by Φ_F the natural mapping

$$\Phi_F : \mathcal{L}(E,F) \hookrightarrow (E \tilde{\otimes}_\pi F')' \xrightarrow{transposition} (F' \tilde{\otimes}_\pi E)' \ .$$

LEMMA: $\Phi_F : \mathcal{L}_{co}(E,F) \to [(F' \tilde{\otimes}_\pi E)', \text{weak-}*\text{-topology}\,]$ *is continuous and therefore*

$$D_F := \Phi'_F : F' \tilde{\otimes}_\pi E \longrightarrow (\mathcal{L}_{co}(E,F))'$$

is defined; if $z = \sum_{n=1}^\infty y'_n \otimes x_n \in F' \tilde{\otimes}_\pi E$ *is an arbitrary expansion, then*

$$\langle D_F(z), T\rangle = \langle z, \Phi_F(T)\rangle = \langle tr_F, id_{F'} \tilde{\otimes}_\pi T(z)\rangle = \sum_{n=1}^\infty \langle y'_n, Tx_n\rangle$$

for all $T \in \mathcal{L}(E,F)$.

PROOF: It has to be shown that for $z = \sum_{n=1}^\infty y'_n \otimes x_n \in F' \tilde{\otimes}_\pi E$ the mapping

$$T \rightsquigarrow \langle \Phi_F(T), z\rangle = \sum_{n=1}^\infty \langle y'_n, Tx_n\rangle$$

is continuous: Choose $\sum_{n=1}^\infty \|y'_n\| < \infty$ and $x_n \to 0$; then

$$|\langle \Phi_F(T), z\rangle| \le \sum_{n=1}^\infty \|y'_n\| \ \sup\left\{\|Tx\|\big| x \in \overline{\Gamma\{x_n\}}\right\}$$

and the latter is a continuous seminorm. Thus D_F is defined and the formula follows from

$$\sum_{n=1}^\infty \langle y'_n, Tx_n\rangle = \langle tr_F, id_{F'} \tilde{\otimes}_\pi T(z)\rangle \ . \ \square$$

If F is reflexive, Φ_F is an isomorphism (onto) and therefore also D_F.

PROPOSITION:

$$D_F : F' \tilde{\otimes}_\pi E \longrightarrow (\mathcal{L}_{co}(E,F))'$$

is surjective. If F *is reflexive, then* D_F *is even an isomorphism (of vector spaces).*

PROOF: Given $\varphi \in (\mathfrak{L}_{co}(E,F))'$, take a compact set $K \subset \overline{\Gamma\{x_n\}} \subset E$ (where (x_n) is a zero sequence, see 3.5.) such that

$$|\langle \varphi, T \rangle| \leq \sup_n \|Tx_n\| \quad .$$

Consider
$$\Psi : \mathfrak{L}_{co}(E,F) \longrightarrow c_o(F)$$
$$T \longmapsto (Tx_n) \; .$$

Then the inequality for φ and the Hahn–Banach theorem give a $\tilde{\varphi} = (y'_n) \in \ell_1(F') = (c_o(F))'$ such that $\tilde{\varphi} \circ \Psi = \varphi$. For $z := \sum_{n=1}^{\infty} y'_n \otimes x_n \in F' \tilde{\otimes}_\pi E$ it follows that

$$\langle D_F(z), T \rangle = \sum_{n=1}^{\infty} \langle y'_n, Tx_n \rangle = \langle \tilde{\varphi}, \Psi(T) \rangle = \langle \varphi, T \rangle \; . \quad \square$$

5.6. For every F there are natural mappings

$$F' \tilde{\otimes}_\pi E \quad \begin{array}{c} \xrightarrow{J_{F'}} \mathfrak{N}(F,E) \hookrightarrow F' \tilde{\otimes}_\varepsilon E \\ \searrow_{D_F} (\mathfrak{L}_{co}(E,F))' \end{array}$$

where $J_{F'}(\sum_n y'_n \otimes x_n) = \sum_n y'_n \underline{\otimes} x_n$ (it is sometimes good to maintain the notation $y' \underline{\otimes} x$ for the operator associated with $y' \otimes x$). It follows that

(*) $$\langle D_F(z), x' \otimes y \rangle = \langle [J_{F'}(z)](y), x' \rangle$$

for all $(x', y) \in E' \times F$ and $z \in F' \tilde{\otimes}_\pi E$. In particular,

(**) $$D_F(z)|_{\mathfrak{F}(E,F)} = 0 \text{ if and only if } z \in \ker J_{F'} \quad .$$

Now E has the approximation property if and only if

$$id_E \in \overline{\mathfrak{F}(E,E)}^{\mathfrak{L}_{co}(E,E)} \quad ;$$

since $D_E : E' \tilde{\otimes}_\pi E \to (\mathfrak{L}_{co}(E,E))'$ is surjective, the Hahn-Banach theorem and (**) imply that this is equivalent to : If $z \in \ker J_{E'}$, then

$$0 = \langle D_E(z), id_E \rangle = \langle tr_E, z \rangle \; ;$$

in other words, E has the approximation property if and only if $\ker J_{E'} \subset \ker tr_E$. This means that the trace : $E' \tilde{\otimes}_\pi E \to \mathbb{K}$ factors through the nuclear operators.

THEOREM: *For a Banach space E the following statements are equivalent:*

(a) *E has the approximation property.*

(b) $\ker J_{E'} \subset \ker tr_E$, *i.e. there is a continuous linear functional $\hat{tr}_E$ on $\mathfrak{N}(E, E)$ with $tr_E = \hat{tr}_E \circ J_{E'}$ (such a functional is called a trace on the space of nuclear operators).*

(c) *The natural mapping*
$$J_{E'} : E' \tilde{\otimes}_\pi E \longrightarrow \mathfrak{N}(E, E)$$
is injective.

(d) *For all Banach spaces F (or only $F = E'$) the natural map*
$$J_F : F \tilde{\otimes}_\pi E \longrightarrow F \tilde{\otimes}_\epsilon E$$
is injective.

Usually the trace $\hat{tr}_E$ on $\mathfrak{N}(E, E)$ will also be denoted by tr_E — if it exists.

PROOF: The equivalence of the first two statements was already shown, clearly (d)$\curvearrowright$ (c) $\curvearrowright$ (b) and it remains to show that (b) implies (d) : For this take $z \in F \tilde{\otimes}_\pi E$ with $J_F(z) = 0$ and $\varphi \in (F \otimes_\pi E)'$ with the associated operator $L_\varphi \in \mathcal{L}(F, E')$; note that

$$\langle \varphi, z \rangle = \langle tr_E, L_\varphi \tilde{\otimes}_\pi id_E(z) \rangle$$

(see 3.7.). A look at the commutative diagram

$$\begin{array}{ccc}
F \tilde{\otimes}_\pi E & \longrightarrow & F \tilde{\otimes}_\epsilon E \\
{\scriptstyle L_\varphi \tilde{\otimes}_\pi id_E} \downarrow & & \downarrow {\scriptstyle L_\varphi \tilde{\otimes}_\epsilon id_E} \\
E' \tilde{\otimes}_\pi E & \longrightarrow \mathfrak{N}(E, E) \hookrightarrow & E' \tilde{\otimes}_\epsilon E \\
{\scriptstyle tr_E} \downarrow & \swarrow {\scriptstyle \hat{tr}_E} & \\
\mathbb{K} & &
\end{array}$$

shows $\langle \varphi, z \rangle = 0$, hence $z = 0$. $\square$

It will be seen later on (21.9.) that it is enough to check (d) for separable reflexive spaces F only. It follows from 5.4. that E has the approximation property if and only if for all Banach spaces F (or only $F = E'$)

$$F \tilde{\otimes}_\pi E \longrightarrow (F' \otimes_\epsilon E')'$$

is injective.

5.7. Other consequences are immediate.

COROLLARY 1: *If E' or F has the approximation property, then*
$$E'\tilde{\otimes}_\pi F = \mathfrak{N}(E, F)$$
holds isometrically (E and F Banach spaces).

If E' has the approximation property, then, by the theorem,
$$E'\tilde{\otimes}_\pi E \longrightarrow E'\tilde{\otimes}_\varepsilon E$$
is injective, hence — again by the theorem — E has the approximation property:

COROLLARY 2: *A Banach space E has the approximation property whenever E' has it.*

The converse is false : It will be seen in Ex 12.9. that the space $\ell_2\tilde{\otimes}_\pi\ell_2$ has a basis but its dual $(\ell_2\tilde{\otimes}_\pi\ell_2)' = \mathfrak{L}(\ell_2, \ell_2)$ does not even have the approximation property. Since it will be shown in 16.3. that corollary 2 has an analogue for the metric approximation property, 5.2.(8) implies that the examples in 5.2.(5) and (6) also have duals without the approximation property.

COROLLARY 3: *Let E and F be Banach spaces. If F is reflexive or: if F' or E has the approximation property, then*
$$F'\tilde{\otimes}_\pi E = (\mathfrak{L}_{co}(E, F))' \quad (via\ D_F) \ .$$

We do not know whether or not D_F is always injective.

PROOF: The reflexive case was already treated in 5.5.. If F' or E has the approximation property, then
$$J_{F'} : F'\tilde{\otimes}_\pi E \longrightarrow F'\tilde{\otimes}_\varepsilon F$$
is injective. Since, by the formula (∗) in 5.6.,
$$\ker D_F \subset \ker J_{F'}$$
the result follows. □

This is a duality result between ε and π : if one defines the injective tensor product for locally convex spaces, then $E'_{co}\tilde{\otimes}_\varepsilon F = \mathfrak{L}_{co}(E, F)$ (if E or F has the approximation property), hence
$$E\tilde{\otimes}_\pi F' = (E'_{co}\tilde{\otimes}_\varepsilon F)'$$
whenever E or F' has the approximation property. The duality between ε and π will be studied in sections 6 and 16.

5.8. If $T_i \in \mathcal{L}(E_i, F_i)$ are injective operators and E_i are Banach spaces, then the operator $T_1 \tilde{\otimes}_\varepsilon T_2 : E_1 \tilde{\otimes}_\varepsilon E_2 \to F_1 \tilde{\otimes}_\varepsilon F_2$ is injective (see 4.3.). What about $T_1 \tilde{\otimes}_\pi T_2$? The commutative diagram

$$\begin{array}{ccc} E_1 \tilde{\otimes}_\pi E_2 & \longrightarrow & E_1 \tilde{\otimes}_\varepsilon E_2 \\ T_1 \tilde{\otimes}_\pi T_2 \downarrow & & \downarrow T_1 \tilde{\otimes}_\varepsilon T_2 \\ F_1 \tilde{\otimes}_\pi F_2 & \longrightarrow & F_1 \tilde{\otimes}_\varepsilon F_2 \end{array}$$

and the theorem lead to

COROLLARY 4: (1) *If the Banach space E_1 or E_2 has the approximation property, then $T_1 \tilde{\otimes}_\pi T_2$ is injective whenever both $T_i \in \mathcal{L}(E_i, F_i)$ are injective.*
(2) *Conversely, if for a Banach space E and its natural metric injection I_E into $\ell_\infty(B_{E'})$ the mapping*

$$id_{E'} \tilde{\otimes}_\pi I_E : E' \tilde{\otimes}_\pi E \longrightarrow E' \tilde{\otimes}_\pi \ell_\infty(B_{E'})$$

is injective, then E has the approximation property.

For (2) use the fact that $\ell_\infty(B_{E'})$ has the approximation property.

5.9. If $T = \sum_{n=1}^\infty x'_n \otimes y_n \in \mathfrak{N}(E, F)$, then

$$T' = \sum_{n=1}^\infty y_n \otimes x'_n \in \mathfrak{N}(F', E')$$

and this shows at the same time that $\mathbf{N}(T') \leq \mathbf{N}(T)$.

PROPOSITION: *Let E and F be Banach spaces, E' with the approximation property and $T \in \mathcal{L}(E, F)$. If T' is nuclear, then T is nuclear and $\mathbf{N}(T) = \mathbf{N}(T')$ holds.*

PROOF: First recall that $F \tilde{\otimes}_\pi E' \hookrightarrow F'' \tilde{\otimes}_\pi E'$ is a metric injection (3.9.). For a functional $\varphi \in (F'' \tilde{\otimes}_\pi E')'$ the following diagram holds (recall that transposition for tensors = dualization of operators):

$$\begin{array}{ccc} E' \tilde{\otimes}_\pi F & \hookrightarrow & \mathcal{L}(E, F) \ni T \\ \downarrow & & \downarrow \wr \\ \mathbb{K} \xleftarrow{\varphi} F'' \tilde{\otimes}_\pi E' & \xhookrightarrow{J} & \mathcal{L}(F', E') \ni T' \\ {}_{tr_{E'}}\nwarrow \quad \downarrow L_\varphi \tilde{\otimes}_\pi id_{E'} & & \downarrow \wr \\ E'' \tilde{\otimes}_\pi E' & \hookrightarrow & \mathcal{L}(E'', E'') \ni L_\varphi \circ T'' \end{array}$$

The three horizontal maps in the middle are injective since E' has the approximation property. If T' is nuclear, there is a $z \in F'' \tilde{\otimes}_\pi E'$ with $J(z) = T'$. It must be shown that $z \in E' \tilde{\otimes}_\pi F$: for this, take, by the Hahn–Banach theorem, a $\varphi \in (F'' \tilde{\otimes}_\pi E')'$ which is zero on $E' \tilde{\otimes}_\pi F$; this means precisely that $L_\varphi : F'' \to E''$ is zero on F. Since T' is nuclear, T is compact and hence $T''(E'') \subset F$, which implies that $L_\varphi \circ T'' = 0$ and the diagram shows that $\langle \varphi, z \rangle = 0$. □

This result will be improved in 22.8.; but in general there are non–nuclear operators the duals of which are nuclear (see 16.8. and 22.9.). For the nuclear norm it may even happen that $N(T') < N(T)$ for finite rank operators (see Ex 16.10.)!

Exercises:

Ex 5.1. $(\mathfrak{L}(E, F), \tau_s)' = E \otimes F'$, where τ_s is the topology of pointwise convergence in $\mathfrak{L}(E, F)$. Hint: Use the Hahn–Banach theorem as in proposition 5.5..

Ex 5.2. (a) The completion $\tilde{E}$ of a normed space E has the λ–bounded approximation property if E has it.

(b) For every normed space E there is a net (T_η) of finite rank operators which converges pointwise to the id_E. Hint: Finite dimensional subspaces are complemented.

(c) Every normed space of countable dimension has the approximation property. Hint: Banach discs are finite dimensional.

(d) The statement (a) is not true for the approximation property.

Ex 5.3. A separable Banach space E has the bounded approximation property if and only if there is a sequence (T_n) in $\mathfrak{F}(E, E)$ converging pointwise to id_E .

Ex 5.4. If E and F are Banach spaces with the approximation property (resp. bounded a.p.; metric a.p.), then $E \tilde{\otimes}_\pi F$ has the approximation property (resp. bounded a.p., metric a.p.). Hint: Lift compact sets.

Ex 5.5. The property of Banach spaces having a basis, the approximation property, the bounded or the metric approximation property is not stable under forming subspaces or quotients. The bounded approximation property and the approximation property are inherited by complemented subspaces. The latter is not true for having a basis — this follows from a result of Pełczyński's [200] together with Szarek's counterexample mentioned in 5.2.(7).

Ex 5.6. (a) Every abstract L–space has the metric approximation property.

(b) Using the fact that E has the metric approximation property if E' has it (this will be proven in 16.3.), show that abstract M–spaces have the metric approximation property.

Ex 5.7. If E has the λ-bounded approximation property for all $\lambda > \lambda_o$, then it has the λ_o-bounded approximation property.

Ex 5.8. Use the Davis–Figiel–Johnson–Pełczyński factorization theorem (see 9.6.) to show that in proposition 5.3. it is enough to test for reflexive spaces F.

Ex 5.9. Let $I : E_2 \overset{1}{\hookrightarrow} G$ be a metric injection, $Q : E \overset{1}{\twoheadrightarrow} E_1$ a metric surjection and $T \in \mathfrak{K}(E_1, E_2)$ such that $I \circ T$ and $T \circ Q$ are approximable. Is T approximable?

Ex 5.10. (a) If $F \subset F''$ is complemented and $T \in \mathfrak{L}(E, F)$ with $T' \in \mathfrak{N}(F', E')$, then T is nuclear as well. What about the nuclear norms?

(b) If F is reflexive, then $T \in \mathfrak{L}(E, F)$ is nuclear if and only if its dual is; moreover, $\mathbf{N}(T) = \mathbf{N}(T')$.

(c) If T' is nuclear and S weakly compact, then $S \circ T$ is nuclear.

Ex 5.11. The natural map $J : E' \tilde{\otimes}_\pi F \to \mathfrak{N}(E, F)$ is a metric surjection. Is it possible to find a finite rank operator $T \in \mathfrak{N}(E, F)$ such that for each n there exists $z_n \in E' \tilde{\otimes}_\pi F$ (or $E' \otimes_\pi F$) with $\pi(z_n; E', F) \geq n$ and $J(z_n) = T$?

Ex 5.12. Take $z \in E \tilde{\otimes}_\pi F$ such that for all $\varphi \in E'$
$$0 = (\varphi \tilde{\otimes}_\pi id_F)(z) \in \mathbb{K} \tilde{\otimes}_\pi F = F \ .$$
Is it true that $z = 0$? Under which conditions?

Ex 5.13. Let E and F be Banach spaces, $G \subset E$ a closed subspace with canonical injection I and assume that F' has the approximation property. Then the following two statements are equivalent:

(a) Each $T \in \mathfrak{L}(F, G)$ is nuclear if $I \circ T$ is nuclear.

(b) $F' \otimes_\pi E$ induces an equivalent norm on $F' \otimes_\pi G$.

Hint: $F'\tilde{\otimes}_\pi G \hookrightarrow F'\tilde{\otimes}_\pi E \overset{J}{\hookrightarrow} \mathfrak{L}(F, E)$ are injective and $M := J^{-1}(\mathfrak{L}(F, G))$ is closed, hence (a)$\sim$(b) is simple. For the converse try to use an argument as in the proof of 5.9. and observe that for $S \in \mathfrak{L}(E, F'') = (F'\tilde{\otimes}_\pi E)'$ with $S|_G = 0$ and $z \in M$ the operator $U \in \mathfrak{L}(F'', F'')$ coming from $id_{F'} \tilde{\otimes}_\pi S(z)$ is just $S \circ J(z)'' = 0$.

Ex 5.14. If E is a Banach space with the approximation property, $S \in \mathfrak{L}(E, F)$ invertible, then
$$\langle tr_E, T \rangle = \langle tr_F, STS^{-1} \rangle$$
for all $T \in \mathfrak{N}(E, E)$.

Ex 5.15. If (e_α) is an orthonormal basis of a Hilbert space, then
$$tr_H(T) = \sum_\alpha (Te_\alpha \mid e_\alpha)_H$$
for each $T \in \mathfrak{N}(H, H)$.

Ex 5.16. Denote by $J : E' \otimes F \to \mathfrak{F}(E,F)$ the canonical isomorphism and define a new norm on $\mathfrak{F}(E,F)$ by

$$\mathbf{N}^\circ(J(z)) := \pi(z; E', F) \ .$$

(a) $(\mathfrak{F}(E,F), \mathbf{N}^\circ)' = \mathfrak{L}(E', F')$ holds isometrically with the duality bracket (trace duality)

$$\langle S, T \rangle_{\mathfrak{F},\mathfrak{L}} := tr_{F'}(T \circ S') = tr_{E'}(S' \circ T)$$

(the trace is defined for finite rank operators).

(b) For $T \in \mathfrak{L}(E,F)$

$$\|T\| = \sup\ \{|tr_E(S \circ T)| \big| S \in \mathfrak{F}(F, E), \mathbf{N}^\circ(S) \leq 1\} \ .$$

(c) If F is reflexive, $S \in \mathfrak{F}(F, E)$, then there is a $T \in \mathfrak{L}(E, F)$ with $\|T\| = 1$ and

$$\mathbf{N}^\circ(S) = tr_E(S \circ T) \ .$$

(d) If E' or F has the approximation property, then $\mathbf{N}^\circ = \mathbf{N}$ on $\mathfrak{F}(E, F)$.

(e) $\mathbf{N} = \mathbf{N}^\circ$ on $\mathfrak{F}(E, E)$ if and only if E has the approximation property.

Ex 5.17. The operators

$$T \in \overline{\mathfrak{F}(E,F)}^{T_{co}} \subset \mathfrak{L}(E,F)$$

are called *compactly approximable*. With the notation of 5.5. and 5.6. observe that $\langle tr_F, T \circ J_{F'}(z) \rangle = \langle D_F(z), T \rangle$ for all $T \in \mathfrak{L}(E, F)$ and $z \in F' \otimes E$ and

$$\|\langle D_F(\cdot), T \rangle\|_{(F' \tilde{\otimes}_\pi E)'} = \|T\|$$

for all $T \in \mathfrak{L}(E, F)$. Show that for $T \in \mathfrak{L}(E, F)$ the following statements are equivalent:

(a) There is a linear functional η on $\mathfrak{N}(F, E)$ such that for all $z \in F' \tilde{\otimes}_\pi E$

$$\langle D_F(z), T \rangle = \langle \eta, J_{F'}(z) \rangle \ .$$

(b) For all $S \in \mathfrak{F}(F, E)$

$$|\langle tr_F, T \circ S \rangle| \leq \mathbf{N}(S) \|T\| \ .$$

(c) T is compactly approximable.

Note that η is necessarily continuous if it exists.

6. Duality of the Projective and Injective Norm

The injective norm ε was defined to be the "dual" of the projective norm π in the sense of trace duality; surprisingly enough, π is not dual to ε in this sense — but nearly. A consequence of this restricted duality is the (weak) principle of local reflexivity which says that the bidual E'' of a normed space looks "locally" like E — a fact which is fundamental for the understanding of modern Banach space theory.

6.1. The fact that $\overset{\circ}{B}_{E'\otimes_\pi F'} = \Gamma(\overset{\circ}{B}_{E'} \otimes \overset{\circ}{B}_{F'})$ implied, by the very definition of the injective norm ε, that

$$E \otimes_\varepsilon F \xrightarrow{\;1\;} (E' \otimes_\pi F')' \ .$$

Using the fact that the ε-norm on $E' \otimes F$ can be calculated on norming subsets of the unit balls instead of $B_{E''} \otimes B_{F'}$ (e.g. $B_E \subset B_{E''}$), it follows in the same way that

$$E' \otimes_\varepsilon F \xrightarrow{\;1\;} (E \otimes_\pi F')'$$
$$E' \otimes_\varepsilon F' \xrightarrow{\;1\;} (E \otimes_\pi F)' \ .$$

In this sense ε is dual to π on all pairs (E, F) of normed spaces.

6.2. Again by the definition of ε, each $x' \otimes y' \in B_{E'} \otimes B_{F'}$ defines a functional on $E \otimes_\varepsilon F$ of "integral" norm ≤ 1 (and $x' \otimes y$ on $E \otimes_\varepsilon F'$, ...). This implies that

$$\begin{aligned} \| E' \otimes_\pi F' &\longrightarrow (E \otimes_\varepsilon F)' \| \leq 1 \\ \| E' \otimes_\pi F &\longrightarrow (E \otimes_\varepsilon F')' \| \leq 1 \\ \| E \otimes_\pi F &\longrightarrow (E' \otimes_\varepsilon F')' \| \leq 1 \ . \end{aligned}$$

Note that this also follows from dualizing the natural embeddings in 6.1.. Since the space $(E' \otimes_\varepsilon F')'$ is complete and continuously embedded into $(E' \otimes_\pi F')' = \mathcal{L}(E', F''')$, the study of the approximation property showed that the extension to the completion

$$E \tilde{\otimes}_\pi F \longrightarrow (E' \otimes_\varepsilon F')'$$

may fail to be injective — in particular, $E \tilde{\otimes}_\pi F \hookrightarrow (E' \otimes_\varepsilon F')'$ may not be an isomorphic embedding. However, it will be shown that these embeddings are even isometries if one of the spaces is finite dimensional. The proof of this needs some preparation.

6.3. The *Banach–Mazur distance* of two normed spaces is defined by

$$d(E, F) := \inf\{\|T\|\,\|T^{-1}\| \mid T \in \mathcal{L}(E, F) \text{ bijective}\} \in [1, \infty] \ .$$

LEMMA: *For each finite dimensional normed space M and $\varepsilon > 0$ there is a subspace G of some ℓ_∞^m such that*

$$d(M, G) \leq 1 + \varepsilon \ .$$

Clearly, by dualization, there is also a metric quotient of ℓ_1^m with this property.

PROOF: Choose an ε-net $(x_1', ..., x_m')$ in the dual unit ball $B_{M'}$ (i.e. $\min_k \|x_k' - x'\| \leq \varepsilon$ for all $x' \in B_{M'}$) and define

$$T : M \longrightarrow \ell_\infty^m$$
$$x \rightsquigarrow (\langle x_k', x \rangle)_k$$

Clearly $\|T\| \leq 1$ and for each $x' \in B_{M'}$ there is a k such that

$$|\langle x', x \rangle| \leq |\langle x' - x_k', x \rangle| + |\langle x_k', x \rangle| \leq \varepsilon \|x\| + \|Tx\| \quad ,$$

hence $\|T^{-1} : TM \to M\| \leq (1-\varepsilon)^{-1}$, which implies the result. $\square$

Every finite dimensional space M (for short $M \in \text{FIN}$) is isometric to $\mathbb{K}^n$ with some norm. Choosing the functionals $x_k' \in (\mathbb{K}^n)^* = \mathbb{K}^n$ in the preceding proof with rational components, it is clear that there is a sequence (G_n) of normed spaces, each a subspace of some ℓ_∞^m such that for every finite dimensional normed space M and $\varepsilon > 0$ there is an n with

$$d(M, G_n) \leq 1 + \varepsilon \quad .$$

If $(G_n)_{n \in \mathbb{N}}$ is such a sequence,

$$C_p := \ell_p(G_n) \oplus_p \ell_p(G_n')$$

is said to be a *Johnson space* $(1 \leq p \leq \infty)$. Every $M \in \text{FIN}$ is $(1+\varepsilon)$–isometric to a 1–complemented subspace of C_p. All the C_p are reflexive for $1 < p < \infty$ and it is easy to show that for $1 \leq p < \infty$ the C_p have the metric approximation property (Ex 6.3.). However, C_∞ does not even have the approximation property since (by Ex 6.5.) every reflexive separable Banach space is 1–complemented in C_∞, hence also those which do not have the approximation property.

6.4. With the lemma the announced duality result can be demonstrated.

THEOREM: *If E and F are normed spaces, one of them finite dimensional, then*

(1) $\qquad E \otimes_\pi F \xhookrightarrow{1} (E' \otimes_\varepsilon F')'$

(2) $\qquad E' \otimes_\pi F' \xhookrightarrow{1} (E \otimes_\varepsilon F)'$

(3) $\qquad E' \otimes_\pi F \xhookrightarrow{1} (E \otimes_\varepsilon F')'$

(canonical embeddings).

If F is finite dimensional, then the mappings in (2) and (3) are surjective.

PROOF: The third isometry follows from (2) and 3.9.(3). For the proof of (1) and (2) assume E to be finite dimensional. Since by 6.1.

$$E' \otimes_\varepsilon F' = (E \otimes_\pi F)'$$

holds isometrically, the isometry (1) follows by dualization. The statement (2) is more difficult to prove — it will be shown in three steps:

(a) Assume first $E = \ell_\infty^n$, then by the examples 3.3. and 4.2.(2)

$$\ell_1^n \otimes_\pi F' = \ell_1^n(F') = (\ell_\infty^n(F))' = (\ell_\infty^n \otimes_\varepsilon F)' \quad \text{isometrically .}$$

(b) If $I : G \hookrightarrow \ell_\infty^n$ is a subspace, then the diagram

$$\begin{array}{ccc}
G' \otimes_\pi F' & \xrightarrow{\;J\;} & (G \otimes_\varepsilon F)' \\
{\scriptstyle I' \otimes id_{F'}} \uparrow & & \uparrow {\scriptstyle (I \otimes \cdot id_F)'} \\
\ell_1^n \otimes_\pi F' & \xrightarrow{\;1\;} & (\ell_\infty^n \otimes_\varepsilon F)'
\end{array} \qquad \text{by (a)}$$

commutes. Since π respects quotients and ε subspaces metrically, the vertical mappings are metric surjections and therefore J is an isometry (clearly, onto as well).

(c) Finally, assume that E is an arbitrary finite dimensional normed space. Choose, by 6.3., a subspace $G \subset \ell_\infty^n$ and a bijection $T : E \to G$ with $\|T\|\,\|T^{-1}\| \leq 1+\varepsilon$. The diagram

$$\begin{array}{ccc}
E' \otimes_\pi F' & \xrightarrow{\;J\;} & (E \otimes_\varepsilon F)' \\
{\scriptstyle T' \otimes id_{F'}} \uparrow & & \downarrow {\scriptstyle (T^{-1} \otimes \cdot id_F)'} \\
G' \otimes_\pi F' & \xrightarrow{\;1\;} & (G \otimes_\varepsilon F)'
\end{array} \qquad \text{by (b)}$$

implies $\|J^{-1}\| \leq \|T^{-1}\|\|T\| \leq 1+\varepsilon$ by the metric mapping property and the conclusion follows. □

Since $\mathfrak{N}(E, F) = E' \tilde\otimes_\pi F$ if E' or F has the approximation property, the following is a simple consequence of (3).

COROLLARY: *If E and F are Banach spaces, one of them finite dimensional, then*

$$\mathfrak{N}(E, F) \xrightarrow{\;1\;} (\mathfrak{L}(F, E))'$$

holds isometrically, the duality given by the trace.

Clearly $(\mathfrak{N}(E, F))' = (E' \otimes_\pi F)' = \mathfrak{L}(E', F')$ in this case. We shall come back to the question of the duality between $\mathfrak{L}$, $\mathfrak{K}$, and $\mathfrak{N}$ in section 16.

6.5. This kind of duality between ε and π produces an important tool for the local techniques in Banach space theory: the "local reflexivity" of all normed spaces; more precisely:

WEAK PRINCIPLE OF LOCAL REFLEXIVITY: *If M is a finite dimensional normed space, F an arbitrary normed space, and $S \in \mathfrak{L}(M, F'')$, then for every $\varepsilon > 0$ and every finite dimensional subspace N of F' there is an $R \in \mathfrak{L}(M, F)$ with $\|R\| \leq (1+\varepsilon)\|S\|$ such that for all $x \in M$ and $y' \in N$*

$$\langle y', Rx \rangle = \langle Sx, y' \rangle \quad .$$

PROOF: It follows from the duality relations that

$$M \otimes_\pi F' \stackrel{1}{=} (M' \otimes_\varepsilon F)' \stackrel{1}{=} (\mathfrak{L}(M, F))'$$
$$(\mathfrak{L}(M, F))'' \stackrel{1}{=} (M \otimes_\pi F')' \stackrel{1}{=} M' \otimes_\varepsilon F'' \stackrel{1}{=} \mathfrak{L}(M, F'') \quad .$$

A check on elementary tensors shows that $T \rightsquigarrow \kappa_F \circ T$

$$\mathfrak{L}(M, F) \hookrightarrow \mathfrak{L}(M, F'') = (\mathfrak{L}(M, F))''$$

is exactly the canonical injection of a normed space into its bidual. The space

$$L := M \otimes N \subset M \otimes F' = (\mathfrak{L}(M, F))'$$

is finite dimensional, hence the result follows from

HELLY'S LEMMA: *If E is a normed space, $L \subset E'$ finite dimensional and $x_0'' \in E''$, then for every $\varepsilon > 0$ there is an $x_o \in E$ with $\|x_o\| \leq (1+\varepsilon)\|x_0''\|$ and*

$$\langle x', x_o \rangle = \langle x_0'', x' \rangle \qquad \text{for all } x' \in L \quad .$$

PROOF: If L^0 is the polar of L in E'' and $^0L := L^0 \cap E$ the polar of L in E, then it follows that $\dim E/^0L = \dim L < \infty$ and hence the natural mapping

$$E/^0L \longrightarrow (E/^0L)'' = L' = E''/L^0$$

is a surjective isometry. This implies the result. $\square$

6.6. Although in most cases the weak principle is sufficient, it is interesting to know that the operator R can be chosen such that $Sx = Rx$ whenever $Sx \in F$, and injective if S is — with control of the norm of the inverse:

PRINCIPLE OF LOCAL REFLEXIVITY: *Let M and F be normed, M finite dimensional, $S \in \mathfrak{L}(M, F'')$ and N a finite dimensional subspace of F'. Then for every $\varepsilon > 0$ there is an $R \in \mathfrak{L}(M, F)$ such that*

(1) $\|R\| \leq (1+\varepsilon)\|S\|$.

(2) $\langle x', Ry \rangle = \langle Sy, x' \rangle$ *for all* $x' \in N$ *and* $y \in M$.

(3) $Ry = Sy$ *whenever* $Sy \in F$.

If S is injective, then R can be chosen to be injective such that additionally

(4) $\|R^{-1}\| \leq (1+\varepsilon)\|S^{-1}\|$.

PROOF: Denote by J the embedding $M_o := \{y \in M \mid Sy \in F\} \hookrightarrow M$ and define $S_o : M_o \longrightarrow F$ by $S_o y := Sy$. Now choose, according to the weak principle, for every finite dimensional $N \subset L \subset F'$ an operator $R_L \in \mathfrak{L}(M, F)$ with $\|R_L\| \leq (1+\varepsilon)\|S\|$ and $\langle x', R_L y \rangle = \langle Sy, x' \rangle$ for all $x' \in L$, $y \in M$; in particular, $\langle x', R_L Jy \rangle = \langle x', S_o y \rangle$ for all $x' \in L$, $y \in M_o$. Since $(\mathfrak{L}(M_o, F))' = (M'_o \otimes_\varepsilon F)' = M_o \otimes F'$, this means that the net $(R_L J)$ converges in $\mathfrak{L}(M_o, F)$ to S_o with respect to the weak topology on $\mathfrak{L}(M_o, F)$; it follows that S_o is in the weak=norm closure of the convex set

$$C := \left\{ UJ \in \mathfrak{L}(M_o, F) \;\middle|\; \begin{array}{l} U \in \mathfrak{L}(M, F) \text{ and } \|U\| \leq (1+\varepsilon)\|S\| \\ \langle x', Uy \rangle = \langle Sy, x' \rangle \text{ for all } x' \in N, y \in M \end{array} \right\} ,$$

hence there is for every $\delta > 0$ a $U_o J \in C$ with

$$\|S_o - U_o J\| \leq \delta \;\; ;$$

clearly, $\langle x', (S_o - U_o J)y \rangle = 0$ for all $x' \in N$ and $y \in M_o$. Now take any projection $Q : M \twoheadrightarrow M_o$ and define

$$R := (S_o - U_o J)Q + U_o \in \mathfrak{L}(M, F) .$$

Then (2) and (3) are obvious and (1) follows from

$$\|R\| \leq \|S_o - U_o J\| \, \|Q\| + \|U_o\| \leq \delta \|Q\| + (1+\varepsilon)\|S\| \;\; .$$

(b) Now suppose that S is injective and put $\lambda := \|S^{-1}\|^{-1}$. Then there is an ε_1-net $(x_i)_{i=1}^k$ in the unit sphere of M and y'_i in the unit sphere of E' such that

$$|\langle Sx_i, y'_i \rangle| \geq \lambda(1 - \varepsilon_1) \;\; .$$

Enlarging N if necessary, it may be assumed that all $y'_i \in N$ — hence the U_o constructed in (a) satisfies

$$\|U_o x_i\| \geq |\langle y'_i, U_o x_i \rangle| = |\langle Sx_i, y'_i \rangle| \geq \lambda(1 - \varepsilon_1) \;\; .$$

If $x \in M$ has norm 1 and $\|x - x_i\| \leq \varepsilon_1$, then

$$\|U_o x\| \geq \|U_o x_i\| - \|U_o(x - x_i)\| \geq \lambda(1 - \varepsilon_1) - (1+\varepsilon)\|S\|\varepsilon_1 > 0 ,$$

hence U_o is invertible and $\|U_o^{-1}\| \leq (1+\varepsilon)\|S^{-1}\|$ if ε_1 is chosen sufficiently small. Now the perturbation
$$W := R - U_o = (S_o - U_o J)Q \in \mathfrak{L}(M, F)$$
which was found in (a) has to be controlled. The operator
$$V := U_o^{-1}(W + U_o) = U_o^{-1}W + id_M$$
is invertible if $\|U_o^{-1}W\| \leq (1+\varepsilon)\,\delta\,\|S^{-1}\|\,\|Q\|$ is small and
$$\|V^{-1}\| \leq (1 - (1+\varepsilon)\,\delta\,\|S^{-1}\|\,\|Q\|)^{-1}$$
by the Neumann series. It follows that
$$\|R^{-1}\| = \|(U_o V)^{-1}\| \leq \|V^{-1}\|\,\|U_o^{-1}\| \leq (1+\varepsilon)^2\|S^{-1}\|$$
if δ was taken sufficiently small. $\square$

An immediate consequence is the

COROLLARY: *For every finite dimensional subspace $M \subset E''$ of a normed space E and $\varepsilon > 0$ there is a subspace $N \subset E$ with $d(M, N) \leq 1 + \varepsilon$.*

This property is called: E'' is *1–represented* in E — and this notion serves the purpose that if F is represented in E, then F shares those properties with E which are determined by the finite dimensional subspaces. Clearly, the bidual E'' is much more than 1–represented in E: The subspace N and the bijection $R : M \to N$ can be chosen such that $M \cap E \subset N$ and $Rx = x$ for all $x \in M \cap E$.

6.7. An application: Every $\varphi \in \mathfrak{Bil}(E, F) = (E \otimes_\pi F)'$ has, by 1.9., a canonical extension $\varphi^\wedge \in \mathfrak{Bil}(E, F'') = (E \otimes_\pi F'')'$ with $\|\varphi\| = \|\varphi^\wedge\|$. What if φ is continuous with respect to the injective norm?

EXTENSION LEMMA: *Let E and F be normed and $\varphi \in (E \otimes_\pi F)'$. Then $\varphi \in (E \otimes_\varepsilon F)'$ if and only if $\varphi^\wedge \in (E \otimes_\varepsilon F'')'$; in this case,*
$$\|\varphi\|_{(E\otimes_\varepsilon F)'} = \|\varphi^\wedge\|_{(E\otimes_\varepsilon F'')'} \quad .$$

PROOF: Since φ is the restriction of $\varphi^\wedge$ and $E \otimes_\varepsilon F \overset{1}{\hookrightarrow} E \otimes_\varepsilon F''$ holds isometrically, it follows that $\|\varphi\|_{...} \leq \|\varphi^\wedge\|_{...}$. Take, conversely, $z \in E \otimes_\varepsilon F''$ and finite dimensional subspaces $N \subset E$ and $M \subset F''$ with $z \in N \otimes M$. Then, by the weak principle of local reflexivity, there is (for every $\varepsilon > 0$) an $R \in \mathfrak{L}(M, F)$ with $\|R\| \leq 1 + \varepsilon$ such that for all $y'' \in M$ and $x \in N$
$$\langle y'', L_\varphi x\rangle_{F'',F'} = \langle L_\varphi x, Ry''\rangle_{F',F}$$

($L_\varphi : E \to F'$ the linear operator associated with φ); this means

$$\langle \varphi^\wedge, x \otimes y'' \rangle = \langle \varphi, (id_E \otimes R)(x \otimes y'') \rangle$$

and therefore $\langle \varphi^\wedge, z \rangle = \langle \varphi, (id_E \otimes R)(z) \rangle$ for all $z \in E \otimes F''$. It follows that

$$|\langle \varphi^\wedge, z \rangle| \leq \|\varphi\|_{...}\varepsilon((id_E \otimes R)(z); E, F) \leq$$
$$\leq \|\varphi\|_{...}(1+\varepsilon)\varepsilon(z; E, M) = \|\varphi\|_{...}(1+\varepsilon)\varepsilon(z; E, F'')$$

and hence $\|\varphi^\wedge\|_{...} \leq \|\varphi\|_{...}$. $\square$

This lemma will be used in section 10 for the investigation of integral operators.

Exercises:

Ex 6.1. Show that $(\mathfrak{L}(E,F))' \stackrel{1}{=} \mathfrak{N}(F,E)$ under trace duality if the Banach spaces E and F are reflexive, and one of them is finite dimensional. In particular,

$$\mathbf{N}(T : F \to E) = \sup\{|tr(S \circ T)| \mid S \in \mathfrak{L}(E,F), \|S\| \leq 1\} .$$

Ex 6.2. Let $T \in \mathfrak{F}(F', E')$, $A \in \mathfrak{L}(E', M')$ and suppose that M is finite dimensional. Then, given $\varepsilon > 0$, there is an operator $R \in \mathfrak{L}(M, E)$ such that $\|R\| \leq (1+\varepsilon)\|A\|$ and $R' \circ T = A \circ T$. Hint: Apply the weak principle of local reflexivity to $S := A'$ and $N := \text{im}(T)$.

Ex 6.3. Check that the Johnson spaces C_p have the metric approximation property if $1 \leq p < \infty$.

Ex 6.4. *Lindenstrauss' compactness argument:* Take Banach spaces E and F, an operator $T \in \mathfrak{L}(E, F)$ and a net (F_α) of subspaces of F, directed by inclusion, such that $F_o := \bigcup_\alpha F_\alpha$ is dense in F. Let $S_\alpha \in \mathfrak{L}(F_\alpha, E)$ be operators such that $TS_\alpha y = y$ for all $y \in F_\alpha$ and $\limsup \|S_\alpha\| = \lambda < \infty$.
(a) Define $\varphi_\alpha : E' \times F_o \to \mathbb{K} \cup \{\infty\}$ (one-point compactification) by

$$\varphi_\alpha(x', y) := \begin{cases} \langle x', S_\alpha y \rangle & \text{if } y \in F_\alpha \\ 0 & \text{if } y \notin F_\alpha \end{cases} .$$

Use the fact that (φ_α) is a net in the compact space $(\mathbb{K} \cup \{\infty\})^{E' \times F_o}$ (hence has a convergent subnet) to show that there is a $\varphi \in \mathfrak{Bil}(E', F_o)$ such that $\|\varphi\| \leq \lambda$ and

$$\varphi(T'y', y) = \langle y', y \rangle$$

for all $y' \in F'$ and $y \in F_o$.

(b) Let $U \in \mathfrak{L}(E', F')$ be the operator associated with φ. Show that $UT' = id_{F'}$.

(c) T' is invertible on $T'(F') \subset E'$ and $\|(T')^{-1}\| \leq \lambda$; the subspace $T'(F')$ is complemented in E' with a projection of norm $\leq \lambda \|T\|$.

This result is quite relevant as the next exercise will show. The technique was refined to the so-called ultraproduct technique; see section 18, in particular Ex 18.9..

Ex 6.5. If F is a separable Banach space, then F' is isometric to a 1-complemented subspace of each Johnson space C_∞ . Hint: Choose an increasing sequence (F_k) of finite dimensional subspaces and $T_n : G_{n(k)} \to F_k$ with $\|T_n\| = 1$ and $\|T_n^{-1}\| \leq 1 + 1/n$; take in the foregoing exercise $T = \sum T_n \in \mathfrak{L}(C_1, F)$ and $S_n := T_n^{-1}$. This result is due to Johnson [130] .

Ex 6.6. Use the principle of local reflexivity to show that a normed space E is an $\mathfrak{L}_{p,\lambda}^g$-space if E'' or the completion $\tilde{E}$ is $\mathfrak{L}_{p,\lambda}^g$. The full duality will be proved in section 23; for the converse of the statement about the completion see Ex 4.7..

Ex 6.7. Every abstract M-space is an $\mathfrak{L}_{\infty,1}^g$-space .

Ex 6.8. (a) Show that there are elements $z_n \in \ell_\infty^n \otimes \ell_\infty^n$ with $\varepsilon(z_n; \ell_\infty^n, \ell_\infty^n) \leq 1$ and $\pi(z_n; \ell_\infty^n, \ell_\infty^n) \geq \sqrt{n}$. Hint: Ex 4.3..

(b) If K and L are infinite compact sets, then there are $f_n \in C(K) \otimes C(L)$ with $\|f_n\|_\infty = 1$ and $\pi(f_n; C(K), C(L)) \geq \sqrt{n}$.

7. The Natural Norm on the p–Integrable Functions

The third type of natural norms for tensor products is the norm induced on $L_p \otimes E$ from $L_p(E)$. It behaves fundamentally different from ε and π in one very important aspect: It does *not* satisfy the mapping property! Therefore, this section studies some additional conditions under which tensor product operators are continuous with respect to Δ_p and gives a first look at the highly interesting consequences of this "failure".

7.1. Let (Ω, μ) be an arbitrary measure space, $1 \leq p \leq \infty$, and E a normed space. Then for $p \neq \infty$ the space of (classes of a.e. equal) Bochner p-integrable functions $\Omega \to \tilde{E}$ is denoted by

$$L_p(\mu, \tilde{E})$$

and the space of (classes of locally a.e. equal) bounded μ-measurable functions $\Omega \to \tilde{E}$

by $L_\infty(\mu, \tilde{E})$, see Appendix B12.. Since the natural mapping

$$L_p(\mu) \otimes E \hookrightarrow L_p(\mu, \tilde{E})$$
$$\tilde{f} \otimes x \rightsquigarrow \widetilde{f(\cdot)x}$$

is injective, the definitions

$$\Delta_p(\tilde{f}; L_p, E) := \left(\int_\Omega \|f(w)\|_E^p \mu(dw) \right)^{1/p}$$

for $1 \leq p < \infty$ and

$$\Delta_\infty(\tilde{f}; L_\infty, E) := \text{ess-sup } \|f(\cdot)\|_E$$

give norms on $L_p \otimes E$; notation: $L_p(\mu) \otimes_{\Delta_p} E$ and $L_p(\mu) \tilde{\otimes}_{\Delta_p} E$ for the completion.

PROPOSITION:
(1) $\quad\quad\quad\quad\quad\quad \Delta_1 = \pi \quad \text{on} \quad L_1(\mu) \otimes E$.
(2) $\quad\quad\quad\quad\quad\quad \Delta_\infty = \varepsilon \quad \text{on} \quad L_\infty(\mu) \otimes E$.
(3) $\quad\quad\quad\quad\quad\quad \varepsilon \leq \Delta_p \leq \pi \quad \text{on} \quad L_p(\mu) \otimes E$.

PROOF: (1) was one of the main examples for the projective norm (see 3.3.) and (2) was Ex 4.9.. For (3) observe first that for $g \otimes x \in L_p \otimes E$ the relation $\Delta_p(g \otimes x) = \|g\|_{L_p} \|x\|_E$ implies $\Delta_p \leq \pi$ by the triangle inequality. To see $\varepsilon \leq \Delta_p$ take $h \in B_{L_{p'}}$ (conjugate exponent), $x' \in B_{E'}$ and fix $f \in L_p \otimes E$; then

$$|\langle h \otimes x', f \rangle| = \left| \int h(w) \langle x', f(w) \rangle \mu(dw) \right| \leq$$
$$\leq \|h\|_{L_{p'}} \left(\int |\langle x', f(w) \rangle|^p \mu(dw) \right)^{1/p} \leq \Delta_p(f; L_p, E) . \quad \Box$$

7.2. For $1 \leq p < \infty$ the step functions

$$S(\mu) \otimes E = \left\{ \sum_{k=1}^n \chi_{A_k} \otimes x_k \mid n \in \mathbb{N}, A_k \text{ integrable, } x_k \in E \right\} / \sim$$

are dense in $L_p(\mu, \tilde{E})$. This implies that $L_p \otimes E$ is dense in $L_p(\tilde{E})$ and therefore

$$L_p(\mu) \tilde{\otimes}_{\Delta_p} E = L_p(\mu) \tilde{\otimes}_{\Delta_p} \tilde{E} = L_p(\mu, \tilde{E})$$

holds isometrically. For the product measure $\mu_1 \otimes \mu_2$ on $\Omega_1 \times \Omega_2$ the Fubini–Tonelli theorem shows that

$$L_p(\mu_1) \tilde{\otimes}_{\Delta_p} L_p(\mu_2) \xrightarrow{\;1\;} L_p(\mu_1 \otimes \mu_2)$$

holds isometrically. Since the step functions in $S(\mu_1) \otimes S(\mu_2)$ are dense in $L_p(\mu_1 \otimes \mu_2)$, it follows that this injection has dense range and therefore isometrically

$$L_p(\mu_1 \otimes \mu_2) \stackrel{1}{=} L_p(\mu_1) \tilde{\otimes}_{\Delta_p} L_p(\mu_2) \stackrel{1}{=} L_p(\mu_1, L_p(\mu_2))$$

($1 \leq p < \infty$). For $p = \infty$ this is false (Ex 4.8. for discrete measures and Ex 7.2. for the general case). Recall from Ex 4.9. that, in general, $L_\infty \otimes E$ is not *dense* in $L_\infty(\tilde{E})$. Finally, it is convenient to define the "transpose" Δ_p^t of Δ_p:

$$\begin{array}{ccc} E \otimes_{\Delta_p^t} L_p(\mu) & :\stackrel{1}{=} & L_p(\mu) \otimes_{\Delta_p} E \\ x \otimes f & \rightsquigarrow & f \otimes x \end{array}$$

— the isometry being the transposition map. Clearly,

$$L_p(\mu) \otimes_{\Delta_p^t} L_p(\nu) \stackrel{1}{=} L_p(\mu) \otimes_{\Delta_p} L_p(\nu) \xhookrightarrow{\;\;1\;\;} L_p(\mu \otimes \nu)$$

holds.

The norm in $L_q(\Omega_1, \mu_1) \otimes_{\Delta_p^t} L_p(\Omega_2, \mu_2)$ is

$$\Delta_p^t(f; L_q, L_p) = \left(\int_{\Omega_2} \left(\int_{\Omega_1} |f(w_1, w_2)|^q \mu_1(dw_1) \right)^{p/q} \mu_2(dw_2) \right)^{1/p}$$

and hence the continuous triangle inequality (see Appendix B5.) gives the

PROPOSITION: *For measures μ_1 and μ_2 (different from zero) and $1 \leq q \leq p \leq \infty$*

$$\|id_{L_q} \otimes id_{L_p} : L_q(\mu_1) \otimes_{\Delta_q} L_p(\mu_2) \longrightarrow L_q(\mu_1) \otimes_{\Delta_p^t} L_p(\mu_2)\| = 1 \ .$$

The case $\Delta_\infty = \varepsilon$ is trivial. This proposition has an interesting consequence for the continuity of operators $S \otimes T : L_q \otimes L_q \to L_p \otimes L_p$, see theorem 7.9..

7.3. Take operators $S \in \mathcal{L}(L_p(\mu), L_p(\nu))$ and $T \in \mathcal{L}(E, F)$. Is it true that

$$S \otimes T : L_p(\mu) \otimes_{\Delta_p} E \longrightarrow L_p(\nu) \otimes_{\Delta_p} F$$

is continuous, in other words: does the mapping property which holds for π and ε (see 3.2. and 4.1.) also hold for Δ_p? There are natural examples showing that this is not true for $1 < p < \infty$ (see 7.5. and 7.6.). But at least in some cases $S \otimes T$ is continuous: Take $S = id_{L_p}$ and $T \in \mathcal{L}(E, F)$. Then

$$\Delta_p \left(\sum_{k=1}^n f_k \otimes Tx_k; L_p, F \right) = \left(\int \left\| \sum_{k=1}^n f_k(w) Tx_k \right\|_F^p \mu(dw) \right)^{1/p} =$$

$$= \left(\int \left\| T \left(\sum_{k=1}^n f_k(w) x_k \right) \right\|_F^p \mu(dw) \right)^{1/p} \leq \|T\| \Delta_p \left(\sum_{k=1}^n f_k \otimes x_k; L_p, E \right)$$

and hence
$$\|id \otimes T : L_p \otimes_{\Delta_p} E \longrightarrow L_p \otimes_{\Delta_p} F\| \leq \|T\| \quad .$$
More generally,

THEOREM: *Let μ and ν be arbitrary measures, $p, q \in [1, \infty]$ and $T \in \mathfrak{L}(E, F)$.*
(1) *If $S \in \mathfrak{L}(L_p(\mu), L_q(\nu))$ is positive, then*
$$\|S \otimes T : L_p(\mu) \otimes_{\Delta_p} E \longrightarrow L_q(\nu) \otimes_{\Delta_q} F\| = \|S\| \, \|T\| \quad .$$
(2) *If K is compact and $S \in \mathfrak{L}(C(K), L_q(\nu))$ positive, then*
$$\|S \otimes T : C(K) \otimes_{\varepsilon} E \longrightarrow L_q(\nu) \otimes_{\Delta_q} F\| = \|S\| \, \|T\| \quad .$$

Since positive operators are involved, it is worthwhile to note that the real and the complex case are treated jointly. Before beginning the proof recall the obvious (see also Ex 3.4.)

DENSITY LEMMA: *Let $\| \; \|_i$ be two norms on a vector space E with $\| \; \|_2 \leq c \| \; \|_1$ for some $c > 0$ and $T : E \to F$ a linear map into a normed space F. If T is $\| \; \|_1$-continuous and for some $\lambda \geq 0$*
$$\|Tx\|_F \leq \lambda \|x\|_2$$
holds for all x in a $\| \; \|_1$-dense subset of E, then
$$\|T : (E, \| \; \|_2) \to F\| \leq \lambda \quad .$$

PROOF of the proposition: Since $S \otimes T = (id_{L_q(\nu)} \otimes T) \circ (S \otimes id_E)$, it is enough to take $T = id_E$. For $G = L_p(\mu)$ or $C(K)$, clearly
$$G \otimes_{\pi} E \xrightarrow{S \otimes id_E} L_q(\nu) \otimes_{\pi} E \hookrightarrow L_q(\nu) \otimes_{\Delta_q} E$$
is continuous, hence, by the density lemma, it suffices to check the continuity-inequality on a π-dense subset of $G \otimes_{\pi} E$. For (1) it is therefore enough to take step functions
$$f = \sum_{m=1}^{n} \chi_{A_m} \otimes x_m$$
with pairwise disjoint A_m : Since $|S\chi_A| = S\chi_A$, it follows that
$$\Delta_q (S \otimes id_E(f); L_q(\nu), E) = \left(\int \| \sum_{m=1}^{n} [S\chi_{A_m}](w) x_m \|_E^q \nu(dw) \right)^{1/q} \leq$$
$$\leq \left(\int \left(\sum_{m=1}^{n} |S\chi_{A_m}| \|x_m\| \right)^q d\nu \right)^{1/q} = \left(\int |S\left(\sum_{m=1}^{n} \|x_m\| \chi_{A_m} \right)|^q d\nu \right)^{1/q} \leq$$
$$\leq \|S\| \left(\int \left(\sum_{m=1}^{n} \|x_m\| \chi_{A_m} \right)^p d\mu \right)^{1/p} = \|S\| \Delta_p(f; L_p(\mu), E)$$

— with obvious modifications if $p = \infty$ (the case $q = \infty$ follows also from $\varepsilon \leq \Delta_p$). For (2) take a partition (φ_m) of unity in K such that there are $t_m \in K$ with $\varphi_m(t_m) = 1$ and put

$$f := \sum_{m=1}^n \varphi_m \otimes x_m \in C(K) \otimes E$$

(these functions form a dense subset of $C(K) \otimes_\pi E$); then

$$\varepsilon(f; C(K), E) = \max_{m=1,\ldots,n} \|x_m\| = \left\| \sum_{m=1}^n \varphi_m \|x_m\| \right\|_{C(K)}$$

and therefore

$$\Delta_q \left(S \otimes id_E(f); L_q(\nu), E \right) = \left(\int \left\| \sum_{m=1}^n S(\varphi_m) x_m \right\|^q d\nu \right)^{1/q} \leq$$

$$\leq \left(\int \left| \sum_{m=1}^n |S\varphi_m| \|x_m\| \right|^q d\nu \right)^{1/q} \leq \left(\int \left| S\left(\sum_{m=1}^n \|x_m\| \varphi_m \right) \right|^q d\nu \right)^{1/q} \leq$$

$$\leq \|S\| \, \varepsilon(f; C(K), E) . \quad \square$$

It is obvious that $S \otimes T$ is also continuous if the operator $S \in \mathfrak{L}(L_p(\mu), L_q(\nu))$ is *regular*, i.e. $S = S_1 - S_2$ in the real case, or $S = S_1 - S_2 + i(S_3 - S_4)$ in the complex case, where the operators S_j are positive. The following result (due to Virot [275]) shows that this is best possible:

PROPOSITION: *Let $p, q \in [1, \infty]$ with $q < \infty$ and $S \in \mathfrak{L}(L_p(\mu), L_q(\nu))$ such that*

$$S \otimes id_{\ell_\infty} : L_p(\mu) \otimes_{\Delta_p} \ell_\infty \longrightarrow L_q(\nu) \otimes_{\Delta_q} \ell_\infty$$

is continuous. Then S is regular.

The analoguous result holds for operators $S \in \mathfrak{L}(C(K), L_q(\nu))$. The proof will show that both results also hold for $q = \infty$ if ν is localizable, i.e. $L_\infty(\nu)$ is order complete. Since $\Delta_1 = \pi$ and π has the metric mapping property, the proposition shows that every operator $L_1(\mu) \longrightarrow L_1(\nu)$ is regular. See also Ex 7.19..

PROOF: Considering $\sum_{k=1}^n f_k \otimes e_k$, it is obvious that continuity means that

$$\left\| \max_{k=1,\ldots,n} |Sf_k| \right\|_{L_q(\nu)} \leq \|S \otimes id_{\ell_\infty} : \ldots\| \left\| \max_{k=1,\ldots,n} |f_k| \right\|_{L_p(\mu)} .$$

The type of order completeness of $L_q(\nu)$ (see Appendix B5.) shows that S throws order bounded sets into order bounded sets. But this implies that S is regular (see e.g. Schaefer [244], p.229 and p.233). $\square$

7.4. With regard to subspaces and quotients the norm Δ_p is well–behaved:

PROPOSITION: *Let μ be a measure and $1 \leq p \leq \infty$. Then*

$$L_p(\mu) \otimes_{\Delta_p} \cdot$$

respects metric injections and metric surjections.

PROOF: For $p = 1$ or ∞ this was shown in 3.8., 3.10., 4.3. and 4.4. (recall that $L_\infty(\mu) = C(K)$ for some compact K). That $L_p(\mu) \otimes_{\Delta_p} \cdot$ respects metric injections follows (for $1 < p < \infty$) directly from the definition of the norm. Now take $Q : E \twoheadrightarrow F$ a metric surjection between normed spaces; by 7.3.

$$id \otimes Q : L_p \otimes_{\Delta_p} E \longrightarrow L_p \otimes_{\Delta_p} F$$

is continuous with norm ≤ 1 and clearly surjective. For every $g \in L_p \otimes_{\Delta_p} F$ and $\varepsilon > 0$ it is necessary to find $f \in L_p \otimes_{\Delta_p} E$ with $id \otimes Q(f) = g$ and

$$\Delta_p(f) \leq (1+\varepsilon)\Delta_p(g) \ .$$

If $g = \sum_{k=1}^n \chi_{A_k} \otimes y_k$ is a step function (the A_k pairwise disjoint), then there are $x_k \in E$ with $Q(x_k) = y_k$ and $\|x_k\| \leq (1+\varepsilon)\|y_k\|$. Hence $f := \sum_{k=1}^n \chi_{A_k} \otimes x_k$ covers g and

$$\Delta_p(f) = \left(\sum_{k=1}^n \int_{A_k} \|x_k\|^p d\mu\right)^{1/p} \leq (1+\varepsilon)\Delta_p(g) \ .$$

In other words,

$$S_p(\mu) \otimes_{\Delta_p} E \xrightarrow{id \otimes Q} S_p(\mu) \otimes_{\Delta_p} F$$

(induced norms) is a metric surjection. The result now follows from the following lemma which is of independent interest (for the condition on the kernels see Ex 7.7.):

QUOTIENT LEMMA: *Let E and F be normed spaces, $Q \in \mathfrak{L}(E, F)$ surjective, $E_o \subset E$ dense and $Q_o := Q|_{E_o} : E_o \to Q(E_o)$ the (surjective) restriction. Then Q_o is a metric surjection if and only if $\overline{\ker Q_o} = \ker Q$ and Q is a metric surjection.*

PROOF: Assume first that Q_o is a metric surjection and take $x \in \ker Q$. Then there are $x_n \in E_o$ with $x_n \to x$ and, by assumption, $z_n \in E_o$ with $Q_o(z_n) = Q(x_n)$ and

$$\|z_n\| \leq 2\|Q(x_n)\| \longrightarrow 2\|Q(x)\| = 0 \ .$$

It follows that $x_n - z_n \in \ker Q_o$ and $x_n - z_n \to x$, hence

$$\ker Q \subset \overline{\ker Q_o} \ .$$

Both implications of the claim now follow from the simple observation that

$$\|Q(x)\|_1 := \inf\{\|x+z\| \,|\, z \in \ker Q_o\} = \inf\{\|x+z\| \,|\, z \in \overline{\ker Q_o}\}$$

for all $x \in E$ and $\|\ \| \leq \|\ \|_1$ on F. □

7.5. As was already mentioned, the operator $S \otimes id_E : L_p \otimes_{\Delta_p} E \to L_p \otimes_{\Delta_p} E$ is not always continuous. To see a concrete example, take S to be the *Fourier transform* $\mathfrak{F} : L_2(\mathbb{R}) \to L_2(\mathbb{R})$ defined by

$$[\mathfrak{F}(f)](t) := (2\pi)^{-1/2} \int_{-\infty}^{\infty} \exp(-ist) f(s) ds$$

if f has compact "support" — and then continuous extension; $\mathfrak{F}$ is an isometry. For normed spaces E, consider the E-valued Fourier transform

$$\mathfrak{F}_E := \mathfrak{F} \otimes id_E : L_2(\mathbb{R}) \otimes_{\Delta_2} E \longrightarrow L_2(\mathbb{R}) \otimes_{\Delta_2} E \ .$$

If H is a Hilbert space, a simple calculation with scalar products shows that $\mathfrak{F}_H$ is an isometry (see Ex 7.8.); considering H as some L_2, this also follows from 7.4.. But if $E = \ell_1$, the operator $\mathfrak{F}_{\ell_1}$ is not even continuous: For the unit vectors e_n in ℓ_1 take

$$f_n := \sum_{m=1}^{n} \chi_{[m,m+1[} \otimes e_m \ ;$$

then

$$\Delta_2(f_n; L_2, \ell_1) = \left(\int_{-\infty}^{\infty} \left\| \sum_{m=1}^{n} \chi_{[m,m+1[}(t) e_m \right\|_{\ell_1}^2 dt \right)^{1/2} = \sqrt{n} \ .$$

However, from

$$[\mathfrak{F}(\chi_{[a,b[})](t) = \frac{i}{\sqrt{2\pi}\, t} (\exp(-ibt) - \exp(-iat))$$

it follows that

$$\Delta_2(\mathfrak{F}_{\ell_1}(f_n); L_2, \ell_1) =$$

$$= \frac{1}{\sqrt{2\pi}} \left(\int_{-\infty}^{\infty} \left\| \frac{1}{t} \sum_{m=1}^{n} [\exp(-i(m+1)t) - \exp(-imt)] e_m \right\|_{\ell_1}^2 dt \right)^{1/2} =$$

$$= \frac{1}{\sqrt{2\pi}} \left(\int_{-\infty}^{\infty} \frac{1}{t^2} \left(\sum_{m=1}^{n} |\exp(-i(m+1)t) - \exp(-imt)| \right)^2 dt \right)^{1/2} =$$

$$= \frac{n}{\sqrt{2\pi}} \left(\int_{-\infty}^{\infty} \left| \frac{\exp(-it) - 1}{t} \right|^2 dt \right)^{1/2} = n \cdot c \ .$$

This shows that the operator $\mathfrak{F}_{\ell_1} = \mathfrak{F} \otimes id_{\ell_1}$ cannot be continuous. For the Fourier transform on $\ell_2(\mathbb{Z})$ see Ex 7.18..

A celebrated result of Kwapień [161] says that a Banach space E is isomorphic to a Hilbert space if and only if $\mathfrak{F}_E$ is continuous; this will be shown in section 30.

7.6. Another example is given by the discrete *Hilbert transform* $\mathfrak{H} : \ell_p(\mathbb{Z}) \to \ell_p(\mathbb{Z})$ for $1 < p < \infty$ which is defined by

$$\mathfrak{H}(e_k) := \sum_{\substack{m \in \mathbb{Z} \\ m \neq k}} \frac{1}{m-k} e_m \ .$$

Due to a result of M. Riesz $\mathfrak{H}$ is continuous (see for example Edwards–Gaudrey [62], theorem 6.7.4.), but $\mathfrak{H} \otimes id_{\ell_1}$ is not: Take $z_n := \sum_{k=1}^n e_k \otimes e_k \in \ell_p(\mathbb{Z}) \otimes \ell_1$; then

$$\Delta_p(z_n; \ell_p(\mathbb{Z}), \ell_1) = n^{1/p}$$

and

$$\mathfrak{H} \otimes id_{\ell_1}(z_n) = \sum_{k=1}^n \sum_{\substack{m \in \mathbb{Z} \\ m \neq k}} \frac{1}{m-k} e_m \otimes e_k = \sum_{m \in \mathbb{Z}} e_m \otimes \sum_{\substack{k=1 \\ k \neq m}}^n \frac{1}{m-k} e_k \ .$$

Therefore

$$\Delta_p(\mathfrak{H} \otimes id_{\ell_1}(z_n); \ell_p(\mathbb{Z}), \ell_1) = \Big(\sum_{m \in \mathbb{Z}} \Big(\sum_{\substack{k=1 \\ k \neq m}}^n |\frac{1}{m-k}|\Big)^p\Big)^{1/p} \geq$$

$$\geq \Big(\sum_{m=3}^n \Big(\sum_{k=1}^{m-1} \frac{1}{m-k}\Big)^p\Big)^{1/p} \geq \Big(\sum_{m=3}^n (\log m)^p\Big)^{1/p} \geq (\log \frac{n!}{2})^{1/p}$$

and since $(1/n) \cdot \log(n!/2) \to \infty$ the operator $\mathfrak{H} \otimes id_{\ell_1}$ cannot be continuous.

7.7. Though at first glance it is disappointing that Δ_p does not satisfy the mapping property, Kwapień's result shows that this behaviour gives new structural insights. It even can be said that this "misbehaviour" is an incredibly rich source for interesting properties of spaces and operators in modern Banach space theory: One of these sources is to study, for a fixed operator $S_o \in \mathfrak{L}(L_p(\mu), L_q(\nu))$, the class of all those operators $T \in \mathfrak{L}(E, F)$ such that

$$S_o \otimes T : L_p(\mu) \otimes_{\Delta_p} E \longrightarrow L_q(\nu) \otimes_{\Delta_q} F$$

is continuous — and those spaces E for which id_E is in this class.

To see an example of this kind, consider the Rademacher functions $r_n \in L_2[0,1]$ defined in 1.5.(2). They form an orthonormal sequence, hence

$$\text{Rad} := \overline{\text{span}}\{r_n\} = \Big\{\sum_{n=1}^\infty \alpha_n r_n \in L_2[0,1] \,\Big|\, (\alpha_n) \in \ell_2\Big\}$$

is a complemented subspace of $L_2[0,1]$. If E is normed, define Δ_2 on $\text{Rad} \otimes E$ to be the norm induced from
$$\text{Rad} \otimes E \hookrightarrow L_2[0,1] \otimes_{\Delta_2} E \ .$$
This means that
$$\Delta_2(f; \text{Rad}, E) := \left(\int_0^1 \left\|\sum_{n=1}^\infty r_n(t)x_n\right\|_E^2 dt\right)^{1/2}$$
is the norm of $f = \sum_{n=1}^\infty r_n x_n$ in $\text{Rad} \otimes_{\Delta_2} E$ and *not* the norm coming from $\ell_2 \otimes E$ via the identification
$$I : \ell_2 \longrightarrow \text{Rad}$$
$$(\alpha_n) \rightsquigarrow \sum \alpha_n r_n$$
of Rad with ℓ_2. Actually it is this difference which counts: An operator $T \in \mathcal{L}(E,F)$ is called *of type p* (for $1 \le p \le 2$, hence $I_p : \ell_p \hookrightarrow \ell_2 \xrightarrow{I} \text{Rad}$ is defined) if
$$\|I_p \otimes T : \ell_p \otimes_{\Delta_p} E \longrightarrow L_2[0,1] \otimes_{\Delta_2} F\| =: \mathbf{T}_p(T) < \infty$$
and *of cotype q* (for $2 \le q \le \infty$, hence $J_q : \text{Rad} \xrightarrow{I^{-1}} \ell_2 \hookrightarrow \ell_q$ is defined) if
$$\|J_q \otimes T : \text{Rad} \otimes_{\Delta_2} E \longrightarrow \ell_q \otimes_{\Delta_q} F\| =: \mathbf{C}_q(T) < \infty \ .$$
A normed space E is called of type p (resp. cotype q) if id_E is of type p (resp. of cotype q) and $\mathbf{T}_p(E) := \mathbf{T}_p(id_E)$ (analogously for the cotype). These notions are due to Maurey [285] and Hoffmann-Jørgensen [106]. An appeal to the density lemma 7.3. shows that the inequalities for the mappings to be continuous need only be verified on π-dense subspaces of the original tensor product — for example, ℓ_p can be replaced by the dense subspace of finite sequences and E or Rad by any dense subspace. In particular, $T \in \mathcal{L}(E,F)$ is of type p if and only if there is some $c \ge 0$ such that
$$\left(\int_0^1 \left\|\sum_{k=1}^n r_k(t) Tx_k\right\|_F^2 dt\right)^{1/2} \le c \left(\sum_{k=1}^n \|x_k\|_E^p\right)^{1/p}$$
for all $x_1, ..., x_n \in E$, and of cotype q if and only if
$$\left(\sum_{k=1}^n \|Tx_k\|_F^q\right)^{1/q} \le c \left(\int_0^1 \left\|\sum_{k=1}^n r_k(t) x_k\right\|_E^2 dt\right)^{1/2} \ .$$
Since $\Delta_1 = \pi$ and $\Delta_\infty = \varepsilon$, every operator/space has type 1 and cotype ∞. In the exercises it will be shown that Hilbert spaces are of type 2 and cotype 2, for the converse see section 30 and also Ex 30.1.(b). Moreover, infinite dimensional $C(K)$ have neither cotype $< \infty$ nor type > 1 (one says: do not have *proper cotype/type*), and infinite dimensional $L_1(\mu)$ do not have proper type. In the next section type and cotype of

L_p–spaces will be studied and some results will be presented which begin to show the importance of these notions. For $p > 2$ (resp. $q < 2$) there are no non–zero operators satisfying the type p (resp. cotype q) estimates; see Ex 7.15..

7.8. For the Hilbert transform $\mathfrak{H} : L_2(\mathbb{R}) \to L_2(\mathbb{R})$ operators (and spaces) such that $\mathfrak{H} \otimes_{\Delta_2} T$ is continuous were studied by Burkholder [19], Bourgain [15] and M. Defant [50], [51]; for the Paley projection $L_2[0, 2\pi] \to L_2[0, 2\pi]$ by Pisier [223], see also the surveys of Figiel [64] and Pełczyński [204] during the IMC 1983. For the projection R of $L_2[0, 1]$ onto the Rademacher functions see Ex 9.16. and section 31.; the operators T such that $R \otimes T$ is continuous will be called K–convex.

Neither the ℓ_1–valued Fourier transform (7.5.) nor and the ℓ_1–valued Hilbert transform (7.6.) was continuous — and ℓ_1 did not have proper type. That the counterexamples were always in ℓ_1 is not by chance:

PROPOSITION: *Let $T \in \mathfrak{L}(L_p(\mu), L_q(\nu))$ such that*

$$T \otimes id_E : L_p(\mu) \otimes_{\Delta_p} E \longrightarrow L_q(\nu) \otimes_{\Delta_q} E$$

is continuous for $E = \ell_1$. Then it is continuous for all Banach spaces E.

PROOF: $T \otimes id_{\ell_1}$ being continuous implies that there is a constant $c \geq 0$ such that

$$\left(\int \left(\sum_{k=1}^n |(Tf_k)(t)| \right)^q \nu(dt) \right)^{1/q} \leq c \left(\int \left(\sum_{k=1}^n |f_k(s)| \right)^p \mu(ds) \right)^{1/p}$$

for all $f_1, ..., f_n \in L_p(\mu)$. The density lemma 7.3. gives that

$$T \otimes id_{\ell_1(\Gamma)} : L_p(\mu) \otimes_{\Delta_p} \ell_1(\Gamma) \longrightarrow L_q(\nu) \otimes_{\Delta_q} \ell_1(\Gamma)$$

has norm $\leq c$ for all Γ. Since every Banach space is a quotient of some $\ell_1(\Gamma)$, the fact that $L_r \otimes_{\Delta_r} \cdot$ respects quotients (theorem 7.4.) yields the result. $\square$

Together with proposition 7.3. this implies for each $T \in \mathfrak{L}(L_p(\mu), L_q(\nu))$ (where ν is supposed to be localizable if $q = \infty$) the following: *If $T \otimes id_E$ is continuous for $E = \ell_1$ or $E = \ell_\infty$, then T is regular and $T \otimes id_E$ is continuous for all Banach spaces E.*

7.9. In particular, $T \otimes S : L_q \otimes_{\Delta_q} E \longrightarrow L_p \otimes_{\Delta_p} F$ is continuous for all $S \in \mathfrak{L}(E, F)$ if $T \otimes id_{\ell_1}$ is continuous. For operators $T \otimes S : L_q \otimes L_q \longrightarrow L_p \otimes L_p$ there is another quite useful positive result in this direction — which even holds for subspaces of L_q. For this denote for a subspace $E_1 \subset L_q(\mu)$ and a normed space E_2 by

$$E_1 \otimes_{\Delta_q} E_2$$

the space $E_1 \otimes E_2$ with the norm inherited from $L_q(\mu) \otimes_{\Delta_q} E_2$ (see Ex 7.17. for more information about this definition). This notation was already used in 7.7. for Rad.

THEOREM: Let μ and ν be measures, $E_1 \subset L_q(\mu)$ a subspace, E_2 and F normed spaces, $T \in \mathfrak{L}(E_1, F)$ and $S \in \mathfrak{L}(E_2, L_p(\nu))$. If $1 \leq q \leq p \leq \infty$, then

$$\|T \otimes S : E_1 \otimes_{\Delta_q} E_2 \longrightarrow F \otimes_{\Delta_p^i} L_p(\nu)\| = \|T\| \, \|S\| \; .$$

For $E_i = L_q(\mu_i)$ this result is due to Beckner [8]; the present form is due to Figiel-Iwaniec-Pełczyński [65]. For another proof of Beckner's result see 15.12.. Note the special case

$$\|T \otimes S : L_q \otimes_{\Delta_q} L_q \longrightarrow L_p \otimes_{\Delta_p} L_p\| = \|T\| \, \|S\|$$

for $q \leq p$. This is false for $q > p$: see 26.3. for an example; some positive results will be presented in 18.9., 26.3. and 32.3..

PROOF: Recall that $id_{L_q} \otimes S$ and $T \otimes id_{L_p}$ are continuous. Now the continuous triangle inequality in the form of proposition 7.2. and the diagram

$$\begin{array}{ccc}
E_1 \otimes_{\Delta_q} E_2 & \xrightarrow{\;1\;} & L_q(\mu) \otimes_{\Delta_q} E_2 \\
{\scriptstyle id_{E_1} \otimes S} \Big\downarrow & & \Big\downarrow {\scriptstyle id_{L_q} \otimes S} \\
 & & L_q(\mu) \otimes_{\Delta_q} L_p(\nu) \\
 & & \Big\downarrow {\scriptstyle id} \\
E_1 \otimes_{\Delta_p^i} L_p(\nu) & \xrightarrow{\;1\;} & L_q(\mu) \otimes_{\Delta_p^i} L_p(\nu) \\
{\scriptstyle T \otimes id_{L_p}} \Big\downarrow & & \\
F \otimes_{\Delta_p^i} L_p(\nu) & &
\end{array}$$

give that $\|T \otimes S\| \leq \|T \otimes id_{L_p}\| \, \|id_{E_1} \otimes S\| \leq \|T\| \, \|S\|$. $\square$

The general question about the continuity of operators $S \otimes T : L_q \otimes_{\Delta_q} E \longrightarrow L_p \otimes_{\Delta_p} F$ is rather involved; the reader will find at various places of this book more information about this problem — in particular, in the sections 18, 26 and 28 – 32. For a survey on this topic see [46].

Exercises:

Ex 7.1. Let $L_p(\mu)$ be infinite dimensional ($1 \leq p \leq \infty$). For a sequence (A_n) of pairwise disjoint integrable sets with positive measure the averaging operator $P : L_p(\mu) \to L_p(\mu)$ is defined by

$$P(f) := \sum_{n=1}^{\infty} \frac{1}{\mu(A_n)} \int_{A_n} f d\mu \; \chi_{A_n} \quad .$$

Show that for any normed space E

$$P \otimes id_E : L_p(\mu) \otimes_{\Delta_p} E \longrightarrow L_p(\mu) \otimes_{\Delta_p} E$$

is a norm–one projection such that im$(P \otimes id_E)$ is isometric to $\ell_p \otimes_{\Delta_p} E$.

Ex 7.2. Use the fact that $L_\infty \tilde{\otimes}_\varepsilon L_\infty = \mathfrak{K}(L_1, L_\infty) \subset (L_1 \tilde{\otimes}_\pi L_1)'$ to show that in general $L_\infty(\mu) \otimes_\varepsilon L_\infty(\nu)$ is not dense in $L_\infty(\mu \otimes \nu)$ for finite measures (see also Ex 4.8.).

Ex 7.3. Let μ be a measure, $1/p = 1/r + 1/q$ and $g \in L_r(\mu)$. Then for every normed space E the multiplication operator

$$\begin{array}{ccc} L_q(\mu, E) & \longrightarrow & L_p(\mu, E) \\ f & \rightsquigarrow & g \cdot f \end{array}$$

is continuous and has norm $\|g\|_r$ (if $E \neq \{0\}$). If $M_g : L_q(\mu) \to L_p(\mu)$ denotes this multiplication operator in the scalar case, then

$$\|M_g \otimes T : L_q \otimes_{\Delta_q} E \longrightarrow L_p \otimes_{\Delta_p} F\| = \|g\|_r \|T\|$$

for all $T \in \mathcal{L}(E, F)$. Hint: Hölder inequality.

Ex 7.4. If $S \in \mathcal{L}(L_p(\mu), G)$ and $T \in \mathcal{L}(E, F)$, then

$$\|S \otimes T : L_p(\mu) \otimes_{\Delta_p} E \longrightarrow G \otimes_\varepsilon F\| = \|S\| \, \|T\|$$

for each $1 \leq p \leq \infty$.

Ex 7.5. If $T \in \mathcal{L}(E, F)$ factors through a positive operator $T_o \in \mathcal{L}(L_p(\mu_1), L_q(\mu_2))$, then for all $S \in \mathcal{L}(L_p(\mu), L_q(\nu))$ the operator

$$S \otimes T : L_p(\mu) \otimes_{\Delta_p} E \longrightarrow L_q(\nu) \otimes_{\Delta_q} F$$

is continuous.

Ex 7.6. If E and F are Banach lattices, then every positive operator $E \to F$ is continuous. Hint: Assume not, then there exist $x_n \geq 0$ such that $\|x_n\| \leq 2^{-n}$ and $\|Tx_n\| \geq 2^n$.

Ex 7.7. Use the metric surjection $E' \tilde{\otimes}_\pi F \twoheadrightarrow \mathfrak{N}(E, F)$ to show that the kernel condition in the quotient lemma 7.4. is indispensable!

Ex 7.8. (a) If $T_i \in \mathcal{L}(L_p(\mu_i), L_p(\nu_i))$, then

$$\|T_1 \otimes T_2 : L_p(\mu_1) \otimes_{\Delta_p} L_p(\mu_2) \longrightarrow L_p(\nu_1) \otimes_{\Delta_p} L_p(\nu_2)\| = \|T_1\| \, \|T_2\| \quad .$$

(b) If $S : L_2(\mu) \to L_2(\nu)$ is a (not necessarily surjective) isometry and H a Hilbert space, then

$$S \otimes id_H : L_2(\mu) \otimes_{\Delta_2} H \longrightarrow L_2(\nu) \otimes_{\Delta_2} H$$

is an isometry. Hint: $(Sf|Sg) = (f|g)$ or look at Δ_2^t.

Ex 7.9. If H is a Hilbert space, then $\mathbf{T}_2(H) = \mathbf{C}_2(H) = 1$.

Ex 7.10. $\mathbf{T}_{p_1}(T) \leq \mathbf{T}_{p_2}(T)$ and $\mathbf{C}_{q_2}(T) \leq \mathbf{C}_{q_1}(T)$ if $1 \leq p_1 \leq p_2 \leq 2 \leq q_1 \leq q_2 \leq \infty$.

Ex 7.11. If $S \circ T \circ U$ is defined for three operators, then

$$\mathbf{C}_q(S \circ T \circ U) \leq \|S\| \mathbf{C}_q(T) \|U\|$$

— and the same holds for the type constants $\mathbf{T}_p$.

Ex 7.12. (a) If F is a subspace of E, then $\mathbf{C}_q(F) \leq \mathbf{C}_q(E)$ and $\mathbf{T}_p(F) \leq \mathbf{T}_p(E)$.

(b) If (E_α) is a family of subspaces of a normed space E with dense union such that for all α, β there is a γ with $E_\alpha \cup E_\beta \subset E_\gamma$, then

$$\mathbf{C}_q(E) = \sup_\alpha \mathbf{C}_q(I_\alpha : E_\alpha \hookrightarrow E) = \sup_\alpha \mathbf{C}_q(E_\alpha).$$

The same holds for the type constants $\mathbf{T}_p$.

Ex 7.13. Use the unit vectors e_k to show that $\mathbf{T}_p(\ell_1^n) \geq n^{1-1/p}$. Deduce that no space E which contains the ℓ_1^n uniformly (i.e. there are subspaces $M_n \subset E$ with $\sup d(M_n, \ell_1^n) < \infty$) does have proper type (Pisier [217] proved the converse; see also Beauzamy [7]). In particular, all infinite dimensional $L_1(\mu)$ and abstract L–spaces do not have proper type; by 23.3. this is also true for $\mathcal{L}_1^g$–spaces.

Ex 7.14. Show that:

(a) $\sup_n \mathbf{T}_p(\ell_\infty^n) = \infty$ for $1 < p \leq 2$. Hint: Ex 7.13. and lemma 6.3..

(b) $\mathbf{C}_q(\ell_\infty^n) \geq n^{1/q}$ for $2 \leq q < \infty$.

(c) A space which contains the ℓ_∞^n uniformly has neither proper type nor proper cotype. In particular, this is true for all infinite dimensional $C(K), L_\infty(\mu)$ and abstract M–spaces (use local reflexivity); by 23.3. this is also the case for arbitrary $\mathcal{L}_\infty^g$–spaces.

Ex 7.15. Show with $x_k = 1$ that the space $\mathbb{R}$ has neither type p for $2 < p \leq \infty$ nor cotype q for $1 \leq q < 2$ in the sense of the inequalities at the end of 7.7..

Ex 7.16. Let E be a normed space.

(a) For each finite dimensional subspace $M \subset E$ and $\varepsilon > 0$ there is a finite codimensional, closed $L \subset E$ such that $(1 - \varepsilon)\|x\| \leq \|Q_L^E(x)\| \leq \|x\|$ for all $x \in M$. L can be chosen to be weak-$*$-closed if E is a dual space. Hint: Proof of 6.3..

(b) Show that for $f \in L_p \otimes E$

$$\Delta_p(f; L_p, E) = \inf\{\Delta_p(f; L_p, M) \mid f \in L_p \otimes M, M \subset E \text{ finite dimensional}\} =$$
$$= \sup\{\Delta_p(id_{L_p} \otimes Q_L^E(f); L_p, E/L) \mid L \subset E \text{ closed, finite codimensional}\}.$$

Ex 7.17. (a) Let $E_{i,j} \subset L_p(\mu_{i,j})$ be subspaces such that $E_{i,1}$ is isometric to $E_{i,2}$ (for $i = 1, 2$). Use the generalization 7.9. of Beckner's result to show that $E_{1,1} \otimes_{\Delta_p} E_{2,1}$ is isometric to $E_{1,2} \otimes_{\Delta_p} E_{2,2}$. This shows that, for fixed p, the definition of $E_1 \otimes_{\Delta_p} E_2$ is independent of the $L_p(\mu_i)$ in which the E_i are embedded. However, $\mathrm{Rad} \otimes_{\Delta_2} E$ is, in general, not even isomorphic to $\ell_2 \otimes_{\Delta_2} E$, hence $E_1 \otimes_{\Delta_p} E$ (where E is an arbitrary Banach space) depends on the embedding $E_1 \stackrel{1}{\hookrightarrow} L_p(\mu)$.

(b) Use the fact that $L_1[0,1]$ has a non–complemented reflexive subspace E (e.g. isomorphic to ℓ_2 via Rademacher functions) to verify that Δ_1 is not equivalent to π on $E \otimes L_1[0,1]$.

Ex 7.18. Consider the Fourier transform $T : \ell_2(\mathbb{Z}) \to L_2[0, 2\pi]$

$$Te_k := (2\pi)^{-1/2} \exp(ik \cdot) \qquad \text{for all } k \in \mathbb{Z}.$$

Show that

$$T \otimes id_{\ell_1} : \ell_2(\mathbb{Z}) \otimes_{\Delta_2} \ell_1 \to L_2[0, 2\pi] \otimes_{\Delta_2} \ell_1$$

is not continuous. Hint: $\sum e_k \otimes e_k$.

Ex 7.19. Recall from 7.3. that an operator between complex Banach lattices is regular if it is of the form $S_1 - S_2 + i(S_1 - S_2)$ with positive operators S_j.

(a) Show that the Fourier transform on $L_2(\mathbb{R})$ and the discrete Hilbert transform on $\ell_p(\mathbb{Z})$ (for $1 < p < \infty$) are not regular.

(b) Let ν be a localizable measure. Then all operators in $\mathcal{L}(C(K)), L_\infty(\nu))$ and all operators in $\mathcal{L}(L_p(\mu), L_\infty(\nu))$ (where μ is an arbitrary measure and $1 \leq p \leq \infty$) are regular. Hint: 7.3. and the mapping property of ε.

8. Absolutely and Weakly p–Summable Series and Averaging Techniques

This section treats various types of series in Banach spaces which are related to ε and Δ_p. Rademacher and Gauss averaging techniques will be introduced, in particular the Khintchine inequality will be proved. As first applications a theorem of Orlicz' on the convergence of series in $L_p(\mu)$ and the calculation of certain nuclear norms are given.

8.1. A sequence (x_n) in a normed space E is called *absolutely p–summable* $(1 \leq p < \infty)$ if

$$\ell_p(x_n; E) := \ell_p(x_n) := \left(\sum_{n=1}^\infty \|x_n\|_E^p \right)^{1/p} < \infty$$

(the notation will be used for finite sequences as well); in other words,

$$(x_n) \in \ell_p(\tilde{E}) = \ell_p \tilde{\otimes}_{\Delta_p} \tilde{E}.$$

$p = \infty$ is just the case of bounded sequences: $\ell_\infty(E)$, and it is known (Ex 4.8., Ex 7.2.) that
$$\ell_\infty \tilde{\otimes}_\varepsilon E = \ell_\infty \tilde{\otimes}_{\Delta_\infty} E \underset{\neq}{\subseteq} \ell_\infty(\tilde{E}) \ .$$

The sequence (x_n) is called *weakly p-summable* if
$$\sum_{n=1}^\infty |\langle x', x_n \rangle|^p < \infty$$
for all $x' \in E'$; this means that the operator
$$T : E' \longrightarrow \ell_p$$
$$x' \rightsquigarrow (\langle x', x_n \rangle)$$
is defined. The closed graph theorem shows that this operator T is bounded and hence $w_p(x_n; E) := w_p(x_n) := \|T\| < \infty$:

$$w_p(x_n; E) = \sup_{x' \in B_{E'}} \left(\sum_{n=1}^\infty |\langle x', x_n \rangle|^p \right)^{1/p} = \sup \left\{ \left\| \sum_{n=1}^N \lambda_n x_n \right\| \, \Big| \, N \in \mathbb{N}, \|(\lambda_n)\|_{\ell_{p'}} \leq 1 \right\}$$

defines a norm on the space $\ell_p^w(E)$ of all weakly p-summable sequences in E. This formula shows that for each norming set $D \subset B_{E'}$

$$w_p(x_n; E) = \sup_{x' \in D} \left(\sum_{n=1}^\infty |\langle x', x_n \rangle|^p \right)^{1/p} \ .$$

Note that $T' = \sum e_n \otimes x_n$ and $T'(\ell_{p'}) \subset \tilde{E}$ if $1 < p < \infty$; for $p = 1$ one has $T'(c_o) \subset \tilde{E}$. The notation $w_p(x_n)$ will also be used for finite sequences. The case $p = \infty$ is, as before, the case of bounded sequences and $w_\infty(x_n; E) = \ell_\infty(x_n; E)$.

PROPOSITION: *If $D \subset B_{E'}$ is norming and $1 \leq p \leq \infty$, then for all $x_1, ..., x_n \in E$*

$$w_p(x_k; E) = \sup_{x' \in D} \|(\langle x', x_k \rangle)\|_{\ell_p^n} = \varepsilon\left(\sum_{k=1}^n e_k \otimes x_k; \ell_p^n, E \right) \ .$$

PROOF: Since
$$w_p(x_k; E) = \sup \left\{ \left| \langle x', \sum_{k=1}^n \lambda_k x_k \rangle \right| \, \Big| \, x' \in B_{E'}, \|(\lambda_k)\|_{\ell_{p'}^n} \leq 1 \right\}$$
and
$$\left\langle x', \sum_{k=1}^n \lambda_k x_k \right\rangle = \left\langle (\lambda_k) \otimes x', \sum_{k=1}^n e_k \otimes x_k \right\rangle,$$
the relation $\varepsilon = w_p$ on $\ell_p^n \otimes E$ follows from the definition of ε. □

The following inequalities are simple but important: If $T \in \mathfrak{L}(E,F)$ and (x_k) in E, then
$$\ell_p(Tx_k; F) \leq \|T\|\ell_p(x_k; E) \quad \text{and} \quad w_p(Tx_k; F) \leq \|T\|w_p(x_k; E).$$

8.2. In order to describe the weak sequence spaces in terms of operators define the closed subspace
$$\ell_p^{w,o}(E) := \{(x_n) \in \ell_p^w(E) \mid w_p((x_n)_{n=N}^\infty; E) \longrightarrow 0 \text{ for } N \to \infty\}$$
and give it the induced norm from $\ell_p^w(E)$. Clearly, $\ell_\infty^{w,o}(E)$ is the space $c_o(E)$ of zero sequences in E. It is easy to see that
$$E' \tilde{\otimes}_\varepsilon \ell_\infty = \mathfrak{K}(E, \ell_\infty) \subset \mathfrak{L}(E, \ell_\infty) = \ell_\infty(E')$$
$$E' \tilde{\otimes}_\varepsilon c_o = \mathfrak{K}(E, c_o) \subset \mathfrak{L}(E, c_o) = \{(x'_n) \in \ell_\infty(E') \mid x'_n \to 0 \text{ weak}*\}$$
$$\ell_\infty \tilde{\otimes}_\varepsilon E = \mathfrak{K}(\ell_1, E) \subset \mathfrak{L}(\ell_1, E) = \ell_\infty(E)$$
hold isometrically.

PROPOSITION: *Let E be a Banach space.*
(1) *The following relations hold isometrically:*
$$\ell_p^w(E) = \mathfrak{L}(\ell_{p'}, E) \supset \mathfrak{K}(\ell_{p'}, E) = \ell_p^{w,o}(E) = \ell_p \tilde{\otimes}_\varepsilon E \quad \text{for } 1 < p < \infty$$
$$\ell_1^w(E) = \mathfrak{L}(c_o, E) \supset \mathfrak{K}(c_o, E) = \ell_1^{w,o}(E) = \ell_1 \tilde{\otimes}_\varepsilon E.$$
The identification is given by $(x_n) \leadsto \sum_{n=1}^\infty e_n \otimes x_n$; *the sum converges pointwise in $\mathfrak{L}$ and with respect to the operator norm if it represents a compact operator.*
(2) *Dually:*
$$\ell_p^w(E') = \mathfrak{L}(E, \ell_p) \supset \mathfrak{K}(E, \ell_p) = \ell_p^{w,o}(E') = E' \tilde{\otimes}_\varepsilon \ell_p$$
hold isometrically for $1 \leq p < \infty$; the identification is given by $(x'_n) \leadsto \sum_{n=1}^\infty x'_n \otimes e_n$, *convergence as in (1).*
(3) *For $p = \infty$ the following equalities hold isometrically:*
$$\{(x_n) \in \ell_\infty(E) \mid \{x_n\} \text{ relatively compact}\} = \ell_\infty \tilde{\otimes}_\varepsilon E = \mathfrak{K}(\ell_1, E)$$
$$c_o(E) = c_o \tilde{\otimes}_\varepsilon E \ .$$

PROOF: That $\ell_p^w(E) \hookrightarrow \mathfrak{L}(\ell_{p'}, E)$ (or $\mathfrak{L}(c_o, E)$) via $S := \sum e_n \otimes x_n$ and
$$\|S\| = w_p(x_n; E)$$
was already shown when defining $w_p(x_n; E)$ in 8.1.. The norm relation also shows that S is compact if $w_p((x_n)_{n=N}^\infty) \to 0$. Conversely, take $S \in \mathfrak{L}(\ell_{p'}, E)$ (or $\mathfrak{L}(c_o, E)$), define $x_n := Se_n$ and observe
$$w_p(x_n; E) = w_p(Se_n; E) \leq \|S\|w_p(e_n; \ell_{p'}) = \|S\| \ .$$

Therefore, $\ell_p^w(E) = \mathfrak{L}(\ell_{p'}, E)$ holds isometrically; moreover,
$$\ell_p \otimes_\varepsilon E \hookrightarrow \ell_p^{w,o}(E) \hookrightarrow \mathfrak{K}(\ell_{p'}, E) = \ell_p \tilde\otimes_\varepsilon E$$
holds isometrically (ℓ_p has the approximation property, 5.3.) and (1) is proven.

For (2) take $(x_n') \in \ell_p^w(E')$. Then
$$w_p(x_n') = \sup_{x \in B_E} \|(\langle x_n', x\rangle)\|_{\ell_p} = \|x \rightsquigarrow (\langle x_n', x\rangle)\| = \left\|\sum_{n=1}^\infty x_n' \otimes e_n\right\|;$$
this shows that
$$\ell_p^w(E') \hookrightarrow \mathfrak{L}(E, \ell_p) \text{ and } \ell_p^{w,o}(E') \hookrightarrow \mathfrak{K}(E, \ell_p)$$
hold isometrically. If $T \in \mathfrak{L}(E, \ell_p)$ and $x_n' := T'e_n'$ (where the e_n' are the coordinate functionals), then, clearly, $Tx = (\langle x_n', x\rangle)$ and $\ell_p^w(E') = \mathfrak{L}(E, \ell_p)$ follows; the density of $E' \otimes \ell_p$ in the space of compact operators gives, as in (1), the result.

The first relation in (3) follows from
$$\ell_\infty \tilde\otimes_\varepsilon E = \mathfrak{K}(\ell_1, E)$$
while the second is a consequence of the facts that
$$c_o \otimes_\varepsilon E \hookrightarrow c_o(E)$$
holds isometrically (since both are isometrically in $\ell_\infty(E)$) and that $c_o \otimes E$ is dense in the Banach space $c_o(E)$: Just cut off the tail of the sequence! □

One might be tempted to think that a description like that in (3) also holds for the space $\ell_p \tilde\otimes_\varepsilon E = \ell_p^{w,o}(E)$ if $1 \le p < \infty$; but this is not the case. Take $E = c_o$ and (x_n) the following zero sequence in c_o:
$$e_1, 2^{-1/p}e_2, 2^{-1/p}e_2, 3^{-1/p}e_3, 3^{-1/p}e_3, 3^{-1/p}e_3, \ldots ;$$
then $(x_n) \in \ell_p^w(c_o)$ but $(x_n) \notin \ell_p^{w,o}(c_o)$. In Ex 8.4. a related question is treated: When is every $(x_n) \in \ell_p^w(E)$ a zero sequence in E?

8.3. A sequence (x_n) in a normed space E is called *unconditionally summable* if for each bijective $\eta : \mathbb{N} \to \mathbb{N}$ the series $\sum(x_{\eta(n)})$ converges; this is easily seen to be equivalent to the convergence of the net of finite sums
$$\sum_{n \in I} x_n,$$

where I runs through the directed set of finite subsets of $\mathbb{N}$.

PROPOSITION: *Let E be a Banach space. Then a sequence (x_n) in E is unconditionally summable if and only if it is in $\ell_1^{w,o}(E) = \ell_1 \tilde{\otimes}_\varepsilon E$.*

PROOF: Observe first the simple fact (decomposition in positive, negative, real, imaginary parts) that if (λ_n) is a sequence of scalars such that $\left|\sum_{n \in I} \lambda_n\right| \leq 1$ for all finite $I \subset \mathbb{N}$, then $\sum_{n=1}^{\infty} |\lambda_n| \leq 4$.

Now assume (x_n) to be unconditionally summable. Then there is an $n_o \in \mathbb{N}$ such that $\left\|\sum_{n \in I} x_n\right\| \leq \varepsilon$ for all finite $I \subset \{n_o, n_o+1, ...\}$ and therefore $\left|\sum_{n \in I} \langle x', x_n \rangle\right| \leq \varepsilon$ for all these I and $x' \in B_{E'}$: it follows that

$$w_1\left((x_n)_{n=n_o}^{\infty}\right) = \sup_{x' \in B_{E'}} \sum_{n=n_o}^{\infty} |\langle x', x_n \rangle| \leq 4\varepsilon .$$

Vice–versa: If $w_1\left((x_n)_{n=n_o}^{\infty}\right) \leq \varepsilon$, then for all finite $I \subset \{n_o, n_o+1, ...\}$

$$\left\|\sum_{n \in I} x_n\right\| = \sup_{x' \in B_{E'}} \left|\sum_{n \in I} \langle x', x_n \rangle\right| \leq \sup_{x' \in B_{E'}} \sum_{n=n_o}^{\infty} |\langle x', x_n \rangle| \leq \varepsilon . \quad \square$$

Bessaga–Pełczyński proved (see Diestel [54], p.45) that a Banach space does not contain c_o isomorphically if and only if each weakly summable sequence is unconditionally summable. Hence, it was not by chance that the example at the end of 8.2. was constructed in the space c_o. See also Ex 8.4..

8.4. The following powerful tool will be used frequently:

Take a probability measure μ on a set Ω and an orthonormal sequence of functions $f_n \in L_2(\mu)$ such that $\text{span}\{f_n\} \subset L_p(\mu)$ for all $1 \leq p < \infty$ and the norms coming from the $L_p(\mu)$ are equivalent:

$$\left(\sum_m |\alpha_m|^2\right)^{1/2} = \left(\int_\Omega \left|\sum_m \alpha_m f_m(w)\right|^2 \mu(dw)\right)^{1/2} \sim \left(\int_\Omega \left|\sum_m \alpha_m f_m(w)\right|^p \mu(dw)\right)^{1/p} .$$

Clearly, when investigating sequence spaces or function spaces these inequalities may be very useful. Additionally, the orthogonality has another quite interesting aspect: If $z = \sum_{m=1}^n x_m \otimes y_m$ is an element in the tensor product of two normed spaces, then

$$z = \sum_{m=1}^n x_m \otimes y_m = \sum_{k,m=1}^n (f_m|f_k)_{L_2(\mu)} x_m \otimes y_k =$$

$$= \int_\Omega \left(\sum_{m=1}^n f_m(w) x_m\right) \otimes \left(\sum_{k=1}^n f_k(w) y_k\right) \mu(dw)$$

— and the element z is represented in a new way as a sum/integral of elementary tensors; note that the Bochner integral in this formula exists in the finite dimensional space $\text{span}\{x_m\} \otimes \text{span}\{y_m\}$. It will be seen that in many cases this new representation, together with the norm equivalences turns out to be extremely fruitful! The method is called the *averaging technique*.

As is already indicated by the definition of type and cotype, *averages* of the form

$$\int \left\| \sum_k f_k(w) x_k \right\|_E^p \mu(dw)$$

for elements $x_1, ..., x_n$ in a normed space E will play an important role. Note that for Hilbert spaces H biorthonormality gives

$$\int \left\| \sum_k f_k(w) x_k \right\|_H^2 \mu(dw) = \sum_k \|x_k\|_H^2$$

(see also Ex 7.8.(b)).

For calculations it is very helpful if the functions f_n are of a simple nature, e.g. if Ω is a cartesian product, μ a product measure and f_n the coordinate functions: The probabilist would say in this case that the (f_n) are "stochastically independent random variables" — and this is why the averaging technique is often labelled a probabilistic method. Anyhow, this observation allows one to use powerful theorems from probability theory. There will be two examples used for averaging: the Rademacher functions 8.5. and the Gauss functions 8.7.. The Rademacher functions attain only values ± 1 and are actually defined on the countable product of a discrete space (as will be seen in a moment); in the case of Gauss functions Ω is $\mathbb{R}^N$ or $\mathbb{C}^N$ with the Gauss measure — and hence invariant under orthogonal transformations.

8.5. Rademacher functions: For every $n \in \mathbb{N}$ define on $D_n := \{-1, 1\}^n$ the probability measure μ_n by $\mu_n(\{w\}) := 2^{-n}$ for every $w \in D_n$; it follows that μ_n is the n-fold product measure of $\mu_1 = 2^{-1}(\delta_{-1} + \delta_1)$ on $D_1 = \{-1, 1\}$. On $D := \{-1, 1\}^{\mathbb{N}}$, the countable product of the μ_1 is a probability measure μ, which has, by definition, the following property: If $f : D \to \mathbb{C}$ depends only on the first n coordinates, i.e. $f : D \xrightarrow{projection} D_n \xrightarrow{f_n} \mathbb{C}$, then

$$\int_D f d\mu = \int_{D_n} f_n d\mu_n \ .$$

The coordinate projections $\varepsilon_k : D \to \{-1, 1\}$ or $\varepsilon_k : D_n \to \{-1, 1\}$ (this will cause no confusion) have norm 1 in $L_2(\mu)$ and $L_2(\mu_n)$ and are biorthonormal: Take $k < \ell \leq n$; then

$$\int_D \varepsilon_k(w) \varepsilon_\ell(w) \mu(dw) = \int_{D_n} w_k w_\ell \mu_n(d(w_1, ..., w_n)) =$$
$$= \int_{D_1} w_\ell \left[\int_{D_1} w_k \mu_1(dw_k) \right] \mu_1(dw_\ell) = 0 \ .$$

8. Absolutely and Weakly p–Summable Series and Averaging Techniques

The functions ε_k are called *discrete Rademacher functions*.

REMARK: If r_k are the Rademacher functions on $[0,1]$ with the Lebesgue measure (see 1.5.(2)), E a normed space and $x_1, ..., x_n \in E$, then for all $1 \le p < \infty$

$$\int_0^1 \Big\|\sum_{k=1}^n r_k(t) x_k\Big\|_E^p dt = \int_{D_n} \Big\|\sum_{k=1}^n \varepsilon_k(w) x_k\Big\|_E^p \mu_n(dw) = \int_D \Big\|\sum_{k=1}^n \varepsilon_k(w) x_k\Big\|_E^p \mu(dw).$$

This follows from the fact that all Rademacher functions r_k are constant on the intervals $[m2^{-n}, (m+1)2^{-n}[$. The same argument gives that (λ the Lebesgue measure)

$$\lambda\Big(\{t \in [0,1] \mid r_k(t) \le \alpha_k \text{ for } k = 1, ..., n\}\Big) = \mu_n\Big(D_n \cap \prod_{k=1}^n]-\infty, \alpha_k]\Big)$$

for all $\alpha_1, ..., \alpha_n \in \mathbb{R}$ and this means, in the language of probability theory, that the measure μ_n on $D_n \subset \mathbb{R}^n$ is the joint distribution of the "random variables" $(r_1, ..., r_n)$.

The basic inequality is the

KHINTCHINE INEQUALITY: *For $1 \le p < \infty$ there are constants a_p and $b_p \ge 1$ such that*

$$a_p^{-1}\Big(\sum_{k=1}^n |\alpha_k|^2\Big)^{1/2} \le \Big(\int_{D_n} \Big|\sum_{k=1}^n \alpha_k \varepsilon_k(w)\Big|^p \mu_n(dw)\Big)^{1/p} \le b_p \Big(\sum_{k=1}^n |\alpha_k|^2\Big)^{1/2}$$

for all $n \in \mathbb{N}$ and $\alpha_1, ..., \alpha_n \in \mathbb{C}$.

This is false for the L_∞–norm, see Ex 8.18. (which also treats the case $0 < p < 1$). The best constants a_p and b_p were calculated by Haagerup [96] in 1982; they are the same if one considers only real α_k (see Ex 26.9.):

$$a_p = 2^{1/p - 1/2} \le a_1 = \sqrt{2} \quad \text{if } 1 \le p \le p_o$$
$$a_p = 2^{-1/2}(\sqrt{\pi}/\Gamma((p+1)/2))^{1/p} \le a_1 \quad \text{if } p_o < p < 2$$
$$a_p = 1 \quad \text{if } 2 \le p < \infty \text{ (obvious)}$$
$$b_p = 1 \quad \text{if } 1 \le p \le 2 \text{ (obvious)}$$
$$b_p = 2^{1/2}(\Gamma((p+1)/2)/\sqrt{\pi})^{1/p} \le \sqrt{p} \quad \text{if } 2 < p < \infty \ .$$

The number $p_o \approx 1,84742$ is the solution of $\Gamma((p+1)/2) = \sqrt{\pi}/2$ in $]1, 2[$.

PROOF: Since the following proof (taken from Stein [260], Appendix D) will not give the best constants anyway, it is enough to check the inequalities for real α_k, the left inequality for $p=1$ and the right one for natural numbers $p > 2$. Take $\sum_{k=1}^{n} \alpha_k^2 = 1$ and put

$$f(w) := \sum_{k=1}^{n} \alpha_k \varepsilon_k(w) \ .$$

Then, for each natural number $p > 2$ and $\lambda > 0$

$$(p!)^{-1} \int_{D_n} \lambda^p |f(w)|^p \mu_n(dw) \leq \int_{D_n} \exp |\lambda f(w)| \mu_n(dw) \leq$$

$$\leq \int_{D_n} [\exp \lambda f(w) + \exp(-\lambda f(w))] \, \mu_n(dw) \ .$$

Since

$$\int_{D_n} \exp \lambda f(w) \mu_n(dw) = \int_{D_1} \cdots \int_{D_1} \prod_{k=1}^{n} \exp(\lambda \alpha_k w_k) \mu_1(dw_1) \ldots \mu_1(dw_n)$$

$$= \prod_{k=1}^{n} \frac{1}{2} [\exp \lambda \alpha_k + \exp(-\lambda \alpha_k)] = \prod_{k=1}^{n} \Big(\sum_{m=0}^{\infty} \frac{(\lambda \alpha_k)^{2m}}{(2m)!} \Big) \leq \prod_{k=1}^{n} \exp \frac{(\lambda \alpha_k)^2}{2} = \exp \frac{\lambda^2}{2} \ ,$$

it follows that $\|f\|_{L_p(\mu_n)} \leq (2p!)^{1/p} \lambda^{-1} \exp(\lambda^2/2p)$. Taking $\lambda := \sqrt{p}$, gives $b_p \leq \sqrt{2ep}$. An estimate for a_1 is a consequence of Hölder's inequality:

$$1 = \int_{D_n} |f|^2 d\mu_n = \int_{D_n} |f|^{2/3} |f|^{4/3} d\mu_n \leq$$

$$\leq \Big(\int_{D_n} |f| d\mu_n \Big)^{2/3} \Big(\int_{D_n} |f|^4 d\mu_n \Big)^{1/3} \ ,$$

hence

$$1 \leq \|f\|_{L_1(\mu_n)} \|f\|_{L_4(\mu_n)}^2 \leq \|f\|_{L_1(\mu_n)} b_4^2$$

and therefore $a_1 \leq b_4^2$. □

8.6. For the Rademacher functions on $[0,1]$ the Khintchine inequality implies that on the subspace $\text{Rad} := \overline{\text{span}}\{r_n\} \subset L_2[0,1]$ all norms coming from $L_p[0,1]$ are equivalent $(1 \leq p < \infty)$:

$$a_p^{-1} \Big(\sum_{k=1}^{\infty} |\alpha_k|^2 \Big)^{1/2} \leq \Big(\int_0^1 \Big| \sum_{k=1}^{\infty} \alpha_k r_k(t) \Big|^p dt \Big)^{1/p} \leq b_p \Big(\sum_{k=1}^{\infty} |\alpha_k|^2 \Big)^{1/2}$$

— a fact which was already used in section 1; for $p = \infty$ see Ex 8.18.. For $p > 1$ Rad is even complemented in $L_p[0,1]$, see Ex 8.17.. It might be interesting to know that the only continuous functions f in Rad are the multiples of the function $1-2t$ (see Chatterji [26]). What happens if the scalars α_k are replaced by x_k in a normed space

E and the modulus by the norm? For $p = 2$ the definitions in 7.7. say that E is of cotype 2 if and only if the left inequality holds (for some constant) — and of type 2 if and only if the right one holds. Since there are spaces which have neither type 2 nor cotype 2 (for example $C[0,1]$, see Ex 7.14.) there is no vector valued Khintchine inequality. However, the *Kahane inequality* says that there are constants K_p such that for all normed spaces E and $x_1, ..., x_n \in E$

$$\int_0^1 \Big\|\sum_{k=1}^n r_k(t)x_k\Big\|_E dt \leq \Big(\int_0^1 \Big\|\sum_{k=1}^n r_k(t)x_k\Big\|_E^p dt\Big)^{1/p} \leq K_p \int_0^1 \Big\|\sum_{k=1}^n r_k(t)x_k\Big\|_E dt$$

(for a proof see e.g. Lindenstrauss–Tzafriri [177], II, p.74). By remark 8.5. the same holds for the discrete Rademacher functions.

So it is important to have more examples of spaces with type and cotype. The Khintchine inequality will help. Moreover, in the following proof repeated use will be made of the "continuous triangle inequality" (see Appendix B5. and 7.2.)

$$\Big(\int_X \Big(\int_Y |f(x,y)|dy\Big)^r dx\Big)^{1/r} \leq \int_Y \Big(\int_X |f(x,y)|^r dx\Big)^{1/r} dy \ .$$

PROPOSITION: *Let (Ω, μ) be a measure space.*
(1) *For $1 \leq p \leq 2$ the space $L_p(\mu)$ is of type p and cotype 2.*
(2) *For $2 \leq q < \infty$ the space $L_q(\mu)$ is of cotype q and type 2.*

If $L_p(\mu)$ is infinite dimensional, then it is of no type larger than $\min\{p,2\}$ and of no cotype smaller than $\max\{p,2\}$, see Ex 8.13..

PROOF: Discrete Rademacher functions will be used. A $*$ will indicate the application of the "continuous triangle inequality". If $1 \leq p \leq 2$ and $f_1, ..., f_n \in L_p(\mu)$, then

$$\Big(\sum_{k=1}^n \|f_k\|_p^2\Big)^{1/2} = \Big(\sum_k \Big(\int_\Omega |f_k(t)|^p dt\Big)^{2/p}\Big)^{1/2} \overset{*}{\leq}$$

$$\leq \Big(\int_\Omega \Big(\sum_k |f_k(t)|^2\Big)^{p/2} dt\Big)^{1/p} \leq a_p \Big(\int_\Omega \int_{D_n} \Big|\sum_k \varepsilon_k(w) f_k(t)\Big|^p dw dt\Big)^{1/p} =$$

$$= a_p \Big(\int_{D_n} \Big\|\sum_k \varepsilon_k(w) f_k\Big\|_p^p dw\Big)^{1/p} \leq a_p \Big(\int_{D_n} \Big\|\sum_k \varepsilon_k(w) f_k\Big\|_p^2 dw\Big)^{1/2} ,$$

which shows that L_p is of cotype 2; for type p

$$\Big(\int_{D_n} \Big(\int_\Omega \Big|\sum_k \varepsilon_k(w) f_k(t)\Big|^p dt\Big)^{2/p} dw\Big)^{1/2} \overset{*}{\leq} \Big(\int_\Omega \Big(\int_{D_n} \Big|\sum_k \varepsilon_k(w) f_k(t)\Big|^2 dw\Big)^{p/2} dt\Big)^{1/p} =$$

$$= \Big(\int_\Omega \Big(\sum_k |f_k(t)|^2\Big)^{p/2} dt\Big)^{1/p} \leq \Big(\int_\Omega \sum_k |f_k(t)|^p dt\Big)^{1/p} = \Big(\sum_k \|f_k\|_p^p\Big)^{1/p} \ .$$

For $2 \leq q < \infty$ the same type of argument applies — but we prefer to write them down as a chain of mappings: Here Rad_p denotes the closed span of the Rademacher functions with the norm induced from $L_p(D)$. The continuous triangle inequality $*$ will be used in the form of 7.2.:

$$\text{Rad}_2 \otimes_{\Delta_2} L_q \xrightarrow{*} \text{Rad}_2 \otimes_{\Delta_q^t} L_q \longrightarrow \ell_2 \otimes_{\Delta_q^t} L_q \longrightarrow \ell_q \otimes_{\Delta_q^t} L_q = \ell_q \otimes_{\Delta_q} L_q$$

and it follows that $\mathbf{C}_q(id_{L_q}) = 1$. That L_q has type 2 follows from

$$\ell_2 \otimes_{\Delta_2} L_q \xrightarrow{*} \ell_2 \otimes_{\Delta_q^t} L_q \xrightarrow{Khintchine} \text{Rad}_q \otimes_{\Delta_q^t} L_q =$$
$$= \text{Rad}_q \otimes_{\Delta_q} L_q \longrightarrow \text{Rad}_2 \otimes_{\Delta_2} L_q$$

and $\mathbf{T}_2(id_{L_q}) \leq b_q$ from the Khintchine inequality. □

Of course, it makes no difference whether one writes the proof as a chain of inequalities or as a chain of maps. This is only a question of mathematical taste — we sometimes prefer maps.

The proof showed that for all $n \in \mathbb{N}$

$$\begin{array}{lll} \mathbf{C}_2(\ell_p^n) \leq a_p \quad \text{and} \quad \mathbf{T}_p(\ell_p^n) = 1 & & \text{for } 1 \leq p \leq 2 \\ \mathbf{C}_q(\ell_q^n) = 1 \quad \text{and} \quad \mathbf{T}_2(\ell_q^n) \leq b_q & & \text{for } 2 \leq q < \infty \,. \end{array}$$

Ex 7.12. implies that for every $\mathfrak{L}_{p,\lambda}^g$-space E

$$\begin{array}{lll} \mathbf{C}_2(E) \leq \lambda a_p \quad \text{and} \quad \mathbf{T}_p(E) \leq \lambda & & \text{for } 1 \leq p \leq 2 \\ \mathbf{C}_p(E) \leq \lambda \quad \text{and} \quad \mathbf{T}_2(E) \leq \lambda b_p & & \text{for } 2 \leq p < \infty \,. \end{array}$$

For the remaining type and cotype constants of ℓ_p^n see Ex 7.13., Ex 7.14. and Ex 8.13.. It is obvious, by the proof, that one could have used other averaging methods instead of the Rademacher one to obtain that L_p (as above) is of some type/cotype with respect to this method; this may improve the final constants in calculations for L_p which use type/cotype constants.

8.7. Gauss functions: The Gauss measure γ_n on $\mathbb{K}^n$ is defined by

$$\int_{\mathbb{K}^n} f(w) \gamma_n(dw) = c_n \int_{\mathbb{K}^n} f(w) \exp(-\frac{1}{2}\|w\|_2^2) dw \,,$$

where $c_n = (2\pi)^{-n/2}$ in the real case and $c_n = (2\pi)^{-n}$ in the complex case (i.e. the Gauss measures on $\mathbb{C}^n$ and $\mathbb{R}^{2n}$ coincide); γ_n is a probability measure — clearly the n–fold product of γ_1. On $\mathbb{K}^{\mathbb{N}}$ the Gauss measure γ is the countable product of γ_1 on $\mathbb{K}$; as before, 8.5., functions on $\mathbb{K}^{\mathbb{N}}$ depending only on the first n coordinates can be integrated with respect to γ_n. The coordinate projections $g_k : \mathbb{K}^{\mathbb{N}} \to \mathbb{K}$ or $\mathbb{K}^n \to \mathbb{K}$ are called Gauss functions; for $\mathbb{K} = \mathbb{R}$ they have norm 1 in $L_2(\gamma)$ or $L_2(\gamma_n)$ and they are biorthogonal: if $1 \leq k < l \leq n$, then

$$\int_{\mathbb{R}^{\mathbb{N}}} g_k g_l \, d\gamma = \int_{\mathbb{R}^n} g_k g_l \, d\gamma_n = \int_{\mathbb{R}} w_k \left(\int_{\mathbb{R}} w_l \gamma_1(dw_l) \right) \gamma_1(dw_k) = 0 .$$

Therefore the (g_k) form an orthonormal system in $L_2(\gamma)$. If $\mathbb{K} = \mathbb{C}$, the g_k are biorthogonal as well, but have norm $\sqrt{2}$ — however, for our purpose there is no need to normalize them. Note that for $\alpha = (\alpha_1, ..., \alpha_n) \in \mathbb{C}^n$

$$\int_{\mathbb{K}^{\mathbb{N}}} \left| \sum_{k=1}^n \alpha_k g_k \right|^p d\gamma = \int_{\mathbb{K}^n} \left| \sum_{k=1}^n \alpha_k g_k \right|^p d\gamma_n = \int_{\mathbb{K}^n} |\langle \alpha, w \rangle_{\mathbb{C}^n}|^p \gamma_n(dw)$$

and, by the transformation formula for the Lebesgue measure,

$$\int_{\mathbb{K}^n} f(Aw)\gamma_n(dw) = \int_{\mathbb{K}^n} f(w)\gamma_n(dw)$$

for all isometries $A : \ell_2^n \to \ell_2^n$: The Gauss measure is invariant under isometries. This is the great advantage of Gauss averaging — as can already be seen in the proof of the following result of Khintchine type.

PROPOSITION: *For $1 \leq p < \infty$ and $\alpha_1, ..., \alpha_n \in \mathbb{K}$*

$$\left(\int_{\mathbb{K}^n} \left| \sum_{k=1}^n \alpha_k g_k \right|^p d\gamma_n \right)^{1/p} = \left(\sum_{k=1}^n |\alpha_k|^2 \right)^{1/2} \cdot \|g_1\|_{p,\mathbb{K}}$$

where $\|g_1\|_{p,\mathbb{K}} := \left(\int_{\mathbb{K}} |w|^p \gamma_1(dw) \right)^{1/p}$ is the p–th moment of the Gauss measure.

PROOF: For every isometry $A : \ell_2^n \to \ell_2^n$ and $\alpha \in \mathbb{K}^n$

$$\int_{\mathbb{K}^n} |\langle \alpha, w \rangle|^p \gamma_n(dw) = \int_{\mathbb{K}^n} |\langle A\alpha, Aw \rangle|^p \gamma_n(dw) = \int_{\mathbb{K}^n} |\langle A\alpha, w \rangle|^p \gamma_n(dw)$$

by the invariance of γ_n under isometries. Therefore it is enough to prove the result for $\alpha = e_1$:

$$\int_{\mathbb{K}^n} |\langle e_1, w \rangle|^p \gamma_n(dw) = \int_{\mathbb{K}} |w|^p \gamma_1(dw) = \|g_1\|_{p,\mathbb{K}}^p . \quad \square$$

It is routine work to calculate the constants:

$$\|g_1\|_{p,\mathbb{R}} = \sqrt{2} \left(\frac{1}{\sqrt{\pi}} \Gamma\left(\frac{p+1}{2} \right) \right)^{1/p} , \quad \|g_1\|_{1,\mathbb{R}} = \sqrt{\frac{2}{\pi}} , \quad \|g_1\|_{2,\mathbb{R}} = 1$$

$$\|g_1\|_{p,\mathbb{C}} = \sqrt{2} \left(\Gamma\left(\frac{p+2}{2} \right) \right)^{1/p} , \quad \|g_1\|_{1,\mathbb{C}} = \sqrt{\frac{\pi}{2}} , \quad \|g_1\|_{2,\mathbb{C}} = \sqrt{2} ;$$

moreover,
$$\|g_1\|_{p,\mathbb{C}} = \left(\frac{\pi}{2}\right)^{1/2p} (\|g_1\|_{p+1,\mathbb{R}})^{1+1/p} .$$

It turns out that the real and complex constants are different — in some sense contrary to the case of the Khintchine inequality for Rademacher functions (but recall also that the g_k are different in the real and complex case).

As in the case of Rademacher functions, there is no vector valued Khintchine result for Gauss functions, and there is a type of Kahane inequality saying that on the subspace $[\text{span } \{g_k\}] \otimes E$ all norms coming from $L_p(\gamma, E)$ are equivalent (see Marcus–Pisier [183], p.44). It can be shown that for all $x_1, ..., x_n \in E$

$$\left(\int_{D_n} \left\|\sum_{k=1}^n \varepsilon_k(w) x_k\right\|_E^p \mu_n(dw)\right)^{1/p} \leq \sqrt{\frac{\pi}{2}} \left(\int_{\mathbb{R}^n} \left\|\sum_{k=1}^n g_k(w) x_k\right\|_E^p \gamma_n(dw)\right)^{1/p} ,$$

but there is no converse estimate (see Ex 8.9.); it is a deep result of Maurey and Pisier [187] that exactly the Banach spaces E which do not contain the ℓ_∞^n uniformly permit the other estimate.

8.8. The averaging technique will now be used to calculate various norms of the element $\sum_{k=1}^n e_k \otimes e_k \in \mathbb{K}^n \otimes \mathbb{K}^n$; considered in $L(\mathbb{K}^n, \mathbb{K}^n)$ this is the identity operator. The Hölder inequality gives

$$\varepsilon\left(\sum_{k=1}^n e_k \otimes e_k; \ell_p^n, \ell_q^n\right) = \|id : \ell_{p'}^n \to \ell_q^n\| = \begin{cases} 1 & \text{if } 1/p + 1/q \leq 1 \\ n^{1/p+1/q-1} & \text{if } 1/p + 1/q \geq 1 \end{cases} .$$

PROPOSITION:
$$\pi\left(\sum_{k=1}^n e_k \otimes e_k; \ell_\infty^n, \ell_\infty^n\right) = N(id : \ell_1^n \to \ell_\infty^n) = 1 .$$

PROOF: $\varepsilon \leq \pi$, hence $\pi(...) \geq 1$. For the converse inequality Rademacher averaging will be used:

$$\sum_{k=1}^n e_k \otimes e_k = \sum_{k,\ell=1}^n (\varepsilon_k|\varepsilon_\ell) e_k \otimes e_\ell = \int_{D_n} \left(\sum_{k=1}^n \varepsilon_k(w) e_k\right) \otimes \left(\sum_{\ell=1}^n \varepsilon_\ell(w) e_\ell\right) \mu_n(dw) .$$

Since $\|\sum_{k=1}^n \varepsilon_k(w) e_k\|_{\ell_\infty^n} = 1$, this gives

$$\pi\left(\sum_{k=1}^n e_k \otimes e_k; \ell_\infty^n, \ell_\infty^n\right) \leq \int_{D_n} \pi(...) \mu_n(dw) = 1 . \quad \square$$

COROLLARY:
$$\pi\left(\sum_{k=1}^n e_k \otimes e_k; \ell_p^n, \ell_q^n\right) = N(id : \ell_{p'}^n \to \ell_q^n) = \begin{cases} n^{1/p+1/q} & \text{if } 1/p + 1/q \leq 1 \\ n & \text{if } 1/p + 1/q \geq 1 \end{cases} .$$

PROOF: The mapping property for π gives
$$\|id \otimes id : \ell_\infty^n \otimes_\pi \ell_\infty^n \longrightarrow \ell_p^n \otimes_\pi \ell_q^n \| \leq n^{1/p+1/q} \quad,$$
hence $\pi(\sum_{k=1}^n e_k \otimes e_k; \ell_p^n, \ell_q^n) \leq n^{1/p+1/q}$. The same argument implies
$$\pi\Big(\sum_{k=1}^n e_k \otimes e_k; \ell_p^n, \ell_q^n\Big) \leq \pi\Big(\sum_{k=1}^n e_k \otimes e_k; \ell_1^n, \ell_1^n\Big) = n \quad.$$
Since $(\ell_p^n \otimes_\pi \ell_q^n)' = \ell_{p'}^n \otimes_\varepsilon \ell_{q'}^n$ holds isometrically,
$$n = \Big\langle \sum_{k=1}^n e_k \otimes e_k, \sum_{l=1}^n e_l \otimes e_l \Big\rangle \leq \pi\Big(\sum_{k=1}^n e_k \otimes e_k; \ell_p^n, \ell_q^n\Big) \varepsilon\Big(\sum_{k=1}^n e_k \otimes e_k; \ell_{p'}^n, \ell_{q'}^n\Big) \leq$$
$$\leq \pi\Big(\sum_{k=1}^n e_k \otimes e_k; \ell_p^n, \ell_q^n\Big) \cdot \begin{cases} n^{1-1/p-1/q} & \text{if } 1/p+1/q \leq 1 \\ 1 & \text{if } 1/p+1/q > 1 \end{cases}.$$
These three estimates give the result. □

8.9. The cotype estimates serve nicely when investigating the following question: In which Banach spaces is every weakly r–summable sequence absolutely s–summable? In other words, when is
$$\ell_r^w(E) \subset \ell_s(E) \quad ?$$
The question goes back to a result of Orlicz 1933 [195] stating that in $L_p[0,1]$ for $1 \leq p \leq 2$ each unconditionally convergent series (see 8.3.) is absolutely 2–summable. A closed graph argument shows that $\ell_1^w(E) \subset \ell_2(E)$ if and only if there is a constant c with
$$\ell_2(x_n; E) \leq c \, w_1(x_n; E)$$
— and clearly it is enough to check this inequality for finite sequences. So the question is whether or not
$$\sup_n \|id \otimes id : \ell_1^n \otimes_\varepsilon E \longrightarrow \ell_2^n \otimes_{\Delta_2} E\| < \infty \quad ?$$
If E is a Hilbert space, a nice direct argument shows (Ex 8.10.) that this supremum is 1. Orlicz used Rademacher functions in his proof for L_p; let us do the same: First observe that
$$\Big(\int_{D_n} \Big\|\sum_{k=1}^n \varepsilon_k(w) x_k\Big\|_E^2 \mu_n(dw)\Big)^{1/2} = \Big(\int_{D_n} \Big(\sup_{\|x'\|\leq 1} \Big|\sum_{k=1}^n \varepsilon_k(w)\langle x', x_k\rangle\Big|\Big)^2 \mu_n(dw)\Big)^{1/2} \leq$$
$$\leq \sup_{\|x'\|\leq 1} \sup_{\|y'\|_{\ell_\infty^n}\leq 1} \Big|\sum_{k=1}^n \langle y', e_k\rangle\langle x', x_k\rangle\Big| = \varepsilon\Big(\sum_{k=1}^n e_k \otimes x_k; \ell_1^n, E\Big) = w_1(x_k; E) \quad.$$

Now if E has cotype 2 the Rademacher average (for r_k or ε_k) dominates, by definition, $\ell_2(x_k; E)$, hence

PROPOSITION(Orlicz): *If E has cotype 2, then every unconditionally convergent series in E is absolutely 2-summable; more precisely, for all $x_1, ..., x_n \in E$*

$$\ell_2(x_k; E) \leq C_2(E) w_1(x_k; E) \ .$$

By 8.6. the $L_p(\mu)$ (for $1 \leq p \leq 2$) are of cotype 2 — hence Orlicz' original theorem was also obtained (the work for the result was done when proving that the L_p have cotype 2!). Maurey and Pisier [187] showed that each Banach space with the "Orlicz property" (stated in the proposition) has cotype $(2+\varepsilon)$ for all $\varepsilon > 0$, but Talagrand [266] recently constructed a Banach lattice with the Orlicz property which does not have cotype 2. Note that, by Ex 8.1.(c) the Orlicz property of E means that there is a constant c satisfying

$$\left(\sum_{k=1}^n \|x_k\|^2\right)^{1/2} \leq c \sup\left\{ \left\|\sum_{k=1}^n \varepsilon_k(w) x_k\right\| \ \Big| \ w \in D_n \right\}$$

for all $x_1, ..., x_n \in E$. In Ex 8.11. the reader finds a nice application of Orlicz' theorem to pointwise convergence of series.

COROLLARY: *If E has cotype 2 and $1/r \leq 1/q - 1/2$, then*

$$\ell_r(x_k; E) \leq C_2(E) w_q(x_k; E) \ ,$$

i.e. every weakly q-summable sequence in E is absolutely r-summable.

In particular, this applies to all L_p with $1 \leq p \leq 2$.

PROOF: For $\lambda = (\lambda_n) \in B_{\ell_{q'}}$ the diagonal operator $D_\lambda : \ell_q \to \ell_1$ has norm ≤ 1; therefore, by the proposition,

$$\|\ell_q \otimes_\varepsilon E \xrightarrow{D_\lambda \otimes id} \ell_1 \otimes_\varepsilon E \xrightarrow{id \otimes id} \ell_2 \otimes_{\Delta_2} E\| \leq C_2(E) \ ,$$

which implies

$$\sup_{\lambda \in B_{\ell_{q'}}} \left(\sum_{k=1}^n |\lambda_k|^2 \|x_k\|^2\right)^{1/2} \leq C_2(E) w_q(x_k; E)$$

for all $x_1, ..., x_n \in E$. For $1/r_o := 1/q - 1/2$ the left supremum is $\ell_{r_o}(x_k; E)$, hence

$$\ell_r(x_k; E) \leq C_2(E) w_q(x_k; E)$$

for all $r \geq r_o$. □

The idea of Orlicz in 1933 to use the Rademacher functions and Khintchine's inequality for the investigation of Banach spaces may be considered to be the beginning of the "modern" averaging techniques.

Exercises:

Ex 8.1. (a) Show that for all $x_1, ..., x_n \in E$

$$w_p(x_k; E) = \sup\left\{ \left\| \sum_{k=1}^n \alpha_k x_k \right\| \mid (\alpha_k) \in B_{\ell_{p'}^n} \right\}.$$

(b) $w_p(f_k; C(K)) = \|(\sum_{k=1}^n |f_k|^p)^{1/p}\|_{C(K)}$ for all $f_1, ..., f_n \in C(K)$.

(c) Show that

$$w_1(x_k; E) = \sup\left\{ \left\| \sum_{k=1}^n \varepsilon_k(w) x_k \right\| \mid w \in D_n \right\} \quad \text{real case},$$

$$w_1(x_k; E) \leq 2 \sup\left\{ \left\| \sum_{k=1}^n \varepsilon_k(w) x_k \right\| \mid w \in D_n \right\} \quad \text{complex case}.$$

Ex 8.2. If (x_n) is weakly p-summable and $(\alpha_n) \in c_o$, then $(\alpha_n x_n) \in \ell_p \tilde\otimes_\varepsilon E = \ell_p^{w,o}(E)$.

Ex 8.3. Give an example of a zero sequence in a Banach space such that

$$\left\{ \sum_{k=1}^n \lambda_k x_k \mid (\lambda_k) \in B_{\ell_2} \text{ and } n \in \mathbb{N} \right\}$$

is bounded but not relatively compact. Hint: 8.2..

Ex 8.4. (a) Use the fact that every continuous operator $c_o \to \ell_1$ is compact to show that $(c_o \otimes_\pi c_o)' = \ell_1 \tilde\otimes_\varepsilon \ell_1$.

(b) *Pitt's theorem* states that $\mathcal{L}(\ell_p, \ell_q) = \mathfrak{K}(\ell_p, \ell_q)$ whenever $1 \leq q < p < \infty$. Use this to show for $p, q \in]1, \infty[$ that $\ell_p^w(\ell_q) = \ell_p^{w,o}(\ell_q)$ if and only if $p' > q$. Hint: 8.2.(1).

(c) Show that $\ell_1^w(E) \subset c_o(E)$ if and only if $\ell_1^w(E) = \ell_1^{w,o}(E)$. Hint: Use the Bessaga–Pełczyński result mentioned at the end of 8.3..

(d) Every block basis sequence (y_n) with respect to the unit vector basis (e_n) in ℓ_p (this means $y_k \in \text{span}\{e_{n_k+1}, ..., e_{n_{k+1}}\}$ for some $0 = n_1 < n_2 < ...$ and $\|y_k\| = 1$) is weakly p'-summable.

(e) *Bessaga–Pełczyński selection principle:* Let (e_n) be a basis of a Banach space E with coefficient functionals (e_n'). If (y_n) is a sequence such that $\|y_n\| \geq c$ for some $c > 0$ and

$$\lim_{n \to \infty} \langle e_m', y_n \rangle = 0$$

for all $m \in \mathbb{N}$, then (y_n) has a subsequence $(y_{\varphi(n)})$ which is equivalent to a block basis sequence (x_n) with respect to (e_n), i.e. $\sum(\alpha_n x_n)$ converges if and only if $\sum(\alpha_n y_{\varphi(n)})$ converges. For a proof see Diestel [54], p.46. Use this to show the following result, which is due to Castillo [25]: Take $1 < p < \infty$. Then $\ell_p^w(E) \subset c_0(E)$ if and only if $\ell_p^w(E) = \ell_p^{w,o}(E)$. Hint: An operator $T: \ell_{p'} \to E$ is compact if and only if $\|Ty_n\| \to 0$ whenever y_n converges weakly to zero.

Ex 8.5. Give examples of unconditionally summable but not absolutely summable sequences.

Ex 8.6. Show that:
$$\varepsilon\left(\sum_{j,k=1}^n t_{j,k} e_j \otimes e_k; \ell_\infty^n, \ell_1^n\right) = \sup_{j=1,\ldots,n} \sum_{k=1}^n |t_{j,k}|\ .$$

Ex 8.7. Use Khintchine's inequality to prove that for all $x_1, \ldots, x_n \in \ell_1 \subset \ell_2$
$$\ell_1(x_k; \ell_2) \leq \sqrt{2}\, w_1(x_k; \ell_1)\ .$$

Ex 8.8. Let $\Phi: D_n \to D_n$ be bijective; then for all $f: D_n \to \mathbb{C}$
$$\int_{D_n} f(\Phi(w))\mu_n(dw) = \int_{D_n} f(w)\mu_n(dw)\ .$$

Ex 8.9. (a) Show that for all normed spaces E and elements $x_1, \ldots, x_n \in E$
$$\left(\int_{D_n} \left\|\sum_{k=1}^n \varepsilon_k(w) x_k\right\|_E^p \mu_n(dw)\right)^{1/p} \leq \sqrt{\frac{\pi}{2}} \left(\int_{\mathbb{R}^n} \left\|\sum_{k=1}^n g_k(v) x_k\right\|_E^p \gamma_n(dv)\right)^{1/p}\ .$$

Hint: Integrate $\|\sum \varepsilon_k(w) g_k(v) x_k\|^p = \|\sum \varepsilon_k(w)\, \mathrm{sign}\, g_k(v)|g_k(v)|x_k\|^p$ with respect to $d\gamma_n d\mu_n$ and $d\mu_n d\gamma_n$; Ex 8.8. and $\|\int \|_E^p \leq \int \| \|_E^p$.

(b) Show by using the unit vectors in ℓ_∞^n that there is no converse inequality for any $1 \leq p < \infty$. Hint: Use the fact that $\max\{|g_k| \mid k \leq n\}$ converges to infinity γ-a.e. in $\mathbb{R}^\mathbb{N}$.

Maurey and Pisier [187] showed that for a Banach space E the converse inequality holds (with an additional constant) if and only if E does not contain the ℓ_∞^n uniformly. Note that, by Kahane's inequalities for the Gauss– and Rademacher functions, this need only be checked for one p.

Ex 8.10. Let H be a Hilbert space and (x_n) a sequence in H. Use the parallelogram identity to construct by induction a sequence (w_n) in $\{-1, 1\}$ such that $\|\sum_{k=1}^n w_k x_k\|^2 \geq \sum_{k=1}^n \|x_k\|^2$ for all n. Deduce that every unconditionally summable sequence in H is absolutely 2–summable.

Ex 8.11. (Orlicz) Show that for every unconditionally convergent series $\sum(f_n)$ in $L_1(\Omega, \mu)$ the series $\sum(|f_n(x)|^2)$ converges for μ–almost all $x \in \Omega$.

Ex 8.12. (a) For $1 \leq s \leq r \leq \infty$ call $T \in \mathcal{L}(E, F)$ absolutely (r, s)–summing if there is a constant $c \geq 0$ such that
$$\ell_r(Tx_k; F) \leq c w_s(x_k; E)\ .$$

for all finite sequences (x_k) in E; the smallest such c is denoted by $\mathbf{P}_{r,s}(T)$. By using the idea of 8.9. show that $\mathbf{P}_{r,s}(T) \leq \mathbf{C}_2(T)$ if $1/r \leq 1/s - 1/2$.

(b) Why does the definition of absolutely (r,s)–summing operators only make sense for $s \leq r$?

(c) Show that $T \in \mathcal{L}(E,F)$ is absolutely (r,s)–summing if $(Tx_k) \in \ell_r(F)$ for all sequences $(x_k) \in \ell_s^{w,o}(E)$.

Ex 8.13. (a) Show with the unit vectors e_k that

$$\mathbf{T}_r(\ell_p^n) \geq \max\left\{1, n^{1/p-1/r}\right\} \quad \text{and} \quad \mathbf{C}_s(\ell_p^n) \geq \max\left\{1, n^{1/s-1/p}\right\}.$$

(b) Deduce that infinite dimensional $L_p(\mu)$ do not have type for any $r > p$ and do not have cotype for any $s < p$.

Ex 8.14. (a) If $1 \leq p \leq 2$ and $T \in \mathcal{L}(E,F)$, then

$$\mathbf{C}_{p'}(T') \leq \mathbf{T}_p(T) \quad \text{and} \quad \mathbf{C}_{p'}(T) \leq \mathbf{T}_p(T').$$

Hint: Use that $\sum \langle T'y_k', x_k \rangle = \int \langle \sum \varepsilon_k y_k', \sum \varepsilon_\ell T x_\ell \rangle d\mu_n$ or dualize the tensor product characterization.

(b) Show that the converse inequalities are false — even with an additional constant. Hint: $\ell_1 = c_o'$. See Pisier [223] and also section 31 for positive results if $T = id_E$.

(c) With the weak principle of local reflexivity show that

$$\mathbf{C}_q(T'') = \mathbf{C}_q(T) \quad \text{and} \quad \mathbf{T}_p(T'') = \mathbf{T}_p(T)$$

for all $T \in \mathcal{L}(E,F)$. Hint: Use $y_k' \in B_{F'}$ with $\langle T''x_k'', y_k' \rangle \approx \|T''x_k''\|$; for the type use discrete Rademacher functions and the same idea.

Ex 8.15. For each finite set $A \subset \mathbb{N}$ define the *Walsh function*

$$\varepsilon_A := \prod_{k \in A} \varepsilon_k : D \longrightarrow \{-1,1\}$$

if $A \neq \emptyset$, and $\varepsilon_\emptyset := 1$. Show that $\{\varepsilon_A \mid A \subset \mathbb{N} \text{ finite }\}$ is an orthonormal basis of $L_2(D,\mu)$. Hint: Do this first for D_n.

Ex 8.16. For $1 \leq p \leq \infty$ and elements $x_1, ..., x_n$ in a normed space E use the convexity of the function

$$[-1,1]^n \ni (\alpha_k) \rightsquigarrow \left(\int_{D_n} \Big\| \sum_{k=1}^n \alpha_k \varepsilon_k(w) x_k \Big\|_E^p \mu_n(dw) \right)^{1/p}$$

(and Ex 8.8.) to show that (with $c_R = 1$ and $c_{\mathcal{C}} = 2$)

$$\left(\int_{D_n}\left\|\sum_{k=1}^n \alpha_k\varepsilon_k(w)x_k\right\|_E^p \mu_n(dw)\right)^{1/p}$$

$$\leq c_K \max_{k=1,\ldots,n}|\alpha_k|\left(\int_{D_n}\left\|\sum_{k=1}^n \varepsilon_k(w)x_k\right\|_E^p \mu_n(dw)\right)^{1/p}$$

for all $\alpha_1, \ldots, \alpha_n \in K$. This is a very special case of Kahane's contraction principle (see e.g. Marcus–Pisier [183], p.45).

Ex 8.17. For $1 < p < \infty$ the space Rad of Rademacher functions is complemented in $L_p[0,1]$. Hint: $L_p \hookrightarrow L_2 \to \ell_2$ if $p > 2$; for $1 < p < 2$ define a projection on $L_p \cap L_2$ by $Pf = \sum (f|r_n)r_n$, estimate $\|Pf\|_p \leq b_p (\sum |(f|r_n)|^2)^{1/2}$ and apply Hölder's and again Khintchine's inequality to

$$\sum_{n=1}^{\infty} \lambda_n(f|r_n) = \int f \sum_{n=1}^{\infty} \lambda_n r_n dt \qquad (\lambda_n) \in \ell_2$$

to see that P is continuous on $L_p \cap L_2 \subset L_p$. The Rademacher functions are not complemented in $L_1[0,1]$, see Appendix C7..

Ex 8.18. (a) Show that $\|\sum_{k=1}^n \alpha_k r_k\|_{L_\infty[0,1]} = \sum_{k=1}^n |\alpha_k|$ for real α_k. In the complex case these two norms are equivalent. Note that (in the real case) this result provides an isometric embedding $\ell_1^n \hookrightarrow \ell_\infty^{2^n}$.

(b) Modify the second part of its proof to show that the Khintchine inequality holds also for $0 < p < 1$.

Ex 8.19. Take $x_1', \ldots, x_n' \in E'$. Show that

$$w_p(x_k'; E') = \left\|\sum_{k=1}^n x_k' \otimes e_k : E \longrightarrow \ell_p^n\right\|$$

$$\ell_p(x_k'; E') = \min \|S : E \longrightarrow \ell_\infty^n\| \cdot \|D_\lambda : \ell_\infty^n \longrightarrow \ell_p^n\|,$$

where the latter minimum is taken over all factorizations

$$D_\lambda \circ S = \sum_{k=1}^n x_k' \otimes e_k : E \longrightarrow \ell_p^n$$

with diagonal operators D_λ. It is enough to take D_λ with non–negative entries.

9. Operator Ideals

The notion of a Banach operator ideal is simple but nevertheless important since it structures the way of thinking and hence also the proofs. It was Pietsch who in 1968 noticed the relevance of this notion, and in the following years he and his school investigated all aspects of the abstract theory of operator ideals — a research which culminated in the incredibly rich monograph "Operator Ideals" in 1978. In this section we collect some (not all) of the fundamental notions of the theory which we will need later on.

9.1. An *operator ideal* $\mathfrak{A}$ is a subclass of the class $\mathfrak{L}$ of all continuous linear operators between Banach spaces such that for all Banach spaces E and F its *components*

$$\mathfrak{A}(E, F) := \mathfrak{L}(E, F) \cap \mathfrak{A}$$

satisfy

(1) $\mathfrak{A}(E, F)$ *is a linear subspace of* $\mathfrak{L}(E, F)$ *which contains the finite rank operators.*
(2) *The ideal property: If* $S \in \mathfrak{A}(E_o, F_o), R \in \mathfrak{L}(E, E_o)$ *and* $T \in \mathfrak{L}(F_o, F)$, *then the composition* TSR *is in* $\mathfrak{A}(E, F)$.

The ideal property of certain classes of operators has already been used quite often. The ideal $\mathfrak{F}$ of finite rank operators is obviously the smallest operator ideal and $\mathfrak{L}$ the largest one. Note that for the theory of operator ideals we are only considering operators between Banach spaces.

9.2. For a good analytic theory of operator ideals one needs convergence in $\mathfrak{A}$. Recall that a function $\|\cdot\| : E \to [0, \infty[$ on a $\mathbb{K}$-vector space E is said to be a *quasinorm* if there is a constant $c \geq 1$ such that

(1) $\quad\quad\quad\quad \|x\| = 0 \text{ implies } x = 0 \quad \text{for each } x \in E$.
(2) $\quad\quad\quad\quad \|\lambda x\| = |\lambda| \|x\| \quad \text{for all } (\lambda, x) \in \mathbb{K} \times E$.
(3) $\quad\quad\quad\quad \|x + y\| \leq c(\|x\| + \|y\|) \quad \text{for all } x, y \in E$.

If E is complete with respect to the uniform structure coming from the quasinorm, it is called a *quasi-Banach space*.
For $0 < p \leq 1$ a *p-norm* is a quasinorm satisfying

(3') $\quad\quad\quad\quad \|x + y\|^p \leq \|x\|^p + \|y\|^p \quad \text{for all } x, y \in E;$

in this case $c := 2^{1/p-1}$ may serve as a constant in the quasi-triangle inequality (3). If $(E, \|\ \|)$ is quasinormed with constant c and $p \in]0, 1]$ such that $c = 2^{1/p-1}$, then

$$\|x\|_o := \inf\left\{ \left(\sum_{k=1}^n \|x_k\|^p\right)^{1/p} \mid x = \sum_{k=1}^n x_k \right\}$$

defines a p–norm on E satisfying

$$\|x\|_o \leq \|x\| \leq 2^{1/p}\|x\|_o \qquad \text{for all } x \in E,$$

hence for every quasinorm there is an equivalent p–norm. We omit the tricky proof since it is not important for our purposes; see e.g. Pietsch [214], p.92.

9.3. A *quasinormed operator ideal* $(\mathfrak{A}, \mathbf{A})$ is an operator ideal $\mathfrak{A}$ together with a function $\mathbf{A}: \mathfrak{A} \to [0, \infty[$ such that

(1) $\mathbf{A}|_{\mathfrak{A}(E,F)}$ is a quasinorm for all Banach spaces E and F.
(2) $\mathbf{A}(id_K) = 1$.
(3) If $S \in \mathfrak{A}(E_o, F_o), T \in \mathcal{L}(F_o, F)$ and $R \in \mathcal{L}(E, E_o)$, then the composition satisfies $\mathbf{A}(TSR) \leq \|T\|\mathbf{A}(S)\|R\|$.

If all the components $\mathfrak{A}(E, F)$ are complete (with respect to $\mathbf{A}$), then $(\mathfrak{A}, \mathbf{A})$ is called a *quasi-Banach operator ideal*. A simple argument (with ℓ_p–sums) shows that the quasi-triangle constant may be chosen independently from (E, F); therefore the construction in 9.2. shows that there is always a $p \in]0, 1]$ and an equivalent $\mathbf{A}_o$ such that $(\mathfrak{A}, \mathbf{A}_o)$ is a quasinormed operator ideal and $\mathbf{A}_o$ a p–norm on each component. The advantage of a p–norm is that it is continuous (for the topology it generates) while a quasinorm need not be! See Ex 9.2. for an example of this "quasinorm catastrophe". If $\mathbf{A}$ is a norm (resp. p–norm) on each component $\mathfrak{A}(E, F)$, then $(\mathfrak{A}, \mathbf{A})$ is called a *normed* (resp. *p–normed*) operator ideal. In the case that the components are complete one speaks of *Banach* and *p-Banach operator ideals*.

PROPOSITION: *The following statements hold for each quasinormed operator ideal $(\mathfrak{A}, \mathbf{A})$:*

(1) $\qquad \|T\| \leq \mathbf{A}(T)$ for all $T \in \mathfrak{A}$.
(2) $\qquad \mathbf{A}(x' \otimes y) = \|x'\| \, \|y\|$ for all $x' \in E', y \in F$.

PROOF: (1) follows from the fact that for all $x \in B_E$ and $y' \in B_{F'}$

$$|\langle y', Tx \rangle| = \mathbf{A}(\langle y', Tx \rangle id_K) = \mathbf{A}(y' \circ T \circ (id_K \otimes x)) \leq \|y'\|\mathbf{A}(T)\|x\|$$

and (2) with the same idea of factorization:

$$\|x'\| \, \|y\| = \|x' \otimes y\| \leq \mathbf{A}(x' \otimes y) = \mathbf{A}((id_K \otimes y) \circ id_K \circ x') \leq$$
$$\leq \|x'\|\mathbf{A}(id_K)\|y\| = \|x'\| \, \|y\| \quad . \qquad \square$$

9.4. It is rather simple to check the following very useful

CRITERION: *Let be given a subclass $\mathfrak{A}$ of $\mathcal{L}$, a function $\mathbf{A}: \mathfrak{A} \to [0, \infty[$ and $0 < p \leq 1$. Then $(\mathfrak{A}, \mathbf{A})$ is a p–Banach operator ideal if (and only if)*

(1) $id_K \in \mathfrak{A}$ and $\mathbf{A}(id_K) = 1$.
(2) If TSR is defined and $S \in \mathfrak{A}$, then $TSR \in \mathfrak{A}$ and $\mathbf{A}(TSR) \leq \|T\|\mathbf{A}(S)\|R\|$.
(3) If $S_n \in \mathfrak{A}(E, F)$ such that $\sum_{n=1}^{\infty} \mathbf{A}(S_n)^p < \infty$, then $S := \sum_{n=1}^{\infty} S_n \in \mathfrak{A}$ and

$$\mathbf{A}(S)^p \leq \sum_{n=1}^{\infty} \mathbf{A}(S_n)^p \quad .$$

Observe that (3) accumulates linearity of the components, the triangle inequality and completeness.

9.5. A norm on a vector space E is good for analytic investigations of this space if it is unique, up to equivalence, under all "reasonable" norms; if E is complete under these norms, the closed graph theorem usually helps. The ideal norm on an operator ideal is always "reasonable" in this sense since it dominates the operator norm:

PROPOSITION: *If $(\mathfrak{A}, \mathbf{A})$ and $(\mathfrak{B}, \mathbf{B})$ are two quasi–Banach operator ideals such that $\mathfrak{A} \subset \mathfrak{B}$, then there is a constant $\rho > 0$ such that for all $T \in \mathfrak{A}$*

$$\mathbf{B}(T) \leq \rho \mathbf{A}(T) \quad .$$

PROOF: Since the embeddings $\mathfrak{A}(E, F)$ and $\mathfrak{B}(E, F)$ into $\mathfrak{L}(E, F)$ are continuous, the closed graph theorem (it holds for quasi–Banach spaces) shows that there is a constant $\rho_{E,F}$ such that $\mathbf{B} \leq \rho_{E,F} \mathbf{A}$ on $\mathfrak{A}(E, F)$. An obvious indirect argument with ℓ_p-sums shows that the $\rho_{E,F}$ can be chosen to be uniformly bounded. □

The reader who is not used to the "obvious" argument with ℓ_p-sums (metric injection of the E_n's into $\ell_p(E_m)$, metric projection, ideal property) should check the details.

9.6. Examples:
(1) The approximable operators $\overline{\mathfrak{F}}$ (see 4.2.(1)) together with the operator norm form a Banach operator ideal $(\overline{\mathfrak{F}}, \|\ \|)$; recall

$$\overline{\mathfrak{F}}(E, F) = E' \tilde{\otimes}_\varepsilon F$$

and that $\overline{\mathfrak{F}}(E, F) = \mathfrak{K}(E, F)$ if E' or F has the approximation property (see 5.3.). The ideal $\mathfrak{K}$ of compact operators and the ideal $\mathfrak{W}$ of weakly compact operators form (again with the operator norm) a Banach operator ideal; ideals of this type are called *closed operator ideals* (=: their components are closed in $\mathfrak{L}(E, F)$). A famous theorem of Calkin states that the only proper (i.e. different from $\mathfrak{L}$ and $\{0\}$) closed ideal in the algebra $\mathfrak{L}(\ell_2, \ell_2)$ is the ideal of compact operators; this is also true for c_o and ℓ_p (if $1 \leq p < \infty$), but not for $L_p[0,1]$ (if $1 < p < \infty$ and $p \neq 2$, see Pietsch [214], p.84).

Weakly compact operators satisfy an important factorization theorem due to Davis, Figiel, Johnson and Pełczyński [37]: If $T \in \mathfrak{W}(E, F)$, then there is a reflexive Banach space G and operators $R \in \mathfrak{L}(E, G)$ and $S \in \mathfrak{L}(G, F)$ such that $T = SR$ and $\|T\| = \|S\| \|R\|$:

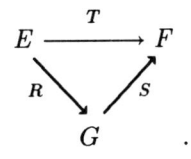

$$\begin{array}{ccc} E & \xrightarrow{T} & F \\ & \searrow R \quad \nearrow S & \\ & G & \end{array}$$

For a proof, including the relation about the norm, see Pietsch [214], p.55.

(2) The class $\mathfrak{N}$ of nuclear operators (see 3.6.) with the norm $\mathbf{N}$ is a Banach operator ideal: Completeness is clear since $\mathbf{N}$ is the quotient norm of

$$E' \tilde{\otimes}_\pi F \xrightarrow{1} \mathfrak{N}(E, F) \quad ,$$

the ideal property is easy to verify (and corresponds to the mapping property of the projective norm) and $\mathbf{N}(id_{I\!\!K}) = 1$ (— it is just a joke to argue for this with $I\!\!K = I\!\!K \otimes_\pi I\!\!K = \mathfrak{N}(I\!\!K)$). By the very definition of the nuclear norm and the fact that

$$\mathbf{A}(x' \otimes y) = \|x'\| \|y\|$$

for all operator ideals, it follows that $(\mathfrak{N}, \mathbf{N})$ is the smallest Banach operator ideal. Recall that the nuclear norm of a $\mathcal{C}$-linear operator may increase if it is considered as an $I\!\!R$-linear operator; this follows from $\mathbf{N}(id_E) = \dim_{I\!\!K} E$ (see 3.7.).

(3) Let (Ω_i, μ_i) be two measure spaces, $p, q \in [1, \infty[, G$ a subspace of $L_p(\mu_1)$ and S an operator in $\mathfrak{L}(G, L_q(\mu_2))$ of norm 1. Take $\mathfrak{L}_S$ to be the class of all operators $T \in \mathfrak{L}(E, F)$ such that

$$S \otimes T : G \otimes_{\Delta_p} E \longrightarrow L_q(\mu_2) \otimes_{\Delta_q} F$$

is continuous. Together with $\mathbf{L}_S(T) := \|S \otimes T : \otimes_{\Delta_p} \to \otimes_{\Delta_q}\|$ the class $\mathfrak{L}_S$ is a Banach operator ideal (this is an exercise). As a special case one gets: The ideal $(\mathfrak{T}_p, \mathbf{T}_p)$ of type p operators and the ideal $(\mathfrak{C}_q, \mathbf{C}_q)$ of cotype q operators are Banach operator ideals. Other ideals of this type and results about $\mathfrak{L}_S$ are treated in Ex 9.15., Ex 9.16. and 17.4..

9.7. An operator ideal $\mathfrak{A}$ is said to be *injective* if for each metric injection $I : F \hookrightarrow G$ an operator $T \in \mathfrak{L}(E, F)$ is in $\mathfrak{A}$ if $I \circ T \in \mathfrak{A}$. A quasinormed ideal $(\mathfrak{A}, \mathbf{A})$ is called injective if moreover, $\mathbf{A}(T) = \mathbf{A}(I \circ T)$. Clearly, the Banach ideals of compact and of weakly compact operators are injective, but the ideals of approximable and of nuclear operators are not, see Ex 3.17., Ex 5.13. and Ex 9.13. for the nuclear operators and Ex 9.12. for the approximable ones.

PROPOSITION: Let $(\mathfrak{A}, \mathbf{A})$ be a quasinormed operator ideal. Then there is a (unique) smallest injective operator ideal $\mathfrak{A}^{inj}$ which contains $\mathfrak{A}$. If I_F denotes the canonical injection $F \hookrightarrow \ell_\infty(B_{F'})$, then $T \in \mathfrak{L}(E, F)$ belongs to $\mathfrak{A}^{inj}$ if and only if $I_F \circ T \in \mathfrak{A}$. With

$$\mathbf{A}^{inj}(T) := \mathbf{A}(I_F T)$$

$(\mathfrak{A}^{inj}, \mathbf{A}^{inj})$ is a quasinormed operator ideal which is quasi-Banach if $(\mathfrak{A}, \mathbf{A})$ is quasi-Banach.

PROOF: Define $\mathfrak{A}^{inj}$ to be the class of all $T \in \mathfrak{L}(E, F)$ such that $I_F \circ T \in \mathfrak{A}$ and $\mathbf{A}^{inj}$ as in the statement. Then it is easily seen that $(\mathfrak{A}^{inj}, \mathbf{A}^{inj})$ is a quasinormed operator ideal (for the ideal property use the metric extension property of $\ell_\infty(\Gamma)$) — with complete components if $\mathfrak{A}$ has. If $I : F \hookrightarrow G$ is a metric injection, the diagram

$$E \xrightarrow{T} F \xhookrightarrow{I} G$$
$$I_F \downarrow \quad\quad \downarrow I_G \quad\quad P \text{ extends } I_F \text{ with}$$
$$\ell_\infty(B_{F'}) \xleftarrow{P} \ell_\infty(B_{G'}) \quad\quad \|P\| = 1$$

shows

$$\mathbf{A}^{inj}(T) = \mathbf{A}(I_F T) = \mathbf{A}(P I_G I T) \leq \|P\| \mathbf{A}^{inj}(IT) \leq \mathbf{A}^{inj}(T) \;\; ,$$

hence $(\mathfrak{A}^{inj}, \mathbf{A}^{inj})$ is injective. If $\mathfrak{B} \supset \mathfrak{A}$ is any injective operator ideal, then clearly $\mathfrak{B} \supset \mathfrak{A}^{inj}$ and the proposition is proven. □

$(\mathfrak{A}^{inj}, \mathbf{A}^{inj})$ is called the *injective hull* of $(\mathfrak{A}, \mathbf{A})$. In Ex 9.13. it will be shown that $\mathfrak{N}^{inj}$ is the class of *quasinuclear operators*.

COROLLARY: If the Banach space F has the λ-extension property and $(\mathfrak{A}, \mathbf{A})$ is a quasinormed operator ideal, then $\mathfrak{A}(E, F) = \mathfrak{A}^{inj}(E, F)$ for each Banach space E and

$$\mathbf{A}(T) \leq \lambda \mathbf{A}^{inj}(T)$$

for all $T \in \mathfrak{A}^{inj}(E, F)$.

This is immediate.

9.8. The dual concept is the following: An operator ideal $\mathfrak{A}$ is called *surjective* if for each metric surjection $Q : G \twoheadrightarrow E$ an operator $T \in \mathfrak{L}(E, F)$ is in $\mathfrak{A}$ if $T \circ Q \in \mathfrak{A}$. A quasinormed operator ideal $(\mathfrak{A}, \mathbf{A})$ is called surjective if moreover, $\mathbf{A}(T) = \mathbf{A}(TQ)$. The Banach ideals $\mathfrak{K}$ and $\mathfrak{W}$ are surjective, $\overline{\mathfrak{F}}$ is not (Ex 9.12.) — and also the nuclear operators do not form a surjective ideal; this may be seen as in Ex 3.17. or by the following argument: Take a metric surjection $Q : G \twoheadrightarrow E$ and F with the approximation property such that $Q' \tilde{\otimes}_\pi id_F$ is not an isomorphic injection, for example (see 1.5.)

$I : \ell_2 \hookrightarrow L_1[0,1]$ the embedding of the Rademacher functions, $Q := I' : L_\infty \to \ell_2$ and $F = \ell_2$. Now

$$E' \tilde\otimes_\pi F = \mathfrak{N}(E,F) \hookrightarrow \mathfrak{L}(E,F) \ni T$$
$$Q' \tilde\otimes_\pi id_F \downarrow \qquad\qquad \downarrow \Phi \quad \}$$
$$G' \tilde\otimes_\pi F = \mathfrak{N}(G,F) \hookrightarrow \mathfrak{L}(G,F) \ni TQ$$

commutes; since the subspace $\mathfrak{N}(G,F) \cap \mathrm{im}\Phi$ is closed in $\mathfrak{N}(G,F)$, it cannot coincide with $\Phi(\mathfrak{N}(E,F))$ by the closed graph theorem; in other words, there is a non–nuclear $T \in \mathfrak{L}(E,F)$ such that $TQ \in \mathfrak{N}(G,F)$: The ideal $\mathfrak{N}$ of nuclear operators is not surjective.

PROPOSITION: *Let* $(\mathfrak{A}, \mathbf{A})$ *be a quasinormed operator ideal. Then there is a (unique) smallest surjective operator ideal* $\mathfrak{A}^{sur}$ *which contains* $\mathfrak{A}$. *If* Q_E *is the canonical surjection* $\ell_1(B_E) \twoheadrightarrow E$, *then* $T \in \mathfrak{L}(E,F)$ *belongs to* $\mathfrak{A}^{sur}$ *if and only if* $T \circ Q_E \in \mathfrak{A}$. *With*

$$\mathbf{A}^{sur}(T) := \mathbf{A}(TQ_E)$$

$(\mathfrak{A}^{sur}, \mathbf{A}^{sur})$ *is a quasinormed operator ideal which is quasi-Banach if* $(\mathfrak{A}, \mathbf{A})$ *is quasi-Banach.*

PROOF: The argument follows the same pattern as that in the case of the injective hull — the metric extension property is replaced by the lifting property of $\ell_1(\Gamma)$, see 3.12.; the central point is to show that $\mathfrak{A}^{sur}$, defined as indicated in the statement of the proposition, is surjective: For every $\varepsilon > 0$ there is a lifting S_ε of Q_E

$$\begin{array}{ccccc} G & \xrightarrow{Q} & E & \xrightarrow{T} & F \\ \uparrow{\scriptstyle Q_G} & & \uparrow{\scriptstyle Q_E} & & \\ \ell_1(B_G) & \xleftarrow{S_\varepsilon} & \ell_1(B_E) & & \end{array} \qquad \|S_\varepsilon\| \leq 1 + \varepsilon$$

hence

$$\mathbf{A}^{sur}(T) = \mathbf{A}(TQ_E) = \mathbf{A}(TQQ_G S_\varepsilon) \leq \mathbf{A}^{sur}(TQ)(1+\varepsilon) \leq \mathbf{A}^{sur}(T)(1+\varepsilon)$$

and this gives $\mathbf{A}^{sur}(T) = \mathbf{A}^{sur}(TQ)$. □

$(\mathfrak{A}^{sur}, \mathbf{A}^{sur})$ is called the *surjective hull* of $(\mathfrak{A}, \mathbf{A})$. The same diagram gives the

COROLLARY: *If* $(\mathfrak{A}, \mathbf{A})$ *is a quasinormed operator ideal, then for all sets* Γ *and all Banach spaces* F

$$\mathfrak{A}(\ell_1(\Gamma), F) = \mathfrak{A}^{sur}(\ell_1(\Gamma), F)$$

holds isometrically.

A description of $\mathfrak{N}^{sur}$ will be given in Ex 9.14..

9.9. If $(\mathfrak{A}, \mathbf{A})$ is a quasinormed operator ideal, then

$$\mathfrak{A}^{dual} := \{T \in \mathfrak{L} \mid T' \in \mathfrak{A}\}$$
$$\mathbf{A}^{dual}(T) := \mathbf{A}(T')$$

defines a quasinormed operator ideal, the dual ideal of $(\mathfrak{A}, \mathbf{A})$. If $(\mathfrak{A}, \mathbf{A})$ is quasi–Banach, then $(\mathfrak{A}^{dual}, \mathbf{A}^{dual})$ is as well. Well–known results state: $\mathfrak{K} = \mathfrak{K}^{dual}$ and $\mathfrak{W} = \mathfrak{W}^{dual}$; ideals $\mathfrak{A}$ with $\mathfrak{A} = \mathfrak{A}^{dual}$ are called *completely symmetric*. $\mathfrak{N} \subset \mathfrak{N}^{dual}$ (see 5.9.), but $\mathfrak{N} \neq \mathfrak{N}^{dual}$ as will be seen in 16.7.. The approximable operators $\overline{\mathfrak{F}}$ are completely symmetric, see Ex 9.12..

9.10. Finally, the notion of the *product* (or the *composition*) of two ideals $\mathfrak{A}$ and $\mathfrak{B}$ will be needed: $T \in \mathfrak{L}(E, F)$ belongs to the product $\mathfrak{A} \circ \mathfrak{B}$ of $\mathfrak{A}$ and $\mathfrak{B}$ if there is a Banach space G and $R \in \mathfrak{B}(E, G)$ and $S \in \mathfrak{A}(G, F)$ such that

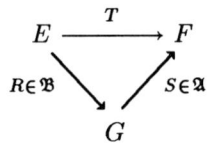

$$E \xrightarrow{T} F$$
$$R \in \mathfrak{B} \searrow \quad \nearrow S \in \mathfrak{A}$$
$$G$$

If $\mathfrak{A}$ and $\mathfrak{B}$ are quasinormed, then

$$\mathbf{A} \circ \mathbf{B}(T) := \inf \mathbf{A}(S)\mathbf{B}(R) ,$$

where the infimum is taken over all such factorizations. It is not difficult to see that $(\mathfrak{A} \circ \mathfrak{B}, \mathbf{A} \circ \mathbf{B})$ is a quasinormed operator ideal. The completeness (if $\mathfrak{A}$ and $\mathfrak{B}$ are) follows from the

PROPOSITION: *If $(\mathfrak{A}, \mathbf{A})$ is a q–Banach and $(\mathfrak{B}, \mathbf{B})$ a p–Banach operator ideal, then $(\mathfrak{A} \circ \mathfrak{B}, \mathbf{A} \circ \mathbf{B})$ is an r–Banach operator ideal where $1/r = 1/p + 1/q$.*

PROOF: Only property (3) of criterion 9.4. needs to be checked: For a sequence (T_n) in $\mathfrak{A} \circ \mathfrak{B}(E, F)$ with

$$\sum_{n=1}^{\infty} [\mathbf{A} \circ \mathbf{B}(T_n)]^r < \infty$$

choose factorizations $T_n = S_n \circ R_n$ through G_n with

$$\mathbf{B}(R_n)^p \leq (1+\varepsilon)\,[\mathbf{A} \circ \mathbf{B}(T_n)]^r$$
$$\mathbf{A}(S_n)^q \leq (1+\varepsilon)\,[\mathbf{A} \circ \mathbf{B}(T_n)]^r$$

and put $G := \ell_2(G_n)$ (denote by I_m and Q_m the canonical injections and projections). Again by the criterion

$$R := \sum_{n=1}^{\infty} I_n R_n \in \mathfrak{B} \quad \text{and} \quad S := \sum_{n=1}^{\infty} S_n Q_n \in \mathfrak{A}$$

with $\sum_{n=1}^{\infty} T_n = SR$ and

$$\mathbf{B}(R)^p \leq (1+\varepsilon) \sum_{n=1}^{\infty} [\mathbf{A} \circ \mathbf{B}(T_n)]^r$$

$$\mathbf{A}(S)^q \leq (1+\varepsilon) \sum_{n=1}^{\infty} [\mathbf{A} \circ \mathbf{B}(T_n)]^r ,$$

hence $(\mathbf{A} \circ \mathbf{B})(\sum_{n=1}^{\infty} T_n) \leq [(1+\varepsilon) \sum_{n=1}^{\infty} [\mathbf{A} \circ \mathbf{B}(T_n)]^r]^{1/r}$ — for all $\varepsilon > 0$. □

Unfortunately, the product of two normed operator ideals need not be normed. For example, if $\mathfrak{I}$ is the ideal of integral operators (to be defined in the next section), the composition ideal $\mathfrak{I} \circ \mathfrak{I}$ is not normed (see 29.8. for an argument showing this). Moreover, $\mathfrak{N} \circ \mathfrak{N}$ is not normed; this follows from the facts that operators in $\mathfrak{N} \circ \mathfrak{N}(E, E)$ have summable eigenvalues, but nuclear operators have, in general, only square–summable eigenvalues (see e.g. König [147], 2.b.14. and 4.a.5.) and $\mathfrak{N}$ is the smallest Banach operator ideal.

Exercises:

Ex 9.1. If $\|\ \|$ is a quasinorm with triangle inequality constant c, then

$$\left\| \sum_{k=1}^{n} x_k \right\| \leq \sum_{k=1}^{n} c^k \|x_k\| \leq c^n \left(\sum_{k=1}^{n} \|x_k\| \right) .$$

Ex 9.2. (a) If $(E, \|\ \|)$ is p–normed, then $\|\ \| : E \to \mathbb{R}$ is continuous.
(b) For $S \in \mathfrak{L}$ define

$$\mathbf{A}(S) := \begin{cases} \|S\| & \text{if } S \in \overline{\mathfrak{F}} \\ 2\|S\| & \text{if } S \notin \overline{\mathfrak{F}} . \end{cases}$$

Then $(\mathfrak{L}, \mathbf{A})$ is a quasi–Banach operator ideal, but $\mathbf{A}$ is not continuous on $\mathfrak{L}(E, E)$ (with respect to the $\mathbf{A}$–convergence) if E is infinite dimensional.

Ex 9.3. Check that the ideals $\mathfrak{V}$ of completely continuous operators $((T x_n)$ is norm convergent if (x_n) converges weakly) and $\mathfrak{X}$ of operators with separable range are closed Banach operator ideals.

Ex 9.4. If Γ has a sufficiently large cardinality, then the algebra $\mathfrak{L}(\ell_2(\Gamma), \ell_2(\Gamma))$ has infinitely many closed ideals.

Ex 9.5. If $T \in \mathfrak{K}(E, F)$ and $n \in \mathbb{N}$, then there are reflexive Banach spaces $E_1, ..., E_n$ and operators $T_o \in \mathfrak{K}(E, E_1), T_k \in \mathfrak{K}(E_k, E_{k+1})$ and $T_n \in \mathfrak{K}(E_n, F)$ such that $T = T_n \circ ... \circ T_o$ and $\|T\| = \|T_n\|...\|T_o\|$. Hint: 3.5., corollary 2.

Ex 9.6. Let $x_1, ..., x_n, y_1, ..., y_n \in \mathbb{K}$ and M the matrix $(x_k y_\ell)_{k,\ell=1}^n$ considered as an operator $\mathbb{K}^n \to \mathbb{K}^n$. If $(\mathfrak{A}, \mathbf{A})$ is an operator ideal, calculate $\mathbf{A}(M : \ell_q^n \to \ell_p^n)$.

Ex 9.7. If $(\mathfrak{A}, \mathbf{A})$ and $(\mathfrak{B}, \mathbf{B})$ are quasi-Banach operator ideals, $R \in \mathfrak{L}(E_o, E)$ and $S \in \mathfrak{L}(F, F_o)$ such that

$$\mathfrak{A}(E, F) \longrightarrow \mathfrak{B}(E_o, F_o)$$
$$T \rightsquigarrow S \circ T \circ R$$

is defined, then this map is continuous.

Ex 9.8. If $\mathfrak{A}$ is a quasinormed operator ideal, the class space($\mathfrak{A}$) of all Banach spaces E with $id_E \in \mathfrak{A}$ is called the *space ideal of* $\mathfrak{A}$. Show that it is stable under isomorphic bijections, complemented subspaces and finite cartesian products. If $\mathfrak{A}$ is injective (resp. surjective), then space($\mathfrak{A}$) is stable under forming subspaces (resp. quotients).

Ex 9.9. For $T \in \mathfrak{L}(E, F)$ put

$$\mathbf{L}_2(T) := \inf \|S\| \, \|R\| \, ,$$

where the infimum is taken over all factorizations $T = SR$ through any Hilbert space. Show that the class $\mathfrak{L}_2$ for all operators with $\mathbf{L}_2(T) < \infty$ is (with $\mathbf{L}_2$) an injective and surjective Banach operator ideal.

Ex 9.10. Let $\mathfrak{A}$ be an operator ideal, $T \in \mathfrak{L}(E, F)$ and G a Banach space.

(a) Let $S \in \mathfrak{A}(G, F)$ such that $\operatorname{im} T \subset \operatorname{im} S$. If S is injective or $\mathfrak{A}$ surjective, then $T \in \mathfrak{A}$.

(b) Let $S \in \mathfrak{A}(E, G)$ such that $\|Tx\| \leq c\|Sx\|$ for some $c \geq 0$ and all $x \in E$. If S has dense range or $\mathfrak{A}$ is injective, then $T \in \mathfrak{A}$.

Ex 9.11. Let $(\mathfrak{A}, \mathbf{A})$ be a quasi-Banach operator ideal such that for every metric injection I the operator T is in $\mathfrak{A}$ if and only if $I \circ T \in \mathfrak{A}$. Then $\mathfrak{A} = \mathfrak{A}^{inj}$ and $\mathbf{A}$ and $\mathbf{A}^{inj}$ are equivalent quasinorms.

Ex 9.12. (a) Use the fact that there exist Banach spaces without the approximation property to show that the ideal $\overline{\mathfrak{F}}$ of all approximable operators is neither injective nor surjective. Moreover, $\overline{\mathfrak{F}}^{inj} = \overline{\mathfrak{F}}^{sur} = \mathfrak{K}$.

(b) Show that $\overline{\mathfrak{F}}$ is completely symmetric: $\overline{\mathfrak{F}}^{dual} = \overline{\mathfrak{F}}$. Hint: If $T \in \mathfrak{K}(E, F)$ and $S \in \mathfrak{F}(E, F'')$ such that $\|T - S\| \leq \varepsilon$, choose an ε-net Tx_k of TB_E and apply the principle of local reflexivity to $\operatorname{span}\{\operatorname{im} S \cup \{Tx_k\}\}$.

(c) Deduce from (b) that $\overline{\mathfrak{F}}$ is regular, i.e.: if $\kappa_F \circ T \in \overline{\mathfrak{F}}(E, F'')$, then $T \in \overline{\mathfrak{F}}(E, F)$.

Ex 9.13. An operator $T \in \mathfrak{L}(E, F)$ is called *quasinuclear* if there exist $(x_n') \in \ell_1(F')$ such that for all $x \in E$

$$\|Tx\| \leq \sum_{n=1}^{\infty} |\langle x_n', x \rangle| \, .$$

(a) With $\mathbf{QN}(T) := \inf\{\sum_{n=1}^{\infty} \|x'_n\| \mid (x'_n) \text{ as above }\}$ the class $\mathfrak{QN}$ of quasinuclear operators is an injective Banach operator ideal.
(b) If T or T' is nuclear, then T is quasinuclear and $\mathbf{QN}(T) \leq \mathbf{N}(T')$.
(c) Give examples of quasinuclear, non–nuclear operators. Hint: Ex 3.17. and Ex 5.13..
(d) If F has the λ–extension property, then $\mathfrak{QN}(E,F) = \mathfrak{N}(E,F)$ and $\mathbf{N}(T) \leq \lambda \mathbf{QN}(T)$. Hint: Factor T through a subspace of ℓ_1 and extend.
(e) $(\mathfrak{N}^{inj}, \mathbf{N}^{inj}) = (\mathfrak{QN}, \mathbf{QN})$.
(f) Quasinuclear operators are compact.
(g) Use the fact that there is a metric surjection $Q: G \twoheadrightarrow F$ such that $Q' \otimes_\pi id_{\ell_\infty}$ is not an isomorphic injection (since ℓ_∞ is no $\mathcal{L}_1^g$–space by 8.6. and Ex 7.14., this will follow from 23.5.) to show that $\mathfrak{QN}$ is not a surjective operator ideal.
(h) $\mathfrak{QN}(c_o, F) = \mathfrak{N}(c_o, F)$ holds isometrically. Hint: Ex 3.13..

Ex 9.14. Show that $T \in \mathcal{L}(E,F)$ is in the surjective hull $\mathfrak{N}^{sur}$ of the nuclear operators if and only if there is a sequence $(y_n) \in \ell_1(F)$ such that

$$TB_E \subset \left\{ \sum_{n=1}^{\infty} \lambda_n y_n \mid |\lambda_n| \leq 1 \right\} .$$

Hint: Ex 9.10..

Ex 9.15. (a) Show that the operator ideal $(\mathcal{L}_S, \mathbf{L}_S)$ defined in 9.6.(3) is always injective. If $G = L_p(\mu_1)$, then $(\mathcal{L}_S, \mathbf{L}_S)$ is even surjective. Hint: 7.4..
(b) What does (a) imply for the ideal $\mathfrak{T}_p$ of type p and the ideal $\mathfrak{C}_q$ of cotype q operators? Show that $\mathfrak{C}_q$ is not surjective; in particular, if $G \subset L_p(\mu)$, then $G \otimes_{\Delta_p} \cdot$ does not in general respect metric surjections.
(c) Let $1 < p < \infty$. If T is an operator which factors through a subspace of a quotient of some $L_p(\nu)$ (or: through a quotient of a subspace of some $L_p(\nu)$ — this is the same, as will be seen later on), then for all $S \in \mathcal{L}(L_p(\mu_1), L_p(\mu_2))$ the operator

$$S \otimes T : L_p(\mu_1) \otimes_{\Delta_p} E \longrightarrow L_p(\mu_2) \otimes_{\Delta_p} F$$

is continuous. Hint: Fix S, consider first $T = id_{L_p}$ and use (a). It is a celebrated result of Kwapień that these T are the only ones which make $S \otimes T$ continuous for all S as above; we shall prove this in Chapter III, 28.4..

Ex 9.16. Denote by R the orthogonal projection of $L_2(D, \mu)$ onto the closed linear span Rad of the discrete Rademacher functions (see 8.5.).
(a) If $E_n := L_1(D_n, \mu_n)$, then

$$\|R \otimes id_{E_n} : L_2(D, \mu) \otimes_{\Delta_2} E_n \longrightarrow L_2(D, \mu) \otimes_{\Delta_2} E_n\| \geq \sqrt{\frac{n}{2}} .$$

Hint: Use the function $f := \prod_{k=1}^n ((1 + \varepsilon_k) \otimes \varepsilon_k) \in L_2(D, \mu) \otimes E_n$ and Khintchine's inequality.

(b) Deduce that
$$R \otimes id_{\ell_1} : L_2(D,\mu) \otimes_{\Delta_2} \ell_1 \longrightarrow L_2(D,\mu) \otimes_{\Delta_2} \ell_1$$
is not continuous.

(c) Following Maurey and Pisier [187], an operator $T \in \mathfrak{L}(E,F)$ is called K-convex if
$$\mathbf{K}(T) := \|R \otimes T : L_2(D,\mu) \otimes_{\Delta_2} E \longrightarrow L_2(D,\mu) \otimes_{\Delta_2} F\| < \infty \;.$$
Show that the class of K-convex operators together with the norm $\mathbf{K}$ is an injective and surjective Banach operator ideal. It follows that a space which contains the ℓ_1^n uniformly is not K-convex; it is a deep result of Pisier [223] that the converse is true. See section 31 for more information about K-convex operators.

Ex 9.17. Show that $\mathfrak{A} \circ (\mathfrak{B} \circ \mathfrak{C}) \stackrel{1}{=} (\mathfrak{A} \circ \mathfrak{B}) \circ \mathfrak{C}$ for quasinormed operator ideals $\mathfrak{A}, \mathfrak{B}$ and $\mathfrak{C}$.

Ex 9.18. Let $(\mathfrak{A}, \mathbf{A})$ be a quasinormed operator ideal. Then $(\mathfrak{A}(E,F), \mathbf{A})$ is complete if E or F is finite dimensional. Hint: $\|T\| \leq \mathbf{A}(T) \leq \|T\|\mathbf{A}(id_F)$.

10. Integral Operators

Grothendieck's integral operators are related to the injective tensor norm in the same way as the continuous operators are related to the projective tensor norm: the associated (bi–)linear form on $E \otimes F'$ is continuous with respect to ε (instead of π). This section gives the basic factorization theorem for integral operators and some applications. Many of the arguments will be the same as those given in Chapter II where the study of maximal operator ideals using tensor norms will be taken up. So the present section is partly a training program for later on.

10.1. Recall that the continuous bilinear forms on $E \times G$ (i.e. the continuous linear forms on $E \otimes_\pi G$) are exactly the continuous linear operators $E \to G'$:
$$\mathfrak{Bil}(E,G) = (E \otimes_\pi G)' = \mathfrak{L}(E,G') \;.$$
$$\varphi \rightsquigarrow L_\varphi$$
$$\beta_T \rightsquigarrow T$$

Every operator $T \in \mathcal{L}(E, F)$ defines a continuous linear functional φ on $E \otimes_\pi F'$ by

$$\langle \varphi, x \otimes y' \rangle := \langle y', Tx \rangle = \langle \kappa_F Tx, y' \rangle$$

and hence $\varphi = \beta_{\kappa_F \circ T} \in (E \otimes_\pi F')'$ is the functional associated with $\kappa_F \circ T : E \to F''$. Recall that $(E \otimes_\varepsilon G)' \subset (E \otimes_\pi G)'$.

An operator $T \in \mathcal{L}(E, F)$ between Banach spaces is called *integral* if its associated linear functional $\beta_{\kappa_F \circ T}$ on $E \otimes F'$ is continuous with respect to the injective norm ε (i.e. is integral in the sense of 4.6.). Define the integral norm by

$$\mathbf{I}(T) := \|\beta_{\kappa_F \circ T} : E \otimes_\varepsilon F' \longrightarrow \mathbb{K}\| \quad .$$

Then the class $\mathfrak{I}$ of integral operators with $\mathbf{I}$ as a norm is a Banach operator ideal; the ideal property of $\mathfrak{I}$ comes from the mapping property of ε, obviously $\mathbf{I}(id_\mathbb{K}) = 1$ and it remains to show the completeness. Since $(E \otimes_\varepsilon F')' \hookrightarrow (E \otimes_\pi F')' = \mathcal{L}(E, F'')$ is continuous and $\mathcal{L}(E, F) \subset \mathcal{L}(E, F'')$ is closed, the space

$$\mathfrak{I}(E, F) = (E \otimes_\varepsilon F')' \cap \mathcal{L}(E, F)$$

(the intersection taken in $\mathcal{L}(E, F'')$) is closed in the Banach space $(E \otimes_\varepsilon F')'$, hence complete (with respect to the induced norm).

PROPOSITION: $\mathfrak{I}(E, F') = (E \otimes_\varepsilon F)'$ *(via $T \leadsto \beta_T$) holds isometrically for all Banach spaces E and F.*

PROOF: $T \in \mathcal{L}(E, F')$ defines a functional $\varphi = \beta_T \in (E \otimes_\pi F)'$. The canonical right–extension $\varphi^\wedge$ of φ to $E \otimes F''$ (see 1.9.) satisfies

$$\langle \varphi^\wedge, x \otimes y'' \rangle = \langle Tx, y'' \rangle_{F', F''} = \langle \kappa_{F'} Tx, y'' \rangle_{F''', F''} \ ,$$

hence $\varphi^\wedge = \beta_{\kappa_{F'} \circ T}$: The extension lemma 6.7. for the injective norm ε says that $\varphi \in (E \otimes_\varepsilon F)'$ if and only if $\varphi^\wedge \in (E \otimes_\varepsilon F'')'$ (with equal norms). This is exactly the statement of the proposition. $\square$

10.2. One should always keep in mind the tensor product representations of $\mathfrak{I}(E, F)$ and $\mathfrak{I}(E, F')$. These representations imply the

COROLLARY 1: $T \in \mathfrak{I}(E, F)$ *if and only if $\kappa_F \circ T \in \mathfrak{I}(E, F'')$; moreover, the norms coincide:* $\mathbf{I}(T) = \mathbf{I}(\kappa_F \circ T)$.

Ideals with this property are called *regular*.

COROLLARY 2: $T \in \mathfrak{I}(E, F)$ *if and only* $T' \in \mathfrak{I}(F', E')$; *moreover,* $\mathbf{I}(T) = \mathbf{I}(T')$.

In other words, $(\mathfrak{J}^{dual}, \mathbf{I}^{dual}) = (\mathfrak{J}, \mathbf{I})$ by the definitions in 9.9..

PROOF: If $T \in \mathcal{L}(E, F)$ and φ is its associated functional on $E \otimes F'$, then

$$\langle \varphi^t, y' \otimes x \rangle = \langle \varphi, x \otimes y' \rangle = \langle y', Tx \rangle = \langle T'y', x \rangle \quad,$$

hence $\varphi^t = \beta_{T'}$. Since, clearly,

$$\|\varphi^t\|_{(F' \otimes_\varepsilon E)'} = \|\varphi\|_{(E \otimes_\varepsilon F')'} \quad,$$

the result follows from the proposition. $\square$

10.3. The definition $\mathfrak{J}(E, F) = (E \otimes_\varepsilon F')' \cap \mathcal{L}(E, F)$ of integral operators, the isometric equality $F' \otimes_\varepsilon E = \mathfrak{F}(F, E)$ and the trace duality 2.6. imply that $T \in \mathcal{L}(E, F)$ is integral if and only if there is a constant $c \geq 0$ such that

$$|tr_E(S \circ T)| \leq c\|S\| \qquad \text{for all } S \in \mathfrak{F}(F, E) \ ;$$

clearly, $\mathbf{I}(T)$ is the minimum of these constants c.

10.4. Some examples: If E or F is finite dimensional, the duality of ε and π (theorem 6.4.) gives that

$$\mathfrak{N}(E, F) = E' \otimes_\pi F \hookrightarrow (E \otimes_\varepsilon F')'$$

holds isometrically. Since $\mathcal{L}(E, F) = \mathfrak{F}(E, F) = E' \otimes F$ in this case, one obtains the

PROPOSITION: *If E and F are Banach spaces one of which is finite dimensional, then for all $T \in \mathcal{L}(E, F)$*

$$\mathfrak{N}(E, F) = \mathfrak{J}(E, F) \text{ and } \mathbf{N}(T) = \mathbf{I}(T) \quad .$$

The results in 8.8. imply

$$\mathbf{I}(id : \ell_p^n \longrightarrow \ell_q^n) = \begin{cases} n^{1-1/p+1/q} & \text{if } 1 \leq p \leq q \leq \infty \\ n & \text{if } 1 \leq q \leq p \leq \infty \end{cases}.$$

EXAMPLE: *The canonical injection $I : \ell_1 \hookrightarrow c_o$ is integral with integral norm 1*.

PROOF: It has to be shown that $\varphi : \ell_1 \otimes_\varepsilon \ell_1 \longrightarrow \mathbb{K}$ defined by

$$\langle \varphi, (\xi_n) \otimes (\eta_n) \rangle := \sum_{n=1}^{\infty} \xi_n \eta_n$$

is continuous with norm 1. Since φ is clearly continuous on $\ell_1 \otimes_\pi \ell_1$ with norm 1, it is enough (by the density lemma 7.3.) to check the ε-continuity of φ on the π-dense subspace $\bigcup_{n=1}^\infty \ell_1^n \otimes \ell_1^n$: Since it was just seen that $\mathbf{I}(id : \ell_1^n \longrightarrow \ell_\infty^n) = 1$, the result follows. $\square$

The ideal property implies that
$$\mathbf{I}(I : \ell_1 \hookrightarrow \ell_\infty) = 1 ,$$
which later in this section will be seen to have interesting consequences.

Which operators are not integral? Integral operators are weakly compact (see Ex 10.1.) — this provides a good number of examples; the product of two integral operators is nuclear (this will follow from the corresponding property for absolutely 2–summing operators) which gives further examples. To see a more concrete example of an operator, recall the Fourier matrices from Ex 4.3.: They provide operators
$$A_n : \ell_\infty^n \longrightarrow \ell_1^n$$
with $\|A_n\| \leq n$ and $\mathbf{I}(A_n) = \mathbf{N}(A_n) = n^{3/2}$. Take $1 < d < \sqrt{2}$ and define
$$T_m := (2d)^{-m} A_{2^m} : \ell_\infty^{2^m} \longrightarrow \ell_1^{2^m} .$$
Then $\|T_m\| \leq d^{-m}$ and $\mathbf{I}(T_m) = (\sqrt{2} \cdot d^{-1})^m \to \infty$; since
$$c_o = c_o(\ell_\infty^{2^m} ; m = 1, 2, ...)$$
$$\ell_1 = \ell_1(\ell_1^{2^m} ; m = 1, 2, ...) ,$$
the operator $T := \oplus T_m : c_o \to \ell_1$ can be defined: it has norm $\leq (d-1)^{-1}$, but is not integral since the integral norms of $T_m = Q_m T I_m$ (canonical injection and projection onto the m-th component) tend to infinity.

10.5. The integrating form $\beta_I : (\tilde{f}, \tilde{g}) \rightsquigarrow \int fg d\mu$ on $L_\infty(\Omega, \mu) \otimes_\varepsilon L_\infty(\Omega, \mu)$ has norm $\mu(\Omega)$ by 4.7., hence, by definition, the canonical embedding
$$I : L_\infty(\mu) \hookrightarrow L_1(\mu)$$
is integral and $\mathbf{I}(I) = \mu(\Omega)$ for all finite measures μ. If μ is a Borel-Radon measure on a compact space Ω, this also implies that
$$I : C(\Omega) \longrightarrow L_1(\Omega, \mu)$$
is integral and $\mathbf{I}(I) \leq \mu(\Omega) = \|I\| \leq \mathbf{I}(I)$, hence this canonical map also has integral norm $\mu(\Omega)$. These are the typical integral mappings — this will follow from the description of the elements of $(E \otimes_\varepsilon F)'$ as integrals.

THEOREM: *For an operator $T \in \mathfrak{L}(E, F)$ between Banach spaces the following statements are equivalent:*

(a) *T is integral.*

(b) *There are a Borel-Radon measure μ on a compact set Ω, operators $R \in \mathfrak{L}(E, C(\Omega))$ and $S \in \mathfrak{L}(L_1(\mu), F'')$ such that*

$$\begin{array}{ccccc} E & \xrightarrow{T} & F & \xrightarrow{\kappa_F} & F'' \\ {\scriptstyle R}\downarrow & & & \nearrow{\scriptstyle S} & \\ C(\Omega) & \xrightarrow{I} & L_1(\mu) & & \end{array}$$

is commutative.

(c) *There are a finite measure space (Ω, μ), an operator $R \in \mathfrak{L}(E, L_\infty(\mu))$ and an operator $S \in \mathfrak{L}(L_1(\mu), F'')$ such that*

$$\begin{array}{ccccc} E & \xrightarrow{T} & F & \xrightarrow{\kappa_F} & F'' \\ {\scriptstyle R}\downarrow & & & \nearrow{\scriptstyle S} & \\ L_\infty(\mu) & \xrightarrow{I} & L_1(\mu) & & \end{array}$$

is commutative.

Moreover, in both cases (b) and (c) the measure μ and the operators can be chosen in such a way that $\|R\| = \|S\| = 1$ (if E and $F \neq \{0\}$) and $\mu(\Omega) = \mathbf{I}(T)$.

In particular, it follows from the ideal property and $\|I\| = \mu(\Omega) = \mathbf{I}(I)$ that

$$\mathbf{I}(T) = \min \|R\| \, \|S\| \, \|I\|$$

with factorizations as in (b) or as in (c).

PROOF: Only (a) $\curvearrowright$ (b) is not obvious: If $\varphi \in (E \otimes_\varepsilon F')'$ is the linear functional associated with $T \in \mathfrak{I}(E, F)$, then corollary 4.7. of the representation theorem for integral bilinear forms gives a Borel-Radon measure μ on a compact Ω with $\mu(\Omega) = \mathbf{I}(T)$, and operators $R_1 \in \mathfrak{L}(E, C(\Omega))$ and $R_2 \in \mathfrak{L}(F', C(\Omega))$ with $\|R_1\| = \|R_2\| = 1$ such that (with $I_o : C(\Omega) \to L_\infty(\Omega, \mu)$ canonical)

$$\varphi : E \otimes_\varepsilon F' \xrightarrow{R_1 \otimes R_2} C(\Omega) \otimes_\varepsilon C(\Omega) \xrightarrow{id \otimes I_o} C(\Omega) \otimes_\varepsilon L_\infty(\mu) \xrightarrow{\beta_I} \mathbb{K} \;,$$

where $\langle \beta_I, f \otimes \tilde{g} \rangle = \int fg \, d\mu$ corresponds to the canonical map $I : C(\Omega) \to L_1(\mu)$. If $R := R_1$ and $S := (I_o R_2)' \kappa_{L_1} : L_1 \to F''$, then $\|S\| \leq 1 = \|R\|$ and

$$\langle \kappa_F T x, y' \rangle_{F'', F'} = \langle \varphi, x \otimes y' \rangle = \langle \beta_I, R_1 x \otimes I_o R_2 y' \rangle =$$
$$= \langle I R_1 x, I_o R_2 y' \rangle = \langle (I_o R_2)' \kappa_{L_1} I R_1 x, y' \rangle = \langle S I R x, y' \rangle \;,$$

hence $\kappa_F T = SIR$ which is the desired factorization. $\square$

This factorization theorem is very important for the understanding of integral operators.

10.6. The canonical maps $L_\infty(\Omega,\mu) \hookrightarrow L_1(\Omega,\mu)$ and $C(\Omega) \to L_1(\Omega,\mu)$ are integral. Even more can be said:

PROPOSITION: *Let K be compact, μ and ν arbitrary measures. If $E = C(K)$ or $L_\infty(\mu)$, then every positive operator $T \in \mathcal{L}(E, L_1(\nu))$ is integral and $\mathbf{I}(T) = \|T\|$.*

This is also true for sublattices of E, see Ex 10.5.. For the proof note first that theorem 7.3. implies that for all Banach spaces G

$$\|T \otimes id_G : E \otimes_\varepsilon G \longrightarrow L_1(\nu) \otimes_\pi G\| = \|T\| \quad .$$

Therefore the conclusion follows from the following extremely useful tensor product characterization of integral operators:

10.7. THEOREM: *An operator $T \in \mathcal{L}(E, F)$ between Banach spaces is integral if and only if for all Banach spaces G (or only $G = F'$)*

$$T \otimes id_G : E \otimes_\varepsilon G \longrightarrow F \otimes_\pi G$$

is continuous. In this case,

$$\mathbf{I}(T) = \|T \otimes id_{F'} : E \otimes_\varepsilon F' \longrightarrow F \otimes_\pi F'\| \geq \|T \otimes id_G : E \otimes_\varepsilon G \longrightarrow F \otimes_\pi G\| \quad .$$

PROOF: If $T \in \mathfrak{I}(E, F)$, then for each $\varphi \in (F \otimes_\pi G)' = \mathcal{L}(F, G')$ and $z \in E \otimes G$ the formula

$$\langle \varphi, T \otimes id_G(z) \rangle = \langle \beta_{L_\varphi \circ T}, z \rangle$$

holds (check for elementary tensors) and hence, since $L_\varphi \circ T$ is integral,

$$|\langle \varphi, T \otimes id_G(z) \rangle| \leq \mathbf{I}(L_\varphi \circ T)\varepsilon(z; E, G) \leq \|\varphi\| \mathbf{I}(T)\varepsilon(z; E, G) \; ,$$

which shows: $\pi(T \otimes id_G(z); F, G) \leq \mathbf{I}(T)\varepsilon(z; E, G)$ — this is one direction of the statement of the theorem. Assume, conversely, that $T \otimes id_{F'}$ is continuous as indicated and let $\varphi_T \in (E \otimes_\pi F')'$ be the functional associated with T: For $z \in E \otimes F'$

$$\langle \varphi_T, z \rangle = \langle tr_F, T \otimes id_{F'}(z) \rangle$$

(again a check for elementary tensors), hence

$$|\langle \varphi_T, z \rangle| \leq \|tr_F\| \pi(T \otimes id_{F'}(z); E, F') \leq c\varepsilon(z; E, F')$$

with $c := \|T \otimes id_{F'} : E \otimes_\varepsilon F' \to F \otimes_\pi F'\|$. It follows that $\mathbf{I}(T) \leq c$ and this is all what was missing from the proof. □

10.8. Another nice application of this characterization is the following

PROPOSITION: *Let E be a Banach space, $(x_n) \in \ell_1^w(E)$ and $(x'_n) \in \ell_1^w(E')$ weakly summable. Then*
$$\sum_{n=1}^{\infty} |\langle x'_n, x_n \rangle| \leq w_1(x'_n; E') w_1(x_n; E) \quad.$$

PROOF: First recall that by 8.1.
$$\varepsilon\left(\sum_{k=1}^{n} e_k \otimes y_k; \ell_1^n, G\right) = w_1((y_k)_{k=1}^n; G)$$

for $y_1, ..., y_n$ in a Banach space G. Since $\mathbf{I}(\ell_1^n \hookrightarrow \ell_\infty^n) = 1$ by 10.4. and hence
$$\|\ell_1^n \otimes_\varepsilon E \xrightarrow{id} \ell_\infty^n \otimes_\pi E\| \leq 1$$

by the theorem, it follows that for $x_1, ..., x_n \in E$
$$\pi\left(\sum_{k=1}^{n} e_k \otimes x_k; \ell_\infty^n, E\right) \leq w_1((x_k)_{k=1}^n; E) \quad.$$

The duality of ε and π shows that $\ell_\infty^n \otimes_\pi E \hookrightarrow (\ell_1^n \otimes_\varepsilon E')'$ holds isometrically, hence
$$\pi\left(\sum_{k=1}^{n} e_k \otimes x_k; \ell_\infty^n, E\right) = \sup\left\{\left|\sum_{k=1}^{n} \langle y'_k, x_k \rangle\right| \mid w_1((y'_k)_{k=1}^n; E') \leq 1\right\} \quad.$$

It follows that for all $x'_1, ..., x'_n \in E'$
$$\left|\sum_{k=1}^{n} \langle x'_k, x_k \rangle\right| \leq w_1((x_k)_{k=1}^n; E) w_1((x'_k)_{k=1}^n; E') ,$$

which implies the result. □

COROLLARY: *If $\varphi \in \mathfrak{Bil}(c_0, c_0)$, then*
$$\sum_{n=1}^{\infty} |\varphi(e_n, e_n)| \leq \|\varphi\| \quad.$$

PROOF: First note that
$$w_1(e_n; c_0) = \sup\left\{\sum_{n=1}^{\infty} |\langle e_n, x \rangle| \mid x \in B_{\ell_1}\right\} = 1 \quad.$$

If $L_\varphi \in \mathfrak{L}(c_o, \ell_1)$ is the operator associated with φ, then

$$w_1(L_\varphi e_n; \ell_1) \leq \|L_\varphi\| w_1(e_n; c_o) = \|\varphi\|$$

and hence the proposition implies that

$$\sum_{n=1}^{\infty} |\varphi(e_n, e_n)| = \sum_{n=1}^{\infty} |\langle L_\varphi e_n, e_n \rangle| \leq \|\varphi\| . \quad \Box$$

These results (also due to Grothendieck [93], p.95; see also Ex 10.7.) are connected with an inequality of Littlewood's [178] from 1930:

$$\left(\sum_{k,\ell=1}^{\infty} |\varphi(e_k, e_\ell)|^{4/3} \right)^{3/4} \leq c \|\varphi\|$$

for all $\varphi \in \mathfrak{Bil}(c_o, c_o)$ where c is a universal constant — and the exponent 4/3 best possible. This will be shown in 34.11. as well as an estimate for the constant c.

10.9. The last corollary was a consequence of the fact that $\mathbf{N}(\ell_1^n \hookrightarrow \ell_\infty^n) = 1$ holds; this was obtained by Rademacher averaging. Using this method directly, one obtains the more general

PROPOSITION: *For $1/p + 1/q + 1/r = 1$ and $\varphi \in \mathfrak{Bil}(\ell_p, \ell_q)$ the following inequality holds:*

$$\|(\varphi(e_k, e_k))_{k \in \mathbb{N}}\|_{\ell_r} \leq \|\varphi\| .$$

One may replace ℓ_∞ by c_o. The result and the following proof are due to Aron (unpublished), see also a recent paper of Zalduendo [278].

PROOF: For $c_k := \varphi(e_k, e_k)$ choose α_k and $\beta_k \in \mathbb{K}$ with

$$\alpha_k \beta_k c_k = |c_k|^r \quad \text{and} \quad |\alpha_k|^p = |\beta_k|^q = |c_k|^r$$

(if $p, q, r \in [1, \infty[$). Rademacher averaging gives

$$\sum_{k=1}^{n} |c_k|^r = \sum_{k=1}^{n} \alpha_k \beta_k \varphi(e_k, e_k) =$$

$$= \int_0^1 \varphi\left(\sum_{k=1}^{n} \alpha_k r_k(t) e_k, \sum_{\ell=1}^{n} \beta_\ell r_\ell(t) e_\ell \right) dt \leq$$

$$\leq \|\varphi\| \left(\sum_{k=1}^{n} |\alpha_k|^p \right)^{1/p} \left(\sum_{\ell=1}^{n} |\beta_\ell|^q \right)^{1/q} = \|\varphi\| \left(\sum_{k=1}^{n} |c_k|^r \right)^{1/p+1/q} .$$

If some of the p, q, r are ∞, some obvious modifications have to be made. □

10.10. The ideal of integral operators is neither injective nor surjective: This can be deduced from the fact that ε does not respect quotients (4.3.) — but the details are omitted since in Chapter II the relationship between ε and $\mathfrak{I}$ will be studied systematically in a more general context. On the other hand the fact that ε respects subspaces implies extension and lifting properties of the ideal $\mathfrak{I}$, see Ex 10.4..

Exercises:

Ex 10.1. Every integral operator is weakly compact, but not necessarily compact (and hence also not necessarily nuclear).

Ex 10.2. For which $1 \leq p \leq q \leq \infty$ is the canonical embedding $\ell_p \hookrightarrow \ell_q$ integral? Hint: 10.4..

Ex 10.3. If F is complemented in F'', then an operator $T \in \mathfrak{L}(E, F)$ is integral if and only if it admits a factorization $E \to L_\infty \hookrightarrow L_1 \to F$; what about the integral norm $\mathbf{I}(T)$? Examples: F reflexive or a dual space, $F = L_1(\mu)$.

Ex 10.4. (a) *Extension property of integral operators:* If G is a closed subspace of a Banach space E and $T \in \mathfrak{I}(G, F)$, then there is a $\tilde{T} \in \mathfrak{I}(E, F'')$ extending $\kappa_F T$ with $\mathbf{I}(\tilde{T}) = \mathbf{I}(T)$. Hint: ε respects subspaces or: the factorization theorem.

(b) *Lifting property of integral operators:* If $Q : G \twoheadrightarrow F$ is a metric surjection and $T \in \mathfrak{I}(E, F)$, then there is a $\tilde{T} \in \mathfrak{I}(E, G'')$ satisfying $Q'' \circ \tilde{T} = \kappa_F \circ T$ and $\mathbf{I}(\tilde{T}) = \mathbf{I}(T)$.

Ex 10.5. Let M be a norm–closed sublattice of an $L_\infty(\mu)$ or $C(K)$ (with K compact) and $T \in \mathfrak{L}(M, L_1(\nu))$ positive, then T is integral and $\mathbf{I}(T) = \|T\|$. Hint: Kakutani's theorem shows that T'' satisfies the assumption of 10.6..

Ex 10.6. Let μ be a finite measure. Then $T \in \mathfrak{L}(E, L_1(\mu))$ is integral if and only if $T B_E$ is lattice bounded. Hint: For one direction use an $h \geq |Tx|$ for all $x \in B_E$ and the measure $d\nu := h d\mu$ to obtain a good factorization; for the other direction recall from 7.4. that every operator $S \in \mathfrak{L}(L_1(\nu), L_1(\mu))$ is regular.

Ex 10.7. Use 10.8. to prove the following result: If (E_n) and (F_n) are two sequences of Banach spaces and $A \in \mathfrak{Bil}(\ell_\infty(E_n), \ell_\infty(F_n))$, then

$$\sum_{n=1}^{\infty} \|A|_{E_n \times F_n}\| \leq \|A\| \ .$$

Ex 10.8. Let $T_i \in \mathfrak{L}(E_i, F_i)$ be two operators one of which is integral. Then the operator $T_1 \otimes T_2 : E_1 \otimes_\varepsilon E_2 \to F_1 \otimes_\pi F_2$ is continuous.

11. Absolutely p–Summing Operators

The Banach ideal of absolutely p–summing operators is in some sense at the heart of Banach space theory. For $p = 1$ these operators are due to Grothendieck ("applications préintégrale droite" in the Résumé or "semi–intégrale droite" in his thesis); they were generalized in the mid sixties (Mityagin, Pełczyński, Pietsch) to $p > 1$; the basic results on factorization and composition are due to Pietsch. The great advantage of absolutely p–summing operators is that — with the aid of various handy characterizations — it is often relatively simple to check whether or not an operator is of this type. Outside Banach space theory these operators had an important impact on the theory of nuclear spaces and the theory of distributions.

11.1. An operator $T \in \mathfrak{L}(E, F)$ between Banach spaces is called *absolutely p–summing* (for $1 \le p < \infty$) if there is a constant $c \ge 0$ such that for all finite sequences $(x_1, ..., x_n)$ in E

$$\ell_p(Tx_k; F) = \left(\sum_{k=1}^n \|Tx_k\|^p\right)^{1/p} \le c\, w_p(x_k; E) = c \sup_{x' \in B_{E'}} \left(\sum_{k=1}^n |\langle x', x_k\rangle|^p\right)^{1/p}.$$

Define $\mathbf{P}_p(T)$ to be the infimum (=minimum) of these constants and write $\mathfrak{P}_p$ for the class of all absolutely p–summing operators; for obvious reasons it is sometimes useful to define $(\mathfrak{P}_\infty, \mathbf{P}_\infty) := (\mathfrak{L}, \|\ \|)$. Clearly, the defining inequality for absolutely p–summing operators also holds for infinite sequences. Recall from section 8 that for the absolutely p–summable sequences

$$\ell_p \tilde{\otimes}_{\Delta_p} F = \ell_p(F) \quad ; \quad \Delta_p\left(\sum_{k=1}^n e_k \otimes x_k; \ell_p, F\right) = \ell_p(x_k; F)$$

and for the weakly p–summable sequences

$$\ell_p \tilde{\otimes}_\varepsilon E \stackrel{1}{=} \ell_p^{w,o}(E) \stackrel{1}{\hookrightarrow} \ell_p^w(E) \quad ; \quad \varepsilon\left(\sum_{k=1}^n e_k \otimes x_k; \ell_p, E\right) = w_p(x_k; E).$$

Now using the definition
$$T^{\mathbb{N}}(x_n) := (Tx_n)$$
and an obvious application of the closed graph theorem, gives the

PROPOSITION: *Let E and F be Banach spaces. For each $T \in \mathfrak{L}(E, F)$ and $1 \le p < \infty$ the following statements are equivalent:*
(a) $T \in \mathfrak{P}_p(E, F)$.

(b) *T maps weakly p-summable sequence sequences into absolutely p-summable sequences:*
$$T^{I\!N}(\ell_p^w(E)) \subset \ell_p(F) .$$

(b') $T^{I\!N} : \ell_p^w(E) \to \ell_p(F)$ *is continuous.*

(c) $T^{I\!N}(\ell_p^{w,o}(E)) \subset \ell_p(F)$.

(c') $T^{I\!N} : \ell_p^{w,o}(E) \to \ell_p(F)$ *is continuous.*

(d) $id \otimes T : \ell_p \otimes_\varepsilon E \to \ell_p \otimes_{\Delta_p} F$ *is continuous.*

(e) *There is a $c \geq 0$ such that for all $n \in I\!N$*
$$\|id \otimes T : \ell_p^n \otimes_\varepsilon E \longrightarrow \ell_p^n \otimes_{\Delta_p} F\| \leq c .$$

In this case,
$$\mathbf{P}_p(T) = \|T^{I\!N} : \ell_p^w(E) \longrightarrow \ell_p(F)\| = \|T^{I\!N} : \ell_p^{w,o}(E) \longrightarrow \ell_p(F)\| =$$
$$= \|id \otimes T : \ell_p \otimes_\varepsilon E \longrightarrow \ell_p \otimes_{\Delta_p} F\| = \sup_n \|id \otimes T : \ell_p^n \otimes_\varepsilon E \longrightarrow \ell_p^n \otimes_{\Delta_p} F\| .$$

Recall that $\Delta_1 = \pi$; Ex 4.3. implies that $\mathbf{P}_1(id_{\ell_1^n}) \geq \sqrt{n}$. The following is an immediate consequence of the definition:

REMARK: $(\mathfrak{P}_p, \mathbf{P}_p)$ *is an injective Banach operator ideal.*

However, it is not surjective, see 11.12.. For $p = 1$ part (c) of the proposition means, by 8.3., that $T \in \mathcal{L}(E, F)$ is absolutely summing (:= absolutely 1–summing) if and only if it maps unconditionally convergent series into absolutely convergent series.

The characterization 10.7. of integral operators by operators of the form $id_G \otimes T$ shows (property (d)) that each integral operator is absolutely summing and
$$\mathbf{P}_1(T) \leq \mathbf{I}(T) .$$

11.2. The following example is basic: If K is compact and μ a (positive) Borel–Radon measure on K, then the canonical map
$$I_p : C(K) \longrightarrow L_p(K, \mu)$$
is absolutely p–summing and $\mathbf{P}_p(I_p) = \|I_p\| = \mu(K)^{1/p}$. This follows from
$$\sum_{k=1}^n \|I_p f_k\|_{L_p}^p = \int_K \sum_{k=1}^n |f_k(x)|^p \mu(dx) \leq \mu(K) \, w_p(f_1, \ldots, f_n; C(K))^p .$$

Much more is true, namely the

COROLLARY: *Let $E = C(K)$ for a compact K or $E = L_\infty(\mu)$ for some measure μ. If ν is any other measure, then every positive operator $T \in \mathcal{L}(E, L_p(\nu))$ is absolutely p-summing and $\mathbf{P}_p(T) = \|T\|$.*

PROOF: It was shown in theorem 7.3. that

$$\|T \otimes \mathrm{id} : E \otimes_\varepsilon \ell_p \longrightarrow L_p(\nu) \otimes_{\Delta_p} \ell_p = L_p(\nu) \otimes_{\Delta_p^!} \ell_p\| = \|T\| \quad ,$$

hence part (d) of the proposition gives the result. □

11.3. To a large extend the theory of absolutely p–summing operators is governed by the following result:

GROTHENDIECK–PIETSCH DOMINATION THEOREM: *Let $1 \le p < \infty$ and $K \subset B_{E'}$ a $\sigma(E', E)$-compact norming subset. Then $T \in \mathcal{L}(E, F)$ is absolutely p-summing if and only if there is a (positive) Borel–Radon measure μ on K with*

$$\|Tx\| \le \left(\int_K |\langle x', x \rangle|^p \mu(dx') \right)^{1/p}$$

for all $x \in E$. In this case, $\mathbf{P}_p(T) = \min \mu(K)^{1/p}$, where the minimum is taken over all such μ.

PROOF: If T satisfies such an estimate, then

$$\sum_{k=1}^n \|Tx_k\|^p \le \int_K \sum_{k=1}^n |\langle x', x_k \rangle|^p \mu(dx') \le \mu(K) w_p(x_k)^p \quad ,$$

hence $\mathbf{P}_p(T) \le \mu(K)^{1/p}$. Conversely, assume $\mathbf{P}_p(T) = 1$, hence

$$\sum_{k=1}^n \|Tx_k\|^p \le \sup_{x' \in B_{E'}} \sum_{k=1}^n |\langle x', x_k \rangle|^p \quad .$$

It is easily seen that on the Banach space $C_\mathbb{R}(K)$ of real-valued continuous functions on K the functional w defined by

$$w(f) := \inf_{x_1, \ldots, x_n \in E} \sup_{x' \in K} \left\{ f(x') + \sum_{k=1}^n [|\langle x', x_k \rangle|^p - \|Tx_k\|^p] \right\}$$

is sublinear and satisfies

$$\inf f(K) \le w(f) \le \sup f(K) \quad .$$

The Hahn–Banach theorem gives a $\mu \in C_{\mathbb{R}}(K)'$ with $\mu \leq w$. Since $\mu(f) \leq w(f) \leq 0$ for $f \leq 0$, it follows that μ is positive, hence a positive Borel–Radon measure by the Riesz representation theorem. For $x \in E$ consider $-|\langle \cdot, x \rangle|^p \in C_{\mathbb{R}}(K)$:

$$\mu(-|\langle \cdot, x \rangle|^p) \leq w(-|\langle \cdot, x \rangle|^p) \leq -\|Tx\|^p,$$

which is exactly the desired estimate since $\mu(K) = \mu(1) \leq 1$. □

The theorem can be rewritten in terms of a factorization — this is why it is sometimes called the *Grothendieck–Pietsch factorization theorem*.

COROLLARY 1: *Let $1 \leq p < \infty$ and $T \in \mathcal{L}(E, F)$. Then the following are equivalent:*

(a) $T \in \mathfrak{P}_p(E, F)$.

(b) *For each norming weak-$*$-compact subset $K \subset B_{E'}$ there are a Borel–Radon measure μ on K, a closed subspace $G \subset L_p(\mu)$, operators $R \in \mathcal{L}(E, C(K))$ and $S \in \mathcal{L}(G, F)$ such that $S I_p R x = T x$ for all $x \in E$:*

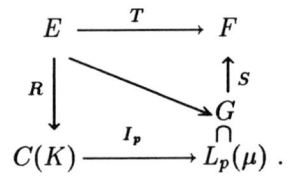

$$\begin{array}{ccc} E & \xrightarrow{T} & F \\ {\scriptstyle R}\downarrow & \nearrow{\scriptstyle S} & \\ & G & \\ & \cap & \\ C(K) & \xrightarrow{I_p} & L_p(\mu). \end{array}$$

(c) *There are a finite measure space (Ω, μ), a closed subspace $G \subset L_p(\mu)$, operators $R \in \mathcal{L}(E, L_\infty(\mu))$ and $S \in \mathcal{L}(G, F)$ such that*

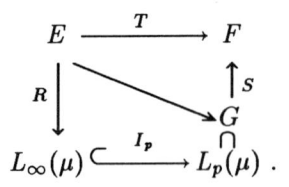

$$\begin{array}{ccc} E & \xrightarrow{T} & F \\ {\scriptstyle R}\downarrow & \nearrow{\scriptstyle S} & \\ & G & \\ & \cap & \\ L_\infty(\mu) & \xrightarrow{I_p} & L_p(\mu). \end{array}$$

In this case, $\mathbf{P}_p(T) = \min \|R\| \, \|S\| \, \|\mu\|^{1/p}$, where the minimum is taken over the factorizations in (b) or over those in (c).

PROOF: Take $T \in \mathfrak{P}_p(E, F)$ and choose μ from the theorem such that

$$\|Tx\|^p \leq \int_K |\langle x', x \rangle|^p \mu(dx') \qquad \text{for all } x \in E$$

and $\mathbf{P}_p(T) = \mu(K)^{1/p}$. Define

$$\begin{aligned} Rx &:= \langle \cdot, x \rangle \in C(K) \\ G &:= \overline{I_p \circ R(E)} \\ S(I_p \circ R(x)) &:= Tx \quad ; \end{aligned}$$

11. Absolutely p–Summing Operators

then S is well-defined by the inequality and both operators R and S have norm ≤ 1, hence (b) is proved. The implication (c) $\sim$ (a) follows from the fact that

$$\mathbf{P}_p(L_\infty(\mu) \hookrightarrow L_p(\mu)) = \mu(\Omega)^{1/p}$$

by the corollary in 11.2. and the injectivity of $\mathfrak{P}_p$. The equalities for the norm come from the fact that $(\mathfrak{P}_p, \mathbf{P}_p)$ is an injective normed operator ideal. $\square$

In special situations the factorization can be improved.

COROLLARY 2: Let $1 \leq p < \infty$ and $T \in \mathfrak{L}(E, F)$.
(1) If $p = 2$ or F has the metric extension property, then $T \in \mathfrak{P}_p(E, F)$ if and only if it admits a factorization

$$\begin{array}{ccc} E & \xrightarrow{T} & F \\ R \downarrow & & \uparrow S \\ C(K) & \xrightarrow{I_p} & L_p(\mu) \end{array}$$

(where, as before, K is any weak-$*$-compact norming subset of $B_{E'}$). In this case,

$$\mathbf{P}_p(T) = \min \|R\| \, \|S\| \, \mu(K)^{1/p} \quad .$$

(2) If $E = C(K)$, then $T \in \mathfrak{P}_p(E, F)$ if and only if there are a Borel–Radon measure μ on K and $S \in \mathfrak{L}(L_p(\mu), F)$ with

$$\begin{array}{ccc} C(K) & \xrightarrow{T} & F \\ & \searrow_{I_p} & \uparrow S \\ & & L_p(\mu) \end{array} \quad .$$

In this case, $\mathbf{P}_p(T) = \min \|S\| \, \mu(K)^{1/p}$.

It follows that in general a $T \in \mathfrak{L}(E, F)$ is absolutely p–summing if and only if it admits a factorization

$$\begin{array}{ccccc} E & \xrightarrow{T} & F & \hookrightarrow & \ell_\infty(B_{F'}) \\ R \downarrow & & & \nearrow S & \\ C(K) & \xrightarrow{I_p} & L_p(\mu) & & \end{array}$$

and $\mathbf{P}_p(T) = \min \|R\| \|S\| \|I_p\|$. The space $C(K)$ can be replaced by $L_\infty(\mu)$.

For $p = 1$ the factorizations show that the "typical" absolutely 1–summing and the "typical" integral map is the same (see 10.5.), only the way of factoring through this mapping is slightly different: The definition of the injective hull of an operator ideal (9.7.) and Ex 10.3. give the first statement of the following result.

COROLLARY 3: (1) $(\mathfrak{J}^{inj}, \mathbf{I}^{inj}) = (\mathfrak{P}_1, \mathbf{P}_1)$ — the absolutely summing operators form the injective hull of the integral operators.

(2) If K is compact, then $\mathfrak{J}(C(K), F) = \mathfrak{P}_1(C(K), F)$ holds isometrically for all Banach spaces F.

Clearly, (2) follows from 10.5. and corollary 2(2). Another immediate consequence of the Grothendieck–Pietsch domination theorem is that:

COROLLARY 4: If $1 \leq p \leq q < \infty$, then $\mathfrak{P}_p \subset \mathfrak{P}_q$ and $\mathbf{P}_q(T) \leq \mathbf{P}_p(T)$.

The inclusion $\mathfrak{P}_p \subset \mathfrak{P}_q$ is strict, see Ex 11.23.(a). One can also show that the canonical mapping $C[0,1] \hookrightarrow L_q[0,1]$ is in no $\mathfrak{P}_p$ for $p < q$ (see Pietsch [214]).

11.4. A first application is the

DVORETZKY–ROGERS THEOREM: *A Banach space E is finite dimensional if (and only if) every unconditionally convergent series is absolutely convergent.*

PROOF: Since, by 8.3., $\ell_1^{w,o}(E)$ is the space of unconditionally convergent series, it follows from 11.1.(c) that the condition implies $id_E \in \mathfrak{P}_1 \subset \mathfrak{P}_2$, hence id_E factors through a Hilbert space by corollary 2:

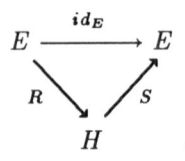

$$\begin{array}{ccc} & id_E & \\ E & \longrightarrow & E \\ & R \searrow \nearrow S & \\ & H & \end{array}$$

Since S must be surjective, E is isomorphic to a Hilbert space. But this implies that E is finite dimensional because $(\frac{1}{n}e_n) \in \ell_1^{w,o}(\ell_2)\setminus \ell_1(\ell_2)$. □

See Ex 11.9. for the Dvoretzky–Rogers theorem for p–summable sequences.

11.5. For many applications Pietsch's multiplication theorem is important:

11. Absolutely p–Summing Operators

THEOREM: If $1 \leq r, p, q < \infty$ with $1/r = 1/p + 1/q$, then $\mathfrak{P}_p \circ \mathfrak{P}_q \subset \mathfrak{P}_r$ and
$$\mathbf{P}_r(ST) \leq \mathbf{P}_p(S)\mathbf{P}_q(T).$$

PROOF: One may assume $\mathbf{P}_q(T : E \to F) = \mathbf{P}_p(S : F \to G) = 1$. It remains to show that $w_r(x_1, ..., x_n; E) \leq 1$ implies that
$$\sum_{k=1}^{n} \|STx_k\|_G^r \leq 1 \ .$$

If μ is a probability measure on $B_{E'}$ with
$$\|Tx\|^q \leq \int_{B_{E'}} |\langle x', x\rangle|^q \mu(dx') \ ,$$

define $c_k := (\int_{B_{E'}} |\langle x', x_k\rangle|^r \mu(dx'))^{1/q}$ and note that $\sum_{k=1}^n c_k^q \leq 1$. Since one may assume $Tx_k \neq 0$, all c_k are different from zero. Now

$$\left(\sum_k \|STx_k\|^r\right)^{1/r} \leq \left(\sum_k \|STc_k^{-1}x_k\|^p\right)^{1/p}\left(\sum_k c_k^q\right)^{1/q} \leq w_p(Tc_k^{-1}x_k; F) =$$

(*)
$$= \sup\left\{\left\|\sum_k \alpha_k Tc_k^{-1}x_k\right\| \ \bigg| \ \sum_k |\alpha_k|^{p'} \leq 1\right\} \ .$$

The Hölder inequality for $1/r' + 1/q + 1/p = 1$ implies for these (α_k) and $x' \in B_{E'}$ that

$$\left|\sum_k \alpha_k c_k^{-1}\langle x', x_k\rangle\right| \leq \sum_k \left[|\alpha_k|^{p'/r'}\right]\left[|\alpha_k|^{p'/q} c_k^{-1}|\langle x', x_k\rangle|^{r/q}\right]\left[|\langle x', x_k\rangle|^{r/p}\right] \leq$$

$$\leq 1 \cdot \left(\sum_k |\alpha_k|^{p'} c_k^{-q}|\langle x', x_k\rangle|^r\right)^{1/q} \cdot 1 \ .$$

It follows from this that

$$\left\|\sum_k \alpha_k Tc_k^{-1}x_k\right\|^q \leq \int_{B_{E'}} \left|\sum_k \alpha_k c_k^{-1}\langle x', x_k\rangle\right|^q \mu(dx') \leq$$

$$\leq \sum_k |\alpha_k|^{p'} c_k^{-q} \int_{B_{E'}} |\langle x', x_k\rangle|^r \mu(dx') \leq 1 \ ;$$

(*) and this inequality conclude the proof. $\square$

This shows in particular that $\mathfrak{P}_2 \circ \mathfrak{P}_2 \subset \mathfrak{P}_1$ and, consequently, that the composition of $2n$ absolutely p–summing mappings is absolutely 1–summing if $2n \geq p$. Clearly, the inclusion $\mathfrak{P}_p \circ \mathfrak{P}_q \subset \mathfrak{P}_s$ is true if $1/s \leq 1/p + 1/q$ but it is false if $1/s > 1/p + 1/q$, see Ex 11.23.(b).

11.6. In order to study the relation between absolutely p–summing operators and other operator ideals more presicely recall that an operator $T \in \mathfrak{L}(H_1, H_2)$ between Hilbert spaces is called a *Hilbert–Schmidt operator* if for some orthonormal basis (e_α) of H_1

$$\mathbf{HS}(T) := \left(\sum_\alpha \|Te_\alpha\|_{H_2}^2 \right)^{1/2} < \infty \; ;$$

notation: $T \in \mathfrak{HS}(H_1, H_2)$. It is easy to see that the definition of the Hilbert–Schmidt norm **HS** is independent of the specific orthonormal basis:

$$\mathbf{HS}(T)^2 = \sum_{\alpha, \beta} |(Te_\alpha \mid f_\beta)_{H_2}|^2 = \sum_{\alpha, \beta} |(e_\alpha \mid T^* f_\beta)_{H_1}|^2 = \mathbf{HS}(T^*)^2$$

where $T^* : H_2 \to H_1$ is the Hilbert space adjoint; in particular, $T \in \mathfrak{HS}$ if and only if $T^* \in \mathfrak{HS}$. The basic facts about Hilbert–Schmidt operators are treated in Ex 11.12., Ex 11.13. and Ex 11.18..

PROPOSITION: $T \in \mathfrak{L}(H_1, H_2)$ is Hilbert–Schmidt if and only if it is absolutely 2-summing. Moreover,

$$\mathbf{HS}(T) = \mathbf{P}_2(T) \; .$$

PROOF: If $T \in \mathfrak{P}_2(H_1, H_2)$, then for all orthonormal $(e_1, ..., e_n)$

$$\left(\sum_{k=1}^n \|Te_k\|^2 \right)^{1/2} \leq \mathbf{P}_2(T) w_2(e_1, ... e_n; H_1) = \mathbf{P}_2(T)$$

and hence $\mathbf{HS}(T) \leq \mathbf{P}_2(T)$. If, conversely, $T \in \mathfrak{HS}(H_1, H_2)$ and $x_1, ..., x_n \in H_1$, then — with an orthonormal basis (e_α) of H_2 —

$$\sum_{k=1}^n \|Tx_k\|^2 = \sum_k \sum_\alpha |(Tx_k \mid e_\alpha)_{H_2}|^2 = \sum_\alpha \sum_k |(x_k \mid T^* e_\alpha)_{H_1}|^2 \leq$$

$$\leq \sum_\alpha \|T^* e_\alpha\|^2 \sum_k |(x_k \mid \|T^* e_\alpha\|^{-1} T^* e_\alpha)_{H_1}|^2 \leq$$

$$\leq \mathbf{HS}(T^*)^2 w_2(x_k; H_1)^2 \; . \; \square$$

Khintchine's inequality actually implies that

$$\mathfrak{P}_p(H_1, H_2) = \mathfrak{HS}(H_1, H_2)$$

for all $1 \leq p < \infty$ (with equivalent norms) — this will be done later in a slightly more general setting (see 26.6.).

11.7. For testing whether or not an operator is Hilbert–Schmidt the following lemma is of great practical importance.

PIETSCH LEMMA: *If H is a Hilbert space, μ a Borel–Radon measure on a compact set K and $S \in \mathfrak{L}(H, C(K))$, then $T := I_2 \circ S : H \xrightarrow{S} C(K) \xrightarrow{I_2} L_2(K,\mu)$ is Hilbert–Schmidt and $\mathbf{HS}(T) \leq \|S\|\mu(K)^{1/2}$.*

This follows from the fact that $\mathbf{P}_2(I_2) = \mu(K)^{1/2}$, the ideal property and the last proposition. But, interestingly enough, this can be proved directly: If $(e_1, ..., e_n)$ is orthonormal, then

$$\sum_k \|Te_k\|^2 = \int \sum_k |(Se_k)(w)|^2 \mu(dw) = \int \sum_k |\langle e_k, S'\delta_w\rangle|^2 \mu(dw)$$

$$\leq \int \|S'\delta_w\|_H^2 \mu(dw) \leq \|S'\|^2 \mu(K)$$

by the Bessel inequality (and $H = H'$).

COROLLARY: *The product of three absolutely 2–summing operators is nuclear.*

PROOF: Consider the factorization

$$\begin{array}{ccccccc}
E_1 & \longrightarrow & E_2 & \longrightarrow & E_3 & \longrightarrow & E_4 \\
\downarrow & & \swarrow\searrow & & \swarrow\searrow & & \uparrow \\
C & \longrightarrow L_2 & C & \longrightarrow L_2 & C & \longrightarrow L_2 &
\end{array}$$

and use the fact that the product of two Hilbert–Schmidt operators is a nuclear operator (see Ex 11.13.). □

Later on this result will be improved to the composition of two operators, thus generalizing the case of Hilbert–Schmidt operators (see also Ex 11.14.). It follows that the composition of $3n$ absolutely p–summing operators is nuclear if $n \geq p/2$. In many concrete cases this is an appropriate tool to show that a given operator is nuclear! In particular, absolutely p–summing operators in a Banach space are power compact and hence have a discrete spectrum; the sequence of eigenvalues of a $T \in \mathfrak{P}_p(E,E)$ is in ℓ_p if $p \geq 2$. This famous result of Johnson–König–Maurey–Retherford [131] is the basis for the recent theory on eigenvalues of operators (see the monographs of König [147] and Pietsch [216]); in 34.10. one step of the proof will be presented.

11.8. A nice consequence of $\mathfrak{P}_2(H,H) = \mathfrak{HS}(H,H)$ is

$$\mathbf{P}_2(id_{\ell_2^n}) = \mathbf{HS}(id_{\ell_2^n}) = \sqrt{n} \ .$$

In order to have this for all n–dimensional spaces the following test, seemingly due to Kwapień, will be useful.

PROPOSITION: *For $T \in \mathfrak{L}(E,F)$ and $1 \le p < \infty$ the following are equivalent:*
(a) $T \in \mathfrak{P}_p(E,F)$.
(b) *There is a $c \ge 0$ such that $\mathbf{P}_p(TS) \le c\|S\|$ for all $S \in \mathfrak{L}(\ell_{p'}^n, E)$ and all $n \in \mathbb{N}$.*
(c) $TS \in \mathfrak{P}_p(\ell_{p'}, F)$ *for any $S \in \mathfrak{L}(\ell_{p'}, E)$.*

Moreover,
$$\mathbf{P}_p(T) := \sup\{\mathbf{P}_p(TS) \mid n \in \mathbb{N}, \ \|S : \ell_{p'}^n \to E\| \le 1\} .$$

PROOF: (a) implies (c) and (c) implies (b) by a closed graph argument. For the remaining implication recall from 8.2. that for $x_1, ..., x_n \in E$
$$w_p(x_k; E) = \left\| S := \sum_{k=1}^n e_k \otimes x_k : \ell_{p'}^n \to E \right\|$$
and therefore
$$\sum_k \|Tx_k\|^p = \sum_k \|TSe_k\|^p \le \mathbf{P}_p(TS)^p w_p(e_k; \ell_{p'}^n)^p \le$$
$$\le c^p \|S\|^p \cdot 1 = c^p w_p(x_k; E)^p . \quad \Box$$

See also Ex 25.14.. Comparing this result with 11.1.(e), one might be tempted to think that $\ell_p^n \otimes_{\Delta_p} F = \mathfrak{P}_p(\ell_{p'}^n, F)$ holds isometrically. This is not true for $p \ne 1$ — a fact which will follow from the results in Chapter II: Δ_p fails to be a "tensor norm".

11.9. The proposition can be used to prove the aforementioned

THEOREM: *If E is a Banach space of dimension n, then $\mathbf{P}_2(id_E) = \sqrt{n}$.*

PROOF: To show $\le \sqrt{n}$, use the foregoing test and take $S \in \mathfrak{L}(\ell_2^m, E)$ with $\|S\| \le 1$. Then
$$S : \ell_2^m \xrightarrow{Q} \ell_2^m/\ker S \xrightarrow{S_o} E ,$$
where $\|Q\| = 1$ and $\|S_o\| = \|S\| \le 1$. Since $\ell_2^m/\ker S$ is a Hilbert space ℓ_2^k with $k \le n$, it follows that
$$\mathbf{P}_2(S) \le \|Q\|\mathbf{P}_2(id_{\ell_2^k})\|S_o\| \le \sqrt{k} \le \sqrt{n} .$$
Conversely, the Grothendieck–Pietsch factorization theorem gives a factorization

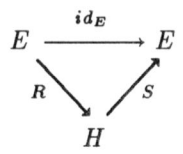

$$\begin{array}{ccc} E & \xrightarrow{id_E} & E \\ & R \searrow \quad \nearrow S & \\ & H & \end{array}$$

through a Hilbert space H with $\|S\| = 1$ and $\mathbf{P}_2(R) \leq \mathbf{P}_2(id_E)$. One may assume that S is bijective, hence

$$\sqrt{n} = \mathbf{P}_2(id_H) = \mathbf{P}_2(RS) \leq \mathbf{P}_2(R)\|S\| \leq \mathbf{P}_2(id_E) \quad . \square$$

This result (due to Garling–Gordon [79]) has interesting consequences for the Banach–Mazur distance between Banach spaces of the same finite dimension and the existence of certain projections on finite dimensional subspaces of Banach spaces, see Ex 11.16..

11.10. Using Gauss averaging it is possible to calculate the p-summing norm of the identity map on ℓ_2^n; the following result is due to Gordon [82], [83] (see also Pietsch [214], 22.1.3.), who also calculated various other $\mathbf{P}_p(id_E)$ for finite dimensional E.

THEOREM:
$$\mathbf{P}_p(id_{\ell_2^n}) = \|g_1\|_{p,K}^{-1} \left(\int_{K^n} \|w\|_2^p \gamma_n(dw) \right)^{1/p} .$$

PROOF: The Khintchine–type equality for the Gauss measure says (8.7.) that for all $x \in \ell_2^n$

$$\|x\|_2 = \left(\int_{K^n} |\langle x, \frac{w}{\|w\|_2} \rangle|^p \nu_n(dw) \right)^{1/p}$$

with $\nu_n(dw) := \|g_1\|_{p,K}^{-p} \|w\|_2^p \gamma_n(dw)$ (which is obviously a finite measure). This equality is the key to the proof (see also Ex 11.19.). The easy part of the Grothendieck–Pietsch domination theorem gives $\mathbf{P}_p(id_{\ell_2^n}) \leq \nu_n(\ell_2^n)^{1/p}$. To obtain the converse, choose — again by the domination theorem — a measure μ on the unit ball B of ℓ_2^n such that

$$\|x\|_2^p \leq \int_B |\langle x, y\rangle|^p \mu(dy) .$$

It follows that

$$\int_{K^n} \|w\|_2^p \gamma_n(dw) \leq \int_{K^n} \int_B |\langle w, y\rangle|^p \mu(dy) \gamma_n(dw) =$$
$$= \int_B \int_{K^n} |\langle w, y\rangle|^p \gamma_n(dw) \mu(dy) = \|g_1\|_{p,K}^p \int_B \|y\|_2^p \mu(dy) \leq \|g_1\|_{p,K}^p \mu(B)$$

by the Khintchine–Gauss equality. This implies that

$$\nu_n(\ell_2^n)^{1/p} = \|g_1\|_{p,K}^{-1} \left(\int_{K^n} \|w\|_2^p \gamma_n(dw) \right)^{1/p} \leq \mu(B)^{1/p} ,$$

which is the claim. $\square$

The explicit calculation for $p = 1$ (for other p see Ex 11.24.) gives the following: If σ_{n-1} is the size of the surface of the euclidean unit ball of $\mathbb{R}^n$, then

$$\int_{\mathbb{R}^n} \|w\|_2 \gamma_n(dw) = (2\pi)^{-n/2} \int_{\mathbb{R}^n} \|w\|_2 \exp\left(-\frac{\|w\|_2^2}{2}\right) dw =$$

$$= (2\pi)^{-n/2}\sigma_{n-1}\int_0^\infty r\exp\left(-\frac{r^2}{2}\right)r^{n-1}dr = (2\pi)^{-n/2}\cdot 2\frac{\pi^{n/2}}{\Gamma(n/2)}2^{(n-1)/2}\Gamma((n+1)/2) =$$

$$= \sqrt{2}\frac{\Gamma((n+1)/2)}{\Gamma(n/2)}.$$

Note that $\int_{\mathbb{C}^n}\|w\|_2\gamma_n(dw) = \int_{\mathbb{R}^{2n}}\|w\|_2\gamma_{2n}(dw)$ and recall that

$$\|g_1\|_{1,\mathbb{R}} = \sqrt{\frac{2}{\pi}} \quad \text{and} \quad \|g_1\|_{1,\mathbb{C}} = \sqrt{\frac{\pi}{2}}$$

from 8.7.. Then (with an obvious notation for ℓ_2^n)

$$\mathbf{P}_1(id_{\mathbb{R}_2^n}) = \sqrt{\pi}\frac{\Gamma((n+1)/2)}{\Gamma(n/2)}$$

and

$$\mathbf{P}_1(id_{\mathbb{C}_2^n}) = \frac{2}{\pi}\mathbf{P}_1(id_{\mathbb{R}_2^{2n}}) = \frac{2}{\sqrt{\pi}}\frac{\Gamma(n+\frac{1}{2})}{(n-1)!} = \prod_{k=1}^{n-1}\left(1+\frac{1}{2k}\right).$$

What is the behaviour of these constants like as $n \to \infty$? Using again

$$\Gamma\left(\frac{2n+1}{2}\right) = 2^{-n}\cdot 1\cdot 3\cdot 5\cdot\ldots\cdot(2n-1)\sqrt{\pi}$$

and Wallis' formula

$$\frac{\pi}{2} = \lim_{n\to\infty}\left[\frac{n!\,2^n}{1\cdot 3\cdot 5\cdot\ldots\cdot(2n-1)}\right]^2\frac{1}{2n},$$

one easily gets

$$\frac{1}{2k}\mathbf{P}_1(id_{\mathbb{R}_2^{2k}})^2 \xrightarrow{k\to\infty} \frac{\pi}{2};$$

since clearly $\mathbf{P}_1(id_{\mathbb{R}_2^n}) \le \mathbf{P}_1(id_{\mathbb{R}_2^{n+1}})$, this is also true for the odd n. It follows that:

COROLLARY:

$$\lim_{n\to\infty} n^{-1/2}\mathbf{P}_1(id_{\mathbb{R}_2^n}) = \sqrt{\frac{\pi}{2}}$$

$$\lim_{n\to\infty} n^{-1/2}\mathbf{P}_1(id_{\mathbb{C}_2^n}) = \frac{2}{\sqrt{\pi}}.$$

11.11. Gauss averages will also give the

11. Absolutely p-Summing Operators

THEOREM: *If H is a Hilbert space and $T \in \mathcal{L}(\ell_\infty^m, H)$, then*

$$\mathbf{P}_2(T) \leq K_{LG}\|T\| \quad,$$

where $K_{LG}^{\mathbb{R}} = \sqrt{\pi/2}$ in the real case and $K_{LG}^{\mathbb{C}} = 2/\sqrt{\pi}$ in the complex case.

It is not by chance that the constants are the same as in the preceding corollary: In 20.19. it will be shown with the aid of this corollary that these constants are actually the best possible ones. The theorem (and the following corollary) is sometimes called the *little Grothendieck theorem* since it already has the flavour of Grothendieck's fundamental theorem of the metric theory of tensor products which will be treated in Chapter II.

PROOF: One may, by the ideal property of $\mathfrak{P}_2$, assume that $H = \ell_2^m$. It is to be shown (proposition 11.1.) that for all $T \in \mathcal{L}(\ell_\infty^m, \ell_2^m)$ and all $n \in \mathbb{N}$

$$\|T \otimes id : \ell_\infty^m \otimes_\varepsilon \ell_2^n \longrightarrow \ell_2^m \otimes_{\Delta_2} \ell_2^n\| \leq K_{LG}\|T\| \quad.$$

Since

$$(\ell_2^m \otimes_{\Delta_2} \ell_2^n)' = (\ell_2^m(\ell_2^n))' = \ell_2^m(\ell_2^n) = \ell_2^m \otimes_{\Delta_2} \ell_2^n$$
$$(\ell_\infty^m \otimes_\varepsilon \ell_2^n)' = \ell_1^m \otimes_\pi \ell_2^n$$

by the duality of ε and π (see section 6), this is the same as showing

$$\|S \otimes id : \ell_2^m \otimes_{\Delta_2} \ell_2^n \longrightarrow \ell_1^m \otimes_\pi \ell_2^n\| \leq K_{LG}\|S\|$$

for all $S \in \mathcal{L}(\ell_2^m, \ell_1^m)$. Fix $S = \sum_{i=1}^m y_i' \otimes e_i$ and take $z = \sum_{k=1}^n x_k \otimes e_k \in \ell_2^m \otimes \ell_2^n$. Then

$$S \otimes id(z) = \sum_{k=1}^n Sx_k \otimes e_k = \sum_{k=1}^n \sum_{i=1}^m \langle y_i', x_k \rangle e_i \otimes e_k = \sum_{i=1}^m e_i \otimes \left(\sum_{k=1}^n \langle y_i', x_k \rangle e_k\right)$$

and hence

$$\pi(S \otimes id(z); \ell_1^m, \ell_2^n) = \sum_{i=1}^m \left\|\sum_{k=1}^n \langle y_i', x_k \rangle e_k\right\|_{\ell_2^n} = \sum_{i=1}^m \left(\sum_{k=1}^n |\langle y_i', x_k \rangle|^2\right)^{1/2} \quad.$$

By the Khintchine–type equality 8.7. for the Gauss measure it follows that

$$\pi(S \otimes id(z); \ell_1^m, \ell_2^n) = \|g_1\|_{1,\mathbb{K}}^{-1} \sum_{i=1}^m \int_{\mathbb{K}^n} \left|\sum_{k=1}^n \langle y_i', x_k \rangle g_k(w)\right| \gamma_n(dw) =$$

$$= \|g_1\|_{1,\mathbb{K}}^{-1} \int_{\mathbb{K}^n} \left\|S\left(\sum_{k=1}^n g_k(w) x_k\right)\right\|_{\ell_1^m} \gamma_n(dw) \leq$$

$$\leq \|S\| \, \|g_1\|_{1,\mathbf{K}}^{-1} \int_{\mathbf{K}^n} \left\| \sum_{k=1}^n g_k(w) x_k \right\|_{\ell_2^m} \gamma_n(dw) =$$

$$\leq \|S\| \, \|g_1\|_{1,\mathbf{K}}^{-1} \left(\int_{\mathbf{K}^n} \left\| \sum_{k=1}^n g_k(w) x_k \right\|_{\ell_2^m}^2 \gamma_n(dw) \right)^{1/2} =$$

$$= \|S\| \, \|g_1\|_{1,\mathbf{K}}^{-1} \|g_1\|_{2,\mathbf{K}} \left(\sum_{k=1}^n \|x_k\|^2 \right)^{1/2}$$

by the biorthogonality of the g_k (see 8.4.). An appeal to 8.7. gives that
$$K_{LG} = \|g_1\|_{1,\mathbf{K}}^{-1} \|g_1\|_{2,\mathbf{K}}$$
is as indicated. $\Box$

COROLLARY ("little Grothendieck theorem"): *If E is an $\mathcal{L}_{\infty,\lambda}^g$- or an $\mathcal{L}_{1,\lambda}^g$-space and H a Hilbert space, then*
$$\mathcal{L}(E,H) = \mathfrak{P}_2(E,H) \quad \text{and} \quad \mathbf{P}_2(T) \leq K_{LG} \lambda \|T\| \quad.$$

In particular, all operators from $C(K)$, L_∞ or L_1 into a Hilbert space are absolutely 2–summing. The "full" Grothendieck theorem will show that all operators $L_1 \to H$ are even 1–summing!

PROOF: (a) Assume first that E is $\mathcal{L}_{\infty,\lambda}^g$ and $T \in \mathcal{L}(E,H)$. To show the relevant $\mathfrak{P}_2$–inequality take $x_1, ..., x_n \in E$; then there is a factorization

$$\text{span}\{x_k\} \hookrightarrow E \qquad \|R\| \, \|S\| \leq \lambda(1+\varepsilon)$$
$$R \searrow \quad \nearrow S$$
$$\ell_\infty^m$$

and hence by the theorem
$$\mathbf{P}_2(TSR) \leq \|R\| \mathbf{P}_2(TS) \leq \|R\| K_{LG} \|TS\| \leq K_{LG} \|T\| \lambda(1+\varepsilon) \quad.$$
(b) If E is an $\mathcal{L}_{1,\lambda}^g$-space, then the same reasoning shows that it is enough to demonstrate that for all $T \in \mathcal{L}(\ell_1^m, H)$
$$\mathbf{P}_2(T) \leq K_{LG} \|T\| \quad.$$
To do this take, by Kwapień's test 11.8., some $S \in \mathcal{L}(\ell_2^k, \ell_1^m)$. Then
$$\mathbf{P}_2(TS) = \mathbf{HS}(TS) = \mathbf{HS}(S'T') = \mathbf{P}_2(S'T') \leq$$
$$\leq \mathbf{P}_2(S' : \ell_\infty^m \to \ell_2^k) \|T'\| \leq K_{LG} \|S'\| \, \|T'\|$$
by the theorem, hence $\mathbf{P}_2(T) \leq K_{LG} \|T\|$. $\Box$

Note the simple dualization argument for $\mathfrak{P}_2(\cdot, H)$ with the aid of Hilbert–Schmidt operators in part (b) of the proof. More systematic dualization arguments for operator ideals will be developped in Chapter II.

One can also prove the corollary for $\mathcal{L}_2^g$-spaces F instead of the Hilbert space H (with an additional constant) just by adding a factorization of $TS: \ell_\infty^m \to F$ through an ℓ_2^k. But Ex 11.17. implies that then F is isomorphic to a Hilbert space: nothing is gained but the

REMARK: *The $\mathcal{L}_2^g$-spaces are exactly the Banach spaces isomorphic to Hilbert spaces.*

This result will also be a trivial consequence of the general characterization of $\mathcal{L}_p^g$-spaces given in section 23.

11.12. The operator ideal $\mathfrak{P}_p$ is injective — but not surjective: The little Grothendieck theorem implies that the canonical quotient map $\ell_1 \twoheadrightarrow \ell_2$ is in $\mathfrak{P}_2$ but clearly id_{ℓ_2} is not since absolutely 2–summing maps are power compact.

The canonical embedding $\ell_1 \hookrightarrow \ell_2$ is in $\mathfrak{P}_2$ (even in $\mathfrak{P}_1$; see Ex 11.5.) but its dual

$$\ell_2 \hookrightarrow c_o \hookrightarrow \ell_\infty$$

is not in $\mathfrak{P}_2$: This follows from the fact that the sequence (e_n) of unit vectors is weakly-2–summable, although $\sum \|e_n\|_{c_o}^2 = \infty$. This shows that the statements "$T \in \mathfrak{P}_2$" and "$T' \in \mathfrak{P}_2$" do not imply each other. Between Hilbert spaces, however, these statements are equivalent since $\mathfrak{P}_2(H_1, H_2) = \mathfrak{HS}(H_1, H_2)$. More generally, if $T \in \mathfrak{P}_2(H, F)$, then $T' \in \mathfrak{P}_2(F', H')$ (where H is a Hilbert and F a Banach space) since, by the factorization theorem, T factors through a Hilbert–Schmidt operator. Cohen [30] and Kwapień [160] proved the following:

PROPOSITION: *A Banach space E is isomorphic to a Hilbert space if and only if the dual $T' \in \mathfrak{P}_2$ whenever $T \in \mathfrak{P}_2(E, \ell_2)$.*

PROOF: One direction was just shown. Conversely, embed

$$I : E \hookrightarrow \ell_\infty(\Gamma) \ .$$

The idea of the proof is to show that the metric surjection I' is in $\mathfrak{P}_2$, hence factors through a Hilbert space and consequently E' is isomorphic to a Hilbert space. To see that $I' \in \mathfrak{P}_2$ take

$$(y_n') \in \ell_2^w(\ell_\infty(\Gamma)') = \mathcal{L}(\ell_\infty(\Gamma), \ell_2) = \mathfrak{P}_2(\ell_\infty(\Gamma), \ell_2)$$

by 8.2. and the little Grothendieck theorem, and consider its associated operator $S := \sum y_n' \otimes e_n$. Since the composition $SI \in \mathfrak{P}_2(E, \ell_2)$, the assumption on E implies that $I'S' \in \mathfrak{P}_2$ as well, therefore

$$\sum_n \|I'y_n'\|^2 = \sum_n \|I'S'e_n\|^2 \le \mathbf{P}_2(I'S')^2 w_2(e_n; \ell_2)^2 < \infty \ ,$$

which shows that $I' \in \mathfrak{P}_2$. □

We shall come back to the relation between T and T' being in $\mathfrak{P}_p$ in 25.9..

Exercises:

Ex 11.1. Show that $\mathfrak{I} \subset \mathfrak{P}_p$ and $\mathbf{P}_p(T) \leq \mathbf{I}(T)$ for all $1 \leq p < \infty$.

Ex 11.2. Prove that for all Banach spaces F
$$\mathfrak{N}(c_o, F) = \mathfrak{I}(c_o, F) = \mathfrak{P}_1(c_o, F) \ .$$
holds isometrically. Hint: $T = \sum e_n \otimes Te_n$. This result is true in many other cases: this can be shown with the Radon–Nikodým property, see Appendix D7. and D8..

Ex 11.3. If a $\mathbb{C}$-linear operator between complex Banach spaces is considered as $\mathbb{R}$-linear, does the p-summing norm change?

Ex 11.4. For $g \in L_p(\mu)$ the multiplication operator
$$M_g : L_\infty(\mu) \longrightarrow L_p(\mu)$$
defined by $M_g(f) := fg$ is absolutely p-summing and $\mathbf{P}_p(M_g) = \|g\|_p$. Hint: 11.2..

Ex 11.5. Use Khintchine's inequality to show that
$$\mathbf{P}_1(\ell_1^n \hookrightarrow \ell_2^n) \leq \sqrt{2}$$
$$\mathbf{P}_2(\ell_1^n \hookrightarrow \ell_2^n) = 1 \ .$$
Deduce that the canonical embedding $I : \ell_1 \hookrightarrow \ell_2$ is in $\mathfrak{P}_1 \subset \mathfrak{P}_2$ with $\mathbf{P}_2(I) = 1 \leq \mathbf{P}_1(I) \leq \sqrt{2}$. Recall from 10.4. that I is not integral!

Ex 11.6. An operator $T \in \mathcal{L}(E, F)$ is absolutely p-summing if there is a finite measure space (Ω, μ) and a $\sigma(E', E)$-measurable $\Phi : \Omega \to B_{E'}$ such that
$$\|Tx\|^p \leq \int_\Omega |\langle \Phi(w), x\rangle|^p \mu(dw) \ .$$
In this case, $\mathbf{P}_p(T) \leq \mu(\Omega)^{1/p}$.

Ex 11.7. For $E = L_\infty(\nu)$ show that $T \in \mathfrak{P}_p(E, F)$ if and only if it admits a factorization $E \to C(K) \to L_p(\mu) \to F$.

Ex 11.8. (a) Every absolutely p-summing operator T is weakly compact and *completely continuous* (this means: if $x_n \to 0$ weakly, then $\|Tx_n\| \to 0$), but in general not compact. Hint: Lebesgue's dominated convergence theorem.

(b) $\mathfrak{P}_p \circ \mathfrak{P}_p \subset \mathfrak{P}_p \circ \mathfrak{W} \subset \mathfrak{K}$, but $\mathfrak{W} \circ \mathfrak{P}_p \not\subset \mathfrak{K}$; moreover, $\mathfrak{P}_p = \mathfrak{W} \circ \mathfrak{P}_p$ if $1 < p < \infty$.

(c) Every quasinuclear operator is absolutely 1–summing, but not conversely.

Ex 11.9. A Banach space E is finite dimensional if and only if $\ell_p^{w,o}(E) = \ell_p(E)$ (as sets) for some (and then for all) $1 \leq p < \infty$. This is the Dvoretzky–Rogers theorem for p–summable sequences.

Ex 11.10. *Extension property for absolutely 2–summing operators:* If $G \subset E$ is a closed subspace of a Banach space E and $T \in \mathfrak{P}_2(G, F)$, then there is an extension $\tilde{T} \in \mathfrak{P}_2(E, F)$ of T with $\mathbf{P}_2(\tilde{T}) = \mathbf{P}_2(T)$. Hint: Factorization through L_∞ and 3.10..

Ex 11.11. If $T \in \mathfrak{P}_2(E, E)$, then there are a Hilbert space H, operators $R \in \mathfrak{L}(E, H)$ and $S \in \mathfrak{L}(H, E)$ such that $SR = T$ and $\hat{T} := RS$ is a Hilbert–Schmidt operator and such that $\mathbf{HS}(\hat{T}) \leq \mathbf{P}_2(T)$. One says that T and $\hat{T}$ are *related operators*, a concept due to Pietsch which is extremely useful since related operators T and $\hat{T}$ have the same non–zero eigenvalues (with the same multiplicity).

Ex 11.12. (a) An operator $T \in \mathfrak{L}(\ell_2, \ell_2)$ is Hilbert–Schmidt if and only if $\mathbf{HS}(T)^2 = \sum_{k,\ell} |(Te_k|e_\ell)|^2 < \infty$.

(b) An operator $T \in \mathfrak{L}(L_2(\Omega_1, \mu_1), L_2(\Omega_2, \mu_2))$ is Hilbert–Schmidt if and only if there is a $k \in L_2(\mu_1 \otimes \mu_2)$ with

$$(Tf|g)_{L_2(\mu_2)} = \int k(w_1, w_2) f(w_1) \overline{g(w_2)} (\mu_1 \otimes \mu_2)(d(w_1, w_2))\ .$$

In this case, $\mathbf{HS}(T) = \|k\|_{L_2}$.

Ex 11.13. (a) $(\mathfrak{H}\mathfrak{S}, \mathbf{HS})$ is a Banach operator ideal in the class of Hilbert spaces:
(1) $\mathfrak{H}\mathfrak{S}(H_1, H_2) \subset \mathfrak{L}(H_1, H_2)$ is a linear subspace containing the finite rank operators.
(2) $\mathbf{HS}(id_K) = 1$.
(3) $\mathbf{HS}(TRS) \leq \|T\|\mathbf{HS}(R)\|S\|$ if the composition TRS is defined.
(4) $(\mathfrak{H}\mathfrak{S}(H_1, H_2), \mathbf{HS})$ is a Banach space.

(b) If (e_α) is an orthonormal basis of H, then

$$(S|T) := \sum_\alpha (Se_\alpha | Te_\alpha)_H$$

defines a scalar product on $\mathfrak{H}\mathfrak{S}(H, H)$ with associated norm $\mathbf{HS}$: It follows that the space $(\mathfrak{H}\mathfrak{S}(H, H), \mathbf{HS})$ is a Hilbert space.

(c) The finite rank operators are $\mathbf{HS}$–dense in $\mathfrak{H}\mathfrak{S}(H_1, H_2)$.

(d) Hilbert–Schmidt operators are compact.

(e) If $T = \sum_{n=1}^\infty \lambda_n e_n \otimes f_n$ with two orthonormal sequences (e_n) and (f_n) and a bounded sequence (λ_n) of scalars, then T is Hilbert–Schmidt if and only if $(\lambda_n) \in \ell_2$. In this case, $\mathbf{HS}(T)^2 = \sum_{n=1}^\infty |\lambda_n|^2$.

(f) Use the spectral decomposition theorem for compact operators (see Ex 3.28.) to show that the composition of two Hilbert–Schmidt operators is nuclear and

$$\mathbf{N}(ST) \leq \mathbf{HS}(S)\mathbf{HS}(T)\ .$$

(g) Show that $(S|T) = tr_H(T^*S)$ if $S, T \in \mathfrak{HS}(H, H)$. Hint: Ex 5.15..

Ex 11.14. $\mathfrak{P}_2 \circ \mathfrak{P}_2(H_1, H_2) \subset \mathfrak{N}(H_1, H_2)$ for Hilbert spaces H_1 and H_2. Hint: Factorization theorem for $\mathfrak{P}_2$, Hilbert–Schmidt operators. This result also holds for arbitrary Banach spaces, see 19.4..

Ex 11.15. $\mathbf{P}_2(T) \leq \|T\|\sqrt{\text{rank}(T)}$ for all $T \in \mathfrak{L}(E, F)$.

Ex 11.16. Prove the following corollaries of theorem 11.9.:

(a) The Banach–Mazur distance from an n–dimensional Banach space E to ℓ_2^n is $\leq \sqrt{n}$. Hint: Use the fact that absolutely 2–summing maps factor through Hilbert spaces.

(b) $d(E, F) \leq n$ if $\dim E = \dim F = n$.

(c) If $G \subset E$ is an n–dimensional subspace of a Banach space E, then there is a projection $P : E \twoheadrightarrow G$ with $\|P\| \leq \sqrt{n}$. Hint: Ex 11.10.. This result is due to Kadec–Snobar [138]; the proof as indicated here follows an idea of Kwapień.

Ex 11.17. A Banach space E is isomorphic to a Hilbert space if and only if every operator $T \in \mathfrak{L}(\ell_1, E)$ is absolutely 2–summing. Hint: $\ell_1(\Gamma) \overset{1}{\twoheadrightarrow} E$.

Ex 11.18. Show that for an operator $T \in \mathfrak{L}(H_1, H_2)$ between Hilbert spaces the following seven statements are equivalent:

(a) T is Hilbert–Schmidt.

(b) T factors through some $C(K)$.

(c) T factors through some space $L_\infty(\mu)$.

(d) T factors through some $L_1(\mu)$.

(e) T factors through ℓ_1.

(f) T factors through c_o.

(g) T factors through ℓ_∞.

Hint: Use T' and little Grothendieck. This is due to Lindenstrauss–Pełczyński [173].

Ex 11.19. (a) Let E be a separable Banach space. Show that $T \in \mathfrak{L}(E, F)$ is absolutely p–summing if and only if, for some $c \geq 0$,

$$\left(\int_E \|Tx\|^p \mu(dx)\right)^{1/p} \leq c \sup_{x' \in B_{E'}} \left(\int_E |\langle x', x\rangle|^p \mu(dx)\right)^{1/p}$$

for all finite Borel measures μ on E; the infimum of these c is $\mathbf{P}_p(T)$. Hint: Choose a partition D_n of E with $|\|Tx\|^p - \|Tx_n\|^p|$ and $\|x - x_n\|$ small on D_n.

(b) Extend (a) to σ–finite Borel measures μ.

(c) If there is a σ–finite Borel measure μ on E with

$$\left(\int_E |\langle x', x\rangle|^p \mu(dx)\right)^{1/p} \leq \|x'\|$$

for all $x' \in E'$, then

$$\left(\int_E \|Tx\|^p \mu(dx)\right)^{1/p} \leq \mathbf{P}_p(T)$$

for each $T \in \mathcal{L}(E, F)$.

(d) Show that for $(x_1, x_2) \in \mathbb{R}^2$

$$\|(x_1, x_2)\|_2 = \frac{1}{4} \int_0^{2\pi} |x_1 \cos t + x_2 \sin t| dt .$$

Use this, (c) and the domination theorem to show that $\mathbf{P}_1(id_{\mathbb{R}^2_2}) = \pi/2$. This method was used by Gordon (see also Jameson [119]) for the calculation of various $\mathbf{P}_p(id_E)$.

Ex 11.20. Show that the class $(\mathfrak{P}_{r,s}, \mathbf{P}_{r,s})$ of absolutely (r, s)-summing operators (see Ex 8.12.) is a Banach operator ideal. Moreover, $T \in \mathcal{L}(E, F)$ is absolutely (r, s)-summing if and only if

$$id \otimes T : \ell_s \otimes_\varepsilon E \longrightarrow \ell_r \otimes_{\Delta_r} F$$

is continuous. In this case the norm of this mapping is $\mathbf{P}_{r,s}(T)$.

Ex 11.21. If $r_1 \leq r_2$ and $s_1 \leq s_2$ such that $1/s_1 - 1/r_1 \leq 1/s_2 - 1/r_2$, then $\mathfrak{P}_{r_1,s_1} \subset \mathfrak{P}_{r_2,s_2}$ and $\mathbf{P}_{r_2,s_2} \leq \mathbf{P}_{r_1,s_1}$. Hint: With $1/r = 1/r_1 - 1/r_2$ and $1/s = 1/s_1 - 1/s_2$ it follows that

$$\ell_{r_1}(\alpha_k\|Tx_k\|) \leq \mathbf{P}_{r_1,s_1}(T)\ell_s(\alpha_k)w_{s_2}(x_k) .$$

Ex 11.22. Use Rademacher functions and Ex 8.1. to show that for $2 \leq p < \infty$ the identity map of any $L_p(\nu)$ is absolutely $(p, 1)$-summing. Hint: $\sum |f_n(t)|^p \leq \int_D |\sum \varepsilon_n(w) f_n(t)|^p \mu(dw)$. For a more general result see 24.7..

Ex 11.23. (a) Take $\lambda \in \ell_\infty$ and consider the diagonal operator $D_\lambda : c_o \to c_o$. Show that D_λ is absolutely p-summing if and only if $\lambda \in \ell_p$. Hint: Ex 11.4. and unit vectors.

(b) Take $p, q, r, s \in [1, \infty[$ with $1/p + 1/q = 1/r < 1/s$. Show with (a) that

$$\mathfrak{P}_q \circ \mathfrak{P}_p \not\subset \mathfrak{P}_s$$

— so the index r in Pietsch's multiplication theorem $\mathfrak{P}_q \circ \mathfrak{P}_p \subset \mathfrak{P}_r$ is best possible. Hint: $D_\lambda \circ D_\eta = D_{\lambda\eta}$.

Ex 11.24. For the Gauss measure γ_n on $\mathbb{K}^n$ and $1 \leq p < \infty$ define

$$c_{2,p}^n := \left(\int_{\mathbb{K}^n} \|w\|_2^p \gamma_n(dw) \right)^{1/p}$$

and observe that $\mathbf{P}_p(id_{\ell_2^n}) = c_{2,p}^n / c_{2,p}^1$ by theorem 11.10..

(a) Using polar coordinates as in 11.10., verify that

$$c_{2,p,\mathbb{R}}^n = \sqrt{2} \left(\frac{\Gamma((n+p)/2)}{\Gamma(n/2)} \right)^{1/p} \quad \text{and} \quad c_{2,p,\mathbb{C}}^n = c_{2,p,\mathbb{R}}^{2n} .$$

(b) Use the Gauss-Khintchine equality $c_{2,p}^1 \|x\|_2 = (\int |\langle x, w \rangle|^p \gamma_n(dw))^{1/p}$, polar coordinates and the calculation for (a) to show that

$$\|x\|_2 = \frac{c_{2,p}^n}{c_{2,p}^1} \left(\int_{S_{n-1}} |\langle x, w \rangle|^p \eta_{n-1}(dw) \right)^{1/p}$$

for every $x \in \mathbb{K}^n$, where η_{n-1} is the normalized Lebesgue measure on the euclidean unit sphere S_{n-1} of $\mathbb{K}^n$. The formula is taken from Pietsch [214], 22.1.2..

Chapter II
Tensor Norms

In this chapter the main topic of this book will be presented: The Grothendieck-Schatten theory of tensor norms, with special emphasis on its interplay with the theory of operator ideals. Sections 12 – 15 and 20 give the basic definitions and properties of tensor norms, including Grothendieck's fundamental theorem of the metric theory of tensor products — nowadays called Grothendieck's inequality. Section 17 deals with the main theorem concerning the relationship between tensor norms and maximal Banach operator ideals — the representation theorem for maximal Banach operator ideals; the representation theorem for minimal Banach operator ideals will be given in section 22.

Sections 18 and 19 study the (p,q)-factorable and (p,q)-dominated operators, the technique of ultraproducts, Maurey's and Kwapień's factorization theorems.

The central tool in all that follows will be the duality theory of tensor norms — and later of operator ideals, in other words: a good understanding of trace duality is basic. Trace duality is "smooth" only for "accessible" tensor norms or under an additional hypothesis such as the metric approximation property. Due to the success of the notion of accessibility for tensor norms, accessible operator ideals will be defined and studied in section 21. It seems to us that the composition formulas for accessible operator ideals given in section 25 (which again are a sort of trace duality) fully justify the introduction of this notion.

At the end of this chapter we hope that the reader is impressed with the power of *also* thinking in terms of tensor norms. In particular, she or he may have learned that it is natural to describe certain phenomena in terms of tensor norms, others in terms of operator ideals.

12. Definition and Examples

In this section the definition and some basic examples of tensor norms are given. Two natural ways will be presented for extending a tensor norm from the class of finite dimensional spaces to all normed spaces — an inductive and a projective procedure. The inductive one ("the finite hull") gives Grothendieck's original definition of a tensor norm, the projective one ("the cofinite hull") — essentially due to Lotz — will turn out to be a useful tool for the understanding of various delicate phenomena related to the approximation property.

12.1. A *tensor norm* α on the class NORM of all normed spaces assigns to each pair (E, F) of normed spaces E and F a norm $\alpha(\,\cdot\,; E, F)$ on the algebraic tensor product $E \otimes F$ (shorthand: $E \otimes_\alpha F$ and $E \tilde{\otimes}_\alpha F$ for the completion) such that the following two conditions are satisfied:

(1) α is reasonable, i.e.: $\varepsilon \leq \alpha \leq \pi$.
(2) α satisfies the metric mapping property: If $T_i \in \mathfrak{L}(E_i, F_i)$, then

$$\|T_1 \otimes T_2 : E_1 \otimes_\alpha E_2 \longrightarrow F_1 \otimes_\alpha F_2\| \leq \|T_1\|\,\|T_2\|\,.$$

This definition goes back to Schatten [245] in 1943 (who called such an α a *uniform crossnorm*) and Grothendieck [93] in 1954. It is clear that it can also be made for subclasses of all normed spaces: for the class FIN of all finite dimensional normed spaces, for the class BAN of all Banach spaces, for the class HILB of all Hilbert spaces or even for classes of pairs of normed spaces such as NORM×BAN or FIN×NORM. A tensor norm α being reasonable means

$$\|E \otimes_\pi F \xrightarrow{id} E \otimes_\alpha F\| \leq 1 \quad \text{and} \quad \|E \otimes_\alpha F \xrightarrow{id} E \otimes_\varepsilon F\| \leq 1\,.$$

It can happen that all tensor norms are equivalent on some $E \otimes F$: the celebrated spaces P of Pisier, mentioned in 4.8., satisfy $P \otimes_\varepsilon P = P \otimes_\pi P$ isomorphically, and hence have this property. ε and π are tensor norms since they have the metric mapping property (3.2. and 4.1.), but Δ_p is not a tensor norm:

PROPOSITION: *For $1 < p < \infty$ there is no tensor norm α such that for all spaces $L_p(\mu)$ and normed spaces F*

$$L_p(\mu) \otimes_{\Delta_p} F = L_p(\mu) \otimes_\alpha F$$

holds isomorphically.

There exist, however, tensor norms "close" to Δ_p (see 15.10. and 15.11.) — and these tensor norms serve well for studying various phenomena related to Δ_p.

PROOF: The reason that no such tensor norm exists is that Δ_p does not satisfy the mapping property: For $p = 2$ it was seen in 7.5. that for the continuous Fourier transform $\mathfrak{F} : L_2(\mathbb{R}) \to L_2(\mathbb{R})$ the operator $\mathfrak{F} \otimes id_{\ell_1} : L_2(\mathbb{R}) \otimes_{\Delta_2} \ell_1 \to L_2(\mathbb{R}) \otimes_{\Delta_2} \ell_1$ is not continuous, but $\mathfrak{F} \otimes id_{\ell_1} : L_2(\mathbb{R}) \otimes_\alpha \ell_1 \to L_2(\mathbb{R}) \otimes_\alpha \ell_1$ is continuous for all tensor norms α. For $1 < p < \infty$ the same argument applies using the continuous Hilbert transform $\ell_p(\mathbb{Z}) \to \ell_p(\mathbb{Z})$ treated in 7.6.. □

On the other hand, as was already shown in 7.1. and 7.3., Δ_p is at least a "one–sided" tensor norm: it is reasonable (i.e. $\varepsilon \leq \Delta_p \leq \pi$) and

$$\|id_{L_p} \otimes T : L_p \otimes_{\Delta_p} E \longrightarrow L_p \otimes_{\Delta_p} F\| = \|T\|$$

for all $T \in \mathfrak{L}(E, F)$.

12.2. The following test (formulated for NORM only) will be quite useful:

CRITERION: α *is a tensor norm on NORM if and only if:*
(1) $\alpha(\,\cdot\,; E, F)$ *is a seminorm on $E \otimes F$ for all pairs (E, F) of normed spaces.*
(2) $\alpha(1 \otimes 1; \mathbb{K}, \mathbb{K}) = 1$.
(3) α *satisfies the metric mapping property.*

Note the similarity with the criterion 9.4. for Banach operator ideals and also that the cases of real and complex scalars will be treated simultaneously.

PROOF: To show that the three properties suffice for α to be a tensor norm, one only has to check that α is reasonable, i.e. $\varepsilon \leq \alpha \leq \pi$. For $x' \in E'$ and $y' \in F'$

$$x' \otimes y' : E \otimes F \longrightarrow \mathbb{K} \otimes \mathbb{K} = \mathbb{K},$$
$$z \rightsquigarrow \langle x' \otimes y', z \rangle$$

hence
$$|\langle x' \otimes y', z \rangle| = \alpha(\langle x' \otimes y', z \rangle; \mathbb{K}, \mathbb{K}) \leq \|x'\| \, \|y'\| \alpha(z; E, F)$$

by (2) and (3), and therefore $\varepsilon \leq \alpha$. To see that π dominates α, observe first that for $x \in E$ the operator $T_x : \mathbb{K} \to E$ defined by $T_x(\lambda) := \lambda x$ has norm $\|x\|$, hence

$$\|T_x \otimes T_y : \mathbb{K} \otimes_\alpha \mathbb{K} \longrightarrow E \otimes_\alpha F\| \leq \|x\| \, \|y\|,$$

and therefore (again by (2) and (3)),

$$\|x\| \, \|y\| = \varepsilon(x \otimes y; E, F) \leq \alpha(x \otimes y; E, F) \leq \|x\| \, \|y\| \alpha(1 \otimes 1; \mathbb{K}, \mathbb{K}) = \|x\| \, \|y\|$$

which means that $\alpha(x \otimes y; E, F) = \|x\| \, \|y\|$. The triangle inequality gives $\alpha \leq \pi$. □

In particular, it was shown that $\alpha(x \otimes y; E, F) = \|x\| \, \|y\|$, and this implies that

$$\|T_1 \otimes T_2 : E_1 \otimes_\alpha E_2 \longrightarrow F_1 \otimes_\alpha F_2\| = \|T_1\| \, \|T_2\|.$$

12.3. In general, tensor norms respect neither subspaces nor quotients isomorphically: this follows from the "bad" properties of ε and π. The mapping property only implies continuity of the respective mappings. Clearly, if the subspace $G \subset E$ is complemented and $P : E \to G$ a projection, then, again by the mapping property,

$$\alpha(z; E, F) \leq \alpha(z; G, F) \leq \|P\| \alpha(z; E, F)$$

for all $z \in G \otimes F$. Recall the natural transposition map

$$\left(\sum_{k=1}^n x_k \otimes y_k \right)^t := \sum_{k=1}^n y_k \otimes x_k \,.$$

If α is a tensor norm, then the *transposed* tensor norm α^t is defined by

$$\alpha^t(z; E, F) := \alpha(z^t; F, E) .$$

It is obvious that α^t is a tensor norm and $\alpha^{tt} = \alpha$. The tensor norms ε and π are *symmetric*, i.e. $\varepsilon^t = \varepsilon$ and $\pi^t = \pi$.

12.4. Given a tensor norm α on the class FIN of all finite dimensional normed spaces, there are two natural ways of extending it to the class of all normed spaces. For this, denote for a normed space E

$$\text{FIN}(E) := \{M \subset E \mid M \in \text{FIN}\} \quad \text{and} \quad \text{COFIN}(E) := \{L \subset E \mid E/L \in \text{FIN}\} ,$$

the sets of finite dimensional subspaces and of finite codimensional, closed subspaces (with induced norms). Now take normed spaces E and F and define for $z \in E \otimes F$

$$\overrightarrow{\alpha}(z; E, F) := \inf\{\alpha(z; M, N) \mid M \in \text{FIN}(E), N \in \text{FIN}(F), z \in M \otimes N\}$$
$$\overleftarrow{\alpha}(z; E, F) := \sup\{\alpha(Q_K^E \otimes Q_L^F(z); E/K, F/L) \mid K \in \text{COFIN}(E), L \in \text{COFIN}(F)\}$$

(where $Q_K^E : E \xrightarrow{1} E/K$ is the canonical mapping); the arrows stem from the fact that the first procedure is an inductive, the second a projective one. By the metric mapping property it is enough to take cofinally many M, N resp. K, L in the definition. With the criterion it is very easy to verify that:

PROPOSITION: *Let α be a tensor norm on* FIN. *Then the "finite hull" $\overrightarrow{\alpha}$ of α and the "cofinite hull" $\overleftarrow{\alpha}$ of α are tensor norms on* NORM *with*

$$\varepsilon \leq \overleftarrow{\alpha} \leq \overrightarrow{\alpha} \leq \pi \quad , \quad \overleftarrow{\alpha}|_{\text{FIN}} = \overrightarrow{\alpha}|_{\text{FIN}} = \alpha \quad .$$

If α is defined on NORM *(and not only on* FIN*), then*

$$\overleftarrow{\alpha} \leq \alpha \leq \overrightarrow{\alpha}$$

holds on NORM.

Clearly, $\overleftarrow{\alpha} \leq c \overleftarrow{\beta}$ and $\overrightarrow{\alpha} \leq c \overrightarrow{\beta}$ whenever $\alpha \leq c\beta$. Since ε respects subspaces, $\varepsilon = \overrightarrow{\varepsilon}$ and hence $\varepsilon = \overleftarrow{\varepsilon}$; the description of the projective norm by an infimum yields that $\pi = \overrightarrow{\pi}$ but it will be shown in 15.6. that $\pi \neq \overleftarrow{\pi}$. A tensor norm α on NORM is called *finitely generated* if $\alpha = \overrightarrow{\alpha}$ and *cofinitely generated* if $\alpha = \overleftarrow{\alpha}$.

The usual tensor norms are finitely generated — but the cofinite hull is natural as well and its consequent use (also for finitely generated tensor norms) helps to structure the theory, leads to a better understanding of various ideas and simplifies many proofs. In his Résumé Grothendieck only considered finitely generated tensor norms — and this is quite understandable: It will be shown in 13.1. that

$$\overrightarrow{\alpha}(\cdot; E, F) = \overleftarrow{\alpha}(\cdot; E, F)$$

whenever E and F have the metric approximation property — and it was not until nearly 20 years after Grothendieck's work when spaces without approximation properties were discovered (see section 5).

It is obvious — but good to always bear in mind — that two finitely generated (or two cofinitely generated) tensor norms are equal if they are equal on finite dimensional spaces. This explains, for example, why these two types of tensor norms are often useful when investigating Banach spaces or operators in terms of finite dimensional subspaces.

12.5. Many of the interesting tensor norms can be obtained from the ones introduced by Lapresté [166], who generalized those of Saphar [237], Chevet [27] and Cohen [31]. For $p, q \in [1, \infty]$ with $1/p + 1/q \geq 1$ take the unique $r \in [1, \infty]$ with

$$1/r = 1/p + 1/q - 1 \quad \text{or equivalently,} \quad 1 = 1/r + 1/p' + 1/q'$$

(where p' is the conjugate of p) and define for normed spaces E and F and $z \in E \otimes F$

$$\alpha_{p,q}(z; E, F) := \inf\left\{\ell_r(\lambda_i) w_{q'}(x_i; E) w_{p'}(y_i; F) \,\Big|\, z = \sum_{i=1}^n \lambda_i x_i \otimes y_i\right\}$$

(with the notation $\ell_p(\cdot)$ and $w_p(\cdot)$ from section 8). Obviously $\alpha_{1,1} = \pi$.

PROPOSITION: (1) $\alpha_{p,q}$ is a finitely generated tensor norm on NORM.
(2) $\alpha_{p_2, q_2} \leq \alpha_{p_1, q_1}$ if $p_1 \leq p_2$ and $q_1 \leq q_2$.
(3) $\alpha_{p,q}^t = \alpha_{q,p}$.

PROOF: (1) By criterion 12.2. only the triangle inequality is not obvious: Take elements $z_1, z_2 \in E \otimes F$ and $\varepsilon > 0$ and choose for $j = 1, 2$ representations

$$z_j = \sum_{i=1}^n \lambda_{ij} x_{ij} \otimes y_{ij}$$

such that

$$\ell_r(\lambda_{ij}) \leq (\alpha_{p,q}(z_j) + \varepsilon)^{1/r}$$
$$w_{q'}(x_{ij}) \leq (\alpha_{p,q}(z_j) + \varepsilon)^{1/q'}$$
$$w_{p'}(y_{ij}) \leq (\alpha_{p,q}(z_j) + \varepsilon)^{1/p'} \quad .$$

It follows that

$$\alpha_{p,q}(z_1+z_2) \leq \ell_r\big((\lambda_{ij})_{i,j}\big) w_{q'}\big((x_{ij})_{i,j}\big) w_{p'}\big((y_{ij})_{i,j}\big) \leq$$
$$\leq \big(\alpha_{p,q}(z_1)+\alpha_{p,q}(z_2)+2\varepsilon\big)^{1/r+1/q'+1/p'}.$$

(2) There is nothing to prove for $1/p_1 + 1/q_1 = 1$, hence assume $r_1 < \infty$ and define

$$1/p := 1/p_1 - 1/p_2 \quad , \quad 1/q := 1/q_1 - 1/q_2$$

which implies that $1/r_1 = 1/r_2 + 1/p + 1/q$. Take $z \in E \otimes F$ and, for $\varepsilon > 0$, a representation

$$z = \sum_i \lambda_i x_i \otimes y_i \qquad \lambda_i \geq 0$$

with

$$\ell_{r_1}(\lambda_i) w_{q'_1}(x_i) w_{p'_1}(y_i) \leq (1+\varepsilon)\alpha_{p_1,q_1}(z) .$$

Now

$$z = \sum_i \lambda_i^{r_1/r_2} (\lambda_i^{r_1/q} x_i) \otimes (\lambda_i^{r_1/p} y_i)$$

and (by Hölder's inequality)

$$\ell_{r_2}(\lambda_i^{r_1/r_2}) = [\ell_{r_1}(\lambda_i)]^{r_1/r_2}$$
$$w_{q'_2}(\lambda_i^{r_1/q} x_i) \leq [\ell_{r_1}(\lambda_i)]^{r_1/q} w_{q'_1}(x_i)$$
$$w_{p'_2}(\lambda_i^{r_1/p} y_i) \leq [\ell_{r_1}(\lambda_i)]^{r_1/p} w_{p'_1}(y_i) ,$$

hence

$$\alpha_{p_2,q_2}(z) \leq \cdots \leq \ell_{r_1}(\lambda_i)^{r_1/r_2+r_1/q+r_1/p} w_{q'_1}(x_i) w_{p'_1}(y_i) \leq$$
$$\leq (1+\varepsilon)\alpha_{p_1,q_1}(z) .$$

(3) is obvious. □

12.6. To describe the completion of the normed space $E \otimes_{\alpha_{p,q}} F$ recall from section 8 the notation
$$\ell_p^w(E) = \{(x_n) \in E^{I\!N} \mid w_p(x_n) < \infty\}$$
$$\ell_p^{w,o}(E) = \{(x_n) \in \ell_p^w(E) \mid w_p((x_n)_{n=N}^\infty) \to 0\} .$$

PROPOSITION: (1) For $(\lambda_n) \in \ell_r$ (in c_o if $r=\infty$), $(x_n) \in \ell_q^w(E)$ and $(y_n) \in \ell_{p'}^w(F)$ the series
$$\sum (\lambda_n x_n \otimes y_n)$$
converges unconditionally in $E \tilde\otimes_{\alpha_{p,q}} F$.

(2) *For every $z \in E\tilde{\otimes}_{\alpha_{p,q}} F$ there are $(\lambda_n) \in \ell_r$ (in c_0 if $r = \infty$), $(x_n) \in \ell_{q'}^{w,\circ}(E)$ and $(y_n) \in \ell_{p'}^{w,\circ}(F)$ with*

$$z = \sum_{n=1}^{\infty} \lambda_n x_n \otimes y_n \ .$$

Moreover,

$$\alpha_{p,q}(z; E, F) = \inf \ell_r(\lambda_n) w_{q'}(x_n) w_{p'}(y_n) \ ,$$

where the infimum is taken over all such (finite or infinite) representations.

PROOF: (1) follows easily from the fact that $(\lambda_n) \in \ell_r$ (or c_0) forces the series to be an $\alpha_{p,q}$–Cauchy series — independent of the ordering. To prove (2) take for $z \in E\tilde{\otimes}_{\alpha_{p,q}} F$ and $\varepsilon > 0$ elements $z_n \in E \otimes F$ with $z = \sum_{n=1}^{\infty} z_n$ and $\sum_{n=1}^{\infty} \alpha_{p,q}(z_n) \leq (1+\varepsilon)\alpha_{p,q}(z)$. Choose $(\lambda_i^n), (x_i^n)$ and (y_i^n) (finite) with $z_n = \sum_i \lambda_i^n x_i^n \otimes y_i^n$ and

$$\ell_r((\lambda_i^n)_i) \leq (\alpha_{p,q}(z_n)(1+\varepsilon))^{1/r}$$
$$w_{q'}((x_i^n)_i) \leq (\alpha_{p,q}(z_n)(1+\varepsilon))^{1/q'}$$
$$w_{p'}((y_i^n)_i) \leq (\alpha_{p,q}(z_n)(1+\varepsilon))^{1/p'} \ .$$

Then

$$\ell_r((\lambda_i^n)_{i,n}) w_{q'}((x_i^n)_{i,n}) w_{p'}((y_i^n)_{i,n}) \leq \alpha_{p,q}(z)(1+\varepsilon)^2$$

and $(\lambda_i^n)_{i,n} \in \ell_r$, $(x_i^n)_{i,n} \in \ell_{q'}^{w,\circ}(E)$ and $(y_i^n)_{i,n} \in \ell_{p'}^{w,\circ}(F)$. Now (1) implies that

$$z = \sum_{n=1}^{\infty} z_n = \sum_{n=1}^{\infty} \sum_i \lambda_i^n x_i^n \otimes y_i^n = \sum_{n,i} \lambda_i^n x_i^n \otimes y_i^n \ .$$

Moreover, if β denotes the seminorm defined by the infimum in the statement of (2), then

$$\beta(z) \leq \alpha_{p,q}(z) \qquad\qquad\qquad \text{for all } z \in E\tilde{\otimes}_{\alpha_{p,q}} F \ .$$

Conversely, if $z = \sum_{n=1}^{\infty} \lambda_n x_n \otimes y_n$ and $z^N := \sum_{n=1}^{N} \lambda_n x_n \otimes y_n$, then

$$\ell_r(\lambda_n) w_{q'}(x_n) w_{p'}(y_n) \geq$$
$$\geq \ell_r((\lambda_n)_{n=1}^N) w_{q'}((x_n)_{n=1}^N) w_{p'}((y_n)_{n=1}^N) \geq \alpha_{p,q}(z^N) \to \alpha_{p,q}(z) \ ;$$

this implies $\beta(z) \geq \alpha_{p,q}(z)$. □

12.7. Special cases of $\alpha_{p,q}$–tensor norms are ($1 \leq p \leq \infty$)

$$\begin{aligned} g_p &:= \alpha_{p,1} & (g \text{ for "gauche"}) \\ d_p &:= \alpha_{1,p} & (d \text{ for "droite"}) \\ w_p &:= \alpha_{p,p'} & (w \text{ for "weak"}) \end{aligned}$$

and therefore

$$\begin{aligned} g_1 &= d_1 = \pi \ , & w_1 &= d_\infty \ , & w_\infty &= g_\infty \\ g_p &= d_p^t \ , & w_p &= w_{p'}^t \\ w_p &\leq g_p \ , & w_{p'} &\leq d_p \ . \end{aligned}$$

For g_p and d_p the index r in the definition is p, for w_p it is ∞; therefore it is easy to see that

$$g_p(z;E,F) = \inf\left\{\ell_p(x_i)w_{p'}(y_i) \mid z = \sum_{i=1}^{n} x_i \otimes y_i\right\}$$

$$d_p(z;E,F) = \inf\left\{w_{p'}(x_i)\ell_p(y_i) \mid z = \sum_{i=1}^{n} x_i \otimes y_i\right\}$$

$$w_p(z;E,F) = \inf\left\{w_p(x_i)w_{p'}(y_i) \mid z = \sum_{i=1}^{n} x_i \otimes y_i\right\} .$$

Clearly, there is also a result for the completions which has the spirit of proposition 12.6.:

COROLLARY: (1) *For* $(x_n) \in \ell_p(E)$ *(in* $c_o(E)$ *if* $p = \infty$*) and* $(y_n) \in \ell_{p'}^w(F)$ *the series* $\sum(x_n \otimes y_n)$ *converges unconditionally in* $E\tilde{\otimes}_{g_p} F$. *If, conversely,* $z \in E\tilde{\otimes}_{g_p} F$, *then there are* $(x_n) \in \ell_p(E)$ *(in* $c_o(E)$ *if* $p = \infty$*) and* $(y_n) \in \ell_{p'}^{w,o}(F)$ *with* $z = \sum_{n=1}^{\infty} x_n \otimes y_n$; *moreover,*

$$g_p(z;E,F) = \inf \ell_p(x_n)w_{p'}(y_n) ,$$

where the infimum is taken over all such representations.

(2) *For* $1 < p < \infty$ *each series* $\sum(x_n \otimes y_n)$ *with* $(x_n) \in \ell_p^{w,o}(E)$ *and* $(y_n) \in \ell_{p'}^w(F)$ *converges unconditionally in* $E\tilde{\otimes}_{w_p} F$. *Conversely, each* $z \in E\tilde{\otimes}_{w_p} F$ *has a representation* $z = \sum_{n=1}^{\infty} x_n \otimes y_n$ *with* $(x_n) \in \ell_p^{w,o}(E)$ *and* $(y_n) \in \ell_{p'}^{w,o}(F)$; *moreover,*

$$w_p(z;E,F) = \inf w_p(x_n)w_{p'}(y_n) ,$$

where the infimum is taken over all such representations.

This follows immediately from the proposition — with the exception of the first part of (2): but this, as in the proof of 12.6., can be checked directly.

12.8. The following picture illustrates the situation:

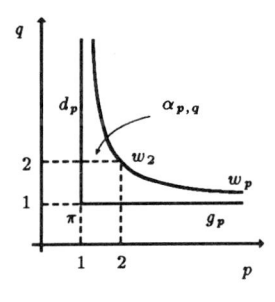

Some of these tensor norms are equivalent (of course, two tensor norms α and β are called *equivalent* if there are $c, d > 0$ with $c\alpha \leq \beta \leq d\alpha$).

PROPOSITION: *For $p,q \in]1,\infty[$ with $1/p + 1/q \geq 1$,*

$$\alpha_{p,q} \leq b_{q'} b_{p'} w_2 \,,$$

where b_r is the constant from the Khintchine inequality. In particular, w_2 and $\alpha_{p,q}$ are equivalent if $p,q \in]1,2]$.

PROOF: As the constants already indicate, Rademacher averaging will be used: For $z = \sum_{i=1}^n x_i \otimes y_i \in E \otimes F$ take the new representation

$$z = \sum_{i,j=1}^n \int_{D_n} \varepsilon_i \varepsilon_j d\mu_n x_i \otimes y_j = \sum_{w \in D_n} \frac{1}{2^n} \left(\sum_{i=1}^n \varepsilon_i(w) x_i \right) \otimes \left(\sum_{j=1}^n \varepsilon_j(w) y_j \right) \,.$$

Now the Khintchine inequality 8.5. gives

$$w_{q'}\left(\left(\sum_{i=1}^n \varepsilon_i(w) x_i \right)_{w \in D_n} \right) = \sup_{\|x'\| \leq 1} \left(\sum_{w \in D_n} \left| \sum_{i=1}^n \varepsilon_i(w) \langle x', x_i \rangle \right|^{q'} \right)^{1/q'} =$$

$$= 2^{n/q'} \sup_{\|x'\| \leq 1} \left(\int_{D_n} \left| \sum_{i=1}^n \langle x', x_i \rangle \varepsilon_i(w) \right|^{q'} \mu_n(dw) \right)^{1/q'} \leq$$

$$\leq 2^{n/q'} b_{q'} w_2((x_i)_{i=1,\cdots,n}) \,.$$

Consequently,

$$\alpha_{p,q}(z; E, F) \leq \left[2^n \frac{1}{2^{nr}} \right]^{1/r} (2^n)^{1/q'+1/p'} b_{q'} b_{p'} w_2(x_i) w_2(y_i)$$

and therefore $\alpha_{p,q} \leq b_{q'} b_{p'} w_2$. □

The tensor norms g_p and d_p cannot be estimated by w_2; otherwise e.g. $w_\infty \leq g_p \leq c w_2$ would follow which, by results to be presented in 23.2., would imply that Hilbert spaces are $\mathcal{L}_\infty^g$-spaces, a contradiction. For the best constants satisfying $\alpha_{p,q} \leq c w_2$ see Ex 24.7..

12.9. It was shown in 8.2. that

$$\ell_p^{w,o}(E) = E \tilde{\otimes}_\varepsilon \ell_p$$

$$(x_n) \rightsquigarrow \sum_{n=1}^\infty x_n \otimes e_n$$

holds isometrically:

$$w_p(x_n; E) = \varepsilon\left(\sum_{n=1}^\infty x_n \otimes e_n; E, \ell_p \right) \,.$$

Since $w_{p'}(e_n; \ell_p) = 1$ and $\varepsilon \leq w_p$, the representation 12.7. of $E \tilde{\otimes}_{w_p} \ell_p$ implies:

REMARK: *The relations*

$$\ell_p^{w,o}(E) = E \tilde{\otimes}_\varepsilon \ell_p = E \tilde{\otimes}_{w_p} \ell_p \qquad \text{for } 1 \leq p < \infty$$
$$c_o(E) = E \tilde{\otimes}_\varepsilon c_o = E \tilde{\otimes}_{w_\infty} c_o$$

hold isometrically. In particular,

$$w_p\left((x_k)_{k=1}^n; E\right) = \varepsilon\left(\sum_{k=1}^n x_k \otimes e_k; E, \ell_p^n\right) = w_p\left(\sum_{k=1}^n x_k \otimes e_k; E, \ell_p^n\right).$$

To obtain another useful description of w_p consider the four-linear map

$$E \times \ell_p \times \ell_{p'} \times F \longrightarrow E \otimes F$$
$$(x, \xi, \eta, y) \rightsquigarrow \langle \xi, \eta \rangle_{\ell_p, \ell_{p'}} x \otimes y$$

which can be linearized to a so-called *tensor contraction*

$$(E \otimes \ell_p) \otimes (\ell_{p'} \otimes F) \longrightarrow E \otimes F \quad .$$

This mapping will turn out to be extremely useful; it mimics the composition of maps (see Ex 12.11.) and this has interesting consequences for theorems about the factorization of operators. A detailed treatment of tensor contractions will be presented in section 29; a first hint of its utility is given by the

COROLLARY: *Let $1 \leq p \leq \infty$. Then the tensor contractions*

$$(E \otimes_\varepsilon \ell_p) \otimes_\pi (\ell_{p'} \otimes_\varepsilon F) \longrightarrow E \otimes_{w_p} F$$
$$(E \tilde{\otimes}_\varepsilon \ell_p) \otimes_\pi (\ell_{p'} \tilde{\otimes}_\varepsilon F) \longrightarrow E \tilde{\otimes}_{w_p} F$$

are metric surjections. Moreover, the norms in the quotients can be calculated by only using elements in $(E \otimes_\varepsilon \ell_p) \times (\ell_{p'} \otimes_\varepsilon F)$ or its completion, respectively. These statements also hold true if ℓ_∞ is replaced by c_o.

PROOF: Denote, for this proof, by $s_p \subset \ell_p$ the subspace of finite sequences with the induced norm. Then, by the definition of w_p and the remark, the map

$$C : (E \otimes_\varepsilon s_p) \times (s_{p'} \otimes_\varepsilon F) = (E \otimes_{w_p} s_p) \times (s_{p'} \otimes_{w_p} F) \longrightarrow E \otimes_{w_p} F$$

maps the cartesian product of the open unit balls onto the open unit ball, hence

$$(E \otimes_\varepsilon s_p) \otimes_\pi (s_{p'} \otimes_\varepsilon F) \longrightarrow E \otimes_{w_p} F$$

is a metric surjection; in particular, the completion of the bilinear mapping C is defined. Note that $(E \otimes_\varepsilon s_p) \otimes_\pi (s_{p'} \otimes_\varepsilon F)$ is a dense subspace of $(E\tilde\otimes_\varepsilon \ell_p) \otimes_\pi (\ell_{p'}\tilde\otimes_\varepsilon F)$ if ℓ_∞ is read as c_0. Since the second mapping in the statement is also surjective by the description 12.7.(2) of $E\tilde\otimes_{w_p} F$ (and the remark), the quotient lemma 7.4. shows that both mappings are metric surjections. If ℓ_∞ appears, note that both mappings have norm ≤ 1 by Ex 12.11.(c); the c_0-case gives the result. □

This result provides a description of the compact sets of $E\tilde\otimes_{w_p} F$, see Ex 12.12.. A similar characterization holds for general $\alpha_{p,q}$, see Ex 35.10..

12.10. The last corollary, easy as it is, has quite interesting consequences. Take a tensor norm α and consider the natural map

$$C(K)\tilde\otimes_\alpha C(L) \longrightarrow C(K)\tilde\otimes_\varepsilon C(L) = C(K \times L)$$

for two compact sets K and L. For $\alpha = \pi$ this map is injective since $C(K)$ has the approximation property (see section 5) and it will be seen in 17.20. that this holds for all finitely generated α. What do the functions coming from $C(K)\tilde\otimes_\alpha C(L)$ look like? For $\alpha = \pi$ a function $f \in C(K \times L)$ is in $C(K)\tilde\otimes_\pi C(L)$ if and only if there are $f_n \in C(K)$ and $g_n \in C(L)$ with

$$f = \sum_{n=1}^\infty f_n \otimes g_n$$

and $\sum_{n=1}^\infty \|f_n\|_\infty \|g_n\|_\infty < \infty$. For $\alpha = w_p$ the corollary above shows that

$$C(K,\ell_p) \times C(L,\ell_{p'}) = (C(K)\tilde\otimes_\varepsilon \ell_p) \times (\ell_{p'}\tilde\otimes_\varepsilon C(L)) \longrightarrow C(K)\tilde\otimes_{w_p} C(L) \subset C(K \times L)$$

is onto — and this map is just $(g,h) \rightsquigarrow \langle g(\cdot), h(\cdot\cdot)\rangle_{\ell_p,\ell_{p'}}$. To see this, calculate first for finite rank functions and then approximate. It follows that:

PROPOSITION: *A function $f \in C(K \times L)$ is in $C(K)\tilde\otimes_{w_p} C(L)$ if and only if there are continuous functions $g : K \to \ell_p$ and $h : L \to \ell_{p'}$ with*

$$f(s,t) = \langle g(s), h(t)\rangle_{\ell_p,\ell_{p'}} \qquad \text{for all } (s,t) \in K \times L \;.$$

Moreover, $w_p(f; C(K), C(L)) \leq \|g\|_\infty \|h\|_\infty$ and for every $\varepsilon > 0$ the functions g and h can be chosen such that $\|g\|_\infty \|h\|_\infty \leq w_p(f) + \varepsilon$.

Exercises:

Ex 12.1. (a) $E \otimes_\alpha \mathbb{K} = E$ holds isometrically for all tensor norms α and normed spaces E.

(b) If $E_o \subset E$ and $F_o \subset F$ are dense subspaces, then $E_o \otimes F_o$ is dense in $E \otimes_\alpha F$ for each tensor norm α.

(c) If $T_i \in \mathfrak{L}(E_i, F_i)$ are isometries (onto), then

$$T_1 \otimes T_2 : E_1 \otimes_\alpha E_2 \longrightarrow F_1 \otimes_\alpha F_2$$

is an isometry for each tensor norm α. Is this true without surjectivity?

Ex 12.2. If α is a tensor norm on FIN, then

$$\overrightarrow{\alpha}(z; E, F) = \inf \left\{ \overrightarrow{\alpha}(z; E, N) \mid N \in \text{FIN}(F), z \in E \otimes N \right\}$$
$$\overleftarrow{\alpha}(z; E, F) = \sup \left\{ \overleftarrow{\alpha}(id_E \otimes Q_L^F(z); E, F/L) \mid L \in \text{COFIN}(F) \right\}.$$

Ex 12.3. Let α be a tensor norm on NORM; show that the "one–sided" finite and cofinite hull (also called right–finite and right–cofinite hull)

$$\alpha^{\rightarrow}(z; E, F) := \inf \{\alpha(z; E, N) \mid N \in \text{FIN}(F), z \in E \otimes N\}$$
$$\alpha^{\leftarrow}(z; E, F) := \sup \{\alpha(id_E \otimes Q_L^F(z); E, F/L) \mid L \in \text{COFIN}(F)\}$$

are tensor norms such that $\overleftarrow{\alpha} \leq \alpha^{\leftarrow} \leq \alpha \leq \alpha^{\rightarrow} \leq \overrightarrow{\alpha}$.

Ex 12.4. Take $S \in \mathfrak{L}(X, Y)$ with $\|S\| = 1$ and α and β tensor norms (on NORM). Show that the class $\mathfrak{L}_S$ of all $T \in \mathfrak{L}(E, F)$ with

$$\mathbf{L}_S(T) := \|S \otimes T : X \otimes_\alpha E \longrightarrow Y \otimes_\beta F\| < \infty$$

is a Banach operator ideal. Check exactly which of the tensor norm–properties are needed for the proof and note that α and β need only satisfy the metric mapping property on the right side of the tensor product. This means that $X \otimes_\alpha \cdot$ or $Y \otimes_\beta \cdot$ could be replaced by $L_p \otimes_{\Delta_p} \cdot$; see also 9.6.(3), Ex 9.15., Ex 9.16. and 17.4..

Ex 12.5. Use Rademacher averaging to show that for all matrices (a_{ij}) and all tensor norms α on FIN the relation

$$\alpha\Big(\sum_{i=1}^n a_{ii} e_i \otimes e_i; \ell_p^n, \ell_q^n\Big) \leq \alpha\Big(\sum_{i,j=1}^n a_{ij} e_i \otimes e_j; \ell_p^n, \ell_q^n\Big)$$

holds. Hint: $I_w(\xi_j) := (\varepsilon_j(w)\xi_j)$ for $w \in D_n$ defines an isometry.

Ex 12.6. For each tensor norm α and $p, q \in [1, \infty]$

$$\alpha\Big(\sum_{i=1}^n e_i \otimes e_i; \ell_p^n, \ell_q^n\Big) \leq n^c$$

where $c := \min\{1/p + 1/q, 1\}$. Hint: 8.8..

Ex 12.7. For $z \in E \otimes F$ define the *rank of* z as follows: $\mathrm{rank}(0) := 0$ and for $z \neq 0$

$$\mathrm{rank}(z) := \min\left\{ n \in \mathbb{N} \ \Big| \ z = \sum_{i=1}^{n} x_i \otimes y_i \right\}.$$

(a) Show that $\mathrm{rank}(z) = \mathrm{rank}(T_z)$, where $T_z : E' \to F$ is the associated linear operator.
(b) For Banach spaces E, F and $T \in \mathfrak{L}(E, F)$ the n-th approximation number $a_n(T)$ of T is defined by

$$a_n(T) := \inf\{\|S - T\| \mid S \in \mathfrak{L}(E, F), \mathrm{rank}(S) < n\}.$$

Show with the aid of these numbers that the set

$$\otimes_n(E, F) := \{z \in E \otimes F \mid \mathrm{rank}(z) \leq n\}$$

is closed in $E \tilde{\otimes}_\varepsilon F$ and in $E \otimes_\alpha F$ for each reasonable norm α on $E \otimes F$. If the map $E \tilde{\otimes}_\alpha F \to E \tilde{\otimes}_\varepsilon F$ is injective, it is also closed in $E \tilde{\otimes}_\alpha F$. Hint: The $\sigma(E', E) - \sigma(F, F')$–continuous operators in $\mathfrak{L}(E', F)$ form a norm–closed subset, a_n is continuous and $a_n(T) = 0$ if and only if $\mathrm{rank}(T) < n$. This result is due to Valdivia [274], see also [72].

Ex 12.8. Let $1/p + 1/q \geq 1$. Then $\beta_{p,q}$ and $\gamma_{p,q}$ defined by

$$\beta_{p,q}(z; E, F) := \inf\left\{ \|(a_{i,j}) : \ell_{q'}^n \to \ell_p^n\| w_{q'}(x_j) w_{p'}(y_i) \ \Big| \ z = \sum_{i,j=1}^{n} a_{i,j} x_j \otimes y_i \right\}$$

$$\gamma_{p,q}(z; E, F) := \inf\left\{ \|(a_{i,j}) : \ell_{q'}^n \to \ell_p^n\| \ell_{q'}(x_j) \ell_{p'}(y_i) \ \Big| \ z = \sum_{i,j=1}^{n} a_{i,j} x_j \otimes y_i \right\}$$

are finitely generated tensor norms with

$$\beta_{p,q} \leq \alpha_{p,q} \quad \text{and} \quad \beta_{p,q} \leq \gamma_{p,q}.$$

Hint: Mimic the proof of 12.5. using the fact that $\|A_1 \oplus A_2 : \ell_{q'}^{n_1+n_2} \to \ell_p^{n_1+n_2}\| \leq (\|A_1\|^r + \|A_2\|^r)^{1/r}$. These tensor norms will be studied in more detail in section 28.

Ex 12.9. If (e_n) and (f_n) are *bases* of Banach spaces E and F, respectively, then $(e_1 \otimes f_1, e_2 \otimes f_1, e_2 \otimes f_2, e_1 \otimes f_2, e_3 \otimes f_1, e_3 \otimes f_2, \ldots)$ — the rectangular ordering — is a basis for $E \tilde{\otimes}_\alpha F$ for each tensor norm α. Hint: Show that the expansion operators are uniformly bounded on $E \otimes_\alpha F$ and determine the coefficient functionals; use the fact that if the expansion operators of a biorthogonal system are uniformly bounded and converge on a dense subspace pointwise to the identity, then this biorthogonal system defines a basis. This result is due to Gelbaum–Gil de Lamadrid [80].

Ex 12.10. Let $(\mathfrak{A}, \mathbf{A})$ be a normed operator ideal and define $\alpha(\cdot; E, F)$ to be the induced norm from the injection $E \otimes F \hookrightarrow \mathfrak{A}(E', \tilde{F})$. Then α is a tensor norm on NORM; moreover, $E \otimes_\alpha F \hookrightarrow E'' \otimes_\alpha F$ holds isometrically. Which property does α have if $(\mathfrak{A}, \mathbf{A})$ is injective? If $(\mathfrak{A}, \mathbf{A})$ is surjective?

Ex 12.11. (a) Show that the tensor contraction

$$(E' \otimes \ell_p) \otimes (\ell_{p'} \otimes F) \longrightarrow E' \otimes F$$

is the same as the composition map

$$\mathfrak{F}(E, \ell_p) \otimes \mathfrak{F}(\ell_p, F) \longrightarrow \mathfrak{F}(E, F)$$

(for $p = \infty$ read c_o instead of ℓ_∞).

(b) Show that $(E \otimes_\varepsilon \ell_{p'}) \otimes_\pi (\ell_p \otimes_{\Delta_p} F) \to E \otimes_{d_p} F$ is a metric surjection (ℓ_∞ can be replaced by c_o). Hint: 12.7. and 12.9..

(c) For normed spaces E, F and G the tensor contraction

$$(E \otimes G') \otimes (G \otimes F) \longrightarrow E \otimes F$$

given by $(x \otimes u') \otimes (u \otimes y) \rightsquigarrow \langle u', u \rangle x \otimes y$ is defined. Show that for an arbitrary tensor norm α

$$\|(E \otimes_\varepsilon G') \otimes_\pi (G \otimes_\alpha F) \longrightarrow E \otimes_\alpha F\| \leq 1 \ .$$

Hint: $\sum_{n,m} \langle u'_n, u_m \rangle x_n \otimes y_m = [(\sum_n u'_n \otimes x_n) \otimes id_F](\sum_m u_m \otimes y_m)$.

Ex 12.12. Use corollary 12.9. and the description of compact subsets of the projective tensor product to show the following: For each compact subset $K \subset E \tilde{\otimes}_{w_p} F$ there are a compact set $D \subset \ell_1$ and zero sequences $(x_n) = ((x_n(m))_m)_n$ in $\ell_p^{w,o}(E)$ and $(y_n) = ((y_n(m))_m)_n$ in $\ell_{p'}^{w,o}(F)$ such that for each $z \in K$ there is a $(\lambda_n) \in D$ with

$$z = \sum_{n=1}^{\infty} \lambda_n \sum_{m=1}^{\infty} x_n(m) \otimes y_n(m) \ .$$

Note the case $p = 1$: Each x_n is an unconditionally convergent series and $y_n \in c_o(F)$.

13. The Five Basic Lemmas

This section contains five relatively simple lemmas which are basic for the understanding and use of tensor norms: the approximation, extension, embedding, density and local technique lemma. The power and importance of these devices will become clear while working with them.

13.1. Recall from section 5 that a Banach space E has the λ–bounded approximation property if there is a net (T_η) of finite rank operators $E \to E$ with $\|T_\eta\| \leq \lambda$ and $T_\eta x \to x$ for all $x \in E$.

APPROXIMATION LEMMA: *Let α and β be tensor norms (on NORM), E, F normed spaces, $c \geq 1$ and*
$$\alpha \leq c\beta \quad \text{on} \quad E \otimes N$$
for cofinally many $N \in \mathrm{FIN}(F)$. If F has the bounded approximation property with constant $\lambda \geq 1$, then
$$\alpha \leq \lambda c \beta \quad \text{on} \quad E \otimes F \ .$$

PROOF: It is easy to see that for each $z \in E \otimes F$
$$id_E \otimes T_\eta(z) \longrightarrow z$$
for the projective norm π and hence for all tensor norms. If η satisfies
$$\alpha(z - id_E \otimes T_\eta(z); E, F) \leq \varepsilon$$
and N is as in the hypothesis with $T_\eta(F) \subset N$, then, by the metric mapping property of tensor norms,
$$\alpha(z; E, F) \leq \alpha(z - id_E \otimes T_\eta(z); E, F) + \alpha(id_E \otimes T_\eta(z); E, F) \leq$$
$$\leq \varepsilon + \alpha(id_E \otimes T_\eta(z); E, N) \leq$$
$$\leq \varepsilon + c\beta(id_E \otimes T_\eta(z); E, N) \leq$$
$$\leq \varepsilon + c\|T_\eta : F \to N\|\beta(z; E, F)$$
which implies the statement. □

For the finite and cofinite hull of a tensor norm this lemma (and its transposed version) implies the following

PROPOSITION: *Let α be a tensor norm (on FIN), E and F Banach spaces with the bounded approximation property with constants λ_E and λ_F, respectively. Then*
$$\overleftarrow{\alpha} \leq \overrightarrow{\alpha} \leq \lambda_E \lambda_F \overleftarrow{\alpha} \quad \text{on} \quad E \otimes F \ .$$

In particular, $\overleftarrow{\alpha} = \overrightarrow{\alpha}$ on $E \otimes F$ if both spaces have the metric approximation property.

13.2. Every $\varphi \in (E \otimes_\pi F)' = \mathfrak{L}(E, F')$ has a canonical extension $\varphi^\wedge \in (E \otimes_\pi F'')'$ which satisfies the relation
$$\langle \varphi^\wedge, x \otimes y'' \rangle = \langle L_\varphi x, y'' \rangle_{F', F''} \ ,$$

where $L_\varphi \in \mathfrak{L}(E, F')$ is the linear operator associated with φ (see 1.9.). What happens if $\varphi \in (E \otimes_\alpha F)'$?

EXTENSION LEMMA: *Let $\varphi \in (E \otimes_\pi F)'$ and let α be a finitely generated tensor norm (on NORM). Then*

$$\varphi \in (E \otimes_\alpha F)' \quad \text{if and only if} \quad \varphi^\wedge \in (E \otimes_\alpha F'')' .$$

In this case, $\|\varphi\|_{(E \otimes_\alpha F)'} = \|\varphi^\wedge\|_{(E \otimes_\alpha F'')'}$.

PROOF: The metric mapping property implies $\|E \otimes_\alpha F \hookrightarrow E \otimes_\alpha F''\| \leq 1$ and hence $\|\varphi\|_\cdots \leq \|\varphi^\wedge\|_\cdots$. Conversely, take $z_o \in E \otimes F''$ and $M \in \text{FIN}(E)$ and $N \in \text{FIN}(F'')$ such that $z_o \in M \otimes N$; then the weak principle of local reflexivity (6.5.) gives for every $\varepsilon > 0$ an $R \in \mathfrak{L}(N, F)$ with $\|R\| \leq 1 + \varepsilon$ such that for all $y'' \in N$ and $x \in M$

$$\langle y'', L_\varphi x \rangle_{F'', F'} = \langle Ry'', L_\varphi x \rangle_{F, F'} .$$

This means that

$$\langle \varphi^\wedge, x \otimes y'' \rangle = \langle \varphi, (id_E \otimes R)(x \otimes y'') \rangle ,$$

therefore

$$\langle \varphi^\wedge, z_o \rangle = \langle \varphi, id_E \otimes R(z_o) \rangle \qquad (*)$$

and hence

$$|\langle \varphi^\wedge, z_o \rangle| \leq \|\varphi\| \, \|R\| \alpha(z_o; E, N) \leq \|\varphi\|(1 + \varepsilon) \alpha(z_o; E, N) ,$$

which implies the result since α is finitely generated (see Ex 12.2.). □

Sometimes the relation $(*)$ is helpful. The reader may have noticed that the proof is nearly identical with the one given for $\alpha = \varepsilon$ in 6.7.. Strangely enough, we do not know *whether the extension lemma holds for cofinitely generated tensor norms.*

There is also a left–canonical extension $^\wedge\varphi \in (E'' \otimes_\pi F)'$ of $\varphi \in (E \otimes_\pi F)'$ and it was shown in 1.9. that $(^\wedge\varphi)^\wedge = {}^\wedge(\varphi^\wedge) \in (E'' \otimes_\pi F'')'$ if and only if L_φ is weakly compact; clearly,

$$\|{}^\wedge(\varphi^\wedge)\|_{(E'' \otimes_\alpha F'')'} = \|(^\wedge\varphi)^\wedge\|_{(E'' \otimes_\alpha F'')'} = \|\varphi\|_{(E \otimes_\alpha F)'}$$

by the lemma. Define two "natural" maps

$$\Phi_j : E'' \otimes F'' \hookrightarrow (E \otimes_\alpha F)''$$

by

$$\langle \Phi_1(z''), \varphi \rangle := \langle {}^\wedge(\varphi^\wedge), z'' \rangle$$
$$\langle \Phi_2(z''), \varphi \rangle := \langle (^\wedge\varphi)^\wedge, z'' \rangle .$$

They obviously have norm 1 on $E'' \otimes_\alpha F'''$. But we do not know *which norms the Φ_j induce on* $E'' \otimes F'''$. If the induced norm were α in reasonable situations, this would easily solve the problem of the bidual mappings which will be treated in Ex 17.16..
Clearly, $\langle E \otimes F''', (E \otimes_\alpha F)' \rangle$ is a separating duality system with the bracket

$$\langle \varphi, z \rangle := \langle \varphi^\wedge, z \rangle .$$

The extension lemma and the bipolar theorem give the

COROLLARY: *If α is a finitely generated tensor norm, then the unit ball $B_{E \otimes_\alpha F}$ is $\sigma(E \otimes F''', (E \otimes_\alpha F)')$-dense in the unit ball $B_{E \otimes_\alpha F''}$.*

13.3. Tensor norms do not respect subspaces but the embedding into the bidual is usually respected:

EMBEDDING LEMMA: *If α is a finitely or cofinitely generated tensor norm (on NORM), then*

$$id_E \otimes \kappa_F : E \otimes_\alpha F \xrightarrow{\;\;1\;\;} E \otimes_\alpha F''$$

is an isometry for all normed spaces E and F.

See Ex 13.4. for a slight refinement of this result.

PROOF: The metric mapping property implies that for all $z \in E \otimes F$

$$\alpha(z; E, F'') \leq \alpha(z; E, F)$$

holds for all tensor norms α (the map $id_E \otimes \kappa_F$ will not always be written).
(1) Let α be finitely generated. Then, by the extension lemma,

$$\begin{aligned}
\alpha(z; E, F) &= \sup\{|\langle \varphi, z \rangle| \mid \varphi \in (E \otimes_\alpha F)', \|\varphi\|_{(E \otimes_\alpha F)'} \leq 1\} \\
&= \sup\{|\langle \varphi^\wedge, z \rangle| \mid \varphi \in (E \otimes_\alpha F)', \|\varphi\|_{(E \otimes_\alpha F)'} \leq 1\} \\
&\leq \sup\{|\langle \psi, z \rangle| \mid \psi \in (E \otimes_\alpha F'')', \|\psi\|_{(E \otimes_\alpha F'')'} \leq 1\} \\
&= \alpha(z; E, F'') ,
\end{aligned}$$

which is the reverse inequality.
(2) If α is cofinitely generated and $L \in \text{COFIN}(F)$, then L^{00} (formed in F'') is in $\text{COFIN}(F'')$ and the map

$$\kappa_{F/L} : F/L \hookrightarrow (F/L)'' = F''/L^{00}$$

is isometric and surjective; moreover, $Q^{F''}_{L^{00}} \circ \kappa_F = \kappa_{F/L} \circ Q^F_L$. Therefore,

$$\begin{aligned}
\alpha(id_E \otimes Q^F_L(z); E, F/L) &= \alpha(id_E \otimes (\kappa_{F/L} \circ Q^F_L)(z); E, (F/L)'') = \\
&= \alpha((id_E \otimes Q^{F''}_{L^{00}}) \circ (id_E \otimes \kappa_F)(z); E, F''/L^{00}) \leq \\
&\leq \overleftarrow{\alpha}(id_E \otimes \kappa_F(z); E, F'') = \overleftarrow{\alpha}(z; E, F'') .
\end{aligned}$$

Taking the supremum over all L gives the missing inequality. □

13.4. In addition, since F and its completion $\tilde{F}$ have the same biduals:

COROLLARY: *If α is finitely or cofinitely generated, then*
$$E \otimes_\alpha F \xhookrightarrow{1} E \otimes_\alpha \tilde{F}$$
is an isometric and dense embedding for all normed spaces E and F.

Density follows from Ex 12.1.. In particular, finitely or cofinitely generated tensor norms respect dense subspaces. This implies that the density lemma 7.3. can be applied to give the more specialized

DENSITY LEMMA: *Let α be a finitely or cofinitely generated tensor norm, E and F normed spaces and E_o and F_o dense subspaces of E and F, respectively. If G is a locally convex space and $T \in \mathfrak{L}(E \otimes_\pi F, G)$ is such that*
$$T|_{E_o \otimes F_o} \in \mathfrak{L}(E_o \otimes_\alpha F_o, G) ,$$
then $T \in \mathfrak{L}(E \otimes_\alpha F, G)$.

Recall that $T : L \to G$ is continuous if each $S \circ T : L \to D$ is continuous, where D is normed and $S \in \mathfrak{L}(G, D)$. A particularly interesting special case is:

COROLLARY: *Let α and β be tensor norms such that α is finitely or cofinitely generated. If $T_i \in \mathfrak{L}(E_i, F_i)$ and $G_i \subset E_i$ are dense subspaces such that*
$$T_1 \otimes T_2|_{G_1 \otimes G_2} \in \mathfrak{L}(G_1 \otimes_\alpha G_2, F_1 \otimes_\beta F_2) ,$$
then $T_1 \otimes T_2 \in \mathfrak{L}(E_1 \otimes_\alpha E_2, F_1 \otimes_\beta F_2)$.

Since
$$T_1 \otimes T_2 : E_1 \otimes_\pi E_2 \longrightarrow F_1 \otimes_\pi F_2 \xrightarrow{id} F_1 \otimes_\beta F_2 =: G$$
is continuous, the proof is obvious.

13.5. The local techniques for $\mathfrak{L}_p^g$-spaces (see 3.13. for the definition) are condensed in the following lemma; for ε and π the argument has already been used several times:

$\mathfrak{L}_p$-LOCAL TECHNIQUE LEMMA: *Let α and β be tensor norms, $c \geq 0$, and E a normed space such that*
$$\alpha \leq c\beta \qquad \text{on} \quad E \otimes \ell_p^n$$

for all $n \in \mathbb{N}$. Then
$$\alpha \leq \lambda c \overrightarrow{\beta} \quad \text{on} \quad E \otimes F$$
for each $\mathcal{L}_{p,\lambda}^{g}$-space F.

For a generalization of this from $T = id_E$ to arbitrary T see Ex 13.7.; see also 23.1. and Ex 23.13..

PROOF: For $N \in \text{FIN}(F)$ take a factorization

$$N \xhookrightarrow{} F, \quad R: N \to \ell_p^m, \quad S: \ell_p^m \to F$$

$$\|R\|\,\|S\| \leq \lambda(1+\varepsilon)$$

$$M := S(\ell_p^m) .$$

Then, for every $z \in E \otimes N$,
$$\alpha(z; E, M) = \alpha(id_E \otimes (S \circ R)(z); E, M) \leq \|S\| \alpha(id_E \otimes R(z); E, \ell_p^m) \leq$$
$$\leq \|S\| c \beta(id_E \otimes R(z); E, \ell_p^m) \leq \|S\|\,\|R\| c \beta(z; E, N) ;$$

this implies the statement. □

Recalling from Ex 12.3. that $\alpha^{\rightarrow}$ is the right–finite hull of α, one notes that actually
$$\alpha^{\rightarrow} \leq \lambda c \beta^{\rightarrow} \quad \text{on} \quad E \otimes F$$
has been shown. Since $w_p = \varepsilon$ on $E \otimes \ell_p^m$ (see 12.9.), it follows that
$$\varepsilon \leq w_p \leq \lambda \varepsilon \quad \text{on} \quad E \otimes F$$
for each $\mathcal{L}_{p,\lambda}^{g}$-space F and arbitrary normed space E. Consequently, $E \tilde{\otimes}_\varepsilon F = E \tilde{\otimes}_{w_p} F$ holds isomorphically whenever F is an $\mathcal{L}_p^g$-space — and, e.g., Ex 12.12. gives a nice description of the compact subsets of $E \tilde{\otimes}_\varepsilon F$, which has an interesting application to compact operators (see Ex 13.6.). In section 23 it will be shown that $\varepsilon \sim w_p$ on $\cdot \otimes F$ actually characterizes the $\mathcal{L}_p^g$-spaces.

Exercises:

Ex 13.1. Let α be a tensor norm and E and F normed spaces with the metric approximation property. Then $E \otimes_\alpha F \hookrightarrow E \otimes_\alpha F''$ holds isometrically. Hint: Use $\overleftarrow{\alpha}$.

Ex 13.2. Let E, F, G be normed spaces, $S \in \mathfrak{L}(E', F')$ and α a finitely generated tensor norm. Then there is a linear $\hat{S} : (G \otimes_\alpha E)' \to (G \otimes_\alpha F)'$ of norm $\|S\|$ (if $G \neq \{0\}$) such that

$$\begin{array}{ccccc}
G' \otimes E' & \hookrightarrow & (G \otimes_\alpha E)' & \hookrightarrow & \mathfrak{L}(G, E') \ni T \\
{\scriptstyle id_{G'} \otimes S} \downarrow & & \downarrow \hat{S} & & \downarrow \} \\
G' \otimes F' & \hookrightarrow & (G \otimes_\alpha F)' & \hookrightarrow & \mathfrak{L}(G, F') \ni S \circ T
\end{array}$$

commutes; $\hat{S}$ is unique. Hint: Extension lemma.

Ex 13.3. If α is finitely or cofinitely generated and $E'' \tilde{\otimes}_\alpha F \to \mathfrak{L}(E', F)$ is injective, then $E \tilde{\otimes}_\alpha F \to \mathfrak{L}(E', F)$ is injective. Is the converse true? Hint: Section 5.

Ex 13.4. Check the proofs of 13.2. and 13.3. in order to show that:

(a) If $\alpha = \alpha^\to$ (i.e. α is finitely generated from the right, see Ex 12.3.), then the extension lemma holds for α.

(b) If $\alpha = \alpha^\to$ or $\alpha = \alpha^\leftarrow$, then the embedding lemma holds for α.

Ex 13.5. Use the special version Ex 13.1. of the embedding lemma and the tensor norm from Ex 12.10. to demonstrate the following statement: If $(\mathfrak{A}, \mathbf{A})$ is a normed operator ideal, E a reflexive and F an arbitrary Banach space such that E' and F have the metric approximation property, then

$$\mathbf{A}(T : E \to F) = \mathbf{A}(\kappa_F \circ T : E \to F'')$$

for all $T \in \mathfrak{F}(E, F)$. This is a sort of "regularity property" of finite rank operators with respect to all normed operator ideals.

Ex 13.6. (a) Let E and F be Banach spaces, F an $\mathfrak{L}_p^g$-space and $K \subset \overline{\mathfrak{F}}(E, F)$ a compact set of approximable operators. Then there is a compact $D \subset \ell_1$, sequences (R_n) in $\mathfrak{F}(E, \ell_p)$ and (S_n) in $\mathfrak{F}(\ell_p, F)$ with $\|R_n\| \to 0$ and $\|S_n\| \to 0$ such that for each $T \in K$ there is a $(\lambda_n) \in D$ with

$$T = \sum_{n=1}^\infty \lambda_n S_n \circ R_n \quad ;$$

for $p = \infty$ replace ℓ_∞ by c_o. Hint: $\varepsilon \sim w_p$ and a modification of Ex 12.12.. It will be shown later on (21.6.) that all $\mathfrak{L}_p^g$-spaces F have the approximation property, hence $\overline{\mathfrak{F}}(E, F) = \mathfrak{K}(E, F)$ holds.

(b) Show that every approximable (=compact) operator with values in an $\mathfrak{L}_p^g$-space factors compactly through ℓ_p (or c_o if $p = \infty$). Hint: 12.9.. See Ex 23.9. for more information.

(c) Use (b) to show the following result of Terzioglu [270] : Every compact operator factors compactly through a subspace of c_o. Hint: $E \to F \hookrightarrow \ell_\infty(\Gamma)$.

Ex 13.7. $\mathcal{L}_p$-*local technique for operators:* Let α and β be tensor norms and α be finitely generated. If $T \in \mathcal{L}(E, F)$ satisfies
$$\|id_{\ell_p^n} \otimes T : \ell_p^n \otimes_\alpha E \longrightarrow \ell_p^n \otimes_\beta F\| \leq c$$
for all $n \in \mathbb{N}$, then
$$\|id_G \otimes T : G \otimes_\alpha E \longrightarrow G \otimes_\beta F\| \leq \lambda c$$
for each $\mathcal{L}_{p,\lambda}^g$-space G.

14. Grothendieck's Inequality

Grothendieck's "théorème fondamental de la théorie métrique des produits tensoriels" ([93], §4, théorème 1) states that the identity operator on a Hilbert space is in the operator ideal $(\mathfrak{J}^{sur})^{inj}$ (see section 20); the operators in this ideal were called "préintégral" by Grothendieck. A remarkable achievement of Lindenstrauss and Pełczyński in their celebrated 1968 paper [173] was to have noticed that this theorem can be reformulated in the form of an inequality concerning $n \times n$ matrices and Hilbert spaces — as well as in the form $\mathcal{L}(\ell_1, \ell_2) = \mathfrak{P}_1(\ell_1, \ell_2)$. This is why the fundamental theorem is nowadays called Grothendieck's inequality — in matrix form, in tensor form, or in operator form in order to highlight its various glamorous facets. Our path of entry to the many equivalent formulations of the inequality is establishing the fact that π and w_2 are equivalent tensor norms on $\ell_\infty \otimes \ell_\infty$; we follow Krivine's proof [155], which was smoothed out by Pisier in his book [225] and polished into its final form by Pełczyński. Simple as it is now, the proof is in some sense dual to Grothendieck's original one, but has still the same ingredients.

14.1. The idea of the proof is to take a function
$$f \in C(K) \otimes_{w_2} C(K)$$
(where $K = \{1, ..., n\}$), factor it as in 12.10. through some Hilbert space ℓ_2^m, i.e.

(∗) $$f = \sum_{k,\ell=1}^n \langle x_k, y_\ell \rangle e_k \otimes e_\ell$$

with $\|x_k\| = \|y_\ell\| = 1$, and use an integral representation of $\langle x, y \rangle$ for elements in the euclidean unit sphere to obtain a new representation
$$f = \int_{\mathbb{R}^n} g(w) \otimes h(w) \gamma_n(dw) \quad \in C(K) \otimes C(K),$$

with the functions g and h being uniformly bounded on $\mathbb{R}^n$ (with values in $C(K)$). However, the formula for $\langle x, y \rangle$ reads $= \sin\left[\int_{\mathbb{R}^n} ...d\gamma_n\right]$, so it is better to represent $\sin f$ instead of f as in $(*)$. In order to be able to define $\sin f$ (with control of the norm) one needs the fact that $C(K) \otimes_{w_2} C(K)$ is a Banach algebra:

LEMMA: *If K and L are compact, $1/p + 1/q \geq 1$, then*

$$C(K) \otimes_{\alpha_{p,q}} C(L)$$

is a normed algebra (under the pointwise multiplication coming from $C(K \times L)$), i.e., if $f_1, f_2 \in C(K) \otimes C(L)$, then

$$\alpha_{p,q}(f_1 \cdot f_2; C(K), C(L)) \leq \alpha_{p,q}(f_1; C(K), C(L))\alpha_{p,q}(f_2; C(K), C(L)) .$$

PROOF: Represent $f_i = \sum_{n=1}^{N} \lambda_n^i g_n^i \otimes h_n^i$. Then

$$f_1 \cdot f_2 = \sum_{n,m=1}^{N} \lambda_n^1 \lambda_m^2 \left(g_n^1 g_m^2\right) \otimes \left(h_n^1 h_m^2\right) \in C(K) \otimes C(L) .$$

Since

$$\ell_r\left((\lambda_n^1 \lambda_m^2)_{n,m}\right) = \ell_r(\lambda_n^1)\ell_r(\lambda_m^2)$$

$$w_{q'}\left((g_n^1 g_m^2)_{n,m}\right) = \sup_{t \in K} \left(\sum_{n,m} |g_n^1(t) g_m^2(t)|^{q'}\right)^{1/q'} \leq w_{q'}(g_n^1) w_{q'}(g_m^2)$$

and likewise for $w_{p'}\left((h_n^1 h_m^2)_{n,m}\right)$, the result is obvious. □

In particular, the space $\ell_\infty^n \otimes_{w_2} \ell_\infty^n$ is a Banach algebra under pointwise multiplication which is nothing more than

$$\langle z_1 \cdot z_2, e_k \otimes e_\ell \rangle = \langle z_1, e_k \otimes e_\ell \rangle \langle z_2, e_k \otimes e_\ell \rangle$$

in tensor notation. If $\sum(a_m t^m)$ is a power series with infinite radius of convergence, then $\sum_{m=0}^{\infty} a_m z^m$ is defined in the Banach algebra $\ell_\infty^n \otimes_{w_2} \ell_\infty^n$ for each $z \in \ell_\infty^n \otimes \ell_\infty^n$ and

$$\left\langle \sum_{m=0}^{\infty} a_m z^m, e_k \otimes e_\ell \right\rangle = \sum_{m=0}^{\infty} a_m (\langle z, e_k \otimes e_\ell \rangle)^m$$

for all $k, \ell = 1, ..., n$ by continuity of the point evaluation $e_k \otimes e_\ell \in (\ell_\infty^n \otimes_{w_2} \ell_\infty^n)'$.

14.2. The second ingredient in the proof will be factorization:

LEMMA: *For $z \in \ell_\infty^n \otimes \ell_\infty^n$ with $w_2(z; \ell_\infty^n, \ell_\infty^n) < 1$ there is an $m \in \mathbb{N}$ and elements $x_1, ..., x_n, y_1, ..., y_n \in \ell_2^m$ of norm 1 such that*

$$z = \sum_{k,\ell=1}^{n} \langle x_k, y_\ell \rangle e_k \otimes e_\ell .$$

PROOF: Each representation $z = \sum_{i=1}^{m} u_i \otimes v_i$ with $w_2(u_i; \ell_\infty^n) < 1$ and $w_2(v_i; \ell_\infty^n) < 1$ allows one to define functions

$$g := \sum_{i=1}^{m} u_i \otimes e_i \text{ and } h := \sum_{i=1}^{m} v_i \otimes e_i \in \ell_\infty^n \otimes_\varepsilon \ell_2^m = C(\{1, ..., n\}, \ell_2^m) \ .$$

It follows from 8.1. that $\|g(k)\|_{\ell_2^m} \leq \|g\|_\infty = w_2(u_i; \ell_\infty^n) < 1$ and also that $\|h(k)\|_{\ell_2^m} < 1$ for all $k = 1, ..., n$. Since

$$z = \sum_{k,\ell=1}^{n} \langle z, e_k \otimes e_\ell \rangle e_k \otimes e_\ell = \sum_{k,\ell=1}^{n} \langle g(k), h(\ell) \rangle e_k \otimes e_\ell \ ,$$

the vectors

$$x_k := g(k) + (1 - \|g(k)\|_2^2)^{1/2} e_{m+1} \ , \ y_\ell := h(\ell) + (1 - \|h(\ell)\|_2^2)^{1/2} e_{m+2}$$

in ℓ_2^{m+2} satisfy the required property. □

The construction in this proof is the same as in 12.9. and 12.10..

14.3. Denote the sign function $\mathbb{R} \to \{-1, 0, +1\}$ by sign. The main tool for the proof of Grothendieck's inequality is the following formula for the scalar product (=duality bracket) of vectors in the *euclidean* unit sphere of $\mathbb{R}^n$.

LEMMA: For $x, y \in \mathbb{R}^n$ with $\|x\|_2 = \|y\|_2 = 1$

$$\langle x, y \rangle = \sin\left[\frac{\pi}{2} \int_{\mathbb{R}^n} \text{sign}\langle x, w \rangle \cdot \text{sign}\langle y, w \rangle \, \gamma_n(dw)\right] \ ,$$

where γ_n is the Gaussian measure on $\mathbb{R}^n$.

PROOF: There is a $\theta \in [0, \pi]$ such that $\langle x, y \rangle = \cos \theta = \sin\left(\frac{\pi}{2} - \theta\right)$. Let $T : \mathbb{R}^n \to \mathbb{R}^n$ be a linear isometry (with respect to the euclidean norm) such that

$$Tx = e_1 \text{ and } Ty = (\cos \theta, \sin \theta, 0, ...0) \ .$$

Since the Gaussian measure is invariant under isometries (8.7.), it follows that

$$\frac{\pi}{2} \int_{\mathbb{R}^n} \text{sign}\langle x, w \rangle \text{sign}\langle y, w \rangle \gamma_n(dw) =$$

$$= \frac{\pi}{2} \int_{\mathbb{R}^n} \text{sign}\langle e_1, w \rangle \text{sign}\langle (\cos \theta, \sin \theta, 0, ...0), w \rangle \gamma_n(dw) =$$

$$= \frac{\pi}{2} \cdot \frac{1}{2\pi} \int_{\mathbb{R}} \int_{\mathbb{R}} \text{sign } s \cdot \text{sign}(s \cos \theta + t \sin \theta) \exp(-\frac{1}{2}(s^2 + t^2)) ds dt =$$

$$= \frac{1}{4} \int_0^{2\pi} \int_0^\infty \text{sign}(r \cos \varphi) \cdot \text{sign}(r \cos \varphi \cos \theta + r \sin \varphi \sin \theta) \exp(-\frac{r^2}{2}) r dr d\varphi =$$

$$= \frac{1}{4} \int_0^{2\pi} \text{sign}(\cos \varphi) \cdot \text{sign}(\cos(\varphi - \theta)) d\varphi \ .$$

The integration of this simple step function gives

$$= \frac{1}{4}[(2\pi - 2\theta) - 2\theta] = \frac{\pi}{2} - \theta ,$$

hence the result. □

14.4. Now everything is prepared for proving

GROTHENDIECK'S INEQUALITY (tensor form): *There is a constant c such that*

$$\pi(\cdot; \ell_\infty^n, \ell_\infty^n) \leq c \cdot w_2(\cdot; \ell_\infty^n, \ell_\infty^n)$$

for all n. If K_G is the best such constant, then

$$K_G^R \leq \frac{\pi}{2\ln(1+\sqrt{2})} \qquad \text{(real theory)}$$

$$K_G^C \leq 2K_G^R \leq \frac{\pi}{\ln(1+\sqrt{2})} \qquad \text{(complex theory)}.$$

PROOF: The complex case will follow from the real one, which will be treated first: Take $a := \ln(1+\sqrt{2}) < 1$ which satisfies $\sinh(a) = \frac{1}{2}(\exp(a) - \exp(-a)) = 1$, and $z \in \ell_\infty^n \otimes \ell_\infty^n$ with $w_2(z; \ell_\infty^n, \ell_\infty^n) < a$. It has to be shown that $\pi(z; \ell_\infty^n, \ell_\infty^n) \leq \pi/2$. The element

$$\sin z := \sum_{k=0}^{\infty}(-1)^k \frac{z^{2k+1}}{(2k+1)!} \in \ell_\infty^n \otimes_{w_2} \ell_\infty^n$$

is defined in the Banach algebra $\ell_\infty^n \otimes_{w_2} \ell_\infty^n$ (by 14.1., recall: pointwise multiplication from $\ell_\infty^{n^2}$) and satisfies

$$w_2(\sin z; \ell_\infty^n, \ell_\infty^n) \leq \sum_{k=0}^{\infty} \frac{1}{(2k+1)!}[w_2(z; \ell_\infty^n, \ell_\infty^n)]^{2k+1} =$$

$$= \sinh[w_2(z; \ell_\infty^n, \ell_\infty^n)] < \sinh(a) = 1 .$$

Lemma 14.2. provides an $m \in \mathbb{N}$ and $x_1, ..., x_n, y_1, ..., y_n \in \ell_2^m$ of norm 1 such that

$$\langle x_k, y_\ell \rangle = \langle \sin z, e_k \otimes e_\ell \rangle = \sin[\langle z, e_k \otimes e_\ell \rangle] .$$

Since, by formula 14.3.,

$$\langle x_k, y_\ell \rangle = \sin\left[\frac{\pi}{2}\int_{\mathbb{R}^n} \text{sign}\langle x_k, w\rangle \cdot \text{sign}\langle y_\ell, w\rangle \, \gamma_n(dw)\right]$$

and

$$|\langle z, e_k \otimes e_\ell \rangle| \leq w_2(z; \ell_\infty^n, \ell_\infty^n) < a < \frac{\pi}{2}$$

$$\left|\frac{\pi}{2}\int_{\mathbb{R}^n} \cdots \gamma_n(dw)\right| \leq \frac{\pi}{2} ,$$

it follows that

$$\langle z, e_k \otimes e_\ell \rangle = \frac{\pi}{2} \int_{\mathbb{R}^n} \text{sign}\langle x_k, w\rangle \cdot \text{sign}\langle y_\ell, w\rangle \, \gamma_n(dw) \ .$$

This implies that

$$z = \sum_{k,\ell=1}^n \langle z, e_k \otimes e_\ell \rangle e_k \otimes e_\ell = \frac{\pi}{2} \int_{\mathbb{R}^n} \left(\sum_{k=1}^n \text{sign}\langle x_k, w\rangle e_k \right) \otimes \left(\sum_{\ell=1}^n \text{sign}\langle y_\ell, w\rangle e_\ell \right) \gamma_n(dw) \ .$$

Since, clearly, $\|\sum_{k=1}^n \text{sign}\langle x_k, w\rangle e_k\|_{\ell_\infty^n} \leq 1$ for all $w \in \mathbb{R}^n$, it follows that

$$\pi(z; \ell_\infty^n, \ell_\infty^n) \leq \frac{\pi}{2} \int_{\mathbb{R}^n} \pi(\cdots; \ell_\infty^n, \ell_\infty^n) \, \gamma_n(dw) \leq \frac{\pi}{2} \ ,$$

which implies the inequality in the real case.

To treat the complex case the notation $\otimes^K, \pi^K$ and w_2^K will be used. It has to be shown that

$$\pi^{\mathbb{C}} \leq 2 K_G^{\mathbb{R}} w_2^{\mathbb{C}} \quad \text{on} \quad \mathbb{C}_\infty^n \otimes^{\mathbb{C}} \mathbb{C}_\infty^n \ ,$$

where $\mathbb{K}_\infty^n$ denotes $\mathbb{K}^n$ with the ℓ_∞-norm. The $\mathbb{R}$-linear identity map $I : \mathbb{C}_\infty^n \to \mathbb{R}_\infty^{2n}$ has norm 1 and its inverse has norm $\sqrt{2}$. The space $(\mathbb{C}_\infty^n)_\mathbb{R}$ is nothing but $\mathbb{C}_\infty^n$ considered as a real space; it was shown (Ex 2.11. and Ex 3.15.) that the natural map

$$\Phi : (\mathbb{C}_\infty^n)_\mathbb{R} \otimes_\pi^\mathbb{R} (\mathbb{C}_\infty^n)_\mathbb{R} \longrightarrow (\mathbb{C}_\infty^n \otimes_\pi^\mathbb{C} \mathbb{C}_\infty^n)_\mathbb{R}$$

has norm one (it is even a metric surjection). For the w_2-norm observe first that for $x_1, ..., x_m \in \mathbb{C}_\infty^n$ Ex 8.1. implies that

$$w_2^\mathbb{R}(x_k; (\mathbb{C}_\infty^n)_\mathbb{R}) = \sup\left\{ \| \sum_{k=1}^m \lambda_k x_k \|_\infty \, \Big| \, \lambda_k \in \mathbb{R}, \sum_{k=1}^m |\lambda_k|^2 \leq 1 \right\} \leq$$

$$\leq \sup\{...|\lambda_k \in \mathbb{C}, ...\} = w_2^\mathbb{C}(x_k; \mathbb{C}_\infty^n) \ .$$

Now take $z_o \in \mathbb{C}_\infty^n \otimes^\mathbb{C} \mathbb{C}_\infty^n$ with $w_2^\mathbb{C}(z_o; \mathbb{C}_\infty^n, \mathbb{C}_\infty^n) < 1$. Then there are elements $x_1, ..., x_m$ and $y_1, ..., y_m \in \mathbb{C}_\infty^n$ with $z_o = \sum_{k=1}^m x_k \otimes^\mathbb{C} y_k$ and $w_2^\mathbb{C}(x_k) \leq 1$ as well as $w_2^\mathbb{C}(y_k) \leq 1$. Therefore

$$z_1 := \sum_{k=1}^m x_k \otimes^\mathbb{R} y_k \in (\mathbb{C}_\infty^n)_\mathbb{R} \otimes^\mathbb{R} (\mathbb{C}_\infty^n)_\mathbb{R}$$

satisfies $\Phi(z_1) = z_o$ and $w_2^\mathbb{R}(z_1; (\mathbb{C}_\infty^n)_\mathbb{R}, (\mathbb{C}_\infty^n)_\mathbb{R}) \leq 1$. It follows that

$$\pi^{\mathbb{C}}(z_0;\mathbb{C}_\infty^n,\mathbb{C}_\infty^n) \leq \pi^{\mathbb{R}}(z_1;(\mathbb{C}_\infty^n)_{\mathbb{R}},(\mathbb{C}_\infty^n)_{\mathbb{R}}) \leq \|I^{-1}\|^2 \pi^{\mathbb{R}}(z_1;\mathbb{R}_\infty^{2n},\mathbb{R}_\infty^{2n}) \leq$$
$$\leq 2K_G^{\mathbb{R}} w_2^{\mathbb{R}}(z_1;\mathbb{R}_\infty^{2n},\mathbb{R}_\infty^{2n}) \leq 2K_G^{\mathbb{R}} w_2^{\mathbb{R}}(z_1;(\mathbb{C}_\infty^n)_{\mathbb{R}},(\mathbb{C}_\infty^n)_{\mathbb{R}}) \leq 2K_G^{\mathbb{R}}$$

which ends the proof. □

14.5. By 1–complementation it is clear that $\pi \leq K_G w_2$ on $\ell_\infty^n \otimes \ell_\infty^m$ for all $n, m \in \mathbb{N}$, hence the $\mathfrak{L}_p$-local technique lemma 13.5. gives that:

COROLLARY 1: *If E is a $\mathfrak{L}_{\infty,\lambda}^g$-space and F a $\mathfrak{L}_{\infty,\mu}^g$-space, then*

$$w_2 \leq \pi \leq K_G \lambda \mu w_2 \qquad \text{on} \quad E \otimes F .$$

In particular,

COROLLARY 2: *For compact spaces K and L*

$$C(K) \tilde{\otimes}_\pi C(L) = C(K) \tilde{\otimes}_{w_2} C(L) \subset C(K \times L) .$$

More precisely, if for $f \in C(K \times L)$ there are functions $g \in C(K, \ell_2)$ and $h \in C(L, \ell_2)$ such that
$$f(s,t) = \langle g(s), h(t) \rangle$$
for all $(s,t) \in K \times L$, then $f \in C(K) \tilde{\otimes}_\pi C(L)$ and

$$\pi(f;C(K),C(L)) \leq K_G \|g\|_\infty \|h\|_\infty .$$

PROOF: Since, by the preceding corollary, $\pi \leq K_G w_2$ on $C(K) \otimes C(L)$, it follows that the natural map

$$C(K) \tilde{\otimes}_\pi C(L) = C(K) \tilde{\otimes}_{w_2} C(L) \hookrightarrow C(K) \tilde{\otimes}_\epsilon C(L) = C(K \times L)$$

is injective. By 12.10. the assumption on the function f means that $f \in C(K) \tilde{\otimes}_{w_2} C(L)$ and $w_2(f;C(K),C(L)) \leq \|g\|_\infty \|h\|_\infty$, which implies the statement. □

Recall that $f \in C(K) \tilde{\otimes}_\pi C(L)$ means that f has an absolutely convergent representation

$$f = \sum_{n=1}^\infty g_n \otimes h_n \in C(K) \tilde{\otimes}_\pi C(L) ,$$

which easily gives a representation $f = \langle g, h \rangle$ as in the corollary.

COROLLARY 3: *If H is a Hilbert space and $x_1, ..., x_n, y_1, ..., y_n \in H$, then*

$$\pi\left(\sum_{i,j=1}^n (x_i|y_j)_H e_i \otimes e_j; \ell_\infty^n, \ell_\infty^n\right) \leq K_G \max_i \|x_i\| \max_j \|y_j\| .$$

For $H = \ell_2$ or L_2 it does not matter whether one writes the scalar product $(\cdot | \cdot\cdot)$ or the duality bracket $\langle \cdot, \cdot\cdot \rangle$, since, clearly, there are elements $\tilde{y}_j$ with $\|\tilde{y}_j\| = \|y_j\|$ and $(x_i|y_j) = \langle x_i, \tilde{y}_j\rangle$.

PROOF: For $H = \ell_2$ define $g(i) := x_i$ and $h(j) := y_j$ and $z := \sum_{i,j}\langle x_i, y_j\rangle e_i \otimes e_j$ in $\ell_\infty^n \otimes \ell_\infty^n$. Then
$$\langle z, e_i \otimes e_j\rangle = \langle g(i), h(j)\rangle ,$$
and hence
$$\pi(z; \ell_\infty^n, \ell_\infty^n) \leq K_G \max_i \|g(i)\| \max_j \|h(j)\|$$
by corollary 2. □

An immediate consequence is the Lindenstrauss–Pełczyński formulation:

GROTHENDIECK'S INEQUALITY (matrix form): *Let $(a_{i,j}) \in \mathcal{L}(\mathbb{K}^n, \mathbb{K}^n)$ be an $n \times n$ matrix. Then for each Hilbert space H*
$$\sup\Big\{\Big|\sum_{i,j=1}^n a_{i,j}(x_i|y_j)\Big|\ \Big|\ x_i, y_j \in B_H\Big\} \leq$$
$$\leq K_G \sup\Big\{\Big|\sum_{i,j=1}^n a_{i,j}s_i t_j\Big|\ \Big|\ s_i, t_j \in B_{\mathbb{K}}\Big\} .$$

PROOF: Define $\varphi \in (\ell_\infty^n \otimes_\pi \ell_\infty^n)'$ by
$$\langle \varphi, e_i \otimes e_j\rangle := a_{i,j} ;$$
then
$$\|\varphi\| = \sup\Big\{\Big|\sum_{i,j=1}^n a_{i,j}s_i t_j\Big|\ \Big|\ (s_1,...,s_n), (t_1,...,t_n) \in B_{\ell_\infty^n}\Big\} .$$
Hence for $x_i, y_j \in B_H$
$$\Big|\sum_{i,j} a_{i,j}(x_i|y_j)\Big| = \Big|\langle \varphi, \sum_{i,j}(x_i|y_j)e_i \otimes e_j\rangle\Big| \leq$$
$$\leq \|\varphi\|\pi\Big(\sum_{i,j}(x_i|y_j)e_i \otimes e_j; \ell_\infty^n, \ell_\infty^n\Big) \leq \|\varphi\|K_G$$

by the preceding corollary. This is the claim. □

14.6. Note that each of the last four statements was deduced directly from the former one with the same constant; hence the best constants in the last four statements are non–increasing. Actually they are the same:

PROPOSITION: *The best constants in*
— *Grothendieck's inequalitiy, tensor form*
— *Grothendieck's inequalitiy, function form (corollary 2)*
— *corollary 3*
— *Grothendieck's inequalitiy, matrix form*
all coincide.

In particular, it will be shown that all of the last four statements are equivalent formulations of the same fact.

PROOF: First observe that for $x_1, ..., x_n, y_1, ..., y_n \in B_{\ell_2}$ and $z := \sum_{i,j} \langle x_i, y_j \rangle e_i \otimes e_j$ in $\ell_\infty^n \otimes \ell_\infty^n$,

$$\pi(z; \ell_\infty^n, \ell_\infty^n) = \sup\left\{ |\langle \varphi, z \rangle| \,\Big|\, \|\varphi\|_{(\ell_\infty^n \otimes_\pi \ell_\infty^n)'} \leq 1 \right\} =$$
$$= \sup\left\{ |\sum_{i,j} \langle x_i, y_j \rangle \langle \varphi, e_i \otimes e_j \rangle| \,\Big|\, \|\varphi\|_{(\ell_\infty^n \otimes_\pi \ell_\infty^n)'} \leq 1 \right\} \leq K_G ,$$

the constant K_G being taken from the matrix form. Now use the tensor contraction

$$(\ell_\infty^n \otimes_\varepsilon \ell_2) \otimes_\pi (\ell_2 \otimes_\varepsilon \ell_\infty^n) \longrightarrow \ell_\infty^n \otimes_{w_2} \ell_\infty^n$$
$$\left(\sum_i e_i \otimes x_i\right) \otimes \left(\sum_j y_j \otimes e_j\right) \rightsquigarrow \sum_{i,j} \langle x_i, y_j \rangle e_i \otimes e_j$$

which is a metric surjection: For $w_2(z; \ell_\infty^n, \ell_\infty^n) < 1$ there are (by 14.2. or 12.9.) $x_1, ..., x_n$ and $y_1, ... y_n \in B_{\ell_2}$ with

$$z = \sum_{i,j} \langle x_i, y_j \rangle e_i \otimes e_j ,$$

hence $\pi(z; \ell_\infty^n, \ell_\infty^n) \leq K_G$, which implies the tensor form with the constant from the matrix form. □

14.7. Considerable effort has been made to find the exact values of the *Grothendieck constants* $K_G^{\mathbb{R}}$ and $K_G^{\mathbb{C}}$ but they are still unknown. A lower estimate (due to Grothendieck) can be obtained as follows. Denote for the moment by K_{LG} the best constant in the little Grothendieck theorem (11.11.); then:

PROPOSITION: $K_{LG}^2 \leq K_G$.

PROOF: For $\varepsilon > 0$ there is an $n \in \mathbb{N}$ and $T \in \mathcal{L}(\ell_\infty^n, \ell_2)$ with $\|T\| = 1$ and $\mathbf{P}_2(T) > K_{LG} - \varepsilon$; this establishes the existence of $x_1, ..., x_m \in \ell_\infty^n$ with

$$\sum_{k=1}^m \|Tx_k\|^2 \geq (K_{LG} - \varepsilon)^2 \text{ and } w_2(x_1, ..., x_m; \ell_\infty^n) = 1 .$$

The functional $\varphi \in (\ell_\infty^n \otimes_\pi \ell_\infty^n)'$ defined by $\langle \varphi, x \otimes y \rangle := \langle Tx, Ty \rangle$ has norm 1 and therefore (with some η_k of modulus 1 in the complex case)

$$(K_{LG} - \varepsilon)^2 \leq \sum_{k=1}^m \|Tx_k\|^2 = \langle \varphi, \sum_{k=1}^m \eta_k x_k \otimes x_k \rangle \leq \pi \Big(\sum_{k=1}^m \eta_k x_k \otimes x_k; \ell_\infty^n, \ell_\infty^n \Big)$$

$$\leq K_G w_2 \Big(\sum_{k=1}^m \eta_k x_k \otimes x_k; \ell_\infty^n, \ell_\infty^n \Big) \leq K_G \left(w_2(x_k; \ell_\infty^n) \right)^2 = K_G \,. \quad \Box$$

It will be shown in 20.19 that the best (real and complex) K_{LG} are actually the ones given in 11.11. : $K_{LG}^{I\!R} = \sqrt{\pi/2}$ and $K_{LG}^{\mathcal{C}} = 2/\sqrt{\pi}$, hence

$$1.57079... = \frac{\pi}{2} \leq K_G^{I\!R} \leq \frac{\pi}{2\ln(1+\sqrt{2})} = 1.78221...$$

$$1.27323... = \frac{4}{\pi} \leq K_G^{\mathcal{C}} \leq \frac{\pi}{\ln(1+\sqrt{2})} = 3.56442...$$

Pisier [220] showed that $K_G^{\mathcal{C}} \leq \exp(1-\gamma) = 1.5262...$ (where γ is Euler's constant) and Haagerup [98] improved this to $K_G^{\mathcal{C}} \leq 1.40491$ using a representation of the scalar product

$$(x|y) = \Phi^{-1}\Big[\int_{\mathcal{C}^n} \text{sign}\langle x, w \rangle \cdot \overline{\text{sign}\langle y, w \rangle} \gamma_n(dw) \Big]$$

for $x, y \in \ell_2$ (complex) with $\|x\|_2 = \|y\|_2 = 1$ (sign denotes the complex sign function) which he deduced from the representation of the real scalar product. The difficulty is that the function

$$\Phi(\lambda) := \lambda \int_0^{\pi/2} \frac{\cos^2 t}{\sqrt{1 - |\lambda|^2 \sin^2 t}} \, dt \qquad \lambda \in \mathcal{C}, |\lambda| \leq 1$$

is much more complicated to handle than the inverse of the sine function, but apart from these tough analytical calculations, the proof runs along the same lines as the proof of the real case since Φ^{-1} can be expressed as a power series (in λ and $\bar{\lambda}$).

In particular, these estimates show that $K_G^{\mathcal{C}} < K_G^{I\!R}$; Krivine [154] showed directly by an estimate of the norm of the complexification of a real operator $\ell_\infty^n \to \ell_1^n$ that $K_G^{\mathcal{C}} \leq \sqrt{2} K_G^{I\!R}$ (see Ex 14.4. for a slightly worse estimate). Concerning the lower bounds it seems that Krivine showed that $K_G^{I\!R} \geq 1.676$ and Davie that $K_G^{\mathcal{C}} \geq 1.3381$, but these results are unpublished (see Pisier [225], p.68, Jameson [119], p.114 and König [148]).

14.8. We resist the temptation of giving the operator form of Grothendieck's inequality now (see 17.14.) and prefer instead to continue with the theory of tensor norms in order to discover the precise connection between operator ideals and tensor norms. Nevertheless, we prove a result, which will be seen to be nothing but $\mathfrak{L}(\ell_1, \ell_2) = \mathfrak{P}_1(\ell_1, \ell_2)$.

PROPOSITION: $\pi \leq K_G d_\infty$ on $\ell_1^n \otimes \ell_2^n$ for all $n \in \mathbb{N}$.

PROOF: For $x_i \in \ell_1^n$ with $x_i = \sum_{j=1}^n a_{i,j} e_j$

$$w_1(x_1, ..., x_m; \ell_1^n) = \sup\left\{\sum_{i=1}^m |\langle x_i, (t_j)\rangle| \,\Big|\, |t_j| \leq 1\right\} =$$

$$= \sup\left\{|\sum_{i=1}^m \sum_{j=1}^n a_{i,j} t_j s_i| \,\Big|\, |t_j| \leq 1, |s_i| \leq 1\right\}.$$

Hence, for every $z = \sum_{i=1}^m x_i \otimes y_i \in \ell_1^n \otimes \ell_2^n$

$$\pi(z; \ell_1^n, \ell_2^n) = \pi\left(\sum_{j=1}^n e_j \otimes \sum_{i=1}^m a_{i,j} y_i; \ell_1^n, \ell_2^n\right) = \sum_{j=1}^n \|\sum_{i=1}^m a_{i,j} y_i\|_2 =$$

$$= \sup\left\{|\sum_{j=1}^n \sum_{i=1}^m a_{i,j} \langle y_i, z_j\rangle| \,\Big|\, z_j \in B_{\ell_2^n}\right\} \leq$$

$$\leq K_G w_1(x_1, ..., x_m; \ell_1^n) \max_i \|y_i\|_2$$

by Grothendieck's inequality (matrix form). Taking the infimum over all representations of z gives $\pi(z; \ell_1^n, \ell_2^n) \leq K_G d_\infty(z; \ell_1^n, \ell_2^n)$. □

It will turn out that the best constant in this relation is also the Grothendieck constant K_G (see 17.14. and 20.17.). For other versions of Grothendieck's inequality see 17.14., 19.5., 20.17., 22.4., 23.10. and Ex 23.18.. Grothendieck's original formulation will be presented in 20.17.. The so-called "finite dimensional" Grothendieck constants will be treated briefly in 20.17. as well. Other proofs of Grothendieck's inequality will be given in Ex 14.1., 31.6. and Ex 32.7..

Exercises:

Ex 14.1. Take elements $x_1, ..., x_n, y_1, ..., y_n \in \mathbb{R}^m$ of euclidean norm 1 and denote by $\Theta(x_i, y_j) \in [0, \pi]$ the angle between x_i and y_j.
(a) Prove that for all $k \in \mathbb{N}$

$$\pi\left(\sum_{i,j=1}^n \left(\frac{\pi}{2} - \Theta(x_i, y_j)\right)^k e_i \otimes e_j; \ell_\infty^n, \ell_\infty^n\right) \leq$$

$$\leq \left[\pi\left(\sum_{i,j=1}^n \left(\frac{\pi}{2} - \Theta(x_i, y_j)\right) e_i \otimes e_j; \ell_\infty^n, \ell_\infty^n\right)\right]^k \leq \left(\frac{\pi}{2}\right)^k.$$

Hint: 14.1. and 14.3..

(b) Check that
$$\langle x_i, y_j \rangle = \sum_{k=0}^{\infty} \frac{(-1)^k}{(2k+1)!} \left(\frac{\pi}{2} - \Theta(x_i, y_j)\right)^{2k+1} \quad i, j = 1, ..., n.$$

(c) Prove that
$$\pi\left(\sum_{i,j=1}^{n} \langle x_i, y_j \rangle e_i \otimes e_j; \ell_{\infty}^n, \ell_{\infty}^n\right) \leq \sinh\left(\frac{\pi}{2}\right).$$

(d) Show that these arguments prove Grothendieck's inequality in tensor form with $K_G^\mathbb{R} \leq \sinh(\pi/2)$. Hint: 14.6.. This is a slight modification of Grothendieck's original proof of the "théorème fondamental".

Ex 14.2. Let K and L be compact sets, $\varphi \in \mathfrak{Bil}(C(K), C(L))$ and $g_k \in C(K), h_k \in C(L)$ be given. Then
$$\left|\sum_{k=1}^{m} \varphi(g_k, h_k)\right| \leq K_G \|\varphi\| \max_{t \in K}\left(\sum_{k=1}^{m} |g_k(t)|^2\right)^{1/2} \max_{s \in L}\left(\sum_{k=1}^{m} |h_k(s)|^2\right)^{1/2}.$$
What happens if φ comes from a signed measure on $K \times L$?

Ex 14.3. Let S_{n-1} be the unit sphere of ℓ_2^n and $s_n(x, y) := (x|y)$ for $x, y \in S_{n-1}$.
(a) Show that $s_n \in C(S_{n-1}) \otimes C(S_{n-1})$ and
$$w_2(s_n; C(S_{n-1}), C(S_{n-1})) \leq 1, \quad \pi(s_n; C(S_{n-1}), C(S_{n-1})) \leq K_G.$$
(b) Prove that for all $x_1, ..., x_n, y_1, ..., y_n \in S_{n-1}$
$$\pi\left(\sum_{i,j=1}^{n} (x_i|y_j) e_i \otimes e_j; \ell_{\infty}^n, \ell_{\infty}^n\right) \leq \pi(s_n; C(S_{n-1}), C(S_{n-1})).$$
Hint: $\pi\left(\sum_{i,j} g(x_i)h(y_j) e_i \otimes e_j; \ell_{\infty}^n, \ell_{\infty}^n\right) \leq \|g\|_\infty \|h\|_\infty$ for all $g, h \in C(S_{n-1})$.
(c) Conclude that $\lim_{n \to \infty} \pi(s_n; C(S_{n-1}), C(S_{n-1})) = K_G$. Hint: Corollary 3. For a more precise statement of this result see Ex 20.16. (and 20.15.).

Ex 14.4. Recall from Ex 11.19.(d) that for every $\lambda \in \mathbb{C}$
$$|\lambda| = \frac{1}{4}\int_0^{2\pi} |\text{Re}(e^{it}\lambda)| dt.$$
(a) If $A: \mathbb{R}_\infty^n \to \mathbb{R}_1^n$ is an $\mathbb{R}$-linear operator and $A^\mathbb{C}$ its complexification $\mathbb{C}_\infty^n \to \mathbb{C}_1^n$, then
$$\|A^\mathbb{C}: \mathbb{C}_\infty^n \to \mathbb{C}_1^n\| \leq \frac{\pi}{2} \|A: \mathbb{R}_\infty^n \to \mathbb{R}_1^n\|.$$
See also 26.3. and Ex 28.14. for the complexification of operators.
(b) Show that $K_G^\mathbb{R} \leq \frac{\pi}{2} K_G^\mathbb{C}$. Hint: Matrix form.
(c) For $A := \begin{pmatrix} 1 & 1 \\ 1 & -1 \end{pmatrix}: \mathbb{R}_\infty^2 \to \mathbb{R}_1^2$ the following holds: $\|A^\mathbb{C}\| = \sqrt{2}\|A\|$. This shows that the best constant in (a) is $\geq \sqrt{2}$. Krivine [154] calculated that it is actually $\sqrt{2}$ and hence obtains $K_G^\mathbb{R} \leq \sqrt{2} K_G^\mathbb{C}$.

15. Dual Tensor Norms

The purpose of this section is to study the norm which is induced on $E' \otimes F'$ by its embedding into $(E \otimes_\beta F)'$. This investigation is quite fundamental for applying tensor product methods to operator ideals, as will be seen shortly. It also gives new insight into the importance of the cofinite hull of a tensor norm. Moreover, some notions ("accessibility") concerning the relationship between $\overrightarrow{\alpha}$ and $\overleftarrow{\alpha}$ are defined and tensor norms close to Δ_p are investigated.

15.1. For two separating dual pairings $\langle E_i, F_i \rangle$ of normed spaces the bilinear mapping

$$(E_1 \otimes E_2) \times (F_1 \otimes F_2) \longrightarrow \mathbb{K}$$

$$\left(\sum_{n=1}^{N} x_n^1 \otimes x_n^2, \sum_{m=1}^{M} y_m^1 \otimes y_m^2 \right) \rightsquigarrow \sum_{n=1}^{N} \sum_{m=1}^{M} \langle x_n^1, y_m^1 \rangle \langle x_n^2, y_m^2 \rangle$$

gives a separating dual pairing $\langle E_1 \otimes E_2, F_1 \otimes F_2 \rangle$ according to Ex 2.4.. This simple and natural pairing is also called *trace duality* since it is the restriction of the trace duality studied in 2.6.: Clearly, the duality bracket induces a mapping Φ

$$\Phi : F_1 \otimes F_2 \longrightarrow (E_1 \otimes E_2)^* = L(E_1, E_2^*)$$

which is injective (since the pairing is separating) and the elements of the range of Φ are certain finite rank operators $E_1 \to E_2^*$. For $u \in E_1 \otimes E_2$ and $v \in F_1 \otimes F_2$ let $T_u \in \mathfrak{F}(E_1', E_2)$ be the operator associated with u and $L_w \in \mathfrak{F}(E_2, E_1')$ the operator associated with $w := \Phi(v)^t \in (E_2 \otimes E_1)^*$. Then

$$\langle u, v \rangle_{E_1 \otimes E_2, F_1 \otimes F_2} = \langle \Phi(v), u \rangle_{(E_1 \otimes E_2)^*, E_1 \otimes E_2} = tr_{E_2}(T_u \circ L_w) = tr_{E_1'}(L_w \circ T_u) \ .$$

Note again that transposing $z \in E \otimes F \hookrightarrow \mathfrak{L}(E', F)$ means passing to the dual of the associated operator T_z — more or less: $T_z' = \kappa_E \circ T_{z^t}$.

15.2. The interesting cases for the (restricted) trace duality are the injections

$$\begin{array}{cccc}
E \otimes F & \hookrightarrow (E' \otimes_\varepsilon F')' & \hookrightarrow (E' \otimes_\beta F')' & \hookrightarrow (E' \otimes_\pi F')' \\
E' \otimes F & \hookrightarrow (E \otimes_\varepsilon F')' & \hookrightarrow (E \otimes_\beta F')' & \hookrightarrow (E \otimes_\pi F')' \\
E' \otimes F' & \hookrightarrow (E \otimes_\varepsilon F)' & \hookrightarrow (E \otimes_\beta F)' & \hookrightarrow (E \otimes_\pi F)'
\end{array}$$

(β any tensor norm). Which norm is induced on $E \otimes F$ by $(E' \otimes_\beta F')'$? To answer this question dual tensor norms will be introduced. Recall that

$$M \otimes N = (M' \otimes_\alpha N')'$$

holds algebraically for all $M, N \in$ FIN.

PROPOSITION: *Let α be a tensor norm on* FIN. *Then α' defined by*
$$\alpha'(z; M, N) := \sup\{|\langle u, z \rangle| \mid \alpha(u; M', N') \leq 1\}$$
(for all $z \in M \otimes N$) is a tensor norm on FIN.

PROOF: To apply the criterion 12.2. (for FIN), observe first that $\alpha'(\cdot; M, N)$ is a norm, (2) is obvious and the mapping property (3) follows from
$$\langle (T_1 \otimes T_2) z, u \rangle = \langle z, (T_1' \otimes T_2') u \rangle \quad . \quad \square$$

In other words,
$$M \otimes_{\alpha'} N := (M' \otimes_\alpha N')'$$
holds isometrically by definition. The finite hull $\vec{\alpha'}$ of α' will be called the *dual tensor norm* α' (on NORM) of the tensor norm α (on FIN or NORM).

15.3. The following properties are obvious:
(1) If $\alpha \leq c\beta$, then $\beta' \leq c\alpha'$.
(2) $\alpha = \alpha''$ on FIN and $\vec{\alpha} = \alpha''$.
(3) $\alpha = \alpha''$ on NORM if and only if α is finitely generated.

By dualization, the relation $\varepsilon \leq \alpha' \leq \pi$ implies for $\alpha = \varepsilon$ that
$$\varepsilon \leq \pi' \leq \varepsilon'' = \varepsilon$$
and hence
$$\pi' = \varepsilon \text{ and } \varepsilon' = \pi \; .$$
This is part of the duality relation between the projective and injective norm which was treated in section 6, where, in fact, much more was achieved. Note that for two finitely generated tensor norms α and β the relation $\beta = \alpha'$ is just a statement about finite dimensional spaces. It is obvious that $(\alpha^t)' = (\alpha')^t$ and this norm is denoted by α^*; it is called the *adjoint* or *contragradient* tensor norm of α. At this point it may not be apparent why one would want to consider α^*. The reason will become clear in 17.9., where the relationship to adjoint operator ideals will be explained.

15.4. There is an interesting connection between Saphar's tensor norms g_p and its duals:

REMARK: $g_p^* = d_p' \leq g_{p'}$ *for* $1 \leq p \leq \infty$ *(where p' is the conjugate exponent of p).*

PROOF: From $d_1 = g_1 = \pi$ it follows that $d'_1 = \varepsilon \leq g_\infty$ and $d'_\infty \leq \pi = g_1$. For $1 < p < \infty$ it has to be shown that

$$\|M \otimes_{g_{p'}} N \xhookrightarrow{} (M' \otimes_{d_p} N')'\| \leq 1$$

for all $M, N \in$ FIN since it is enough to check $d'_p \leq g_{p'}$ on finite dimensional Banach spaces. In other words, it has to be shown that

$$|\langle u, z \rangle| \leq d_p(u; M', N') g_{p'}(z; M, N)$$

for all $u \in M' \otimes N'$ and $z \in M \otimes N$. Take $u = \sum_n x'_n \otimes y'_n$ and $z = \sum_m x_m \otimes y_m$ (finite representations with all entries different from zero); then Hölder's inequality gives

$$|\langle u, z \rangle| = \left| \sum_{n,m} \langle x'_n, x_m \rangle \langle \frac{y'_n}{\|y'_n\|}, y_m \rangle \|y'_n\| \right| \leq$$

$$\leq \left(\sum_n \left| \sum_m \langle x'_n, x_m \rangle \langle \frac{y'_n}{\|y'_n\|}, y_m \rangle \right|^{p'} \right)^{1/p'} \ell_p(y'_n) \leq$$

$$\leq \left(\sum_n \sum_m |\langle x'_n, x_m \rangle|^{p'} w_p(y_k)^{p'} \right)^{1/p'} \ell_p(y'_n) =$$

$$= \left(\sum_m \|x_m\|^{p'} \sum_n |\langle x'_n, \frac{x_m}{\|x_m\|} \rangle|^{p'} \right)^{1/p'} w_p(y_m) \ell_p(y'_n) \leq$$

$$\leq \ell_{p'}(x_m) w_{p'}(x'_n) w_p(y_m) \ell_p(y'_n) \;.$$

Passing to the infimum over all representations of z and u yields, by the representation 12.7. of d_p and $g_{p'}$, the desired inequality. □

At the end of this section this remark will be used for determining the tensor norms close to Δ_p. The inequality $g^*_{p'} \leq g_p$ will be improved in a certain sense in 20.14..

15.5. It is clearly of fundamental importance to know in which sense the isometric embedding for finite dimensional spaces

$$M' \otimes_\alpha N' \xhookrightarrow{1} (M \otimes_{\alpha'} N)'$$

extends to infinite dimensional spaces. The answer is given by the

DUALITY THEOREM: *Let α be a tensor norm (on FIN). Then for all normed spaces E and F the following natural mappings are isometries:*

(1) $$E' \otimes_{\frac{1}{\alpha}} F' \xhookrightarrow{1} (E \otimes_{\alpha'} F)'$$

(2) $$E' \otimes_{\frac{1}{\alpha}} F \xhookrightarrow{1} (E \otimes_{\alpha'} F')'$$

(3) $$E \otimes_{\frac{1}{\alpha}} F \xhookrightarrow{1} (E' \otimes_{\alpha'} F')' \;.$$

PROOF: To prove (3), observe first that

$$\mathrm{FIN}(E') = \{K^0 | K \in \mathrm{COFIN}(E)\}$$

and that, for $(K, L) \in \mathrm{COFIN}(E) \times \mathrm{COFIN}(F)$

$$\langle z, u \rangle = \langle Q_K^E \otimes Q_L^F(z), u \rangle$$

whenever $z \in E \otimes F$ and $u \in K^0 \otimes L^0 \subset E' \otimes F'$. Now, by the valid duality relation for finite dimensional spaces

$$\overleftarrow{\alpha}(z; E, F) = \sup_{K,L} \alpha(Q_K^E \otimes Q_L^F(z); E/K, F/L) =$$

$$= \sup_{K,L} \sup\{|\langle Q_K^E \otimes Q_L^F(z), u \rangle| \mid \alpha'(u; K^0, L^0) < 1\} =$$

$$= \sup\{|\langle z, u \rangle| \mid \overrightarrow{\alpha'}(u; E', F') < 1\}$$

and this is exactly (3). The extension lemma 13.2. and the commutative diagram

$$\begin{array}{c} E' \otimes_{\overleftarrow{\alpha}} F \xrightarrow{1} (E'' \otimes_{\alpha'} F')' \ni \hat{\varphi} \\ \searrow \downarrow 1 \uparrow \\ (E \otimes_{\alpha'} F')' \ni \varphi \end{array}$$

imply (2) — and (1) follows in the same way. □

The proof showed that the duality theorem is nothing but a reformulation of the definition of the cofinite hull. The duality theorem indicates that the cofinite hull $\overleftarrow{\alpha}$ of a tensor norm α will be an extremely helpful device. Since $\overleftarrow{\alpha} \leq \alpha$, the theorem implies in particular that all mappings $\otimes_\alpha \to \cdots$ in the theorem ($\otimes_{\overleftarrow{\alpha}}$ replaced by $\otimes_\alpha$) are continuous and of norm 1 (if E and F are different from $\{0\}$). The approximation lemma 13.1. gives $\overleftarrow{\alpha} = \alpha$ on $E \otimes F$ if both spaces have the metric approximation property, hence $E \otimes_\alpha F \xrightarrow{1} (E' \otimes_{\alpha'} F')'$ in this case. But what is the situation in general?

15.6. To get information in this direction recall first from section 13 that $\overleftarrow{\alpha} \leq \alpha \leq \overrightarrow{\alpha}$ holds for all tensor norms α on NORM. A tensor norm α is called *right–accessible* if

$$\overleftarrow{\alpha}(\cdot; M, F) = \overrightarrow{\alpha}(\cdot; M, F)$$

for all $(M, F) \in \mathrm{FIN} \times \mathrm{NORM}$, *left–accessible* if α^t is right–accessible, and *accessible* if it is right– and left–accessible. α is called *totally accessible* if $\overleftarrow{\alpha} = \overrightarrow{\alpha}$, i.e. if α is finitely and cofinitely generated. The injective norm ε is totally accessible (this was already mentioned in 12.4.), the projective norm π is accessible by theorem 6.4. (duality of π and ε) and the duality theorem — but it is not totally accessible: The mapping

$$E' \tilde{\otimes}_\pi F' \twoheadrightarrow \mathfrak{N}(E, F') \hookrightarrow \mathfrak{J}(E, F') = (E \otimes_\varepsilon F)'$$

is in general not injective (see section 5 about the approximation property and also 6.2.), hence $\pi \neq \overleftarrow{\pi}$ by the duality theorem. It will be shown in section 21 that all $\alpha_{p,q}$ are accessible and that all $\alpha'_{p,q}$ are even totally accessible.

Is every finitely generated tensor norm accessible? This problem seems to be quite hard since — by the approximation lemma — the non-accessibility of a tensor norm can only appear on spaces without the metric approximation property. In section 31 Pisier's space P (which does not have the approximation property, see 5.2.(1)) will be used to construct a tensor norm which is neither left- nor right-accessible.

It is obvious from the definitions that the presence of accessibility facilitates the investigation of questions involving trace duality. It will turn out that in some situations an accessibility assumption is even necessary, see 25.4., proposition 1. In section 21 the accessibility properties of tensor norms will be "translated" to conditions on operator ideals. Concerning the distinction between *right-* and *left-*accessible it is worthwhile to note that in specific situations it is sometimes enough (and easier) to only check that a tensor norm is accessible from one side — and that there are even tensor norms which are only accessible from one side (see 31.6., corollary).

PROPOSITION: *Let α be a tensor norm on* NORM.
(1) α *is right-accessible if and only if α' is right-accessible.*
(2) *If α is accessible, then the transposed tensor norm α^t, the dual tensor norm α' and the adjoint tensor norm α^* are accessible.*

If α is totally accessible, α' is accessible (by (2)) but in general not totally accessible: take $\alpha = \varepsilon$ as an example.

PROOF: Clearly, only (1) needs a proof. Take α right-accessible; then the duality theorem gives

$$M' \otimes_{\alpha''} F' \overset{1}{=} M' \otimes_{\overrightarrow{\alpha}} F' \overset{1}{=} M' \otimes_{\overleftarrow{\alpha}} F' \overset{1}{=} (M \otimes_{\alpha'} F)'$$

for finite dimensional M, hence

$$M \otimes_{\alpha'} F \overset{1}{\hookrightarrow} (M \otimes_{\alpha'} F)'' \overset{1}{=} (M' \otimes_{\alpha''} F')'$$

which means that $\overleftarrow{\alpha'} = \alpha' = \overrightarrow{\alpha'}$ on $M \otimes F$ (again by the duality theorem). This shows that α' is right-accessible. Conversely, if α' is right-accessible, then $\alpha'' = \overrightarrow{\alpha}$ is right-accessible and this is the same as α being right-accessible. $\square$

15.7. Recall that it is an immediate consequence of the approximation lemma 13.1. that $\alpha = \beta$ on $E \otimes F$ whenever $\alpha = \beta$ on $M \otimes F$ for all $M \in \mathrm{FIN}(E)$ and E has the metric approximation property. This observation and the duality theorem show that *for a tensor norm α the relations*

$$E \otimes_\alpha F \stackrel{1}{=} E \otimes_{\overleftarrow{\alpha}} F \text{ and } E \tilde{\otimes}_\alpha F \stackrel{1}{\hookrightarrow} (E' \otimes_{\alpha'} F')'$$

hold in each of the following three cases:
(1) E and F have the metric approximation property.
(2) α is right-accessible and E has the metric approximation property.
(2') α is left-accessible and F has the metric approximation property.
(3) α is totally accessible.

So, "two ingredients" are necessary in order for a "good" relationship to hold between α and α'. For the bounded approximation property the relations hold isomorphically. It will be seen in the next section that for $\alpha = \pi$ the metric approximation property is in some sense even necessary for the "good" duality. Note also that each of the conditions (1) – (3) (with "metric" replaced by "bounded") implies that the natural map $E \tilde{\otimes}_\alpha F \longrightarrow E \tilde{\otimes}_\varepsilon F$ is injective, but this observation will be improved in 17.20..

The definition of the dual tensor norm (as the finite hull of what one might have expected) and the duality theorem (explaining the role of the cofinite hull) appear innocent enough. However, in the forties Schatten defined the dual norm by using the embedding $E \otimes F \subset (E' \otimes_\alpha F')'$ and ran into severe difficulties; it was Grothendieck in his Résumé who used the definition of the dual tensor norm as a finite hull — and it turned out that this is the appropriate notion for the investigations he had in mind, for example, for the study of operator ideals. By the way: the cofinite hull $\overleftarrow{\alpha}$ of a tensor norm α is, by the duality theorem, Grothendieck's norm $\|\cdot\|_\alpha$ (see [93], p.11). Nowadays, when it seems natural to study Banach spaces and operators in terms of finite dimensional spaces, the definitions of $\overrightarrow{\alpha}, \overleftarrow{\alpha}$ and α' fit easily into the accepted philosophy — a successful philosophy whose beginning was marked by Grothendieck's definitions (and results) in his Résumé 1954.

15.8. Two tensor norms being equivalent on some $E \otimes F$ is always a fruitful situation. Up to now, Khintchine's and Grothendieck's inequalities have produced the most powerful equivalences. In 13.5. the relation $\varepsilon \leq w_p \leq \lambda \varepsilon$ on $E \otimes G$ was proven to hold whenever G is an $\mathcal{L}^g_{p,\lambda}$-space. Dually:

PROPOSITION: *Let E and G be normed spaces, G an $\mathcal{L}^g_{p,\lambda}$-space. Then*

$$w'_p \leq \pi \leq \lambda w'_p \qquad \text{on } G \otimes E.$$

PROOF: For $N \in \mathrm{FIN}$ it follows from $\varepsilon = w_{p'}$ on $N' \otimes \ell^n_{p'}$ that

$$\ell^n_p \otimes_\pi N = (\ell^n_{p'} \otimes_\varepsilon N')' = (\ell^n_{p'} \otimes_{w^t_{p'}} N')' = (\ell^n_{p'} \otimes_{w_p} N')' = \ell^n_p \otimes_{w'_p} N \ .$$

The $\mathfrak{L}_p$-local technique lemma 13.5. implies the result since w'_p is, by definition, finitely generated. □

The trace on $E' \otimes E$ is always continuous with respect to the projective norm π. Even more holds in the above situation: The proposition implies that the traces

$$L_p(\mu) \otimes_{w'_p} L_{p'}(\mu) \longrightarrow \mathbb{K}$$
$$f \otimes g \rightsquigarrow \int fg\, d\mu$$

and

$$H \otimes_{w'_2} H' \longrightarrow \mathbb{K}$$
$$x \otimes x' \rightsquigarrow \langle x, x' \rangle$$

(for a Hilbert space H) are continuous. It will be seen in section 23 that the statement of the proposition is actually characteristic for $\mathfrak{L}_p^g$-spaces.

15.9. A nice analytic application is the following: In 8.8. the values of

$$\pi\left(\sum_{k=1}^n e_k \otimes e_k; \ell_p^n, \ell_q^n\right) = \mathbf{N}(id : \ell_{p'}^n \to \ell_q^n)$$

were calculated explicitly by using some averaging and the knowledge of the ε-norm of $\sum_{k=1}^n e_k \otimes e_k$ (which is the operator norm of $id : \ell_{p'}^n \to \ell_q^n$). To obtain a result of this type for general tensor norms α define for the moment

$$u^{(n)} := \sum_{k=1}^n e_k \otimes e_k \in \mathbb{K}^n \otimes \mathbb{K}^n$$

and the operators

$$S(x_1, ..., x_n) := (x_2, x_3, ..., x_n, x_1)$$
$$M_w(x_1, ..., x_n) := (\varepsilon_1(w)x_1, ..., \varepsilon_n(w)x_n)$$

for $w \in D_n = \{-1, +1\}^n$, where the ε_k are the Rademacher functions from 8.5.. It is easy to see that

(1) $$\langle z, u^{(n)} \rangle = tr(z)$$

for all $z \in \mathbb{K}^n \otimes \mathbb{K}^n$ (duality in $\langle \mathbb{K}^n \otimes \mathbb{K}^n, \mathbb{K}^n \otimes \mathbb{K}^n \rangle$). If $z = \sum_{i,j=1}^n \alpha_{ij} e_i \otimes e_j$ and $z_o := \sum_{k=1}^n \alpha_{kk} e_k \otimes e_k$ is the associated "diagonal" element, then

(2) $$\int_{D_n} M_w \otimes M_w(z)\, \mu_n(dw) = z_o$$

(where μ_n is the "Rademacher" measure on D_n) and

(3) $$\sum_{k=1}^n S^k \otimes S^k(z_o) = tr(z) \cdot u^{(n)} = \langle z, u^{(n)}\rangle u^{(n)}.$$

This is obvious for $z = e_i \otimes e_j$ and the general case follows from linearity. Now everything is prepared for demonstrating the following result whose idea is due to Garling and Gordon [79] (see also Jameson [119], p.72):

PROPOSITION: *Let E and F be $\mathbb{K}^n$ equipped with norms such that $\|M_w\| = \|S\| = 1$ (in $\mathcal{L}(E, E)$ and $\mathcal{L}(F, F)$) for all $w \in D_n$. Then*

$$\alpha\left(\sum_{k=1}^n e_k \otimes e_k; E, F\right) \cdot \alpha'\left(\sum_{k=1}^n e_k \otimes e_k; E', F'\right) = n$$

for all tensor norms α.

The conditions on E and F are satisfied for the ℓ_p–norms and more generally for Orlicz–norms on $\mathbb{K}^n$. Note further that the result implies a relationship between certain norms of the identity operators $E' \to F$ and $E \to F'$ (see section 17). A result which is a bit more general will be treated in Ex 15.4..

PROOF: Since $n = \langle u^{(n)}, u^{(n)}\rangle \leq \alpha(u^{(n)}; E, F)\alpha'(u^{(n)}; E', F')$, one inequality is obvious. To see the other one, observe first that $(E' \otimes_{\alpha'} F')' = E \otimes_\alpha F$ holds isometrically; therefore there is a $z \in E \otimes F$ with $\alpha(z; E, F) = 1$ and $\langle z, u^{(n)}\rangle = \alpha'(u^{(n)}; E', F')$. Then, by (2) and (3)

$$\alpha(u^{(n)}; E, F)\alpha'(u^{(n)}; E', F') = \langle z, u^{(n)}\rangle \cdot \alpha(u^{(n)}; E, F) =$$

$$= \alpha\left(\sum_{k=1}^n S^k \otimes S^k(z_o); E, F\right) \leq n\,\alpha(z_o; E, F) \leq$$

$$\leq n \int_{D_n} \alpha(M_w \otimes M_w(z); E, F)\mu_n(dw) \leq n$$

by the mapping property of α and $\mu_n(D_n) = 1$. $\square$

15.10. The duality relation for the "one–sided" tensor norm Δ_p is exactly what one would expect:

PROPOSITION: *For $1 \leq p \leq \infty$, any normed space E and any measure μ*

$$L_p(\mu) \otimes_{\Delta_p} E \xrightarrow{\;\;1\;\;} (L_{p'}(\mu) \otimes_{\Delta_{p'}} E')' \text{ and } L_p(\mu) \otimes_{\Delta_p} E' \xrightarrow{\;\;1\;\;} (L_{p'}(\mu) \otimes_{\Delta_{p'}} E)'$$

hold isometrically.

PROOF: Since $\Delta_1 = \pi$ and $\Delta_\infty = \varepsilon$ (see 7.1.), the claim for $p = 1, \infty$ follows from the duality theorem because ε and π are accessible and L_1 and L_∞ have the metric approximation property. For $1 < p < \infty$ the assertion is an immediate consequence of the isometric embedding

$$L_p(\mu) \tilde{\otimes}_{\Delta_p} E' = L_p(\mu, E') \overset{1}{\hookrightarrow} (L_{p'}(\mu, E))' = (L_{p'}(\mu) \otimes_{\Delta_{p'}} E)'$$

(see Appendix B12.). □

COROLLARY 1: *For every measure μ, every normed space E and every $1 \le p \le \infty$ the following inequalities hold:*

$$d_p \le \Delta_p \le g_{p'}^* \qquad\qquad \text{on } L_p(\mu) \otimes E \ .$$

PROOF: Since $d_\infty = w_1 = \varepsilon$ on $L_\infty \otimes E$ and $\pi = w_1' = g_\infty^*$ on $L_1 \otimes E$ (see 13.5. and 15.8.), the result holds for $p = 1$ and $p = \infty$. Let $1 < p < \infty$. Since d_p and Δ_p are smaller than the projective norm, the relation $d_p \le \Delta_p$ only needs to be verified for step functions

$$f = \sum_{m=1}^n \chi_{A_m} \otimes x_m \ ,$$

the (A_m) being pairwise disjoint and of positive measure (by the density lemma 7.4.). Obviously,

$$\Delta_p(f) = \ell_p(\mu(A_m)^{1/p} x_m; E)$$

and

$$d_p(f; L_p, E) \le w_{p'}(\mu(A_m)^{-1/p} \chi_{A_m}; L_p) \ell_p(\mu(A_m)^{1/p} x_m; E) \ .$$

For $g \in B_{L_{p'}}$

$$\sum_{m=1}^n |\langle \mu(A_m)^{-1/p} \chi_{A_m}, g \rangle|^{p'} = \sum_{m=1}^n \mu(A_m)^{-p'/p} |\int_{A_m} g d\mu|^{p'} \le$$
$$\le \sum_{m=1}^n \mu(A_m)^{-p'/p} \int_{A_m} |g|^{p'} d\mu \ \mu(A_m)^{p'/p} \le 1$$

and it follows that $w_{p'}(\mu(A_m)^{-1/p} \chi_{A_m}) \le 1$. This shows that $d_p \le \Delta_p$. To see that the other inequality holds use duality: By the proposition and the duality theorem there are isometries

$$\begin{array}{ccc} L_p \otimes_{\Delta_p} E & \overset{1}{\hookrightarrow} & (L_{p'} \otimes_{\Delta_{p'}} E')' \\ & & \uparrow \\ L_p \otimes_{d'_{p'}} E & \overset{1}{\hookrightarrow} & (L_{p'} \otimes_{d_{p'}} E')' \end{array}$$

and the vertical mapping has norm ≤ 1 by what was already shown. Therefore,
$$\Delta_p \leq \overleftarrow{d'_{p'}} \leq d'_{p'} = g^*_{p'}$$
and the result is proven. □

Using $g^*_{p'} \leq g_p$ from 15.4., the *Chevet–Persson–Saphar inequalities*
$$d^*_{p'} \leq d_p \leq \Delta_p \leq g^*_{p'} \leq g_p \qquad \text{on } L_p \otimes E$$
are obtained (due to [27], [208], [242]), where E is an arbitrary normed space.

COROLLARY 2: $d^*_{p'} = d_p = \Delta_p = \Delta^t_p = g^*_{p'} = g_p$ on $L_p(\Omega_1, \mu_1) \otimes L_p(\Omega_2, \mu_2)$ for arbitrary measures μ_i.

PROOF: Since $d^*_{p'} \leq d_p \leq \Delta_p \leq g^*_{p'} \leq g_p$ on $L_p(\mu_1) \otimes E$ and $g_p \leq \Delta^t_p \leq d^*_{p'}$ on $F \otimes L_p(\mu_2)$ by the Chevet–Persson–Saphar inequalities, the result follows. □

Later on (25.10.) the spaces E such that $\Delta_p = d_p$ (and so on) on $L_p \otimes E$ will be characterized.

15.11. To finish this section it will be shown that d_p and $g^*_{p'}$ are actually the tensor norms closest to Δ_p, a result of Gordon–Saphar [87].

PROPOSITION: *Let α be a tensor norm, μ a measure such that $L_p(\mu)$ is infinite dimensional and $c \geq 1$.*
(1) *If $\alpha \leq c\Delta_p$ on $L_p(\mu) \otimes E$ for all normed spaces E, then $\alpha \leq c d_p$.*
(2) *If α is finitely generated and $\Delta_p \leq c\alpha$ on $L_p(\mu) \otimes E$ for all normed spaces E, then $g^*_{p'} \leq c\alpha$.*

Actually, the condition that α is finitely generated in (2) is superfluous, see Ex 15.6.. See also 25.10. and 29.11. to understand how close d_p and Δ_p are.

PROOF: Since ℓ_p is "nicely" 1-complemented in $L_p(\mu)$ (see Ex 7.1.), it is enough to consider ℓ_p instead of $L_p(\mu)$.
(1) For the tensor contractions C_i (see 12.9.) the following diagram commutes

$$\begin{array}{ccc} (G \otimes_\varepsilon \ell_{p'}) \otimes_\pi (\ell_p \otimes_{\Delta_p} E) & \xrightarrow{C_1} & G \otimes_{d_p} E \\ {\scriptstyle id} \downarrow & & \downarrow {\scriptstyle id} \\ (G \otimes_\varepsilon \ell_{p'}) \otimes_\pi (\ell_p \otimes_\alpha E) & \xrightarrow{C_2} & G \otimes_\alpha E \end{array}.$$

Ex 12.11. showed that C_1 is a metric surjection and $\|C_2\| \leq 1$; since the left vertical map has, by assumption, norm $\leq c$, it follows that the right one also has norm $\leq c$. This means that $\alpha \leq cd_p$ on $G \otimes E$. Note that E could be fixed.

(2) If $\Delta_p \leq c\alpha$ on $\ell_p \otimes E$, then (by the duality theorem for α and the duality 15.10. for Δ_p)

$$\ell_{p'} \otimes_{\Delta_{p'}} E' \xrightarrow{\;\;1\;\;} (\ell_p \otimes_{\Delta_p} E)' \text{ and } \ell_{p'} \otimes_{\overleftarrow{\alpha'}} E' \xrightarrow{\;\;1\;\;} (\ell_p \otimes_\alpha E)'$$

and therefore $\overleftarrow{\alpha'} \leq c\Delta_{p'}$ on $\ell_{p'} \otimes E'$. From the first part of the proof it follows that $\overleftarrow{\alpha'} \leq cd_{p'}$ on $M' \otimes E'$ for all $M \in \text{FIN}$. Dualizing again yields

$$\overleftarrow{g^*_{p'}} = \overleftarrow{d'_{p'}} \leq c\overleftarrow{(\overleftarrow{\alpha'})'} = c\overleftarrow{\alpha''} = c\overleftarrow{\alpha} \leq c\alpha$$

on $M \otimes E$ for all $M \in \text{FIN}$. Since $g^*_{p'}$ and α are finitely generated, this implies that $g^*_{p'} \leq c\alpha$. □

For a fixed normed space E the proof shows that $\alpha \leq cd_p$ on all $G \otimes E$ if $\alpha \leq c\Delta_p$ on $L_p(\mu) \otimes E$ for some infinite dimensional $L_p(\mu)$; using the fact that $g^*_{p'}$ is totally accessible (this will be proven in section 21), statement (2) reads: $g^*_{p'} \leq c\alpha$ on all $G \otimes E$ if $\Delta_p \leq c\alpha$ on $L_p(\mu) \otimes E$ (where α is finitely generated).

15.12. A neat application of the fact that there are tensor norms close to Δ_p is the following alternative proof of the extension of Beckner's result given in 7.9. (we omit the case of subspaces $E_1 \otimes_{\Delta^t_q} E_2$) :

PROPOSITION: *Let $1 \leq q \leq p \leq \infty$ and $T \in \mathcal{L}(E, L_p(\mu))$ and $S \in \mathcal{L}(L_q(\nu), F)$ for some measures μ and ν. Then*

$$\|T \otimes S : E \otimes_{\Delta^t_q} L_q(\nu) \longrightarrow L_p(\mu) \otimes_{\Delta_p} F\| = \|T\| \, \|S\| \, .$$

PROOF: For $z \in E \otimes L_q$ the Chevet–Persson–Saphar inequalities and 15.4. imply that

$$\Delta_p(T \otimes S(z); L_p, F) \leq g^*_{p'}(T \otimes S(z); L_p, F) \leq g^*_{q'}(T \otimes S(z); L_p, F) \leq$$
$$\leq g_q(T \otimes S(z); L_p, F) \leq \|T\| \, \|S\| g_q(z; E, L_q) \leq$$
$$\leq \|T\| \, \|S\| \Delta^t_q(z; E, L_q) \quad . \square$$

This result does *not* hold for $p < q$: This is the content of Maurey's factorization theorem which will be treated in 18.8.. In 26.3. the reader will find an explicit counterexample and other positive results in this direction. See also section 32.

Exercises:

Ex 15.1. Let α be a tensor norm on NORM and $k \in \mathbb{N}$. Then $\alpha^{(k)}$, defined by

$$\alpha^{(k)}(z; E, F) := \sup\{|\langle u, z\rangle| \mid u \in E' \otimes F', \text{rank}(u) \leq k, \alpha(u; E', F') \leq 1\},$$

is a tensor norm on NORM. If α is finitely generated, then

$$\lim_{k \to \infty} \alpha^{(k)}(z; E, F) = \overleftarrow{\alpha'}(z; E, F)$$

for all normed E, F and $z \in E \otimes F$. Hint: Duality theorem.

Ex 15.2. Recall that there is a natural continuous mapping

$$J : E \tilde{\otimes}_\alpha F \to E \tilde{\otimes}_\varepsilon F \hookrightarrow \mathfrak{L}(E', F).$$

(a) Take $z \in E \tilde{\otimes}_\alpha F$. Then $\varphi \tilde{\otimes}_\alpha id_F(z) = 0 \in \mathbb{K} \otimes F = F$ for all $\varphi \in E'$ if and only if the associated operator $J(z) : E' \to F$ is zero.

(b) Formulate conditions on the spaces E, F and the tensor norm α in terms of accessibility and the bounded approximation property such that the following holds for each $z \in E \tilde{\otimes}_\alpha F$: If $\varphi \tilde{\otimes}_\alpha id_F(z) = 0$ for all $\varphi \in E'$, then $z = 0$. Hint: 15.7.. For another weaker condition see Ex 17.11..

Ex 15.3. (a) Show for $E = \ell_p^n$ that for every $\varepsilon > 0$ there are $R \in \mathfrak{L}(E, \ell_2^n)$ and $S \in \mathfrak{L}(\ell_2^n, E)$ with $S \circ R = id_E$ and

$$\|S\|\|R\| \leq w_2\left(\sum_{k=1}^n e_k \otimes e_k; E', E\right)(1 + \varepsilon).$$

Hint: 12.9..

(b) Show that the trace on $\ell_p \otimes \ell_{p'}$ is *not* continuous with respect to w_2' if $p \neq 2$. Hint: With $\langle tr, z\rangle = \langle u^{(n)}, z\rangle$ this would imply $w_2(u^{(n)}; \ell_{p'}^n, \ell_p^n) \leq c$ for some c and hence, by (a), $d(\ell_p^n, \ell_2^n) \leq 2c$; but $d(\ell_p^n, \ell_2^n) \to \infty$ (a well-known fact, see also Ex 26.4.). This is a proof by local technique. Later on (23.2.) a more natural result will be given: if the trace on E is w_2'-continuous, then E is isomorphic to a Hilbert space.

Ex 15.4. Using the ideas of 15.9. show that: If E and F are $\mathbb{K}^n$ equipped with some norms, then

$$n \leq \alpha\left(\sum_{k=1}^n e_k \otimes e_k; E, F\right) \alpha'\left(\sum_{k=1}^n e_k \otimes e_k; E', F'\right) \leq nc_E^2 c_F^2,$$

where

$$c_E := \sup\{\|\sum_{k=1}^n \varepsilon_k(w)\alpha_k e_{\pi(k)}\|_E \mid \|\sum_{k=1}^n \alpha_k e_k\|_E \leq 1, w \in D_n, \pi \text{ permutation}\}$$

is the *symmetry constant* of the canonical basis (e_k) of E.

Ex 15.5. Show that $L_2(\mu; H) = L_2(\mu) \tilde{\otimes}_{d_2} H = L_2(\mu) \tilde{\otimes}_{g_2} H$ holds isometrically for all Hilbert spaces. Hint: 7.2..

Ex 15.6. Use the fact that g_p^* is totally accessible (to be proven in section 21) to show that for *each* tensor norm α the inequality $\Delta_p \leq c\alpha$ on all $\ell_p \otimes E$ implies that $g_{p'}^* \leq c\alpha$.

Ex 15.7. Is the natural duality $\langle E\tilde{\otimes}_\alpha F, E' \otimes F' \rangle$ separating?

Ex 15.8. Show by using the duality theorem that

$$\overleftarrow{\alpha}(z; E', F) = \sup\left\{\alpha(Q_K^{E'} \otimes Q_L^F(z); E'/K, F/L) \;\middle|\; \begin{array}{l} K, L \in \text{COFIN} \\ K \; \sigma(E', E)\text{-closed} \end{array}\right\}$$

$$\overleftarrow{\alpha}(z; E', F') = \sup\left\{\alpha(Q_F^{E'} \otimes Q_L^{F'}(z); E'/K, F'/L) \;\middle|\; \begin{array}{l} K, L \in \text{COFIN} \\ \text{both weak}-*-\text{closed} \end{array}\right\}.$$

Ex 15.9. If E is an $\mathcal{L}_{p,\lambda}^g$-space, then $d_p \leq \lambda g_{p'}^*$ on $E \otimes F$ for all normed spaces F. Hint: 13.5..

Ex 15.10. Let α be a tensor norm.
(a) If $A \subset E\tilde{\otimes}_\alpha F$ is bounded, then A is $\sigma(E\tilde{\otimes}_\alpha F, E' \otimes F')$-bounded.
(b) The converse of (a) is true for $\alpha = \varepsilon$, but not for $\alpha = \pi$. Hint: Uniform boundedness principle.
(c) If $\alpha = \varepsilon$ or $E' \otimes F'$ is norm-dense in $(E\tilde{\otimes}_\alpha F)'$, then every $\sigma(E\tilde{\otimes}_\alpha F, E' \otimes F')$-convergent norm-bounded sequence in $E\tilde{\otimes}_\alpha F$ is weakly convergent. Hint: 4.6..

Note that if for two locally convex topologies on a vector space the convergent sequences coincide, then the Cauchy sequences also coincide (proof?). A subset A of a normed space is called *weakly conditionally compact* if every sequence in A has a weakly Cauchy subsequence. Assume now for (d), (e) and (f) that E and F are Banach spaces, F separable and α is such that every norm-bounded, $\sigma(E\tilde{\otimes}_\alpha F, E' \otimes F')$-convergent sequence in $E\tilde{\otimes}_\alpha F$ is weakly convergent. This implies, in particular, that the duality system $\langle E\tilde{\otimes}_\alpha F, E' \otimes F' \rangle$ is separating (and so $E\tilde{\otimes}_\alpha F \longrightarrow \mathcal{L}(E', F)$ is injective).
(d) A bounded subset $A \subset E\tilde{\otimes}_\alpha F$ is weakly conditionally compact if it satisfies (denote $J_1 : E\tilde{\otimes}_\alpha F \longrightarrow \mathcal{L}(E', F)$ and $J_2 : F\tilde{\otimes}_{\alpha^t} E \longrightarrow \mathcal{L}(F', E)$ the canonical maps)
  (1) $\{[J_1 z](x') | z \in A\} \subset F$ is weakly relatively compact for each $x' \in E'$.
  (2) $\{[J_2 z^t](y') | z \in A\} \subset E$ is weakly conditionally compact for each $y' \in F'$.

Hint: Take a countable $D \subset F'$ which is $\sigma(F', F)$-dense, hence dense for the Mackey topology $\mu(F', F)$. For a given sequence in A choose a subsequence (z_n) for which $([J_2 z_n^t](y'))$ converges for all $y' \in D$. For $x'_o \otimes y'_o \in E' \otimes F'$ use (1) and the $\mu(F', F)$-density to show that $\langle z_n, x'_o \otimes y'_o \rangle = \langle [J_1 z_n](x'_o), y'_o \rangle = \langle [J_2 z_n^t](y'_o), x'_o \rangle$ converges.
(e) If, additionally, E and F are reflexive, $E\tilde{\otimes}_\alpha F$ is reflexive if and only if it is weakly sequentially complete.
(f) Rosenthal's theorem (see Diestel [54]) states that ℓ_1 is not contained isomorphically in E if and only if every bounded subset of E is weakly conditionally compact. Use this to show that, if $F \neq \{0\}$ is in addition reflexive, then ℓ_1 is not contained isomorphically in $E\tilde{\otimes}_\alpha F$ if and only if it is not in E.

(g) Using only the fact that $E \otimes_\varepsilon \cdot$ respects subspaces, show that (d) and (e) hold for $\alpha = \varepsilon$ if F is not necessarily separable.

These results are due to Lewis [167] and Rivera Ortun [232]. There are many interesting situations in which $E' \tilde{\otimes}_\alpha F' = (E \tilde{\otimes}_\alpha F)'$ holds isometrically, hence the condition on $E \tilde{\otimes}_\alpha F$ is satisfied, see section 33.

16. The Bounded Approximation Property

This section resumes the study of the bounded approximation property introduced in section 5. A characterization in terms of $\pi \approx {}^t\pi$ will be given. With the aid of the Radon–Nikodým property (see Appendix D) the full duality of ε and π will be investigated — as well as the duality of the spaces of compact, nuclear, and all continuous operators.

16.1. According to definition 5.1. a normed space E has the λ-bounded approximation property (for $\lambda \geq 1$) whenever $id_E \in \lambda \overline{B_{\mathfrak{F}(E,E)}}^{\tau_{co}}$, where τ_{co} stands for the topology of uniform convergence on the absolutely convex, compact subsets of E; in other words, there is a net (T_η) of finite rank operators of norm $\leq \lambda$ such that $T_\eta x \to x$ uniformly on absolutely convex compact sets. There are two other locally convex topologies on $\mathfrak{L}(E, F)$ which will be needed:

$\tau_s :=$ the topology of pointwise convergence

$\tau_w := \sigma(\mathfrak{L}(E, F), E \otimes F')$, the topology of pointwise weak convergence.

LEMMA: *Let E and F be normed spaces.*
(1) *On bounded subsets of $\mathfrak{L}(E, F)$ the topologies τ_{co} and τ_s coincide.*
(2) $(\mathfrak{L}(E, F), \tau_s)' = E \otimes F'$ *with the obvious duality bracket*

$$\left\langle T, \sum_{k=1}^n x_k \otimes y'_k \right\rangle := \sum_{k=1}^n \langle y'_k, T x_k \rangle .$$

(3) *For bounded, convex subsets $A \subset \mathfrak{L}(E, F)$ the closures with respect to τ_{co}, τ_s and τ_w coincide:*

$$\overline{A}^{\tau_{co}} = \overline{A}^{\tau_s} = \overline{A}^{\tau_w} \subset \mathfrak{L}(E, F) .$$

PROOF: (1) is simple and was already used in 5.1.(2). To see (2) take $\varphi \in (\mathfrak{L}(E,F), \tau_s)'$; then there are $x_1, ..., x_n \in E$ with

$$|\langle \varphi, T \rangle| \leq \max_{k=1,...,n} \|Tx_k\|,$$

and hence, by the Hahn–Banach theorem, φ factors through

$$\Phi : \mathfrak{L}(E,F) \longrightarrow F^n$$
$$T \rightsquigarrow (Tx_k)_{k=1}^n .$$

Since $(F^n)' = (F')^n$, the result follows (this was Ex 5.1.). But now $\overline{A}^{\tau_w} = \overline{A}^{\tau_s}$ (for convex A) is an immediate consequence of the following well-known fact: if $\langle G, H \rangle$ is a separating dual system, then the closures of a convex set $A \subset G$ coincide with respect to all locally convex topologies τ on G satisfying $(G, \tau)' = H$. This, together with (1) gives (3). □

REMARK: *A normed space E has the λ-bounded approximation property if and only if $id_E \in \lambda \overline{B_{\mathfrak{F}(E,E)}}^{\tau_w}$.*

16.2. Since π is accessible, it follows that $\pi = \overleftarrow{\pi}$ holds on $E \otimes F$ whenever one of the spaces is finite dimensional. The approximation lemma 13.1. gives $\pi \leq \lambda \overleftarrow{\pi}$ on $E \otimes F$ if E or F has the λ-bounded approximation property. The key for proving a sort of a converse to this observation is the following lemma — which will even be stated for arbitrary tensor norms. It characterizes $\alpha \leq \lambda \overleftarrow{\alpha}$ in terms of approximating operators in $(E \otimes_\alpha F')' \subset \mathfrak{L}(E, F'')$ by finite rank operators $E \to F$:

LEMMA: *Let α be a tensor norm, E and F normed spaces. Then $\alpha \leq \lambda \overleftarrow{\alpha}$ on $E \otimes F'$ if and only if*

$$B_{(E \otimes_\alpha F')'} \subset \lambda \overline{B_{E' \otimes_{\alpha'} F}}^\sigma \subset \mathfrak{L}(E, F''),$$

where $\sigma := \sigma(\mathfrak{L}(E, F''), E \otimes F')$.

PROOF: This sounds more complicated than it is. Clearly, both $E' \otimes F$ and $(E \otimes_\alpha F')'$ are regarded as subspaces of $\mathfrak{L}(E, F'')$. For each $z \in E \otimes F'$ the duality theorem 15.5. gives

$$\overleftarrow{\alpha}(z; E, F') = \sup \left\{ |\langle T, z \rangle| \;\middle|\; T \in B_{E' \otimes_{\alpha'} F} \right\}$$

and it is straightforward that

$$\alpha(z; E, F') = \sup \left\{ |\langle T, z \rangle| \;\middle|\; T \in B_{(E \otimes_\alpha F')'} \right\}.$$

Now the result is obvious if one applies the following simple fact to the separating dual system $\langle \mathfrak{L}(E, F''), E \otimes F' \rangle$: If $\langle G, H \rangle$ is a separating dual system, A and B are subsets of G, then $A \subset \lambda B^{00}$ if and only if $B^0 \subset \lambda A^0$. Just take $A := B_{(E \otimes_\alpha F')'}$ and $B := B_{E' \otimes_{\alpha'} F}$ and apply the bipolar theorem. □

The proof shows that the result also holds with the topology σ replaced by $\tilde{\sigma} := \sigma((E \otimes_\alpha F')', E \tilde{\otimes}_\alpha F')$.

In section 21 this lemma will be used to investigate the so-called bounded α-approximation property. For the moment only $\alpha = \pi$ is interesting. From $\pi' = \varepsilon$ it follows that the relevant spaces in the lemma are

$$E' \otimes_\varepsilon F = \mathfrak{F}(E, F) \quad \text{and} \quad (E \otimes_\pi F')' = \mathfrak{L}(E, F'')$$

(operator norm in both cases), and hence $\pi \leq \lambda \overleftarrow{\pi}$ on $E \otimes F'$ if and only if

$$B_{\mathfrak{L}(E, F'')} \subset \overline{\lambda B_{\mathfrak{F}(E, F)}}^\sigma$$

(with $\sigma := \sigma(\mathfrak{L}(E, F''), E \otimes F')$). In particular, $\pi \leq \lambda \overleftarrow{\pi}$ on $E \otimes E'$ implies

$$id_E \in B_{\mathfrak{L}(E, E)} \subset \overline{\lambda B_{\mathfrak{F}(E, E)}}^\sigma \cap \mathfrak{L}(E, E) = \overline{\lambda B_{\mathfrak{F}(E, E)}}^{T_w} = \overline{\lambda B_{\mathfrak{F}(E, E)}}^{T_{co}}$$

(by 16.1.), hence E has the λ-bounded approximation property. Collecting all this, the following result has been proved:

THEOREM: *A normed space E has the λ-bounded approximation property if and only if*

$$\pi \leq \lambda \overleftarrow{\pi} \quad \text{on } E \otimes F$$

for all normed spaces F (or only $F = E'$). In particular, E has the metric approximation property if and only if

$$\pi = \overleftarrow{\pi} \quad \text{on } E \otimes F$$

for all normed spaces F (or only $F = E'$).

16.3. From the structural point of view this result is very important. Some immediate consequences are:

COROLLARY 1: *If for a normed space E the dual E' has the bounded approximation property, then E has the bounded approximation property as well, with the same constant.*

PROOF: By the easy part of the theorem (which was just the approximation lemma) $\pi \leq \lambda \overleftarrow{\pi}$ on $E' \otimes E$. Since $\pi = \pi^t$ the other part of the theorem gives the result. □

In the following statement the constants for the bounded approximation property are omitted — since they will be clear anyhow.

COROLLARY 2: *Let E be a normed space.*
(1) *E has the metric (resp. bounded) approximation property if and only if*

$$E \otimes_\pi F \hookrightarrow (E' \otimes_\varepsilon F')'$$

holds isometrically (resp. isomorphically) for all normed spaces F (or only $F = E'$); moreover, this is equivalent to saying that

$$E \otimes_\pi F' \hookrightarrow (E' \otimes_\varepsilon F)'$$

holds isometrically (resp. isomorphically) for all normed spaces F (or only $F = E$).
(2) *E' has the metric (resp. bounded) approximation property if and only if*

$$E' \otimes_\pi F \hookrightarrow (E \otimes_\varepsilon F')'$$

holds isometrically (resp. isomorphically) for all F (or only $F = E''$); this is also equivalent to the statement:

$$E' \otimes_\pi F' \hookrightarrow (E \otimes_\varepsilon F)'$$

holds isometrically (resp. isomorphically) for all F (or only $F = E'$).

PROOF: This is an immediate consequence of the theorem and the duality theorem 15.5.. □

Clearly, these injections extend to the completions isometrically, resp. isomorphically. For the embedding of the nuclear into the integral operators

$$E' \tilde{\otimes}_\pi F \overset{1}{\twoheadrightarrow} \mathfrak{N}(E,F) \hookrightarrow \mathfrak{I}(E,F) \overset{1}{\hookrightarrow} (E \otimes_\varepsilon F')'$$

this implies: If E' or F has the λ-bounded approximation property, then

$$\mathbf{I}(T) \leq \mathbf{N}(T) \leq \lambda \mathbf{I}(T)$$

for all $T \in \mathfrak{N}(E,F)$. Under what conditions is $\mathfrak{N}(E,F) = \mathfrak{I}(E,F)$? To answer this question the Radon–Nikodým property is the appropriate tool.

16.4. A Banach space E possesses the *Radon–Nikodým property* if for every finite measure μ every operator $T : L_1(\Omega, \mu) \to E$ is *representable*, i.e. there exists a bounded μ-measurable function $g : \Omega \to E$ with

$$Tf = \int fg \, d\mu \qquad \text{for all } f \in L_1(\Omega, \mu) \,.$$

The results needed concerning representable operators and the Radon–Nikodým property can be found in Appendix C and D, as well as the notion of Pietsch integral operators. The Radon–Nikodým property has powerful consequences for the questions treated in this section. For tensor norms a variant of the Radon–Nikodým property will be investigated in Chapter III, section 33.

COROLLARY 3: *Let E be a Banach space with the Radon–Nikodým property which is λ-complemented in its bidual (i.e. there is a projection $P : E'' \twoheadrightarrow E$ with $\|P\| \leq \lambda$). If E has the approximation property, then it has the bounded approximation property with constant λ.*

PROOF: It follows from Appendix D7. that $\mathfrak{N}(E, E)$ coincides isometrically with the space $\mathfrak{PI}(E, E)$ of Pietsch integral operators. The factorization properties of integral and Pietsch integral operators and the existence of a projection $E'' \to E$ imply that $\mathfrak{I}(E, E) = \mathfrak{PI}(E, E)$ and $\mathbf{PI}(T) \leq \lambda \mathbf{I}(T)$. Since E has the approximation property, it follows that

$$E' \tilde{\otimes}_\pi E \stackrel{1}{=} \mathfrak{PI}(E, E) = \mathfrak{I}(E, E) \stackrel{1}{\hookrightarrow} (E \otimes_\varepsilon E')'$$

and hence, by the duality theorem,

$$\pi(z; E', E) = \mathbf{PI}(T_z) \leq \lambda \mathbf{I}(T_z) = \lambda \overleftarrow{\pi}(z; E', E) ;$$

therefore, the theorem implies corollary 3. □

Reflexive spaces and separable dual spaces have the Radon–Nikodým property and are clearly 1–complemented in their biduals, hence:

COROLLARY 4: *If E is reflexive or a dual space which is separable, then E has the metric approximation property if it has the approximation property.*

16.5. The Radon–Nikodým property is intimately connected with the duality of the tensor norms ε and π.

PROPOSITION: *Let F be a Banach space. Then F' has the Radon–Nikodým property if and only if the canonical mapping*

$$J_E : E' \tilde{\otimes}_\pi F' \longrightarrow (E \otimes_\varepsilon F)'$$

is surjective for all Banach spaces E (or only $E = C(K)$ with K compact, or only $L_\infty(\mu)$ for arbitrary finite measures). In this case, J_E is even a metric surjection for all Banach spaces E.

PROOF: Surjectivity of J_E means that $\mathfrak{N}(E, F') = \mathfrak{I}(E, F')$ by the very definitions of nuclear and integral operators, so the result is nothing but a reformulation of corollary 1 in Appendix D7.. □

The result cited from the appendix even implies that in order for F' to have the Radon–Nikodým property it is enough to take only $E = C[0,1]$ or only $E = L_\infty[0,1]$ (Lebesgue measure) in the proposition.

16.6. Combining the results of 16.4. and 16.5. easily gives the following result on the duality of the projective and injective norm:

THEOREM: *Let E and F be Banach spaces.*
(1) *If E' or F' has the approximation property and: E' or F' has the Radon–Nikodým property, then*
$$E' \tilde\otimes_\pi F' = (E \otimes_\varepsilon F)' \qquad \text{isometrically.}$$
(2) *Conversely: If $E' \tilde\otimes_\pi G' = (E \otimes_\varepsilon G)'$ (as vector spaces) for all Banach spaces G, then E' has the Radon–Nikodým property and metric approximation property.*

PROOF: Since E' or F' has the approximation property, the canonical map
$$J : E' \tilde\otimes_\pi F' \longrightarrow (E \otimes_\varepsilon F)'$$
is injective (see 5.6.), hence proposition 16.5. gives (1). For the converse, proposition 16.5. again shows that E' has the Radon–Nikodým property. The injectivity of the mapping $E' \tilde\otimes_\pi G' \longrightarrow (E \otimes_\varepsilon G)'$ for all G implies that E' has the approximation property (again by theorem 5.6.) and hence the metric approximation property by corollary 3 from 16.4.. □

16.7. These results have nice applications for spaces of operators. From section 5 on the approximation property it is known that
$$\mathfrak{N}(E,F) \stackrel{1}{=} E' \tilde\otimes_\pi F \quad \text{and} \quad \mathfrak{K}(E,F) \stackrel{1}{=} E' \tilde\otimes_\varepsilon F$$
if E' or F have the approximation property. Proposition 16.5. gives
$$\mathfrak{N}(E,F') = (E \tilde\otimes_\varepsilon F)'$$
if E' or F' has the Radon–Nikodým property and, clearly,
$$\mathfrak{L}(E,F') = (E \tilde\otimes_\pi F)'.$$
The duality is always the natural trace duality.

PROPOSITION: *Let E and F be Banach spaces such that E'' or F' has the approximation property and: E'' or F' has the Radon–Nikodým property. Then*
$$(\mathfrak{K}(E,F))' = \mathfrak{N}(E',F') \qquad \text{isometrically}$$
$$(\mathfrak{K}(E,F))'' = (\mathfrak{N}(E',F'))' = \mathfrak{L}(E'',F'') \qquad \text{isometrically.}$$

The duality brackets are given by

$$\langle S, T \rangle_{\mathfrak{K},\mathfrak{N}} = \sum_{n=1}^{\infty} \langle x_n'', S' y_n' \rangle_{E'',E'}$$

$$\langle T, S \rangle_{\mathfrak{N},\mathfrak{L}} = \sum_{n=1}^{\infty} \langle S x_n'', y_n' \rangle_{E'',E'}$$

if $T = \sum_{n=1}^{\infty} x_n'' \otimes y_n' \in \mathfrak{N}(E', F') = E'' \tilde{\otimes}_\pi F'$. The natural embedding of $\mathfrak{K}(E, F)$ into its bidual

$$\mathfrak{K}(E, F) \hookrightarrow (\mathfrak{K}(E, F))'' = \mathfrak{L}(E'', F'')$$

is passing to the bidual: $S \rightsquigarrow S''$.

PROOF: The statements $\mathfrak{K}' \stackrel{1}{=} \mathfrak{N}$ and $\mathfrak{N}' \stackrel{1}{=} \mathfrak{L}$ follow from what was said before. The duality brackets originate from the natural embedding

$$\mathfrak{N}(E', F') = E'' \tilde{\otimes}_\pi F' \xrightarrow{=} (E' \tilde{\otimes}_\varepsilon F)' = (\mathfrak{K}(E, F))'$$

and the identification

$$(\mathfrak{N}(E', F'))' = (E'' \tilde{\otimes}_\pi F')' = \mathfrak{L}(E'', F'') .$$

It also follows (check for elementary tensors) that for $S \in \mathfrak{K}$ and $T \in \mathfrak{N}$

$$\langle S, T \rangle_{\mathfrak{K},\mathfrak{N}} = \langle S'', T \rangle_{\mathfrak{L},\mathfrak{N}}$$

which is nothing but the statement that $\mathfrak{K} \hookrightarrow \mathfrak{K}'' = \mathfrak{L}$. $\square$

If E'' has the approximation property, then the duality brackets are

$$\langle S, T \rangle_{\mathfrak{K},\mathfrak{N}} = tr_{E'}(S' \circ T)$$
$$\langle T, S \rangle_{\mathfrak{N},\mathfrak{L}} = tr_{E''}(T' \circ S) .$$

An immediate consequence is

COROLLARY: *Let E and F be Banach spaces ($\neq \{0\}$), one of which has the approximation property. Then $E \tilde{\otimes}_\pi F$ is reflexive if and only if E and F are reflexive and every $T \in \mathfrak{L}(E, F')$ is compact.*

PROOF: If E and F are reflexive and $\mathfrak{L}(E, F') = \mathfrak{K}(E, F')$, then

$$(\mathfrak{K}(E, F'))'' = \mathfrak{L}(E, F') = \mathfrak{K}(E, F')$$

by the proposition, therefore $\mathfrak{L}(E, F') = (E \tilde{\otimes}_\pi F)'$ is reflexive, hence also $E \tilde{\otimes}_\pi F$. Conversely, if $E \tilde{\otimes}_\pi F$ is reflexive, then E and F are reflexive (as complemented subspaces of $E \tilde{\otimes}_\pi F$), moreover $\mathfrak{L}(E, F') = (E \tilde{\otimes}_\pi F)'$ and its subspace $\mathfrak{K}(E, F')$ are also reflexive. The proposition gives

$$\mathfrak{K}(E, F') = (\mathfrak{K}(E, F'))'' = \mathfrak{L}(E, F')$$

and the result is proven. □

For an example, recall that Pitt's theorem (see e.g. Köthe [152], II, p.208) says that

$$\mathfrak{L}(\ell_p, \ell_q) = \mathfrak{K}(\ell_p, \ell_q)$$

if $1 \leq q < p < \infty$. It follows that for $1 < q < p < \infty$ the space

$$\ell_p \tilde{\otimes}_\pi \ell_{q'}$$

is reflexive. Recall that $\ell_2 \tilde{\otimes}_\pi \ell_2$ is not reflexive (Ex 3.9.). For some more details in this direction, see Ex 16.4. – 16.8.; in section 33 the reflexivity of other operator ideals will be treated.

16.8. Finally an answer to the following question concerning dual nuclear mappings is obtained: If $T \in \mathfrak{L}(E, F)$ is such that T' is nuclear, T is nuclear as well? In 5.9. it was shown that the answer is positive if E' has the approximation property. But in general this is false; this can be seen as follows: Take a Banach space E with the approximation property such that E' has the Radon–Nikodým property (e.g.: E' is separable). Clearly

$$\begin{array}{ccc} E' \tilde{\otimes}_\pi E \stackrel{1}{=} \mathfrak{N}(E, E) & \stackrel{J}{\longrightarrow} & (E \otimes_\varepsilon E')' = \mathfrak{I}(E', E') \\ \downarrow I_1 & & \uparrow I_2 \\ G := \mathfrak{N}(E', E') \cap \{T' | T \in \mathfrak{L}(E, E)\} & \stackrel{1}{\hookrightarrow} & \mathfrak{N}(E', E') \end{array}$$

commutes. The Radon–Nikodým property (16.5. or Appendix D7.) implies that I_2 is an isometry. Since the subspace of dual operators is closed in $\mathfrak{L}(E', E')$, the space G is closed in $\mathfrak{N}(E', E')$. So *if I_1 is surjective*, the closed graph theorem yields that I_1 is a topological isomorphism and hence J is as well: Corollary 2 in 16.3. now implies that *E has the bounded approximation property.*

However, the Figiel–Johnson example 5.2.(5) is a Banach space E with the approximation property and such that E' has the Radon–Nikodým property (since E' is separable), but without the bounded approximation property. It follows that:

PROPOSITION: *There is a non-nuclear operator the dual of which is nuclear.*

Note that the counterexample is an operator $T \in \mathcal{L}(E, E)$, where E has the approximation property and E' is separable. For another counterexample, see 22.9..

16.9. There is a simple characterization of the bounded approximation property which has interesting consequences for the factorization of finite rank operators:

PROPOSITION: *A normed space E has the λ-bounded approximation property if and only if for every $M \in FIN(E)$ and $\varepsilon > 0$ there is an $R \in \mathfrak{F}(E, E)$ with $\|R\| \leq \lambda(1 + \varepsilon)$ and $Rx = x$ for all $x \in M$.*

PROOF: Clearly, the condition is sufficient (see also Ex 5.7.). Conversely, take a basis $(e_1, ..., e_n)$ of M with coefficient functionals $e'_k \in E'$ and choose $T \in \mathfrak{F}(E, E)$ with $\|T\| \leq \lambda$ and $\|Te_k - e_k\| \leq \delta$ for all $k = 1, ..., n$. Then

$$R := T + \sum_{k=1}^{n} e'_k \otimes (e_k - Te_k)$$

satisfies $Re_k = e_k$ and $\|R\| \leq \lambda + n\delta \max \|e'_k\|$. □

COROLLARY: *Let E and F be normed spaces, $T \in \mathfrak{F}(E, F)$ and $\varepsilon > 0$.*
(1) If F has the λ-bounded approximation property, then there is an $R \in \mathfrak{F}(F, F)$ with $\|R\| \leq \lambda(1 + \varepsilon)$ and $R \circ T = T$.
(2) If E' has the λ-bounded approximation property, then there is an $S \in \mathfrak{F}(E, E)$ with $\|S\| \leq \lambda(1 + \varepsilon)$ and $T \circ S = T$.

PROOF: (1) is immediate. To see (2) observe first that, by (1), there is an $R \in \mathfrak{F}(E', E')$ with $\|R\| \leq \lambda(1+\varepsilon)$ and $R \circ T' = T'$. The weak principle of local reflexivity 6.5. applied to the finite rank operator $R' \circ \kappa_E \in \mathcal{L}(E, E'')$ and the finite dimensional subspace $T'(F') \subset E'$ gives an $S \in \mathfrak{F}(E, E)$ with $\|S\| \leq \|R' \circ \kappa_E\|(1 + \varepsilon) \leq \lambda(1 + \varepsilon)^2$ and

$$\langle (R' \circ \kappa_E)x, T'y' \rangle = \langle Sx, T'y' \rangle$$

for all $x \in E$ and $y' \in F'$. This implies $T \circ S = T$. □

Exercises:

Ex 16.1. A Banach space F has the λ-bounded approximation property if and only if for every Banach space E each operator $T \in \mathcal{L}(E, F)$ can be approximated with respect to the topology of pointwise weak convergence by finite rank operators of norm $\leq \lambda$.

Ex 16.2. (a) If G_o and G are normed spaces and $I \in \mathcal{L}(G_o, G)$ is of norm ≤ 1 such that $I(B_{G_o})$ is norming for G', then $I' : G' \to G'_o$ is an isometry, hence $\tilde{I} : \tilde{G}_o \to \tilde{G}$ is a metric surjection.
(b) Use (a) to decide whether or not $B_{E'} \otimes B_{F'} \subset (E \otimes_\varepsilon F)'$ is norming for $(E \otimes_\varepsilon F)''$.

Ex 16.3. (a) If K is compact and F a Banach space the dual of which has the Radon–Nikodým property, then

$$(C(K,F))' = M(K) \tilde{\otimes}_\pi F',$$

where $M(K)$ is the space of signed regular Borel measures on K. In particular, for every $\varphi \in (C(K,F))'$ there are measures $\mu_n \in B_{M(K)}$, elements $y'_n \in B_{F'}$ and $(\lambda_n) \in \ell_1$ such that

$$\langle \varphi, f \rangle = \sum_{n=1}^{\infty} \lambda_n \int_K \langle x'_n, f(w) \rangle \mu_n(dw)$$

for all $f \in C(K,F)$. Hint: Recall that $M(K)$ has the metric approximation property by Ex 5.6..
(b) Is statement (a) true for all Banach spaces F? More explicitly, show that

$$C([0,1])' \tilde{\otimes}_\pi C([0,1])' \subsetneq C([0,1]^2)'.$$

Hint: $C([0,1])'$ does not have the Radon–Nikodým property and $C[0,1]$ is a "test–space" for this property, see Appendix D7., proposition (2).

The following results (Ex 16.4. – Ex 16.8.) are due to Holub [114]; in section 33 the reflexivity of other operator ideals will be treated.

Ex 16.4. Let E and F be Banach spaces one of which has the approximation property. Then $E \tilde{\otimes}_\varepsilon F$ is reflexive if and only if $E' \tilde{\otimes}_\pi F'$ is reflexive.

Ex 16.5. Show that for $p, q \in [1, \infty]$:
(a) $\ell_p \tilde{\otimes}_\pi \ell_q$ is reflexive if and only if $1 < q' < p < \infty$.
(b) $\ell_p \tilde{\otimes}_\varepsilon \ell_q$ is reflexive if and only if $1 < q < p' < \infty$.

Ex 16.6. If E is an infinite dimensional Banach space with the approximation property, then neither $E \tilde{\otimes}_\varepsilon E'$ nor $E \tilde{\otimes}_\pi E'$ is reflexive.

Ex 16.7. $L_p[0,1] \tilde{\otimes}_\alpha L_q[0,1]$ is not reflexive if $p, q \in [1, \infty]$ and $\alpha = \pi$ or ε. Hint: The Rademacher functions are complemented in $L_p[0,1]$ if $1 < p < \infty$, see Ex 8.17..

Ex 16.8. Use the fact (to be proved in 26.1.) that every operator from an $\mathcal{L}_p^g$-space into an $\mathcal{L}_q^g$-space factors through a Hilbert space (if $1 \leq p \leq 2 \leq q \leq \infty$) and Pitt's theorem to show that

$$\mathfrak{L}(\ell_q, L_p(\mu))$$

is reflexive if $1 < p \leq 2 < q < \infty$, and

$$\mathfrak{L}(L_q(\mu), \ell_p)$$

is reflexive if $1 < p < 2 \leq q < \infty$. What does this imply for the reflexivity of $L_q \tilde{\otimes}_\pi \ell_p$ and $L_q \tilde{\otimes}_\varepsilon \ell_p$?

Ex 16.9. If $T \in \mathfrak{L}(E, F)$ has a nuclear dual, then it is integral.

Ex 16.10. It may happen that $\mathbf{N}(T') < \mathbf{N}(T)$ even for finite rank operators. Hint: Look at the diagram in 16.8..

Ex 16.11. An operator ideal $\mathfrak{A}$ is called *regular* if $T \in \mathfrak{A}(E, F)$ whenever the operator $\kappa_F \circ T$ is in $\mathfrak{A}(E, F'')$. Using 16.8., show that the ideal $\mathfrak{N}$ of nuclear operators is not regular. For positive results in this direction see 22.8..

Ex 16.12. If $\mathfrak{L}(\ell_\infty, F) = \mathfrak{P}_1(\ell_\infty, F)$, then F is finite dimensional. Hint: The assumption implies that $\ell_1 \tilde{\otimes}_\varepsilon F \hookrightarrow \mathfrak{I}(\ell_\infty, F) \overset{1}{\hookrightarrow} (\ell_\infty \otimes_\varepsilon F')'$; 11.3., corollary 3.

17. The Representation Theorem for Maximal Operator Ideals

This section deals with the one–to–one correspondence between finitely generated tensor norms and maximal Banach operator ideals — a simple, but extremely powerful correspondence, which will be central to the investigations which follow. It links together thinking in terms of operators with "tensorial" thinking — and as a result both theories, the theory of tensor products and the theory of Banach operator ideals, are incredibly enriched (and very often simplified). There will be two types of characterizations for operators T which are in a maximal operator ideal: firstly, in terms of the associated bilinear form and, secondly, in terms of the continuity behaviour of $T \otimes id : \otimes_\alpha \to \otimes_\beta$ for appropriate tensor norms.

17.1. If $(\mathfrak{A}, \mathbf{A})$ is a normed operator ideal, then

$$M \otimes_\alpha N : \overset{1}{=} \mathfrak{A}(M', N)$$

defines a tensor norm α on FIN; in other words, if $z \in M \otimes N$ and $T_z \in \mathfrak{L}(M', N)$ is the associated operator, then

$$\alpha(z; M, N) := \mathbf{A}(T_z : M' \to N) \ .$$

That α is a tensor norm is an immediate consequence of the criterion 12.2.: *the metric mapping property of α is the same as the ideal property for $\mathfrak{A}$* (for finite dimensional spaces) since for $z \in M \otimes N$

$$R \circ T_z \circ S' = T_{S \otimes R(z)} \qquad (*)$$

holds for all $S \in \mathfrak{L}(M, M_1)$ and $R \in \mathfrak{L}(N, N_1)$. Note that the criteria 12.2. and 9.4. for tensor norms and operator ideals are quite similar.

17.2. Conversely, let α be a tensor norm on FIN. Define, for finite dimensional Banach spaces M and N, a norm $\mathbf{A}$ on $\mathfrak{L}(M, N)$ by

$$\mathbf{A}(T : M \to N) := \alpha(z_T; M', N) ,$$

where $z_T \in M' \otimes N = \mathfrak{L}(M, N)$ is the associated element in the tensor product. It follows from formula (∗) that in this way a normed operator ideal $(\mathfrak{A}, \mathbf{A})$ for finite dimensional spaces is defined. How to extend it to all Banach spaces? (Recall that operator ideals were only defined for Banach spaces, not for general normed spaces.) Of course there are several different ways to do this — for our purposes the smallest and the largest extensions will be the most interesting ones. The smallest will be treated in section 22; for the largest take the following

DEFINITION: *Let $(\mathfrak{A}, \mathbf{A})$ be a (quasi-) normed operator ideal.*
(1) *For $T \in \mathfrak{L}(E, F)$ define*

$$\mathbf{A}^{max}(T) := \sup \left\{ \mathbf{A}(Q_L^F \circ T \circ I_M^E) \,\bigg|\, \begin{array}{l} M \in \mathrm{FIN}(E) \\ L \in \mathrm{COFIN}(F) \end{array} \right\} \in [0, \infty]$$

$$\mathfrak{A}^{max} := \{T \in \mathfrak{L} \mid \mathbf{A}^{max}(T) < \infty\}$$

and call $(\mathfrak{A}, \mathbf{A})^{max} := (\mathfrak{A}^{max}, \mathbf{A}^{max})$ the maximal hull of $(\mathfrak{A}, \mathbf{A})$.
(2) *$(\mathfrak{A}, \mathbf{A})$ is called maximal if $(\mathfrak{A}, \mathbf{A}) = (\mathfrak{A}^{max}, \mathbf{A}^{max})$.*

It is clear that $(\mathfrak{A}^{max}, \mathbf{A}^{max})$ is always a maximal (quasi-) normed operator ideal. It is also immediate from criterion 9.4. that maximal (quasi-) normed ideals are necessarily (quasi-) Banach ideals. The following observation is simple, but useful.

REMARK: *Let $(\mathfrak{A}, \mathbf{A})$ be a maximal quasinormed operator ideal and $(\mathfrak{B}, \mathbf{B})$ a quasinormed operator ideal which coincides with $(\mathfrak{A}, \mathbf{A})$ for finite dimensional spaces. Then $\mathfrak{B} \subset \mathfrak{A}$ and $\mathbf{A} \leq \mathbf{B}$.*

In particular, two maximal ideals coincide if they coincide on finite dimensional spaces.

Returning to the tensor norm α on FIN, for $T \in \mathfrak{L}(E, F)$ define

$$\mathbf{A}(T : E \to F) := \sup \left\{ \alpha(z_{Q_L^F \circ T \circ I_M^E}; M', F/L) \,\bigg|\, \begin{array}{l} M \in \mathrm{FIN}(E) \\ L \in \mathrm{COFIN}(F) \end{array} \right\} \in [0, \infty]$$

and $\mathfrak{A} := \{T \in \mathfrak{L} \mid \mathbf{A}(T) < \infty\}$. It is clear that $(\mathfrak{A}, \mathbf{A})$ is a normed operator ideal, that $\mathfrak{A}(M, N) = M' \otimes_\alpha N$ holds isometrically for $M, N \in \mathrm{FIN}$, and hence that $(\mathfrak{A}, \mathbf{A})$ is even a maximal normed operator ideal.

17.3. Both finitely generated tensor norms and maximal Banach operator ideals are, by their definition, uniquely determined by their behaviour on FIN.

DEFINITION: *A finitely generated tensornom α (on NORM) and a maximal Banach operator ideal $(\mathfrak{A}, \mathbf{A})$ are said to be associated, notation:*
$$(\mathfrak{A}, \mathbf{A}) \sim \alpha$$
if for all $M, N \in$ FIN
$$\mathfrak{A}(M, N) = M' \otimes_\alpha N$$
holds isometrically: $\mathbf{A}(T_z : M \to N) = \alpha(z; M', N)$.

The constructions in 17.1. and 17.2. show that in this way a *one–to–one correspondence between maximal Banach operator ideals $(\mathfrak{A}, \mathbf{A})$ and finitely generated tensor norms α* is established. This correspondence is basic to the understanding of all that follows.

17.4. Before examining this correspondence more precisely, the maximality of various operator ideals will be checked. Obviously, $(\mathfrak{L}, \|\ \|)$ is maximal and $(\mathfrak{K}, \|\ \|)$ is not. A quick look at the definition shows that the absolutely p-summing operators $\mathfrak{P}_p$ are maximal — but there is a nice, rather general result which gives this and a great variety of other maximal Banach operator ideals.

For this, take a normed space G and a tensor norm α. Then
(1) $\varepsilon \leq \alpha \leq \pi$ on $G \otimes E$ for all normed E;
(2) $\alpha(id_G \otimes T(z); G, F) \leq \|T\| \alpha(z; G, E)$ for all $T \in \mathfrak{L}(E, F)$.

If G is a subspace of some $L_p(\mu)$ and $\alpha = \Delta_p$ is the norm induced on $G \otimes E$ by $L_p(\mu) \otimes_{\Delta_p} E$, then (1) and (2) also hold by the results in 7.3.. This is why we adopt a more general notion and call α a *right–tensor norm (with respect to a fixed normed space G)* if it is defined on all $G \otimes E$ with $E \in$ NORM, and satisfies (1) and (2). We repeat: we do not want to make a fuss over this definition — it is just more practical to treat α and Δ_p jointly. A right–tensor norm is *finitely generated (from the right)* if
$$\alpha(z; G, E) = \inf\{\alpha(z; G, N) \mid N \in \text{FIN}(E), z \in G \otimes N\}$$
for all $E \in$ NORM and *cofinitely generated (from the right)* if
$$\alpha(z; G, E) = \sup\{\alpha(id_G \otimes Q_L^E(z); G, E/L) \mid L \in \text{COFIN}(E)\}.$$

Clearly, finitely/cofinitely generated tensor norms are finitely/cofinitely generated from the right (see Ex 12.3.) and it was shown in Ex 7.16. that Δ_p is finitely and cofinitely generated from the right with respect to every subspace $G \subset L_p(\mu)$.

PROPOSITION: *Let G, H be normed spaces, $S \in \mathfrak{L}(G, H)$ with $\|S\| = 1$, α a finitely generated right–tensor norm with respect to G and β a cofinitely generated right–tensor norm with respect to H. Then the ideal $\left(\mathfrak{L}_S^{\alpha, \beta}, \mathbf{L}_S^{\alpha, \beta} \right)$ defined by*
$$\left\{ T \in \mathfrak{L}(E, F) \,\middle|\, \mathbf{L}_S^{\alpha, \beta}(T) := \|S \otimes T : G \otimes_\alpha E \longrightarrow H \otimes_\beta F\| < \infty \right\}$$

is a maximal Banach operator ideal.

The superscripts α and β will be omitted if they are clear from the context.

PROOF: Clearly, $\mathbf{L}_S$ is a norm on each $\mathcal{L}_S(E,F)$; the ideal property and the relation $\mathbf{L}_S(id_K) = \|S\| = 1$ are obvious as well. Maximality follows from

$$\mathbf{L}_S(T) = \sup\left\{\beta(S \otimes T(z); H, F) \mid \alpha(z; G, E) < 1\right\} =$$
$$= \sup\left\{\beta\left((id_H \otimes Q_L^F)(S \otimes T)(z); H, F/L\right) \middle| \begin{array}{l}\alpha(z; G, M) < 1, \; z \in G \otimes M \\ M \in \mathrm{FIN}(E), L \in \mathrm{COFIN}(F)\end{array}\right\}$$
$$= \sup\left\{\mathbf{L}_S(Q_L^F \circ T \circ I_M^E : M \to F/L) \mid M \in \mathrm{FIN}(E), \; L \in \mathrm{COFIN}(F)\right\}$$

— and this completes the proof. □

Note that $H \otimes_\pi \cdot$ is cofinitely generated if H has the metric approximation property (since π is accessible) and hence the proposition implies that the following ideals are maximal:
— the ideal $\mathfrak{P}_p$ of absolutely p–summing operators (see 11.1.(d))
— the ideal $\mathfrak{P}_{(r,s)}$ of absolutely (r,s)–summing operators (see Ex 11.20.)
— the ideal $\mathfrak{I}$ of integral operators (see Ex 17.1.)
— the ideal $\mathfrak{T}_p$ of type p operators (see the definition 7.7.)
— the ideal $\mathfrak{C}_q$ of cotype q operators (see the definition 7.7.)
— the ideal of K–convex operators (see Ex 9.16.)
— the ideal $\mathcal{L}_\mathfrak{F}^{\Delta_2,\Delta_2}$, where $\mathfrak{F} : L_2(\mathbb{R}) \to L_2(\mathbb{R})$ is the Fourier transform (see 7.5.)
— the ideal $\mathcal{L}_\mathfrak{H}^{\Delta_p,\Delta_p}$, where $\mathfrak{H} : \ell_p(\mathbb{Z}) \to \ell_p(\mathbb{Z}))$ is the Hilbert transform and $1 < p < \infty$ (see 7.6.).

Actually, it is much easier (and more natural) to deduce that the integral operators form a maximal ideal by using the following representation theorem.

17.5. If a maximal operator ideal $(\mathfrak{A}, \mathbf{A})$ is associated with a finitely generated tensor norm α, then
$$\mathfrak{A}(M, N) = M' \otimes_\alpha N = (M \otimes_{\alpha'} N')'$$
holds isometrically for all $M, N \in \mathrm{FIN}$. The extension of this to infinite dimensional spaces is the important

REPRESENTATION THEOREM FOR MAXIMAL OPERATOR IDEALS: Let $(\mathfrak{A}, \mathbf{A})$ be a maximal normed ideal and α a finitely generated tensor norm which are associated with each other: $(\mathfrak{A}, \mathbf{A}) \sim \alpha$. Then for all Banach spaces E and F the relations
$$\mathfrak{A}(E, F') = (E \otimes_{\alpha'} F)'$$

and
$$\mathfrak{A}(E,F) = (E \otimes_{\alpha'} F')' \cap \mathfrak{L}(E,F)$$

hold isometrically.

By definition 10.1. of the integral operators, $\mathfrak{I}(E,F) = (E \otimes_\varepsilon F')' \cap \mathfrak{L}(E,F)$ holds isometrically. In particular, $\mathfrak{I}(M,N) \stackrel{1}{=} (M \otimes_\varepsilon N')' \stackrel{1}{=} M' \otimes_\pi N$ and hence $\pi \sim \mathfrak{I}^{max}$; the representation theorem gives

$$\mathfrak{I}^{max}(E,F) \stackrel{1}{=} (E \otimes_\varepsilon F')' \cap \mathfrak{L}(E,F) \stackrel{1}{=} \mathfrak{I}(E,F)$$

and therefore $\mathfrak{I} = \mathfrak{I}^{max}$ and $\pi \sim \mathfrak{I}$ (see also Ex 17.2.). This example explains why operators in $\mathfrak{A}$ are sometimes called *α-integral operators* whenever $\mathfrak{A} \sim \alpha$.

This theorem is due to Lotz [180]. His approach to tensor norms was different from ours and was very influential in the development of the theory of operator ideals: Roughly speaking, he took the representation theorem as a definition for tensor norms and consequently emphasized the one–to–one correspondence between maximal normed operator ideals and tensor norms; see also Ex 17.2..

PROOF: The second formula will be proven first: it needs to be shown for $T \in \mathfrak{L}(E,F)$ that $T \in \mathfrak{A}(E,F)$ if and only if the associated bilinear form $\beta_{\kappa_F \circ T}$ is α'-continuous:

$$\beta_{\kappa_F \circ T} \in (E \otimes_{\alpha'} F')'$$

(with equal norms). But this is easy: $T \in \mathfrak{A}(E,F)$ and $\mathbf{A}(T) \leq c$ if and only if

$$\mathbf{A}\left(Q_L^F \circ T \circ I_M^E\right) \leq c$$

for all $(M,L) \in \text{FIN}(E) \times \text{COFIN}(F)$; by $\mathfrak{A}(M,F/L) = (M \otimes_{\alpha'} L^0)'$ this is equivalent to
$$|\langle \beta_{\kappa_F \circ T}, z \rangle| = \left|\langle \beta_{Q_L^F \circ T \circ I_M^E}, z \rangle\right| \leq c\alpha'(z; M, L^0) \text{ for all } z \in M \otimes L^0 .$$

Therefore, the result follows since α' is finitely generated. To see that the first formula holds just look at the diagram

$$\begin{array}{ccc} \varphi \in (E \otimes_{\alpha'} F)' & \hookrightarrow & (E \otimes_\pi F)' = \mathfrak{L}(E,F') \\ \downarrow{\scriptstyle 1} & & \uparrow \\ \varphi^\wedge \in (E \otimes_{\alpha'} F'')' & \hookrightarrow & (E \otimes_\pi F'')' \end{array}$$

and use the extension lemma. □

17.6. An obvious combination of the representation theorem and the duality theorem 15.5. gives the

EMBEDDING THEOREM: *If* $(\mathfrak{A}, \mathbf{A}) \sim \alpha$ *are associated, then the relations*

$$E' \otimes_{\overleftarrow{\alpha}} F' \xhookrightarrow{1} \mathfrak{A}(E, F')$$
$$E \otimes_{\overleftarrow{\alpha}} F \xhookrightarrow{1} \mathfrak{A}(E', F)$$
$$E' \otimes_{\overleftarrow{\alpha}} F \xhookrightarrow{1} \mathfrak{A}(E, F)$$

hold isometrically.

In particular, the extension

$$E' \tilde{\otimes}_\alpha F \longrightarrow \mathfrak{A}(E, F)$$

of $E' \otimes_\alpha F \to \mathfrak{A}(E, F)$ is well–defined and has norm 1 (if E and F are not trivial). In 16.3. this map was studied for the case $\alpha = \pi$ and was useful in understanding the relationship between nuclear and integral operators. More generally, in section 22 the embedding theorem will be crucial for the investigation of minimal operator ideals.

17.7. An essential goal of the theory is to compare different tensor norms/maximal operator ideals. The very definition of $(\mathfrak{A}, \mathbf{A}) \sim \alpha$ using finite dimensional spaces shows the

REMARK: *Let* $(\mathfrak{A}, \mathbf{A}) \sim \alpha$ *and* $(\mathfrak{B}, \mathbf{B}) \sim \beta$ *be associated*, $c \geq 1$. *Then* $\alpha \leq c\beta$ *if and only if* $\mathbf{A} \leq c\mathbf{B}$. *In this case*, $\mathfrak{B} \subset \mathfrak{A}$.

Very interesting phenomena can be deduced from estimates on special Banach spaces. The representation and embedding theorems

$$E' \otimes_{\overleftarrow{\alpha}} F' \xhookrightarrow{1} \mathfrak{A}(E, F') \stackrel{1}{=} (E \otimes_{\alpha'} F)'$$

imply the following:

COROLLARY 1: *Let* $(\mathfrak{A}, \mathbf{A}) \sim \alpha$ *and* $(\mathfrak{B}, \mathbf{B}) \sim \beta$ *be associated*, $c \geq 1$ *and* E *and* F *Banach spaces. Consider the following conditions:*
(a) $\beta' \leq c\alpha'$ *on* $E \otimes F$.
(b) $\mathfrak{B}(E, F') \subset \mathfrak{A}(E, F')$ *and* $\mathbf{A} \leq c\mathbf{B}$.
(c) $\overleftarrow{\alpha} \leq c\overleftarrow{\beta}$ *on* $E' \otimes F'$.
Then
(1) (a)$\leftrightarrow$(b)$\to$(c).
(2) *If* E' *and* F' *have the metric approximation property, or:* α *and* β *are accessible and* E' *or* F' *has the metric approximation property or:* α *and* β' *are totally accessible, then* (a), (b) *and* (c) *are equivalent.*

PROOF: (1) follows from the observation above. So it remains to show that (c) implies (a) under the conditions of (2). But this follows from the diagram

$$
\begin{array}{ccccc}
E \otimes_{\alpha'} F & \xrightarrow{\leq 1} & (E' \otimes_{\alpha} F')' & \xrightarrow{?} & (E' \otimes_{\overleftarrow{\alpha}} F')' \\
\downarrow & & & & \downarrow \leq c \text{ by (c)} \\
E \otimes_{\overleftarrow{\beta'}} F & \xrightarrow{1} & (E' \otimes_{\beta} F')' & \xleftarrow{\leq 1} & (E' \otimes_{\beta} F')'
\end{array}
$$

since $\overleftarrow{\alpha} = \alpha$ on $E' \otimes F'$ and $\overleftarrow{\beta'} = \beta'$ on $E \otimes F$ under the assumptions of (2) (use 15.7., 15.6. and 16.3.). □

This result, although simple, is often quite useful and will be fundamental in the understanding of the interplay between maximal ideals and tensor norms. Since it transfers statements about operator ideals to inequalities involving tensor norms and vice–versa, it will be referred to as the *transfer argument*. Note that (2) gives conditions under which: $\alpha \leq c\beta$ on $E' \otimes F'$ if and only if $\beta' \leq c\alpha'$ on $E \otimes F$.

17.8. The representation theorem has simple, but important consequences for the behaviour of maximal operator ideals. Looking at

$$\mathfrak{A}(E, F) \xrightarrow{1} (E \otimes_{\alpha'} F')' \stackrel{1}{=} \mathfrak{A}(E, F'')\ ,$$

gives the

COROLLARY 2: *Maximal operator ideals* $(\mathfrak{A}, \mathbf{A})$ *are regular, which means:* $T \in \mathfrak{L}(E, F)$ *is in* $\mathfrak{A}$ *if (and only if)* $\kappa_F \circ T \in \mathfrak{A}(E, F'')$; *moreover,*

$$\mathbf{A}(T : E \to F) = \mathbf{A}(\kappa_F \circ T : E \to F'')\ .$$

Recall from 9.9. that $\mathfrak{A}^{dual} := \{T \in \mathfrak{L} \mid T' \in \mathfrak{A}\}$ and $\mathbf{A}^{dual}(T) := \mathbf{A}(T')$.

COROLLARY 3: *If* $(\mathfrak{A}, \mathbf{A})$ *is a maximal normed operator ideal associated with a finitely generated tensor norm* α, *then* $(\mathfrak{A}^{dual}, \mathbf{A}^{dual})$ *is maximal and is associated with the transposed tensor norm* α^t:

$$(\mathfrak{A}^{dual}, \mathbf{A}^{dual}) \sim \alpha^t\ .$$

PROOF: Denote by $(\mathfrak{B}, \mathbf{B})$ the maximal normed operator ideal associated with α^t. Then, by the representation theorem,

$$\mathfrak{B}(E, F) = (E \otimes_{(\alpha^t)'} F')' \cap \mathfrak{L}(E, F) =$$
$$= \left\{ T \in \mathfrak{L}(E, F) \mid \beta_{T'} \in (F' \otimes_{\alpha'} E)' \right\} =$$
$$= \left\{ T \in \mathfrak{L}(E, F) \mid T' \in \mathfrak{A}(F', E') \right\}$$

with $\mathbf{B}(T) = \mathbf{A}(T')$. In other words, $(\mathfrak{B}, \mathbf{B}) = (\mathfrak{A}^{dual}, \mathbf{A}^{dual})$; this was the claim. □

Since $(\mathfrak{A}^{dual})^{dual} \sim \alpha^{tt} = \alpha \sim \mathfrak{A}$, one obtains

COROLLARY 4: *Let $(\mathfrak{A}, \mathbf{A})$ be a maximal normed operator ideal. Then $(\mathfrak{A}^{dual})^{dual} = \mathfrak{A}$ holds isometrically. In other words, $T \in \mathfrak{A}$ if and only if $T'' \in \mathfrak{A}$; in this case,*

$$\mathbf{A}(T) = \mathbf{A}(T'') .$$

Another way to formulate this is: T is α–integral if and only if T' is α^t–integral.

17.9. Let $(\mathfrak{A}, \mathbf{A}) \sim \alpha$ be associated. Then for $M, N \in \text{FIN}$

$$(\mathfrak{A}(M, N))' = (M' \otimes_\alpha N)' = (N \otimes_{\alpha^t} M')' = N' \otimes_{(\alpha^t)'} M$$

holds isometrically. Hence, if $(\mathfrak{A}^*, \mathbf{A}^*)$ is the maximal normed operator ideal associated with $(\alpha^t)' = \alpha^*$, one obtains

$$(\mathfrak{A}(M, N))' = \mathfrak{A}^*(N, M)$$

— the duality bracket given by

$$\langle T, S \rangle_{\mathfrak{A}^*, \mathfrak{A}} = tr_N(S \circ T) = tr_M(T \circ S)$$

whenever $T \in \mathfrak{A}^*(N, M)$ and $S \in \mathfrak{A}(M, N)$. This is why $(\mathfrak{A}^*, \mathbf{A}^*)$ is called the *adjoint maximal normed operator ideal* of $(\mathfrak{A}, \mathbf{A})$. In section 22 it will be seen how the duality relation is extended to infinite dimensional spaces. When is an operator $T \in \mathcal{L}(E, F)$ in $\mathfrak{A}^*$? Observe first that, by duality,

$$\mathbf{A}^*(T : M \to N) = \sup \left\{ |tr_M(S \circ T)| \,\Big|\, \mathbf{A}(S : N \to M) \leq 1 \right\}$$

for finite dimensional spaces. But now it is clear from the very definition of maximal operator ideals that $T \in \mathcal{L}(E, F)$ is in $\mathfrak{A}^*$ if and only if

$$\sup \left\{ |tr_M(S \circ Q_L^F \circ T \circ I_M^E)| \,\Bigg|\, \begin{array}{c} M \in \text{FIN}(E) \\ L \in \text{COFIN}(F) \\ \mathbf{A}(S : F/L \to M) \leq 1 \end{array} \right\} < \infty$$

— and this number is $\mathbf{A}^*(T : E \to F)$:

$$\begin{array}{ccc} E & \xrightarrow{T} & F \\ {\scriptstyle I_M^E}\uparrow & & \downarrow{\scriptstyle Q_L^F} \\ M & \xleftarrow{S} & F/L \end{array}.$$

This justifies the following definition: For every quasinormed ideal $(\mathfrak{A}, \mathbf{A})$ the *adjoint operator ideal* $(\mathfrak{A}^*, \mathbf{A}^*)$ is the class of all T such that

$$\mathbf{A}^*(T : E \to F) := \sup \left\{ |tr_M(S \circ Q_L^F \circ T \circ I_M^E)| \;\middle|\; \begin{array}{l} M \in \mathrm{FIN}(E) \\ L \in \mathrm{COFIN}(F) \\ \mathbf{A}(S : F/L \to M) \le 1 \end{array} \right\} < \infty.$$

$(\mathfrak{A}^*, \mathbf{A}^*)$ is always a Banach operator ideal which for $M, N \in \mathrm{FIN}$ satisfies the relation $(\mathfrak{A}(M,N))' = \mathfrak{A}^*(N,M)$ isometrically. Note that $tr_M(S \circ \cdots) = tr_{F/L}(\cdots \circ S)$ in the formula for $\mathbf{A}^*$. Since

$$|tr(T \circ S)| \le \mathbf{A}^*(T) \mathbf{A}(S)$$

for operators T, S between finite dimensional spaces, it follows that $\mathfrak{A} \subset \mathfrak{A}^{**}$ and $\mathbf{A}^{**} \le \mathbf{A}$. Under certain circumstances the description of $\mathfrak{A}^*$ is simpler, see Ex 17.4..

17.10. The maximal normed operator ideal associated with the tensor norm $\alpha_{p,q}$ of Lapresté (see 12.5.) is the ideal $(\mathfrak{L}_{p,q}, \mathbf{L}_{p,q})$ of (p,q)-*factorable operators*; recall that $1/p + 1/q \ge 1$ and $1/r := 1/p + 1/q - 1$. If for finite dimensional M, N

$$T \in \mathfrak{L}_{p,q}(M, N) = M' \otimes_{\alpha_{p,q}} N,$$

then, by definition of the norm, $\mathbf{L}_{p,q}(T) = \alpha_{p,q}(z_T; M', N)$, there is for every $\varepsilon > 0$ a representation

$$T = \sum_{k=1}^{n} \lambda_k x'_k \otimes y_k,$$

with $\ell_r(\lambda_k) w_{q'}(x'_k; M') w_{p'}(y_k; N) \le \mathbf{L}_{p,q}(T)(1 + \varepsilon)$. Since

$$\varepsilon \Big(\sum_{k=1}^{n} x'_k \otimes e_k; M', \ell_{q'}^n\Big) = w_{q'}(x'_k; M') \quad \text{on} \quad M' \otimes_\varepsilon \ell_{q'}^n = \mathfrak{L}(M, \ell_{q'}^n)$$

$$\varepsilon \Big(\sum_{k=1}^{n} e_k \otimes y_k; \ell_{p'}^n, N\Big) = w_{p'}(y_k; N) \quad \text{on} \quad \ell_{p'}^n \otimes_\varepsilon N = \mathfrak{L}(\ell_p^n, N)$$

(by 12.9.) and $\|D_\lambda : \ell_{q'}^n \to \ell_p^n\| = \ell_r(\lambda_k)$ for the diagonal operator associated with $\lambda := (\lambda_k)$, it follows that T admits a factorization

$$\begin{array}{ccc} M & \xrightarrow{T} & N \\ {\scriptstyle R := \sum_{k=1}^{n} x'_k \otimes e_k} \downarrow & & \uparrow {\scriptstyle S := \sum_{k=1}^{n} e_k \otimes y_k} \\ \ell_{q'}^n & \xrightarrow{D_\lambda} & \ell_p^n \end{array}$$

with $\|R\| \|D_\lambda\| \|S\| \le \mathbf{L}_{p,q}(T)(1 + \varepsilon)$. Reading this argument backwards, it follows that

$$\mathbf{L}_{p,q}(T) = \inf \|R\| \|D_\lambda\| \|S\|,$$

where the infimum is taken over all such factorizations.
An example: Since $\|id : \ell_{q'}^n \to \ell_p^n\| = n^{1/r}$, it follows that

$$\mathbf{L}_{p,q}(id : \ell_{q'}^n \to \ell_p^n) = n^{1/r} .$$

These factorizations help to explain the name of the ideal; it will be proved in section 18 that an operator T is in $\mathfrak{L}_{p,q}(E, F)$ if and only if it factors

$$\begin{array}{ccc} E & \xrightarrow{T} & F \hookrightarrow F'' \\ \downarrow & & \uparrow \\ L_{q'}(\mu) & \xrightarrow{I} & L_p(\mu) \end{array} ;$$

note that $q' \geq p$: if $q' > p$, then the measure μ is understood to be finite and I is the canonical inclusion, whereas in the case $q' = p$ the measure may be arbitrary and I is just the identity map. There are several interesting special cases: the ideal $(\mathfrak{L}_p, \mathbf{L}_p)$ associated with $w_p = \alpha_{p,p'}$ is the ideal of all p-factorable operators. By what was just said these are the operators T such that $\kappa_F \circ T$ factors through some $L_p(\mu)$. Since $\pi = \alpha_{1,1}$, the $(1,1)$-factorable operators are exactly the integral operators and the factorization theorem mentioned above is already known from 10.5.; more generally, the ideal $(\mathfrak{J}_p, \mathbf{I}_p) := (\mathfrak{L}_{p,1}, \mathbf{L}_{p,1})$ associated with $g_p = \alpha_{p,1}$ is called the ideal of p-integral operators. Note that $\mathfrak{J}_\infty = \mathfrak{L}_\infty$ since $w_\infty = g_\infty$.

It is important to reformulate the factorization of $T \in \mathfrak{L}_{p,q}(M, N') = (M \otimes_{\alpha'_{p,q}} N)'$ in terms of bilinear forms: The diagonal operator $D_\lambda \in \mathfrak{L}(\ell_{q'}^n, \ell_p^n) = (\ell_{q'}^n \otimes_\pi \ell_{p'}^n)'$ corresponds to the "diagonal" bilinear form

$$\delta_\lambda := \sum_{k=1}^n \lambda_k e_k \otimes e_k \in (\ell_{q'}^n \otimes_\pi \ell_{p'}^n)' .$$

Therefore, for every $\varphi \in (M \otimes_\pi N)'$,

$$\|\varphi\|_{(M \otimes_{\alpha'_{p,q}} N)'} = \inf \|R\| \, \|S\| \, \|\delta_\lambda\| ,$$

where the infimum is taken over all factorizations

$$\varphi : M \otimes N \xrightarrow{R \otimes S} \ell_{q'}^n \otimes \ell_{p'}^n \xrightarrow{\delta_\lambda} \mathbb{K} ,$$

where $R \in \mathfrak{L}(M, \ell_{q'}^n), S \in \mathfrak{L}(N, \ell_{p'}^n)$ and $\lambda \in \mathbb{K}^n$. Clearly, it is enough to take only $\lambda \in \mathbb{K}^n$ with non–negative components.

17.11. Take $p, q \in [1, \infty]$ such that $1/p + 1/q \leq 1$. Then $1/p' + 1/q' \geq 1$ and $\alpha_{q',p'}$ is defined. The ideal $(\mathfrak{D}_{p,q}, \mathbf{D}_{p,q})$ associated with $\alpha'_{q',p'} = \alpha^*_{p',q'}$ is the maximal normed

operator ideal of so-called (p,q)-*dominated operators*. To describe the operators in $\mathfrak{D}_{p,q}$, define $r \in [1, \infty]$ by
$$1/r := 1/p + 1/q \,,$$
so that $1/p + 1/q + 1/r' = 1$. The representation theorem for maximal operator ideals shows that $T \in \mathcal{L}(E, F)$ is in $\mathfrak{D}_{p,q}$ if and only if the associated bilinear form is $\alpha_{q',p'}$-continuous:
$$\beta_{\kappa_F \circ T} \in (E \otimes_{\alpha_{q',p'}} F')'$$
— with equal norms. This means that $\mathbf{D}_{p,q}(T) \le c$ if and only if for every element $z = \sum_{k=1}^n \lambda_k x_k \otimes y'_k \in E \otimes F'$
$$|\langle \beta_{\kappa_F \circ T}, z \rangle| = \left| \sum_{k=1}^n \lambda_k \langle y'_k, Tx_k \rangle \right| \le c \, \ell_{r'}(\lambda_k) w_p(x_k; E) w_q(y'_k; F') \,,$$
which is equivalent to
$$\ell_r(\langle y'_k, Tx_k \rangle) \le c \, w_p(x_k; E) w_q(y'_k; F')$$
for all $x_1, ..., x_n \in E$ and $y'_1, ..., y'_n \in F'$; clearly, $\mathbf{D}_{p,q}(T)$ is the minimum of all such $c \ge 0$.

Special cases: $(\mathfrak{D}_p, \mathbf{D}_p) \sim w^*_{p'} = \alpha^*_{p',p}$ is the maximal operator ideal of p-*dominated* operators; the operators $T \in \mathfrak{D}_p(E, F) = \mathfrak{D}_{p,p'}(E, F)$ are characterized by
$$\ell_1(\langle y'_k, Tx_k \rangle) \le c \, w_p(x_k; E) w_{p'}(y'_k; F') \,.$$
For $q = \infty$ one obtains $r = p$ and the operators in $\mathfrak{D}_{p,\infty}(E, F)$ are those satisfying
$$\ell_p(\langle y'_k, Tx_k \rangle) \le c \, w_p(x_k; E) w_\infty(y'_k; F')$$
or equivalently (choose appropriate $y'_k \in B_{F'}$)
$$\ell_p(\|Tx_k\|) \le c \, w_p(x_k; E)$$
for all $x_1, ..., x_n \in E$. These are exactly the *absolutely p-summing* operators from section 11:
$$(\mathfrak{P}_p, \mathbf{P}_p) = (\mathfrak{D}_{p,\infty}, \mathbf{D}_{p,\infty}) \sim \alpha^*_{p',1} = g^*_{p'} \,.$$
In section 19 the Grothendieck–Pietsch domination theorem for $\mathfrak{P}_p$ will be generalized to $\mathfrak{D}_{p,q}$, thus explaining the name (p,q)-dominated.

17.12. Before going into some examples showing the interplay between maximal operator ideals and tensor norms, it might be helpful to have a list of the operator ideals connected with $\alpha_{p,q}$. It is clear from the definition of the adjoint operator ideal and $\alpha^{**} = \alpha$ (if α is finitely generated) that
$$\mathfrak{L}^*_{p,q} = \mathfrak{D}_{p',q'} \quad \text{and} \quad \mathfrak{D}^*_{p',q'} = \mathfrak{L}_{p,q}$$
hold isometrically for $p, q \in [1, \infty]$ with $1/p + 1/q \ge 1$. Hence

(1) $\quad \varepsilon \sim \mathfrak{L}$ $\quad\quad\quad\quad\quad\quad\quad\quad\quad\quad$ all operators
$\quad\quad \pi \sim \mathfrak{I} = \mathfrak{I}_1 = \mathfrak{L}_{1,1} = \mathfrak{L}^*$ $\quad$ integral operators

(2) $\quad \alpha_{p,q} \sim \mathfrak{L}_{p,q}$ $\quad\quad\quad\quad\quad\quad\quad$ (p,q)–factorable operators
$\quad\quad \alpha_{p,q}^* \sim \mathfrak{D}_{p',q'} = \mathfrak{L}_{p,q}^*$ $\quad\quad$ (p',q')–dominated operators

(3) $\quad w_p \sim \mathfrak{L}_p = \mathfrak{L}_{p,p'}$ $\quad\quad\quad\quad$ p–factorable operators
$\quad\quad w_p^* \sim \mathfrak{D}_{p'} = \mathfrak{D}_{p',p} = \mathfrak{L}_p^*$ $\quad$ p'–dominated operators

(4) $\quad g_p \sim \mathfrak{I}_p = \mathfrak{L}_{p,1}$ $\quad\quad\quad\quad\quad$ p–integral operators
$\quad\quad g_p^* \sim \mathfrak{P}_{p'} = \mathfrak{D}_{p',\infty} = \mathfrak{I}_p^*$ $\quad$ absolutely p'–summing operators ,

where $\mathfrak{P}_\infty := \mathfrak{L}$, which is a reasonable definition from several points of view.

17.13. Transfering results from tensor norms to operator ideals and vice–versa, is a sort of "game" whose value stems from the fact that it is quite often easier (or much more natural) to prove a result in one form than in the other. We are going to illustrate this now, although not always with the intention of getting the best result; this will be done later.

The Khintchine equality gives that $\alpha_{p,q} \leq b_{q'} b_{p'} w_2$ whenever $p, q \in]1, \infty[$ (see 12.8.). From the very definition of $\mathfrak{A}$ being associated with α it follows that

$$\mathfrak{A}(M, N) \stackrel{1}{=} M' \otimes_\alpha N ,$$

hence

$$\mathfrak{L}_2(E, F) \subset \mathfrak{L}_{p,q}(E, F) \text{ and } \mathbf{L}_{p,q}(T) \leq b_{q'} b_{p'} \mathbf{L}_2(T) \text{ for } p, q \in]1, \infty[$$

for all Banach spaces E and F. Since $w_2 \leq \alpha_{p,q}$, whenever $p, q \in]1, 2]$, it follows even that

$$\mathfrak{L}_2(E, F) = \mathfrak{L}_{p,q}(E, F) \text{ and } \mathbf{L}_2(T) \leq \mathbf{L}_{p,q}(T) \leq b_{q'} b_{p'} \mathbf{L}_2(T) \text{ for } p, q \in]1, 2] .$$

The relevance of these results stems from the fact that the 2–factorable operators are exactly those which factor through a Hilbert space (to be shown in the next section), thus emphasizing the very special role of Hilbert spaces in the theory. What does $g_{p'}^* \leq g_p$ (see 15.4.) imply? Since

$$\mathfrak{P}_p \sim g_{p'}^* \leq g_p \sim \mathfrak{I}_p ,$$

this clearly means (again by the definition of $\mathfrak{A} \sim \alpha$ and extension to infinite dimensional spaces) that

$$\mathfrak{I}_p(E, F) \subset \mathfrak{P}_p(E, F) \text{ and } \mathbf{P}_p(T) \leq \mathbf{I}_p(T)$$

(this inclusion will also follow easily from the factorization behaviour of p–integral and absolutely p–summing operators). These results were obtained by using the simple *transfer argument* : $\alpha \leq c\beta$ if and only if $\mathbf{A} \leq c\mathbf{B}$.

For special pairs of spaces the method of arguing with the representation and embedding theorem is the transfer argument from corollary 1 in 17.7.. An example: From the Chevet–Persson–Saphar inequalities (see 15.10.) it is known that $d_p \leq g_{p'}^*$ on $L_p \otimes \cdot$ which implies

$$\|(L_p \otimes_{d_p} E')' \hookrightarrow (L_p \otimes_{g_{p'}^*} E')'\| \leq 1 \ .$$

Since $d_p' = g_p^* \sim \mathfrak{P}_{p'}$ and $(g_{p'}^*)' = g_{p'}^t \sim \mathfrak{I}_{p'}^{dual}$ (by the description of the dual operator ideal in 17.8.), it follows from the representation theorem for maximal operator ideals that

$$\mathfrak{P}_{p'}(L_p, E) \subset \mathfrak{I}_{p'}^{dual}(L_p, E) \text{ and } \mathbf{I}_{p'}(T') = \mathbf{I}_{p'}^{dual}(T) \leq \mathbf{P}_{p'}(T) \ .$$

For a "transposed" result see Ex 17.7., for $p = 2$ see Ex 17.8.. The spaces F which satisfy this property of $F = L_p$ will be characterized in 25.9..

17.14. As has already been mentioned, Grothendieck's fundamental theorem of the metric theory of tensor products (section 14) has powerful consequences for operators; at this point we could examine the general versions in terms of $\mathcal{L}_p^g$-spaces, but we will postpone this task to section 23 and restrict ourselves for the moment to considering the sequence spaces ℓ_p for $p = 1, 2$, and ∞.

Since $\pi \leq K_G w_2$ on $\ell_\infty \otimes c_o$ and $w_2 \leq d_2$ (see 12.5.), it follows that

$$\mathfrak{L}(\ell_\infty, \ell_1) = (\ell_\infty \otimes_\pi c_o)' = (\ell_\infty \otimes_{w_2} c_o)' = (\ell_\infty \otimes_{d_2} c_o)' =$$
$$= \mathfrak{D}_2(\ell_\infty, \ell_1) = \mathfrak{P}_2(\ell_\infty, \ell_1)$$

by the representation theorem, and therefore:

THEOREM: *Every operator $T \in \mathfrak{L}(\ell_\infty, \ell_1)$ is 2-dominated and (hence) absolutely 2-summing; moreover,*

$$\mathbf{P}_2(T) \leq \mathbf{D}_2(T) \leq K_G \|T\| \ .$$

In 20.18. and Ex 20.14. it will be shown that this result is also equivalent to Grothendieck's inequality: For $\mathfrak{L} = \mathfrak{D}_2$ with the same constant (this is obvious), for $\mathfrak{L} = \mathfrak{P}_2$ with another one; see also Ex 25.5.. It is customary nowadays to call another operator form of Grothendieck's inequality

GROTHENDIECK'S THEOREM: *Every $T \in \mathfrak{L}(\ell_1, \ell_2)$ is absolutely 1-summing and*

$$\mathbf{P}_1(T) \leq K_G \|T\| \ .$$

PROOF: It was shown in 14.8. that $\pi \leq K_G d_\infty$ on $\ell_1^n \otimes \ell_2^n$ and hence, by 1-complementation, on $\ell_1^m \otimes \ell_2^n$ for all m, n. The density lemma (or the $\mathcal{L}_p^g$-local technique lemma) implies that this holds on $\ell_1 \otimes \ell_2$, hence $\mathfrak{L}(\ell_1, \ell_2) = (\ell_1 \otimes_\pi \ell_2)' = (\ell_1 \otimes_{d_\infty} \ell_2)' = \mathfrak{P}_1(\ell_1, \ell_2)$ since $\mathfrak{P}_1 \sim g_\infty^* = d_\infty'$. □

See 20.17. concerning the fact that the best constant in Grothendieck's theorem is the Grothendieck constant K_G. The little Grothendieck theorem (11.11.) says that

$$\mathcal{L}(E, \ell_2) = \mathfrak{P}_2(E, \ell_2) \text{ and } \mathbf{P}_2(T) \leq K_{LG}\|T\|,$$

whenever $E = \ell_\infty$ or ℓ_1. Since

$$(E \otimes_{d_2} \ell_2)' = \mathfrak{P}_2(E, \ell_2) = \mathcal{L}(E, \ell_2) = (E \otimes_\pi \ell_2)',$$

it follows:

PROPOSITION: *If $E = \ell_1$ or ℓ_∞, then*

$$d_2 \leq \pi \leq K_{LG}d_2 \qquad \text{on } E \otimes \ell_2.$$

17.15. Apart from the representation theorem there are other extremely useful characterizations of α–integral operators $T \in \mathcal{L}(E, F)$ in terms of tensor product mappings

$$id_G \otimes T : G \otimes_\beta E \longrightarrow G \otimes_\gamma F$$

with appropriate tensor norms β and γ — as has already been seen in the characterization 10.7. of integral operators and 11.1. of absolutely p–summing operators. Actually the following theorem will be proved along the lines of the proof of 10.7.; the key to that proof was to use two simple formulas (check for elementary tensors) which connect $T \in \mathcal{L}(E, F)$ and $T \otimes id_G$:
(1) For $\varphi \in (F \otimes_\pi G)'$ and $z \in E \otimes G$

$$\langle \beta_{L_\varphi \circ T}, z \rangle = \langle \varphi, T \otimes id_G(z) \rangle.$$

(2) For $z \in E \otimes F'$

$$\langle \beta_{\kappa_F \circ T}, z \rangle = \langle tr_F, T \otimes id_{F'}(z) \rangle.$$

THEOREM: *Let $(\mathfrak{A}, \mathbf{A}) \sim \alpha$ be associated and $T \in \mathcal{L}(E, F)$. Then the following two statements are equivalent:*
(a) $T \in \mathfrak{A}(E, F)$.
(b) *For all Banach spaces G (or only $G = F'$ or $G = L$ if $L' = F$ isometrically)*

$$id_G \otimes T : G \otimes_{\alpha^*} E \longrightarrow G \otimes_\pi F$$

is continuous.
In this case,

$$\mathbf{A}(T) = \|id_{F'} \otimes T : F' \otimes_{\alpha^*} E \longrightarrow F' \otimes_\pi F\| \geq \|id_G \otimes T : G \otimes_{\alpha^*} E \longrightarrow G \otimes_\pi F\|.$$

We prefer to formulate the theorem for $id_G \otimes T$ instead of $T \otimes id_G$, even though the proof will be for $T \otimes id_G : \otimes_{\alpha'} \longrightarrow \otimes_\pi$.

PROOF: If $T \in \mathfrak{A}(E, F)$ and $\varphi \in (F \otimes_\pi G)' = \mathfrak{L}(F, G')$, then
$$L_\varphi \circ T \in \mathfrak{A}(E, G') = (E \otimes_{\alpha'} G)'$$
and formula (1) implies that
$$|\langle \varphi, T \otimes id_G(z) \rangle| = |\langle \beta_{L_\varphi \circ T}, z \rangle| \leq \mathbf{A}(L_\varphi \circ T)\alpha'(z; E, G) \leq$$
$$\leq \|\varphi\| \mathbf{A}(T)\alpha'(z; E, G)$$
which shows that
$$\pi(T \otimes id_G(z); F, G) \leq \mathbf{A}(T)\alpha'(z; E, G) \ .$$
Conversely, assume (b) to be satisfied for $G = F'$. Since $\mathfrak{A}$ is regular (17.8.), it is enough to show that
$$\beta_{\kappa_F \circ T} \in \mathfrak{A}(E, F'') = (E \otimes_{\alpha'} F')' \ .$$
For $z \in E \otimes F'$ formula (2) implies that
$$|\langle \beta_{\kappa_F \circ T}, z \rangle| = |\langle tr_F, T \otimes id_{F'}(z) \rangle| \leq \pi(T \otimes id_{F'}(z); F, F') \leq$$
$$\leq \|T \otimes id_{F'} : \otimes_{\alpha'} \longrightarrow \otimes_\pi\| \alpha'(z; E, F')$$
which gives $T \in \mathfrak{A}(E, F)$ with the norm estimate. For $F = L'$ the proof is nearly the same for checking that $\beta_T \in (E \otimes_{\alpha'} L)' \ . \ \square$

It is important to note that (b) is actually a statement about the composition of operators
$$\mathfrak{F}(G, E) = G' \otimes E \xrightarrow{id \otimes T} G' \otimes F = \mathfrak{F}(G, F)$$
$$S \rightsquigarrow T \circ S$$
(see also 17.19. and Ex 17.6.).

17.16. In order to obtain characterizations involving ε — which will turn out to be a sort of a dualization of the last theorem — the following lemma will be needed; the reader will find it to be a simple consequence of the metric mapping property of tensor norms and the fact that for every $M \in \mathbf{FIN}$ and $\varepsilon > 0$ there is a 1–complemented subspace N of the Johnson space C_p (see 6.3.) with $d(M, N) \leq 1 + \varepsilon$.

LEMMA: Let β and γ be tensor norms, β finitely generated, $c \geq 0$ and $T \in \mathfrak{L}(E, F)$.
(1) If for a normed space G
$$\|id_M \otimes T : M \otimes_\beta E \longrightarrow M \otimes_\gamma F\| \leq c$$

for cofinally many $M \in \text{FIN}(G)$, then
$$\|id_G \otimes T : G \otimes_\beta E \longrightarrow G \otimes_\gamma F\| \leq c \,.$$

(2) *If for some Johnson space* C_p
$$\|id_{C_p} \otimes T : C_p \otimes_\beta E \longrightarrow C_p \otimes_\gamma F\| \leq c \,,$$

then
$$\|id_G \otimes T : G \otimes_\beta E \longrightarrow G \otimes_\gamma F\| \leq c$$

for all normed spaces G.

Everything is now prepared for the

COROLLARY: *Let* $(\mathfrak{A}, \mathbf{A}) \sim \alpha$ *be associated and* $T \in \mathfrak{L}(E, F)$. *If* α *is a right-accessible and finitely generated tensor norm or if* F *has the metric approximation property, then the following statements are equivalent:*
(a) $T \in \mathfrak{A}(E, F)$.
(b) *For all Banach spaces* G *(or only* G *a Johnson space* C_p *for some* $1 \leq p \leq \infty$*)*
$$id_G \otimes T : G \otimes_\varepsilon E \longrightarrow G \otimes_\alpha F$$

is continuous.
In this case,
$$\mathbf{A}(T) = \|id_{C_p} \otimes T : C_p \otimes_\varepsilon E \longrightarrow C_p \otimes_\alpha F\| \geq \|id_G \otimes T : G \otimes_\varepsilon E \longrightarrow G \otimes_\alpha F\| \,.$$

PROOF: $T \in \mathfrak{A}$ if and only if $T' \in \mathfrak{A}^{dual} \sim \alpha^t$ and in this case $\mathbf{A}(T) = \mathbf{A}^{dual}(T')$ holds. So the theorem and the lemma imply that $\mathbf{A}(T) \leq c$ if and only if

$$\|M' \otimes_{\alpha'} F' \xrightarrow{id \otimes T'} M' \otimes_\pi E'\| \leq c$$

for all $M \in \text{FIN}$. Since the map (see 6.4. for the duality of ε and π)

$$M' \otimes_{\alpha'} F' \xrightarrow{id \otimes T'} M' \otimes_\pi E' \stackrel{1}{=} (M \otimes_\varepsilon E)'$$

is the restriction to $M' \otimes_{\alpha'} F'$ of the dual of

$$M \otimes_\varepsilon E \xrightarrow{id \otimes T} M \otimes_\alpha F \xrightarrow{\leq 1} M \otimes_{\frac{1}{\alpha}} F \xrightarrow{1} (M' \otimes_{\alpha'} F')' \,,$$

it is clear from the lemma that (b) implies (a), and the converse is true if the relation $M \otimes_\alpha F \stackrel{1}{=} M \otimes_{\frac{1}{\alpha}} F$ holds for all $M \in \text{FIN}$; but this is true if α is right–accessible or if F has the metric approximation property. $\square$

Note that the assumption about the accessibility of α or the assumption on F was only used for the proof that (a) implies (b).

17.17. An example: *For the ideal* $(\mathfrak{I}_p, \mathbf{I}_p) \sim g_p$ *of p–integral operators the following three statements are equivalent:*
(a) $T \in \mathfrak{L}(E, F)$ *is p–integral.*
(b) *For all Banach spaces G (or only $G = F'$)*

$$id_G \otimes T : G \otimes_{g_p^*} E \longrightarrow G \otimes_\pi F$$

is continuous.
(c) *For all Banach spaces G*

$$id_G \otimes T : G \otimes_\varepsilon E \longrightarrow G \otimes_{g_p} F$$

is continuous.

For (a) or (b) implies (c) the fact that g_p is accessible is used — a fact which will be proved in section 21. However, this will not be needed in the following nice application of this result.

PROPOSITION: *Every positive operator $T \in \mathfrak{L}(E, L_p(\mu))$, where E is a $C(K)$ or some $L_\infty(\nu)$, is p–integral and $\mathbf{I}_p(T) = \|T\|$.*

It has already been seen in 11.2. that these operators are absolutely p–summing and $\mathbf{P}_p(T) = \|T\|$; see also Ex 18.3..

PROOF: It is known from theorem 7.3. that for all Banach spaces G

$$\|id_G \otimes T : G \otimes_\varepsilon E \longrightarrow G \otimes_{\Delta_p^t} L_p\| = \|id_G\| \, \|T\| \, .$$

Since $g_p \leq \Delta_p^t$ by the Chevet–Persson–Saphar inequality 15.10., it follows from the corollary (since L_p has the metric approximation property) that T is p–integral and $\mathbf{I}_p(T) = \|T\|$. $\square$

17.18. Sometimes it is possible to find better test spaces than $G = C_p$ in corollary 17.16.. Using the fact that $g_{p'}^*$ is accessible (this will be proved in section 21), it is clear that for $T \in \mathfrak{P}_p(E, F)$ and $G = \ell_p$ the map

$$id_{\ell_p} \otimes T : \ell_p \otimes_\varepsilon E \longrightarrow \ell_p \otimes_{g_{p'}^*} F$$

is continuous by the corollary. Since $\Delta_p \leq g_{p'}^*$ on $\ell_p \otimes F$, it follows from proposition 11.1. that this actually characterizes p–summing operators:

REMARK: $T \in \mathcal{L}(E,F)$ is absolutely p-summing if and only if
$$id_{\ell_p} \otimes T : \ell_p \otimes_\varepsilon E \longrightarrow \ell_p \otimes_{g_{p'}^*} F$$
is continuous. In this case, $\mathbf{P}_p(T) = \|id_{\ell_p} \otimes T : \ell_p \otimes_\varepsilon E \to \ell_p \otimes_{g_{p'}^*} F\|$.

Another description of $\mathfrak{P}_p(E,F)$ is given in Ex 17.9. with the help of Kwapień's characterization 11.8..

17.19. The theorem (and its corollary) also has structural consequences.

PROPOSITION 1: *Let $(\mathfrak{A}, \mathbf{A})$ be a maximal normed operator ideal such that the associated tensor norm α is accessible. Then*
$$\mathfrak{A}^* \circ \mathfrak{A} \subset \mathfrak{I} \quad \text{and} \quad \mathbf{I}(T \circ S) \leq \mathbf{A}^*(T)\mathbf{A}(S) ;$$
the composition $T \circ S$ of operators $S \in \mathfrak{A}$ and $T \in \mathfrak{A}^*$ is integral.

In 25.4. this result will be generalized to the so-called "cyclic composition theorem"; it will also be shown in 25.4. that $\mathfrak{A}^* \circ \mathfrak{A} \subset \mathfrak{I}$ implies that α is necessarily right-accessible. In section 21 the accessibility of α will be explained in terms of the operator ideal $\mathfrak{A}$. If α is not accessible (recall that there is an example known), the proof will also show that
$$\mathfrak{A}^*(F,G) \circ \mathfrak{A}(E,F) \subset \mathfrak{I}(E,G) ,$$
with the norm inequality whenever F has the metric approximation property. In 19.4. the reader will find examples in which $\mathfrak{A}^* \circ \mathfrak{A}$ consists of nuclear operators only. For another "converse" of the proposition, see Ex 17.6., Ex 17.15. and Ex 25.10..

PROOF: Since $\mathfrak{A} \sim \alpha$ and $\mathfrak{A}^* \sim \alpha^*$, the theorem and its corollary show that for $S \in \mathfrak{A}(E,F)$ and $T \in \mathfrak{A}^*(F,G)$ the composition
$$id_{G'} \otimes (T \circ S) : G' \otimes_\varepsilon E \xrightarrow{id \otimes S} G' \otimes_\alpha F \xrightarrow{id \otimes T} G' \otimes_\pi G$$
has norm $\leq \mathbf{A}^*(T)\mathbf{A}(S)$; since $\pi \sim \mathfrak{I}$ and $\pi^* = \varepsilon$, the theorem implies that $T \circ S$ is integral and $\mathbf{I}(T \circ S) \leq \mathbf{A}^*(T)\mathbf{A}(S)$. □

If $\dim E = n$, then $\mathbf{I}(id_E) = \mathbf{N}(id_E) = n$ (see 10.4. and 3.7.). It follows that for each isomorphism $T \in \mathcal{L}(E,F)$
$$\mathbf{A}(T : E \to F)\mathbf{A}^*(T^{-1} : F \to E) \geq n$$
for each normed operator ideal $(\mathfrak{A}, \mathbf{A})$ (maximal or not: E and F are finite dimensional). For special cases there is equality: Recall the investigation 15.9. of certain

norms on $\mathbb{K}^n$ which are invariant under the isomorphism $\mathbf{M}_w$ (multiplication of the components by ± 1) and the cyclic shift S. In terms of operators, 15.9. reads as follows:

PROPOSITION 2: *Let E and F be $\mathbb{K}^n$ equipped with norms such that $\|M_w\| = \|S\| = 1$ in $\mathfrak{L}(E', E')$ and $\mathfrak{L}(F, F)$ for all $w \in D_n$. Then*

$$\mathbf{A}(id_{\mathbb{K}^n} : E \to F)\mathbf{A}^*(id_{\mathbb{K}^n} : F \to E) = n$$

for all normed operator ideals $(\mathfrak{A}, \mathbf{A})$.

PROOF: The element $u^{(n)} := \sum_{k=1}^{n} e_k \otimes e_k$ corresponds to the identity operator. If α is the tensor norm associated with $\mathfrak{A}^{max}$, then

$$\mathbf{A}(id : E \to F) = \alpha(u^{(n)}; E', F)$$
$$\mathbf{A}^*(id : F \to E) = \alpha^*(u^{(n)}; F', E) = \alpha'(u^{(n)}; E, F')$$

— and the result follows from $\alpha(u^{(n)}; E', F)\alpha'(u^{(n)}; E, F') = n$ which was shown in 15.9.. □

This implies, for example, that the results in 8.8. about the nuclear norm of the identity $id : \ell_p^n \to \ell_q^n$ can be directly calculated from the operator norm of $id : \ell_q^n \to \ell_p^n$ since $\mathfrak{L} \sim \varepsilon$ and $\varepsilon^* = \pi \sim \mathfrak{J}$, which is $\mathfrak{N}$ on finite dimensional spaces. The exact calculation of $\mathbf{P}_1(id_{\ell_2^n})$ in 11.10. gives the exact values of the norm of $id_{\ell_2^n}$ in $\mathfrak{P}_1^* = \mathfrak{J}_\infty = \mathfrak{L}_\infty$, that is, the exact values of the ∞–factorable norm:

$$\mathbf{L}_\infty(id_{\ell_2^n}) = n \left(\mathbf{P}_1(id_{\ell_2^n})\right)^{-1}.$$

It follows from corollary 11.10. that

$$\lim_{n \to \infty} n^{-1/2} \mathbf{L}_\infty(id_{\ell_2^n}) = \begin{cases} \sqrt{2/\pi} & \text{real case} \\ \sqrt{\pi}/2 & \text{complex case} \end{cases}.$$

These results are quite interesting since it turns out that $\mathbf{L}_\infty(id_E)$ is the projection constant of E whenever E is finite dimensional, see 23.6..

17.20. The final topic of this section is the following question: Under what conditions is the natural map

$$I : E \tilde{\otimes}_\alpha F \longrightarrow E \tilde{\otimes}_\varepsilon F \xhookrightarrow{1} \mathfrak{L}(E', F)$$

injective? If α is totally accessible, then the duality theorem 15.5. for tensor norms implies that

17. The Representation Theorem for Maximal Operator Ideals

$$E \tilde{\otimes}_\alpha F = E \tilde{\otimes}_{\overset{1}{\alpha}} F \overset{1}{\hookrightarrow} (E' \tilde{\otimes}_{\alpha'} F')' \hookrightarrow \mathcal{L}(E', F'')$$

with the diagonal maps going through $E \tilde{\otimes}_\varepsilon F$ (labels I and 1),

hence I is injective.

PROPOSITION: *If α is a finitely generated tensor norm and E and F are Banach spaces, one of which has the approximation property, then the natural map*

$$I : E \tilde{\otimes}_\alpha F \longrightarrow E \tilde{\otimes}_\varepsilon F$$

is injective.

PROOF: Assume that F has the approximation property, $z \in E \tilde{\otimes}_\alpha F$ and $I(z) = 0$. It has to be shown that $\langle \varphi, z \rangle = 0$ for all

$$\varphi \in (E \tilde{\otimes}_\alpha F)' \hookrightarrow \mathcal{L}(E, F') .$$

By theorem 17.15. (and obviously also by the correspondence between maximal operator ideals and tensor norms)

$$L_\varphi \otimes id_F : E \tilde{\otimes}_\alpha F \longrightarrow F' \tilde{\otimes}_\pi F$$

is continuous. The lower map in the diagram

$$\begin{array}{ccc} E \tilde{\otimes}_\alpha F & \overset{I}{\longrightarrow} & E \tilde{\otimes}_\varepsilon F \\ L_\varphi \tilde{\otimes}_{\alpha, \pi} id_F \downarrow & & \downarrow L_\varphi \tilde{\otimes}_\varepsilon id_F \\ F' \tilde{\otimes}_\pi F & \longrightarrow & F' \tilde{\otimes}_\varepsilon F \end{array}$$

is injective by the approximation property of F (see 5.6.), hence

$$L_\varphi \tilde{\otimes}_{\alpha, \pi} id_F(z) = 0 \in F' \tilde{\otimes}_\pi F .$$

Formula (2) in 17.15. implies that

$$\langle \varphi, z \rangle = \langle tr_F, L_\varphi \tilde{\otimes}_{\alpha, \pi} id_F(z) \rangle = 0 ,$$

which was to be shown. □

The diagram

$$E'\tilde{\otimes}_\alpha F \rightleftarrows \begin{array}{c} E'\tilde{\otimes}_\varepsilon F \\ \mathfrak{A}(E,F) \end{array} \rightrightarrows \mathfrak{L}(E,F)$$

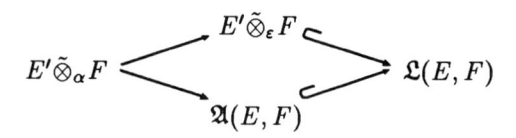

shows that:

COROLLARY: *If $(\mathfrak{A}, \mathbf{A}) \sim \alpha$, then the natural map*

$$E'\tilde{\otimes}_\alpha F \longrightarrow \mathfrak{A}(E,F)$$

is injective if α is totally accessible or if E' or F has the approximation property.

See Ex 25.4. and Ex 17.11. – Ex 17.14. for more results connected with the injectivity of $\tilde{\otimes}_\alpha \to \tilde{\otimes}_\varepsilon$.

17.21. The components $\mathfrak{A}(E,F)$ of a maximal operator ideal are not in general closed in $\mathfrak{L}(E,F)$ (with the operator norm) (why?) — however the unit balls are. Even more is true:

PROPOSITION: *Let $(\mathfrak{A}, \mathbf{A})$ be a maximal normed operator ideal. Then the closed unit ball of $\mathfrak{A}(E,F)$ is closed in the space $\mathfrak{L}(E,F)$ with respect to the weak operator topology $\sigma(\mathfrak{L}(E,F), E \otimes F')$. In other words: If (T_ι) is a net in $\mathfrak{A}(E,F)$ which converges pointwise weakly to $T \in \mathfrak{L}(E,F)$ and $\sup_\iota \mathbf{A}(T_\iota) < \infty$, then $T \in \mathfrak{A}$ and*

$$\mathbf{A}(T) \leq \liminf \mathbf{A}(T_\iota) .$$

PROOF: Since the unit ball of $\mathfrak{A}(M,N)$ is closed for $\sigma(..., M \otimes N')$ (finite dimensional spaces), the second assertion follows from the definition of $\mathbf{A}^{max}$. □

Exercises:

Ex 17.1. (a) Let $\{(\mathfrak{A}_\lambda, \mathbf{A}_\lambda) | \lambda \in \Lambda\}$ be a family of maximal normed operator ideals. Then the class of all $T \in \mathfrak{L}$ with

$$\mathbf{A}(T) := \sup_{\lambda \in \Lambda} \mathbf{A}_\lambda(T) < \infty$$

is a maximal normed operator ideal.
(b) Deduce from (a) and 10.7. that the ideal $(\mathfrak{I}, \mathbf{I})$ of integral operators is maximal.

Ex 17.2. A normed operator ideal $(\mathfrak{A}, \mathbf{A})$ is maximal if and only if there is a finitely generated tensor norm β such that

$$\mathfrak{A}(E,F) \stackrel{1}{=} (E \otimes_\beta F')' \cap \mathfrak{L}(E,F)$$

for all Banach spaces E and F. In this case,
$$\mathfrak{A}(E, F') \stackrel{1}{=} (E \otimes_\beta F)' .$$

Ex 17.3. Deduce directly from the representation theorem that $T \in \mathfrak{A}$ if and only if $T'' \in \mathfrak{A}$ whenever $\mathfrak{A}$ is maximal. Hint: Embedding lemma and $^\wedge \varphi$.

Ex 17.4. Let $(\mathfrak{A}, \mathbf{A}) \sim \alpha$ and E, F be Banach spaces. Assume that α is totally accessible or: α accessible and E or F' has the metric approximation property or: E and F' have the metric approximation property. Then an operator $T \in \mathcal{L}(E, F)$ is in the adjoint operator ideal $\mathfrak{A}^*$ if and only if
$$\mathbf{A}^\triangle(T) := \sup\{|tr_E(S \circ T)| \,\big|\, S \in \mathfrak{F}(F, E), \mathbf{A}(S) \le 1\} < \infty .$$
In this case this number is $\mathbf{A}^*(T)$. Hint: Embedding theorem.

Gordon–Lewis–Retherford [85] studied the ideal $\mathfrak{A}^\triangle$ of all T with $\mathbf{A}^\triangle(T) < \infty$ under the name *conjugate* ideal of $\mathfrak{A}$.

Ex 17.5. Let $(\mathfrak{A}, \mathbf{A})$ be a normed operator ideal. Show that $T \in \mathcal{L}(E, F)$ is in $\mathfrak{A}^*$ if and only if there is a $c \ge 0$ such that for all $M, N \in$ FIN
$$|tr_N(U \circ T \circ S \circ R)| \le c\|U\|\,\|S\|\,\mathbf{A}(R)$$
for all $R \in \mathcal{L}(N, M), S \in \mathcal{L}(M, E)$ and $U \in \mathcal{L}(F, N)$.

Ex 17.6. Use the embedding theorem and
$$\|T' \otimes id_E : F' \otimes_{\alpha^*} E \longrightarrow E' \otimes_\pi E\| = \mathbf{A}(T)$$
to show that: If $(\mathfrak{A}, \mathbf{A}) \sim \alpha$ are associated, α is accessible, E and F are Banach spaces and E has the metric approximation property, then $T \in \mathfrak{A}(E, F)$ if and only if there is a $c \ge 0$ such that
$$\mathbf{I}(S \circ T) \le c\mathbf{A}^*(S)$$
for all $S \in \mathfrak{F}(F, E)$. In this case, $\mathbf{A}(T)$ is the minimum of these constants c. See also Ex 17.15..

Ex 17.7. Show that $\mathfrak{P}_p^{dual}(E, L_p) \subset \mathfrak{I}_p(E, L_p)$. Hint: Chevet–Persson–Saphar inequality. See 25.9. for more information about this relation.

Ex 17.8. Use the result in 11.12. to show that:
(a) There are no constants c_1 and c_2 with $d_2 \le c_1 g_2$ and $g_2 \le c_2 d_2$.
(b) A Banach space E is a Hilbert space if and only if $d_2 \le cg_2$ on $E \otimes \ell_2$ for some $c \ge 1$.

Ex 17.9. Use Kwapień's characterization 11.8. of absolutely p–summing operators to show that $T \in \mathfrak{P}_p(E, F)$ if and only if
$$id_{\ell_{p'}} \otimes T' : \ell_{p'} \otimes_{d_{p'}} F' \longrightarrow \ell_{p'} \otimes_\pi E'$$

is continuous.

Ex 17.10. If $(\mathfrak{A}, \mathbf{A}) \sim \alpha$ and α is accessible, then the following three statements are equivalent for each Banach space E and $c \geq 0$:

(a) $E \in \text{space}(\mathfrak{A})$ and $\mathbf{A}(id_E) \leq c$.

(b) $\pi \leq c\alpha^*$ on $G \otimes E$ for all Banach spaces G (or only $G = E'$).

(c) $\alpha \leq c\varepsilon$ on $G \otimes E$ for all Banach spaces G (or only some Johnson space C_p).

Ex 17.11. Let E or F have the approximation property and $z \in E \tilde{\otimes}_\alpha F$ for some finitely generated tensor norm. If $\varphi \tilde{\otimes}_\alpha id_F(z) = 0$ for all $\varphi \in E'$, then $z = 0$. Hint: 17.20. and Ex 15.2..

Ex 17.12. Let $T_i \in \mathfrak{L}(E_i, F_i)$ be two injective operators between Banach spaces and α a finitely generated tensor norm. If E_1 or E_2 has the approximation property, then $T_1 \tilde{\otimes}_\alpha T_2 \in \mathfrak{L}(E_1 \tilde{\otimes}_\alpha E_2, F_1 \tilde{\otimes}_\alpha F_2)$ is injective. Hint: 17.20. and this property for $\alpha = \varepsilon$.

Ex 17.13. The canonical map $L_p(\mu) \tilde{\otimes}_{\Delta_p} E \to L_p(\mu) \tilde{\otimes}_\varepsilon E$ is injective. Hint: Use 15.10. — or prove it directly for $p < \infty$ using the fact that every $f \in L_p(E)$ has separable range.

Ex 17.14. If α is a finitely generated tensor norm and $\alpha \leq \Delta_p$ on $L_p(\mu) \otimes E$, then $L_p(\mu) \tilde{\otimes}_{\Delta_p} E \to L_p(\mu) \tilde{\otimes}_\alpha E$ is injective. Also: if $\alpha \geq \Delta_p$ on $L_p(\mu) \otimes E$, the mapping $L_p(\mu) \tilde{\otimes}_\alpha E \to L_p(\mu) \tilde{\otimes}_{\Delta_p} E$ is injective.

Ex 17.15. Use the representation theorem and corollary 17.16. to prove the converse of proposition 1 in 17.19.: Let $(\mathfrak{A}, \mathbf{A})$ be a maximal normed operator ideal the associated tensor norm of which is accessible. Then $T \in \mathfrak{L}(E, F)$ is in $\mathfrak{A}$ if and only if for all Banach spaces G and operators $S \in \mathfrak{A}^*(F, G)$ the composition $S \circ T$ is integral. See also Ex 25.10..

Ex 17.16. Let $S \in \mathfrak{L}(X, Y)$ and $T \in \mathfrak{L}(E, F)$ be continuous operators between Banach spaces and α, β two tensor norms such that

$$S \otimes T : X \otimes_\alpha E \longrightarrow Y \otimes_\beta F$$

is continuous. Question: Is $S \otimes T'' : X \otimes_\alpha E'' \to Y \otimes_\beta F''$ also continuous? Unfortunately, we do not have a complete answer, even in the case where $\beta = \pi$. Assume α to be finitely generated. Show that the answer is positive under each of the following conditions:

(a) T is weakly compact. Hint: $[(S \otimes T)'\varphi]^\wedge = \varphi \circ (S \otimes T^*)$ for all $\varphi \in (Y \otimes_\pi F)'$ if $T^* : E'' \to F$ is the astriction of T''.

(b) $\beta \leq c\beta^{\leftarrow}$ on $Y \otimes F''$ for some $c \geq 0$ (the right–cofinite hull, see Ex 12.3.) Hint: Proposition 17.4..

(c) β is totally accessible.

(d) β is accessible and: Y or F'' has the bounded approximation property.

(e) Y and F'' have the bounded approximation property.

Ex 17.17. *Density lemma for maximal normed operator ideals:* Let $(\mathfrak{A}, \mathbf{A})$ be a normed operator ideal, E and F Banach spaces, $E_o \subset E$ and $G_o \subset F'$ norm–dense subspaces, $C \subset \text{FIN}(E_o)$ and $D \subset \text{FIN}(G_o)$ cofinal subsets. Then
$$\mathbf{A}^{max}(T : E \longrightarrow F) = \sup\{\mathbf{A}(Q_L^F \circ T \circ I_M^E) \mid M \in C, L^0 \in D\} \ .$$
Hint: Representation theorem, density lemma 13.4. and α' finitely generated.

18. (p,q)–Factorable Operators

The purpose of this section is to show that every $\alpha'_{p,q}$–continuous functional on $E \otimes F$ factors through the integrating functional on $L_{q'}(\mu) \otimes L_{p'}(\mu)$ for some measure μ. In terms of operators this means that every (p, q)–factorable operator $T \in \mathfrak{L}(E, F)$ has the property that $\kappa_F \circ T$ factors through the embedding $L_{q'}(\mu) \hookrightarrow L_p(\mu)$. The most important special cases are the p–factorable and p–integral operators. In these cases, the proofs are actually simpler than in the general case. A central tool for the demonstrations will be the ultraproduct technique, the basic properties of which will be presented as well.

18.1. The maximal normed operator ideal $(\mathfrak{L}_{p,q}, \mathbf{L}_{p,q})$ of (p, q)–factorable operators is associated with the tensor norm $\alpha_{p,q}$. Using the definition of $\alpha_{p,q}$ directly, it was shown in 17.10. that for finite dimensional Banach spaces M and N the $\mathbf{L}_{p,q}$–norm of $T \in \mathfrak{L}(M, N)$ is given by the factorizations

$$\begin{array}{ccc} M & \xrightarrow{T} & N \\ R \downarrow & & \uparrow S \\ \ell_{q'}^n & \xrightarrow{D_\lambda} & \ell_p^n \end{array},$$

where D_λ is a diagonal operator the entries of which can clearly be assumed to be non–negative:
$$\mathbf{L}_{p,q}(T) = \inf \|S\| \, \|D_\lambda\| \, \|R\| \ .$$
It is a simple matter of concentration (and this will be done in 18.10.) to show that D_λ factors through an embedding $L_{q'}(\mu) \hookrightarrow L_p(\mu)$ for a finite measure with control of the

norms. Since the operator ideal $\mathfrak{L}_{p,q}$ is maximal, every $T \in \mathfrak{L}_{p,q}$ is determined by its finite dimensional components — and a tricky "fitting together" technique (ultraproducts), representation theorems for abstract L_p–spaces and a characterization of positive operators $L_{q'}(\mu) \longrightarrow L_p(\nu)$ will give the desired characterization: $T \in \mathfrak{L}_{p,q}(E,F)$ if and only if $\kappa_F \circ T$ factors through some embedding $L_{q'}(\mu) \hookrightarrow L_p(\mu)$ (see 18.11.); recall from 17.8. that $T \in \mathfrak{L}_{p,q}$ if and only if $\kappa_F \circ T \in \mathfrak{L}_{p,q}$.

18.2. Not $T : E \to F$ but $\kappa_F \circ T : E \to F''$ can be factored. Therefore it is more reasonable to treat functionals in $(E \otimes_{\alpha'_{p,q}} F)'$ rather than operators.

Using the fact that $\varepsilon = w_{p'}$ on $N' \otimes \ell^n_{p'}$, it was shown in 15.8. that $\pi = w'_p$ on $L_p(\mu) \otimes F$ for every measure μ (not necessarily finite), and hence the integrating functional

$$\beta_I : L_p(\mu) \otimes L_{p'}(\mu) \longrightarrow \mathbb{K}$$
$$f \otimes g \rightsquigarrow \int fg\, d\mu$$

(which is nothing but the trace) is w'_p-continuous. In other words, for every measure μ the identity map on $L_p(\mu)$ is in the ideal $\mathfrak{L}_p$ of p-factorable operators associated with w_p. For the ideal $\mathfrak{L}_{p,q} \sim \alpha_{p,q}$ of (p,q)-factorable operators the following result holds (the integrating functional is defined for $1/p + 1/q > 1$ only if μ is a finite measure!):

PROPOSITION: Let $p, q, r \in [1, \infty]$ with $1/p + 1/q = 1 + 1/r$ and μ a finite measure on a set Ω. Then the integrating functional

$$\beta_I : L_{q'}(\mu) \otimes L_{p'}(\mu) \longrightarrow \mathbb{K}$$
$$f \otimes g \rightsquigarrow \int fg\, d\mu$$

is $\alpha'_{p,q}$-continuous. In other words, the embedding $I : L_{q'}(\mu) \hookrightarrow L_p(\mu)$ is (p,q)-factorable; moreover, $\mathbf{L}_{p,q}(I) = \|I\| = \mu(\Omega)^{1/r} = \|\beta_I\|_{(L_{q'} \otimes_{\alpha'_{p,q}} L_{p'})'}$.

Clearly, $\mu(\Omega)^{1/\infty}$ is understood as being 1 if $\mu(\Omega) > 0$.

PROOF: Since $q' \geq p$, the embedding I is defined and continuous (the norm is $\mu(\Omega)^{1/r}$) and so is the integrating functional β_I. Therefore, by the density lemma 13.4., it is enough to check the $\alpha'_{p,q}$-continuity of β_I on the dense subspace of step functions. Since $\alpha'_{p,q}$ is finitely generated, it is enough to take pairwise disjoint integrable sets $C_1, ..., C_m \subset \Omega$, the subspace

$$M := \left\{ \sum_{k=1}^{m} \alpha_k \chi_{C_k} \,\Big|\, \alpha_k \in \mathbb{K} \right\}$$

and $M_s := (M, \|\ \|_{L_s})$. Then it has to be shown that

$$\beta_I|_{M \otimes M} \in (M_{q'} \otimes_{\alpha'_{p,q}} M_{p'})' \stackrel{1}{=} M'_{q'} \otimes_{\alpha_{p,q}} M'_{p'}$$

has norm $\leq \mu(\Omega)^{1/r}$. For this define the functionals φ_k on M by

$$\langle \varphi_k, f \rangle := \mu(C_k)^{-1} \int_{C_k} f d\mu ;$$

it follows that

$$\langle \beta_I, f \otimes g \rangle = \int fg d\mu = \sum_{k=1}^{m} \mu(C_k) \langle \varphi_k, f \rangle \langle \varphi_k, g \rangle =$$

$$= \langle \sum_{k=1}^{m} \mu(C_k) \varphi_k \otimes \varphi_k, f \otimes g \rangle$$

for all $f, g \in M$ and therefore

$$z_M := \sum_{k=1}^{m} \mu(C_k)^{1/r} (\mu(C_k)^{1/q'} \varphi_k) \otimes (\mu(C_k)^{1/p'} \varphi_k) \in M'_{q'} \otimes_{\alpha_{p,q}} M'_{p'}$$

corresponds to $\beta_I|_{M \otimes M}$ under the isometry $(M_{q'} \otimes_{\alpha'_{p,q}} M_{p'})' = M'_{q'} \otimes_{\alpha_{p,q}} M'_{p'}$. Since it is immediate that

$$w_s(\mu(C_k)^{1/s} \varphi_k; M'_s) = 1 \text{ and } \ell_r(\mu(C_k)^{1/r}) \leq \mu(\Omega)^{1/r} ,$$

it follows from the definition of $\alpha_{p,q}$ that

$$\|\beta_I|_{M \otimes M}\|_{(M_{q'} \otimes_{\alpha'_{p,q}} M_{p'})'} \leq \mu(\Omega)^{1/r} . \quad \square$$

Thus, the first step towards the factorization theorem has been completed: the integrating functional (or the embedding), which will turn out to be the typical $\alpha'_{p,q}$-continuous functional, actually is $\alpha'_{p,q}$-continuous.

18.3. For finite dimensional M and N the typical factorization of

$$\varphi \in (M \otimes_{\alpha'_{p,q}} N)' = M' \otimes_{\alpha_{p,q}} N'$$

goes as follows: the functional φ has $\alpha'_{p,q}$-norm $< c$ if and only if there exist operators $R \in \mathcal{L}(M, \ell_{q'}^n)$, $S \in \mathcal{L}(N, \ell_{p'}^n)$ and $\lambda \in [0, \infty[^n$ such that

$$\varphi : M \otimes N \xrightarrow{R \otimes S} \ell_{q'}^n \otimes \ell_{p'}^n \xrightarrow{\delta_\lambda} \mathbb{K}$$

where δ_λ is the bilinear form $\sum_{k=1}^{n} \lambda_k e_k \otimes e_k$ and $\|R\| \|S\| \|\delta_\lambda\| < c$; this was already mentioned in 17.10.. The functional δ_λ is *positive* (i.e. $\langle \delta_\lambda, f \otimes g \rangle \geq 0$ whenever $f \geq 0$ and $g \geq 0$). Since $\alpha'_{p,q}$ is finitely generated, $\varphi \in (E \otimes F)^*$ is $\alpha'_{p,q}$-continuous if and only if the restrictions $\varphi|_{M \otimes N}$ have uniformly bounded $\alpha'_{p,q}$-norm (this is another formulation of the maximality of $\mathfrak{L}_{p,q}$). The ultraproduct technique (which will be explained now) allows one to study φ in terms of its finite dimensional parts.

18.4. Let I be a non–empty set. If $\mathfrak{V}$ is any filter on I, Zorn's lemma implies that there is an ultrafilter $\mathfrak{U}$ finer than $\mathfrak{V}$. Recall that a topological Hausdorff space X is compact if and only if every ultrafilter on X converges; in particular, if $f : I \to \mathbb{R}$ is a bounded map, then the family $(f(\iota))_{\iota \in I}$ converges along every ultrafilter $\mathfrak{U}$; the limit is usually denoted by $\lim_{\mathfrak{U}} f(\iota)$.

Now fix an ultrafilter $\mathfrak{U}$ in I. If $(E_\iota)_{\iota \in I}$ is a family of Banach spaces, consider in the Banach space

$$\ell_\infty(E_\iota) := \left\{ x = (x_\iota)_{\iota \in I} \in \prod_{\iota \in I} E_\iota \;\middle|\; \|x\|_\infty := \sup_{\iota \in I} \|x_\iota\| < \infty \right\}$$

the closed subspace $c_o^{\mathfrak{U}}(E_\iota)$ of all (x_ι) with $\lim_{\mathfrak{U}} \|x_\iota\| = 0$.

The *ultraproduct of* $(E_\iota)_{\iota \in I}$ *along* $\mathfrak{U}$ is defined to be

$$(E_\iota)_{\mathfrak{U}} := \ell_\infty(E_\iota)/c_o^{\mathfrak{U}}(E_\iota) .$$

With the quotient norm this is a Banach space, the elements of which are denoted by $(x_\iota)_{\mathfrak{U}}$ (whenever $(x_\iota)_{\iota \in I} \in \ell_\infty(E_\iota)$), and after a moment's reflection one sees that

$$\|(x_\iota)_{\mathfrak{U}}\| = \lim_{\mathfrak{U}} \|x_\iota\| .$$

In the case that each $E_\iota = E$ the ultraproduct $(E_\iota)_{\mathfrak{U}}$ is called *the ultrapower of E along $\mathfrak{U}$*. Perhaps the reader is tempted to picture the ultraproduct as being something like ℓ_∞/c_o — this is erroneous: The right analogue is c/c_o (where c is the space of convergent sequences). This indicates that ultraproducts are much smaller than one might at first guess; see in particular Ex 18.5. and Ex 18.10..

If $(E_\iota)_{\iota \in I}$ and $(F_\iota)_{\iota \in I}$ are two families of Banach spaces and $T_\iota \in \mathfrak{L}(E_\iota, F_\iota)$ with $\|T_\iota\| \leq c < \infty$ for all $\iota \in I$, then

$$T^I : \ell_\infty(E_\iota) \longrightarrow \ell_\infty(F_\iota)$$
$$(x_\iota) \rightsquigarrow (T_\iota x_\iota)$$

is defined and $\|T^I\| \leq c$. Since $\|T_\iota x_\iota\| \leq c\|x_\iota\|$ holds, T^I maps $c_o^{\mathfrak{U}}(E_\iota)$ into $c_o^{\mathfrak{U}}(F_\iota)$, hence there is a unique map $T \in \mathfrak{L}((E_\iota)_{\mathfrak{U}}, (F_\iota)_{\mathfrak{U}})$ such that

$$\begin{array}{ccc} \ell_\infty(E_\iota) & \xrightarrow{T^I} & \ell_\infty(F_\iota) \\ \downarrow & & \downarrow \\ (E_\iota)_{\mathfrak{U}} & \xrightarrow{T} & (F_\iota)_{\mathfrak{U}} \end{array}$$

commutes; T is called the *ultraproduct* $(T_\iota)_\mathfrak{U}$ of the operators T_ι. It is clear from the description of the norm in the ultraproduct that

$$\|(T_\iota)\|_\mathfrak{U} = \lim_\mathfrak{U} \|T_\iota\| \ .$$

LEMMA: *If $\psi_\iota \in (E_\iota \otimes_\pi F_\iota)'$ are functionals with $\sup_{\iota \in I} \|\psi_\iota\| < \infty$, then*

$$\langle \psi, (x_\iota)_\mathfrak{U} \otimes (y_\iota)_\mathfrak{U} \rangle := \lim_\mathfrak{U} \langle \psi_\iota, x_\iota \otimes y_\iota \rangle$$

defines a continuous linear functional on $(E_\iota)_\mathfrak{U} \otimes_\pi (F_\iota)_\mathfrak{U}$ with $\|\psi\| = \lim_\mathfrak{U} \|\psi_\iota\|$. Notation: $\psi = \lim_\mathfrak{U} \psi_\iota$.

PROOF: The mapping

$$\tilde\psi : \ell_\infty(E_\iota) \times \ell_\infty(F_\iota) \longrightarrow \mathbb{K}$$
$$((x_\iota), (y_\iota)) \rightsquigarrow \lim_\mathfrak{U} \langle \psi_\iota, x_\iota \otimes y_\iota \rangle$$

is bilinear. With $c := \sup \|\psi_\iota\|$, it follows from

$$|\langle \psi_\iota, x_\iota \otimes y_\iota \rangle - \langle \psi_\iota, \overline{x}_\iota \otimes \overline{y}_\iota \rangle| \le c(\|x_\iota - \overline{x}_\iota\| \, \|y_\iota\| + \|\overline{x}_\iota\| \, \|y_\iota - \overline{y}_\iota\|)$$

that $\tilde\psi$ factors through the map ψ of the statement. Since

$$|\lim_\mathfrak{U} \langle \psi_\iota, x_\iota \otimes y_\iota \rangle| \le \lim_\mathfrak{U} \|\psi_\iota\| \, \|x_\iota\| \, \|y_\iota\| \ ,$$

one obtains $\|\psi\| \le \lim_\mathfrak{U} \|\psi_\iota\|$ and even equality by choosing appropriate x_ι and y_ι. $\square$

If each E_ι is a Banach lattice, then $\ell_\infty(E_\iota)$ is also a Banach lattice under componentwise ordering. Since $c_o^\mathfrak{U}(E_\iota)$ is a closed, solid subspace (since the norms are lattice norms), it follows (see Appendix A6.) that $(E_\iota)_\mathfrak{U}$ is a Banach lattice under its natural quotient ordering. If each E_ι is an abstract L_p-spaces ($1 \le p < \infty$), i.e.

$$\|x + y\|^p = \|x\|^p + \|y\|^p \qquad \qquad \text{if } x \wedge y = 0 \ ,$$

then their ultraproduct is also an abstract L_p-space: take $x = (x_\iota)_\mathfrak{U}$ and $y = (y_\iota)_\mathfrak{U}$ with $x \wedge y = 0$; then $(x_\iota \wedge y_\iota) \in c_o^\mathfrak{U}(E_\iota)$ and $(x_\iota - x_\iota \wedge y_\iota) \wedge (y_\iota - x_\iota \wedge y_\iota) = 0$, hence

$$\|x + y\|^p = \lim_\mathfrak{U} \|(x_\iota - x_\iota \wedge y_\iota) + (y_\iota - x_\iota \wedge y_\iota)\|^p =$$
$$= \lim_\mathfrak{U} \|x_\iota - x_\iota \wedge y_\iota\|^p + \lim_\mathfrak{U} \|y_\iota - x_\iota \wedge y_\iota\|^p = \|x\|^p + \|y\|^p \ .$$

The same holds for abstract M-spaces ($\|x + y\| = \max\{\|x\|, \|y\|\}$ if $x \wedge y = 0$).

The ultraproduct techniques in Banach space theory were developed around 1970 by various authors; recent references are Heinrich [104] and Sims [257].

18.5. To apply the ultraproduct technique for the description of a $\psi \in (E \otimes_{\alpha'_{p,q}} F)'$, with the help of its finite dimensional restrictions, define on the index set

$$I := \text{FIN}(E) \times \text{FIN}(F) \times \,]0,1]$$

an ordering by

$$(M_1, N_1, \varepsilon_1) \leq (M_2, N_2, \varepsilon_2) \text{ whenever } M_1 \subset M_2, N_1 \subset N_2 \text{ and } \varepsilon_1 \geq \varepsilon_2 ,$$

and take $\mathfrak{W}$ the order filter which has as basis the sets $\{\iota \in I | \iota \geq \iota_o\}$ and $\mathfrak{U}$ any ultrafilter finer than $\mathfrak{W}$; note that $\mathfrak{U}$ is not unique. For $x \in E$ and $\iota = (M, N, \varepsilon)$ define

$$x_\iota := \begin{cases} x & \text{if } x \in M \\ 0 & \text{if } x \notin M \end{cases}$$

and $E_\iota := M$. It follows that

$$J_E : E \longrightarrow (E_\iota)_\mathfrak{U}$$
$$x \rightsquigarrow (x_\iota)_\mathfrak{U}$$

is a linear map — which is an isometry by the definition of the norm in the ultraproduct. With $F_\iota := N$ for $\iota = (M, N, \varepsilon)$ there is also an isometry $J_F : F \to (F_\iota)_\mathfrak{U}$. Everything is now prepared for proving the

THEOREM: Let $1/p + 1/q \geq 1$ and $\varphi \in (E \otimes_{\alpha'_{p,q}} F)'$. Then there are strictly localizable measures μ and ν, operators $R \in \mathfrak{L}(E, L_{q'}(\mu))$, $S \in \mathfrak{L}(F, L_{p'}(\nu))$ and a positive functional $\delta \in (L_{q'}(\mu) \otimes_\pi L_{p'}(\nu))'$ with $\langle \varphi, x \otimes y \rangle = \langle \delta, Rx \otimes Sy \rangle$, i.e.

$$\varphi : E \otimes F \xrightarrow{R \otimes S} L_{q'}(\mu) \otimes_\pi L_{p'}(\nu) \xrightarrow{\delta} \mathbb{K}$$

and $\|\varphi\|_{(E \otimes_{\alpha'_{p,q}} F)'} \geq \|R\| \, \|S\| \, \|\delta\|$.

It will follow from corollary 18.10. below that $\|\varphi\|_{...} = \|R\| \, \|S\| \, \|\delta\|$ actually holds. (A linear functional $\varphi : E \otimes F \longrightarrow \mathbb{K}$ on the tensor product of two Banach lattices is called *positive* if $\langle \varphi, x \otimes y \rangle \geq 0$ whenever $x \geq 0$ and $y \geq 0$.)

PROOF: For $\iota = (M, N, \varepsilon) \in I$ consider $\varphi_\iota := \varphi|_{M \otimes N}$ and factor it, as was done in 18.3.,

$$\varphi_\iota : M \otimes N \xrightarrow{R_\iota \otimes S_\iota} \ell^{n_\iota}_{q'} \otimes \ell^{n_\iota}_{p'} \xrightarrow{\delta_\iota} \mathbb{K}$$

such that δ_ι is positive, $\|\delta_\iota\| = \|S_\iota\| = 1$ and

$$\|R_\iota\| \leq \|\varphi_\iota\|_{(M \otimes_{\alpha'_{p,q}} N)'}(1+\varepsilon) \leq \|\varphi\|_{(E \otimes_{\alpha'_{p,q}} F)'}(1+\varepsilon) .$$

Now consider (see lemma 18.4.)

$$\tilde{\varphi}: E \otimes F \xrightarrow{J_E \otimes J_F} (E_\iota)_{\mathfrak{U}} \otimes (F_\iota)_{\mathfrak{U}} \xrightarrow{(R_\iota)_{\mathfrak{U}} \otimes (S_\iota)_{\mathfrak{U}}} (\ell_{q'}^{n_\iota})_{\mathfrak{U}} \otimes_\pi (\ell_{p'}^{n_\iota})_{\mathfrak{U}} \xrightarrow{\lim_{\mathfrak{U}} \delta_\iota} \mathbb{K} \ .$$

It follows, by the very definitions (recall the definition of x_ι for $x \in E$ and similarly y_ι) that for $x \otimes y \in E \otimes F$

$$[((R_\iota)_{\mathfrak{U}} \circ J_E) \otimes ((S_\iota)_{\mathfrak{U}} \circ J_F)] (x \otimes y) = (R_\iota x_\iota)_{\mathfrak{U}} \otimes (S_\iota y_\iota)_{\mathfrak{U}}$$

and therefore

$$\langle \tilde{\varphi}, x \otimes y \rangle = \langle \lim_{\mathfrak{U}} \delta_\iota, (R_\iota x_\iota)_{\mathfrak{U}} \otimes (S_\iota y_\iota)_{\mathfrak{U}} \rangle = \lim_{\mathfrak{U}} \langle \delta_\iota, R_\iota x_\iota \otimes S_\iota y_\iota \rangle = \langle \varphi, x \otimes y \rangle \ ,$$

hence $\varphi = \tilde{\varphi}$. For $R := (R_\iota)_{\mathfrak{U}} \circ J_E$, $S := (S_\iota)_{\mathfrak{U}} \circ J_F$ and $\delta := \lim_{\mathfrak{U}} \delta_\iota$ it follows that

$$\|R\| \leq \lim_{\mathfrak{U}} \|R_\iota\| \ \|J_E\| \leq \|\varphi\|_{(E \otimes_{\alpha'_{p,q}} F)'}$$

$$\|S\| \leq \lim_{\mathfrak{U}} \|S_\iota\| \ \|J_F\| = 1$$

$$\|\delta\| = \lim_{\mathfrak{U}} \|\delta_\iota\| = 1 \ .$$

Since all δ_ι are positive, it is clear from the definition of the ordering in the ultraproducts $(\ell_{q'}^{n_\iota})_{\mathfrak{U}}$ and $(\ell_{p'}^{n_\iota})_{\mathfrak{U}}$ that $\delta = \lim_{\mathfrak{U}} \delta_\iota$ is positive. By what was said above $G_s := (\ell_s^{n_\iota})_{\mathfrak{U}}$ is an abstract L_s-space if $s < \infty$ and an abstract M-space if $s = \infty$. The representation theorems (see Appendix A7.) show that $G_s = L_s(\mu)$ (with norm and order) for some strictly localizable μ if $s < \infty$ and that for $s = \infty$ the bidual G''_∞ is an $L_\infty(\mu)$. Therefore, if $p', q' \in [1, \infty[$, the theorem is completely proven. If, for example, $q' = \infty$ then δ factors

$$\delta: G_\infty \otimes_\pi G_{p'} \xhookrightarrow{1} G''_\infty \otimes_\pi G_{p'} = L_\infty(\mu) \otimes_\pi L_{p'}(\nu) \xrightarrow{\wedge \delta} \mathbb{K}$$

through its left canonical extension $\wedge \delta$ which, clearly, is also positive and has the same norm as δ. $\square$

The structure theorems imply that μ and ν can be chosen to be Borel-Radon measures on locally compact spaces. In terms of operators the theorem implies that for each $T \in \mathfrak{L}_{p,q}(E, F)$ there are strictly localizable measures μ and ν, operators $R \in \mathfrak{L}(E, L_{q'}(\mu)), S \in \mathfrak{L}(L_p(\nu), F'')$ and a *positive* operator $U \in \mathfrak{L}(L_{q'}(\mu), L_p(\nu))$ such that

$$\begin{array}{ccccc} E & \xrightarrow{T} & F & \xhookrightarrow{\kappa_F} & F'' \\ R \downarrow & & & \nearrow S & \\ L_{q'}(\mu) & \xrightarrow{U} & L_p(\nu) & & \end{array}$$

and $\mathbf{L}_{pq}(T) \geq \|R\| \|U\| \|S\|$. See also Ex 18.8..

18.6. $\mathbf{L}_p(id_{L_p}) = 1$ by what was said at the beginning of 18.2.. Therefore this factorization and $\mathbf{L}_p(T) \geq \|R\| \|U\| \|S\| \geq \|R\| \|U \circ S\|$ immediately imply:

COROLLARY 1: *An operator $T \in \mathfrak{L}(E,F)$ is p-factorable if and only if $\kappa_F \circ T$ factors through some $L_p(\mu)$* :

$$\begin{array}{ccccc} E & \xrightarrow{T} & F & \xrightarrow{\kappa_F} & F'' \\ & R \searrow & & \nearrow S & \\ & & L_p(\mu) & & \end{array}$$

Moreover, $\mathbf{L}_p(T) = \min \|R\| \|S\|$, the minimum taken over all such factorizations.

The measure can clearly be chosen to be a Borel–Radon measure. For $p = 2$ the fact that every subspace of a Hilbert space is 1–complemented gives the following:

COROLLARY 2: *$T \in \mathfrak{L}(E,F)$ is 2-factorable if and only if it factors through a Hilbert space. Moreover, $\mathbf{L}_2(T) = \min \|R\| \|S\|$, where the minimum is taken over all operators $R \in \mathfrak{L}(E,H)$ and $S \in \mathfrak{L}(H,F)$ with $T = S \circ R$ (and H an arbitrary Hilbert space).*

In general, the factorization cannot be chosen to be through an $L_p(\mu)$ with a finite measure! Take $T = id_{\ell_1(\Gamma)}$ and a factorization

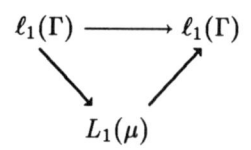

$$\begin{array}{ccc} \ell_1(\Gamma) & \longrightarrow & \ell_1(\Gamma) \\ & \searrow \quad \nearrow & \\ & L_1(\mu) & \end{array}$$

($\ell_1(\Gamma)$ is complemented in its bidual), so $\ell_1(\Gamma)$ is isomorphic to a subspace of $L_1(\mu)$. If Γ is uncountable, μ cannot be σ–finite by a result of Pełczyński [199], p.244.

18.7. For the ideal $\mathfrak{I}_p \sim g_p = \alpha_{p,1}$ of p–integral operators ($1 \leq p < \infty$) the factorization 18.5. reads

$$\begin{array}{ccccc} E & \xrightarrow{T} & F & \xrightarrow{\kappa_F} & F'' \\ R \downarrow & & & \nearrow S & \\ L_\infty(\mu) & \xrightarrow{U} & L_p(\nu) & & \end{array}$$

Since, by 17.17., the positive operator U is even p–integral and absolutely p–summing with $\mathbf{P}_p(U) = \mathbf{I}_p(U) = \|U\|$, it follows from the Grothendieck–Pietsch factorization theorem (recall that $L_\infty(\mu) \stackrel{1}{=} C(K)$ for some compact K and use 11.3., corollary 2) that U factors

$$\begin{array}{ccc} L_\infty(\mu) & \xrightarrow{U} & L_p(\nu) \\ \| & & \uparrow S_o \\ C(K) & \xrightarrow{I} & L_p(\eta) \end{array}$$

with $\|U\| = \|S_o\| \, \|I\|$. This implies:

COROLLARY 3: *For an operator $T \in \mathcal{L}(E,F)$ and $1 \leq p < \infty$ the following three statements are equivalent:*

(a) T is p–integral.

(b) *There are a Borel–Radon measure μ on a compact K, operators $R \in \mathcal{L}(E, C(K))$ and $S \in \mathcal{L}(L_p(\mu), F'')$ such that*

$$\begin{array}{ccc} E \xrightarrow{T} & F \xhookrightarrow{\kappa_F} & F'' \\ R \downarrow & & \nearrow S \\ C(K) & \xrightarrow{I} & L_p(\mu) \end{array}$$

commutes; I the canonical map $C(K) \to L_p(\mu)$.

(c) *There are a finite measure μ and operators $R \in \mathcal{L}(E, L_\infty(\mu))$ and $S \in \mathcal{L}(L_p(\mu), F'')$ such that $\kappa_F \circ T = S \circ I \circ R$.*

In this case, $\mathbf{I}_p(T) = \min \|S\| \, \|I\| \, \|R\|$, where the minimum is taken over all factorizations as in (b) or as in (c).

Thus, the natural map $C(K) \xrightarrow{I} L_p(K,\mu)$ is not only the typical absolutely p–summing but also the typical p–integral operator — only the form of the factorization is different: for absolutely p–summing operators into $\ell_\infty(B_{F'})$, for p–integral operators into F'' (for $p=1$ this is theorem 10.5.). For $p = \infty$ the ideals $\mathfrak{I}_\infty$ and $\mathcal{L}_\infty$ coincide.

The factorization cannot be into F (instead of F'') since there are $(1-)$integral operators which are not Pietsch integral (see Appendix D9.). However, for $p = 2$, it follows as in 18.6. (corollary 2) that the operator S in the factorization can be chosen to have values in F, hence, $T \in \mathfrak{J}_2(E,F)$ if and only if there is a factorization

$$\begin{array}{ccc} E & \xrightarrow{T} & F \\ R \downarrow & & \uparrow S \\ C(K) & \xrightarrow{I} & L_2(\mu) \end{array}$$

Moreover, $\mathbf{I}_2(T) = \min \|R\| \|I\| \|S\|$. But this is exactly the factorization of the absolutely 2–summing operators (see 11.3., corollary 2), hence

$$(\mathfrak{P}_2, \mathbf{P}_2) = (\mathfrak{I}_2, \mathbf{I}_2) .$$

Since $g_2^* \sim \mathfrak{P}_2 = \mathfrak{I}_2 \sim g_2$, it follows that:

COROLLARY 4: $g_2 = g_2^*$.

In Ex 18.3. it will be shown that $\mathfrak{P}_p(E, F) = \mathfrak{I}_p(E, F)$ holds isometrically if $E = C(K)$ or L_∞. If F is 1–complemented in F''', then obviously $T \in \mathfrak{I}_p(E, F)$ if and only if T factors $E \xrightarrow{R} C \xrightarrow{I} L_p \xrightarrow{S} F$ and $\mathbf{I}_p(T) = \min \|R\| \|I\| \|S\|$, the minimum taken over these factorizations. In particular, this holds if F is a dual space or if F has the metric extension property; example: $F = \ell_\infty(\Gamma)$. It follows from the Grothendieck–Pietsch factorization theorem that

$$\mathfrak{P}_p(E, \ell_\infty(\Gamma)) \stackrel{1}{=} \mathfrak{I}_p(E, \ell_\infty(\Gamma))$$

holds isometrically. Since $(\mathfrak{P}_p, \mathbf{P}_p)$ is an injective operator ideal, this implies:

COROLLARY 5: *For $1 \leq p < \infty$ the injective hull of the ideal $\mathfrak{I}_p$ of p–integral operators is the ideal $\mathfrak{P}_p$ of absolutely p–summing operators:*

$$(\mathfrak{P}_p, \mathbf{P}_p) = (\mathfrak{I}_p^{inj}, \mathbf{I}_p^{inj}) .$$

For the associated tensor norm one obtains

$$N \otimes_{g_{p'}^*} \ell_\infty^n \stackrel{1}{=} \mathfrak{P}_p(N', \ell_\infty^n) \stackrel{1}{=} \mathfrak{I}_p(N', \ell_\infty^n) \stackrel{1}{=} N \otimes_{g_p} \ell_\infty^n .$$

Since $g_{p'}^*$ and g_p are finitely generated, it follows that:

COROLLARY 6: $g_{p'}^* = g_p$ *on $E \otimes \ell_\infty^n$ for all normed spaces E.*

We shall come back to this result in 20.14..

18.8. For p–factorable operators the factorization theorem followed directly from theorem 18.5., for p–integral operators the Grothendieck–Pietsch factorization theorem was used. This suggests that the general case of (p, q)–factorable operators is even more complicated — and it is. Starting from the fact that for $1/p + 1/q > 1$ every

(p,q)-factorable operator factors (into F'') through a positive operator $L_{q'}(\mu) \to L_p(\nu)$ for some measures μ and ν (this is a consequence of theorem 18.5.), we shall proceed as follows:

(1) Every positive $L_{q'}(\mu) \to L_p(\nu)$ factors through a multiplication operator

$$M_g : L_{q'}(\nu) \longrightarrow L_p(\nu) \; ; \; M_g(f) := fg \; .$$

(2) Multiplication operators $M_g : L_{q'}(\mu) \to L_p(\mu)$ factor through an embedding $L_{q'}(\mu_o) \hookrightarrow L_p(\mu_o)$ for some finite measure μ_o, hence are (p,q)-factorable by what was said in 18.2.; moreover, $\mathbf{L}_{p,q}(M_g) = \|M_g\| = \|g\|_r$. Note that $q' \geq p$.

The fact (2) will be shown in 18.10. and (1) follows from Maurey's factorization theorem, which will be proven now.

18.9. Every positive operator $T : L_q(\mu_1) \to L_p(\mu_2)$ has the property (see 7.3.) that for all $S \in \mathcal{L}(E, F)$

$$\|T \otimes S : L_q(\mu_1) \otimes_{\Delta_q} E \longrightarrow L_p(\mu_2) \otimes_{\Delta_p} F\| = \|T\| \, \|S\|$$

($p, q \in [1, \infty]$ arbitrary). In particular,

$$\|T \otimes S : L_q(\mu_1) \otimes_{\Delta_q^t} L_q(\nu) \longrightarrow L_p(\mu_2) \otimes_{\Delta_p} F\| = \|T\| \, \|S\|$$

for each measure ν and each $S \in \mathcal{L}(L_q(\nu), F)$. So T satisfies condition (a) of

MAUREY'S FACTORIZATION THEOREM: *Let $1 \leq p \leq q \leq \infty$ and define $r \in [1, \infty]$ by $1/p = 1/q + 1/r$. For $T \in \mathcal{L}(E, L_p(\mu))$ the following two statements are equivalent:*
(a) *For every measure ν and every $S \in \mathcal{L}(L_q(\nu), F)$ (or only: $S := id_{\ell_q}$) the operator*

$$T \otimes S : E \otimes_{\Delta_q^t} L_q(\nu) \longrightarrow L_p(\mu) \otimes_{\Delta_p} F$$

is continuous.
(b) *There are $g \in L_r(\mu)$ and $R \in \mathcal{L}(E, L_q(\mu))$ with $T = M_g \circ R$.*
In this case,

$$\|T \otimes id_{\ell_q} : E \otimes_{\Delta_q^t} \ell_q \longrightarrow L_p(\mu) \otimes_{\Delta_p} \ell_q\| = \min \|R\| \, \|g\|_r \; ,$$

where the minimum is taken over all factorizations as in (b).

Recall that $\|M_g : L_q(\mu) \to L_p(\mu)\| = \|g\|_r$.

PROOF: Assume first (b). Then it was proven in Ex 7.3. with the help of Hölder's inequality, that

$$\|M_g \otimes S : L_q(\mu) \otimes_{\Delta_q^t} L_q(\nu) \longrightarrow L_p(\mu) \otimes_{\Delta_p} F\| = \|g\|_r \|S\| \; .$$

It follows that

$$T \otimes S : E \otimes_{\Delta_q^t} L_q(\nu) \xrightarrow{R \otimes id} L_q(\mu) \otimes_{\Delta_q^t} L_q(\nu) \xrightarrow{M_g \otimes S} L_p(\mu) \otimes_{\Delta_p} F$$

has norm $\leq \|R \otimes id : \cdots\| \, \|M_g \otimes S : \cdots\| = \|R\| \, \|g\|_r \|S\|$.

Conversely, assume (a). If $p = q$, the function $g = 1$ gives the desired factorization. So assume (a) with $S = id_{\ell_q}$ for $r < \infty$; this implies that $c := \|T \otimes id_{\ell_q} : \cdots\|$ is a constant satisfying

(*) $$\left(\int \left(\sum_{k=1}^n |Tx_k|^q\right)^{p/q} d\mu\right)^{1/p} = \Delta_p\left(\sum_{k=1}^n Tx_k \otimes e_k; L_p(\mu), \ell_q\right) \leq$$

$$\leq c\, \Delta_q^t\left(\sum_{k=1}^n x_k \otimes e_k; E, \ell_q\right) = c \left(\sum_{k=1}^n \|x_k\|^q\right)^{1/q}$$

for all finite families $(x_1, ..., x_n)$ in E.

For $q = \infty$ (and hence $p = r < \infty$) the sum has to be replaced by a maximum, which means, in particular, that the increasing net (f_A)

$$f_A := \max\{|Tx| \mid x \in A\} \subset L_p(\mu)$$

(where $A \subset B_E$ is finite) is norm-bounded by c. The type of order completeness which $L_r(\mu)$ has (see Appendix B5.) gives a non-negative $g \in L_r(\mu)$ with $\|g\|_r \leq c$ and $|Tx| \leq g\|x\|$ for all $x \in E$. This shows that $R \in \mathfrak{L}(E, L_\infty(\mu))$ defined by

$$Rx := \frac{Tx}{g} \qquad \left(\frac{0}{0} := 0\right)$$

has norm 1 and that $T = M_g \circ R$ is a factorization as in (b) with $\|R\| \, \|M_g\| \leq c$.

For $1 \leq p < q < \infty$ (this is the remaining case) one has to find a $g \in L_r(\mu)$ with $\|g\|_r \leq 1$ and

(**) $$\left(\int \left(\frac{|Tx|}{g}\right)^q d\mu\right)^{1/q} \leq c\|x\|.$$

Then $Rx := g^{-1} \cdot Tx$ and M_g again give the desired factorization $T = M_g \circ R$ with $\|R\| \, \|M_g\| \leq c\|g\|_r \leq c$. For this, define $s := q/p$ so $s' = r/p$. To apply Ky Fan's lemma (see Appendix A3.) observe that

$$K := \{f \in L_{s'}(\mu) \mid f \geq 0, \|f\|_{s'} \leq 1\}$$

is weakly compact and convex. For every μ-measurable $h \geq 0$ the function $\Phi_h : K \to [0, \infty]$ defined by

$$\Phi_h(f) := \begin{cases} \int h f^{-s} d\mu & \text{if this integral exists} \\ \infty & \text{if not} \end{cases}$$

is convex. The convex set $\{f \in K \mid \Phi_h(f) \leq \alpha\}$ is norm–closed by Fatou's lemma from integration theory and hence weakly closed; this means that Φ_h is lower semi–continuous. Therefore the set of functions $\Phi_{x_1,...,x_n} : K \to [-\infty, \infty[$ for $x_1, ..., x_n \in E$ given by

$$\Phi_{x_1,...,x_n}(f) := c^q \sum_{k=1}^n \|x_k\|^q - \int \sum_{k=1}^n |Tx_k|^q f^{-s} d\mu$$

is a convex set of concave, upper semi–continuous functions. For $x_1, ..., x_n \in E$ the function f defined by

$$\alpha := \int \left(\sum_{k=1}^n |Tx_k|^q\right)^{p/q} d\mu \quad ; \quad f := \alpha^{-1/s'} \left(\sum_{k=1}^n |Tx_k|^q\right)^{1/ss'}$$

is in K and satisfies, by inequality (*),

$$\Phi_{x_1,...,x_n}(f) \geq 0$$

(the case $\alpha = 0$ is trivial). Ky Fan's lemma gives a function $f_o \in K$ such that for all $x_1, ..., x_n \in E$ the inequality

$$\Phi_{x_1,...,x_n}(f_o) \geq 0$$

holds. In particular, $\Phi_x(f_o) \geq 0$, which means that

$$\int |Tx|^q f_o^{-s} d\mu \leq c^q \|x\|^q ,$$

for all $x \in E$. The function $g := f_o^{1/p}$ satisfies $\|g\|_r \leq 1$ and (**). □

The proof shows that T factors through a multiplication operator (as in (b)) if and only if (*) holds and this is how Maurey's factorization theorem was originally stated [186]. Recall from 7.9. that $T \otimes S$ is always continuous if $q \leq p$. Other factorization results in the spirit of this theorem will be given in Ex 18.14. and Ex 18.15..

18.10. For positive operators $L_{q'} \to L_p$ this implies:

COROLLARY: *Let $1/p + 1/q \geq 1$. If $T : L_{q'}(\mu) \to L_p(\nu)$ is a positive operator, then there are a finite measure μ_o and operators $R \in \mathcal{L}(L_{q'}(\mu), L_{q'}(\mu_o))$ and $S \in \mathcal{L}(L_p(\mu_o), L_p(\nu))$ such that $T = S \circ I \circ R$*

$$\begin{array}{ccc} L_{q'}(\mu) & \xrightarrow{T} & L_p(\nu) \\ R \downarrow & & \uparrow S \\ L_{q'}(\mu_o) & \xrightarrow{I} & L_p(\mu_o) \end{array}$$

and $\|T\| = \|S\|\,\|I\|\,\|R\|$. In particular, T is (p,q)-factorable and $\mathbf{L}_{p,q}(T) = \|T\|$.

PROOF: Since $q' \geq p$ and T satisfies condition (a) of Maurey's factorization theorem (this was already noted at the beginning of 18.9.), T decomposes into $T = M_g \circ R_o$ with $\|T\| = \|g\|_r \|R_o\|$. It remains to factor the multiplication operator $M_g : L_{q'}(\nu) \to L_p(\nu)$; recall that $1/p = 1/q' + 1/r$.

For this, take
$$g_1 := \operatorname{sign}(g)|g|^{-r/q'} \quad \text{and} \quad g_2 := |g|^{r/p}$$
and the probability measure $d\mu_o := |g|^r \|g\|_r^{-r} d\nu$ (recall $0/0 := 0$). Then
$$\|M_{g_1} : L_{q'}(\nu) \longrightarrow L_{q'}(\mu_o)\| = \|g\|_r^{-r/q'}$$
$$\|M_{g_2} : L_p(\mu_o) \longrightarrow L_p(\nu)\| = \|g\|_r^{r/p}$$
and $M_g = M_{g_2} \circ [I : L_{q'}(\mu_o) \hookrightarrow L_p(\mu_o)] \circ M_{g_1}$. It follows that
$$T = M_{g_2} \circ I \circ M_{g_1} \circ R_o .$$
Since $\|I\| = \mathbf{L}_{p,q}(I) = 1$ by 18.2., this factorization satisfies
$$\mathbf{L}_{p,q}(T) \geq \|T\| = \|g\|_r \|R_o\| = \|M_{g_2}\|\,\|M_{g_1}\|\,\|R_o\| \geq$$
$$\geq \|M_{g_2}\|\,\|I\|\,\|M_{g_1} \circ R_o\| = \|M_{g_2}\|\,\mathbf{L}_{p,q}(I)\,\|M_{g_1} \circ R_o\| \geq \mathbf{L}_{p,q}(T)$$
and everything has been proven. □

18.11. Summarizing these results, one obtains the general factorization theorem for (p,q)-factorable operators.

THEOREM: Let $p, q \in [1, \infty[$ such that $1/p + 1/q > 1$. For each operator $T \in \mathcal{L}(E, F)$ the following two statements are equivalent:

(a) $T \in \mathfrak{L}_{p,q}(E, F)$.

(b) There are a finite measure μ, operators $R \in \mathcal{L}(E, L_{q'}(\mu))$ and $S \in \mathcal{L}(L_p(\mu), F'')$ such that $\kappa_F \circ T = S \circ I \circ R$

$$\begin{array}{ccccc}
E & \xrightarrow{T} & F & \hookrightarrow & F'' \\
{\scriptstyle R}\downarrow & & & \nearrow{\scriptstyle S} & \\
L_{q'}(\mu) & \xrightarrow{I} & L_p(\mu) & &
\end{array}.$$

In this case, $\mathbf{L}_{p,q}(T) = \min \|S\|\,\|I\|\,\|R\|$ over all such factorizations.

Recall that the case $1/p + 1/q = 1$ of p-factorable operators was already treated in 18.6..

PROOF: If T has such a factorization, it follows from proposition 18.2. (and the regularity of maximal operator ideals, 17.8.) that

$$\mathbf{L}_{p,q}(T) = \mathbf{L}_{p,q}(\kappa_F \circ T) \leq \|S\| \, \mathbf{L}_{p,q}(I) \, \|R\| = \|S\| \, \|I\| \, \|R\| \ .$$

Conversely, if $T \in \mathfrak{L}_{p,q}(E, F)$ the ultraproduct theorem 18.5. gives a factorization

$$\begin{array}{ccccc} E & \xrightarrow{T} & F & \xhookrightarrow{\kappa_F} & F'' \\ {\scriptstyle R}\downarrow & & & \nearrow {\scriptstyle S} & \\ L_{q'}(\mu) & \xrightarrow{U} & L_p(\nu) & & \end{array}$$

with some measures μ and ν, a positive operator U and

$$\mathbf{L}_{p,q}(T) \geq \|S\| \, \|U\| \, \|R\| \ .$$

By corollary 18.10. the operator U can be decomposed into $U = S_o \circ I \circ R_o$ with $\mathbf{L}_{p,q}(U) = \|U\| = \|S_o\| \, \|I\| \, \|R_o\|$ and $I: L_{q'}(\mu_o) \hookrightarrow L_p(\mu_o)$ for some finite measure μ_o. It follows that

$$\kappa_F \circ T = S \circ S_o \circ I \circ R_o \circ R$$

and

$$\mathbf{L}_{p,q}(T) \geq \|S\| \, \|U\| \, \|R\| \geq \|S \circ S_o\| \, \|I\| \, \|R_o \circ R\| \geq \mathbf{L}_{p,q}(T)$$

— the last inequality by the first part of the proof. $\square$

By examining the way in which the measure μ is found it becomes clear that μ can be chosen to be a finite Borel–Radon measure on a locally compact space Ω — and hence, even on a compact space (by passing, e.g., to the one–point compactification of Ω). If the range space F is 1–complemented in F'' (or $p = 2$), then the factorization can be made into F. Simple manipulations show that:

COROLLARY: *Take $1/p + 1/q > 1$. For each $\varphi \in (E \otimes_\pi F)'$ the following statements are equivalent:*
(a) *$\varphi \in (E \otimes_{\alpha'_{p,q}} F)'$.*
(b) *There are a finite measure μ, operators $R_1 \in \mathfrak{L}(E, L_{q'}(\mu))$ and $R_2 \in \mathfrak{L}(F, L_{p'}(\mu))$ such that*

$$\langle \varphi, x \otimes y \rangle = \int_\Omega (R_1 x) \cdot (R_2 y) d\mu \ .$$

In this case, $\|\varphi\|_{(E \otimes_{\alpha'_{p,q}} F)'} = \min \|R_1\| \, \|R_2\| \mu(\Omega)^{1/r}$, the minimum taken over all such factorizations.

Exercises:

Ex 18.1. Use the factorization 18.1. of $T \in \mathfrak{L}_{p,q}(M, \ell_2^n)$ and the little Grothendieck theorem to show that for all $1 \leq q \leq \infty$

$$g_2 \leq K_{LG} d_q \quad \text{and} \quad g_2 \leq K_{LG} g_\infty \leq K_{LG} g_q \qquad \text{on } E \otimes H$$

whenever H is a Hilbert space and E an arbitrary normed space.

Ex 18.2. (a) An operator $T \in \mathfrak{L}(E, F)$ is in the injective hull $\mathfrak{L}_p^{inj}$ of the p–factorable operators if and only if T factors through a subspace of some $L_p(\mu)$. Moreover,

$$\mathbf{L}_p^{inj}(T) = \min \|R\| \, \|S\| \, ,$$

where the minimum is taken over all $R \in \mathfrak{L}(E, G)$ and $S \in \mathfrak{L}(G, F)$ with $T = SR$ and $G \subset L_p(\mu)$ for some measure μ.

(b) $T \in \mathfrak{L}(E, F)$ is in the surjective hull $\mathfrak{L}_p^{sur}$ of $\mathfrak{L}_p$ if and only if $\kappa_F \circ T$ factors through a quotient of some $L_p(\mu)$. Moreover,

$$\mathbf{L}_p^{sur}(T) = \min \|R\| \, \|S\| \, ,$$

the minimum taken over all such factorizations.

Ex 18.3. If $E = C(K)$ or $L_\infty(\mu)$, then $\mathfrak{P}_p(E, F) = \mathfrak{I}_p(E, F)$ holds isometrically and the operator S in the factorization 18.7. has necessarily values in F (instead of the bidual F'').

Ex 18.4. (a) Let $\mathfrak{U}$ be the ultrafilter of all sets containing a fixed point $\iota_o \in I$. What is the ultraproduct along $\mathfrak{U}$ of a family $(E_\iota)_\mathfrak{U}$ of Banach spaces?

(b) Describe the relationship between ultrafilters on I and points of the Stone–Čech–compactification βI of I (discrete topology).

Ex 18.5. If $\mathfrak{U}$ is an ultrafilter on I and $E_\iota := \mathbb{K}$ for all $\iota \in I$, then $(E_\iota)_\mathfrak{U} = \mathbb{K}$. Hint: $\lim_\mathfrak{U} f(\iota)$ defines a linear functional on $\ell_\infty(I)$. See also Ex 18.10..

Ex 18.6. (a) If $T_\iota \in \mathfrak{L}(E_\iota, F_\iota)$ are metric injections, then $(T_\iota)_\mathfrak{U}$ is a metric injection.

(b) Show that the construction of $(T_\iota)_\mathfrak{U}$ establishes an isometry

$$(\mathfrak{L}(E_\iota, F_\iota))_\mathfrak{U} \xhookrightarrow{1} \mathfrak{L}((E_\iota)_\mathfrak{U}, (F_\iota)_\mathfrak{U}) \, .$$

Ex 18.7. (a) The ultraproduct of C^*–algebras is a C^*–algebra.

(b) The ultraproduct of $C(K)$–spaces is a $C(K)$–space.

(c) For every Hilbert space H and $1 \leq p < \infty$ there exist a strictly localizable measure μ and an isometry $I : H \hookrightarrow L_p(\mu)$. For $1 < p < \infty$ the Hilbert space is isomorphic to a complemented subspace of some $L_p(\mu)$. Hint: $\ell_2^m \xhookrightarrow{1} L_p(\mu_n)$ for some μ_n via Gauss functions, see 8.7.; for the second part use Rademacher functions, see Ex 8.17..

Ex 18.8. (a) If $\varphi_\iota \in E'_\iota$ is a uniformly bounded family of functionals, then $\langle \varphi, (x_\iota)_\mathfrak{U} \rangle := \lim_\mathfrak{U} \langle \varphi_\iota, x_\iota \rangle$ defines a linear functional φ on $(E_\iota)_\mathfrak{U}$ with $\|\varphi\| = \lim_\mathfrak{U} \|\varphi_\iota\|$. It follows that this defines an isometric injection

$$(E'_\iota)_\mathfrak{U} \xhookrightarrow{1} ((E_\iota)_\mathfrak{U})' \ .$$

(b) Take an operator $T \in \mathcal{L}(E, G)$ and a special ultrafilter $\mathfrak{U}$ defined on the index set $I := \mathrm{FIN}(E) \times \mathrm{FIN}(G') \times \,]0,1]$ as in 18.5.. For $\iota = (M, N, \varepsilon) \in I$ define $E_\iota := M$, $G_\iota := G/N^0$ and

$$T_\iota := Q_{N^0}^G \circ T \circ I_M^E : E_\iota = M \hookrightarrow E \to G \twoheadrightarrow G/N^0 = G_\iota$$

— the finite parts of T. Then there is a factorization as follows:

$$\begin{array}{ccccc}
E & \xrightarrow{T} & G & \xrightarrow{\kappa_G} & G'' \\
{\scriptstyle J_E} \downarrow & & {\scriptstyle Q_G} \nearrow & & \uparrow {\scriptstyle J'_{G'}} \\
(E_\iota)_\mathfrak{U} & \xrightarrow{(T_\iota)_\mathfrak{U}} & (G_\iota)_\mathfrak{U} & \xhookrightarrow{1} & (G'_\iota)'_\mathfrak{U}
\end{array}$$

In the index set one can drop the interval $]0, 1]$ for this construction, but (as one saw in the proof of 18.5.) it is sometimes quite useful to have it. The decomposition

$$\kappa_G \circ T = Q_G \circ (T_\iota)_\mathfrak{U} \circ J_E$$

of T is sometimes called the *canonical decomposition of T into its finite parts* (with respect to $\mathfrak{U}$). Note that Q_G and J_E are independent of T.

(c) Use (b) to prove the factorization result 18.5. for operators $T \in \mathcal{L}_{p,q}(E, F)$.

Ex 18.9. (a) Let F be a normed space and $\mathfrak{U}$ an ultrafilter on $\mathrm{FIN}(F)$ finer than the order filter. Define $F_\mathfrak{U}$ to be the ultraproduct of all $N \in \mathrm{FIN}(F)$ along $\mathfrak{U}$. Then there is an isometry $J_F : F \hookrightarrow F_\mathfrak{U}$.

(b) Let $T \in \mathcal{L}(F', E')$ and $\varphi_N \in (N \otimes_\pi E')'$ satisfy

$$\langle \varphi_N, y \otimes Ty' \rangle = \langle y, y' \rangle \qquad \text{for all } y \otimes y' \in N \otimes F'$$

and $\sup_N \|\varphi_N\| < \infty$. Then there is a $\varphi \in (F_\mathfrak{U} \otimes_\pi E')'$ with

$$\langle \varphi, J_F y \otimes Ty' \rangle = \langle y, y' \rangle \qquad \text{for all } y \otimes y' \in F \otimes F'$$

and $\|\varphi\| = \lim_\mathfrak{U} \|\varphi_N\|$. If $S \in \mathcal{L}(E', F'_\mathfrak{U})$ is the operator associated with the functional φ, then the relation $J'_F \circ S \circ T = \mathrm{id}_{F'}$ holds. In particular, $\mathrm{im}(T)$ is complemented in F' and $T : F' \to \mathrm{im}(T)$ invertible. Hint: Lemma 18.4..

This result is just "Lindenstrauss' compactness argument" treated in Ex 6.4. in its ultraproduct version.

Ex 18.10. (a) If dim $E_\iota \leq n$ for a family (E_ι) of Banach spaces, then its ultraproduct $(E_\iota)_{\mathfrak{U}}$ has dimension $\leq n$ as well. Hint: For dim $E_\iota = n$ take Auerbach bases $(x_{\iota,k}, \varphi_{\iota,k})_{k=1,\ldots,n}$ and define $\varphi_k := \lim_{\mathfrak{U}} \varphi_{\iota,k}$ (Ex 18.8.). Then $\langle \varphi_k, (x_\iota)_{\mathfrak{U}} \rangle = 0$ for all k implies $(x_\iota)_{\mathfrak{U}} = 0$.

(b) If $T_\iota \in \mathfrak{L}(E_\iota, F_\iota)$ all have rank $\leq n$, then $(T_\iota)_{\mathfrak{U}}$ has rank $\leq n$.

Ex 18.11. Let $T \in \mathfrak{L}(E, F)$ be such that for all $\varepsilon > 0$ there is an $n \in \mathbb{N}$ such that for all $M \in \text{FIN}(E)$ there is an $S \in \mathfrak{L}(M, F)$ of rank $\leq n$ with $\|T|_M - S\| \leq \varepsilon$. Use the factorization of T into its finite parts (Ex 18.8.(b)) and Ex 18.10.(b) to show that $\kappa_F \circ T : E \to F''$ is approximable. Since the ideal $\mathfrak{F}$ of approximable operators is regular (Ex 9.12.(c)), it follows that T is approximable.

Ex 18.12. Use the previous exercise to prove the following result of Heinrich [104]: Let $T \in \mathfrak{L}(E, F)$ be such that for all separable subspaces $E_o \subset E$ the restriction $T|_{E_o} \in \mathfrak{L}(E_o, F)$ is approximable, then T is approximable. Hint: Negation and Ex 18.11..

Ex 18.13. For $1 \leq p \leq q < \infty$ and $T \in \mathfrak{L}(E, \ell_p)$ the following statements are equivalent $(1/p = 1/q + 1/r)$:

(a) For all $x_1, \ldots, x_n \in E$

$$\left(\sum_{\ell=1}^{\infty} \left(\sum_{k=1}^{n} |\langle Tx_k, e_\ell \rangle|^q \right)^{p/q} \right)^{1/p} \leq \left(\sum_{k=1}^{n} \|x_k\|^q \right)^{1/q}.$$

(b) There are an $R \in \mathfrak{L}(E, \ell_q)$ with $\|R\| \leq 1$ and $\lambda \in B_{\ell_r}$ such that $T = D_\lambda \circ R$, where $D_\lambda : \ell_q \longrightarrow \ell_p$ is the diagonal operator associated with λ.

Ex 18.14. Let $1 \leq q \leq p \leq \infty$ and $1/q = 1/p + 1/r$. For $T \in \mathfrak{L}(L_p(\mu), F)$ the following statements are equivalent:

(a) For each $S \in \mathfrak{L}(E, L_q(\nu))$ (or only $S = id_{\ell_q}$)

$$T \otimes S : L_p(\mu) \otimes_{\Delta_p} E \longrightarrow F \otimes_{\Delta_q^i} L_q(\nu)$$

is continuous.

(b) There are a $g \in L_r(\mu)$ and $R \in \mathfrak{L}(L_q(\mu), F'')$ such that $\kappa_F \circ T = R \circ M_g$. Hint: Dualize Maurey's factorization theorem.

Ex 18.15. The following variation of Maurey's factorization theorem 18.9. was given to us by Pisier; according to him, it "has been known in the neighbourhood of Maurey for a long time". The idea of proof is exactly as in 18.9..

Take $1 \leq p < r < q < \infty$ and $T \in \mathfrak{L}(L_q(\mu), L_p(\nu))$ and define s, t by $1/s + 1/q = 1/r$ and $1/t + 1/r = 1/p$.

(a) If there are non-negative functions $g_o \in L_s(\mu)$ and $h_o \in L_t(\nu)$ such that

$$\|h_o^{-1} Tf\|_r \leq c \|g_o f\|_r$$

for all $f \in L_q(\mu)$, then there are multiplication operators M_{g_o} and M_{h_o} and an operator $\tilde{T} \in \mathfrak{L}(L_r(\mu), L_r(\nu))$ such that $T = M_{h_o} \circ \tilde{T} \circ M_{g_o}$ and $\|\tilde{T}\| \leq c$. Hint: The closure of $\{g_o f \mid f \in L_q\}$ is 1-complemented in L_r.

(b) Assume that T satisfies

$$\left(\int \left(\sum_{k=1}^{n} |Tf_k|^r\right)^{p/r} d\nu\right)^{r/p} \leq c^r \left(\int \left(\sum_{k=1}^{n} |f_k|^r\right)^{q/r} d\mu\right)^{r/q}$$

for all $f_1, ..., f_n \in L_q(\mu)$ and define $u := q/r$ and $v := r/p$. Use the function

$$\Phi_{f_1,...,f_n}(g, h) := c^r \int \sum_{k=1}^{n} |f_k|^r g^{u'} d\mu - \int \sum_{k=1}^{n} |Tf_k|^r h^{-v} d\nu$$

(for non-negative g and h in the unit balls of $L_{(u')^2}(\mu)$ and $L_{v'}(\nu)$, respectively) to deduce from the lemma of Ky Fan that there are functions g_o and h_o as in (a). Hint: The first term is norm-continuous since

$$\left|\left(\int fg_n^{u'} d\mu\right)^{1/u'} - \left(\int fg^{u'} d\mu\right)^{1/u'}\right| \leq \left(\int |g_n - g|^{u'} f d\mu\right)^{1/u'} \leq \|g_n - g\|_{(u')^2} \|f\|_u^{1/u'}.$$

(c) Verify from these results the following theorem for $1 \leq p < r < q < \infty$:

$$\|T \otimes id_{\ell_r} : L_q(\mu) \otimes_{\Delta_q} \ell_r \longrightarrow L_p(\nu) \otimes_{\Delta_p} \ell_r\| \leq c$$

if and only if there are functions $g \in L_s(\mu)$ and $h \in L_t(\nu)$ and $\tilde{T} \in \mathcal{L}(L_r(\mu), L_r(\nu))$ such that $T = M_h \circ \tilde{T} \circ M_g$ and $\|g\|_s \|\tilde{T}\| \|h\|_t \leq c$.

Note that the case $r = q$ is a special case of Maurey's factorization theorem and $r = p$ a special case of Ex 18.14..

Ex 18.16. The ideal $\mathfrak{L}_{p,q}$ of (p, q)-factorable operators is contained in the ideal $\mathfrak{W}$ of weakly compact operators if and only if $q' < \infty$ or $p < \infty$.

19. (p,q)–Dominated Operators

In this section the Grothendieck–Pietsch domination theorem for absolutely p–summing operators will be appropriately generalized to the ideal $\mathfrak{D}_{p,q} = \mathfrak{L}^*_{p',q'}$ of (p, q)-dominated operators. The interpretation in terms of factorizations will give Kwapień's factorization theorem: $\mathfrak{D}_{p,q} = \mathfrak{P}_q^{dual} \circ \mathfrak{P}_p$ — a result which shows the central role of the absolutely p–summing operators in the theory. At the end of this section some comments about the so-called "non-commutative" Grothendieck inequality will be made.

19.1. For $p, q \in [1, \infty]$, with $1/p + 1/q \leq 1$ (or equivalently: $1/p' + 1/q' \geq 1$), the ideal $\mathfrak{D}_{p,q}$ of (p,q)-dominated operators was defined to be the ideal associated with the tensor norm $\alpha_{p',q'}^* = \alpha_{q',p'}'$. In other words (see 17.11.), $T \in \mathfrak{D}_{p,q}(E, F)$ if and only if
$$\ell_r(\langle y_k', Tx_k \rangle) \leq c \, w_p(x_k; E) w_q(y_k'; F')$$
where $1/p + 1/q + 1/r' = 1$, the norm $\mathbf{D}_{p,q}(T)$ being the minimum of these c. This estimate was a consequence of the obvious characterization of the $\alpha_{q',p'}$-continuous bilinear forms: Take $1/p + 1/q = 1 + 1/r$, then $\varphi \in (E \otimes F)^*$ is $\alpha_{p,q}$-continuous if and only if
$$\ell_{r'}(\langle \varphi, x_k \otimes y_k \rangle) \leq c \, w_{q'}(x_k; E) w_{p'}(y_k; F) \, .$$

19.2. Since $\mathfrak{P}_p \sim g_{p'}^* = d_{p'}' = \alpha_{1,p'}'$ (see 17.12.), the Grothendieck–Pietsch domination theorem implies that
$$|\langle \varphi, x \otimes y \rangle| \leq \|L_\varphi x\| \, \|y\| \leq \mathbf{P}_p(L_\varphi) \left(\int_{B_{E'}} |\langle x', x \rangle|^p \mu(dx') \right)^{1/p} \|y\|$$
for each $\varphi \in (E \otimes_{d_{p'}} F)' = \mathfrak{P}_p(E, F')$ and vice-versa. For arbitrary $\alpha_{p,q}$-continuous forms (or, equivalently, for (p,q)-dominated operators, see the corollary below) there is also a powerful characterization of this type — a result which is essentially due to Kwapień [160]:

THEOREM: Let $p, q \in [1, \infty]$ with $1/p + 1/q \geq 1$ and $K \subset B_{E'}$ and $L \subset B_{F'}$ weak-*-compact norming sets for E and F, respectively. For $\varphi \in (E \otimes F)^*$ the following two statements are equivalent:
(a) $\varphi \in (E \otimes_{\alpha_{p,q}} F)'$.
(b) There are a constant $c \geq 0$ and normalized Borel-Radon measures μ on K and ν on L such that for all $x \in E$ and $y \in F$
$$|\langle \varphi, x \otimes y \rangle| \leq c \left(\int_K |\langle x', x \rangle|^{q'} \mu(dx') \right)^{1/q'} \left(\int_L |\langle y', y \rangle|^{p'} \nu(dy') \right)^{1/p'}$$
(where the integral must be replaced by $\| \, \|$ if the exponent is ∞).
In this case, $\|\varphi\|_{(E \otimes_{\alpha_{p,q}} F)'} = \min\{c \mid c \text{ as in (b)}\}$.

PROOF: For $p' = \infty$ (or, by symmetry, for $q' = \infty$) this is just the Grothendieck–Pietsch domination theorem 11.3.; so it may be assumed that p and q are > 1, hence $r' < \infty$. If (b) is satisfied, take $x_1, ..., x_n \in E$ and $y_1, ..., y_n \in F$. Since $r'/q' + r'/p' = 1$, it follows from Hölder's inequality that
$$\sum_{k=1}^n |\langle \varphi, x_k \otimes y_k \rangle|^{r'} \leq c^{r'} \sum_{k=1}^n \left(\int_K |\langle x', x_k \rangle|^{q'} \mu(dx') \right)^{r'/q'} \left(\int_L |\langle y', y_k \rangle|^{p'} \nu(dy') \right)^{r'/p'}$$
$$\leq c^{r'} \left(\int_K \sum_{k=1}^n |\langle x', x_k \rangle|^{q'} \mu(dx') \right)^{r'/q'} \left(\int_L \sum_{k=1}^n |\langle y', y_k \rangle|^{p'} \nu(dy') \right)^{r'/p'} \leq$$
$$\leq [c w_{q'}(x_k) w_{p'}(y_k)]^{r'} \, .$$

The converse will be proven using Ky Fan's lemma (see Appendix A3.). For this assume $\|\varphi\|_{(E\otimes_{\alpha_{p,q}}F)'} = 1$, take the convex weak-*-compact set $M_1^+(K) \subset C(K)'$ of probability measures on K and define

$$C := M_1^+(K) \times M_1^+(L) \subset C(K)' \times C(L)' .$$

Consider the set $\mathfrak{F}$ of all functions $f : C \to \mathbb{R}$ for which there exist $x_1, ..., x_n \in E$ and $y_1, ..., y_n \in F$ such that

$$f(\mu, \nu) := \sum_{k=1}^{n} \left[\frac{r'}{q'} \int_K |\langle x', x_k\rangle|^{q'} \mu(dx') + \frac{r'}{p'} \int_L |\langle y', y_k\rangle|^{p'} \nu(dy') - |\langle \varphi, x_k \otimes y_k\rangle|^{r'} \right]$$

for all $(\mu, \nu) \in C$. It is clear that all such f are continuous and affine and that the set $\mathfrak{F}$ is a convex cone. Since K and L are weak-*-compact and norming, there exist for $f \in \mathfrak{F}$ (coming from (x_k) in E and (y_k) in F) elements $x'_o \in K$ and $y'_o \in L$ such that

$$w_{q'}(x_k; E)^{q'} = \sum_{k=1}^{n} |\langle x'_o, x_k\rangle|^{q'} \text{ and } w_{p'}(y_k; F)^{p'} = \sum_{k=1}^{n} |\langle y'_o, y_k\rangle|^{p'} .$$

The inequality $a/s + b/s' \geq a^{1/s} b^{1/s'}$ (for $s \in]1, \infty[$ and $a, b \geq 0$), implies for the Dirac measures

$$f(\delta_{x'_o}, \delta_{y'_o}) = \frac{r'}{q'} w_{q'}(x_k)^{q'} + \frac{r'}{p'} w_{p'}(y_k)^{p'} - \sum_{k=1}^{n} |\langle \varphi, x_k \otimes y_k\rangle|^{r'} \geq$$

$$\geq [w_{q'}(x_k) w_{p'}(y_k)]^{r'} - \sum_{k=1}^{n} |\langle \varphi, x_k \otimes y_k\rangle|^{r'} \geq 0 ,$$

since $\|\varphi\|_{(E\otimes_{\alpha_{p,q}}F)'} = 1$: All the assumptions for Ky Fan's lemma are satisfied! It follows that there is a $(\mu, \nu) \in C$ with $f(\mu, \nu) \geq 0$ for all $f \in \mathfrak{F}$; in particular,

$$|\langle \varphi, x \otimes y\rangle| = \alpha\beta |\langle \varphi, \alpha^{-1} x \otimes \beta^{-1} y\rangle| \leq$$

$$\leq \alpha\beta \left[\frac{r'}{q'} \alpha^{-q'} \int_K |\langle x', x\rangle|^{q'} \mu(dx') + \frac{r'}{p'} \beta^{-p'} \int_L |\langle y', y\rangle|^{p'} \nu(dy') \right]^{1/r'} \leq \alpha\beta$$

for $\alpha := (\int_K |\langle x', x\rangle|^{q'} \mu(dx'))^{1/q'}$ and $\beta := (\int_L |\langle y', y\rangle|^{p'} \nu(dy'))^{1/p'}$ (if $\alpha = 0$, then $|\langle \varphi, \lambda x \otimes y\rangle|^{r'} \leq r'/p' \int_L |\langle y', y\rangle|^{p'} \nu(dy')$ for all λ and hence $\langle \varphi, x \otimes y\rangle = 0$). This is the inequality in (b). $\square$

Since $\mathfrak{D}_{p,q} \sim \alpha'_{q',p'}$ and

$$\mathfrak{D}_{p,q}(E, F) = (E \otimes_{\alpha_{q',p'}} F')' \cap \mathcal{L}(E, F)$$

(isometrically) by the representation theorem for maximal operator ideals, the theorem has the following consequence:

COROLLARY: *Take $p, q \in [1, \infty]$ with $1/p + 1/q \leq 1$ and $K \subset B_{E'}$ and $L \subset B_{F''}$ weak-$*$-compact norming sets. For $T \in \mathfrak{L}(E, F)$ the following statements are equivalent:*
(a) $T \in \mathfrak{D}_{p,q}(E, F)$.
(b) *There are a $c \geq 0$ and probability measures μ and ν such that*

$$|\langle y', Tx \rangle| \leq c \left(\int_K |\langle x', x \rangle|^p \mu(dx') \right)^{1/p} \left(\int_L |\langle y'', y' \rangle|^q \nu(dy'') \right)^{1/q}$$

holds for all $x \in E$ and $y' \in F'$, (replace the integral by $\| \ \|$ if the exponent is ∞). In this case, $\mathbf{D}_{p,q}(T) = \min\{c \,|\, c$ as in (b)$\}$.

19.3. The proof will show that the following result is just a reformulation of the characterization of the (p, q)-dominated operators which was just obtained.

KWAPIEŃ'S FACTORIZATION THEOREM: *Let $p, q \in [1, \infty]$ such that $1/p + 1/q \leq 1$. Then*

$$\mathfrak{D}_{p,q} = \mathfrak{P}_q^{dual} \circ \mathfrak{P}_p$$

holds isometrically. In other words, $T \in \mathfrak{L}(E, F)$ is (p,q)-dominated if and only if T admits a factorization

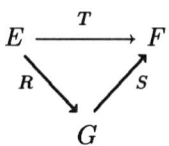

$$\begin{array}{ccc} E & \xrightarrow{T} & F \\ & R \searrow \quad \nearrow S & \\ & G & \end{array}$$

such that $R \in \mathfrak{P}_p$ and $S' \in \mathfrak{P}_q$. Moreover,

$$\mathbf{D}_{p,q}(T) = \min\{\mathbf{P}_q^{dual}(S) \mathbf{P}_p(R) \,|\, T = S \circ R\}.$$

Recall that $\mathfrak{P}_\infty := \mathfrak{L}$, so the result is nothing new if p or q is ∞.

PROOF: If T has such a factorization and $1/r = 1/p + 1/q$, then

$$\ell_r(\langle y'_k, Tx_k \rangle) = \ell_r(\langle S' y'_k, Rx_k \rangle) \leq \ell_p(Rx_k) \ell_q(S' y'_k) \leq$$
$$\leq \mathbf{P}_p(R) w_p(x_k) \mathbf{P}_q(S') w_q(y'_k)$$

and therefore, $\mathbf{D}_{p,q}(T) \leq \mathbf{P}_q^{dual}(S) \mathbf{P}_p(R)$ (by what was repeated in 19.1.). Conversely, take $T \in \mathfrak{D}_{p,q}(E, F)$. Then, by the corollary above, there are probability measures μ on $B_{E'}$ and ν on $B_{F''}$ such that

$$(*) \quad |\langle y', Tx \rangle| \leq \mathbf{D}_{p,q}(T) \left(\int_{B_{E'}} |\langle x', x \rangle|^p \mu(dx') \right)^{1/p} \left(\int_{B_{F''}} |\langle y'', y' \rangle|^q \nu(dy'') \right)^{1/q}$$

(the cases p or $q = \infty$ need not be treated). Define $R_o x := \langle \cdot, x \rangle \in C(B_{E'})$ and consider the diagram

$$\begin{array}{ccc} E & \xrightarrow{\ T\ } & F \\ {\scriptstyle R_o}\downarrow & {\scriptstyle R}\searrow \quad {\scriptstyle \nearrow S} & \\ & G := \overline{I \circ R_o(E)} & \\ & \cap & \\ C(B_{E'}) & \xrightarrow[\ I\ ]{} L_p(B_{E'}, \mu) & , \end{array}$$

where $Rx := I(R_o x)$. Since $\mathbf{P}_p(I) = 1$ and $\mathfrak{P}_p$ is injective (see 11.1. and 11.2.), it follows that $\mathbf{P}_p(R) \leq 1$. The operator S is defined on $R(E)$ by

$$SRx := Tx$$

— and this definition makes sense because

$$|\langle y', SRx \rangle| \leq \mathbf{D}_{p,q}(T) \|Rx\|_G \left(\int_{B_{F''}} |\langle y'', y' \rangle|^q \nu(dy'') \right)^{1/q} .$$

It follows that S is continuous on $R(E)$ and has a unique extension to $\overline{R(E)} = G$; moreover, the inequality implies that

$$\|S'y'\| = \sup\{ |\langle S'y', Rx \rangle| \mid \|Rx\|_G \leq 1 \} \leq$$

$$\leq \mathbf{D}_{p,q}(T) \left(\int_{B_{F''}} |\langle y'', y' \rangle|^q \nu(dy'') \right)^{1/q} ,$$

which means that S' is absolutely q-summing and

$$\mathbf{P}_q^{dual}(S) = \mathbf{P}_q(S') \leq \mathbf{D}_{p,q}(T) .$$

This ends the proof. $\square$

19.4. Since $\mathfrak{L}_{p,q} \sim \alpha_{p,q}$ and $\mathfrak{D}^*_{p',q'} \sim \alpha^*_{p,q}$, it follows that the ideals $\mathfrak{L}_{p,q}$ of (p,q)-factorable operators and $\mathfrak{D}_{p',q'}$ of (p',q')-dominated operators are adjoint to each other (see 17.9.). Anticipating the fact that $\alpha_{p,q}$ and $\alpha^*_{p,q}$ are accessible tensor norms (to be proven in section 21), it follows from the inclusion $\mathfrak{A}^* \circ \mathfrak{A} \subset \mathfrak{I}$ (see 17.19.) that

$$\mathfrak{L}_{p,q} \circ \mathfrak{D}_{p',q'} \subset \mathfrak{I} \quad \text{and} \quad \mathbf{I}(TS) \leq \mathbf{L}_{p,q}(T)\mathbf{D}_{p',q'}(S)$$
$$\mathfrak{D}_{p',q'} \circ \mathfrak{L}_{p,q} \subset \mathfrak{I} \quad \text{and} \quad \mathbf{I}(TS) \leq \mathbf{D}_{p',q'}(T)\mathbf{L}_{p,q}(S)$$

(for $1/p + 1/q \geq 1$). The composition is even nuclear in most cases:

PROPOSITION *If $1/p + 1/q \geq 1$ and $(p,q) \notin \{(1,1), (1,\infty), (\infty,1)\}$, then*

$$\mathfrak{D}_{p',q'} \circ \mathfrak{L}_{p,q} \subset \mathfrak{N} \quad \text{and} \quad \mathbf{N}(TS) \leq \mathbf{D}_{p',q'}(T)\mathbf{L}_{p,q}(S) .$$

This relation is false for the excluded pairs (p, q).

For the nuclearity of operators in $\mathfrak{L}_{p,q} \circ \mathfrak{D}_{p',q'}$ see Ex 19.3..

PROOF: The best way to check that an integral operator is nuclear, is to use the Radon–Nikodým property. Take $S \in \mathfrak{L}_{p,q}(E, F)$ and $T \in \mathfrak{D}_{p',q'}(F, G)$ and assume that $1 < q < \infty$; then

$$\mathfrak{D}_{p',q'} \subset \mathfrak{P}_q^{dual} \subset \mathfrak{W} \qquad \text{(weakly compact operators)}$$

by Kwapień's factorization theorem (or more simply: since $d_q^* \leq \alpha_{p,q}^*$). Therefore $T''(F'') \subset G$ which implies that the astriction $T^\pi : F'' \to G$ of T'' is also (p', q')-dominated and $\mathbf{D}_{p',q'}(T^\pi) = \mathbf{D}_{p',q'}(T)$ (since T'' is, and maximal ideals are regular, see 17.8.). The factorization theorem 18.11. for (p, q)-factorable operators gives

$$T \circ S : E \xhookrightarrow{\kappa_E} E'' \xrightarrow{S''} F'' \xrightarrow{T^\pi} G$$

with diagram $U : E'' \to L_{q'}$, $I : L_{q'} \hookrightarrow L_p$, $V : L_p \to F''$,

with $\mathbf{L}_{p,q}(S) = \mathbf{L}_{p,q}(S'') = \|U\| \, \|I\| \, \|V\|$. Since $I \in \mathfrak{L}_{p,q}$, it follows that $T^\pi \circ V \circ I$ is integral and hence nuclear because $L_{q'}$ is reflexive and has the approximation property (see Appendix D8., corollary); moreover,

$$\mathbf{N}(T^\pi \circ V \circ I) = \mathbf{I}(T^\pi \circ V \circ I) \leq \mathbf{D}_{p',q'}(T)\|V\| \, \|I\| \ .$$

It follows that $T \circ S$ is nuclear with the desired estimate.

If $q = 1$ and $1 < p < \infty$, observe first that the Grothendieck–Pietsch factorization theorem (11.3., corollary 1, (c)) implies that $\mathfrak{P}_{p'} = \mathfrak{W} \circ \mathfrak{P}_{p'}$ holds isometrically; hence,

$$\mathfrak{D}_{p',\infty} \circ \mathfrak{L}_{p,1} = \mathfrak{D}_{p',\infty} \circ \mathfrak{D}_{p',\infty}^* = \mathfrak{P}_{p'} \circ \mathfrak{P}_{p'}^* \subset \mathfrak{W} \circ \mathfrak{I} \subset \mathfrak{N}$$

and $\mathbf{N} \leq \| \ \| \circ \mathbf{I}$, where the latter is one of Grothendieck's results related to the Radon–Nikodým property (see Appendix C8.). To see the negative results, first take $(p, q) = (1, 1)$:

$$\mathfrak{D}_{\infty,\infty} \circ \mathfrak{L}_{1,1} = \mathfrak{L} \circ \mathfrak{I} = \mathfrak{I} \not\subset \mathfrak{N} \ .$$

For the remaining two cases $(p, q) = (1, \infty)$ or $(\infty, 1)$ recall that c_o does not have the Radon–Nikodým property (Appendix D3.), hence there is a $T : C[0, 1] \to c_o$ which is absolutely summing but not nuclear (Appendix D7.). Since $C[0, 1]'$ has the approximation property (it is an abstract L–space), T' is also not nuclear by 5.9.; since, clearly, $id_{C[0,1]} \in \mathfrak{L}_\infty$ and $id_{\ell_1} \in \mathfrak{L}_1$ (by 18.6., for example), it follows that

$$T \in \mathfrak{P}_1 \circ \mathfrak{L}_\infty = \mathfrak{D}_{1,\infty} \circ \mathfrak{L}_{\infty,1}$$
$$T' \in \mathfrak{P}_1^{dual} \circ \mathfrak{L}_1 = \mathfrak{D}_{\infty,1} \circ \mathfrak{L}_{1,\infty}$$

— and this completes the proof. □

The following important result of Grothendieck is a special case:

COROLLARY: *The composition of two absolutely 2-summing operators is nuclear; moreover,* $\mathbf{N}(TS) \leq \mathbf{P}_2(T)\mathbf{P}_2(S)$.

Since $\mathfrak{I}_2 = \mathfrak{P}_2$ (see 18.7.), this follows from

$$\mathfrak{P}_2 \circ \mathfrak{P}_2 = \mathfrak{P}_2 \circ \mathfrak{I}_2 = \mathfrak{D}_{2,\infty} \circ \mathfrak{L}_{2,1} \subset \mathfrak{N}.$$

19.5. If K and L are two compact topological spaces, then

$$\pi \leq K_G w_2 \quad \text{on } C(K) \otimes C(L)$$

by Grothendieck's inequality (see 14.5.). Since $(C(K) \otimes_\pi C(L))' = \mathfrak{Bil}(C(K), C(L))$, theorem 19.2. implies:

PROPOSITION: *For each $\varphi \in \mathfrak{Bil}(C(K), C(L))$ there are normalized Borel–Radon measures μ on K and ν on L such that for all $(f, g) \in C(K) \times C(L)$*

$$|\varphi(f,g)| \leq K_G \|\varphi\| \left(\int_K |f|^2 d\mu \right)^{1/2} \left(\int_L |g|^2 d\nu \right)^{1/2}.$$

Obviously, the smallest constant having this property (for all K and L) is the Grothendieck constant (recall proposition 14.6.).

19.6. A *state* λ on a C^*-algebra A is a linear functional $\lambda \in A'$ with $\|\lambda\| = 1$ which is positive, i.e. $\lambda(xx^*) \geq 0$ for all $x \in A$. Since every commutative C^*-algebra is isometric (as a C^*-algebra) to some $C(K)$ and the probability measures on K are exactly the states of $C(K)$, it follows: Let A and B be commutative C^*-algebras and $\varphi \in \mathfrak{Bil}(A, B)$. Then there are states λ_1 on A and λ_2 on B such that

$$|\varphi(x,y)| \leq K_G^{\mathbb{C}} \|\varphi\| \, \lambda_1(|x|^2)^{1/2} \lambda_2(|y|^2)^{1/2}.$$

This is another formulation of Grothendieck's inequality which (with another constant) also holds for arbitrary C^*-algebras A and B if (different from the usual definition of the modulus) $|x|^2 := 1/2(xx^* + x^*x)$; this deep result was conjectured by Grothendieck (Résumé, problème 4), first proven by Pisier [220], assuming the approximation property, and then by Haagerup [97] in the general case. The result is sometimes called the *Haagerup–Pisier–Grothendieck inequality* or the *non-commutative Grothendieck inequality*. If K_* denotes the best possible constant, then a simple example in $A = B = \mathfrak{K}(\ell_2, \ell_2)$ shows that $K_* \geq 2$ (see Pisier [225], p.119), which implies, in particular, that $K_G^{\mathbb{C}} < K_*$. Actually, Haagerup [97] proved the following stronger version, which easily implies that $K_* \leq 4$, and in which the constant 1 cannot be improved:

If A and B are C^*-algebras, $\varphi \in \mathfrak{Bil}(A,B)$, then there are states λ_1, λ_2 on A and μ_1, μ_2 on B such that

$$|\varphi(x,y)| \leq \|\varphi\|(\lambda_1(xx^*) + \lambda_2(x^*x))^{1/2}(\mu_1(yy^*) + \mu_2(y^*y))^{1/2} .$$

Unfortunately, it is not true (as it is in the commutative case) that the tensor norms π and w_2 are equivalent on the tensor product $A \otimes B$ of two C^*-algebras: Since ℓ_2 is 1-complemented in the C^*-algebra $\mathfrak{L}(\ell_2, \ell_2)$, this would imply that π and w_2 were equivalent on $\ell_2 \otimes \ell_2$, but, clearly, id_{ℓ_2} is not 2-dominated, i.e. $tr \notin (\ell_2 \otimes_{w_2} \ell_2)'$; this observation is due to Pisier. Since $w_2 \leq w_2^*$ (see Ex 19.5.), the following result is weaker; recall that $w_2 \sim \mathfrak{L}_2$, the ideal of operators factoring through a Hilbert space (18.6.).

COROLLARY (Haagerup [97]): *If A and B are C^*-algebras, then every continuous operator $T : A \to B'$ factors through a Hilbert space; moreover, $\mathbf{L}_2(T) \leq 2\|T\|$. In other words,*

$$\pi \leq 2w_2^* \quad \text{on } A \otimes B .$$

PROOF: The non-commutative Grothendieck inequality implies that

$$|\langle Tx, y\rangle| \leq \|T\|\sqrt{2}\left[2^{-1}(\lambda_1(xx^*) + \lambda_2(x^*x))\right]^{1/2} [\|yy^*\| + \|y^*y\|]^{1/2} \leq$$
$$\leq 2\|T\|\left[2^{-1}(\lambda_1(xx^*) + \lambda_2(x^*x))\right]^{1/2} \|y\| .$$

The definition

$$(x|z) := 2^{-1}(\lambda_1(xz^*) + \lambda_2(z^*x))$$

gives a non-negative sesquilinear form on $A \times A$ such that the associated seminorm p satisfies $p(x) := (x|x)^{1/2} \leq \|x\|$. It follows that $A/\ker p$ (with its natural norm coming from p) is a pre-Hilbert space and hence its completion H is a Hilbert space. The canonical map $R : A \to H$ has norm ≤ 1. The inequality at the beginning implies that

$$\|Tx\| \leq 2\|T\| \|Rx\|_H .$$

If $S_o : R(A) \to B'$ is defined by $S_o(Rx) := Tx$, then S_o is well-defined, has norm $\leq 2\|T\|$ and extends to an operator $S : H \to B'$ with $\|S\| = \|S_o\|$. Clearly, $T = S \circ R$ and therefore,

$$\mathbf{L}_2(T) \leq \|S\| \|R\| \leq 2\|T\| .$$

Since

$$(A \otimes_\pi B)' \stackrel{1}{=} \mathfrak{L}(A, B') = \mathfrak{L}_2(A, B') \stackrel{1}{=} (A \otimes_{w_2^*} B)'$$

by the representation theorem for maximal operator ideals, the result is proven. □

For a nice application of this see Ex 19.9. – Ex 19.11.. The reader will find more information about the Haagerup–Pisier–Grothendieck inequality in Pisier's book [225]. Using the ideas of Pisier and Haagerup, Grothendieck's inequality for C^*-algebras was extended by Barton–Friedman [6] to continuous bilinear forms on the larger class of so–called JB^*–triples (see also Chu–Iochum–Loupias [29]).

Exercises:

Ex 19.1. If $id_E \in \mathfrak{D}_{p,q}$ (where p or q are different from ∞), then E is finite dimensional.

Ex 19.2. Using the Grothendieck–Pietsch domination theorem, show directly that every operator $T \in \mathfrak{P}_q^{dual} \circ \mathfrak{P}_p$ satisfies an integral estimate as in corollary 19.2..

Ex 19.3. If $(p,q) \notin \{(1,1),(1,\infty),(\infty,1)\}$ and $T \in (\mathfrak{L}_{p,q} \circ \mathfrak{D}_{p',q'})(E,F)$, then T' is nuclear and $\mathbf{N}(T') \leq \mathbf{L}_{p,q}(R)\mathbf{D}_{p',q'}(S)$ whenever $T = R \circ S$ is such a factorization. In particular, $\kappa_F \circ T$ is nuclear and, if E' has the approximation property, then T is nuclear as well. If $(p,q) \in \{(1,1),(1,\infty),\infty,1)\}$, then it may happen that none of the operators T, T' or $\kappa_F \circ T$ is nuclear. Hint for the negative part: T at the end of 19.4. has a bidual which is not nuclear.

Ex 19.4. Deduce from 12.5. and 12.8. the following results:
(a) If $p_1 \leq p_2$ and $q_1 \leq q_2$, then $(\mathfrak{D}_{p_1,q_1}, \mathbf{D}_{p_1,q_1}) \subset (\mathfrak{D}_{p_2,q_2}, \mathbf{D}_{p_2,q_2})$.
(b) If $p,q \in]1,\infty[$, then $\mathfrak{D}_{p,q} \subset \mathfrak{D}_2$.
(c) If $p,q \in [2,\infty[$, then $\mathfrak{D}_{p,q} = \mathfrak{D}_2$.

Ex 19.5. Show that $w_2 \leq g_2^* \leq w_2^*$. Hint: Operator ideals.

Ex 19.6. If p or q is $< \infty$, then $\mathfrak{D}_{p,q} \subset \mathfrak{W}$.

Ex 19.7. (a) Every operator $C(K) \to C(L)'$ is in $\mathfrak{D}_2$. Hint: Grothendieck's inequality.
(b) $\mathfrak{D}_2 \not\subset \mathfrak{K}$. Hint: Consider the operator $T : L_\infty[0,1] \ni f \rightsquigarrow \int f d\lambda \in L_\infty[0,1]'$ for the Lebesgue measure λ.

Ex 19.8. (a) Show with the little Grothendieck theorem that $\mathfrak{P}_2^{dual}(E,F) \subset \mathfrak{D}_2(E,F)$ whenever E is an $\mathcal{L}_1^g$–space. Hint: Prove this first for ℓ_1 and then use the $\mathcal{L}_p$–local technique lemma.
(b) There is a set Γ and a Banach space F such that

$$\mathfrak{P}_2(F, \ell_\infty(\Gamma)) \not\subset (\mathfrak{P}_2 \circ \mathfrak{W})(F, \ell_\infty(\Gamma)).$$

Hint: $\mathfrak{P}_2$ and $\mathfrak{P}_2 \circ \mathfrak{W}$ are injective and $\mathfrak{P}_2 \not\subset \mathfrak{K}$, but $\mathfrak{P}_2 \circ \mathfrak{W} \subset \mathfrak{K}$.
(c) Deduce from (a) and (b) that $\mathfrak{D}_2 \not\subset \mathfrak{W} \circ \mathfrak{D}_2$.

Ex 19.9. Use corollary 19.6. and the fact that w_2^* and w_2 are totally accessible (to be proven in section 21) to show the following results for arbitrary C^*–algebras A and B.
(a) The canonical map $A \tilde{\otimes}_\pi B \to A \tilde{\otimes}_\varepsilon B$ is injective.
(b) $w_2 \leq 2\varepsilon$ on $A' \otimes B'$.
(c) The natural embedding $A \otimes_\pi B \hookrightarrow (A' \otimes_\varepsilon B')'$ is an isomorphic embedding; more precisely, $\pi \leq 2\overleftarrow{\pi}$ on $A \otimes B$.

Clearly, (c) improves (a). Recall that the C^*-algebra $\mathfrak{L}(\ell_2, \ell_2)$ does not have the approximation property.

Ex 19.10. Let H be a Hilbert space. Then every operator $\mathfrak{L}(H, H) \to \mathfrak{N}(H, H)$ factors through a Hilbert space.

Ex 19.11. If A is a C^*-algebra such that A' is isomorphic to a C^*-algebra, then A is finite dimensional. Hint: Use the fact that infinite dimensional C^*-algebras are not reflexive (this follows, e.g., from Exercise 4.6.12 in Kadison–Ringrose [139]).

20. Projective and Injective Tensor Norms

The projective norm π respects metric surjections and the injective norm ε respects metric injections. In this section these properties are studied systematically for arbitrary tensor norms. In particular, injective and projective norms (i.e. those respecting metric injections or surjections, respectively) closest to a given tensor norm are constructed. The associated operator ideals will be described. Finally, some more aspects of Grothendieck's inequality will be treated.

20.1. A tensor norm α on NORM (or on FIN, or other classes) is called *right–injective on* NORM (or *on* FIN, ...) if for all metric injections $I : F \hookrightarrow G$ the operator

$$id_E \otimes I : E \otimes_\alpha F \longrightarrow E \otimes_\alpha G$$

is a metric injection for $E, F, G \in$ NORM (or $\in$ FIN, ...) and *right–projective on* NORM (or *on* FIN, ...) if for all metric surjections $Q : F \twoheadrightarrow G$ the operator

$$id_E \otimes Q : E \otimes_\alpha F \longrightarrow E \otimes_\alpha G$$

is a metric surjection for $E, F, G \in$ NORM (or $\in$ FIN, ...). If α^t is right–injective (resp. right–projective), then the tensor norm α is called *left–injective* (resp. *left–projective*) on NORM/FIN/...; if α is right– and left–injective (resp. right– and left–projective), it is called *injective* (resp. *projective*) on NORM/FIN/... . Clearly, ε is injective and π is projective.

Observe that $id_E \tilde\otimes I : E \tilde\otimes_\alpha F \hookrightarrow E \tilde\otimes_\alpha G$ is a metric injection whenever $id_E \otimes I$ is — and it follows from the quotient lemma 7.4. that $id_E \tilde\otimes Q : E \tilde\otimes_\alpha F \twoheadrightarrow E \tilde\otimes_\alpha G$ is a metric surjection whenever $id_E \otimes Q$ is.

The duality
$$M \otimes_{\alpha'} N \stackrel{1}{=} (M' \otimes_\alpha N')' \qquad M, N \in \text{FIN}$$
implies the

REMARK: α is right–injective on FIN if and only if α' is right–projective on FIN.

20.2. This remark will be extended to tensor norms on NORM. The right–injective tensor norms are relatively easy to handle, but when treating the right–projective ones some problems occur when passing from Banach to arbitrary normed spaces. So their study will be prepared by a precise investigation of their behaviour with respect to dense subspaces. For this, let β be a tensor norm on NORM × C (where C is either the class of all Banach spaces or the class of all normed spaces), and recall from Ex 12.3. the definition of the right–finite hull $\beta^\rightarrow$

$$\beta^\rightarrow(z; E, F) = \inf\{\beta(z; E, N) \mid N \in \text{FIN}(F), z \in E \otimes N\} .$$

$\beta^\rightarrow$ is a tensor norm and $\beta \leq \beta^\rightarrow \leq \overrightarrow{\beta}$.

LEMMA:
(1) If β is right–projective on NORM × C, then $\beta = \beta^\rightarrow$ on NORM × C.
(2) If β is a tensor norm on NORM such that $\beta = \beta^\rightarrow$ on NORM × BAN, then $\beta = \beta^\rightarrow$ on NORM × NORM and
$$E \otimes_\beta F \xhookrightarrow{1} E \otimes_\beta \tilde{F}$$
for all $(E, F) \in$ NORM × NORM.
(3) If β is a tensor norm on NORM, right–projective on NORM × BAN, then it is right–projective on NORM × NORM.

PROOF: (1) If $G \in C$, then there is a metric surjection
$$Q : F \xtwoheadrightarrow{1} G$$
such that F has the metric approximation property (if G is complete, take $F := \ell_1(B_G)$ and in the general case an appropriate dense subspace of $\ell_1(B_{\tilde{G}})$ which necessarily has the metric approximation property by 16.3., for example); then, for each normed space E the approximation lemma 13.1. implies that
$$\beta(\cdot; E, F) = \beta^\rightarrow(\cdot; E, F) .$$
It follows that for each element $z \in E \otimes G$ there is an $N \in \text{FIN}(F)$ and a $\hat{z} \in E \otimes N$ with $id_E \otimes Q(\hat{z}) = z$ and
$$\beta(\hat{z}; E, N) \leq (1+\varepsilon)\beta(z; E, G)$$

and therefore,
$$\beta(z; E, G) \le \beta^\to(z; E, G) \le \beta(z; E, QN) \le \beta(\hat{z}; E, N) \le (1+\varepsilon)\beta(z; E, G) \ .$$

(2) Take $z \in E \otimes F$. Then the metric mapping property gives
$$\beta^\to(z; E, \tilde{F}) = \beta(z; E, \tilde{F}) \le \beta(z; E, F) \le \beta^\to(z; E, F) \ .$$

For $N \in \mathrm{FIN}(\tilde{F})$, with $z \in E \otimes N$ and
$$\beta(z; E, N) \le (1+\varepsilon)\beta^\to(z; E, \tilde{F}) \ ,$$

use the principle of local reflexivity 6.6. to find an operator $R : N \to F$ with $\|R\| \le 1+\varepsilon$ and $Ry = y$ for all $y \in N \cap F$. The operator $T : E' \to F$ associated with z has values in $N \cap F$, which implies that
$$id_E \otimes R(z) = z \in E \otimes RN \ .$$

Consequently,
$$\beta^\to(z; E, F) \le \beta(id_E \otimes R(z); E, RN) \le \|R\|\beta(z; E, N) \le$$
$$\le (1+\varepsilon)^2 \beta^\to(z; E, \tilde{F}) \ ,$$

which proves (2).

To see (3), recall again the quotient lemma in 7.4. and take $Q : F \twoheadrightarrow G$ a metric surjection between normed spaces; then the completion $\tilde{Q} : \tilde{F} \to \tilde{G}$ is a metric surjection with $\ker \tilde{Q} = \overline{\ker Q}$ and, by assumption,
$$id_E \otimes \tilde{Q} : E \otimes_\beta \tilde{F} \longrightarrow E \otimes_\beta \tilde{G}$$

is also a metric surjection. Since by (1) and (2)
$$E \otimes_\beta F \hookrightarrow E \otimes_\beta \tilde{F} \quad \text{and} \quad E \otimes_\beta G \hookrightarrow E \otimes_\beta \tilde{G}$$

are metric injections and, clearly $(id_E \otimes Q)(E \otimes F) = E \otimes G$, it follows from the quotient lemma that
$$id_E \otimes Q : E \otimes_\beta F \longrightarrow E \otimes_\beta G$$

is a metric surjection if
$$\ker(id_E \otimes \tilde{Q}) = E \otimes \ker \tilde{Q} \overset{!}{\subset} \overline{\ker(id_E \otimes Q)}^{E \otimes_\beta \tilde{F}}$$

(see 2.7.). But this is obvious since $\ker \tilde{Q} = \overline{\ker Q}$. $\square$

20.3. Now the announced duality between right–injective and right–projective tensor norms can be proved. At the same time, and this is in some sense natural, a preliminary observation concerning the accessibility of these tensor norms will be made (a more careful investigation will be given in section 21).

PROPOSITION: *Let α be a tensor norm on* NORM.

(1) *If α is right–injective on* FIN, *then $\overleftarrow{\alpha}$ and $\overrightarrow{\alpha}$ are right–injective on* NORM.

(2) *If α is right–projective on* FIN, *then $\overrightarrow{\alpha}$ is right–projective on* NORM.

(3) *If α is finitely or cofinitely generated, then: α is right–injective on* NORM *if and only if α' is right–projective on* NORM.

(4) *If α is right–injective or right–projective on* FIN, *then α is right–accessible.*

PROOF: (1) and (4): If α is right–injective on FIN, then for $F \overset{1}{\hookrightarrow} G$ and $z \in E \otimes F$

$$\overrightarrow{\alpha}(z; E, G) \leq \overrightarrow{\alpha}(z; E, F) =$$
$$= \inf\{\alpha(z; M, N \cap F) | M \in \text{FIN}(E), N \in \text{FIN}(G), z \in M \otimes N\} =$$
$$= \inf\{\alpha(z; M, N) | ...\} = \overrightarrow{\alpha}(z; E, G) \,,$$

so $\overrightarrow{\alpha}$ is right–injective. To treat the cofinite hull, first (4) will be shown: For this, take $(N, F) \in \text{FIN} \times \text{NORM}$ and $z \in N \otimes F$ and assume α is right–injective on FIN. Then, by what was already shown and by the approximation lemma, it follows that

$$\overrightarrow{\alpha}(z; N, F) = \overrightarrow{\alpha}(z; N, \ell_\infty(B_{F'})) = \overleftarrow{\alpha}(z; N, \ell_\infty(B_{F'})) \leq \overleftarrow{\alpha}(z; N, F) \,,$$

hence α is right–accessible. Now recall that α is right–accessible if α' is (see 15.6.). Hence, if α is right–projective on FIN, the dual α' is right–injective on FIN, therefore α' is right–accessible and so is α.

Now it is possible to show that $\overleftarrow{\alpha}$ is right–injective on NORM if α is right–injective on FIN: For $F \overset{1}{\hookrightarrow} G$ and $z \in E \otimes F$ the following holds by the two results which were already shown:

$$\overleftarrow{\alpha}(z; E, F) = \sup\left\{\overleftarrow{\alpha}(Q_K^E \otimes id_F(z); E/K, F) \mid K \in \text{COFIN}(E)\right\} =$$
$$= \sup\left\{\overrightarrow{\alpha}(Q_K^E \otimes id_F(z); E/K, F) \mid K \in \text{COFIN}(E)\right\} =$$
$$= \sup\left\{\overrightarrow{\alpha}(Q_K^E \otimes id_G(z); E/K, G) \mid K \in \text{COFIN}(E)\right\} =$$
$$= \sup\left\{\overleftarrow{\alpha}(Q_K^E \otimes id_G(z); E/K, G) \mid K \in \text{COFIN}(E)\right\} =$$
$$= \overleftarrow{\alpha}(z; E, G) \,.$$

(2) Using lemma 20.2.(3), it is enough to consider a metric surjection $Q : F \xrightarrow{1} G$ between Banach spaces. By (4) the tensor norm α' is right–accessible, hence for every $N \in \text{FIN}$ result (1) implies that

$$\left(N \otimes_{\overrightarrow{\alpha}} G\right)' \stackrel{1}{=} N' \otimes_{\alpha'} G' \xhookrightarrow{1} N' \otimes_{\alpha'} F' \stackrel{1}{=} \left(N \otimes_{\overrightarrow{\alpha}} F\right)'$$

and therefore

$$id_N \otimes Q : N \otimes_{\overrightarrow{\alpha}} F \longrightarrow N \otimes_{\overrightarrow{\alpha}} G$$

is a metric surjection. Now take E an arbitrary normed space:

$$\overrightarrow{\alpha}(z; E, G) = \inf\left\{\overrightarrow{\alpha}(z; N, G) \mid N \in \text{FIN}(E), z \in N \otimes G\right\} =$$
$$= \inf\left\{\overrightarrow{\alpha}(w; N, F) \mid N \in \text{FIN}(E), id_N \otimes Q(w) = z\right\} =$$
$$= \inf\left\{\overrightarrow{\alpha}(w; E, F) \mid id_E \otimes Q(w) = z\right\}.$$

Finally, statement (3) follows from (1), (2) and the remark at the end of 20.1.. □

It is *not true* that the cofinite hull $\overleftarrow{\alpha}$ is right–projective on BAN if α is right–projective on FIN; to see an example take $\alpha = \pi$ and $\ell_1(B_F) \xrightarrow{1} F$ for a Banach space F without the metric approximation property. Then

$$F' \otimes_{\overleftarrow{\pi}} \ell_1(B_F) = F' \otimes_\pi \ell_1(B_F) \xrightarrow{1} F' \otimes_\pi F \neq F' \otimes_{\overleftarrow{\pi}} F.$$

Since there is no Hahn–Banach theorem for general operators between Banach spaces (1.5.), π is neither right– nor left–injective. Dually (this means using (3) of the last proposition), ε is neither right– nor left–projective; see also 20.20..

20.4. For the $\alpha_{p,q}$-tensor norms the following result holds:

PROPOSITION: *Let $1 \leq p \leq \infty$. Then*

(1) d_p *is right–projective; consequently, g_p is left–projective and $g_p^* = d_p'$ is right–injective.*

(2) $\alpha_{2,p}$ *is right–injective, $\alpha_{p,2}$ left–injective and $\alpha'_{2,p}$ right–projective. In particular, w_2 is injective and $w_2^* = w_2'$ is projective.*

PROOF: Since

$$d_p(z; E, F) = \inf\left\{w_{p'}(x_i)\ell_p(y_i) \mid z = \sum_{i=1}^n x_i \otimes y_i\right\},$$

result (1) is a direct consequence of the following observation: If $Q : F \twoheadrightarrow G$ is a metric surjection, $\varepsilon > 0$ and $y_1, ..., y_n \in G$, then there are $\hat{y}_i \in F$ with $Q(\hat{y}_i) = y_i$ and

$$\ell_p(y_i) \leq \ell_p(\hat{y}_i) \leq (1+\varepsilon)\ell_p(y_i) .$$

To see that $\alpha_{2,p}$ is right–injective, take a metric injection $I : F \hookrightarrow G$, an element $z \in E \otimes F$ and $\varepsilon > 0$. Choose a representation of z in $E \otimes G$ with

$$\ell_r(\lambda_i)w_{p'}(x_i)w_2(y_i) \leq (1+\varepsilon)\alpha_{2,p}(z; E, G) .$$

Then (by what was explained in 17.10.) the associated operator $T_z : E' \to F$ admits a factorization

$$\begin{array}{ccccc} E' & \xrightarrow{T_z} & F & \xhookrightarrow{I} & G \\ {\scriptstyle R}\downarrow & & & \nearrow {\scriptstyle S} & \\ \ell_{p'}^n & \xrightarrow{D_\lambda} & \ell_2^n & & \end{array}$$

with $\|R\| \|D_\lambda\| \|S\| \leq (1+\varepsilon)\alpha_{2,p}(z; E, G)$. The subspace $S^{-1}(I(F))$ is 1–complemented in ℓ_2^n, hence there is an $S_o : \ell_2^n \to F$ with $\|S_o\| \leq \|S\|$ and $T_z = S_o \circ D_\lambda \circ R$. It follows that

$$z = \sum_{k=1}^{n} \lambda_k x_k \otimes S_o e_k \in E \otimes F$$

and

$$\alpha_{2,p}(z; E, G) \leq \alpha_{2,p}(z; E, F) \leq \ell_r(\lambda_k)w_{p'}(x_k)\|S_o\| \leq (1+\varepsilon)\alpha_{2,p}(z; E, G) ,$$

which shows that $\alpha_{2,p}$ is right–injective.

The other statements in (1) and (2) follow easily by dualization and transposition. $\square$

20.5. To see an application of this, recall from 15.10. the Chevet–Persson–Saphar inequalities

$$d_p \leq \Delta_p \leq g_{p'}^* \qquad \text{on } L_p(\mu) \otimes E$$

and also that d_p and $g_{p'}^*$ are the closest tensor norms to Δ_p (see 15.11.). It was shown in 15.10. that $d_p = \Delta_p = g_{p'}^*$ if $E = L_p(\nu)$. More generally,

PROPOSITION: *If E is a Banach space which can be obtained by starting with some $L_p(\nu)$ and then taking metric quotients and metric subspaces finitely many times, then*

$$d_p = \Delta_p = g_{p'}^* \qquad \text{on } L_p(\mu) \otimes E$$

for every measure μ.

PROOF: Fix μ and consider the maximal normed operator ideal $\mathfrak{L}_\mu$ of all $T \in \mathfrak{L}(E,F)$ such that

$$\mathbf{L}_\mu(T) := \|id_{L_p(\mu)} \otimes T : L_p(\mu) \otimes_{d_p} E \longrightarrow L_p(\mu) \otimes_{\Delta_p} F)\| < \infty$$

(see 17.4.). Since d_p is right–projective and Δ_p right–injective (7.4.), this operator ideal is injective and surjective. This implies that the class of spaces E such that $\mathbf{L}_\mu(id_E) = 1$ is stable under forming metric subspaces and metric quotients (both $\neq \{0\}$, clearly). Since $\mathbf{L}_\mu(id_{L_p(\nu)}) = 1$ by what was said earlier, it follows that $d_p = \Delta_p$ for those spaces satisfying the assumption. For $g_{p'}^*$, consider the ideal of those T such that

$$\|id_{L_p(\mu)} \otimes T : L_p(\mu) \otimes_{\Delta_p} E \longrightarrow L_p(\mu) \otimes_{g_{p'}^*} F\| < \infty .$$

Observe that it is again injective and surjective since $g_{p'}^*$ is right–injective and Δ_p is right–projective (7.4.), and argue in exactly the same way. □

In Ex 20.8. and Ex 20.9. it will be shown that the spaces in the assumption of the proposition are precisely the subspaces of quotients of some L_p (equivalently: quotients of subspaces of some L_p). It will turn out in 25.10. that actually these are exactly the spaces E with $d_p = \Delta_p = g_{p'}^*$ on $L_p \otimes E$. For $\Delta_p = d_p^*$ and $\Delta_p = g_p$ see also 25.10..

20.6. Every tensor norm α is less than or equal to π and π is projective. Hence it is reasonable to search for a smallest tensor norm $\beta \geq \alpha$ which is projective.

THEOREM: *Let α be a tensor norm on* NORM. *Then there is a unique right–projective tensor norm $\alpha/ \geq \alpha$ on* NORM *with the following property: If $\beta \geq \alpha$ is right–projective, then $\beta \geq \alpha/$. Moreover, $\alpha/$ is finitely generated from the right, i.e. $\alpha/ = (\alpha/)^{\rightarrow}$.*

The *right–projective associate* $\alpha/$ of α can be calculated using the following property:

If E is normed and F a Banach space, then

(∗) $\qquad E \otimes_\alpha \ell_1(B_F) \xrightarrow{1} E \otimes_{\alpha/} F$

is a metric surjection. If E and F are arbitrary normed spaces and $z \in E \otimes F$, then

(∗) $\qquad \alpha/(z; E, F) = \inf\{\alpha/(z; E, N) \mid N \in \text{FIN}(F), z \in E \otimes N\} .$

The fact that $\alpha/$ respects quotient mappings $F \xrightarrow{1} F/G$ is of course responsible for the use of the symbol $\alpha/$.

PROOF: Uniqueness presents no problem. Existence: $\alpha/$ will be constructed first on NORM × BAN and then extended, using the introductory lemma 20.2..

(a) If $(E, F) \in \text{NORM} \times \text{BAN}$, define $\alpha/$ to be the quotient seminorm on $E \otimes F$ given by the mapping
$$E \otimes_\alpha \ell_1(B_F) \longrightarrow E \otimes F.$$
Since $\varepsilon \leq \alpha \leq \pi$ and π is right–projective, it follows that $\varepsilon \leq \alpha/ \leq \pi$ on the class NORM $\times$ BAN. To see the metric mapping property on NORM $\times$ BAN, take operators $T_i \in \mathfrak{L}(E_i, F_i)$ and use the lifting property of the spaces $\ell_1(\Gamma)$ (see 3.12.) to obtain a $\hat{T}_2$ with

$$\begin{array}{ccc} \ell_1(B_{E_2}) & \xrightarrow{\hat{T}_2} & \ell_1(B_{F_2}) \\ {\scriptstyle 1}\downarrow & & \downarrow{\scriptstyle 1} \\ E_2 & \xrightarrow{T_2} & F_2 \end{array} \qquad \|\hat{T}_2\| \leq (1+\varepsilon)\|T_2\|.$$

It follows that $\|T_1 \otimes_\alpha \hat{T}_2\| \leq \|T_1\| \, \|\hat{T}_2\| \leq \|T_1\|(1+\varepsilon)\|T_2\|$, and hence (by the definition of $\alpha/$) that

$$\|T_1 \otimes T_2 : E_1 \otimes_{\alpha/} E_2 \longrightarrow F_1 \otimes_{\alpha/} F_2\| \leq \|T_1\|(1+\varepsilon)\|T_2\|$$

for all $\varepsilon > 0$. Thus, $\alpha/$ is a tensor norm on NORM $\times$ BAN.

(b) To show that $\alpha/$ is right–projective on NORM $\times$ BAN take a metric surjection $Q : F \twoheadrightarrow G$ between Banach spaces and again use the lifting property:

$$\begin{array}{ccc} \ell_1(B_F) & \xleftarrow{\hat{Q}_\varepsilon} & \ell_1(B_G) \\ {\scriptstyle 1}\downarrow & & \downarrow{\scriptstyle 1} \\ F & \xrightarrow{Q} & G \end{array} \qquad \|\hat{Q}_\varepsilon\| \leq 1 + \varepsilon.$$

The diagram

$$\begin{array}{ccc} E \otimes_\alpha \ell_1(B_F) & \xleftarrow{\mathrm{id}_E \otimes \hat{Q}_\varepsilon} & E \otimes_\alpha \ell_1(B_G) \\ {\scriptstyle 1}\downarrow & & \downarrow{\scriptstyle 1} \\ E \otimes_{\alpha/} F & \xrightarrow{\mathrm{id}_E \otimes Q} & E \otimes_{\alpha/} G \end{array}$$

implies that $\mathrm{id}_E \otimes Q$ is a metric surjection.

(c) Lemma 20.2.(1) now implies that $\alpha/ = (\alpha/)^\rightarrow$ on NORM $\times$ BAN. This means that the definition
$$\alpha/ := (\alpha/)^\rightarrow \qquad \text{on NORM} \times \text{NORM}$$
gives a tensor norm which extends $\alpha/$ from NORM $\times$ BAN to NORM $\times$ NORM. Lemma 20.2.(3) shows that $\alpha/$ is right–projective on NORM; moreover, by the definition $\alpha \leq \alpha/$ on NORM $\times$ FIN, and therefore, $\alpha \leq \alpha^\rightarrow \leq (\alpha/)^\rightarrow = \alpha/$.

(d) Finally, take any right–projective tensor norm $\beta \geq \alpha$; it follows that $\alpha/ \leq \beta$ on NORM $\times$ FIN by the definition of $\alpha/$. Since $\beta = \beta^{\rightarrow}$ on NORM by lemma 20.2.(1), it follows that $\alpha/ = (\alpha/)^{\rightarrow} \leq \beta^{\rightarrow} = \beta$. This ends the proof. $\square$

Consider the natural metric surjection

$$Q : \ell_1(B_{\ell_1(\Gamma)}) \xrightarrow{1} \ell_1(\Gamma) ;$$

then there is a lifting $T : \ell_1(\Gamma) \to \ell_1(B_{\ell_1(\Gamma)})$ of $id_{\ell_1(\Gamma)}$ (i.e. $Q \circ T = id_{\ell_1(\Gamma)}$) with $\|T\| \leq 1 + \varepsilon$. Therefore, for any normed space E, the diagram

$$\begin{array}{ccc}
E \otimes_\alpha \ell_1(\Gamma) & \xrightarrow{\;\;id\;\;} & E \otimes_{\alpha/} \ell_1(\Gamma) \\
& \searrow {\scriptstyle id_E \otimes T} & \uparrow {\scriptstyle id_E \otimes Q} \\
& & E \otimes_\alpha \ell_1(B_{\ell_1(\Gamma)})
\end{array}$$

shows that $\alpha/ \leq (1 + \varepsilon)\alpha$ on $E \otimes \ell_1(\Gamma)$. Since $\alpha \leq \alpha/$ always holds, it follows:

COROLLARY 1: *If E is a normed space, then*

$$\alpha = \alpha/ \quad \text{on } E \otimes \ell_1(\Gamma)$$

for every set Γ.

For obvious reasons

$$\backslash \alpha := ((\alpha^t)/)^t$$

is called the *left–projective associate* of α.

COROLLARY 2: *Let α be a tensor norm. Then*

$$\backslash(\alpha/) = (\backslash\alpha)/ =: \backslash\alpha/$$

is called the projective associate *of α; it is the unique smallest projective tensor norm $\geq \alpha$, is finitely generated and*

$$\ell_1(B_E) \otimes_\alpha \ell_1(B_F) \xrightarrow{1} E \otimes_{\backslash\alpha/} F$$

is a metric surjection whenever E and F are Banach spaces.

The proof follows easily from the "transitivity of metric surjections" and the theorem.

20.7. Fortunately, the injective case is simpler.

THEOREM: *Let α be a tensor norm on* NORM. *Then there is a unique right–injective tensor norm $\alpha\backslash \le \alpha$ on* NORM *such that $\beta \le \alpha\backslash$ for all right–injective tensor norms $\beta \le \alpha$. For all normed spaces E and F*

(*)
$$E \otimes_{\alpha\backslash} F \xrightarrow{\quad 1 \quad} E \otimes_\alpha \ell_\infty(B_{F'})$$

is a metric injection. $\alpha\backslash$ is finitely generated from the right.

The tensor norm $\alpha\backslash$ is called the *right–injective associate* of α.

PROOF: Define $\alpha\backslash$ on $E \otimes F$ to be the norm induced by
$$E \otimes F \hookrightarrow E \otimes_\alpha \ell_\infty(B_{F'}) \ .$$
Then $\varepsilon \le \alpha\backslash \le \pi$ on NORM since ε is right–injective. To see the metric mapping property, take $S \in \mathcal{L}(E_1, E_2)$ and $T \in \mathcal{L}(F_1, F_2)$. By the metric extension property of $\ell_\infty(\Gamma)$ there is an extension $\hat{T}$ of T such that

$$\begin{array}{ccc} \ell_\infty(B_{F_1'}) & \xrightarrow{\hat{T}} & \ell_\infty(B_{F_2'}) \\ \uparrow & & \uparrow \\ F_1 & \xrightarrow{T} & F_2 \end{array} \qquad \|\hat{T}\| = \|T\|$$

commutes. This easily implies the conclusion. $\alpha\backslash$ is right–injective: If $I : F \hookrightarrow G$ is a metric injection, then the metric extension property again gives a mapping $\hat{I}$ such that

$$\begin{array}{ccc} \ell_\infty(B_{F'}) & \xleftarrow{\hat{I}} & \ell_\infty(B_{G'}) \\ \uparrow & & \uparrow \\ F & \hookrightarrow & G \end{array} \qquad \|\hat{I}\| = \|I\| = 1$$

commutes; it follows that

$$\alpha\backslash(z; E, G) \le \alpha\backslash(z; E, F) = \alpha(z; E, \ell_\infty(B_{F'})) = \alpha(id_E \otimes \hat{I}(z); E, \ell_\infty(B_{F'})) \le$$
$$\le \|\hat{I}\| \alpha(z; E, \ell_\infty(B_{G'})) = \alpha\backslash(z; E, G) \ .$$

The definition of $\alpha\backslash$ immediately implies that $\beta \le \alpha\backslash$ whenever $\beta \le \alpha$ is right–injective. Clearly, $\beta = \beta^\rightarrow$ for every right–injective tensor norm. $\square$

As in the projective case,
$$/\alpha := ((\alpha^t)\backslash)^t$$

is called the *left-injective associate* of α and

$$/\alpha\backslash := (/\alpha)\backslash = /(\alpha\backslash)$$

is the *injective associate* which is the unique largest injective tensor norm smaller than α. It follows that

$$E \otimes_{/\alpha\backslash} F \xrightarrow{1} \ell_\infty(B_{E'}) \otimes_\alpha \ell_\infty(B_{F'}) .$$

Clearly, every injective tensor norm is finitely generated.

The notation, with all the various slashes, may seem confusing at first, but it is really quite natural and not that difficult to memorize: Start with the right–projective associate $\alpha/$ (coming from quotients F/G), then take its counterpart $\alpha\backslash$ for the right–injective associate — and finally the rest by symmetry.

COROLLARY: *If the Banach space F has the λ-extension property, then*

$$\alpha\backslash \leq \alpha \leq \lambda\alpha\backslash \qquad \text{on } E \otimes F$$

for all normed spaces E.

PROOF: F is λ-complemented in $\ell_\infty(B_{F'})$. □

20.8. The following is a special case.

PROPOSITION: *For every tensor norm α, normed space E and $n \in \mathbb{N}$*

$$E \otimes_\alpha \ell_1^n = E \otimes_{\alpha/} \ell_1^n \quad \text{isometrically}$$
$$E \otimes_\alpha \ell_\infty^n = E \otimes_{\alpha\backslash} \ell_\infty^n \quad \text{isometrically} .$$

Now the $\mathcal{L}_p$-local technique lemma 13.5. can be applied.

COROLLARY: *Let α be a tensor norm and E a normed space.*
(1) *If F is an $\mathcal{L}^g_{1,\lambda}$-space, then*

$$\alpha \leq \alpha/ \leq \lambda\alpha^{\rightarrow} \qquad \text{on } E \otimes F .$$

(2) *If F is an $\mathcal{L}^g_{\infty,\lambda}$-space, then*

$$\alpha\backslash \leq \alpha \leq \lambda\alpha\backslash \qquad \text{on } E \otimes F .$$

Note that $\alpha^{\rightarrow} \leq \mu\alpha$ on $E \otimes F$ if F has the μ-bounded approximation property by the approximation lemma ($\mathcal{L}^g_p$-spaces have the bounded approximation property, see 21.6.) and $\alpha = \alpha^{\rightarrow}$ if α is finitely generated.

PROOF: The local technique lemma showed, in case (1), that the proposition implies that $\alpha/ \leq (\alpha/)^{\rightarrow} \leq \lambda \alpha^{\rightarrow}$ on $E \otimes F$ (see the remark after the proof in 13.5.). In case (2) it follows that $\alpha^{\rightarrow} \leq \lambda(\alpha\backslash)^{\rightarrow} = \lambda \alpha\backslash$. □

20.9. The following gives a simple test for recognizing whether or not a tensor norm β is the projective or injective associate of a given tensor norm α:

PROPOSITION: *Let α and β be tensor norms.*
(1) *If β is right-projective, then the following statements are equivalent:*
  (a) $\beta = \alpha/$.
  (b) *For all $E \in$ NORM and $n \in \mathbb{N}$*
  $$E \otimes_\beta \ell_1^n = E \otimes_\alpha \ell_1^n \quad \text{isometrically.}$$
(2) *If β is right-injective, then the following statements are equivalent:*
  (a) $\beta = \alpha\backslash$.
  (b) *For all $E \in$ NORM and $n \in \mathbb{N}$*
  $$E \otimes_\beta \ell_\infty^n = E \otimes_\alpha \ell_\infty^n \quad \text{isometrically.}$$
(3) *If α and β are finitely generated, then in both cases it is enough to test only for finite dimensional E.*
(4) *If β is projective, then $\beta = \backslash\alpha/$ if and only if for all $n \in \mathbb{N}$*
  $$\ell_1^n \otimes_\beta \ell_1^n = \ell_1^n \otimes_\alpha \ell_1^n \quad \text{isometrically.}$$
(5) *If β is injective, then $\beta = /\alpha\backslash$ if and only if for all $n \in \mathbb{N}$*
  $$\ell_\infty^n \otimes_\beta \ell_\infty^n = \ell_\infty^n \otimes_\alpha \ell_\infty^n \quad \text{isometrically.}$$

PROOF: That (a) implies (b) (in (1) and (2)) follows from the last proposition. Assume (1)(b). Then by the local technique lemma, $\beta^{\rightarrow} = \alpha^{\rightarrow}$ on $E \otimes \ell_1(\Gamma)$ and hence $\beta = \alpha$ on $E \otimes \ell_1(\Gamma)$ by the approximation lemma: the fact that β is right-projective along with property $(*)$ in theorem 20.6. gives (a). The remaining implication in (2), as well as (4) and (5), follow in the same fashion. Statement (3) is obvious. □

It is worth noting that these characterizations (with the same proofs) also hold if constants are involved — and this will be used. For example: *If α and β are tensor norms, β injective, then $c\beta \leq /\alpha\backslash \leq d\beta$ if and only if for all $n \in \mathbb{N}$*
$$c\beta \leq \alpha \leq d\beta \quad \text{on } \ell_\infty^n \otimes \ell_\infty^n.$$

20.10. The following formulas contain many of the fundamental properties of projective/injective associates and finite/cofinite hulls; they create a type of "calculus" which will be helpful when dealing with accessibility in the next section.

PROPOSITION: Let α be a tensor norm on NORM.
(1) $(\overrightarrow{\alpha})\backslash = \overrightarrow{\alpha\backslash}$ and $(\overrightarrow{\alpha})/ = \overrightarrow{\alpha/}$.
(2) $(\overleftarrow{\alpha})\backslash = \overleftarrow{\alpha\backslash}$, but, in general, $(\overleftarrow{\alpha})/ \neq \overleftarrow{\alpha/}$.
(3) $(\alpha/)' = (\alpha')\backslash$ and $(\alpha\backslash)' = (\alpha')/$.
(4) $(\alpha/)^* = /(\alpha^*)$ and $(\alpha\backslash)^* = \backslash(\alpha^*)$.

PROOF: (1) By 20.8. it follows that
$$\overrightarrow{\alpha} = \alpha = \alpha\backslash = (\overrightarrow{\alpha})\backslash \quad \text{on} \quad N \otimes \ell_\infty^n .$$
Since $\beta := \overrightarrow{\alpha\backslash}$ is right–injective by proposition 20.3.(1), the test 20.9. gives
$$\overrightarrow{\alpha\backslash} = (\overrightarrow{\alpha})\backslash .$$
The same argument applies to the right–projective associate.
(3) and (4) also follow from the preceding test: α' and $(\alpha/)'$ are finitely generated and clearly,
$$N \otimes_{\alpha'} \ell_\infty^n = (N' \otimes_\alpha \ell_1^n)' = (N' \otimes_{\alpha/} \ell_1^n)' = N \otimes_{(\alpha/)'} \ell_\infty^n ,$$
hence, $(\alpha')\backslash = (\alpha/)'$, which implies all the formulas in (3) and (4).
(2) Note first that $\overleftarrow{\alpha\backslash}$ is right–injective by proposition 20.3.(1). Since (by (3) and 20.6., corollary 1)
$$E' \otimes_{\alpha'} \ell_1(B_{F'}) = E' \otimes_{(\alpha\backslash)'} \ell_1(B_{F'})$$
and (by the duality theorem 15.5.)
$$E \otimes_{(\overleftarrow{\alpha})\backslash} F \overset{1}{\hookrightarrow} E \otimes_{\overleftarrow{\alpha}} \ell_\infty(B_{F'}) \overset{1}{\hookrightarrow} (E' \otimes_{\alpha'} \ell_1(B_{F'}))'$$
$$E \otimes_{\overleftarrow{\alpha\backslash}} F \overset{1}{\hookrightarrow} E \otimes_{\overleftarrow{\alpha\backslash}} \ell_\infty(B_{F'}) \overset{1}{\hookrightarrow} (E' \otimes_{(\alpha\backslash)'} \ell_1(B_{F'}))' ,$$
one obtains $(\overleftarrow{\alpha})\backslash = \overleftarrow{\alpha\backslash}$. The related formula for the right–projective associate fails to hold: In fact, $\overleftarrow{\pi} = \pi$ on $E \otimes \ell_1(\Gamma)$ and hence,
$$(\overleftarrow{\pi})/ = \pi/ = \pi \neq \overleftarrow{\pi} = \overleftarrow{(\pi/)}$$
since π is right–projective. □

The first statement of the proposition implies the

COROLLARY: If α is a finitely generated tensor norm on NORM, then all of its associates $\alpha/, \alpha\backslash, \backslash\alpha, /\alpha, /\alpha\backslash$ and $\backslash\alpha/$ are finitely generated as well.

20.11. If α is a finitely generated tensor norm and $(\mathfrak{A}, \mathbf{A})$ its associated maximal operator ideal, what do the ideals associated with $\alpha/, \alpha\backslash, /\alpha, \backslash\alpha, \ldots$ look like? Finding an answer to this question will be very fruitful for the study of both tensor norms and ideals. Start with the right-injective associate $\alpha\backslash$ of α, denote by $(\mathfrak{B}, \mathbf{B})$ the ideal associated with $\alpha\backslash$ and fix $T \in \mathfrak{L}(E, F)$. Since $(\ell_\infty(B_{F'}))'$ is an $L_1(\mu)$ (by Kakutani's representation theorem, Appendix A7.) and hence an $\mathcal{L}^g_{1,1}$-space, it follows that

$$E \otimes_{\alpha'} (\ell_\infty(B_{F'}))' \stackrel{1}{=} E \otimes_{\alpha'/} (\ell_\infty(B_{F'}))' \, .$$

If β_T and $\beta_{I \circ T}$ are the functionals associated with T and $I \circ T$ (where $I : F \hookrightarrow \ell_\infty(B_{F'})$), then

$$\begin{array}{c} E \otimes_{\alpha'} (\ell_\infty(B_{F'}))' \xrightarrow{\beta_{I \circ T}} \mathbb{K} \\ {\scriptstyle id \otimes I'} \downarrow 1 \quad \nearrow {\scriptstyle \beta_T} \\ E \otimes_{(\alpha\backslash)'} F' = E \otimes_{\alpha'/} F' \end{array}$$

commutes. Since $id \otimes I'$ is a metric surjection, β_T is continuous if and only if $\beta_{I \circ T}$ is continuous (with the same norms); the representation theorem for maximal operator ideals yields that $T \in \mathfrak{B}$ if and only if $I \circ T \in \mathfrak{A}$; moreover, $\mathbf{B}(T) = \mathbf{A}(I \circ T)$. Hence, the definition 9.7. of the injective hull $\mathfrak{A}^{inj}$ of an operator ideal gives the first part of the

THEOREM: *Let $\alpha \sim (\mathfrak{A}, \mathbf{A})$ be associated.*
(1) The injective hull $(\mathfrak{A}^{inj}, \mathbf{A}^{inj})$ of $(\mathfrak{A}, \mathbf{A})$ is maximal and associated with $\alpha\backslash$. In particular, α is right-injective if and only if $(\mathfrak{A}, \mathbf{A})$ is an injective operator ideal.
(2) The surjective hull $(\mathfrak{A}^{sur}, \mathbf{A}^{sur})$ of $(\mathfrak{A}, \mathbf{A})$ is maximal and associated with $/\alpha$. In particular, α is left-injective if and only if $(\mathfrak{A}, \mathbf{A})$ is a surjective operator ideal.

PROOF: To see (2) take the canonical quotient map $Q : \ell_1(B_E) \stackrel{1}{\twoheadrightarrow} E$ and $T \in \mathfrak{L}(E, F)$; look at the diagram

$$\begin{array}{c} \ell_1(B_E) \otimes_{\alpha'} F' \xrightarrow{\beta_{T \circ Q}} \mathbb{K} \\ {\scriptstyle Q \otimes id} \downarrow 1 \quad \nearrow {\scriptstyle \beta_T} \\ E \otimes_{(/\alpha)'} F' = E \otimes_{\backslash \alpha'} F' \end{array}$$

to obtain the result, as before. $\square$

To quickly draw one consequence from these relationships:

COROLLARY: *If* $(\mathfrak{A}, \mathbf{A})$ *is a maximal normed operator ideal, then*

$$(\mathfrak{A}^{dual})^{inj} = (\mathfrak{A}^{sur})^{dual} \quad \text{isometrically.}$$

PROOF: This is just $(\alpha^t)\backslash = (/\alpha)^t$. □

Clearly, $/\alpha\backslash = (/\alpha)\backslash = /(\alpha\backslash) \sim (\mathfrak{A}^{sur})^{inj} = (\mathfrak{A}^{inj})^{sur}$.

20.12. The projective associates give rise to factorization theorems for the corresponding operator ideals. Denote by $(\mathfrak{A}/, \mathbf{A}/)$ the ideal associated with $\alpha/$ and take $T \in \mathfrak{L}(E, F)$. Then

$$\begin{array}{c} E \otimes_{(\alpha/)'} F' = E \otimes_{\alpha'\backslash} F' \xrightarrow{\beta_T} \mathbb{K} \\ {\scriptstyle id \otimes I} \downarrow \quad 1 \qquad \nearrow \varphi \\ E \otimes_{\alpha'} \ell_\infty(B_{F''}) \end{array}$$

If β_T is continuous, then the Hahn–Banach theorem gives an extension φ with the same norm; it follows that $\kappa_F \circ T = I' \circ L_\varphi$ and, by the representation theorem for maximal operator ideals, $\mathbf{A}(L_\varphi) = \mathbf{A}/(T)$. So part of the following result has already been shown.

PROPOSITION: *Let* $\alpha \sim (\mathfrak{A}, \mathbf{A})$ *and denote by* $(\mathfrak{A}/, \mathbf{A}/) \sim \alpha/$ *and* $(\backslash\mathfrak{A}, \backslash\mathbf{A}) \sim \backslash\alpha$ *the maximal operator ideals associated with the projective associates; take* $T \in \mathfrak{L}(E, F)$:

(1) $T \in \mathfrak{A}/(E, F)$ *if and only if there exist a measure* μ, *operators* $R \in \mathfrak{A}$ *and* $S \in \mathfrak{L}$ *such that*

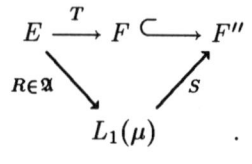

$$\begin{array}{c} E \xrightarrow{T} F \hookrightarrow F'' \\ {\scriptstyle R \in \mathfrak{A}} \searrow \quad \nearrow {\scriptstyle S} \\ L_1(\mu) \end{array} \quad .$$

In this case, $\mathbf{A}/(T) = \min \mathbf{A}(R)\|S\|$, *where the minimum is taken over all such factorizations and is attained for a metric surjection* $S : L_1(\mu) \to F''$ *for a Borel-Radon measure* μ.

(2) $T \in \backslash\mathfrak{A}(E, F)$ *if and only if there exist a compact set* K, *operators* $R \in \mathfrak{L}$ *and* $S \in \mathfrak{A}$ *such that*

$$E \xrightarrow{T} F \hookrightarrow F''$$
$$R \searrow \quad \nearrow S \in \mathfrak{A}$$
$$C(K)$$

In this case, $\backslash \mathbf{A}(T) = \min \|R\| \mathbf{A}(S)$, where the minimum is taken over all such factorizations and is attained for a metric injection R.

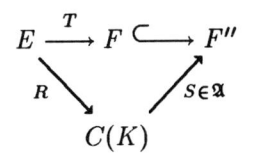

The space $C(K)$ in (2) can be replaced by $L_\infty(\mu)$.

PROOF: For $T \in \mathfrak{A}/(E, F)$,
$$\|I'\| \mathbf{A}(L_\varphi) = \mathbf{A}/(T) ,$$
and it remains only to use Kakutani's representation theorem Appendix A7., which gives that $\ell_\infty(B_{F''})' = L_1(\mu)$ for some Borel–Radon measure.
For the converse, note first that $L_1(\mu) \stackrel{1}{=} L_1(\mu_o)$ for some Borel–Radon measure μ_o (by Kakutani's representation theorem), therefore, $L_1(\mu)' \stackrel{1}{=} L_\infty(\mu_o)$. Now 20.8. and 20.10. imply that $E \otimes_{(\alpha/)'} L_1' = E \otimes_{\alpha'} L_1'$ and hence $\mathbf{A}(R) = \mathbf{A}/(R)$ by the representation theorem for maximal operator ideals. To see (2) apply the same type of argumentation using the diagram

$$E \otimes_{(\backslash\alpha)'} F' \xrightarrow{\beta_T} \mathbb{K}$$
$$\downarrow \quad \nearrow \varphi$$
$$\ell_\infty(B_{E'}) \otimes_{\alpha'} F'$$
$\square$

In Ex 20.7. the associated operator ideal for the projective associate $\backslash \alpha/$ is described. Ex 20.6. discusses extension and lifting properties related to $\mathfrak{A}/$ and $\backslash \mathfrak{A}$.

20.13. The characterization of $T \in \mathfrak{A}^{inj} \sim \alpha\backslash$ and $T \in \mathfrak{A}^{sur} \sim /\alpha$ in terms of $id_G \otimes T$ given in 17.15. and 17.16. can be modified to be in terms of α (instead of $\alpha\backslash$ and $/\alpha$). We omit stating and proving this now: In 29.10. it will be a very easy consequence of the calculus of traced tensor norms.

20.14. The simple relationship between the injective associates of α and surjective and injective operator ideals provides nice information about the associates of $\alpha_{p,q}$:

PROPOSITION: *Let* $1 \leq p \leq \infty$. *Then*
(1) $g_p\backslash = g_{p'}^* = d_{p'}'$.
(2) $\backslash g_p^* = g_{p'}$ and $d_p^*/ = d_{p'}$.
(3) $\backslash(g_p\backslash) = g_p$ and $(/d_p)/ = d_p$.
(4) $\pi\backslash = g_\infty^* = w_\infty^* = w_1' = d_\infty'$ and $\varepsilon/ = d_\infty = w_1$.

(5) $g_\infty\backslash = /d_\infty = \varepsilon$.
(6) $g_2^* = g_2$ and $d_2^* = d_2$.

PROOF: Corollary 5 in 18.7. states that $\mathfrak{J}_p^{inj} = \mathfrak{P}_p$ holds isometrically whenever $1 \leq p < \infty$; since $g_p \sim \mathfrak{J}_p$ and $g_{p'}^* \sim \mathfrak{P}_p$, theorem 20.11. ($\mathfrak{A}^{inj} \sim \alpha\backslash$) implies that $g_p\backslash = g_{p'}^*$.
For $p = \infty$ it is known from 12.9. that

$$g_\infty = w_\infty = \varepsilon \qquad \text{on } E \otimes \ell_\infty^n$$

and therefore 20.9. implies that $g_\infty\backslash = \varepsilon$. Since $g_2 = \alpha_{2,1}$ is right–injective (20.4.), the last statement follows from the first (note that $g_2 = g_2^*$ was already shown in 18.7.). All the other statements come from (1) as well, using transposition and dualization (see 20.10.). □

20.15. The behaviour of the $\alpha_{p,q}$ is particularly interesting when one of the involved indices is 2: in such a case, the discussion is about tensor norms/operator ideals connected with Hilbert spaces. The following list exhibits some of the most important examples:

α	$/\alpha$	$\alpha\backslash$	$\backslash\alpha$	$\alpha/$	$/\alpha\backslash$	$\backslash\alpha/$
d_2	d_2	$\sim w_2$	$\sim w_2^*$	d_2	$\sim w_2$	$\sim w_2^*$
g_2	$\sim w_2$	g_2	g_2	$\sim w_2^*$	$\sim w_2$	$\sim w_2^*$
w_2	w_2	w_2	$\sim g_2$	$\sim d_2$	w_2	$\sim w_2^*$
w_2^*	$\sim d_2$	$\sim g_2$	w_2^*	w_2^*	$\sim w_2$	w_2^*

The symbol $\sim$ indicates that the norms are equivalent.

PROOF: Only the statements for g_2 and w_2 will be proven — the others follow by transposition, dualization, and from $g_2 = g_2^*$.
Since g_2 is left–projective and right–injective, it follows that $g_2 = \backslash g_2 = g_2\backslash$. Moreover, w_2 is injective, hence $w_2 = /w_2\backslash = w_2\backslash = /w_2$.
The remaining statements are a consequence of the little Grothendieck theorem: In Ex 20.10. the reader is asked to show (with the factorizations coming from the characterization of $\backslash\mathfrak{A}, \mathfrak{A}/$ and $\backslash\mathfrak{A}/$) that

$$\backslash w_2 \leq g_2 \leq K_{LG}\backslash w_2$$
$$\backslash w_2/ \leq w_2^* \leq K_{LG}^2(\backslash w_2/)$$
$$g_2/ \leq w_2^* \leq K_{LG}g_2/ .$$

Moreover, this implies that

$$w_2/ = (\backslash w_2)^t \leq d_2 \leq K_{LG}w_2/$$
$$K_{LG}w_2 = K_{LG}w_2^{**} \geq (g_2/)^* = /(g_2^*) = /g_2 \geq w_2 .$$

Since $\backslash g_2/ = g_2/$ and $/g_2\backslash = /g_2$, all the relations have been demonstrated. □

20.16. The injective associate of $\alpha_{p,q}$ and, dually, $\backslash\alpha_{p,q}^*/$ have an interesting description. Recall from Ex 12.8. the tensor norm

$$\gamma_{p,q}(z;E,F) := \inf\left\{\|(a_{k,\ell})\| : \ell_{q'}^n \to \ell_p^n \|\ell_{q'}(x_\ell)\ell_{p'}(y_k)\ \middle|\ z = \sum_{k,\ell=1}^n a_{k,\ell}x_\ell \otimes y_k\right\}$$

(for $1/p + 1/q \geq 1$); as in 20.4., it is easily seen that $\gamma_{p,q}$ is projective (Ex 20.2.) — and it turns out that
$$\backslash\alpha_{p,q}^*/ = \gamma_{q,p}$$
— in particular, $w_2^* = \gamma_{2,2}$. Everything has been prepared for giving a proof of this result but we prefer to postpone it to section 28.

20.17. Certainly the most exciting relationship between tensor norms is

GROTHENDIECK'S INEQUALITY (tensor form):

$$w_2 \leq /\pi\backslash \leq K_G w_2 \quad \text{and} \quad \backslash\varepsilon/ \leq w_2^* \leq K_G\backslash\varepsilon/\ .$$

PROOF: Since w_2 is injective, it follows from the test 20.9. that the first relation is nothing but Grothendieck's inequality 14.4.. The inequality with the projective associate $\backslash\varepsilon/$ follows by duality. □

Since $\pi \sim \mathfrak{I}$ (the integral operators), $w_2 \sim \mathfrak{L}_2$ (the operators factoring through a Hilbert space, see 18.6.), $\pi\backslash \sim \mathfrak{I}^{inj} \stackrel{1}{=} \mathfrak{P}_1$ (the absolutely summing operators, see 11.3. or 18.7.) and $/\pi\backslash = /(\pi\backslash) = (/\pi)\backslash$, this can be reformulated as

GROTHENDIECK'S INEQUALITY (operator form):

$$\mathfrak{P}_1^{sur} = (\mathfrak{I}^{sur})^{inj} = \mathfrak{L}_2$$
$$\mathbf{L}_2(T) \leq \mathbf{P}_1^{sur}(T) = (\mathbf{I}^{sur})^{inj}(T) \leq K_G \mathbf{L}_2(T)\ .$$

Moreover, Grothendieck's constant K_G is the best constant in this inequality.

Or — as Grothendieck formulated it orginally in the Résumé:

COROLLARY: *If H is a Hilbert space, then $\mathbf{P}_1^{sur}(id_H) \leq K_G$.*

Since for $T \in \mathfrak{L}(\ell_1,\ell_2)$ the relation

$$\mathbf{P}_1(T) = \mathbf{P}_1^{sur}(T) \leq \mathbf{P}_1^{sur}(id_{\ell_2})\|T\| \leq K_G\|T\|$$

holds, it follows what was called "Grothendieck's theorem" in 17.14.:

$$\mathfrak{L}(\ell_1, \ell_2) = \mathfrak{P}_1(\ell_1, \ell_2) \text{ and } \mathbf{P}_1(T) \leq K_G \|T\| \ .$$

In particular, the best constant c in this inequality is $\leq \mathbf{P}_1^{sur}(id_{\ell_2}) \leq K_G$; for an operator $T \in \mathfrak{L}_2$ factorization shows that $\mathbf{P}_1^{sur}(T) \leq c\mathbf{L}_2(T)$ and hence $K_G \leq c$. It follows that $c = \mathbf{P}_1^{sur}(id_{\ell_2}) = K_G$. Moreover, the maximality of the operator ideal $\mathfrak{P}_1^{sur}$ (and an obvious complementation argument) implies that

$$K_G = \mathbf{P}_1^{sur}(id_{\ell_2}) = \sup_n \mathbf{P}_1^{sur}(id_{\ell_2^n})$$

and $\mathbf{P}_1^{sur}(id_H) = K_G$ for all infinite dimensional Hilbert spaces H (see also Ex 20.12.).

PROPOSITION: *The number* $K_G(n) := \mathbf{P}_1^{sur}(id_{\ell_2^n})$ *is the smallest constant* $c \geq 1$ *such that Grothendieck's inequality in matrix form is satisfied for* ℓ_2^n: *Let* $(a_{i,j})_{i,j=1}^m$ *be a* $m \times m$ *matrix* $(m \in \mathbb{N}$ *arbitrary) such that*

$$\sup\left\{ |\sum_{i,j=1}^m a_{i,j} s_i t_j| \ \bigg| \ |s_i| \leq 1, |t_j| \leq 1 \right\} \leq 1 \ .$$

Then

$$\left| \sum_{i,j=1}^m a_{i,j} \langle x_i, y_j \rangle \right| \leq c$$

for all $x_i, y_j \in B_{\ell_2^n}$.

PROOF: For the moment, call the smallest constant c in the matrix inequality $M_G(n)$. As in the proof of 14.8., it follows that $\pi \leq M_G(n) d_\infty$ on $\ell_1 \otimes \ell_2^n$, which means that $\mathbf{P}_1(T : \ell_1 \to \ell_2^n) \leq M_G(n) \|T\|$. In particular,

$$K_G(n) = \mathbf{P}_1^{sur}(id_{\ell_2^n}) \leq M_G(n) \ .$$

Conversely, define for $n \in \mathbb{N}$ and $z \in M \otimes N = L(M', N)$ (finite dimensional Banach spaces)

$$w_2^{(n)}(z; M, N) = \inf\left\{ w_2(x_i) w_2(y_i) \ \bigg| \ z = \sum_{i=1}^n x_i \otimes y_i \right\} =$$
$$= \inf\{\|R\| \, \|S\| \ | \ R \in \mathfrak{L}(M', \ell_2^n), S \in \mathfrak{L}(\ell_2^n, N), T_z = S \circ R\} \in [0, \infty]$$

(this is not a tensor norm!). Since

$$\ell_\infty^m \otimes_\pi \ell_\infty^m = \mathfrak{N}(\ell_1^m, \ell_\infty^m) = \mathfrak{I}(\ell_1^m, \ell_\infty^m) = \mathfrak{P}_1^{sur}(\ell_1^m, \ell_\infty^m)$$

20. Projective and Injective Tensor Norms

holds isometrically, it follows that $\pi \leq K_G(n)w_2^{(n)}$ on $\ell_\infty^m \otimes \ell_\infty^m$ for all $m \in \mathbb{N}$. Moreover, if x_i and y_j are in the unit ball of ℓ_2^n, then

$$R := \sum_{i=1}^m e_i \otimes x_i : \ell_1^m \longrightarrow \ell_2^n \quad \text{and} \quad S := \sum_{j=1}^m y_j \otimes e_j : \ell_2^n \longrightarrow \ell_\infty^m$$

have norm ≤ 1 and factor $z_o := \sum_{i,j=1}^m \langle x_i, y_j \rangle e_i \otimes e_j \in \ell_\infty^m \otimes \ell_\infty^m$, which means that

$$\pi(z_o; \ell_\infty^m, \ell_\infty^m) \leq K_G(n) w_2^{(n)}(z_o; \ell_\infty^m, \ell_\infty^m) \leq K_G(n) .$$

If $(a_{i,j})$ is an $m \times m$ matrix satisfying the assumption, then φ defined by

$$\langle \varphi, z \rangle := \sum_{i,j=1}^m a_{i,j} \langle z, e_i \otimes e_j \rangle \qquad \text{for } z \in \ell_\infty^m \otimes \ell_\infty^m$$

has norm ≤ 1 in $(\ell_\infty^m \otimes_\pi \ell_\infty^m)'$, and hence

$$\left| \sum_{i,j=1}^m a_{i,j} \langle x_i, y_j \rangle \right| = \left| \langle \varphi, \sum_{i,j=1}^m \langle x_i, y_j \rangle e_i \otimes e_j \rangle \right| \leq K_G(n) . \quad \Box$$

In the real case Krivine [154] proved:

$$K_G^\mathbb{R}(2) = \sqrt{2}$$
$$K_G^\mathbb{R}(3) < 1.517$$
$$K_G^\mathbb{R}(4) \leq \frac{\pi}{2} .$$

Recall that $\pi/2 \leq K_G^\mathbb{R}$ from 14.7.. Using Krivine's method and its complex modification by Haagerup (see 14.7.), König [148] estimated some "finite dimensional" Grothendieck constants in the complex case

$1.1526 \leq$	$K_G^\mathbb{C}(2)$	≤ 1.2157
$1.2108 \leq$	$K_G^\mathbb{C}(3)$	≤ 1.2744
$1.2413 \leq$	$K_G^\mathbb{C}(4)$	≤ 1.3048
$1.2600 \leq$	$K_G^\mathbb{C}(5)$	≤ 1.3236
$1.2984 \leq$	$K_G^\mathbb{C}(10)$	≤ 1.3628
$1.3181 \leq$	$K_G^\mathbb{C}(20)$	≤ 1.3834
$1.3300 \leq$	$K_G^\mathbb{C}(50)$	≤ 1.3962
$1.3381 \leq$	$K_G^\mathbb{C}$	≤ 1.4049

and conjectured slightly better upper bounds. For other descriptions of $K_G(n)$ see Ex 20.16., Ex 28.6., Ex 28.13. and Ex 28.14..

20.18. There are more Banach spaces E which *satisfy Grothendieck's theorem*

$$\mathfrak{L}(E, \ell_2) = \mathfrak{P}_1(E, \ell_2)$$

(they are sometimes called *G.T. spaces*). In fact, in 23.10. it will be shown that all $\mathfrak{L}_1^g$-spaces do — this is rather simple. For a long time it was not known whether or not there are any other spaces satisfying Grothendieck's theorem. In 1976, Kisliakov [142] and Pisier [219] independently constructed spaces which do: for example, quotients $L_1(\mu)/R$ where R is a reflexive subspace. Bourgain [16] showed that L_1/H^1 (where H^1 is the Hardy space) satisfies Grothendieck's theorem. We refer the reader to Pisier's book [225], ch.6, for more details on this topic and give only the following characterization, which appeared first in [225] in a slightly different form (see Ex 20.14.), but is in some sense also contained in Lindenstrauss–Pełczyński's famous 1968 analysis of Grothendieck's inequality; see also Ex 20.13. and Ex 20.14..

PROPOSITION: *For every Banach space E the following two statements are equivalent:*
(a) $\mathfrak{L}(E', \ell_2) = \mathfrak{P}_1(E', \ell_2)$.
(b) $\mathfrak{L}(E, \ell_1) = \mathfrak{D}_2(E, \ell_1)$.
The constants in (a) and (b) are the same.

That (b) holds for $E = c_o$ can be shown as in 17.14. using $\pi \leq K_G w_2$ on $c_o \otimes c_o$; see also section 23 on $\mathfrak{L}_p^g$-spaces.

PROOF: Observe first (by the representation theorem 17.5. for maximal ideals and the $\mathfrak{L}_p$-local technique lemma 13.5.) that (a) means that there is a best constant $c_1 < \infty$ such that $\mathbf{P}_1(T) \leq c_1 \|T\|$ for all $T \in \mathfrak{L}(E', \ell_2^n)$, and (b) that there is a best $c_2 < \infty$ such that $\mathbf{D}_2(T) \leq c_2 \|T\|$ for all $T \in \mathfrak{L}(E, \ell_1^n)$. Assume (a) and take $T \in \mathfrak{L}(E, \ell_1^n)$. Then, by 17.9. and since $\mathfrak{D}_2^* = \mathfrak{L}_2$,

$$\mathbf{D}_2(T) = \sup\{|tr_M(S \circ T \circ I_M^E)| \mid M \in \mathrm{FIN}(E), \mathbf{L}_2(S : \ell_1^n \longrightarrow M) < 1\} .$$

Hence the factorization of $\mathbf{L}_2$ from 18.1., $T = \hat{T} \circ Q_{\ker T}^E$, the fact that $\mathfrak{P}_1^{dual} \sim g_\infty^{*t} = w_1^* \sim \mathfrak{L}_1^*$ and (a) give

$$\mathbf{D}_2(T) = \sup \left\{ |tr_M(V \circ U \circ \hat{T} \circ Q_{\ker T}^E \circ I_M^E)| \,\Big|\, \begin{array}{l} M \in \mathrm{FIN}(E), \|U : \ell_1^n \to \ell_2^m\| \leq 1 \\ \|V : \ell_2^m \to M\| \leq 1 \end{array} \right\}$$
$$\leq \sup\{\mathbf{P}_1^{dual}(Q_{\ker T}^E \circ I_M^E \circ V) \mathbf{L}_1(U \circ \hat{T}) \mid ...\} \leq c_1 \|T\|$$

since $|tr(R \circ S)| \leq \mathbf{A}^*(R) \mathbf{A}(S)$ for operators between finite dimensional spaces. Conversely, take $T \in \mathfrak{L}(E', \ell_2^n)$ and, in order to apply Kwapień's characterization 11.8., $S \in \mathfrak{L}(\ell_\infty^k, E')$. Then $\mathfrak{L}^* = \mathfrak{J}$ implies that

$$\mathbf{P}_1(T \circ S) \leq \mathbf{I}(T \circ S) = \sup\{|tr_{\ell_\infty^k}(U \circ T \circ S)| \mid \|U : \ell_2^n \longrightarrow \ell_\infty^k\| \leq 1\} .$$

Now $\mathfrak{D}_2 = \mathfrak{D}_2^{dual} \sim w_2^* \sim \mathfrak{L}_2^*$ and (b) give

$$\mathbf{P}_1(T \circ S) \leq \sup\{\mathbf{L}_2(U)\mathbf{D}_2^{dual}(T \circ S)| \cdots\} \leq \|T\|c_2\|S\| ,$$

which is the conclusion. □

In Ex 20.14. it will be seen that $\mathbf{P}_2 \leq \mathfrak{D}_2 \leq K_{LG}\mathbf{P}_2$ on $\mathfrak{D}_2(E, \ell_1) = \mathfrak{P}_2(E, \ell_1)$. Therefore, one can replace $\mathfrak{D}_2$ by $\mathfrak{P}_2$ in this proposition — at the price of an additional constant in one implication.

COROLLARY: *Let E be a Banach space and $c \geq 1$. Then the following are equivalent:*
(a) $\pi \leq c\,d_\infty$ *on* $E' \otimes \ell_2$.
(b) $\pi \leq c\,w_2$ *on* $E \otimes c_o$.

20.19. There is a quite useful result due to Saphar [242] which is in the same spirit as the previous one:

PROPOSITION: *Let E be a Banach space, $c \geq 0$ and $1 \leq p \leq \infty$. Then the following three statements are equivalent:*
(a) *For every $M \in $ FIN and $T \in \mathfrak{L}(E, M)$*

$$\mathbf{P}_1(T) \leq c\mathbf{I}_{p'}(T) .$$

(b) *For every $M \in$ FIN and $T \in \mathfrak{L}(E, M)$*

$$\mathbf{P}_1(T) \leq c\mathbf{P}_{p'}(T) .$$

(c) *For every $n \in \mathbb{N}$ and $T \in \mathfrak{L}(\ell_\infty^n, E)$*

$$\mathbf{P}_p(T) \leq c\|T\| .$$

PROOF: Observe first, that (by 15.7. and the embedding theorem 17.6.)

$$\ell_1^n \otimes_{g_{p'}^*} E \xrightarrow{\ 1\ } \mathfrak{P}_p(\ell_\infty^n, E)$$

since $g_{p'}^*$ is right–injective and hence right–accessible by 20.3.. This implies that (c) is equivalent to $g_{p'}^* \leq c\varepsilon$ on $\ell_1^n \otimes E$ and therefore, by 20.9. (recall the remark at the end of 20.9. concerning constants) and 20.14.

$$\backslash g_{p'}^* \leq c(\backslash \varepsilon) = cg_\infty \qquad\qquad \text{on } M \otimes E.$$

By the representation theorem for maximal ideals, (a) is equivalent to

$$g^*_{p'} \leq cg_\infty \qquad \text{on } M \otimes E.$$

Since g_∞ is left–projective, it follows again from 20.9. that $\backslash g^*_{p'} \leq cg_\infty$ on $M \otimes E$. Hence, the equivalence of (a) and (c) is shown. Since $g_p = \backslash g^*_{p'}$ by 20.14., the same arguments give the equivalence with (b). $\square$

In 25.6. a much more general statement of this type will be given. For $p = 2$ the proposition shows that

$$\mathbf{P}_1(T) \leq c\mathbf{P}_2(T)$$

for every $T \in \mathfrak{L}(E, F)$ whenever E satisfies (c) and F is an arbitrary Banach space. The little Grothendieck theorem 11.11. shows the following:

COROLLARY: *Every absolutely 2-summing operator $T : \ell_2 \to F$ (where F is an arbitrary Banach space) is absolutely 1-summing and*

$$\mathbf{P}_1(T) \leq K_{LG}\mathbf{P}_2(T) .$$

In 11.9. it was shown that $\mathbf{P}_2(id_{\ell_2^n}) = \sqrt{n}$. It follows that

$$n^{-1/2}\mathbf{P}_1(id_{\ell_2^n}) \leq K_{LG}$$

for all $n \in \mathbb{N}$. But

$$\lim_{n \to \infty} n^{-1/2}\mathbf{P}_1(id_{\mathbb{R}_2^n}) = \sqrt{\frac{\pi}{2}}$$

$$\lim_{n \to \infty} n^{-1/2}\mathbf{P}_1(id_{\mathbb{C}_2^n}) = \frac{2}{\sqrt{\pi}}$$

by 11.10., hence the best constants in the little Grothendieck theorem are actually those stated in 11.11..

20.20. One may ask whether or not there exists a tensor norm α which is at the same time injective and projective! Existence would imply, by 20.17.,

$$g^*_2 \leq w^*_2 \sim \backslash\varepsilon/ \leq \alpha \leq /\pi\backslash \sim w_2 ,$$

hence $\mathfrak{L}_2 \subset \mathfrak{P}_2$ which is impossible since id_{ℓ_2} is not power compact. More generally, and at a much deeper level:

PROPOSITION: *There is no tensor norm which is right–injective and right–projective.*

PROOF: This would imply that

$$w_1 = \varepsilon/ \leq \alpha \leq \pi\backslash = w'_1 = g^*_\infty$$

and hence that $\mathfrak{P}_1 \subset \mathcal{L}_1$. But this is not true by a result of Gordon and Lewis [84] which solved an old problem from Grothendieck's Résumé ([93], p.72, question 2). □

Exercises:

Ex 20.1. Given a normed space E and a tensor norm α such that for all $G \in$ NORM and $F \subset G$ the space $E \otimes_\alpha F$ is an *isomorphic* subspace of $E \otimes_\alpha G$, then there is a constant $\lambda \geq 1$ such that

$$\alpha(z; E, G) \leq \alpha(z; E, F) \leq \lambda \alpha(z; E, G)$$

for all $F \subset G$ and $z \in E \otimes F$.

Ex 20.2. Use the method of proof from 20.4. to show that the tensor norms $\gamma_{p,q}$ from Ex 12.8. are projective and the tensor norms $\beta_{2,p}$ are right–injective.

Ex 20.3. Show that $\alpha = \alpha\backslash = /\alpha = /\alpha\backslash$ on $C(K) \otimes C(L)$ for all compact K and L and all tensor norms α.

Ex 20.4. (a) Show that the ideal $\mathfrak{I}$ of integral operators is neither injective nor surjective.

(b) Show that $\mathfrak{P}_p^{dual} = (\mathfrak{I}_p^{dual})^{sur}$ holds isometrically.

Ex 20.5. (a) $(\alpha/)\backslash \geq \alpha$ if α is right–injective. In particular, $((\pi\backslash)/)\backslash = \pi\backslash$.

(b) Show that $(\backslash\alpha)\backslash \leq \backslash(\alpha\backslash)$ for every tensor norm α. Hint: Use the description in terms of $\ell_1(B_E)$ and $\ell_\infty(B_{F'})$.

Ex 20.6. Assume $(\mathfrak{A}, \mathbf{A}) \sim \alpha$.

(a) α is left–projective if and only if for all $E, G \in$ BAN, each subspace $F \subset E$ and each $T \in \mathfrak{A}(F, G)$ there exists an *extension* $\tilde{T} \in \mathfrak{A}(E, G'')$ of $\kappa_G \circ T$ with $\mathbf{A}(T) = \mathbf{A}(\tilde{T})$.

(b) α is right–projective if and only if for all Banach spaces E, metric surjections $Q : F \to G$ and all $T \in \mathfrak{A}(E, G)$ there exists a *lifting* $\hat{T} \in \mathfrak{A}(E, F'')$ of $\kappa_G \circ T$ (i.e. $Q'' \circ \hat{T} = \kappa_G \circ T$) such that $\mathbf{A}(\hat{T}) = \mathbf{A}(T)$. Hint: It is enough to show that α is right–projective on FIN.

Ex 20.7. For a given tensor norm α denote by $(\backslash\mathfrak{A}/, \backslash\mathbf{A}/)$ the maximal normed operator ideal associated with $\backslash\alpha/$. Show that $T \in \mathcal{L}(E, F)$ is in $\backslash\mathfrak{A}/$ if and only if there are a compact K, a strictly localizable measure μ on some set Ω, operators $R, S \in \mathcal{L}$ and $U \in \mathfrak{A}$ such that

$$\begin{array}{ccccc} & T & & & \\ E & \longrightarrow & F & \hookrightarrow & F'' \\ R \downarrow & & & \nearrow S & \\ & & U \in \mathfrak{A} & & \\ C(K) & \longrightarrow & L_1(\Omega, \mu) & & \end{array}.$$

Moreover, $\backslash \mathbf{A}/(T) = \min \|R\| \mathbf{A}(U) \|S\|$, the minimum being taken over all such factorizations.

Ex 20.8. Let E be a Banach space and $1 \leq p \leq \infty$. The following three statements are equivalent:

(a) $id_E \in \mathcal{L}_p^{sur\ inj}$.

(b) E is isomorphic to a subspace of a quotient of some $L_p(\mu)$.

(c) E is isomorphic to a quotient of a subspace of some $L_p(\mu)$. Hint: $\mathfrak{A}^{sur\ inj} = \mathfrak{A}^{inj\ sur}$.

Ex 20.9. The class of all Banach spaces E such that $\mathbf{L}_p^{inj\ sur}(id_E) = 1$ is stable under formation of isometric subspaces and quotients. Hint: $(/w_p\backslash)\backslash = /w_p\backslash\ldots$

Ex 20.10. Use the little Grothendieck theorem to show that $\backslash \mathcal{L}_2 = \mathfrak{J}_2, \backslash \mathcal{L}_2/ = \mathfrak{P}_2^{dual} \circ \mathfrak{P}_2 (= \mathfrak{D}_2)$ and $\mathfrak{J}_2/ = \mathfrak{D}_2$ with equivalent norms. More precisely,

$$\backslash w_2 \leq g_2 \leq K_{LG} \backslash w_2$$
$$\backslash w_2/ \leq w_2^* \leq K_{LG}^2(\backslash w_2/)$$
$$g_2/ \leq w_2^* \leq K_{LG} g_2/ \ .$$

Ex 20.11. Show that $(\backslash w_2)\backslash = \backslash w_2$ and $/(w_2/) = w_2/$. Hint: Consider the associate ideals, 20.11. and 20.12..

Ex 20.12. Show that $\|tr_{\ell_2} : \ell_2 \otimes_{(/\pi\backslash)'} \ell_2 \longrightarrow \mathbb{K}\| = K_G$.

Ex 20.13. (a) If α is left-accessible, E and N are normed spaces, N finite dimensional, then

$$(E \otimes_\alpha N)'' = E'' \otimes_\alpha N$$

holds isometrically.

(b) Use (a) and the fact that g_∞^* is accessible (to be proven in the next section) to verify that the bidual E'' of a Banach space satisfies Grothendieck's theorem if and only if E does.

Ex 20.14. (a) Show that $\mathbf{P}_2 \leq \mathbf{D}_2 \leq K_{LG} \mathbf{P}_2$ on $\mathfrak{D}_2(E, \ell_1) = \mathfrak{P}_2(E, \ell_1)$ for every Banach space E.

(b) Formulate (a) in terms of tensor norms on $E \otimes c_o$.

Ex 20.15. (a) Show that $\sqrt{n} \leq \mathbf{P}_1(id_{\ell_1^n}) \leq K_G \sqrt{n}$. Hint: 11.9..

(b) Deduce from this that $\|id \otimes id : \ell_1^n \otimes_\varepsilon \ell_1^n \longrightarrow \ell_1^n \otimes_\pi \ell_1^n \| \leq K_G \sqrt{n}$.

Ex 20.16. Let S_{n-1} be the unit sphere of ℓ_2^n and s_n the scalar product restricted to $S_{n-1} \times S_{n-1}$. If $\alpha \sim (\mathfrak{A}, \mathbf{A})$, then

$$\mathbf{A}^{inj\ sur}(id_{\ell_2^n}) = \alpha(s_n; C(S_{n-1}), C(S_{n-1})) \ .$$

Hint: Use the isometries $x \rightsquigarrow (x\,|\,\cdot)$ and $x \rightsquigarrow (x\,|\,\cdot)$ of ℓ_2^n into $C(S_{n-1})$. Note the special case $K_G(n) = \pi(s_n; C(S_{n-1}), C(S_{n-1}))$.

21. Accessible Tensor Norms and Operator Ideals

The natural mapping
$$I : E \tilde{\otimes}_\alpha F \longrightarrow (E' \otimes_{\alpha'} F')'$$
is a metric injection if E and F have the metric approximation property (see section 15). In this section other conditions are investigated under which I is an isomorphic injection — or at least injective. Some criteria will be presented in order to show that a given tensor norm is totally accessible. With the help of certain "accessibility" notions for operator ideals it will turn out that all $\alpha_{p,q}^*$ are totally accessible. In the second part of this section some more general approximation properties will be analyzed.

21.1. The duality theorem 15.5. for a tensor norm α, which is nothing else but the embedding theorem 17.6. for the associated maximal operator ideal $\mathfrak{A}$, states that
$$E \tilde{\otimes}_{\overleftarrow{\alpha}} F \xhookrightarrow{1} (E' \otimes_{\alpha'} F')' = \mathfrak{A}(E', F'') .$$

Since $\overleftarrow{\alpha} \leq \alpha \leq \overrightarrow{\alpha}$ always holds for the finite and cofinite hull α, the problem under investigation is to find conditions which ensure that $\alpha \leq \overleftarrow{\alpha}$ on $E \otimes F$ for a given finitely generated tensor norm and Banach spaces E and F. Recall from 15.6. that a tensor norm α is called right-accessible if $\overrightarrow{\alpha}(\cdot; M, F) = \overleftarrow{\alpha}(\cdot; M, F)$ for all $M \in \text{FIN}$ and $F \in \text{NORM}$ (or only $\in \text{BAN}$ by 13.4.), left–accessible if its transpose α^t is right–accessible, accessible if it is both right– and left–accessible and totally accessible if $\overleftarrow{\alpha} = \overrightarrow{\alpha}$.

With the notions of injectivity and projectivity of tensor norms it is possible to verify accessibility in many cases. The following result is basic to what will be done; it is due to Saphar [241].

PROPOSITION: *Let α be a tensor norm.*
(1) *The right–projective associate $\alpha/$ and the right–injective associate $\alpha\backslash$ are right–accessible.*
(2) *If α is left–accessible, then $\alpha\backslash$ is totally accessible.*
(3) *The tensor norms $(\backslash\alpha)\backslash, /(\alpha/)$, and the injective associate $/\alpha\backslash$ are totally accessible. In particular, injective tensor norms are totally accessible.*

PROOF: The first statement was already proven in 20.3.(4). For the proof of (2) take normed spaces E and F. Since $\ell_\infty(B_{F'})$ has the metric approximation property and α

is left–accessible, the approximation lemma shows that $\overleftarrow{\alpha} = \overrightarrow{\alpha}$ on $E \otimes \ell_\infty(B_{F'})$. Now the formulas from 20.10. imply that for $z \in E \otimes F$

$$\overrightarrow{\alpha\backslash}(z; E, F) = \overrightarrow{\alpha}\backslash(z; E, F) = \overrightarrow{\alpha}(z; E, \ell_\infty(B_{F'})) =$$
$$= \overleftarrow{\alpha}(z; E, \ell_\infty(B_{F'})) = \overleftarrow{\alpha}\backslash(z; E, F) = \overleftarrow{\alpha\backslash}(z; E, F).$$

(3) is a simple consequence of (1) and (2). □

There is no statement like (2) for the projective associate $\alpha/$: Just take the projective norm $\pi = \pi/$ as an example. But even if α is totally accessible, $\alpha/$ need not be; this follows from the following two facts which will be shown in a moment: w'_p is totally accessible (see 21.5.) but $w'_p/$ is not (see 21.6.).

COROLLARY: *The tensor norms g_p^*, w_2 and g_2 are totally accessible.*

PROOF: Since $g_{p'}$ is left–projective and $g_p^* = g_{p'}\backslash$ (see 20.14.), it follows that

$$g_p^* = g_{p'}\backslash = (\backslash g_{p'})\backslash$$

is totally accessible. Moreover, $g_2^* = g_2$ and w_2 is injective. □

21.2. It turns out that it is sometimes easier to check the accessibility of a given finitely generated tensor norm through its associated maximal operator ideal. A quasi–Banach operator ideal $(\mathfrak{A}, \mathbf{A})$ is called *right–accessible*, if for all $(M, F) \in$ FIN × BAN, operators $T \in \mathfrak{L}(M, F)$ and $\varepsilon > 0$ there are $N \in$ FIN(F) and $S \in \mathfrak{L}(M, N)$ such that

$$\begin{array}{ccc} M & \xrightarrow{T} & F \\ & S \searrow & \uparrow I_N^F \\ & & N \end{array} \quad \text{and } \mathbf{A}(S) \leq (1+\varepsilon)\mathbf{A}(T).$$

It is called *left–accessible*, if for all $(E, N) \in$ BAN × FIN, operators $T \in \mathfrak{L}(E, N)$ and $\varepsilon > 0$ there are $L \in$ COFIN(E) and $S \in \mathfrak{L}(E/L, N)$ such that

$$\begin{array}{ccc} E & \xrightarrow{T} & N \\ Q_L^E \downarrow & \nearrow S & \\ E/L & & \end{array} \quad \text{and } \mathbf{A}(S) \leq (1+\varepsilon)\mathbf{A}(T).$$

A left– and right–accessible ideal is called *accessible*. Moreover, $(\mathfrak{A}, \mathbf{A})$ is *totally accessible*, if for every finite rank operator $T \in \mathfrak{F}(E, F)$ between Banach spaces and $\varepsilon > 0$ there are $(L, N) \in$ COFIN$(E) \times$ FIN(F) and $S \in \mathfrak{L}(E/L, N)$ such that

$$\begin{array}{ccc} E & \xrightarrow{T} & F \\ Q_L^E \downarrow & & \uparrow I_N^F \\ E/L & \xrightarrow{S} & N \end{array} \quad \text{and } \mathbf{A}(S) \le (1+\varepsilon)\mathbf{A}(T) \ .$$

Obviously, every injective quasi-Banach ideal is right-accessible and every surjective ideal is left-accessible. The canonical factorization

$$\begin{array}{ccc} E & \xrightarrow{T} & F \\ \downarrow & & \uparrow \\ E/\ker T & \longrightarrow & \operatorname{im} T \end{array} \quad T \in \mathfrak{F}(E,F)$$

shows that a surjective and injective quasi-Banach ideal is even totally accessible. For example, the injective and surjective ideal $\mathfrak{T}_p$ of type p operators (see Ex 9.15.) is totally accessible and the injective ideal $\mathfrak{C}_q$ of cotype q operators is right-accessible. We do not know whether or not $\mathfrak{C}_q$ is left-accessible (being injective it would even be totally accessible — by 21.1.(2) and the following proposition, or by a direct proof).

In Ex 21.1. it will be checked that the factorizations in the definitions of accessibility also hold when only the metric approximation property is assumed. Reisner [230] called accessible operator ideals "semi-tensorial".

21.3. For a maximal Banach operator ideal $\mathfrak{A}$ associated with a (finitely generated) tensor norm α the notions of accessibility coincide; this will follow from the embedding theorem
$$E' \otimes_{\overleftarrow{\alpha}} F \xhookrightarrow{1} \mathfrak{A}(E,F) \ .$$

PROPOSITION: *A finitely generated tensor norm is right-accessible (resp. left-accessible, accessible, totally accessible) if and only if its associated maximal Banach ideal is.*

PROOF: It will be shown that α is totally accessible if and only if $(\mathfrak{A}, \mathbf{A})$ has this property; all other proofs are similar. Assume that α is totally accessible and let $T \in \mathfrak{F}(E,F)$. Then
$$\overrightarrow{\alpha}(z_T; E', F) = \overleftarrow{\alpha}(z_T; E', F) = \mathbf{A}(T) \ ,$$
which implies that there are $(M,N) \in \mathrm{FIN}(E') \times \mathrm{FIN}(F)$ and $u \in M \otimes N$ with
$$\alpha(u; M, N) \le (1+\varepsilon)\mathbf{A}(T) \quad \text{and} \quad I_M^{E'} \otimes I_N^F(u) = z_T \ .$$
Hence, $T_u \in \mathcal{L}(E/M^0, N)$ satisfies $\mathbf{A}(T_u) \le (1+\varepsilon)\mathbf{A}(T)$ and $I_N^F \circ T_u \circ Q_{M^0}^E = T$.
Conversely, let $(\mathfrak{A}, \mathbf{A})$ be totally accessible. By the embedding lemma 13.3. it suffices to check that
$$\alpha(\cdot; E', F) \le \overleftarrow{\alpha}(\cdot; E', F)$$

for all Banach spaces E and F. For $z \in E' \otimes F$ and $\varepsilon > 0$ there are $L \in \mathrm{COFIN}(E)$ and $N \in \mathrm{FIN}(F)$ as well as an operator $S \in \mathfrak{L}(E/L, N)$ such that

$$\mathbf{A}(S) \leq (1+\varepsilon)\mathbf{A}(T_z) \quad \text{and} \quad I_N^F \circ S \circ Q_L^E = T_z \, .$$

For $z_S \in L^0 \otimes_\alpha N = (E/L)' \otimes_\alpha N \doteq \mathfrak{A}(E/L, N)$ it follows that

$$\alpha(z_S; L^0, N) = \mathbf{A}(S) \leq (1+\varepsilon)\overleftarrow{\alpha}(z; E', F) \quad \text{and} \quad I_{L^0}^{E'} \otimes I_N^F(z_S) = z \, ,$$

which completes the proof. □

Since $\mathfrak{A}^{dual} \sim \alpha^t$ and $\mathfrak{A}^* \sim \alpha^*$ (see 17.8. and 17.9.) for the dual and adjoint operator ideals, the proposition 15.6. gives the

COROLLARY: *Let $(\mathfrak{A}, A)$ be a maximal normed operator ideal.*
(1) *$(\mathfrak{A}^{dual}, \mathbf{A}^{dual})$ is right-accessible (resp. left-accessible, totally accessible) if and only if $(\mathfrak{A}, A)$ is left-accessible (resp. right-accessible, totally accessible).*
(2) *$(\mathfrak{A}^*, \mathbf{A}^*)$ is right-accessible (resp. left-accessible) if and only if $(\mathfrak{A}, A)$ is left-accessible (resp. right-accessible).*

21.4. The following result will be quite useful:

PROPOSITION: *Let $(\mathfrak{A}, \mathbf{A})$ and $(\mathfrak{B}, \mathbf{B})$ be quasi-Banach ideals, $(\mathfrak{A}, \mathbf{A})$ injective and left-accessible, $(\mathfrak{B}, \mathbf{B})$ totally accessible. Then $(\mathfrak{B} \circ \mathfrak{A}, \mathbf{B} \circ \mathbf{A})$ is totally accessible.*

PROOF: Take $T \in \mathfrak{F}(E, F)$ and $\varepsilon > 0$. Then there are $R \in \mathfrak{A}(E, G)$ and $S \in \mathfrak{B}(G, F)$ such that $T = S \circ R$ and $\mathbf{B}(S)\mathbf{A}(R) \leq (1+\varepsilon)(\mathbf{B} \circ \mathbf{A})(T)$. Since $\mathfrak{A}$ is injective, one can choose this factorization such that $\overline{R(E)} = G$, hence $S(G) \subset T(E)$ and S is also a finite rank operator. Since $\mathfrak{B}$ is totally accessible and $\mathfrak{A}$ left-accessible, T factors as follows:

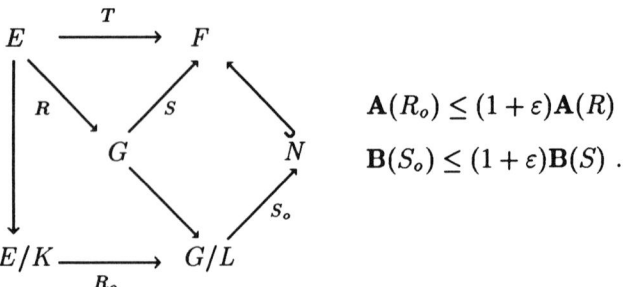

$$\begin{array}{c} E \xrightarrow{T} F \\ {\scriptstyle R} \searrow \nearrow {\scriptstyle S} \\ G \\ \downarrow \searrow {\scriptstyle S_o} \\ E/K \xrightarrow[R_o]{} G/L \end{array} \qquad \begin{array}{l} \mathbf{A}(R_o) \leq (1+\varepsilon)\mathbf{A}(R) \\ \mathbf{B}(S_o) \leq (1+\varepsilon)\mathbf{B}(S) \, . \end{array}$$

Consequently,

$$\mathbf{B} \circ \mathbf{A}(S_o \circ R_o) \leq \mathbf{B}(S_o)\mathbf{A}(R_o) \leq (1+\varepsilon)\mathbf{B}(S)(1+\varepsilon)\mathbf{A}(R) \leq (1+\varepsilon)^3 \mathbf{B} \circ \mathbf{A}(T) \, ,$$

which proves the result. □

For more results on the accessibility of composition ideals see Ex 21.2. and 28.6..

21.5. Everything is prepared for giving an easy proof of the following important result, apparently due to Gilbert–Leih [81]:

THEOREM: *Let $p, q \in [1, \infty]$ satisfy $1/p + 1/q \geq 1$.*
(1) *$\alpha_{p,q}$ and $(\mathfrak{L}_{p,q}, \mathbf{L}_{p,q})$ are accessible.*
(2) *$\alpha_{p,q}^*$ and $(\mathfrak{D}_{p',q'}, \mathbf{D}_{p',q'})$ are totally accessible.*

PROOF: Since the tensor norms and operator ideals in question are associated and α is accessible if α^* is (by 15.6.), it is enough to show that $\mathfrak{D}_{p',q'}$ is totally accessible. Kwapień's factorization theorem 19.3. states that

$$\mathfrak{D}_{p',q'} = \mathfrak{P}_{q'}^{dual} \circ \mathfrak{P}_{p'}.$$

Since $\mathfrak{P}_{p'} \sim g_p^*$ and $\mathfrak{P}_{q'}^{dual} \sim g_q^{*t}$ are totally accessible (by corollary 21.1. and 21.3.) and the operator ideal $\mathfrak{P}_{p'}$ is injective, the preceding proposition shows that $\mathfrak{D}_{p',q'}$ is totally accessible. □

21.6. A structurally interesting application of the total accessibility of certain tensor norms is the

PROPOSITION: *Let $(\mathfrak{A}, \mathbf{A})$ be a maximal normed operator ideal with associated tensor norm α. If α^* is totally accessible, then every Banach space E with $id_E \in \mathfrak{A}$ has the bounded approximation property with constant $\mathbf{A}(id_E)$.*

PROOF: In order to use the characterization theorem 16.2. (concerning $\pi \leq \overleftarrow{\lambda \pi}$) for the bounded approximation property, observe that $\pi \leq \mathbf{A}(id_E)\alpha^*$ on $E' \otimes E$ by 17.15.. Consequently,

$$\pi \leq \mathbf{A}(id_E)\alpha^* = \mathbf{A}(id_E)\overleftarrow{\alpha^*} \leq \mathbf{A}(id_E)\overleftarrow{\pi} \qquad \text{on } E' \otimes E,$$

which gives the result by 16.2.. □

COROLLARY 1: *For $1 \leq p \leq \infty$ and $\lambda \geq 1$ every $\mathcal{L}_{p,\lambda}^g$-space E has the bounded approximation property with constant λ.*

PROOF: Since $\mathfrak{L}_p \sim w_p$ is accessible and $\varepsilon \leq w_p \leq \lambda\varepsilon$ on $G \otimes E$ for every Banach space G by the local technique lemma (see 13.5.), it follows from the characterization 17.16. that $\mathbf{L}_p(id_E) \leq \lambda$. The fact that w_p^* is totally accessible implies the result. □

This was a "positive" application of the proposition. But the proposition also has a negative flavour: if there is an $E \in \text{space}(\mathfrak{A})$ (this means, by definition $id_E \in \mathfrak{A}$)

without the bounded approximation property, then α^* is *not* totally accessible! An example: Each subspace $E \subset \ell_p$ is contained in space($\mathfrak{L}_p^{inj}$); since there are subspaces of ℓ_p without the approximation property (if $p \neq 2$, see 5.2.), the adjoint of $w_p \backslash \sim \mathfrak{L}_p^{inj}$ cannot be totally accessible:

COROLLARY 2: *For $1 \leq p \leq \infty$ and $p \neq 2$ the tensor norm $w_p'/$ is not totally accessible.*

21.7. Let α be a finitely generated tensor norm. A Banach space E is said to have the α-*approximation property* if for all Banach spaces F the natural mapping

$$F \tilde{\otimes}_\alpha E \longrightarrow F \tilde{\otimes}_\varepsilon E \qquad [\hookrightarrow \mathfrak{L}(F', E'')]$$

is injective. E has the *bounded α-approximation property with constant* $\lambda \geq 1$ if for all Banach spaces F the natural mapping

$$I : F \otimes_\alpha E \longrightarrow (F' \otimes_{\alpha'} E')' \qquad [\hookrightarrow \mathfrak{L}(F', E'')]$$

satisfies $\alpha(z; F, E) \leq \lambda \|I(z)\|_{...}$; if $\lambda = 1$ one speaks of the *metric α-approximation property*. For $\alpha = \pi$ the usual approximation properties are obtained. For $\alpha = g_p$ this notation was introduced by Saphar [243] under the name "approximation property of order p" (and generalized to $\alpha_{p,q}$ by Diaz, López Molina and Rivera Ortun [52]).

PROPOSITION:
(1) *A Banach space with the approximation property has the α-approximation property for all tensor norms α.*
(2) *E has the bounded α-approximation property with constant λ if and only if $\alpha \leq \lambda \overleftarrow{\alpha}$ on $F \otimes E$ for all Banach spaces F. In particular, if α is totally accessible, then each Banach space has the metric α-approximation property.*
(3) *If α is accessible, then E has the bounded α-approximation property if it has the bounded approximation property (same constant).*
(4) *It is enough to check the definitions for the (bounded) α-approximation property for dual spaces F.*
(5) *If E'' has the (bounded/metric) α-approximation property, then E has it.*

PROOF: (1) is a reformulation of 17.20. and (2) follows from the duality theorem 15.5.. The approximation lemma and (2) imply (3) and the rest is a consequence of the embedding lemma. □

Thus, every Banach space has the metric g_2-approximation property since $g_2 = g_2^*$ is totally accessible. It will be shown in 31.7. that Banach spaces E of cotype 2 satisfy

$$F \tilde{\otimes}_{g_2} E = F \tilde{\otimes}_{g_p} E \qquad \text{for all } p \in [2, \infty],$$

hence they have the g_p-approximation property for these p. There are spaces of cotype 2 without the approximation property: Szankowski's subspaces of ℓ_q (where $1 \leq q < 2$, see 8.6. and 5.2.(1)). In 1982 Reinov [228] showed for $1 \leq p < \infty$ and $p \neq 2$ that

(1) There is a Banach E such that $E' \tilde{\otimes}_{g_p} E \to E' \tilde{\otimes}_\varepsilon E$ is not injective. In particular, E does not have the g_p-approximation property and the tensor norm g_p is not totally accessible.

(2) There is a Banach space with the g_p-approximation property but failing the bounded g_p-approximation property.

(3) There is a Banach space E with the approximation property which does not have the bounded g_p-approximation property for any p as above.

21.8. That the (bounded) α-approximation property is really a property of "approximation" of certain operators can be seen as follows. If $\mathfrak{A}$ is the maximal Banach operator ideal associated with α, the adjoint operator ideal satisfies

$$\mathfrak{A}^*(E, F'') = (E \otimes_{\alpha^t} F')' = (E \tilde{\otimes}_{\alpha^t} F')'$$

isometrically. For the weak topologies

$$\sigma := \sigma(\mathfrak{A}^*(E, F''), E \otimes F') \text{ and } \tilde{\sigma} := \sigma(\mathfrak{A}^*(E, F''), E \tilde{\otimes}_{\alpha^t} F')$$

lemma 16.2. reads as follows:

PROPOSITION: *Let E and F be Banach spaces, α a tensor norm and $\lambda \geq 1$. Then $\alpha \leq \lambda \overleftarrow{\alpha}$ on $F' \otimes E$ is equivalent to each of the following two statements about the unit balls:*

(a) $\quad B_{\mathfrak{A}^*(E,F'')} \subset \lambda \overline{B_{E' \otimes_\alpha \cdot F}}^\sigma$.

(b) $\quad B_{\mathfrak{A}^*(E,F'')} \subset \lambda \overline{B_{E' \otimes_\alpha \cdot F}}^{\tilde{\sigma}}$.

In particular, E has the bounded α-approximation property (with constant λ) if and only if (a) *or* (b) *is satisfied for all Banach spaces F.*
The latter follows from proposition 21.7.(4).

21.9. For the bounded g_p-approximation property $(1 < p < \infty)$ it is enough to check the definition for reflexive spaces. This will now be shown.

LEMMA: *Let $1 \leq p \leq \infty$ and E, F arbitrary Banach spaces. For every $z \in F \tilde{\otimes}_{g_p} E$ there are a separable reflexive Banach space G, an injective operator $S \in \mathfrak{L}(G, F)$ and $w \in G \tilde{\otimes}_{g_p} E$ with $S \otimes id_E(w) = z$.*

PROOF: For $p = \infty$ read c_o instead of ℓ_∞ in this proof. By 12.7. the element $z \in F \tilde{\otimes}_{g_p} E$ has a representation

$$z = \sum_{n=1}^{\infty} \lambda_n x_n \otimes y_n$$

with $(\lambda_n) \in \ell_p$, a zero sequence (x_n) in F and $w_{p'}(y_n; E) \leq 1$. Define K to be the absolutely convex, closed hull of $\{x_n\}$. Then $[\![K]\!]$ is a Banach space such that the embedding $[\![K]\!] \hookrightarrow F$ is compact, and hence factors injectively through a reflexive space G_1 by the Davis-Figiel-Johnson-Pełczyński factorization theorem (see 9.6.); take G to be the closed linear span of $\{x_n\}$ in G_1. Since $\{x_n\}$ is bounded in $[\![K]\!]$, it is bounded in G, and $w := \sum_{n=1}^{\infty} \lambda_n x_n \otimes y_n \in G \tilde{\otimes}_{g_p} E$ gives the result. $\square$

In Ex 21.5. there is a metric version of this lemma.

PROPOSITION 1: *For $1 \leq p \leq \infty$ a Banach space E has the g_p-approximation property if (and only if)*
$$I_F : F \tilde{\otimes}_{g_p} E \longrightarrow F \tilde{\otimes}_{\varepsilon} E$$
is injective for all separable, reflexive spaces F.

Recall that the g_1-approximation property is the usual approximation property.

PROOF: Take F to be an arbitrary Banach space, $z \in F \tilde{\otimes}_{g_p} E$ and choose G, S and $w \in G \tilde{\otimes}_{g_p} E$ as in the lemma. Then the diagram

$$\begin{array}{ccc} F \tilde{\otimes}_{g_p} E & \xrightarrow{I_F} & F \tilde{\otimes}_{\varepsilon} E \\ {\scriptstyle S \tilde{\otimes}_{g_p} id_E} \uparrow & & \uparrow {\scriptstyle S \tilde{\otimes}_{\varepsilon} id_E} \\ G \tilde{\otimes}_{g_p} E & \xrightarrow{I_G} & G \tilde{\otimes}_{\varepsilon} E \end{array}$$

implies the result since I_G and $S \tilde{\otimes}_{\varepsilon} id_E$ are injective. $\square$

For the bounded g_p-approximation property the analogous result holds at least for $1 < p < \infty$:

PROPOSITION 2: *For $1 < p < \infty$ a Banach space E has the bounded g_p-approximation property if (and only if) $g_p \leq \lambda \overleftarrow{g_p}$ holds on $F \otimes E$ for all reflexive Banach spaces F.*

PROOF: Take an arbitrary Banach space F and $z \in F \otimes_{g_p} E$. One has to show that for every $\varphi \in (F \otimes_{g_p} E)' = \mathfrak{P}_{p'}(E, F')$ of $\mathbf{P}_{p'}$-norm one the inequality
$$|\langle \varphi, z \rangle| \leq \lambda \overleftarrow{g_p}(z; F, E)$$
holds. The Grothendieck-Pietsch factorization theorem 11.3. shows that the operator $T \in \mathfrak{P}_{p'}(E, F')$ associated with φ factors through operators $R \in \mathfrak{P}_{p'}(E, G)$ and $S \in \mathcal{L}(G, F')$, where G is a subspace of some $L_{p'}$, hence reflexive, $\|S\| = 1$ and $\mathbf{P}_{p'}(T) = \mathbf{P}_{p'}(R)$. It follows that there is a $\varphi_o \in (G' \otimes_{g_p} E)'$ of norm one with
$$\langle \varphi, z \rangle = \langle \varphi_o, [(S' \circ \kappa_F) \otimes id_E](z) \rangle .$$

Therefore,

$$|\langle \varphi, z \rangle| \leq g_p\left([(S' \circ \kappa_F) \otimes id_E](z); G', E\right) \leq \lambda \overleftarrow{g_p}(\cdots; G', E) \leq$$
$$\leq \lambda \|S' \circ \kappa_F\| \, \|id_E\| \overleftarrow{g_p}(z; F, E) \leq \lambda \overleftarrow{g_p}(z; F, E)$$

by the metric mapping property. □

The proof works neither for $p = 1$ nor for $p = \infty$; it is not clear to us whether the statement holds for $p = 1$, which is the usual bounded approximation property (probably not, by the following corollary).

21.10. For $1 \leq p < \infty$ an operator $T \in \mathcal{L}(E, F)$ is called *Pietsch p–integral* if it factors

$$\begin{array}{ccc} E & \xrightarrow{T} & F \\ R \downarrow & & \uparrow S \\ C(K) & \xrightarrow{I} & L_p(\mu) \end{array}$$

for some compact K and Borel–Radon measure μ on K. With

$$\mathbf{PI}_p(T) := \inf \|R\| \, \|I\| \, \|S\|$$

the class $\mathfrak{PI}_p$ of all Pietsch p–integral operators becomes a Banach operator ideal. Clearly, $\mathfrak{PI}_1 = \mathfrak{PI}$ and $\mathfrak{PI}_2 = \mathfrak{I}_2 = \mathfrak{P}_2$ hold isometrically (see 18.7.). Moreover,

$$\mathfrak{PI}_p(E, F') \stackrel{1}{=} \mathfrak{I}_p(E, F')$$

by the factorization theorem for p–integral operators. As for $p = 1$ (see Appendix D9.), the fact that there are (for $p \neq 2$) Banach spaces without the g_p–approximation property was used by Reinov [228] in order to show that there are p–integral operators which are not Pietsch p–integral. It will be shown in 33.6. that

$$F' \tilde{\otimes}_{g_p} E = \mathfrak{PI}_p(F, E)$$

holds isometrically if F' has the Radon–Nikodym property and $F' \tilde{\otimes}_{g_p} E \to F' \tilde{\otimes}_\varepsilon E$ is injective. This will be used to prove the

COROLLARY: *Let $1 < p < \infty$ and E a Banach space which is λ–complemented in its bidual. If E has the g_p–approximation property, then it has the bounded g_p–approximation property with constant λ.*

Note that for $p = 1$ only the weaker statement 16.4. is known.

PROOF: By proposition 2 it is enough to show that $g_p \leq \lambda \overleftarrow{g_p}$ on $F' \otimes E$ for reflexive spaces F. Take $z \in F' \otimes E$, denote by $T_z : F \to E$ the associated operator and let $P : E'' \to E$ be a projection. Since, as a reflexive space, F has the Radon–Nikodym property, it follows that

$$g_p(z; F', E) = \mathbf{PI}_p(P \circ \kappa_E \circ T_z) \leq \|P\| \mathbf{PI}_p(\kappa_E \circ T_z) = \|P\| \mathbf{I}_p(\kappa_E \circ T_z) =$$
$$= \|P\| \mathbf{I}_p(T_z) = \|P\| \overleftarrow{g_p}(z; F', E)$$

by the regularity of $\mathfrak{I}_p$ and the embedding theorem. □

21.11. It is good to be able to verify the α–approximation property with the help of the coincidence of operator ideals:

PROPOSITION: *Let E be a Banach space and α and β finitely generated, accessible tensor norms with associated operator ideals $\mathfrak{A}$ and $\mathfrak{B}$ such that $\mathfrak{A}(E, F) = \mathfrak{B}(E, F)$ for all reflexive Banach spaces F.*

(1) *Then for all even duals $E^{(2k)}$ of E and all Banach spaces F*

$$\mathfrak{A}(E^{(2k)}, F) = \mathfrak{B}(E^{(2k)}, F) \ .$$

(2) *If β^* is totally accessible, then the space E and all its even dual spaces $E^{(2k)}$ have the α^*-approximation property.*

(3) *If α is totally accessible, then all the odd duals of E have the β^t-approximation property.*

PROOF: Since

$$(E \tilde{\otimes}_{\alpha'} F)' = \mathfrak{A}(E, F') = \mathfrak{B}(E, F') = (E \tilde{\otimes}_{\beta'} F)' \ ,$$

it follows that α' and β' are equivalent norms on $E \otimes F$ for all reflexive spaces F. Taking the Johnson space C_2 for F, lemma 17.16. shows that there are $c_1 \geq 0$ and $c_2 \geq 0$ satisfying $c_1 \alpha' \leq \beta' \leq c_2 \alpha'$ on $E \otimes M$ for all $M \in \mathrm{FIN}$. Since $(E \otimes_\gamma N)'' = E'' \otimes_\gamma N$ if γ is accessible (see Ex 20.13.), the fact that the tensor norms are finitely generated implies that

$$E^{(2k)} \tilde{\otimes}_{\alpha'} F = E^{(2k)} \tilde{\otimes}_{\beta'} F$$

for all Banach spaces F; in particular, (2) holds. The regularity of $\mathfrak{A}$ and $\mathfrak{B}$ implies (1). To see (3), observe first that the embedding theorem implies that $\overleftarrow{\alpha}$ and $\overleftarrow{\beta}$ are equivalent norms on all $E' \otimes F'$ for reflexive F. Since α and β are accessible, it follows that $\alpha = \overleftarrow{\alpha}$ and $\beta = \overleftarrow{\beta}$ on $E' \otimes M$ for all $M \in \mathrm{FIN}$ — and the same arguments as before show that

$$E^{(2k+1)} \tilde{\otimes}_\alpha F = E^{(2k+1)} \tilde{\otimes}_\beta F$$

for all Banach spaces F; this implies (3). □

Taking $\alpha = g_p^*$ and $\beta = g_{p'}$, the fact that $g_{p'}^*$ is totally accessible, corollary 21.10. and proposition 21.7.(5) give the

THEOREM: *Let $1 < p < \infty$ and E a Banach space such that*
$$\mathfrak{P}_{p'}(E, F) = \mathfrak{I}_{p'}(E, F)$$
for all reflexive Banach spaces F. Then:
(1) *For all Banach spaces F and all even duals of E*
$$\mathfrak{P}_{p'}(E^{(2k)}, F) = \mathfrak{I}_{p'}(E^{(2k)}, F) \ .$$
(2) *E and all its even duals have the metric g_p-approximation property.*
(3) *The odd duals of E have the $d_{p'}$-approximation property.*

(1) is due to Kisliakov [143] and (2) to Bourgain–Reinov [17]. H^∞ satisfies the assumption of the theorem for all $1 < p < \infty$: This was shown by Kisliakov [143] and Gordon–Reisner [86] — both using the fact that H^∞ is complemented in the bidual of the disk algebra A and a result of Mityagin–Pełczyński (see [202], p.16) stating that $\mathfrak{P}_p(A, \cdot) = \mathfrak{I}_p(A, \cdot)$.

It follows in particular that H^∞ has the metric g_p-approximation property (Bourgain–Reinov); the reader is reminded that it is not known whether or not H^∞ has the approximation property.

21.12. Here is a summary of the results which were obtained in this section about the accessibility of Lapresté's tensor norms $\alpha_{p,q}$:

(1) All $\alpha_{p,q}$ and all $\alpha_{p,q}^*$ as well as all their projective and injective associates are accessible (21.5. and 21.1.).

(2) Totally accessible are: $\alpha_{p,q}^*, w_2, d_2, g_2, \alpha_{2,q}$ and $\alpha_{p,2}$ (by 21.5. and 21.1. since $\alpha_{2,q}$ is right–injective).

(3) Not totally accessible are: d_p and g_p for $1 \leq p < \infty$ and $p \neq 2$ (by Reinov's counterexample, 21.7.), all $w_p'/$ for $p \neq 2$ (see 21.6.).

We do not know whether or not g_∞ and w_p (for $1 \leq p \leq \infty$ and $p \neq 2$) are totally accessible.

Exercises:

Ex 21.1. Let $(\mathfrak{A}, \mathbf{A})$ be a quasi-Banach operator ideal. Use the factorization of finite rank operators proved in 16.9. to show the following results:

(a) If $M \in$ FIN and F has the bounded approximation property with constant $\lambda \geq 1$, then for every $T \in \mathfrak{A}(M, F)$ and $\varepsilon > 0$ there is a subspace $N \in \text{FIN}(F)$ such that $\mathbf{A}(T : M \to N) \leq \lambda(1+\varepsilon)\mathbf{A}(T : M \to F)$.

(b) If $N \in$ FIN and E' has the bounded approximation property with constant $\lambda \geq 1$, then there is an $L \in \text{COFIN}(E)$ and $T_o \in \mathfrak{A}(E/L, N)$ such that $T = T_o \circ Q_L^E$ and $\mathbf{A}(T_o) \leq \lambda(1+\varepsilon)\mathbf{A}(T)$.

(c) If $\mathfrak{A}$ is left–accessible and F has the bounded approximation property with constant $\lambda \geq 1$ (or: $\mathfrak{A}$ is right–accessible and E' has the bounded approximation property with constant $\lambda \geq 1$), then for every $T \in \mathfrak{F}(E, F)$ and $\varepsilon > 0$ there are subspaces $L \in \text{COFIN}(E)$ and $N \in \text{FIN}(F)$ and $T_o \in \mathfrak{L}(E/L, N)$ such that $T = I_N^F \circ T_o \circ Q_L^E$ and $\mathbf{A}(T_o) \leq \lambda(1+\varepsilon)\mathbf{A}(T)$.

(d) If $\mathfrak{A}$ is left–accessible, then $\mathfrak{A}^{inj}$ is totally accessible.

(e) If $\mathfrak{A}$ is right–accessible, then $\mathfrak{A}^{sur}$ is totally accessible.

If $(\mathfrak{A}, \mathbf{A})$ is maximal normed, note that all these statements follow from 21.3., the embedding theorem, the approximation lemma and the respective properties for tensor norms.

Ex 21.2. (a) If $(\mathfrak{A}, \mathbf{A})$ and $(\mathfrak{B}, \mathbf{B})$ are both right–accessible (resp: both left–accessible), then $(\mathfrak{B} \circ \mathfrak{A}, \mathbf{B} \circ \mathbf{A})$ is right–accessible (resp: left–accessible).

(b) Let $(\mathfrak{A}, \mathbf{A})$ and $(\mathfrak{B}, \mathbf{B})$ be two quasi–Banach operator ideals with $\mathfrak{A} \subset \mathfrak{B}$ and $\mathbf{B} \leq \mathbf{A}$ such that $(\mathfrak{A}, \mathbf{A}) = (\mathfrak{B}, \mathbf{B})$ on FIN. Then $(\mathfrak{A}, \mathbf{A})$ is right–accessible (resp. left–accessible, totally accessible) whenever $(\mathfrak{B}, \mathbf{B})$ is.

Ex 21.3. *Principle of local reflexivity for operator ideals:* Let $(\mathfrak{A}, \mathbf{A})$ be a right–accessible quasi–Banach operator ideal, F a Banach space, $M \in$ FIN and $N \in \text{FIN}(F')$. Then for every $T \in \mathfrak{L}(M, F'')$ and $\varepsilon > 0$ there is an operator $R \in \mathfrak{L}(M, F)$ such that

(1) $\mathbf{A}(R) \leq (1+\varepsilon)\mathbf{A}(T)$,

(2) $\langle y', Rx \rangle = \langle Tx, y' \rangle$ for all $(x, y') \in M \times N$,

(3) $Rx = Tx$ whenever $Tx \in E$,

(4) If T is injective, R can be chosen to be injective and satisfy $\|R^{-1}\| \leq (1+\varepsilon)\|T^{-1}\|$.

Hint: Use 6.6.. If $(\mathfrak{A}, \mathbf{A})$ is maximal normed, the weak principle (i.e. only (1) and (2)) follows from $\mathfrak{A}(M, E)'' = \mathfrak{A}(M, E'')$ (see Ex 20.13.), by Helly's lemma.

Ex 21.4. (a) Given $\varepsilon > 0$, every finite rank operator $T : E \to \ell_\infty(\Gamma)$ has the form $T = \sum_{m=1}^n x'_m \otimes y_m$, with $\ell_p(x'_m) \leq (1+\varepsilon)\mathbf{P}_p(T)$ and $w_{p'}(y_m) \leq 1$. Hint: $g_{p'}^* = g_p \backslash$ and accessibility.

(b) For each $T \in \mathfrak{F}(E, F)$ with $\mathbf{P}_p(T) < 1$ there are $x'_1, ..., x'_n \in E'$ with $\ell_p(x'_m) \leq 1$ and

$$\|Tx\| \leq \left(\sum_{m=1}^n |\langle x'_m, x \rangle|^p \right)^{1/p} \qquad \text{for all } x \in E.$$

Ex 21.5. Check that in lemma 21.9. for every $\varepsilon > 0$ the operator $S \in \mathfrak{L}(G, F)$ and $w \in G \tilde{\otimes}_{g_p} E$ can be chosen so that $\|S\| \leq 1$ and $g_p(w; G, E) \leq (1+\varepsilon)g_p(z; F, E)$.

Ex 21.6. If $T \in \mathfrak{N}(E, F)$ and $\varepsilon > 0$, then there are reflexive spaces G_1 and G_2, operators $R \in \mathfrak{L}(E, G_1)$, $T_o \in \mathfrak{N}(G_1, G_2)$ and $S \in \mathfrak{L}(G_2, F)$ such that $T = S \circ T_o \circ R$ and $\|S\|\mathbf{N}(T_o)\|R\| \leq (1+\varepsilon)\mathbf{N}(T)$. Hint: Ex 21.5..

Ex 21.7. (a) Take $1/p + 1/q \geq 1$. Then for each operator $T \in \mathfrak{P}_{q'}(F, G)$ and each Banach space E

$$\|T \otimes id_E : F \tilde{\otimes}_{\alpha_{p,q}} E \longrightarrow G \tilde{\otimes}_{g_p} E\| \leq \mathbf{P}_{q'}(T) \, .$$

(b) Use Kwapień's factorization theorem to show that a Banach space E has the $\alpha_{p,q}$-approximation property if it has the g_p-approximation property. Hint: For each functional $\varphi \in (F \otimes_{\alpha_{p,q}} E)'$ and $L_\varphi = U \circ S$ one has $\langle \varphi, z \rangle = tr_E(U \otimes id_E)(S \otimes id_E)(z)$. This result is due to Diaz, López Molina and Rivera Ortun [52].

22. Minimal Operator Ideals

The smallest Banach operator ideal which coincides with a given Banach operator ideal $\mathfrak{A}$ for finite dimensional spaces is called its minimal kernel $\mathfrak{A}^{min}$. The main result of this section is the representation theorem of minimal Banach operator ideals in terms of the tensor norm which is associated with the maximal hull $\mathfrak{A}^{max}$ of $\mathfrak{A}$.

22.1. If $(\mathfrak{A}, \mathbf{A})$ is a quasi–Banach operator ideal, then its *minimal kernel* is defined to be the quasi–Banach operator ideal

$$(\mathfrak{A}, \mathbf{A})^{min} := (\mathfrak{A}^{min}, \mathbf{A}^{min}) := (\overline{\mathfrak{F}}, \| \ \|) \circ (\mathfrak{A}, \mathbf{A}) \circ (\overline{\mathfrak{F}}, \| \ \|) \, ,$$

where $\overline{\mathfrak{F}}$ is the ideal of all approximable operators; in other words (see 9.10. for the definition of the composition of operator ideals),

$$\mathbf{A}^{min}(T) = \inf\{ \, \|S\|\mathbf{A}(T_o)\|R\| \mid T = S \circ T_o \circ R \text{ with } S, R \in \overline{\mathfrak{F}} \text{ and } T_o \in \mathfrak{A} \} \, .$$

$(\mathfrak{A}, \mathbf{A})$ is called *minimal* if $(\mathfrak{A}, \mathbf{A}) = (\mathfrak{A}, \mathbf{A})^{min}$.

REMARK: $\overline{\mathfrak{F}} \circ \overline{\mathfrak{F}} = \overline{\mathfrak{F}}$ holds isometrically.

PROOF: Take $T \in \overline{\mathfrak{F}}(E, F)$ and choose $T_n \in \mathfrak{F}(E, F)$ and $1 \leq \lambda_n \uparrow \infty$ with $T = \sum_{n=1}^\infty T_n$ (with respect to the operator norm $\| \ \|$) and $\sum_{n=1}^\infty \lambda_n \|T_n\| \leq \|T\|(1+\varepsilon)$. If

$N_n := \text{im}(T_n)$, then

$$R : E \longrightarrow c_o(N_n) \qquad \text{and} \qquad S : c_o(N_n) \longrightarrow F$$

$$x \rightsquigarrow ((\lambda_n \|T_n\|)^{-1} T_n x)_{n \in \mathbb{N}} \qquad\qquad (y_n) \rightsquigarrow \sum_{n=1}^{\infty} \lambda_n \|T_n\| y_n$$

give a factorization $T = S \circ R \in \overline{\mathfrak{F}} \circ \overline{\mathfrak{F}}$ with $\|S\| \|R\| \leq \|T\|(1+\varepsilon)$. □

This implies that $\overline{\mathfrak{F}} = \overline{\mathfrak{F}} \circ \overline{\mathfrak{F}} = \overline{\mathfrak{F}} \circ \mathfrak{L} \circ \overline{\mathfrak{F}}$, hence $\overline{\mathfrak{F}}$ is the minimal kernel of all operators: $\mathfrak{L}^{min} = \overline{\mathfrak{F}}$. Moreover,

$$\mathfrak{A}^{min} = \overline{\mathfrak{F}} \circ \mathfrak{A} \circ \overline{\mathfrak{F}} = \overline{\mathfrak{F}} \circ \overline{\mathfrak{F}} \circ \mathfrak{A} \circ \overline{\mathfrak{F}} \circ \overline{\mathfrak{F}} = \mathfrak{A}^{min\ min}$$

(isometrically) — in other words, $\mathfrak{A}^{min}$ is a minimal operator ideal. Since, obviously, $\mathfrak{A}^{min} \subset \mathfrak{B}^{min}$ if $\mathfrak{A} \subset \mathfrak{B}$, these facts show that $\overline{\mathfrak{F}}$ is the largest minimal operator ideal. If $\mathfrak{A}$ is p-Banach, then $\mathfrak{A}^{min}$ is q-Banach for some q by 9.10.. To avoid the quasinorm catastrophe the following result will be formulated only for p-Banach operator ideals; but recall from 9.3. that every quasi-Banach ideal has an equivalent p-norm.

PROPOSITION: *Let $(\mathfrak{A}, \mathbf{A})$ be a p-Banach operator ideal.*
(1) $\mathfrak{A}^{min}(M,N) = \mathfrak{A}^{max}(M,N) = \mathfrak{A}(M,N)$ *holds isometrically for all $M, N \in$ FIN.*
(2) *For each $T \in \mathfrak{A}^{min}$ there are $T_n \in \mathfrak{F} \circ \mathfrak{A} \circ \mathfrak{F}$ with $\mathbf{A}^{min}(T_n - T) \to 0$.*
(3) $(\mathfrak{A}, \mathbf{A})^{max\ min} = (\mathfrak{A}, \mathbf{A})^{min}$.
(4) $(\mathfrak{A}, \mathbf{A})^{min\ max} = (\mathfrak{A}, \mathbf{A})^{max}$.
(5) *If $(\mathfrak{B}, \mathbf{B})$ is a q-Banach operator ideal with $\mathbf{B} \leq c\mathbf{A}$ on FIN for some $c \geq 1$, then $\mathfrak{A}^{min} \subset \mathfrak{B}^{min}$ and $\mathbf{B}^{min} \leq c\mathbf{A}^{min}$.*

In particular, (5) shows that $(\mathfrak{A}, \mathbf{A})^{min}$ is the smallest q-Banach operator ideal (where q is arbitrary) which induces the norm $\mathbf{A}$ on FIN.

PROOF: The relation for $\mathfrak{A}^{max}$ in (1) is immediate from the definition 17.2. of maximal operator ideals. $\mathbf{A} \leq \mathbf{A}^{min}$ is always true and $\mathbf{A}^{min} \leq \mathbf{A}$ on $\mathfrak{L}(M,N)$ follows from $T = id_N \circ T \circ id_M$. Clearly, (1) implies (4). To see (2), take $T \in \mathfrak{A}^{min}(E, F)$, an $\varepsilon > 0$ and a factorization $T = S \circ T_o \circ R \in \overline{\mathfrak{F}} \circ \mathfrak{A} \circ \overline{\mathfrak{F}}$ with $\|S\| \mathbf{A}(T_o) \|R\| \leq \mathbf{A}^{min}(T)(1+\varepsilon)$. For $S_n, R_n \in \mathfrak{F}$ with $\|R - R_n\| \to 0$ and $\|S - S_n\| \to 0$ define the operators $T_n := S_n \circ T_o \circ R_n$. Since $\mathbf{A}^{min}$ is an r-norm for some r, (2) follows from

$$\mathbf{A}^{min}(T_n - T) \leq [(\|S_n - S\| \mathbf{A}(T_o) \|R_n\|)^r + (\|S\| \mathbf{A}(T_o) \|R_n - R\|)^r]^{1/r} \longrightarrow 0 .$$

To see (3), take $T \in \mathfrak{A}^{max\ min}$. Then this construction (applied to $\mathfrak{A}^{max}$ instead of $\mathfrak{A}$) gives the factorization

$$\begin{array}{ccc}
E & \xrightarrow{T_n} & F \\
\downarrow & & \uparrow \\
M_n := \operatorname{im}(R_n) & \xdashrightarrow{T_{o,n}} & \operatorname{im}(S_n) =: N_n \\
\uparrow & & \uparrow \\
G_1 & \xrightarrow{T_o} & G_2
\end{array}$$

where $T_{o,n} \in \mathfrak{A}^{max}(M_n, N_n) \stackrel{1}{=} \mathfrak{A}(M_n, N_n)$. It follows that

$$\mathbf{A}(T_n) \leq \|S_n\| \mathbf{A}^{max}(T_o) \|R_n\|$$

$$\mathbf{A}(T_n - T_m) \leq \left[(\|S_n - S_m\| \mathbf{A}^{max}(T_o) \|R_n\|)^p + (\|S_m\| \mathbf{A}^{max}(T_o) \|R_n - R_m\|)^p \right]^{1/p},$$

hence (T_n) is an $\mathbf{A}$–Cauchy sequence. It converges in $\mathfrak{A}(E, F)$ and its limit is clearly T. Since $\mathbf{A}$ is a p-norm, it follows that

$$\mathbf{A}(T) = \lim_{n \to \infty} \mathbf{A}(T_n) \leq \lim_{n \to \infty} \|S_n\| \mathbf{A}^{max}(T_o) \|R_n\| \leq \mathbf{A}^{max\ min}(T)(1 + \varepsilon).$$

This means that $\mathfrak{A}^{max\ min} \subset \mathfrak{A}$ and $\mathbf{A} \leq \mathbf{A}^{max\ min}$, which implies that

$$\mathfrak{A}^{max\ min} \subset \mathfrak{A}^{min} \text{ and } \mathbf{A}^{min} \leq \mathbf{A}^{max\ min}.$$

The reverse, however, is obvious. The remaining statement (5) follows from (1) and (3). □

The proposition implies that for the study of the maximal and minimal operator ideals $\mathfrak{A}^{max}$ and $\mathfrak{A}^{min}$ one may always assume that $\mathfrak{A}$ is maximal.

22.2. Let $(\mathfrak{A}, \mathbf{A})$ be a maximal *normed* operator ideal and α its associated finitely generated tensor norm, i.e. $\mathfrak{A}(M, N) = M' \otimes_\alpha N$ for all finite dimensional M, N (see section 17), and denote by

$$J : E' \otimes F \longrightarrow \mathfrak{A}^{min}(E, F)$$

the natural embedding $J(z) := T_z$. For $M \in \operatorname{FIN}(E')$ and $N \in \operatorname{FIN}(F)$ the diagram

$$\begin{array}{ccc}
E' \otimes_\alpha F & \xrightarrow{J} & \mathfrak{A}^{min}(E, F) \ni I_N^F \circ T \circ Q_{M^0}^E \\
I_M^{E'} \otimes I_N^F \uparrow & & \uparrow \updownarrow \\
M \otimes_\alpha N & \stackrel{1}{=} & \mathfrak{A}(E/M^0, N) \ni T
\end{array}$$

obviously commutes. If $z \in E' \otimes F$ and $u \in M \otimes N$ with $I_M^{E'} \otimes I_N^F(u) = z$, then

$$\mathbf{A}^{min}(J(z)) = \mathbf{A}^{min}(I_N^F \circ T_u \circ Q_{M^0}^E) \leq \mathbf{A}(T_u) = \alpha(u; M, N),$$

which implies that $\mathbf{A}^{min}(\tilde{J}(z)) \leq \alpha(z; E', F)$. Even more is true:

REPRESENTATION THEOREM FOR MINIMAL OPERATOR IDEALS: *Let $(\mathfrak{A}, \mathbf{A})$ be a Banach operator ideal and α the tensor norm associated with $(\mathfrak{A}, \mathbf{A})^{max}$. Then the canonical map*

$$\tilde{J}: E' \tilde{\otimes}_\alpha F \longrightarrow \mathfrak{A}^{min}(E, F)$$

is a metric surjection for all Banach spaces E and F. In particular, the minimal kernel of a Banach operator ideal is normed.

PROOF: One may assume that $\mathfrak{A}$ is maximal. By what was said above $\tilde{J}$ is defined and has norm ≤ 1. By definition, $M' \otimes_\alpha N = \mathfrak{A}(M, N)$ holds isometrically whenever $M, N \in$ FIN. Take operators $R \in \mathfrak{F}(E, G_1), T_o \in \mathfrak{A}(G_1, G_2)$ and $S \in \mathfrak{F}(G_2, F)$; if $z_{S \circ T_o \circ R} := J^{-1}(S \circ T_o \circ R)$, then

$$\alpha(z_{S \circ T_o \circ R}; E', F) \leq \|S\|\mathbf{A}(T_o)\|R\| :$$

Indeed, take (as in the proof of 22.1.) the factorization

$$\begin{array}{ccc}
E & \xrightarrow{S \circ T_o \circ R} & F \\
R \downarrow & & \uparrow I_2 \\
M := \text{im}(R) & \xdashrightarrow{T_1} & \text{im}(S) =: N \\
I_1 \downarrow & & \uparrow S \\
G_1 & \xrightarrow{T_o} & G_2
\end{array}$$

Then

$$\alpha(z_{S \circ T_o \circ R}; E', F) = \alpha(\overline{R}' \otimes I_2(z_{T_1}); E', F) \leq \|\overline{R}\| \, \|I_2\| \alpha(z_{T_1}; M', N) =$$
$$= \|\overline{R}\|\mathbf{A}(T_1) \leq \|\overline{R}\| \, \|I_1\|\mathbf{A}(T_o)\|\overline{S}\| = \|R\|\mathbf{A}(T_o)\|S\| \, .$$

This inequality is the key to the following reasoning. Take $T \in \mathfrak{A}^{min}(E, F)$, factor $T = S \circ T_o \circ R$ with $\|S\|\mathbf{A}(T_o)\|R\| \leq (1+\varepsilon)\mathbf{A}^{min}(T)$ and choose $R_n, S_n \in \mathfrak{F}$ with $\|R - R_n\| \to 0, \|S - S_n\| \to 0$. The proof of 22.1.(2) showed that $T_n := S_n \circ T_o \circ R_n$ converges to T in $\mathfrak{A}^{min}(E, F)$. Denote $z_n := J^{-1}(S_n \circ T_o \circ R_n) \in E' \otimes F$; then

$$z_n - z_m = J^{-1}((S_n - S_m) \circ T_o \circ R_n) + J^{-1}(S_m \circ T_o \circ (R_n - R_m))$$

and therefore by the inequality,

$$\alpha(z_n - z_m; E', F) \leq \|S_n - S_m\|\mathbf{A}(T_o)\|R_n\| + \|S_m\|\mathbf{A}(T_o)\|R_n - R_m\| \, .$$

It follows that (z_n) is a Cauchy sequence in $E'\tilde{\otimes}_\alpha F$. Its limit $z \in E'\tilde{\otimes}_\alpha F$ satisfies

$$\tilde{J}(z) = \lim_{n\to\infty} J(z_n) = \lim_{n\to\infty} T_n = T \quad \text{in } \mathfrak{A}^{min}$$

since $\tilde{J}$ is continuous. Moreover,

$$\alpha(z; E', F) = \lim_{n\to\infty} \alpha(z_n; E', F) \le \lim_{n\to\infty} \|S_n\| \mathbf{A}(T_o) \|R_n\| \le \mathbf{A}^{min}(T)(1+\varepsilon) \ .$$

This, together with $\|\tilde{J}\| \le 1$, shows that $\tilde{J}$ is a metric surjection. $\square$

A special case: The projective norm π is associated with the ideal $\mathfrak{I}$ of integral operators, hence the description 3.6. of nuclear operators gives

$$(\mathfrak{I}, \mathbf{I})^{min} = (\mathfrak{N}, \mathbf{N}) \ .$$

This is the reason why the operators in $\mathfrak{A}^{min}(E, F)$ are called α-*nuclear* if $\mathfrak{A}^{max} \sim \alpha$. Moreover, it follows that the ideal $\mathfrak{N}$ is the smallest *normed* minimal operator ideal.

COROLLARY 1: *Let α be associated with the maximal hull of the Banach operator ideal $\mathfrak{A}$. If E' or F has the approximation property or if α is totally accessible, then*

$$E'\tilde{\otimes}_\alpha F = \mathfrak{A}^{min}(E, F)$$

holds isometrically.

This follows from corollary 17.20.; clearly, it would be enough to assume that F had the α-approximation property or E' the α^t-approximation property (see 21.7.).

22.3. Examples:

$(\mathfrak{K}_{p,q}, \mathbf{K}_{p,q}) := (\mathfrak{L}_{p,q}, \mathbf{L}_{p,q})^{min}$ the ideal of (p,q) – compact operators.
$(\mathfrak{N}_p, \mathbf{N}_p) := (\mathfrak{I}_p, \mathbf{I}_p)^{min}$ the ideal of p – nuclear operators.
$(\mathfrak{K}_p, \mathbf{K}_p) := (\mathfrak{L}_p, \mathbf{L}_p)^{min}$ the ideal of p – compact operators.

The associated tensor norms are $\alpha_{p,q}$, $g_p = \alpha_{p,1}$ and $w_p = \alpha_{p,p'}$. The representation theorems 12.6. and 12.7. for $E'\tilde{\otimes}_{\alpha_{p,q}} F$ and the representation theorem for minimal operator ideals give

$$\mathbf{K}_{p,q}(T) := \inf\left\{\ell_r(\lambda_n) w_{q'}(x'_n) w_{p'}(y_n) \,\Big|\, T = \sum_{n=1}^\infty \lambda_n x'_n \otimes y_n\right\}$$

$$\mathbf{N}_p(T) := \inf\left\{\ell_p(x'_n) w_{p'}(y_n) \,\Big|\, T = \sum_{n=1}^\infty x'_n \otimes y_n\right\}$$

$$\mathbf{K}_p(T) := \inf\left\{w_p(x'_n) w_{p'}(y_n) \,\Big|\, T = \sum_{n=1}^\infty x'_n \otimes y_n\right\} \ .$$

The formula for $\mathbf{K}_{p,q}$ also shows that the (p,q)–compact operators are those which factor

$$\begin{array}{ccc} E & \xrightarrow{T} & F \\ R \downarrow & & \uparrow S \\ \ell_{q'} & \xrightarrow{D_\lambda} & \ell_p \end{array} \qquad \begin{array}{l} D_\lambda \text{ a diagonal operator} \\ (\ell_\infty \text{ can be replaced by } c_o), \end{array}$$

and $\mathbf{K}_{p,q}(T) = \inf \|S\| \|D_\lambda\| \|R\|$. Since $\mathfrak{A}^{min} = \overline{\mathfrak{F}} \circ \mathfrak{A}^{min} \circ \overline{\mathfrak{F}}$, the operators R and S can be taken to be compact. Note the special cases $\mathfrak{N}_p = \mathfrak{K}_{p,1}$ and $\mathfrak{K}_p = \mathfrak{K}_{p,p'}$; it follows that the p–compact operators are just those which factor compactly through the sequence space ℓ_p.

22.4. The following consequence of the representation theorem for minimal operator ideals is trivial, but useful:

COROLLARY 2: *If E and F are Banach spaces, $\alpha \sim \mathfrak{A}$ and $\beta \sim \mathfrak{B}$ such that $\alpha \leq c\beta$ on $E' \otimes F$, then $\mathfrak{B}^{min}(E,F) \subset \mathfrak{A}^{min}(E,F)$ and*

$$\mathbf{A}^{min}(T) \leq c\, \mathbf{B}^{min}(T)$$

for all $T \in \mathfrak{B}^{min}(E,F)$.

By the embedding theorem (and 15.7.) the next result (again for maximal normed ideals and fixed Banach spaces E and F) appears as a special case of this corollary: If $\mathfrak{B}(E,F) \subset \mathfrak{A}(E,F)$ and $\mathfrak{A}$ is totally accessible (or: $\mathfrak{A}$ accessible and E' or F has the bounded approximation property or: E' and F have the bounded approximation property), then

$$\mathfrak{B}^{min}(E,F) \subset \mathfrak{A}^{min}(E,F).$$

Since $\pi \leq K_G g_\infty$ on $\ell_2^m \otimes \ell_1^m$ (see 14.8.), the local technique lemma implies that

$$\pi \leq \lambda K_G g_\infty \leq \lambda K_G g_p \qquad \text{on } H \otimes F$$

for every Hilbert space H, every $\mathcal{L}_{1,\lambda}^g$–space F and all $1 \leq p \leq \infty$. Since $g_p \sim \mathfrak{I}_p$ and $\mathfrak{I}_p^{min} = \mathfrak{N}_p$, it follows that:

PROPOSITION: $\mathfrak{N}(H,F) = \mathfrak{N}_p(H,F)$ *and* $\mathbf{N}(T) \leq \lambda K_G \mathbf{N}_p(T)$ *for every Hilbert space H, $\mathcal{L}_{1,\lambda}^g$–space F and $p \in [1,\infty]$.*

This is in some sense a "nuclear" form of Grothendieck's inequality.

22.5. The representation theorem for minimal operator ideals represents the third of the three basic links between the metric theory of tensor products and the theory of

Banach operator ideals: If the maximal Banach operator ideal $(\mathfrak{A}, \mathbf{A})$ and the finitely generated tensor norm α are associated, i.e. if

$$M' \otimes_\alpha N = \mathfrak{A}(M, N)$$

holds isometrically for all $M, N \in \text{FIN}$, then for all Banach spaces E and F the following theorems are valid (17.5., 17.6. and 22.2.):

(1) *The representation theorem for maximal Banach operator ideals:*

$$\mathfrak{A}(E, F') \stackrel{1}{=} (E \otimes_{\alpha'} F)'$$

and

$$\mathfrak{A}(E, F) \stackrel{1}{=} (E \otimes_{\alpha'} F')' \cap \mathfrak{L}(E, F) .$$

(2) *The embedding theorem:*

$$E' \tilde{\otimes}_{\overset{\sim}{\alpha}} F \stackrel{1}{\hookrightarrow} \mathfrak{A}(E, F) .$$

(3) *The representation theorem for minimal Banach operator ideals:*

$$E' \tilde{\otimes}_\alpha F \stackrel{1}{\twoheadrightarrow} \mathfrak{A}^{min}(E, F) .$$

This troika of results is a powerful tool for investigating maximal and minimal *normed* operator ideals. A first example is the

COROLLARY: *Let $(\mathfrak{A}, \mathbf{A})$ be an accessible maximal Banach operator ideal. If E' or F has the metric approximation property or if $(\mathfrak{A}, \mathbf{A})$ is totally accessible, then*

$$\mathfrak{A}^{min}(E, F) = \overline{\mathfrak{F}(E, F)}^{\mathbf{A}} \stackrel{1}{\hookrightarrow} \mathfrak{A}(E, F) .$$

The proof follows from the fact that, in this case, $\alpha = \overleftarrow{\alpha}$ on $E' \otimes F$ (see 15.7. and 21.3.). As before, it would be enough to assume that, for example, F has the metric α-approximation property. A consequence of this corollary is:

$$\mathfrak{P}_p^{min}(E, F) = \overline{\mathfrak{F}(E, F)}^{\mathbf{P}_p} \stackrel{1}{\hookrightarrow} \mathfrak{P}_p(E, F)$$

since $g_{p'}^* \sim \mathfrak{P}_p$ is totally accessible.

22.6. Another nice example of the interplay between the three basic theorems is the following slight extension of a result due to Schwarz [254].

PROPOSITION: *Let $(\mathfrak{A}, \mathbf{A})$ be a maximal Banach operator ideal. If $(\mathfrak{A}, \mathbf{A})$ is totally accessible or if E or F' has the approximation property, then*

$$(\mathfrak{A}^{min}(F, E))' = \mathfrak{A}^*(E, F'')$$

holds isometrically.

PROOF: If α is the tensor norm associated with $\mathfrak{A}$, then $\mathfrak{A}^* \sim \alpha^*$ and therefore
$$\mathfrak{A}^*(E, F'') \stackrel{1}{=} (E \otimes_{\alpha^t} F')' \stackrel{1}{=} (F' \tilde{\otimes}_\alpha E)'.$$
Since $F' \tilde{\otimes}_\alpha E = \mathfrak{A}^{min}(F, E)$ by 22.2., corollary 1, the result follows. $\square$

The duality bracket can be calculated by using the trace: Use 17.15. to see (first for elementary tensors) that for $T \in \mathfrak{A}^*(E, F'')$

$$\mathfrak{A}^{min}(F, E) = F' \tilde{\otimes}_\alpha E \xrightarrow{id_{F'} \tilde{\otimes} T} F' \tilde{\otimes}_\pi F'' \longrightarrow \mathfrak{N}(F', F')$$
$$S \rightsquigarrow \qquad\qquad\qquad \rightsquigarrow S' \circ T' \circ \kappa_{F'}$$
$$\downarrow tr_{F'}$$
$$\rightsquigarrow \langle T, S \rangle \in \mathbb{K}$$

and

$$\mathfrak{A}^{min}(F, E) = F' \tilde{\otimes}_\alpha E \hookrightarrow F''' \tilde{\otimes}_\alpha E \xrightarrow{T' \tilde{\otimes} id_E} E' \tilde{\otimes}_\pi E \longrightarrow \mathfrak{N}(E, E)$$
$$S \rightsquigarrow \qquad\qquad\qquad\qquad \rightsquigarrow S^\pi \circ T$$
$$\downarrow tr_E$$
$$\rightsquigarrow \langle T, S \rangle \in \mathbb{K}$$

where $S^\pi : F'' \to E$ is the astriction of S''; it follows that

$$\langle T, S \rangle = \begin{cases} tr_{F'}(S' \circ T' \circ \kappa_{F'}) & \text{if } F' \text{ has a.p.} \\ tr_E(S^\pi \circ T) & \text{if } E \text{ has a.p.} \end{cases}$$

In the case of α being totally accessible, the duality bracket cannot always be calculated with the trace *on operators*. This can be seen as follows: The diagram shows that $S' \circ T' \circ \kappa_{F'}$ and $S^\pi \circ T$ are nuclear operators — and it may happen that all nuclear operators in $\mathfrak{N}(F', F')$ [or $\mathfrak{N}(E, E)$] appear this way; therefore, if F' [or E] does not have the approximation property, the trace is not defined. For an example, take $\alpha = \varepsilon$, hence $\mathfrak{A}^* = \mathfrak{J}$ and $\mathfrak{A}^{min} = \overline{\mathfrak{F}}$, and G a reflexive space without the approximation property. Then $\mathfrak{N}(G, G) = \mathfrak{N}(c_o, G) \circ \overline{\mathfrak{F}}(G, c_o)$ and $\mathfrak{N}(G, G) = \overline{\mathfrak{F}}(\ell_1, G) \circ \mathfrak{N}(G, \ell_1)$. Since ℓ_1 and G' have the Radon–Nikodým property, it follows from Appendix D8. that

$$\mathfrak{N}(G, G) = \mathfrak{J}(c_o, G) \circ \overline{\mathfrak{F}}(G, c_o) = \overline{\mathfrak{F}}(\ell_1, G) \circ \mathfrak{J}(G, \ell_1).$$

Therefore $(E, F) = (c_o, G')$ and $(E, F) = (G, \ell_1)$ show that for $\alpha = \varepsilon$ none of the two operator trace formulas holds *for all* pairs (E, F) of Banach spaces. Even more: Taking the direct sum of these two counterexamples one can verify that there is no quasi–Banach operator ideal $\mathfrak{A}$ with a continuous trace such that for each (E, F) either all $S' \circ T' \circ \kappa_{F'} \in \mathfrak{A}(F', F')$ or all $S^\pi \circ T \in \mathfrak{A}(E, E)$; see Ex 22.9. for the definition of such ideals and the main argument for this.

22.7. If T is nuclear, then T' is — but the converse is not true by the example given in 16.8.. For other *normed* minimal operator ideals the question of duality reads as follows: If $(\mathfrak{A}, \mathbf{A})$ is a maximal normed operator ideal with associated tensor norm α, then the transposed tensor norm α^t is associated to $\mathfrak{A}^{dual}$, the ideal of all T such that $T' \in \mathfrak{A}$ (see 17.8.). The diagram

$$\begin{array}{ccc} E' \tilde{\otimes}_{\alpha^t} F & \xrightarrow{1} & \mathfrak{A}^{dual\ min}(E,F) \ni T \\ \| & & \downarrow\ \wr \\ F \tilde{\otimes}_\alpha E' & \xrightarrow{1} F'' \tilde{\otimes}_\alpha E' \xrightarrow{1} & \mathfrak{A}^{min}(F', E') \ni T' \end{array}$$

shows that T' is α-nuclear if T is α^t-nuclear and $\mathbf{A}^{min}(T') \leq \mathbf{A}^{dual\ min}(T)$ holds; see also Ex 22.5.. Under which circumstances is the converse true? In other words, when does

$$\mathfrak{A}^{min\ dual}(E, F) \stackrel{?}{\subset} \mathfrak{A}^{dual\ min}(E, F) \subset \mathfrak{A}^{min\ dual}(E, F)$$

hold? If F is a dual space, then it is easy to see that this is true (Ex 22.5.). As for further situations where the answer is positive an appropriate description of the $\sigma(F', F)-\sigma(E, E')$-continuous operators in $\mathfrak{A}^{min}(F', E)$ will be helpful:

PROPOSITION: *Let $\mathfrak{A} \sim \alpha$ be associated and E, F Banach spaces. Then the canonical mapping*

$$F \tilde{\otimes}_\alpha E \longrightarrow \{T \in \mathfrak{A}^{min}(F', E) \mid T\ \sigma(F', F)-\sigma(E, E')-\text{continuous}\}$$

is defined and has norm ≤ 1. It is a metric surjection under each of the following conditions:

(1) F *is 1-complemented in* F''.

(2) F''/F *has the approximation property.*

(3) E *has the approximation property.*

PROOF: The first statement is clear since, by 4.2., all operators coming from $F \tilde{\otimes}_\varepsilon E$ are $\sigma(F', F) - \sigma(E, E')$ continuous. The representation theorem also gives that, for a given $\sigma(F', F) - \sigma(E, E')$ continuous $T_o \in \mathfrak{A}^{min}(F', E)$, there is an $z_o \in F'' \tilde{\otimes}_\alpha E$ with $\tilde{J}(z_o) = T_o$ and

$$\alpha(z_o; F'', E) \leq (1+\varepsilon)\mathbf{A}^{min}(T_o)\ .$$

It remains to prove that z_o is (or can be taken to be) in $F \tilde{\otimes}_\alpha E$.

(1) If $P : F'' \to F$ is a projection with $\|P\| = 1$, then the diagram

$$\begin{array}{ccc}
F''\tilde{\otimes}_\alpha E \xrightarrow{\tilde{j}} \mathfrak{A}^{min}(F',E) \hookrightarrow \mathcal{L}(E',F''') \ni S \\
P\tilde{\otimes}id_E \downarrow \uparrow & \downarrow \downarrow & \} \\
F\tilde{\otimes}_\alpha E \longrightarrow & \mathcal{L}(E',F) \ni P\circ S
\end{array}$$

commutes. Since $\kappa_F \circ P \circ T'_o = T'_o$ by assumption, it follows that

$$T_o = \tilde{J}\left((\kappa_F \tilde{\otimes} id_E)(P \tilde{\otimes} id_E)(z_o)\right)$$

and hence $P\tilde{\otimes}id_E(z_o) \in F\tilde{\otimes}_\alpha E$ represents T_o.

(3) If the space E has the approximation property, it will be shown that z_o is already an element of $F\tilde{\otimes}_\alpha E$. For this, take $\varphi \in (F''\tilde{\otimes}_\alpha E)'$ with $\varphi_{|F\otimes E} = 0$; it has to be shown that $\langle \varphi, z_o \rangle = 0$. If $L_\varphi \in \mathcal{L}(F'', E')$ is the operator associated with φ, then, by 17.15.,

$$L_\varphi \tilde{\otimes} id_E : F''\tilde{\otimes}_\alpha E \longrightarrow E'\tilde{\otimes}_\pi E$$

is continuous and $\langle \varphi, z_o \rangle = tr_E(L_\varphi \tilde{\otimes} id_E(z_o))$:

$$\begin{array}{ccc}
F''\tilde{\otimes}_\alpha E \longrightarrow & \mathcal{L}(F',E) \ni S \\
\varphi \nearrow \quad \downarrow L_\varphi \tilde{\otimes} id_E & \downarrow & \} \\
\mathbb{K} \quad tr_E \nwarrow \quad E'\tilde{\otimes}_\pi E \xrightarrow{I} & \mathcal{L}(E',E') \ni L_\varphi \circ S' & .
\end{array}$$

Since $L_{\varphi|F} = 0$ and $T'_o(E') \subset F$, it follows that $L_\varphi \circ T'_o = 0$. The space E has the approximation property, hence the lower map I is injective; therefore $L_\varphi \circ T'_o = 0$ implies that $L_\varphi \tilde{\otimes} id_E(z_o) = 0$ which gives $\langle \varphi, z_o \rangle = 0$.

(2) Finally, assume that F''/F has the approximation property. As before, take a functional $\varphi \in (F''\tilde{\otimes}_\alpha E)'$ with $\varphi_{|F\otimes E} = 0$. Denote by $U \in \mathcal{L}(E, F''')$ the operator associated with $\varphi^t \in (E\tilde{\otimes}_{\alpha^t} F'')' = \mathfrak{A}^*(E, F''')$. Since

$$im(U) \subset F^0 \hookrightarrow F'''$$

(the polar taken in F''') and F^0 is complemented in F''' (by $x''' \rightsquigarrow x''' - \kappa_{F'}(x'''|_F)$), it follows that the astriction $U_o : E \to F^0$ of U is in $\mathfrak{A}^*$ as well. Note that $F^0 = (F''/F)'$. With the canonical surjection $Q : F'' \to F''/F$, it follows from 17.15. that

$$Q\tilde{\otimes}U_o : F''\tilde{\otimes}_\alpha E \longrightarrow F''\tilde{\otimes}_\pi(F''/F)' \longrightarrow F''/F\tilde{\otimes}_\pi(F''/F)'$$

is continuous. A check on elementary tensors shows that the diagram

$$\begin{array}{ccc}
F''\tilde{\otimes}_\alpha E \longrightarrow & \mathcal{L}(F',E) \ni S \\
\varphi \nearrow \quad \downarrow Q\tilde{\otimes}U_o & \downarrow & \} \\
\mathbb{K} \quad tr \nwarrow \quad F''/F\tilde{\otimes}_\pi(F''/F)' \xrightarrow{I} & \mathcal{L}((F''/F)'',F''/F) \ni Q\circ S'\circ U'_o
\end{array}$$

commutes. Since $T'_o(E') \subset F$, it follows that $Q \circ T'_o \circ U'_o = 0$ and the same argument as before gives the result. □

The space $(F''/F)'$ is complemented in F''' (this was just shown in the proof), hence F''/F has the approximation property if F''' has it. It will be shown in a moment (22.9.) that the statement of the proposition *does not hold* if only F'' is required to have the approximation property.

22.8. Coming back to the question under which conditions $\mathfrak{A}^{min\ dual} \subset \mathfrak{A}^{dual\ min}$ holds, the proposition gives the

COROLLARY 1: *Let $(\mathfrak{A}, \mathbf{A})$ be a Banach operator ideal and E, F Banach spaces. If F is 1-complemented in F'' or if E' or F''/F has the approximation property, then*

$$\mathfrak{A}^{min\ dual}(E,F) = \mathfrak{A}^{dual\ min}(E,F)$$

holds isometrically. If $\alpha \sim (\mathfrak{A}^{max}, \mathbf{A}^{max})$ this means that: $T \in \mathcal{L}(E, F)$ is α^t-nuclear if and only if T' is α-nuclear; the respective norms coincide.

PROOF: Since clearly, $(\mathfrak{A}, \mathbf{A})^{max\ dual} = (\mathfrak{A}, \mathbf{A})^{dual\ max}$, one may assume $(\mathfrak{A}, \mathbf{A})$ to be maximal. A weakly compact operator $S \in \mathcal{L}(F', E')$ is $\sigma(F', F) - \sigma(E', E'')$ continuous if and only if it is a dual operator. Since all operators in $\mathfrak{A}^{min}$ are compact, the proposition implies that the canonical map

$$F \tilde{\otimes}_\alpha E' \longrightarrow \{T' \in \mathfrak{A}^{min}(F', E') \mid T \in \mathcal{L}(E, F)\}$$

is a metric surjection under each of the hypotheses of the corollary. Since

$$\begin{array}{ccc} F \tilde{\otimes}_\alpha E' & \longrightarrow & \mathcal{L}(F', E') \ni T' \\ \| & & \updownarrow \\ E' \tilde{\otimes}_{\alpha^t} F & \longrightarrow & \mathcal{L}(E, F) \ni T \end{array}$$

commutes, this (and $E' \tilde{\otimes}_{\alpha^t} F \twoheadrightarrow \mathfrak{A}^{dual\ min}(E, F)$) shows that T is α^t-nuclear if T' is α-nuclear; moreover,

$$\mathbf{A}^{dual\ min}(T) \leq \mathbf{A}^{min}(T') = \mathbf{A}^{min\ dual}(T) .$$

The converse is always true. □

The fact that S' is β-nuclear if S is β^t-nuclear easily implies the following: *The dual T' of $T \in \mathcal{L}(E, F)$ is α^t-nuclear if and only if $\kappa_F \circ T = T'' \circ \kappa_E : E \to F''$ is α-nuclear.* Therefore, corollary 1 (applied to α^t) gives the

COROLLARY 2: *Let* $(\mathfrak{A}, \mathbf{A})$ *be a Banach operator ideal and* E, F *Banach spaces. If* F *is 1–complemented in* F'' *or if* E' *or* F''/F *has the approximation property, then* $T \in \mathfrak{A}^{min}(E, F)$ *if and only if* $\kappa_F \circ T \in \mathfrak{A}^{min}(E, F'')$; *moreover,*

$$\mathbf{A}^{min}(T) = \mathbf{A}^{min}(\kappa_F \circ T) .$$

In other words, T is α–nuclear if and only if $\kappa_F \circ T$ is α–nuclear (where α is the tensor norm associated with $\mathfrak{A}^{max}$). Clearly, the result is trivial under the first assumption that F is complemented in F''; for more information about these questions, see 25.11., Ex 22.6. and Ex 25.12..

22.9. For $\alpha = \pi$ these results were already proven if F is complemented in F'' (see Ex 3.34.) and if E' has the approximation property (see 5.9.) — but not yet under the hypotheses on F''/F. The example in 16.8. showed that it is not enough to assume that E has the approximation property. It is also not enough that F'' has the approximation property; the key to a counterexample is the following result due to Lindenstrauss [172]: If G is a separable Banach space, then there is a Banach space F such that F'' has a basis and F''/F is isomorphic to G. If one takes G to be a space without the approximation property, one obtains a space F'' with a basis such that F''/F does not have the approximation property.

PROPOSITION: *There are Banach spaces* E *and* F *such that* F'' *has a basis and there is a non–nuclear* $T \in \mathfrak{L}(E, F)$ *with nuclear dual.*

PROOF: For $E := F''/F$, as before, it is enough to find a non–nuclear $T \in \mathfrak{L}(E, F)$ such that $\kappa_F \circ T$ is nuclear. For the quotient map $Q : F'' \to F''/F$ consider the commutative diagram

$$\begin{array}{ccccc}
(F''/F)' \tilde{\otimes}_\pi F''/F & \xrightarrow{J_1} & \mathfrak{N}(F''/F, F''/F) & \ni & Q \circ S \\
{\scriptstyle id \tilde{\otimes} Q} \uparrow & & \uparrow & & \\
(F''/F)' \tilde{\otimes}_\pi F'' & \xrightarrow{J_2} & \mathfrak{N}(F''/F, F'') & \ni & S \\
 & & & \ni & \kappa_F \circ U \\
{\scriptstyle id \tilde{\otimes} \kappa_F} \uparrow & & \uparrow & & \\
(F''/F)' \tilde{\otimes}_\pi F & \xrightarrow{J_3} & \mathfrak{N}(F''/F, F) & \ni & U .
\end{array}$$

The construction of F implies that J_2 and J_3 are injective, but J_1 is not! Therefore there is a $z_o \in (F''/F)' \tilde{\otimes}_\pi F''$ such that $w_o := id \tilde{\otimes} Q(z_o) \neq 0$, but its associated operator $J_1(w_o)$ is zero. If $T_o := J_2(z_o)$, then $Q \circ T_o = 0$, hence $\operatorname{im}(T_o) \subset F$ and $T_o = \kappa_F \circ T$ for some $T \in \mathfrak{L}(F''/F, F)$. If T would be nuclear, then the diagram would imply that $T_o = 0$. $\square$

The results of 22.7.–22.9. are taken from Losert–Michor's exposition [179] of Grothendieck's Résumé. A counterexample similar to the above one was also recently constructed by Oya–Reinov [196]. The investigation of $\mathfrak{A}^{sur\ min}$ and $\mathfrak{A}^{inj\ min}$ will be postponed to 25.11. since it requires some more flexible descriptions of $\mathfrak{A}^{min}$ for accessible operator ideals $\mathfrak{A}$.

Exercises:

Ex 22.1. Use the Davis–Figiel–Johnson–Pełczyński-factorization theorem (see 9.6.) to show that for every $T \in \mathfrak{A}^{min}(E,F)$

$$\mathbf{A}^{min}(T) = \inf \|S\| \mathbf{A}(T_o) \|R\| \ ,$$

where $T_o \in \mathfrak{A}(G_1, G_2)$ for *reflexive* G_1 and G_2 and, obviously, $R, S \in \overline{\mathfrak{F}}$ and $T = S \circ T_o \circ R$.

Ex 22.2. Let E and F be Banach spaces, one of which has the approximation property. Then
$$E \tilde{\otimes}_\varepsilon F = \{T \in \mathfrak{K}(E',F) \mid T \ \sigma(E',E)-\sigma(F,F')-\text{continuous}\} \ .$$
Hint: 22.7. if F has the approximation property; dualize and approximate T' in the other case.

Ex 22.3. Give conditions (like the approximation property, accessibility) which imply that $\mathbf{A}^{min}(T) = \mathbf{A}^{max}(T)$ on $\mathfrak{F}(E,F)$. Check this for the ideals: $\mathfrak{K}, \mathfrak{N}, \mathfrak{L}_{p,q}, \mathfrak{L}_p, \mathfrak{I}_p, \mathfrak{D}_{p,q}$, and $\mathfrak{P}_p$.

Ex 22.4. If $(\mathfrak{A}, \mathbf{A})$ is a maximal Banach operator ideal, F an $\mathcal{L}^g_\infty$-space, then
$$\mathfrak{A}^{inj\ min}(E,F) = \mathfrak{A}^{min}(E,F)$$
for all Banach spaces E. What about the norm? Hint: 20.8.. For nuclear operators the converse is true, see 23.7..

Ex 22.5. Use factorization to show that
$$\mathfrak{A}^{dual\ min} \subset \mathfrak{A}^{min\ dual} \quad \text{and} \quad \mathbf{A}^{min}(T') \leq \mathbf{A}^{dual\ min}(T)$$
for all quasi-Banach operator ideals. Moreover,
$$\mathfrak{A}^{dual\ min}(E,F') = \mathfrak{A}^{min\ dual}(E,F')$$
with equal norms. Hint: Ex 22.1..

Ex 22.6. Let $(\mathfrak{A}, \mathbf{A})$ be a quasi-Banach operator ideal. The *regular hull* $(\mathfrak{A}, \mathbf{A})^{reg}$ of $(\mathfrak{A}, \mathbf{A})$ is defined to be the class of all $T \in \mathcal{L}(E,F)$ such that $\kappa_F \circ T \in \mathfrak{A}$ and $\mathbf{A}^{reg}(T) := \mathbf{A}(\kappa_F \circ T)$.

(a) Show that $(\mathfrak{A}, \mathbf{A})^{reg}$ is a quasi-Banach operator ideal containing $\mathfrak{A}$.

(b) $(\mathfrak{A}, \mathbf{A})^{dual\ reg} = (\mathfrak{A}, \mathbf{A})^{dual}$; in other words, $(\mathfrak{A}, \mathbf{A})^{dual}$ is regular.

(c) $(\mathfrak{A}, \mathbf{A})^{dual\ min\ reg} = (\mathfrak{A}, \mathbf{A})^{min\ dual}$. Hint: Ex 22.5..

Ex 22.7. *Extension and lifting properties of minimal operator ideals:* Let α be a tensor norm with associated maximal normed operator ideal $(\mathfrak{A}, \mathbf{A})$.

(a) α is left–projective if and only if: Whenever $I : G \overset{1}{\hookrightarrow} E$ and $T \in \mathfrak{A}^{min}(G, F)$, then for every $\varepsilon > 0$ there is an *extension* $\tilde{T} \in \mathfrak{A}^{min}(E, F)$ with $\mathbf{A}^{min}(\tilde{T}) \leq (1+\varepsilon)\mathbf{A}^{min}(T)$ and $T = \tilde{T} \circ I$.

(b) α is right–projective if and only if: Whenever $Q : G \overset{1}{\twoheadrightarrow} F$ and $T \in \mathfrak{A}^{min}(E, F)$, then for every $\varepsilon > 0$ there is a *lifting* $\tilde{T} \in \mathfrak{A}^{min}(E, G)$ with $\mathbf{A}^{min}(\tilde{T}) \leq (1+\varepsilon)\mathbf{A}^{min}(T)$ and $T = Q \circ \tilde{T}$. Hint: Representation theorem and 20.3.(1) and (2).

Ex 22.8. Let $(\mathfrak{A}, \mathbf{A})$ be a maximal normed operator ideal with associated tensor norm α. Use the characterization 17.15. to show that

$$\mathfrak{A}^* \circ \mathfrak{A}^{min} \subset \mathfrak{N}$$

— in other words, $T \circ S$ is nuclear if S is α–nuclear and T is α^*–integral. Conversely, $T \in \mathfrak{A}^*(E, F)$ if for all $S \in \mathfrak{A}^{min}(C_2, E)$ the operator $T \circ S$ is nuclear. Hint: Lemma 17.16. and 22.2., corollary 1.

Ex 22.9. A *continuous trace* τ on a quasi-Banach operator ideal $(\mathfrak{A}, \mathbf{A})$ assigns to each $T \in \mathfrak{A}$ a number $\tau(T) \in \mathbb{K}$ such that

(1) τ restricted to each $\mathfrak{A}(E, E)$ is linear and continuous.

(2) $\tau(x' \tilde{\otimes} x) = \langle x', x \rangle$ for $x' \in E'$ and $x \in E$.

As for the existence of ideals with a continuous trace see the books of König [147] and Pietsch [216]. Show the following: If the Banach space G does not have the approximation property, then $\mathfrak{N}(G, G) \not\subset \mathfrak{A}(G, G)$ for each quasi-Banach operator ideal $\mathfrak{A}$ which admits a continuous trace. Hint: Closed graph theorem, 5.6. and $tr(z) = \tau(J_{G'}(z))$ for all $z \in G' \tilde{\otimes}_\pi G$.

23. $\mathcal{L}_p^g$–Spaces

This section gives a systematic presentation of the class of all $\mathcal{L}_p^g$–spaces. These spaces were introduced in Chapter I and have already been used at various times. Locally they look like ℓ_p^m and hence share many properties of ℓ_p. They are very closely related to the class of $\mathcal{L}_p$–spaces investigated by Lindenstrauss and Pełczyński in their famous 1968–paper; in order to make them easier to use we modified their definition slightly and as a result they behave better under dualization and complementation. The main results about $\mathcal{L}_p$- and hence $\mathcal{L}_p^g$-spaces presented in this section are, at least in spirit, due to Lindenstrauss, Pełczyński and Rosenthal.

23.1. Recall from 3.13. that for $1 \leq p \leq \infty$ and $1 \leq \lambda < \infty$ a normed space E is called an $\mathcal{L}_{p,\lambda}^g$-space, if for each $M \in \text{FIN}(E)$ and $\varepsilon > 0$ there are $R \in \mathcal{L}(M, \ell_p^m)$ and $S \in \mathcal{L}(\ell_p^m, E)$ for some $m \in \mathbb{N}$ factoring the embedding I_M^E such that $\|S\|\,\|R\| \leq \lambda + \varepsilon$:

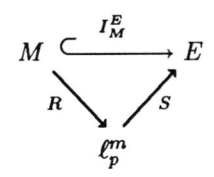

$$\begin{array}{ccc} M & \xrightarrow{I_M^E} & E \\ & R \searrow \quad \nearrow S & \\ & \ell_p^m & \end{array}$$

E is called an $\mathcal{L}_p^g$-space if it is an $\mathcal{L}_{p,\lambda}^g$ for some $\lambda \geq 1$. It has already been shown that all $C(K)$ and $L_\infty(\mu)$ are $\mathcal{L}_{\infty,1}^g$-spaces (see lemma 4.4.) and that all $L_p(\mu)$ are $\mathcal{L}_{p,1}^g$-spaces (Ex 4.7.). It is easy to see (Ex 4.7. and Ex 6.6.) that E is an $\mathcal{L}_{p,\lambda}^g$-space if and only if its completion $\tilde{E}$ is an $\mathcal{L}_{p,\lambda}^g$-space.

The local techniques for $\mathcal{L}_p^g$-spaces, i.e. passing from statements about the ℓ_p^n to all $\mathcal{L}_p^g$-spaces, are summed-up in

— the $\mathcal{L}_p$-local technique lemma about inequalities between tensor norms (see 13.5.).
— its generalization to the continuity of operators $id_{\mathcal{L}_p^g} \otimes T$ (see Ex 13.7.).
— the $\mathcal{L}_p$-local technique for minimal operator ideals (Ex 23.13.).
— and the following lemma for maximal operator ideals (which by corollary 2 of 22.3. also has a consequence for minimal ideals):

$\mathcal{L}_p$-LOCAL TECHNIQUE LEMMA FOR OPERATOR IDEALS: *Let $(\mathfrak{A}, \mathbf{A})$ and $(\mathfrak{B}, \mathbf{B})$ be quasi-Banach operator ideals, $(\mathfrak{B}, \mathbf{B})$ maximal, $c \geq 0$ and E a Banach space.*
(1) *If $\mathbf{B} \leq c\mathbf{A}$ on $\mathcal{L}(\ell_p^n, E)$ for all n, then $\mathfrak{A}(F, E) \subset \mathfrak{B}(F, E)$ and*

$$\mathbf{B}(T) \leq c\lambda \mathbf{A}(T)$$

for all $\mathcal{L}_{p,\lambda}^g$-spaces F and $T \in \mathfrak{A}(F, E)$.
(2) *If $\mathbf{B} \leq c\mathbf{A}$ on $\mathcal{L}(E, \ell_p^n)$ for all n, then $\mathfrak{A}(E, F) \subset \mathfrak{B}(E, F)$ and*

$$\mathbf{B}(T) \leq c\lambda \mathbf{A}(T)$$

for all $\mathcal{L}_{p,\lambda}^g$-spaces F and $T \in \mathfrak{A}(E, F)$.
(3) *If $\mathfrak{A}(\ell_p, \ell_q) \subset \mathfrak{B}(\ell_p, \ell_q)$ and $\mathbf{B} \leq c\mathbf{A}$ on $\mathfrak{A}(\ell_p, \ell_q)$, then $\mathfrak{A}(E, F) \subset \mathfrak{B}(E, F)$ and*

$$\mathbf{B}(T) \leq c\lambda\mu \mathbf{A}(T)$$

for all $\mathcal{L}_{p,\lambda}^g$-spaces E, $\mathcal{L}_{q,\mu}^g$-spaces F and $T \in \mathfrak{A}(E, F)$.

PROOF: Since $\mathfrak{B}$ is maximal, consider for $M \in \text{FIN}(F)$

$$M \xrightarrow{I} F \xrightarrow{T} E$$
$$R \searrow \quad \nearrow S \qquad \|R\|\,\|S\| \leq \lambda + \varepsilon ,$$
$$\ell_p^n$$

which gives (1). For (2) anticipate the dual characterization in corollary 4 below (23.4.):

$$E \xrightarrow{T} F \xrightarrow{Q} F/L$$
$$R \searrow \quad \nearrow S \qquad \|R\|\,\|S\| \leq \lambda + \varepsilon$$
$$\ell_p^n$$

for $L \in \text{COFIN}(F)$ — and again maximality gives the result. (3) is a consequence of (1) and (2). □

The ideal $(\mathcal{L}_p, \mathbf{L}_p)$ of p-factorable operators turned out to be the class of all operators $T \in \mathcal{L}(E, F)$ such that $\kappa_F \circ T$ factors through some $L_p(\mu)$; moreover,

$$\mathbf{L}_p(T) = \min\{\|R\|\,\|S\| \mid \kappa_F \circ T : E \xrightarrow{R} L_p \xrightarrow{S} F''\} .$$

This was obtained in 18.6. by ultraproduct techniques.

COROLLARY: *The assumptions of (3) imply the formulas*
$$\mathcal{L}_q \circ \mathfrak{A} \circ \mathcal{L}_p \subset \mathfrak{B} \quad \text{and} \quad \mathbf{B} \leq c\,(\mathbf{L}_q \circ \mathbf{A} \circ \mathbf{L}_p) .$$

PROOF: If $U \in \mathcal{L}_r$, then U'' factors through some $L_r(\mu)$. Take $T \circ S \circ R \in \mathcal{L}_q \circ \mathfrak{A} \circ \mathcal{L}_p$; then

$$E_1'' \xrightarrow{R''} E_2'' \xrightarrow{S''} E_3'' \xrightarrow{T''} E_4''$$
$$R_1 \searrow \nearrow R_2 \quad T_1 \searrow \nearrow T_2 \qquad \begin{array}{l}\|R_1\|\,\|R_2\| \leq \mathbf{L}_p(R)(1+\varepsilon)\\ \|T_2\|\,\|T_1\| \leq \mathbf{L}_q(T)(1+\varepsilon)\end{array}$$
$$L_p \qquad\qquad L_q$$

Since $\mathbf{B}(T_1 S'' R_2) \leq c\mathbf{A}(T_1 S'' R_2)$ by (3) of the proposition, it follows that $(TSR)''$, and hence TSR is in $\mathfrak{B}$ — with the desired norm estimate. □

In the spirit of this corollary, statement (1) of the proposition reads $\mathfrak{A} \circ \mathcal{L}_p(\cdot, E) \subset \mathfrak{B}(\cdot, E)$ and (2) $\mathcal{L}_p \circ \mathfrak{A}(E, \cdot) \subset \mathfrak{B}(E, \cdot)$ with the norm estimates.

23.2. The ideal $(\mathfrak{L}_p, \mathbf{L}_p)$ of p-factorable operators is associated with the accessible tensor norm w_p, the adjoint w_p^* of which is even totally accessible (21.5.). This implied that every $\mathfrak{L}_{p,\lambda}^g$-spaces has the λ-bounded approximation property (see 21.6. and also Ex 23.1.). Recall from Ex 9.8. that the space ideal space($\mathfrak{L}_p$) is the class of all Banach spaces such that $id_E \in \mathfrak{L}_p$.

THEOREM: *Let $1 \leq p \leq \infty$ and $1 \leq \lambda < \infty$. For each normed space E the following statements are equivalent:*

(a) *E is an $\mathfrak{L}_{p,\lambda}^g$-space.*
(b) *$id_{\tilde{E}}$ is p-factorable, i.e. $\tilde{E} \in$ space($\mathfrak{L}_p$), and $\mathbf{L}_p(id_{\tilde{E}}) \leq \lambda$.*
(c) *For all Banach spaces G (or only $G = E'$ or G some predual of E)*

$$w_p^* \leq \pi \leq \lambda w_p^* \qquad \text{on } G \otimes E.$$

(d) *For all Banach spaces G (or only G some Johnson space C_q)*

$$\varepsilon \leq w_p \leq \lambda \varepsilon \qquad \text{on } G \otimes E.$$

(e) *There is a factorization $id_{E''} = S \circ R$ through some $L_p(\mu)$ (where μ is strictly localizable) with $\|S\| \|R\| \leq \lambda$.*

PROOF: By what was already said in 23.1. and 13.4. one may assume E to be complete. The characterization of the accessible operator ideal $\mathfrak{L}_p$ in section 17 (see 17.15. and 17.16.) applied to $T = id_E$ gives the equivalences between (b), (c) and (d). The last statement means that $\mathbf{L}_p(id_{E''}) \leq \lambda$ (note that $\mathbf{L}_p(T) = min...$), so (e) is equivalent to (b) since $T \in \mathfrak{A}$ if and only if $T'' \in \mathfrak{A}$ for each maximal normed operator ideal $\mathfrak{A}$ (see 17.8.).
Since $G \otimes_\varepsilon \ell_p^n = G \otimes_{w_p} \ell_p^n$ (see 12.9.), the $\mathfrak{L}_p$-local technique lemma 13.5. shows that (a) implies (d). To see the converse take $M \in \text{FIN}(E)$ and denote by $z_M \in M' \otimes E$ the element associated with $I_M^E \in \mathfrak{F}(M, E)$. Then

$$w_p(z_M; M', E) \leq \lambda \varepsilon(z_M; M', E) = \lambda \|I_M^E\| = \lambda,$$

which, by the definition of w_p (see 17.10.), gives the desired factorization $I_M^E = S \circ R$ through some ℓ_p^m with $\|S\| \|R\| \leq \lambda + \varepsilon$. $\square$

The last calculation was in some sense the crucial point of the proof of the theorem.

COROLLARY 1: (1) *A normed space E is an $\mathfrak{L}_{p,\lambda}^g$-space if and only if its dual E' is an $\mathfrak{L}_{p',\lambda}^g$-space.*
(2) *Complemented subspaces of $\mathfrak{L}_p^g$-spaces are $\mathfrak{L}_p^g$-spaces; more precisely, if $G \subset E$ is ρ-complemented and E an $\mathfrak{L}_{p,\lambda}^g$-space, then G is an $\mathfrak{L}_{p,\rho\lambda}^g$-space.*

(3) E is an $\mathcal{L}_p^g$-space if and only if its bidual E'' is isomorphic to a complemented subspace of some $L_p(\mu)$. The bidual of an $\mathcal{L}_{p,1}^g$-space is isometric to a 1-complemented subspace of some $L_p(\mu)$.

PROOF: The relation $\mathcal{L}_{p'}^{dual} \sim w_{p'}^t = w_p \sim \mathcal{L}_p$ means: $id_E \in \mathcal{L}_p$ if and only if $(id_E)' = id_{E'} \in \mathcal{L}_{p'}$ (with the same norms, see 17.8.). Statement (2) is obvious from (c) or (d) of the theorem and the last statement follows from this and (e). □

In particular, the class of $\mathcal{L}_2^g$-spaces is exactly the class of spaces isomorphic to a Hilbert space (this is already known from 11.11. where this result was shown using the little Grothendieck theorem). Since $w_p \leq c_p w_2$ for all $1 < p < \infty$ for some $c_p \geq 1$ (see 12.8.), it follows that

COROLLARY 2: *Hilbert spaces are $\mathcal{L}_p^g$-spaces for all $1 < p < \infty$.*

However, they are neither $\mathcal{L}_\infty^g$- nor $\mathcal{L}_1^g$-spaces (if they are infinite dimensional) by the little Grothendieck theorem (11.11., corollary).

23.3. Lindenstrauss–Pełczyński [173] called a Banach space E an $\mathcal{L}_{p,\lambda}$-space if for each $M \in \text{FIN}(E)$ there is an $N \in \text{FIN}(E)$ with $M \subset N$ and the Banach–Mazur distance

$$d(N, \ell_p^{dimN}) \leq \lambda \ .$$

Obviously, $\mathcal{L}_{p,\lambda}$-spaces are $\mathcal{L}_{p,\lambda}^g$-spaces. Infinite dimensional Hilbert spaces fail to be $\mathcal{L}_{p,\lambda}$-spaces (for $p \neq 2$) since $d(\ell_2^n, \ell_p^n) = n^{|1/2 - 1/p|} \to \infty$ for $n \to \infty$ (see Ex 26.4. for a proof of this well-known fact); but this is the only difference: Lindenstrauss–Rosenthal [174], theorem 4.3., showed exactly the non-trivial implications of the following two statements:

$1 < p < \infty$: A Banach space is an $\mathcal{L}_p^g$-space if and only if it is an $\mathcal{L}_p$-space or isomorphic to a Hilbert space.

$p = 1$ or ∞ : A Banach space is an $\mathcal{L}_p^g$-space if and only if it is an $\mathcal{L}_p$-space.

However, the constants may vary! $\mathcal{L}_p^g$-spaces are much handier than $\mathcal{L}_p$-spaces; for example, it is not true that the $\mathcal{L}_{p'}$-constant of the dual of an $\mathcal{L}_{p,\lambda}$-space is $\leq \lambda$ since there are infinite dimensional $\mathcal{L}_{\infty,1}$-spaces, but no infinite dimensional $\mathcal{L}_{1,1}$-spaces! See Lindenstrauss–Tzafriri [176], p.199. It seems to be unknown whether or not the dual of an $\mathcal{L}_{p,\lambda}$ is an $\mathcal{L}_{p',\lambda+\varepsilon}$ for every $\varepsilon > 0$.

The Lindenstrauss–Rosenthal characterization of $\mathcal{L}_p^g$-spaces implies that $\mathcal{L}_p^g$-spaces E which are not isomorphic to a Hilbert space contain the ℓ_p^n uniformly: There are $M_n \in \text{FIN}(E)$ with $\sup d(M_n, \ell_p^n) < \infty$. For $p = 1, \infty$ this can be seen as follows: By local reflexivity (see explicitly corollary 6.6.) and theorem 23.2.(e) one may assume that E is complemented in some $L_p(\mu)$. If $p = \infty$, it follows from a result of Pełczyński

[197] that E contains a subspace isomorphic to c_o, which implies the result. But this observation is helpful also when $p = 1$: In this case it follows that the $\mathcal{L}_\infty^g$-space E' contains c_o, hence there is a surjection $E'' \twoheadrightarrow \ell_1$; this easily implies (by lifting the unit vectors) that E'' has a subspace isomorphic to ℓ_1.

COROLLARY 3: (1) *Infinite dimensional $\mathcal{L}_\infty^g$-spaces have neither proper type nor proper cotype.*

(2) *Infinite dimensional $\mathcal{L}_1^g$-spaces have no proper type, but do have cotype 2.*

(3) *$\mathcal{L}_p^g$-spaces have type $\min\{p,2\}$ and cotype $\max\{p,2\}$ for $1 \leq p < \infty$.*

(4) *space($\mathcal{L}_p$) $\neq$ space($\mathcal{L}_q$) if $p \neq q$ and $p,q \in [1,\infty]$.*

PROOF: By what was said above (1) follows from Ex 7.14. and (2) from Ex 7.13. and 8.6. — this latter reference also gives (3). These statements and Ex 8.13. give (4). □

In particular, no infinite dimensional space can be simultaneously an $\mathcal{L}_\infty^g$- and an $\mathcal{L}_1^g$-space.

23.4. A dual characterization of $\mathcal{L}_p^g$-spaces is the

COROLLARY 4: *A Banach space E is an $\mathcal{L}_{p,\lambda}^g$-space if and only if for every subspace $L \in \mathrm{COFIN}(E)$ and $\varepsilon > 0$ there is a factorization $Q_L^E = S \circ R$ through some ℓ_p^m with $\|S\| \|R\| \leq \lambda + \varepsilon$:*

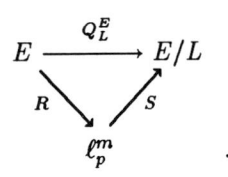

$$\begin{array}{ccc} E & \xrightarrow{Q_L^E} & E/L \\ & {}_R \searrow \quad \nearrow {}_S & \\ & \ell_p^m & \end{array}$$

PROOF: If E is an $\mathcal{L}_{p,\lambda}^g$, then its dual is an $\mathcal{L}_{p',\lambda}^g$. Thus, if $L \in \mathrm{COFIN}(E)$, there is a factorization

$$\begin{array}{ccc} (E/L)' = L^0 & \hookrightarrow & E' \\ & {}_{R_o} \searrow \quad \nearrow {}_{S_o} & \\ & \ell_{p'}^m & \end{array} \qquad \|R_o\| \|S_o\| \leq \lambda+\varepsilon,$$

hence $Q_L^E = R_o' \circ S_o' \circ \kappa_E$. Conversely, and in essentially the same way, the condition implies that E' is an $\mathcal{L}_{p',\lambda}^g$-space. □

23.5. In 20.14. it was shown that $\pi\backslash = w_\infty^*$ and $\varepsilon/ = w_1$; transposition yields

$$/\pi = w_1^* \quad \text{and} \quad \backslash\varepsilon = w_\infty .$$

Thus, theorem 23.2. gives a characterization of $\mathcal{L}_1^g$- and $\mathcal{L}_\infty^g$-spaces in terms of respecting subspaces and quotients; for $\mathcal{L}_1^g$ the result is due to Stegall-Retherford [259].

COROLLARY 5: *Let E be a Banach space.*
(1) E is an $\mathcal{L}_1^g$-space if and only if $E \otimes_\pi \cdot$ respects subspaces isomorphically. More precisely, E is an $\mathcal{L}_{1,\lambda}^g$-space if and only if for each $G \overset{1}{\hookrightarrow} F$ and $z \in E \otimes G$

$$\pi(z; E, G) \leq \lambda \pi(z; E, F) .$$

(2) E is an $\mathcal{L}_\infty^g$-space if and only if $E \otimes_\varepsilon \cdot$ respects quotients isomorphically. More precisely, E is an $\mathcal{L}_{\infty,\lambda}^g$-space if and only if for each $Q : F \twoheadrightarrow G$

$$\mathring{B}_{E\otimes_\varepsilon G} \subset \lambda \left[id \otimes Q(\mathring{B}_{E\otimes_\varepsilon F})\right] .$$

PROOF: That the conditions are necessary follows from the properties of $\pi\backslash$ and $\varepsilon/$. Conversely, observe that if $E \otimes_\varepsilon \cdot$ respects quotients, then, using $\ell_1(B_G) \twoheadrightarrow G$ (recall theorem 20.6.), there is a λ_G with $\varepsilon/ \leq \lambda_G \varepsilon$ on $E \otimes G$ — and a simple ℓ_p-sum-argument shows that the constant can be chosen independently from G. The inclusion about the unit balls gives as well $\varepsilon/ \leq \lambda \varepsilon$ on $E \otimes G$ for all G. Since $\varepsilon/ = w_\infty^t$, this implies (2). Similar arguments give the result for (1). □

Grothendieck's theorem 3.11. says that the $\mathcal{L}_{1,1}^g$-spaces are exactly the spaces of the form $L_1(\mu)$ for some strictly localizable measure μ. Moreover, it follows that $\mathcal{L}_1^g$-spaces share the interesting lifting property expressed in 3.12., corollary 1 — in particular, the lifting of compact operators, see Ex 3.31.. The result in 3.10. now reads that E is an $\mathcal{L}_1^g$-space (resp. an $\mathcal{L}_{1,\lambda}^g$-space) if and only if E' is an injective Banach space (resp. has the λ-extension property).

For $\mathcal{L}_\infty^g$-spaces this means (by the duality result in corollary 1) that *a dual space is $\mathcal{L}_{\infty,\lambda}^g$ if and only if it has the λ-extension property.* Moreover, as in 4.5., the fact that for an $\mathcal{L}_{\infty,\lambda}^g$-space E

$$\mathfrak{K}(G, E) = G' \tilde{\otimes}_\varepsilon E = G' \tilde{\otimes}_{\backslash\varepsilon} E$$

($\mathcal{L}_\infty^g$-spaces have the approximation property) shows that these spaces E have the *compact extension property* with constant λ : For every isometric embedding $G \hookrightarrow F$, every operator $T \in \mathfrak{K}(G, E)$ and $\varepsilon > 0$ there is an extension $\tilde{T} \in \mathfrak{K}(F, E)$ of T with $\|\tilde{T}\| \leq (\lambda+\varepsilon)\|T\|$. This, as well as its converse (see Ex 23.4.), is — in its spirit — due to Lindenstrauss [171].

Going to the completion, the corollary gives the following result due to Kaballo [134], which can be used for the lifting of vector valued functions, see also 4.5., corollary 1.

COROLLARY 6: *A Banach space E is an $\mathcal{L}^g_\infty$-space if and only if the completion $E\tilde{\otimes}_\varepsilon \cdot$ respects quotients isomorphically.*

PROOF: The quotient lemma 7.4. implies that $E\tilde{\otimes}_{\varepsilon/} \cdot$ respects quotients; so the condition is necessary. Conversely, take a metric surjection $\ell_1(\Gamma) \twoheadrightarrow C_2$ onto a Johnson space C_2. Then the condition implies that $E\tilde{\otimes}_\varepsilon / C_2 = E\tilde{\otimes}_\varepsilon C_2$ holds isomorphically and hence $\varepsilon/ \leq \lambda\varepsilon$ for all $E \otimes G$ by the very properties of C_2, see lemma 17.16.. □

23.6. It was shown that dual $\mathcal{L}^g_\infty$-spaces are injective. This is not true for arbitrary $\mathcal{L}^g_\infty$-spaces; take the $\mathcal{L}^g_\infty$-space c_o which is not complemented in ℓ_∞. Every injective space E has the λ-extension property with $\lambda = \|P\|$, where $P : \ell_\infty(B_{E'}) \to E$ is a projection (Ex 1.7.). Define the *projection constant* of a Banach space E to be

$$\lambda(E) := \inf\{\|P\| \mid P \in \mathcal{L}(\ell_\infty(B_{E'}), E) \text{ projection onto } E\} \in [0, \infty] .$$

PROPOSITION: $\mathbf{L}_\infty(id_{E'}) = \lambda(E')$ *for each Banach space E.*

PROOF: If P is such a projection, then $id_{E'} = P \circ I_{E'}^{\ell_\infty(B_{E''})}$ is a factorization through an $L_\infty(\mu)$, hence $\mathbf{L}_\infty(id_{E'}) \leq \|P\|$. Conversely, if $\lambda = \mathbf{L}_\infty(id_{E'})$, then, by what was said above, E' has the λ-extension property and therefore $\lambda(E') \leq \lambda$. □

Note that it was shown that E' has the $\lambda(E')$-extension property, in particular, $\lambda(E') = \min\{\|P\| | ...\}$.

Since the adjoint operator ideal $\mathcal{L}^*_\infty \sim w^*_\infty = g^*_\infty$ is the ideal $\mathfrak{P}_1$ of all absolutely 1-summing operators, it follows from 17.19. that for each $M \in$ FIN

$$\lambda(M)\mathbf{P}_1(id_M) \geq \dim M$$

and there is even equality if the norm on M satisfies certain invariance conditions (e.g. $M = \ell_p^n$). This gives a possibility of calculating projection constants; for the euclidean space ℓ_2^n see 11.10.. See also Ex 23.8. for the study of the extension of operators.

23.7. In order to obtain some operator characterizations of $\mathcal{L}^g_\infty$-spaces recall $\mathfrak{I} \sim \pi$ for the integral operators, $\mathfrak{I}^{min} = \mathfrak{N}$ and the representation theorem for minimal operator ideals. The diagram

$$\begin{array}{ccc} E'\tilde{\otimes}_{\pi\backslash}F & \xrightarrow{\;\;1\;\;} & E'\tilde{\otimes}_\pi \ell_\infty(B_{F'}) \\ \downarrow & & \downarrow = \\ (\mathfrak{I}^{inj})^{min}(E,F) & \hookrightarrow & \mathfrak{N}(E,\ell_\infty(B_{F'})) \end{array}$$

commutes, and also the left vertical map is injective since $\pi\backslash$ is totally accessible by 21.1.(2). Arguing as in 5.9. or Ex 5.13. with the Hahn-Banach theorem, one obtains

$$(\mathfrak{J}^{inj})^{min}(E,F) = \mathfrak{N}(E, \ell_\infty(B_{F'})) \cap \mathfrak{L}(E,F) = \mathfrak{N}^{inj}(E,F)$$

if E' has the approximation property. In 25.11. a much more general result will be obtained, showing that this also holds if F has the approximation property. By the characterization of quasinuclear operators (see Ex 9.13.) this means exactly that:

LEMMA: *If E' or F has the approximation property, then*

$$E' \tilde{\otimes}_{\pi\backslash} F = \mathfrak{QN}(E,F)$$

holds isometrically.

The following characterization is due to Stegall-Retherford [259].

PROPOSITION: *For each Banach space E the following statements are equivalent:*
(a) *E is an $\mathfrak{L}_\infty^g$-space.*
(b) *$\mathfrak{QN}(E,F) = \mathfrak{N}(E,F)$ for all Banach spaces F (or only C_2).*
(c) *$\mathfrak{QN}(F,E) = \mathfrak{N}(F,E)$ for all Banach spaces F (or only C_2).*

PROOF: If E is an $\mathfrak{L}_\infty^g$, its dual E' is an $\mathfrak{L}_1^g$-space; since $\pi\backslash = w_1^{*t}$, theorem 23.2. implies that $\pi\backslash$ and π are equivalent on $E' \otimes F$, which implies (b) by the lemma. Conversely, (b) gives $\pi\backslash \approx \pi$ on $E' \otimes C_2$, which shows (17.16.) that $\pi \leq \lambda\pi\backslash$ for some λ on all $E' \otimes F$. This implies that E' is an $\mathfrak{L}_1^g$-space (by the theorem) and E is an $\mathfrak{L}_\infty^g$-space.
That (a) implies (c) is a consequence of the fact that $\alpha \approx \alpha\backslash$ on $F' \otimes E$ for all tensor norms α if E is an $\mathfrak{L}_\infty^g$-space (see 20.8.). For the converse, it follows that (again by the lemma) $\pi \approx \pi\backslash$ on $E' \otimes E$. The theorem implies that E' is an $\mathfrak{L}_1^g$-space. □

In Ex 23.9. the reader will find a characterization of $\mathfrak{L}_p^g$-spaces in terms of p-compact operators $\mathfrak{K}_p = \mathfrak{L}_p^{min}$.

23.8. The preceding characterizations were in terms of minimal operator ideals. To obtain a characterization with p'-dominated operators (recall $\mathfrak{L}_p^* = \mathfrak{D}_{p'}$) the following lemma, which is a special case of the much more general quotient formula to be proven in section 25, will help.

LEMMA: *Let $\mathfrak{A}$ be an accessible maximal normed operator ideal. Then for each Banach space E the following statements are equivalent:*
(a) *$E \in \text{space}(\mathfrak{A})$.*

(b) $\mathfrak{A}^*(E,F) = \mathfrak{I}(E,F)$ *for all Banach spaces F.*
(c) $\mathfrak{A}^*(F,E) = \mathfrak{I}(F,E)$ *for all Banach spaces F.*

PROOF: The associated tensor norm α of $\mathfrak{A}$ is accessible. Since $\mathfrak{A}^* \circ \mathfrak{A} \subset \mathfrak{I}$ and $\mathfrak{A} \circ \mathfrak{A}^* \subset \mathfrak{I}$ by 17.19., statement (a) implies the others. Assume (b), then

$$(G \otimes_\alpha E)' = \mathfrak{A}^*(E, G') \hookrightarrow \mathfrak{I}(E, G') = (G \otimes_\varepsilon E)'$$

is continuous and hence $G \otimes_\varepsilon E \to G \otimes_\alpha E$, which implies that $id_E \in \mathfrak{A}$, see 17.16.. If (c) is satisfied, use the fact that α^* is accessible and C_2 has the metric approximation property, to see that the diagram

$$\begin{array}{ccc} C_2 \otimes_{\alpha^*} E & \xrightarrow{1} & \mathfrak{A}^*(C_2', E) \\ \downarrow & & \downarrow \\ C_2 \otimes_\pi E & \xrightarrow{1} & \mathfrak{I}(C_2', E) \end{array}$$

proves the continuity of $C_2 \otimes_{\alpha^*} E \to C_2 \otimes_\pi E$, which implies $id_E \in \mathfrak{A}$ by lemma 17.16. and theorem 17.15.. $\square$

It is clear from 17.19. that, for example, $\mathbf{I}(T) \leq \mathbf{A}(id_E)\mathbf{A}^*(T)$ in (b). Since $\mathfrak{L}_p^* = \mathfrak{D}_{p'}$, the p'-dominated operators, it follows that:

PROPOSITION: *For a Banach space E and $1 \leq p \leq \infty$ the following are equivalent:*

(a) *E is an $\mathfrak{L}_p^g$-space.*
(b) $\mathfrak{D}_{p'}(E,F) = \mathfrak{I}(E,F)$ *for all Banach spaces F.*
(c) $\mathfrak{D}_{p'}(F,E) = \mathfrak{I}(F,E)$ *for all Banach spaces F.*

Recall that $\mathfrak{D}_1 = \mathfrak{P}_1$ and $\mathfrak{D}_\infty = \mathfrak{P}_1^{dual}$.

23.9. $\mathfrak{L}_1^g$-spaces enjoying the Radon–Nikodým property are rare:

PROPOSITION: *Let the $\mathfrak{L}_1^g$-space E be complemented in E''. Then E has the Radon–Nikodým property if and only if it is isomorphic to some $\ell_1(\Gamma)$.*

PROOF: Observe first that $\ell_1(\Gamma)$ has the Radon–Nikodým property (see Appendix D3.). Conversely, theorem 23.2.(e) implies that E is isomorphic to a complemented subspace G of some $L_1(\mu)$. By proposition 2 of Appendix D2.(Lewis–Stegall theorem), the projection $P : L_1(\mu) \to G$ factors through some $\ell_1(\Gamma_o)$:

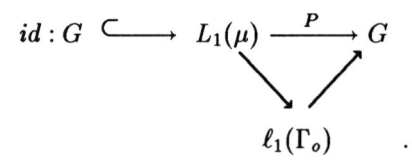

$$id: G \hookrightarrow L_1(\mu) \xrightarrow{P} G$$
$$\searrow \quad \nearrow$$
$$\ell_1(\Gamma_o)$$

This shows that G (and hence E) is isomorphic to a complemented subspace of $\ell_1(\Gamma_o)$ and, as a consequence, has the lifting property (see 3.12.). It follows that E is isomorphic to some $\ell_1(\Gamma)$ by Köthe's result mentioned in 3.12.. □

This result and the following interesting consequence are due to Lewis and Stegall [170].

COROLLARY 1: *The dual E' of a Banach space E is isomorphic to some $\ell_1(\Gamma)$ if and only if*
$$\mathfrak{P}_1(E, F) = \mathfrak{N}(E, F)$$
for all Banach spaces F.

PROOF: It follows from 23.8. and 23.2. that $\mathfrak{P}_1(E, \cdot) = \mathfrak{I}(E, \cdot)$ if and only if E' is an $\mathcal{L}_1^g$-space. Appendix D8. gives that $\mathfrak{I}(E, F') = \mathfrak{N}(E, F')$ for all F if and only if E' has the Radon–Nikodým property. This, together with the proposition and 5.9., easily implies the result. □

The Hardy spaces H^p are $\mathcal{L}_p^g$-spaces for $1 < p < \infty$ since they are complemented in $L_p[0, 2\pi]$ (see Ex 4.7.). Pełczyński [201] showed that H^∞ is not a quotient of any $C(K)$. Since H^∞ is a dual space ([202], p.11), it cannot be $\mathcal{L}_\infty^g$ by theorem 23.2.(e).

COROLLARY 2: *The Hardy space H^∞ is not an $\mathcal{L}_\infty^g$-space, the Hardy space H^1 is not an $\mathcal{L}_1^g$-space.*

PROOF: It is known that H^1 is a dual space ([202], p.12) and separable, and hence has the Radon–Nikodým property. If it were an $\mathcal{L}_1^g$-space, it would be isomorphic to ℓ_1 (by the proposition), hence weak and strong convergence of sequences would coincide: The sequence $(\exp(in\cdot))$ converges weakly to zero by the Riemann–Lebesgue lemma, but each function has norm 1, a contradiction. □

23.10. Recall from 17.14. Grothendieck's theorem
$$\mathfrak{L}(\ell_1, \ell_2) = \mathfrak{P}_1(\ell_1, \ell_2) \quad \text{and} \quad \mathbf{P}_1(T) \leq K_G \|T\| \ .$$
By the $\mathcal{L}_p$-local technique lemma 23.1. for operator ideals this can easily be extended:

GROTHENDIECK'S THEOREM: *If E is an $\mathcal{L}_{1,\lambda}^g$-space and H a Hilbert space, then every $T \in \mathfrak{L}(E, H)$ is absolutely 1-summing and*
$$\mathbf{P}_1(T) \leq K_G \lambda \|T\| \ .$$

This improves the little Grothendieck theorem, which stated that $T \in \mathfrak{P}_2(E, H)$. The same argument as before shows that the first theorem in 17.14. reads as follows for $\mathcal{L}_p^g$-spaces:

THEOREM: *Let E be an $\mathcal{L}_{\infty,\lambda}^g$- and F an $\mathcal{L}_{1,\mu}^g$-space. Then every $T \in \mathcal{L}(E, F)$ is 2-dominated and (hence) absolutely 2-summing; moreover,*

$$\mathbf{P}_2(T) \leq \mathbf{D}_2(T) \leq K_G \lambda \mu \|T\| .$$

In the next section it will be shown that for $1 \leq p \leq 2$ the sequence space ℓ_p is isometrically contained in some $L_1(\mu)$ — the Lévy embedding. Since $\mathfrak{P}_2$ is injective,

$$T: \ell_\infty \longrightarrow \ell_p \hookrightarrow L_1(\mu)$$

implies that $\mathfrak{L}(\ell_\infty, \ell_p) = \mathfrak{P}_2(\ell_\infty, \ell_p)$ and hence, by the $\mathcal{L}_p$-local technique for operators:

COROLLARY 1: *If E is an $\mathcal{L}_{\infty,\lambda}^g$- and F an $\mathcal{L}_{p,\mu}^g$-space for $1 \leq p \leq 2$, then*

$$\mathcal{L}(E, F) = \mathfrak{P}_2(E, F) \quad \text{and} \quad \mathbf{P}_2(T) \leq K_G \lambda \mu \|T\| .$$

Clearly, the result also has a tensor norm formulation, see Ex 23.15.. Note that $\mathfrak{D}_2^{inj} = \mathfrak{P}_2$ (with equivalent norms) since $w_2^* \backslash \sim g_2 = g_2^*$ (see the table in 20.15.). See Ex 25.5. and Ex 25.6. for more information about this result. Proposition 20.19. gives the "adjoint" form:

COROLLARY 2: *If E and F are Banach spaces, E an $\mathcal{L}_{p,\lambda}^g$-space for $1 \leq p \leq 2$, then*

$$\mathfrak{P}_2(E, F) = \mathfrak{P}_1(E, F) \quad \text{and} \quad \mathbf{P}_1(T) \leq K_G \lambda \mathbf{P}_2(T) .$$

Corollary 1 implies, in particular, that every operator $\mathcal{L}_\infty^g \to \mathcal{L}_p^g$ (for $1 \leq p \leq 2$) factors through a Hilbert space. More generally, it will be shown in 26.1. that every operator $\mathcal{L}_q^g \to \mathcal{L}_p^g$ (with $1 \leq p \leq 2 \leq q \leq \infty$) has this property. A nice consequence of this result is a characterization, due to Grothendieck, of Hilbert spaces.

COROLLARY 3: *Let $1 \leq p \leq 2 \leq q \leq \infty$. A Banach space E is isomorphic to a Hilbert space if and only if it is isomorphic to a quotient of an $\mathcal{L}_q^g$-space and isomorphic to a subspace of an $\mathcal{L}_p^g$-space.*

PROOF: If $T: \mathcal{L}_q^g \twoheadrightarrow E \hookrightarrow \mathcal{L}_p^g$, then T factors through a Hilbert space; this easily implies that E is isomorphic to a Hilbert-space. Conversely, since ℓ_2^n is an isometric subspace of some $L_p(\mu_n)$ (for example with Gauss functions), the ultraproduct technique gives that every Hilbert space H is isometric to a subspace of an ultraproduct of L_p, which is an L_p by 18.4. (this was Ex 18.7.(c)). Dualizing the embedding $H \hookrightarrow L_{q'}(\mu)$ gives $L_q(\mu) \twoheadrightarrow H$. □

Exercises:

Ex 23.1. $\mathcal{L}_p^g$-spaces have the metric approximation property if $1 < p < \infty$. Hint: Reflexivity.

Ex 23.2. There is no constant c such that $w_\infty \leq c w_2$. Hint: End of 23.2..

Ex 23.3. A Banach space is an $\mathcal{L}_\infty^g$-space if and only if its bidual is isomorphic to a complemented subspace of some $C(K)$, K compact.

Ex 23.4. A Banach space E is an $\mathcal{L}_{\infty,\lambda}^g$-space if (and only if) it has the compact extension property with the constant λ (see 23.5.). Hint: $C_2 \hookrightarrow \ell_\infty(\Gamma)$, proof of corollary 6.

Ex 23.5. (a) Every injective Banach space is an $\mathcal{L}_\infty^g$-space. Hint: Ex 1.7..

(b) A Banach space is an $\mathcal{L}_\infty^g$-space if and only if its bidual is injective.

(c) E is an $\mathcal{L}_\infty^g$-space if and only if for every metric injection $G \xrightarrow{1} F$ and $T \in \mathcal{L}(G, E)$ there is an extension $\tilde{T} \in \mathcal{L}(F, E'')$ of $\kappa_E \circ T$. See also Ex 3.20..

Ex 23.6. Let E and F be normed spaces and

$$\sigma_1(z; E, F) := \inf \sup \left\{ \sum_{k=1}^n |\langle x' \otimes y', x_k \otimes y_k \rangle| \mid x' \in B_{E'}, y' \in B_{F'} \right\},$$

the infimum being taken over representations $z = \sum_{k=1}^n x_k \otimes y_k$. Show that σ_1 is a finitely generated tensor norm $\leq w_1$ and that $\sigma_1 \leq \lambda \varepsilon$ on $E \otimes F$ if E is an $\mathcal{L}_{\infty,\lambda}^g$-space. This tensor norm is due to Matos (see also [2]). The associated maximal operator ideal consists of those operators $T : E \longrightarrow F$ such that $\kappa_F \circ T$ factors through a Banach lattice (see Pietsch [214], 23.2.9. and 23.3.4.).

Ex 23.7. E is an $\mathcal{L}_p^g$-space if and only if $\mathcal{L}_p(E, \cdot) = \mathcal{L}(E, \cdot)$ and if and only if $\mathcal{L}_p(\cdot, E) = \mathcal{L}(\cdot, E)$.

Ex 23.8. Take E and F finite dimensional Banach spaces and $I : E \hookrightarrow \ell_\infty(\Gamma)$ a metric injection. For $T \in \mathcal{L}(E, F) = E' \otimes F$ define the *extension norm* $\lambda(T)$ of T by

$$\lambda(T) := \inf \left\{ \|\tilde{T}\| \mid \tilde{T} \in \mathcal{L}(\ell_\infty(\Gamma), F), \tilde{T} \circ I = T \right\}.$$

(a) $\lambda(T)$ is independent of the special injection I and Γ; moreover,

$$\|T\| \leq \lambda(T) \leq \lambda(id_E)\|T\|.$$

The number $\lambda(E) := \lambda(id_E)$ is the *projection constant* of E, defined in 23.6..

(b) $\lambda(T) = \backslash\varepsilon(z_T; E', F)$ if $z_T \in E' \otimes F$ represents T.

(c) The following trace duality holds:

$$\lambda(T) = \sup \left\{ |tr (S \circ T)| \mid \mathbf{P}_1(S : F \to E) \leq 1 \right\}$$
$$\mathbf{P}_1(T) = \sup \left\{ |tr (S \circ T)| \mid \lambda(S : F \to E) \leq 1 \right\}.$$

These results are due to Garling–Gordon [79].

Ex 23.9. Recall from 22.3. that $E' \tilde{\otimes}_{w_p} F \twoheadrightarrow \mathfrak{K}_p(E,F) = \mathfrak{L}_p^{min}(E,F)$ is a representation of the p–compact operators. Show that for each Banach space E the following statements are equivalent:

(a) E is $\mathfrak{L}_p^g$–space.
(b) $\mathfrak{K}_p(E,F) = \overline{\mathfrak{F}}(E,F)$ for all Banach spaces F.
(c) $\mathfrak{K}_p(F,E) = \overline{\mathfrak{F}}(F,E)$ for all Banach spaces F.

In both statements $\overline{\mathfrak{F}}$ can be replaced by $\mathfrak{K}$. Hint: $w_{p'}^t = w_p$ and the ideas of 23.7.. The result is due to Johnson [129].

Ex 23.10. Every separable dual space isomorhic to a complemented subspace of an $L_1(\mu)$ is isomorphic to ℓ_1. Hint: 23.9..

Ex 23.11. The dual E' of a Banach space E is isomorphic to some $\ell_1(\Gamma)$ if and only if the following two conditions are satisfied:

(a) For all metric surjections $Q: F \to G$ the mapping

$$id_E \otimes Q : E \otimes_\varepsilon F \longrightarrow E \otimes_\varepsilon G$$

is a quotient mapping.

(b) For all Banach spaces F the canonical mapping

$$E' \tilde{\otimes}_\pi F' \longrightarrow (E \otimes_\varepsilon F)'$$

is surjective. Hint: $\backslash \varepsilon = w_\infty$, 16.5., and 23.9..

Ex 23.12. Apply the Chevet–Persson–Saphar inequalities 15.10. to show that for Banach spaces E and F the following holds:

(a) $\mathfrak{P}_{p'}(E,F) \subset \mathfrak{J}_{p'}^{dual}(E,F)$ if E is an $\mathfrak{L}_p^g$–space
(b) $\mathfrak{P}_p^{dual}(E,F) \subset \mathfrak{J}_p(E,F)$ if F is an $\mathfrak{L}_p^g$–space.

For a converse see 25.9..

Ex 23.13. $\mathfrak{L}_p$–*local technique for minimal operator ideals:* Let $(\mathfrak{A}, \mathbf{A})$ and $(\mathfrak{B}, \mathbf{B})$ be quasi–Banach operator ideals, $(\mathfrak{A}, \mathbf{A})$ minimal and normed. Then the statements (1), (2) and (3) of the $\mathfrak{L}_p$–local technique lemma 23.1. hold. Note that $\mathfrak{B}$ was not supposed to be maximal. Hint: Use the representation theorem for minimal operator ideals with $\alpha \sim \mathfrak{A}^{max}$.

Ex 23.14. For $S \in \mathfrak{L}(L_p(\mu), L_p(\nu))$ (with $\|S\| = 1$) consider the maximal operator ideal $\mathfrak{L}_S := \mathfrak{L}_S^{\Delta_p, \Delta_p}$; see 17.4.. Show that every $\mathfrak{L}_p^g$–space is in space($\mathfrak{L}_S$). In particular, if $1 < p < \infty$, each Hilbert space is in space($\mathfrak{L}_S$).
Hint: $L_p(\mu) \in$ space($\mathfrak{L}_S$), see Ex 7.8.. For a more general result see 26.3..

Ex 23.15. Show that $\pi \leq K_G \lambda \mu d_2$ on $E \otimes F$ if E is an $\mathfrak{L}_{\infty,\lambda}^g$– and F an $\mathfrak{L}_{p,\mu}^g$–space, $2 \leq p \leq \infty$. Hint: 23.10..

Ex 23.16. Let α be a tensor norm.

(a) If E and F are $\mathcal{L}_1^g$-spaces with constants λ and μ, then
$$\alpha \leq \backslash\alpha/ \leq \lambda\mu\alpha \qquad \text{on } E \otimes F.$$

(b) If E and F are $\mathcal{L}_\infty^g$-spaces with constants λ and μ, then
$$/\alpha\backslash \leq \alpha \leq \lambda\mu/\alpha\backslash \qquad \text{on } E \otimes F.$$

(c) If E_i are $\mathcal{L}_1^g$- and F_i are $\mathcal{L}_\infty^g$-spaces, $T_i \in \mathcal{L}(E_i, F_i)$, then
$$T_1 \otimes T_2 : E_1 \otimes_\varepsilon E_2 \longrightarrow F_1 \otimes_\pi F_2$$

is continuous. Hint: $w_2 \leq w_2^*$.

Ex 23.17. (a) $L_p(\mu) \otimes_{\Delta_p} L_p(\nu)$ is an $\mathcal{L}_{p,1}^g$-space. Hint: $\mu \otimes \nu$.
(b) E is an $\mathcal{L}_p^g$-space if and only if $E \hookrightarrow E''$ factors through an $\mathcal{L}_p^g$-space.
(c) $E_1 \otimes_{g_p} E_2$ is an $\mathcal{L}_p^g$-space if the E_i are. Hint: Use Δ_p and 13.2.. See also Ex 34.2..
(d) If E is an $\mathcal{L}_{p'}^g$- and F an $\mathcal{L}_{p'}^g$-space, then the space $\mathfrak{P}_p(E, F')$ of absolutely p–summing operators is an $\mathcal{L}_p^g$-space.

Ex 23.18. Show that $w_2^* \leq K_G \lambda\mu\varepsilon$ on the tensor product $E \otimes F$ of an $\mathcal{L}_{1,\lambda}$-space and an $\mathcal{L}_{1,\mu}^g$-space. Deduce that every 2-factorable operator from an $\mathcal{L}_1^g$ into an $\mathcal{L}_\infty^g$-space is integral; for this, see also the operator form 20.17. of Grothendieck's inequality.

24. Stable Measures

In this section embeddings $\ell_p \hookrightarrow L_q(\mu)$ for an appropriate measure μ will be studied. These embeddings exist only if $p = q$ (trivial) or $p = 2$ or $1 \leq q < p < 2$. The p-stable measures/distributions due to P. Lévy, which are so important for the central limit theorems in probability theory, will be used for the embeddings; Marcinkiewicz–Zygmund in 1939, and later on Kadec, noted that p-stable measures may serve for the construction of the desired embeddings. As was already seen in the improvement of Grothendieck's inequality in 23.10., these Lévy embeddings may be useful; in the final part of this section their importance will become self–evident through results of Saphar, Kwapień and Maurey about absolutely q-summing maps.

24.1. If γ is the Gaussian measure on $\mathbb{K}^\infty$, studied in 8.7., the "Gauss–Khintchine" equality in proposition 8.7. says that for all $0 < q < \infty$ (the same proof for $0 < q \leq 1$)

and all n

$$\left(\int_{K^\infty}\left|\sum_{k=1}^n \alpha_k g_k\right|^q d\gamma\right)^{1/q} = \left(\int_{K^n}\left|\sum_{k=1}^n \alpha_k g_k\right|^q d\gamma_n\right)^{1/q} = \|g_1\|_{q,K}\left(\sum_{k=1}^n |\alpha_k|^2\right)^{1/2}.$$

This means that $I(\alpha_k) := \sum_{k=1}^\infty \alpha_k g_k$ defines a map I from ℓ_2 into the γ-measurable functions such that

$$\|Ix\|_{L_q(\gamma)} = c_{2,q}\|x\|_{\ell_2} \qquad\qquad 1 \le q < \infty$$

for all $x \in \ell_2$. In particular, for every $q \in [1, \infty[$ there is a measure μ (a multiple of γ) and an isometric embedding $\ell_2 \hookrightarrow L_q(\mu)$. Note that the Rademacher functions give an isomorphic (not isometric) embedding of ℓ_2 into $L_q(\mu)$ via the Khintchine inequality. What about the embedding of ℓ_p into $L_q(\mu)$ for some measure μ if $p \in [1,\infty]\setminus\{2,q\}$? If the separable space ℓ_p is isomorphic to a subspace of $L_q(\mu)$, one may clearly assume that μ is separable and σ-finite — and, hence, even finite. Using the Carathéodory–Halmos–von Neumann theorem (every non-atomic separable finite measure is isomorphic to the Lebesgue measure on some interval, see Brown–Pearcy [18], p.181), one obtains that ℓ_p is isomorphic to a subspace of $L_q[0,1] \oplus_q \ell_q \overset{1}{\hookrightarrow} L_q[0,1]$.

PROPOSITION: *Let $p, q \in [1, \infty[$ and $q \ne p \ne 2$. If $q > p$ or $p > 2$, then the sequence space ℓ_p is not isomorphic to a subspace of $L_q(\mu)$ for any measure μ.*

PROOF: Since type and cotype respect subspaces and ℓ_p does not have type $\min\{2, q\}$ if $p < 2$ and $p < q$, and does not have cotype $\max\{2, q\}$ if $p > 2$ and $q < p$ (see 8.6. and Ex 8.13.), only the case $2 < p < q$ remains. An old result of Banach ([4], p.185) says that ℓ_r is not isomorphic to a subspace of ℓ_s for different $r, s \in]1, \infty[$. But Kadec–Pełczyński [137] showed for $2 < q < \infty$ that a closed subspace E of $L_q[0,1]$ is either isomorphic to ℓ_2 or contains ℓ_q isomorphically — so the last case is also impossible. □

Clearly, this proposition concerns the comparability of the linear dimension (in the sense of Banach) of ℓ_p and $L_q[0,1]$.

24.2. In the remaining case $1 \le q < p < 2$ there is even an isometric embedding of ℓ_p into some $L_q(\mu)$! It seems that the only way to show this is using Lévy's stable measures, which can be obtained as Fourier transforms of certain functions. The result which will be presented is implicitly due to Lévy; the interpretation of Lévy's result in terms of embeddings of ℓ_p into L_q was observed by Marcinkiewicz–Zygmund [182] in 1939 — and, more explicitly, by Kadec [136] in 1958 (see also the survey of Pełczyński–Bessaga [205], §12, for more information about this type of question). A continuous function $f : \mathbb{R}^m \to \mathbb{C}$ is called *positive definite* if

$$\sum_{i,j=1}^\ell \xi_i \bar{\xi}_j f(x_i - x_j) \ge 0$$

for all $\ell \in \mathbb{N}$, scalars $\xi_i \in \mathbb{C}$ and $x_i \in \mathbb{R}^m$. If $f_w, w \in \Omega$, are positive definite functions and ν is a (positive) measure on Ω, then it is obvious that the function f defined by

$$f(x) = \int_\Omega f_w(x)\nu(dw) \qquad \text{for } x \in \mathbb{R}^m$$

is positive definite if these integrals exist. In particular, the Fourier transform of a finite measure μ on $\mathbb{R}^m$

$$\mathfrak{F}(\mu)[x] := \int_{\mathbb{R}^m} \exp(i\langle w, x \rangle)\mu(dw)$$

is positive definite (as usual in this context, we omit the constant $(2\pi)^{-m/2}$ in the Fourier transform). Using Choquet's representation theory one can show the converse (see Choquet [28], vol.II, p.260):

BOCHNER'S THEOREM: *Every positive definite function on $\mathbb{R}^m$ is the Fourier transform of a finite Borel-Radon measure on $\mathbb{R}^m$.*

The set $\mathfrak{P}$ of positive definite functions on $\mathbb{R}^m$ has excellent stability properties:
(1) As was already mentioned: $\int_\Omega f_w d\nu \in \mathfrak{P}$ if all $f_w \in \mathfrak{P}$; in particular, $\mathfrak{P}$ is stable under forming positive linear combinations.
(2) If $f_n \in \mathfrak{P}$ and $f_n \to f$ pointwise, then $f \in \mathfrak{P}$.
(3) If $f, g \in \mathfrak{P}$, then $f \cdot g \in \mathfrak{P}$.
The last statement follows from the fact that the Schur product $(a_{i,j} b_{i,j})$ of two positive definite matrices $(a_{i,j})$ and $(b_{i,j})$ is positive definite (see Ex 24.1.). Series expansion and (1) – (3) imply:
(4) If $f \in \mathfrak{P}$, then $\exp \circ f \in \mathfrak{P}$.
The aim is to show that the function

$$f_p(\cdot) := \exp(-\|\cdot\|_2^p)$$

on $\mathbb{R}^m$ is positive definite for $0 < p \leq 2$, where $\|\ \|_2$ is the euclidean norm on $\mathbb{R}^m$; Bochner's theorem will then supply the measures to be used. For $p = 2$

$$\exp(-\|x\|_2^2) = \mathfrak{F}\left((4\pi)^{-m/2} \exp(-\frac{\|\cdot\|_2^2}{4})\right)[x]$$

so $\exp(-\|\cdot\|_2^2)$ is the Fourier transform of the measure having Lebesgue density $(4\pi)^{-m/2} \exp(-\|\cdot\|_2^2/4)$, which is (up to scaling) the Gaussian measure. For $p < 2$ the argument is not as direct: First observe that

$$\cos\langle w, x\rangle = \frac{1}{2}[\exp(i\langle w, x\rangle) + \exp(-i\langle w, x\rangle)] = \mathfrak{F}\left(\frac{\delta_w + \delta_{-w}}{2}\right)[x]$$

(Dirac measures) implies that $\cos\langle w, \cdot\rangle \in \mathfrak{P}$ for all $w \in \mathbb{R}^m$.

LEMMA 1: *For all $x \in \mathbb{R}^m$ and $0 < q < 2$*

$$\|x\|_2^q = K_{q,m}^{-1} \int_{\mathbb{R}^m} \frac{1 - \cos\langle w, x\rangle}{\|w\|_2^{q+m}} dw,$$

where

$$K_{q,1} := \int_{-\infty}^{+\infty} \frac{1 - \cos s}{|s|^{q+1}} ds$$

and, for $m \geq 2$,

$$K_{q,m} := K_{q,1} \cdot \sigma_{m-2} \int_0^\infty (1+r^2)^{-(q+m)/2} r^{m-2} dr$$

(σ_{m-2} is the area of the unit sphere in $\mathbb{R}^{m-1}$).

This is an easy calculation which begins with an isometry T satisfying $x = \|x\| T e_1$. Note that the formula does not hold for $q = 2$. The result (it will also be used for the next lemma) is crucial for the

PROPOSITION: *The function $f_p(\cdot) := \exp(-\|\cdot\|_2^p)$ on $\mathbb{R}^m$ is positive definite for each $0 < p \leq 2$.*

PROOF: For $p = 2$ this was already shown, hence take $0 < p < 2$. Since

$$-\|x\|_2^p = \lim_{\varepsilon \to 0} \left[K_{p,m}^{-1} \int_{\|w\|_2 \geq \varepsilon} \frac{\cos\langle w, x\rangle}{\|w\|_2^{p+m}} dw - K_{p,m}^{-1} \int_{\|w\|_2 \geq \varepsilon} \|w\|_2^{-p-m} dw \right],$$

properties (1), (4) and (2) of $\mathfrak{P}$ imply the result — the last term just gives a non-negative multiplicative constant. □

Obviously the function $-\|\cdot\|_2^p$ fails to be positive definite.

24.3. Bochner's theorem gives that, for $0 < p \leq 2$, there is a finite measure ν_p^m on $\mathbb{R}^m$ with $\mathfrak{F}(\nu_p^m) = f_p$. Since f_p and all its products with polynomials are Lebesgue integrable, it follows that $\mathfrak{F}^{-1}(f_p)$ is an infinitely differentiable function — the density of ν_p^m with respect to the Lebesgue measure on $\mathbb{R}^m$; moreover, the density $\mathfrak{F}^{-1}(f_p)$ is symmetric and satisfies

$$1 = f_p(0) = \int_{\mathbb{R}^m} \mathfrak{F}^{-1}(f_p)(w) dw = \nu_p^m(\mathbb{R}^m).$$

It follows that ν_p^m is a symmetric probability measure on $\mathbb{R}^m$.

LEMMA 2: For $0 < q < p < 2$ the q-th moment of ν_p^m exists; more precisely,
$$c_{p,q,m} := \int_{\mathbb{R}^m} \|x\|_2^q \nu_p^m(dx) = \frac{\sigma_{m-1}}{p \cdot K_{q,m}} \int_0^\infty (1 - \exp(-r)) r^{-(q+p)/p} dr < \infty.$$

PROOF: By lemma 1
$$\int_{\mathbb{R}^m} \|x\|_2^q \nu_p^m(dx) = \int_{\mathbb{R}^m} K_{q,m}^{-1} \int_{\mathbb{R}^m} \frac{1 - \cos\langle w, x \rangle}{\|w\|_2^{q+m}} dw \, \nu_p^m(dx) =$$
$$= K_{q,m}^{-1} \int_{\mathbb{R}^m} \frac{1}{\|w\|_2^{q+m}} \int_{\mathbb{R}^m} (1 - \cos\langle w, x \rangle) \nu_p^m(dx) dw = K_{q,m}^{-1} \int_{\mathbb{R}^m} \frac{1 - \exp(-\|w\|_2^p)}{\|w\|_2^{q+m}} dw$$

since ν_p^m is a probability measure and $\exp(-\|w\|_2^p) = \operatorname{Re}(\mathfrak{F}(\nu_p^m)[x])$. The rest is a standard calculation. □

Note that the calculation shows that the p-th moment of ν_p^m does not exist!

24.4. Only the cases $m = 1, 2$ will be needed from now on (but see Ex 24.3.) and $\mathbb{R}^2$ will be interpreted as $\mathbb{C}$. In this way the real and complex case will be treated jointly: for $0 < p < 2$ there is a probability measure μ_p^1 on $\mathbb{K}$ with Fourier transform $\exp(-|\cdot|^p)$ whose q-th moments exist for $0 < q < p$. By μ_p^n the product measure on $\mathbb{K}^n$ and by μ_p the product measure on $\mathbb{K}^\infty$ will be denoted — as was done in the case of the Rademacher and Gauss functions in section 8. The k-th projection $\mathbb{K}^\infty \to \mathbb{K}$ will be called h_k (to avoid the g_k from section 8 which referred to Gaussian stochastic variables).

LEMMA 3: Let $\alpha_1, ..., \alpha_n \in \mathbb{K}$ and $\sum_{k=1}^n |\alpha_k|^p = 1$. Then the image measure of μ_p via
$$\sum_{k=1}^n \alpha_k h_k : \mathbb{K}^\infty \longrightarrow \mathbb{K}$$
is the measure μ_p^1.

In the language of probability theory this means: The stochastic variables h_k on the probability space $(\mathbb{K}^\infty, \mu_p)$ (with values in $\mathbb{K}$) are stochastically independent, identically distributed (their distribution is μ_p^1) and $\sum_{k=1}^n \alpha_k h_k$ has the same distribution μ_p^1 if $\sum_{k=1}^n |\alpha_k|^p = 1$. This is why the measure μ_p^1 is called p-stable. For the present purposes it will be convenient to call the infinite product μ_p the p-stable Lévy measure on $\mathbb{K}^\infty$.

PROOF: Call ν the image measure of μ_p via $\sum_{k=1}^n \alpha_k h_k$; then
$$\int_{\mathbb{K}} f(t) \nu(dt) = \int_{\mathbb{K}^\infty} f\left(\sum_{k=1}^n \alpha_k h_k\right) d\mu_p = \int_{\mathbb{K}^n} f\left(\sum_{k=1}^n \alpha_k w_k\right) \mu_p^n(dw).$$

By the injectivity of the Fourier transform one has to show that $\mathfrak{F}(\nu) = \mathfrak{F}(\mu_p^1)$. Take $z \in \mathbb{K}$ (and note that $\operatorname{Re}(z_1\bar{z}_2)$ is the scalar product of $\mathbb{C} = \mathbb{R}^2$). Then

$$\int_K \exp[i \operatorname{Re}(tz)]\nu(dt) = \int_{K^n} \exp\left[i \operatorname{Re}\left(z\sum_{k=1}^n \alpha_k w_k\right)\right]\mu_p^n(dw) =$$

$$= \int_K \cdots \int_K \prod_{k=1}^n \exp[i \operatorname{Re}(z\alpha_k w_k)]\mu_p^1(dw_1)\ldots\mu_p^1(dw_n) = \prod_{k=1}^n \exp(-|z\alpha_k|^p) =$$

$$= \exp\left(-\sum_{k=1}^n |\alpha_k z|^p\right) = \exp(-|z|^p) = \int_K \exp[i \operatorname{Re}(zt)]\mu_p^1(dt)$$

by the defining property of μ_p^1. $\square$

Everything is prepared for proving the main result of this section:

THEOREM (Lévy): *For $0 < p < 2$ let μ_p be the p-stable Lévy measure on $\mathbb{K}^\infty$. Then the operator I on ℓ_p defined by $I((\alpha_k)) := \sum_{k=1}^\infty \alpha_k h_k$ maps ℓ_p into $L_q(\mathbb{K}^\infty, \mu_p)$ for all $0 < q < p$ and there are constants $c_{p,q}$ such that*

$$c_{p,q}\|\alpha\|_{\ell_p} = \|I(\alpha)\|_{L_q(\mathbb{K}^\infty,\mu_p)} = \left(\int_{\mathbb{K}^\infty} |\langle\alpha,w\rangle|^q \mu_p(dw)\right)^{1/q}$$

for all $\alpha \in \ell_p$. The constant

$$c_{p,q} := \left(\int_K |x|^q \mu_p^1(dx)\right)^{1/q}$$

is $(c_{p,q,1})^{1/q}$ in the real and $(c_{p,q,2})^{1/q}$ in the complex case, where $c_{p,q,m}$ is the q-th moment of ν_p^m.

The mapping I will be called *Lévy embedding*.

PROOF: Take $\alpha = (\alpha_1, \ldots, \alpha_n) \in \mathbb{K}^n$ with $\|\alpha\|_{\ell_p} = 1$; by lemma 3 the measure μ_p^1 is the image measure of μ_p via $\sum_{k=1}^n \alpha_k h_k$. Therefore, lemma 2 gives

$$c_{p,q}^q = \int_K |t|^q \mu_p^1(dt) = \int_{\mathbb{K}^\infty} \left|\sum_{k=1}^n \alpha_k h_k\right|^q d\mu_p = \|I(\alpha)\|_{L_q(\mu_p)}^q.$$

This relation obviously implies the result. $\square$

24.5. The following consequence is obvious:

COROLLARY 1: *For $0 < q < p < 2$ or: $p = 2$ and $0 < q < \infty$ there are a finite measure μ and an isometry $I : \ell_p \hookrightarrow L_q(\mu)$.*

For $p = 2$ this was already stated at the beginning of this section with the help of the Gauss measure. It is interesting to note that the map I cannot be positive, see Ex 24.2.. For the tensor norms w_p one obtains the

COROLLARY 2: *If* $1 \leq q < p < 2$ *or:* $p = 2$ *and* $1 \leq q < \infty$, *then* $w_q \backslash \leq w_p$.

PROOF: The isometry $\ell_p^n \hookrightarrow L_q$ implies $\mathbf{L}_q^{inj}(id_{\ell_p^n}) = 1$. Recall that $\mathfrak{L}_q^{inj} \sim w_q \backslash$, take $z \in M \otimes N$ and factor

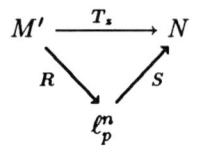

$$M' \xrightarrow{T_z} N$$
$$R \searrow \quad \nearrow S$$
$$\ell_p^n$$

Then $w_q \backslash (z; M, N) = \mathbf{L}_q^{inj}(T_z) \leq \|R\| \|S\| \mathbf{L}_q^{inj}(id_{\ell_p^n}) \leq \|R\| \|S\|$. Passing to the infimum over all such factorizations gives the result by 17.10.. □

COROLLARY 3: *If* E *is an* $\mathfrak{L}_{p,\lambda}^g$*-space and* $1 \leq q < p < 2$ *or:* $p = 2$ *and* $1 \leq q < \infty$, *then there are a Borel-Radon measure* μ *and an operator* $I : E \longrightarrow L_q(\mu)$ *such that*

$$\|x\|_E \leq \|Ix\|_{L_q} \leq \lambda \|x\|_E .$$

PROOF: Since $\mathbf{L}_q^{inj} \leq \mathbf{L}_p$ by corollary 2, it follows that, by theorem 23.2. on the characterization of $\mathfrak{L}_p^g$-spaces, $\mathbf{L}_q^{inj}(id_E) \leq \mathbf{L}_p(id_E) \leq \lambda$. By the definition of $\mathfrak{L}_q^{inj}$ and the factorization theorem 18.6. for q-factorable operators there is a factorization

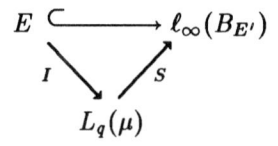

$$E \hookrightarrow \ell_\infty(B_{E'})$$
$$I \searrow \quad \nearrow S$$
$$L_q(\mu)$$

with $\|I\| \leq \lambda$ and $\|S\| = 1$. This gives the result. □

In particular, for every measure ν and $1 \leq q < p < 2$ (or $p = 2$ and $1 \leq q < \infty$) there is a measure μ such that $L_p(\nu)$ is isometrically embedded into $L_q(\mu)$. It is easy to derive this result directly from the Lévy theorem with the canonical embedding of a space into the ultraproduct of its finite dimensional subspaces (see 18.5.). The case $p = 2$ (Hilbert spaces) was already settled in Ex 18.7.(c).

24.6. Lévy embeddings have interesting consequences for tensor norms and operator ideals. The following result is due to Saphar [242].

PROPOSITION: *For $1 \leq q < p < 2$ or: $p = 2$ and $1 \leq q < \infty$ the relation*

$$\pi\backslash \leq c_{p,1}^{-1} c_{p,q} g'_{q'}$$

holds on $E \otimes \ell_p$ for all Banach spaces E.

PROOF: If I denotes the Lévy embedding $\ell_p \hookrightarrow L_q := L_q(\mu_p)$, then $c_{p,q}^{-1} I$ is an isometry. This means that for $z \in E \otimes \ell_p \hookrightarrow E \otimes L_q$

$$\alpha\backslash(z; E, \ell_p) = c_{p,q}^{-1} \alpha\backslash(z; E, L_q)$$
$$\alpha(z; E, L_q) \leq c_{p,q} \alpha(z; E, \ell_p)$$

for all tensor norms α. Since $\Delta_q \leq g_{q'}^*$ (Chevet–Persson–Saphar inequality 15.10.) and μ_p is a probability measure, it follows that

$$\pi\backslash(z; E, \ell_p) = c_{p,1}^{-1} \pi\backslash(z; E, L_1) \leq c_{p,1}^{-1} \pi(z; E, L_1) = c_{p,1}^{-1} \Delta_1^t(z; E, L_1) \leq$$
$$\leq c_{p,1}^{-1} \Delta_q^t(z; E, L_q) \leq c_{p,1}^{-1} g'_{q'}(z; E, L_q) \leq c_{p,1}^{-1} c_{p,q} g'_{q'}(z; E, \ell_p) . \square$$

Note that the proof not only used embeddings $\ell_p \hookrightarrow L_q$ but also that, for fixed p, the *same* embedding can be used for all $q < p$! Since $\pi = \pi\backslash$ on $\ell_1 \otimes \ell_p$, it follows that

$$\pi \leq c_{p,1}^{-1} c_{p,q} g'_{q'} \quad \text{on } \ell_1 \otimes \ell_p .$$

Recall the "little Grothendieck theorem" 11.11.: $\mathfrak{L}(\ell_1, \ell_2) = \mathfrak{P}_2(\ell_1, \ell_2) = \mathfrak{I}_2(\ell_1, \ell_2)$. The proposition implies the following generalization:

COROLLARY: *If $2 < p < q \leq \infty$ or: $p = 2$ and $1 < q \leq \infty$, then*

$$\mathfrak{L}(\ell_1, \ell_p) = \mathfrak{P}_q(\ell_1, \ell_p) = \mathfrak{I}_q(\ell_1, \ell_p) .$$

PROOF: Since $\mathfrak{I}_q \subset \mathfrak{P}_q$ and $\mathfrak{I}_q \sim g_q$, it follows that

$$\mathfrak{I}_q(\ell_1, \ell_p) = (\ell_1 \otimes_{g'_q} \ell_{p'})' = (\ell_1 \otimes_\pi \ell_{p'})' = \mathfrak{L}(\ell_1, \ell_p) . \square$$

The $\mathfrak{L}_p$-local technique 23.1. clearly implies (for p, q as in the corollary) that

$$\mathfrak{L}(E, F) = \mathfrak{P}_q(E, F) = \mathfrak{I}_q(E, F)$$

whenever E is an $\mathfrak{L}_{1,\lambda}^g$- and F an $\mathfrak{L}_{p,\mu}^g$-space, and in this case

$$\mathbf{P}_q(T) \leq \mathbf{I}_q(T) \leq c_{p',1}^{-1} c_{p',q} \lambda \mu \|T\| .$$

Note that $\mathfrak{L}(\ell_1, \ell_2) = \mathfrak{P}_1(\ell_1, \ell_2)$ by Grothendieck's theorem, but $\mathfrak{I}(\ell_1, \ell_2) = \mathfrak{N}(\ell_1, \ell_2) \neq \mathfrak{L}(\ell_1, \ell_2)$. See Ex 24.6. and Ex 25.7. for some consequences of this result.

24.7. As it is the case in the little Grothendieck theorem (i.e. $\mathfrak{L}(E,\ell_2) = \mathfrak{P}_2(E,\ell_2)$ for $E = \ell_1$ or ℓ_∞) the last corollary also holds for ℓ_∞ instead of ℓ_1. This will also be shown using Lévy embeddings; the range space will be even more general than ℓ_p, namely of cotype p. Recall from Ex 8.12. and Ex 11.20. the definition of the Banach ideal $(\mathfrak{P}_{r,s}, \mathbf{P}_{r,s})$ of absolutely (r,s)–summing maps ($1 \leq s \leq r < \infty$). In 8.9. it was shown that for all Banach spaces F and elements $x_1, \ldots, x_n \in F$

$$\left(\int_D \Big\|\sum_{k=1}^n \varepsilon_k x_k\Big\|_F^2 d\mu\right)^{1/2} \leq w_1(x_k; F) .$$

Hence, the definition of cotype q gives the

REMARK: *The identity map id_F of every Banach space F of cotype $q \in [2, \infty[$ is absolutely $(q,1)$–summing, and $\mathbf{P}_{q,1}(id_F) \leq \mathbf{C}_q(F)$.*

The crucial step is to show that $\mathfrak{P}_{q,1}(\ell_\infty, \cdot) \subset \mathfrak{P}_{q+\varepsilon}(\ell_\infty, \cdot)$.

LEMMA: *For $1 < s \leq r < \infty$ and $T \in \mathfrak{L}(\ell_\infty^n, F)$*

$$\mathbf{P}_{r,s}(T) = \sup\Big\{\sum_{j=1}^n \Big(\sum_{k=1}^m |\langle y_k', Te_j\rangle|^{s'}\Big)^{1/s'} \,\Big|\, \ell_{r'}(y_k'; F') \leq 1 \Big\} .$$

PROOF: By definition (see Ex 11.20.) and duality

$$\mathbf{P}_{r,s}(T) = \|id \otimes T : \ell_s \otimes_\varepsilon \ell_\infty^n \longrightarrow \ell_r \otimes_{\Delta_r} F\|$$
$$= \|id \otimes T' : \ell_{r'} \otimes_{\Delta_{r'}} F' \longrightarrow \ell_{s'} \otimes_\pi \ell_1^n\|$$
$$= \sup\Big\{\pi\Big(\sum_{k=1}^m e_k \otimes T'y_k'; \ell_{s'}, \ell_1^n\Big) \,\Big|\, \ell_{r'}(y_k') \leq 1\Big\} .$$

Hence, the conclusion follows from

$$\sum_{k=1}^m e_k \otimes T'y_k' = \sum_{k=1}^m e_k \otimes \sum_{j=1}^n \langle e_j, T'y_k'\rangle e_j = \sum_{j=1}^n \Big(\sum_{k=1}^m \langle Te_j, y_k'\rangle e_k\Big) \otimes e_j$$

and the calculation of the projective norm in $\cdot \otimes \ell_1^n$. $\square$

Now cleverly chosen Lévy embeddings (in the form of Ex 24.8.(b)) give the

PROPOSITION: *Let $1 < s < r < p < \infty$. Then for every Banach space F*

$$\mathfrak{P}_{r,s}(\ell_\infty, F) \subset \mathfrak{P}_p(\ell_\infty, F) .$$

As a consequence, this inclusion holds for every $\mathcal{L}^g_\infty$-space E instead of ℓ_∞.

PROOF: It is enough to show that $\mathbf{P}_p(T) \leq c\mathbf{P}_{r,s}(T)$ for all $T \in \mathfrak{L}(\ell^n_\infty, F)$ — with a constant independent of n. Fix $0 < q < 2/s'$ and define

$$0 < q < p_o := qp' < r_o := qr' < s_o := qs' < 2 \ .$$

To apply the lemma take $y'_1, ..., y'_m \in F'$. Then

$$\sum_{j=1}^n \left(\sum_{k=1}^m |\langle y'_k, Te_j\rangle|^{p'}\right)^{1/p'} = \sum_{j=1}^n \left(\sum_{k=1}^m |\langle y'_k, Te_j\rangle|^{p_o/q}\right)^{q/p_o} \leq$$

(using Ex 24.8.(b) with $q < p_o < s_o$)

$$\leq \sum_{j=1}^n c_{p_o,q}^{-q} \int_{K^\infty} \left(\sum_{k=1}^m \||\langle y'_k, Te_j\rangle|^{1/q} h_k|^{s_o}\right)^{q/s_o} d\mu_{p_o} =$$

$$= c_{p_o,q}^{-q} \int_{K^\infty} \sum_{j=1}^n \left(\sum_{k=1}^m |\langle |h_k(w)|^q y'_k, Te_j\rangle|^{s'}\right)^{1/s'} \mu_{p_o}(dw) =$$

$$\leq c_{p_o,q}^{-q} \mathbf{P}_{r,s}(T) \int_{K^\infty} \ell_{r'}\left(|h_k(w)|^q y'_k; F'\right) \mu_{p_o}(dw) =$$

$$= \mathbf{P}_{r,s}(T) c_{p_o,q}^{-q} \int_{K^\infty} \left(\sum_{k=1}^m (|h_k(w)| \|y'_k\|^{1/q})^{r_o}\right)^{q/r_o} \mu_{p_o}(dw) \ .$$

Applying Ex 24.8.(b) once more, this time with $q < p_o < r_o$, it follows that the last line is

$$\leq \mathbf{P}_{r,s}(T) c_{r_o,p_o}^q c_{r_o,q}^{-q} \ell_{p_o}(\|y'_k\|^{1/q})^q =$$

$$\leq \mathbf{P}_{r,s}(T) c_{r_o,p_o}^q c_{r_o,q}^{-q} \left(\sum_{k=1}^m \|y'_k\|^{p_o/q}\right)^{q/p_o} \ .$$

Since $p_o/q = p'$, the foregoing lemma implies

$$\mathbf{P}_p(T) \leq c\mathbf{P}_{r,s}(T) \ ,$$

with a constant c, depending only on (s, r, p). □

One can show (see Pietsch [214], 22.6.4., where the proof just presented was taken from) that

$$\bar{c}_{p,r} := \lim_{q \downarrow 0} c^q_{r_o,p_o} \cdot c^{-q}_{r_o,q}$$

exists and obviously can be taken to be the constant c in the last inequality. Everything is prepared for the following result due to Maurey [186] — who in this way generalized a result of Kwapień's [159]:

THEOREM: *If F is a Banach space of cotype q and $\varepsilon > 0$, then*

$$\mathfrak{L}(\ell_\infty, F) = \mathfrak{P}_{q+\varepsilon}(\ell_\infty, F)$$

and for every Banach space E

$$\mathfrak{P}_{(q+\varepsilon)'}(F, E) = \mathfrak{P}_1(F, E) .$$

PROOF: By Saphar's duality result in 20.19. it is enough to prove the first statement. Take $\varepsilon > 0$ and choose $1 < s < q < r < q+\varepsilon$ which satisfy $1 - 1/q \leq 1/s - 1/r$. Then, by Ex 11.21. and the proposition

$$\mathfrak{P}_{q,1}(\ell_\infty, F) \subset \mathfrak{P}_{r,s}(\ell_\infty, F) \subset \mathfrak{P}_{q+\varepsilon}(\ell_\infty, F) ;$$

now $id_F \in \mathfrak{P}_{q,1}$ (by the remark) concludes the proof. □

Note that for $q \in [2, \infty[$ the space ℓ_q has cotype q (see 8.6.), hence

$$\mathfrak{L}(\ell_\infty, \ell_q) = \mathfrak{P}_{q+\varepsilon}(\ell_\infty, \ell_q) = \mathfrak{I}_{q+\varepsilon}(\ell_\infty, \ell_q)$$

for all $\varepsilon > 0$. For $q = 2$ the little Grothendieck theorem allows even $\varepsilon = 0$; that this is also true for arbitrary range spaces of cotype 2 will be shown in section 31.

24.8. We end this section with a remark concerning averaging. It is not too difficult to see that the average with respect to the Lévy measure dominates the Rademacher average: "stable" type implies Rademacher type (see Ex 24.5.). The converse is true up to an ε :

PROPOSITION: *Let $1 \leq s < q < p \leq 2$ and E be a Banach space of type p. Then*

$$\left(\int_{K^\infty} \Big\|\sum_{k=1}^n h_k x_k\Big\|_E^s d\mu_q\right)^{1/s} \leq c\, \ell_q(x_k)$$

for all $x_1, ..., x_n \in E$ and $c := \mathbf{T}_p(E) c_{p,s}^{-1} c_{q,s} c_{p,q}$.

PROOF: Take μ the Rademacher measure on $D = \{-1, 1\}^N$ and ε_k the projection (see 8.5.). Since the Lévy measure μ_q^1 is symmetric, one has

$$\int_{K^\infty} \Big\|\sum_{k=1}^n h_k x_k\Big\|^s d\mu_q = \int_D \int_{K^\infty} \Big\|\sum_{k=1}^n \varepsilon_k h_k x_k\Big\|^s d\mu_q d\mu .$$

Using the type p inequality and the Lévy embedding, one obtains

$$\left(\int_{K^\infty}\left\|\sum_{k=1}^n h_k x_k\right\|^s d\mu_q\right)^{1/s} = \left(\int_{K^\infty}\int_D\left\|\sum_{k=1}^n \varepsilon_k h_k x_k\right\|^s d\mu d\mu_q\right)^{1/s} \leq$$

$$\leq \left(\int_{K^\infty}\left(\int_D\left\|\sum_{k=1}^n \varepsilon_k h_k x_k\right\|^2 d\mu\right)^{s/2} d\mu_q\right)^{1/s} \leq$$

$$\leq T_p(E)\left(\int_{K^\infty}\left(\sum_{k=1}^n |h_k(w)|\|x_k\|^p\right)^{s/p} d\mu_q(w)\right)^{1/s} =$$

$$= T_p(E)c_{p,s}^{-1}\left(\int_{K^\infty}\int_{K^\infty} |\langle v,(\|h_k(w)x_k\|)\rangle|^s \mu_p(dv)\mu_q(dw)\right)^{1/s}.$$

Changing the order of integration and applying again the Lévy theorem, the relation $\langle v,(h_k(w)\|x_k\|)\rangle = \langle w,(h_k(v)\|x_k\|)\rangle$ shows that the last line is

$$\leq T_p(E)c_{p,s}^{-1}c_{q,s}\left(\int_{K^\infty}\left(\sum_{k=1}^n |h_k(v)|\|x_k\|^q\right)^{s/q} \mu_p(dv)\right)^{1/s} \leq$$

$$\leq T_p(E)c_{p,s}^{-1}c_{q,s}\left(\sum_{k=1}^n \int_{K^\infty} |h_k(v)|^q \mu_p(dv)\|x_k\|^q\right)^{1/q} =$$

$$\leq T_p(E)c_{p,s}^{-1}c_{q,s}c_{p,q}\left(\sum_{k=1}^n \|x_k\|^q\right)^{1/q}$$

by the definition of $c_{p,q}$. $\square$

Exercises:

Ex 24.1. A matrix $A \in \mathcal{L}(\mathbb{C}^n,\mathbb{C}^n)$ is called positive definite if $(Ax|x) \geq 0$ for all $x \in \mathbb{C}^n$. Use diagonalization of A to show that the *Schur product* $(a_{i,j}b_{i,j})$ of two positive definite matrices $(a_{i,j}),(b_{i,j}) \in \mathcal{L}(\mathbb{C}^n,\mathbb{C}^n)$ is positive definite. Hint: $a_{i,j} = \sum_{\ell=1}^n \lambda_\ell(e_i|y_\ell)\overline{(e_j|y_\ell)}$.

Ex 24.2. If $1 \leq q < p < \infty$, then there does not exist any *positive* isomorphism $\ell_p \longrightarrow L_q(\mu)$. Hint: 11.2. and $\mathbf{P}_q(\ell_\infty^n \hookrightarrow \ell_p^n) \geq n^{1/q}$.

Ex 24.3. Show that for every $m \in \mathbb{N}$ and $0 < q < p < 2$ there is a probability measure μ and an isometric embedding $\ell_p(\ell_2^m) \hookrightarrow L_q(\mu)$. Hint: Follow 24.4..

Ex 24.4. Formulate and prove proposition 24.8. for operators of type p.

Ex 24.5. (a) Show for $p, s \in]0, 2]$ that for the Rademacher measure μ and the p-stable Lévy measure μ_p the averages satisfy

$$\left(\int_D \Big\|\sum_{k=1}^n \varepsilon_k x_k\Big\|_E^s d\mu\right)^{1/s} \leq c_{p,1}^{-1}\left(\int_{K^\infty} \Big\|\sum_{k=1}^n h_k x_k\Big\|_E^s d\mu_p\right)^{1/s}$$

for all $x_1, ..., x_n \in E$. Hint: As in Ex 8.9..

(b) A Banach space has *stable type* $q \in [1, 2]$ if it satisfies the inequalitiy of proposition 24.8. for some $s \in [1, 2]$. Show with Kahane's inequality 8.6. for the Rademacher averages that stable type q implies type q.

Ex 24.6. (a) If $2 < p < q < \infty$ or: $p = 2$ and $1 < q < \infty$, then every compact operator from an $\mathcal{L}_1^g$-space into an $\mathcal{L}_p^g$-space is q-nuclear. Hint: 24.6. and 22.4..

(b) Let $2 < p < q < \infty$. Then every compact operator from an $\mathcal{L}_\infty^g$-space into an $\mathcal{L}_p^g$-space is q-nuclear. Hint: 24.7. and 22.4..

Ex 24.7. Let $I : \ell_2 \hookrightarrow L_p(\gamma)$ be the embedding $Ie_k := g_k$ and denote for the complex case $I_o e_k := \bar{g}_k$. Take $p, q \in]1, \infty[$ with $1/p + 1/q \geq 1$.

(a) Show that the mapping $\ell_2 \xrightarrow{I} L_{q'}(\gamma) \xrightarrow{id} L_p(\gamma) \xrightarrow{I_o'} \ell_2$ is just $\|g_1\|_{2,K}^2 id_{\ell_2}$.

(b) $d_{p,q} := \|g_1\|_{2,K}^{-2} \|g_1\|_{q',K} \|g_1\|_{p',K}$ satisfies

$$\mathbf{L}_{p,q}(id_{\ell_2}) \leq d_{p,q} \quad \text{and} \quad \alpha_{p,q} \leq d_{p,q} w_2 .$$

It can be shown that this constant is actually the best one (see Pietsch [214], 22.1.4. and 22.1.1.) and that $d_{p,q} = b_{p'} b_{q'}$ (see 12.8.) in the real case. With the formulas given in Ex 11.24.(a) for the moments of the Gauss measure one can show, however, that in the complex case in general $d_{p,q} < b_{p'} b_{q'}$ (take $p = 2$ and $q = 4/3$ for an example).

Ex 24.8. (a) Take $p, q \in]0, 2[$ and $r < \min\{p, q\}$. Then for all $\alpha_1, ..., \alpha_n \in \mathbb{K}$

$$c_{q,r}\left(\int_{K^\infty} \Big(\sum_{k=1}^n |\alpha_k h_k|^q\Big)^{r/q} d\mu_p\right)^{1/r} = c_{p,r}\left(\int_{K^\infty} \Big(\sum_{k=1}^n |\alpha_k h_k|^p\Big)^{r/p} d\mu_q\right)^{1/r} .$$

Hint: Express the first integrand in terms of μ_q, the second in terms of μ_p and use Fubini's theorem.

(b) For $0 < r < q < p < 2$ and $\alpha = (\alpha_1, ..., \alpha_n) \in \mathbb{K}^n$ prove the inequalities

$$\|\alpha\|_{\ell_q} \leq c_{q,r}^{-1}\left(\int_{K^\infty} \Big(\sum_{k=1}^n |\alpha_k h_k|^p\Big)^{r/p} d\mu_q\right)^{1/r} \leq c_{p,q} c_{p,r}^{-1} \|\alpha\|_{\ell_q} .$$

Hint: $\|\alpha\|_{\ell_q} = (\sum_k (\int |\alpha_k h_k|^r d\mu_p)^{q/r})^{1/q} c_{p,r}^{-1}$ and the continuous triangle inequality.

Ex 24.9. Let $q \in [2, \infty[$ and $\varepsilon > 0$. Use theorem 24.7. to show that $\mathfrak{L}(\ell_1, \ell_q) = \mathfrak{P}_{q+\varepsilon}(\ell_1, \ell_q)$ (which is the statement of corollary 24.6.). Hint: The dual of the Lévy embedding into L_1 and lifting.

25. Composition of Accessible Operator Ideals

The main topic in this section is an extremely useful result concerning the composition of operator ideals: If $\mathfrak{A} \circ \mathfrak{B} \subset \mathfrak{C}$, then — under certain accessibility conditions — $\mathfrak{C}^* \circ \mathfrak{A} \subset \mathfrak{B}^*$ holds. Various applications of this result are given. It turns out that the theory of accessible operator ideals is quite smooth.

25.1. It will be crucial to exploit the accessibility of operator ideals, defined in section 21: A quasi–Banach operator ideal was called left-accessible if (in brief) each operator $T \in \mathfrak{L}(E, N)$ factors

$$E \xrightarrow{T} N, \quad S: E \to E/L \to N \qquad \mathbf{A}(S) \leq (1+\varepsilon)\mathbf{A}(T)$$

i.e. $\mathfrak{A}(E, N) \stackrel{1}{=} \mathfrak{A} \circ \mathfrak{F}(E, N)$, and right-accessible if

$$M \xrightarrow{T} F, \quad S: M \to N \to F \qquad \mathbf{A}(S) \leq (1+\varepsilon)\mathbf{A}(T)$$

i.e. $\mathfrak{A}(M, F) \stackrel{1}{=} \mathfrak{F} \circ \mathfrak{A}(M, F)$. The aim of the first part of this section is to give a characterization of accessibility in terms of the minimal kernel $\mathfrak{A}^{min}$ of $\mathfrak{A}$.

LEMMA: *Let $(\mathfrak{A}, \mathbf{A})$ be a p–Banach operator ideal.*
(1) *For each Banach space E the following are equivalent:*
  (a) $\mathfrak{F} \circ \mathfrak{A}(E, F) \stackrel{1}{=} \mathfrak{A}^{min}(E, F)$ *for all Banach spaces F.*
  (b) $\mathfrak{A} \circ \mathfrak{F}(E, N) \stackrel{1}{=} \mathfrak{A}(E, N)$ *for all $N \in$ FIN.*
(2) *For each Banach space F the following are equivalent:*
  (a) $\mathfrak{A} \circ \mathfrak{F}(E, F) \stackrel{1}{=} \mathfrak{A}^{min}(E, F)$ *for all Banach spaces E.*
  (b) $\mathfrak{F} \circ \mathfrak{A}(M, F) \stackrel{1}{=} \mathfrak{A}(M, F)$ *for all $M \in$ FIN.*

PROOF: Only (1) will be proven, (2) follows along the same lines. Assume (a); then

$$\mathfrak{A} \circ \mathfrak{F}(E, N) \stackrel{1}{=} \mathfrak{F} \circ \mathfrak{A} \circ \mathfrak{F}(E, N) \stackrel{1}{=} \mathfrak{A}^{min}(E, N) \stackrel{1}{=} \mathfrak{F} \circ \mathfrak{A}(E, N) \stackrel{1}{=} \mathfrak{A}(E, N)$$

— the first and last equalities being obvious consequences of $N \in \text{FIN}$. Conversely, take $T \in \overline{\mathfrak{F}} \circ \mathfrak{A}(E,F)$; it is necessary to show that $T \in \mathfrak{A}^{min}$ with the same norm. There is a factorization

$$E \xrightarrow{T} F \qquad R \searrow \nearrow S \qquad G \qquad \mathbf{A}(R)\|S\| \leq (1+\varepsilon)[\|\ \|\circ \mathbf{A}](T)$$

with $S \in \overline{\mathfrak{F}}$. For $S_n \in \mathfrak{F}(G,F)$ with $\|S-S_n\| \to 0$ define $T_n := S_n \circ R$ and fix $n, m \in \mathbb{N}$. For $N := (S_n - S_m)(G) \in \text{FIN}(F)$, by (b), $\mathfrak{A}^{min}(E,N) \stackrel{1}{=} \mathfrak{A} \circ \overline{\mathfrak{F}}(E,N) \stackrel{1}{=} \mathfrak{A}(E,N)$ and therefore,

$$\mathbf{A}^{min}(T_n - T_m : E \longrightarrow F) \leq \mathbf{A}^{min}((S_n - S_m)R : E \longrightarrow N) =$$
$$\stackrel{1}{=} \mathbf{A}((S_n - S_m)R : E \longrightarrow N) \leq \|S_n - S_m\|\mathbf{A}(R) \ .$$

It follows that (T_n) is an $\mathbf{A}^{min}$-Cauchy sequence, which clearly converges in $\mathfrak{L}$ to T and hence in $\mathbf{A}^{min}$; moreover, since $\mathfrak{A}^{min}$ is q-normed for some q (see 9.10.),

$$\mathbf{A}^{min}(T) = \lim_{n \to \infty} \mathbf{A}^{min}(T_n) \leq \lim_{n \to \infty} \|S_n\|\mathbf{A}(R) = \|S\|\mathbf{A}(R) \leq$$
$$\leq (1+\varepsilon)[\|\ \|\circ \mathbf{A}](T) \ ,$$

which shows that $\mathbf{A}^{min} \leq [\|\ \|\circ \mathbf{A}]$. The converse inequality is trivial. □

25.2. Corollary 16.9. about the factorization of finite rank operators in spaces with the metric approximation property gives immediately the

COROLLARY: *Let $(\mathfrak{A}, \mathbf{A})$ be a p-Banach ideal and E, F Banach spaces.*
(1) *If E' has the metric approximation property, then*

$$\mathfrak{A}^{min}(E,F) = \overline{\mathfrak{F}} \circ \mathfrak{A}(E,F) \qquad\qquad \text{isometrically.}$$

(2) *If F has the metric approximation property, then*

$$\mathfrak{A}^{min}(E,F) = \mathfrak{A} \circ \overline{\mathfrak{F}}(E,F) \qquad\qquad \text{isometrically.}$$

In other words, in $\mathfrak{A}^{min} = \overline{\mathfrak{F}} \circ \mathfrak{A} \circ \overline{\mathfrak{F}}(E,F)$ one $\overline{\mathfrak{F}}$ can be cancelled if E' or F has the metric approximation property. If $\mathfrak{A}$ is accessible, the same is true — and even more:

PROPOSITION: *Let $(\mathfrak{A}, \mathbf{A})$ be a p-Banach operator ideal.*
(1) *The following statements are equivalent:*
   (a) *$(\mathfrak{A}, \mathbf{A})$ is left-accessible.*

(b) $\mathfrak{A}^{min} = \overline{\mathfrak{F}} \circ \mathfrak{A}$ isometrically.

(c) $\mathfrak{A}^{min}(E, N) \stackrel{1}{=} \mathfrak{A}(E, N)$ for all $(E, N) \in$ BAN $\times$ FIN.

(2) For right–accessibility the following characterizations hold:

(a) $(\mathfrak{A}, \mathbf{A})$ is right–accessible.

(b) $\mathfrak{A}^{min} = \mathfrak{A} \circ \overline{\mathfrak{F}}$ isometrically.

(c) $\mathfrak{A}^{min}(M, F) \stackrel{1}{=} \mathfrak{A}(M, F)$ for all $(M, F) \in$ FIN $\times$ BAN.

This result was motivated by an idea of Lewis (see [169], lemma 1).

PROOF: Again, only (1) will be proved. Lemma 25.1. gives the equivalence of (b) and (c) — and (a) implies (c) by the very definition of $\mathfrak{A}$ being left–accessible. Assume (c); in order to show that (a) holds, observe that

$$\mathfrak{A}(E, N) \stackrel{1}{=} \mathfrak{A}^{min}(E, N) \stackrel{1}{=} \mathfrak{A} \circ \overline{\mathfrak{F}}(E, N) \, .$$

It follows that it is enough to show that $\mathfrak{A} \circ \overline{\mathfrak{F}}(E, N) \stackrel{1}{=} \mathfrak{A} \circ \mathfrak{F}(E, N)$. Denote, for the moment, by $\mathbf{F}$ and $\overline{\mathbf{F}}$ the operator norm $\| \ \|$ on $\mathfrak{F}$ and $\overline{\mathfrak{F}}$. Clearly, $\mathbf{A} \circ \overline{\mathbf{F}} \leq \mathbf{A} \circ \mathbf{F}$. Conversely, take $T \in \mathcal{L}(E, N)$ with $\mathbf{A} \circ \overline{\mathbf{F}}(T) < 1$, choose a factorization $T = S \circ R$ with $\mathbf{A}(S)\overline{\mathbf{F}}(R) < 1$ and $R_n \in \mathfrak{F}(E, G)$ such that $\|R - R_n\| \to 0$. Since

$$\mathbf{A} \circ \mathbf{F}(S \circ R_n - S \circ R_m) \leq \mathbf{A}(S)\|R_n - R_m\| \longrightarrow 0 \, ,$$

it follows that $(S \circ R_n)$ is Cauchy in the complete (see Ex 9.18.) space $\mathfrak{A} \circ \mathfrak{F}(E, N)$ — with limit T. Since $\mathfrak{A} \circ \mathfrak{F}$ is q–normed for some q (this has the same proof as 9.10.),

$$\mathbf{A} \circ \mathbf{F}(T) = \lim_{n \to \infty} \mathbf{A} \circ \mathbf{F}(S \circ R_n) \leq \lim_{n \to \infty} \mathbf{A}(S)\mathbf{F}(R_n) = \mathbf{A}(S)\overline{\mathbf{F}}(R) < 1 \, . \ \Box$$

In particular, if $(\mathfrak{A}, \mathbf{A})$ is accessible, then

$$\mathfrak{A}^{min} = \mathfrak{A} \circ \overline{\mathfrak{F}} = \overline{\mathfrak{F}} \circ \mathfrak{A}$$

holds isometrically (see also Ex 25.1.). This is a very useful fact, see for example Ex 25.2. for an application to p–compact operators.

25.3. Since $\overline{\mathfrak{F}} \circ \overline{\mathfrak{F}} \stackrel{1}{=} \overline{\mathfrak{F}}$ (shown in 22.1.), the relations

$$\mathfrak{A}^{min} \circ \overline{\mathfrak{F}} = \overline{\mathfrak{F}} \circ \mathfrak{A} \circ \overline{\mathfrak{F}} \circ \overline{\mathfrak{F}} = \overline{\mathfrak{F}} \circ \overline{\mathfrak{F}} \circ \mathfrak{A} \circ \overline{\mathfrak{F}} \circ \overline{\mathfrak{F}} = \mathfrak{A}^{minmin}$$

hold isometrically, and hence

COROLLARY: *Minimal p–Banach operator ideals are accessible.*

This is a rather interesting result — since it will be shown in section 31 that there are non-accessible maximal normed ideals. It is easy to see (Ex 25.4.) that the ideal $\mathfrak{N}$ of nuclear operators is not totally accessible.

25.4. The following result allows one to obtain various composition formulas from a given formula of the form $\mathfrak{A} \circ \mathfrak{B} \subset \mathfrak{C}$. The proof will reveal that the "cyclic composition theorem" is nothing but a condensation of trace duality: this is why it is natural that some accessibility assumptions have to be made. Recall from 17.9. that

$$|tr(S \circ T)| \leq \mathbf{A}(S)\mathbf{A}^*(T)$$

whenever $S \in \mathfrak{L}(M, N)$ and $T \in \mathfrak{L}(N, M)$ for finite dimensional spaces.

CYCLIC COMPOSITION THEOREM: *Let* $(\mathfrak{A}, \mathbf{A}), (\mathfrak{B}, \mathbf{B})$ *and* $(\mathfrak{C}, \mathbf{C})$ *be quasi-Banach operator ideals with*

$$\mathfrak{A} \circ \mathfrak{B} \subset \mathfrak{C} \quad \text{and} \quad \mathbf{C} \leq c[\mathbf{A} \circ \mathbf{B}]$$

for some $c \geq 0$.
(1) *If* $\mathfrak{A}$ *is right-accessible (or* $\mathfrak{C}^*$ *left-accessible), then*

$$\mathfrak{C}^* \circ \mathfrak{A} \subset \mathfrak{B}^* \quad \text{and} \quad \mathbf{B}^* \leq c[\mathbf{C}^* \circ \mathbf{A}] .$$

(2) *If* $\mathfrak{B}$ *is left-accessible (or* $\mathfrak{C}^*$ *right-accessible), then*

$$\mathfrak{B} \circ \mathfrak{C}^* \subset \mathfrak{A}^* \quad \text{and} \quad \mathbf{A}^* \leq c[\mathbf{B} \circ \mathbf{C}^*] .$$

PROOF: To see (1) fix $R \in \mathfrak{A}(E, F)$ and $S \in \mathfrak{C}^*(F, G)$. To calculate $\mathbf{B}^*(S \circ R)$ take $(M, L) \in \text{FIN}(E) \times \text{COFIN}(G)$ and $U \in \mathfrak{L}(G/L, M)$ with $\mathbf{B}(U) \leq 1$:

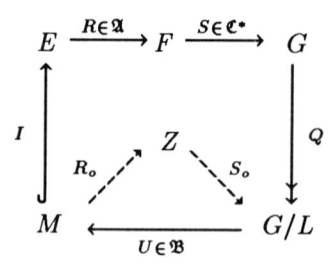

$$\begin{array}{ccccc}
 & R\in\mathfrak{A} & & S\in\mathfrak{C}^* & \\
E & \longrightarrow & F & \longrightarrow & G \\
\uparrow & & & & \downarrow \\
I & \quad R_o \nearrow & Z & \searrow S_o & Q \\
\downarrow & & & & \downarrow \\
M & \longleftarrow & & & G/L \\
 & & U\in\mathfrak{B} & &
\end{array}$$

If $\mathfrak{A}$ is right-accessible, take $Z := N \in \text{FIN}(F)$ and $R_o \in \mathfrak{L}(M, Z)$ such that $\mathbf{A}(R_o) \leq (1 + \varepsilon)\mathbf{A}(R)$ and define $S_o := Q \circ S \circ I_N^F$. If $\mathfrak{C}^*$ is left-accessible, choose $Z := F/K$ with $K \in \text{COFIN}(F)$ and $S_o \in \mathfrak{L}(Z, G/L)$ with $\mathbf{C}^*(S_o) \leq (1 + \varepsilon)\mathbf{C}^*(S)$. In either case there are R_o and S_o with

$$\mathbf{A}(R_o) \leq (1 + \varepsilon)\mathbf{A}(R) , \quad \mathbf{C}^*(S_o) \leq (1 + \varepsilon)\mathbf{C}^*(S)$$

and $QSRI = S_o R_o$. It follows that

$$|tr(QSRIU)| = |tr(S_o R_o U)| \leq \mathbf{C}(R_o U)\mathbf{C}^*(S_o) \leq$$
$$\leq c\mathbf{A}(R_o)\mathbf{B}(U)\mathbf{C}^*(S_o) \leq c(1+\varepsilon)^2 \mathbf{A}(R)\mathbf{C}^*(S) ,$$

which shows that $\mathbf{B}^* \leq c[\mathbf{C}^* \circ \mathbf{A}]$. The proof of (2) is much the same. □

Note that applying the cyclic argument (1) twice gives the statement (2) — but under stronger hypotheses than stated. We shall also use the term "rotating" for applying the cyclic argument. If $\mathfrak{A}$ is right–accessible, then $\mathfrak{A} \circ \mathfrak{L} \subset \mathfrak{A}$ implies

$$\mathfrak{A}^* \circ \mathfrak{A} \subset \mathfrak{J} \quad \text{and} \quad \mathbf{I} \leq \mathbf{A}^* \circ \mathbf{A}$$

— a formula which generalizes the one from 17.19. (proposition 1) to arbitrary quasi–Banach ideals. Oertel [194] showed that right–accessibility is necessary:

PROPOSITION 1: *Let* $(\mathfrak{A}, \mathbf{A})$ *be a maximal normed operator ideal. Then* $(\mathfrak{A}, \mathbf{A})$ *is right–accessible if (and only if)*

$$\mathfrak{A}^* \circ \mathfrak{A} \subset \mathfrak{J} \quad \text{and} \quad \mathbf{I} \leq \mathbf{A}^* \circ \mathbf{A} .$$

PROOF: It will be shown that the condition implies that $\mathfrak{A} \circ \overline{\mathfrak{F}} = \mathfrak{A}^{min}$, which is enough by proposition 25.2.; for this fix $T \in \mathfrak{A}(E, F)$. Then for every Banach space G the map

$$(F\tilde{\otimes}_{\alpha^t} G')' = \mathfrak{A}^*(F, G'') \longrightarrow \mathfrak{J}(E, G'') = (E\tilde{\otimes}_\varepsilon G')'$$
$$S \rightsquigarrow S \circ T$$

has norm $\leq \mathbf{A}(T)$, and hence

$$\|G'\tilde{\otimes}_\varepsilon E \xrightarrow{id\tilde{\otimes}T} G'\tilde{\otimes}_\alpha F\| \leq \mathbf{A}(T) ,$$

which implies that

$$\Phi : \overline{\mathfrak{F}}(G, E) \xrightarrow{1} G'\tilde{\otimes}_\varepsilon E \xrightarrow{id\tilde{\otimes}T} G'\tilde{\otimes}_\alpha F \xrightarrow{1} \mathfrak{A}^{min}(G, F)$$

has norm $\leq \mathbf{A}(T)$ as well. But $\Phi(S) = T \circ S$, which shows that $\mathfrak{A} \circ \overline{\mathfrak{F}} \subset \mathfrak{A}^{min}$ and that the norm inequality holds; the other inclusion is obvious. □

The cyclic composition theorem also gives results for the composition with minimal kernels; this is a consequence of the following easy

PROPOSITION 2: *If* $(\mathfrak{A}, \mathbf{A}), (\mathfrak{B}, \mathbf{B})$ *and* $(\mathfrak{C}, \mathbf{C})$ *are quasi–Banach operator ideals with*

$$\mathfrak{A} \circ \mathfrak{B} \subset \mathfrak{C} \quad \text{and} \quad \mathbf{C} \leq c[\mathbf{A} \circ \mathbf{B}]$$

for some $c \geq 0$ and $(\mathfrak{C}, C)$ is accessible, then

$$\mathfrak{A}^{min} \circ \mathfrak{B} \subset \mathfrak{C}^{min} \quad \text{and} \quad C^{min} \leq c[A^{min} \circ B]$$
$$\mathfrak{A} \circ \mathfrak{B}^{min} \subset \mathfrak{C}^{min} \quad \text{and} \quad C^{min} \leq c[A \circ B^{min}].$$

PROOF: If $(\mathfrak{C}, C)$ is left–accessible, then $\mathfrak{C}^{min} = \overline{\mathfrak{F}} \circ \mathfrak{C}$ by 25.2., and hence,

$$\mathfrak{A}^{min} \circ \mathfrak{B} = \overline{\mathfrak{F}} \circ \mathfrak{A} \circ \overline{\mathfrak{F}} \circ \mathfrak{B} \subset \overline{\mathfrak{F}} \circ \mathfrak{A} \circ \mathfrak{B} \subset \overline{\mathfrak{F}} \circ \mathfrak{C} = \mathfrak{C}^{min}.$$

The other relation follows in the same way. □

25.5. An immediate application is the multiplication table for absolutely p–summing and p–nuclear operators: In 11.5. it was shown that

$$\mathfrak{P}_p \circ \mathfrak{P}_q \subset \mathfrak{P}_r \quad \text{and} \quad P_r \leq P_p \circ P_q$$

if $r, p, q \in [1, \infty]$ and $1/r = 1/p + 1/q$. Using the facts that $\mathfrak{P}_p \sim g_{p'}^*$ is totally accessible, $\mathfrak{P}_p^* = \mathfrak{I}_{p'} \sim g_{p'}$ is accessible and $\mathfrak{I}_p \subset \mathfrak{P}_p$ and $\mathfrak{I}_p^{min} = \mathfrak{N}_p$, one obtains by cyclic composition (and the proposition about minimal operator ideals) the Persson–Pietsch multiplication table from [209]: For $1/r = 1/p + 1/q$

○	$\mathfrak{P}_q$	$\mathfrak{I}_q$	$\mathfrak{N}_q$
$\mathfrak{P}_p$	$\mathfrak{P}_r$	$\mathfrak{I}_r$	$\mathfrak{N}_r$
$\mathfrak{I}_p$	$\mathfrak{I}_r$	$\mathfrak{I}_r$	$\mathfrak{N}_r$
$\mathfrak{N}_p$	$\mathfrak{N}_r$	$\mathfrak{N}_r$	$\mathfrak{N}_r$

(read: $\mathfrak{N}_p \circ \mathfrak{P}_q \subset \mathfrak{N}_r$ ...).

25.6. There have already been examples given of operators T having the property that $S \in \mathfrak{A}$ implies TS or $ST \in \mathfrak{B}$ for some ideals; Grothendieck's inequality (in operator form 23.10.) had this flavour: For $\mathfrak{A} = \mathfrak{L}_p$ (for some $1 \leq p \leq 2$) and $T = id_{\ell_\infty}$ the operator $S = ST$ was in $\mathfrak{B} = \mathfrak{P}_2$. For quasi–Banach operator ideals $(\mathfrak{A}, A)$ and $(\mathfrak{B}, B)$ define for $T \in \mathfrak{L}$

$$\mathbf{A}^{-1} \circ \mathbf{B}(T) := \sup \{B(ST) \mid S \in \mathfrak{A}, A(S) \leq 1 \text{ and } ST \text{ defined}\} \in [0, \infty]$$
$$\mathbf{B} \circ \mathbf{A}^{-1}(T) := \sup \{B(TS) \mid S \in \mathfrak{A}, A(S) \leq 1 \text{ and } TS \text{ defined}\} \in [0, \infty]$$

(with the convention $B(U) = \infty$ whenever $U \notin \mathfrak{B}$).
The *quotient ideals* $\mathfrak{A}^{-1} \circ \mathfrak{B}$ and $\mathfrak{B} \circ \mathfrak{A}^{-1}$ consist, by definition, of all operators T such that $\mathbf{A}^{-1} \circ \mathbf{B}(T) < \infty$ and $\mathbf{B} \circ \mathbf{A}^{-1}(T) < \infty$, respectively.

It is obvious that $(\mathfrak{A}^{-1} \circ \mathfrak{B}, \mathbf{A}^{-1} \circ \mathbf{B})$ and $(\mathfrak{B} \circ \mathfrak{A}^{-1}, \mathbf{B} \circ \mathbf{A}^{-1})$ are quasinormed operator ideals — and the completeness is also not difficult to show. Note (Ex 25.9.) that $T \in \mathfrak{A}^{-1} \circ \mathfrak{B}$ if (and only if) $ST \in \mathfrak{B}$ for all $S \in \mathfrak{A}$. In other words, $\mathfrak{A}^{-1} \circ \mathfrak{B}$ is the largest ideal $\mathfrak{C}$ with the property that $\mathfrak{A} \circ \mathfrak{C} \subset \mathfrak{B}$. If $\mathfrak{B}$ is normed, then $\mathfrak{B} \circ \mathfrak{A}^{-1}$ and $\mathfrak{A}^{-1} \circ \mathfrak{B}$ are normed as well.

Since, obviously,
$$\mathfrak{A} \circ (\mathfrak{A}^{-1} \circ \mathfrak{B}) \subset \mathfrak{B} \quad \text{and} \quad (\mathfrak{B} \circ \mathfrak{A}^{-1}) \circ \mathfrak{A} \subset \mathfrak{B},$$

the cyclic composition theorem gives

$$(\mathfrak{A}^{-1} \circ \mathfrak{B}) \circ \mathfrak{B}^* \subset \mathfrak{A}^* \qquad \text{if } \mathfrak{B}^* \text{ is right-accessible}$$
$$\mathfrak{B}^* \circ (\mathfrak{B} \circ \mathfrak{A}^{-1}) \subset \mathfrak{A}^* \qquad \text{if } \mathfrak{B}^* \text{ is left-accessible,}$$

and hence the

COROLLARY: *Let $\mathfrak{A}$ and $\mathfrak{B}$ be quasi-Banach operator ideals, $\mathfrak{B}^*$ accessible. Then*

$$\mathfrak{A}^{-1} \circ \mathfrak{B} \subset \mathfrak{A}^* \circ (\mathfrak{B}^*)^{-1} \quad \text{and} \quad \mathfrak{B} \circ \mathfrak{A}^{-1} \subset (\mathfrak{B}^*)^{-1} \circ \mathfrak{A}^*.$$

Saphar's result in 20.19. is a special case: for example,

$$\mathfrak{J}_{p'}^{-1} \circ \mathfrak{P}_1 \subset \mathfrak{J}_{p'}^* \circ (\mathfrak{P}_1^*)^{-1} = \mathfrak{P}_p \circ (\mathfrak{L}_\infty)^{-1} \subset (\mathfrak{P}_p^*)^{-1} \circ \mathfrak{L}_\infty^* = \mathfrak{J}_{p'}^{-1} \circ \mathfrak{P}_1$$

applied to $T = id_E$ gives the equivalence of (a) and (c); the other one follows using the fact that $\mathfrak{P}_p(\ell_\infty, \cdot) = \mathfrak{J}_p(\ell_\infty, \cdot)$.

25.7. This already shows the power of these formulas concerning quotient ideals. Another quite useful result is the

QUOTIENT FORMULA: *Let $(\mathfrak{A}, \mathbf{A})$ and $(\mathfrak{B}, \mathbf{B})$ be quasi-Banach operator ideals.*
(1) *If $\mathfrak{A}$ is accessible (or: $\mathfrak{A}$ and $\mathfrak{B}$ right-accessible), then*

$$(\mathfrak{A} \circ \mathfrak{B})^* = \mathfrak{B}^* \circ \mathfrak{A}^{-1} \qquad \text{isometrically.}$$

(2) *If $\mathfrak{B}$ is accessible (or: $\mathfrak{A}$ and $\mathfrak{B}$ left-accessible), then*

$$(\mathfrak{A} \circ \mathfrak{B})^* = \mathfrak{B}^{-1} \circ \mathfrak{A}^* \qquad \text{isometrically.}$$

Recall that $\mathfrak{A}$ is right-accessible if it is injective — and left-accessible if it is surjective.

PROOF: The formulas will be a consequence of various applications of the cyclic composition theorem. Assume first that $\mathfrak{A}$ and $\mathfrak{B}$ are accessible. Rotating $\mathfrak{A} \circ \mathfrak{B} \subset \mathfrak{A} \circ \mathfrak{B}$ with the cyclic composition theorem gives

$$(\mathfrak{A} \circ \mathfrak{B})^* \circ \mathfrak{A} \subset \mathfrak{B}^* \quad \text{and} \quad \mathfrak{B} \circ (\mathfrak{A} \circ \mathfrak{B})^* \subset \mathfrak{A}^*,$$

which means $(\mathfrak{A} \circ \mathfrak{B})^* \subset \mathfrak{B}^* \circ \mathfrak{A}^{-1}$ and $(\mathfrak{A} \circ \mathfrak{B})^* \subset \mathfrak{B}^{-1} \circ \mathfrak{A}^*$ (the norm–inequalities will not be written). Conversely, it is obvious that

$$(\mathfrak{B}^* \circ \mathfrak{A}^{-1}) \circ \mathfrak{A} \subset \mathfrak{B}^* \text{ and } \mathfrak{B} \circ (\mathfrak{B}^{-1} \circ \mathfrak{A}^*) \subset \mathfrak{A}^* .$$

Rotating this yields

$$\mathfrak{A} \circ \mathfrak{B}^{**} \subset (\mathfrak{B}^* \circ \mathfrak{A}^{-1})^* \text{ and } \mathfrak{A}^{**} \circ \mathfrak{B} \subset (\mathfrak{B}^{-1} \circ \mathfrak{A}^*)^* .$$

Since $\mathfrak{C} \subset \mathfrak{C}^{**}$, this easily gives the other inclusions.

Now the assumptions will be weakened, using the facts that $\mathfrak{A}^{min}$ is always accessible and that $\mathfrak{C}^* = (\mathfrak{C}^{min})^*$ — the latter since $\mathfrak{C}^*$ is determined by the components $\mathfrak{C}(M,N) = \mathfrak{C}^{min}(M,N)$. Note first that

$$(\mathfrak{A} \circ \mathfrak{B})^{min} = \overline{\mathfrak{F}} \circ \mathfrak{A} \circ \mathfrak{B} \circ \overline{\mathfrak{F}} \stackrel{1}{=} \overline{\mathfrak{F}} \circ \mathfrak{A} \circ \overline{\mathfrak{F}} \circ \mathfrak{B} \circ \overline{\mathfrak{F}} = \overline{\mathfrak{F}} \circ \mathfrak{A} \circ \overline{\mathfrak{F}} \circ \overline{\mathfrak{F}} \circ \mathfrak{B} \circ \overline{\mathfrak{F}} =$$
$$= \mathfrak{A}^{min} \circ \mathfrak{B}^{min}$$

if $\mathfrak{A}$ is left–accessible or $\mathfrak{B}$ is right–accessible by proposition 25.2.. Therefore,

$$(\mathfrak{A} \circ \mathfrak{B})^* = ((\mathfrak{A} \circ \mathfrak{B})^{min})^* = (\mathfrak{A}^{min} \circ \mathfrak{B}^{min})^* = \mathfrak{B}^* \circ (\mathfrak{A}^{min})^{-1} = (\mathfrak{B}^{min})^{-1} \circ \mathfrak{A}^*$$

by the first part of the proof. Clearly,

$$\mathfrak{B}^* \circ (\mathfrak{A}^{min})^{-1} \supset \mathfrak{B}^* \circ \mathfrak{A}^{-1} \text{ and } (\mathfrak{B}^{min})^{-1} \circ \mathfrak{A}^* \supset \mathfrak{B}^{-1} \circ \mathfrak{A}^* .$$

To show equality take $T \in \mathfrak{B}^* \circ (\mathfrak{A}^{min})^{-1}$ and $S \in \mathfrak{A}$ with $\mathbf{A}(S) \leq 1$. Consider

$$\begin{array}{ccccc} E & \xrightarrow{S \in \mathfrak{A}} & F & \xrightarrow{T} & G \\ I \uparrow & \nearrow{S_o} & & & \downarrow Q \\ M & & \xleftarrow{U \in \mathfrak{B}} & & G/L \end{array} \qquad \mathbf{B}(U) \leq 1 .$$

Then $\mathbf{A}^{min}(S_o) = \mathbf{A}(S_o) \leq 1$ if $\mathfrak{A}$ is right–accessible and hence the inequality

$$|tr(UQTSI)| = |tr(UQTS_o)| \leq \mathbf{B}(U)\mathbf{B}^*(QTS_o) \leq \mathbf{B}^* \circ (\mathbf{A}^{min})^{-1}(T)$$

shows that $\mathfrak{B}^* \circ (\mathfrak{A}^{min})^{-1} \subset \mathfrak{B}^* \circ \mathfrak{A}^{-1}$. The same argument implies that

$$(\mathfrak{B}^{min})^{-1} \circ \mathfrak{A}^* \subset \mathfrak{B}^{-1} \circ \mathfrak{A}^*$$

if $\mathfrak{B}$ is left–accessible. $\square$

Note that the proof showed that

$$(\mathfrak{A} \circ \mathfrak{B})^* \stackrel{1}{=} \mathfrak{B}^* \circ (\mathfrak{A}^{min})^{-1} \stackrel{1}{=} (\mathfrak{B}^{min})^{-1} \circ \mathfrak{A}^*$$

whenever $\mathfrak{A}$ is left–accessible or $\mathfrak{B}$ right–accessible. Moreover, it can easily be seen from proposition 25.2. that

$$(\mathfrak{B}^{min})^{-1} \circ \mathfrak{A}^* \stackrel{1}{=} (\mathfrak{B}^{min})^{-1} \circ (\mathfrak{A}^*)^{min} \quad \text{if } \mathfrak{A}^* \text{ is left–accessible}$$
$$\mathfrak{B}^* \circ (\mathfrak{A}^{min})^{-1} \stackrel{1}{=} (\mathfrak{B}^*)^{min} \circ (\mathfrak{A}^{min})^{-1} \quad \text{if } \mathfrak{B}^* \text{ is right–accessible.}$$

25.8. The quotient formula as well as the following interesting characterization of $(\mathfrak{A} \circ \mathfrak{B})^*$ in terms of the associated tensor norms appeared first in [41].

THEOREM: *Let $(\mathfrak{A}, \mathbf{A})$ and $(\mathfrak{B}, \mathbf{B})$ be maximal normed operator ideals with associated tensor norms α and β, respectively. If β is accessible, then the following statements are equivalent for each operator $T \in \mathcal{L}(E, F)$:*
(a) $T \in (\mathfrak{A} \circ \mathfrak{B})^*(E, F)$.
(b) $id_{C_p} \otimes T : C_p \otimes_\alpha E \longrightarrow C_p \otimes_{\beta^\bullet} F$ *is continuous for some Johnson space C_p.*
(c) $id_G \otimes T : G \otimes_\alpha E \longrightarrow G \otimes_{\beta^\bullet} F$ *is continuous for all Banach spaces G.*
In this case,

$$(\mathbf{A} \circ \mathbf{B})^*(T) = \|id_{C_p} \otimes T : C_p \otimes_\alpha E \longrightarrow C_p \otimes_{\beta^\bullet} F\| \geq \|id_G \otimes T : G \otimes_\alpha E \longrightarrow G \otimes_{\beta^\bullet} F\|.$$

PROOF: Lemma 17.16. shows the equivalence of (b) and (c). The representation theorem for minimal operator ideals gives the following diagram

$$\begin{array}{ccc} G' \tilde{\otimes}_\alpha E & \xrightarrow{id_G \tilde{\otimes} T} & G' \tilde{\otimes}_{\beta^\bullet} F \\ \downarrow & & \downarrow \\ \mathfrak{A}^{min}(G, E) & \xrightarrow{\Phi_G} & (\mathfrak{B}^*)^{min}(G, F) \\ S & \rightsquigarrow & T \circ S \end{array},$$

which clearly commutes. Hence, if (c) holds, then

$$T \in (\mathfrak{B}^*)^{min} \circ (\mathfrak{A}^{min})^{-1} = (\mathfrak{A} \circ \mathfrak{B})^*$$

by the quotient formula (since $\mathfrak{B} \sim \beta$ and $\mathfrak{B}^* \sim \beta^*$ are accessible). Conversely, assume (a) and take $G = C_2'$ which has the approximation property. Then the vertical maps are isometries, and hence (b) also follows from the quotient formula. Moreover,

$$\sup_G \|\Phi_G\| = (\mathbf{B}^*)^{min} \circ (\mathbf{A}^{min})^{-1}(T) = (\mathbf{A} \circ \mathbf{B})^*(T)$$
$$(\mathbf{A} \circ \mathbf{B})^*(T) \leq \sup_G \|id_G \otimes T\| = \|id_{C_p} \otimes T\| \leq \|\Phi_{C_2}\|,$$

which gives the relations about the norm. □

This result generalizes the two tensor product characterizations of $\mathfrak{A}$ given in section 17 since $\mathfrak{A} = (\mathcal{L} \circ \mathfrak{A}^*)^* = (\mathfrak{A}^* \circ \mathcal{L})^*$ and $\mathcal{L} \sim \varepsilon$.

Checking the proofs of the cyclic composition theorem, the quotient formula and this last characterization, reveals that accessibility was always used to put in an additional finite rank operator at an appropriate place. The results at the end of section 16 about the factorization of finite rank operators show that this can be also done if some metric approximation property is at hand. It becomes clear — for example — that for fixed Banach spaces E and F the assumption of accessibility in the last characterization of $T \in (\mathfrak{A} \circ \mathfrak{B})^*(E, F)$ can be replaced by the assumption that E has the metric approximation property. We omit the details.

25.9. Some results of Kwapień's concerning subspaces/quotients of L_p-spaces can be proved using what was developed so far. Recall from 20.12. that for a maximal normed operator ideal $\mathfrak{A}$ with associated tensor norm α, the ideals associated with the projective associates $\alpha/$ and $\backslash\alpha$ were denoted by $\mathfrak{A}/$ and $\backslash\mathfrak{A}$, respectively. They were characterized by factorization (into the bidual) by

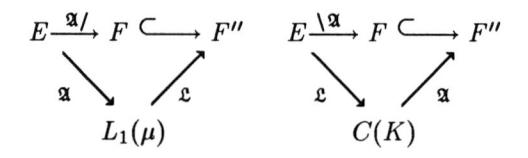

$$E \xrightarrow{\mathfrak{A}/} F \hookrightarrow F'' \qquad E \xrightarrow{\backslash\mathfrak{A}} F \hookrightarrow F''$$
$$\mathfrak{A} \searrow \quad \nearrow \mathcal{L} \qquad \mathcal{L} \searrow \quad \nearrow \mathfrak{A}$$
$$L_1(\mu) \qquad\qquad C(K)$$

To formulate the following lemma in an adequate way recall from Ex 22.6. the *regular hull* $(\mathfrak{A}, \mathbf{A})^{reg}$ of a quasi-Banach operator ideal $(\mathfrak{A}, \mathbf{A})$:

$$\mathfrak{A}^{reg}(E, F) := \{T \in \mathcal{L}(E, F) \mid \kappa_F \circ T \in \mathfrak{A}(E, F'')\}$$
$$\mathbf{A}^{reg}(T : E \to F) := \mathbf{A}(\kappa_F \circ T : E \to F'') .$$

It is straightforward to show that this is a quasi-Banach operator ideal, that $(\mathfrak{A}, \mathbf{A})$ is regular if $(\mathfrak{A}^{reg}, \mathbf{A}^{reg}) = (\mathfrak{A}, \mathbf{A})$ and that $\mathfrak{A}^{reg}(M, N) \stackrel{1}{=} \mathfrak{A}(M, N)$ for finite dimensional spaces. In particular,

$$((\mathfrak{A}, \mathbf{A})^{reg})^* \stackrel{1}{=} (\mathfrak{A}, \mathbf{A})^*$$

for the adjoints. Since $F/L = F''/L^{00}$ if $L \in \text{COFIN}(F)$, it is easy to see that maximal quasi-Banach operator ideals are regular (see also 17.8.).

LEMMA: *Let $(\mathfrak{A}, \mathbf{A})$ and $(\mathfrak{B}, \mathbf{B})$ be maximal normed operator ideals such that the composition ideal $(\mathfrak{A} \circ \mathfrak{B}, \mathbf{A} \circ \mathbf{B})$ is maximal and normed as well. Then the following relations hold isometrically:*

(1) $\backslash(\mathfrak{A} \circ \mathfrak{B}) = (\mathfrak{A} \circ (\backslash\mathfrak{B}))^{reg}$.
(2) $(\mathfrak{A} \circ \mathfrak{B})/ = ((\mathfrak{A}/) \circ \mathfrak{B})^{reg}$.
(3) $\backslash(\mathfrak{A} \circ \mathfrak{B})/ = ((\mathfrak{A}/) \circ (\backslash\mathfrak{B}))^{reg}$.

PROOF: For (1) it is enough to show that $\backslash(\mathfrak{A} \circ \mathfrak{B})(E, F'') = \mathfrak{A} \circ (\backslash\mathfrak{B})(E, F'')$ — but this is immediate from the factorization. (2) and (3) follow in the same way. □

To apply these formulas to Kwapień's factorization theorem $\mathfrak{D}_{p,q} = \mathfrak{P}_q^{dual} \circ \mathfrak{P}_p$ from 19.3., observe first that 20.14. implies that

$$\mathfrak{I}_p^{inj} \sim g_p \backslash = g_{p'}^* \sim \mathfrak{P}_p = \mathfrak{I}_{p'}^*$$
$$\backslash \mathfrak{P}_p \sim \backslash g_{p'}^* = (g_{p'}\backslash)^* = g_p^{**} = g_p \sim \mathfrak{I}_p \ .$$

An immediate application of the quotient formula now gives a result on the injective and surjective hulls of the ideal $\mathfrak{L}_{p,q}$ of (p,q)-factorable operators — a result which at first glance might terrify the reader.

PROPOSITION: *The following relations hold isometrically:*
(1) $\mathfrak{L}_{p,q} = (\mathfrak{P}_{q'}^{dual} \circ \mathfrak{P}_{p'})^* = \mathfrak{I}_p \circ (\mathfrak{P}_{q'}^{dual})^{-1} = \mathfrak{P}_{p'}^{-1} \circ \mathfrak{I}_q^{dual}$.
(2) $\mathfrak{L}_{p,q}^{inj} = (\mathfrak{P}_{q'}^{dual} \circ \mathfrak{I}_{p'})^* = \mathfrak{P}_p \circ (\mathfrak{P}_{q'}^{dual})^{-1} = \mathfrak{I}_{p'}^{-1} \circ \mathfrak{I}_q^{dual}$.
(3) $\mathfrak{L}_{p,q}^{sur} = (\mathfrak{I}_{q'}^{dual} \circ \mathfrak{P}_{p'})^* = \mathfrak{I}_p \circ (\mathfrak{I}_{q'}^{dual})^{-1} = \mathfrak{P}_{p'}^{-1} \circ \mathfrak{P}_q^{dual}$.
(4) $\mathfrak{L}_{p,q}^{inj\ sur} = (\mathfrak{I}_{q'}^{dual} \circ \mathfrak{I}_{p'})^* = \mathfrak{P}_p \circ (\mathfrak{I}_{q'}^{dual})^{-1} = \mathfrak{I}_{p'}^{-1} \circ \mathfrak{P}_q^{dual}$.

PROOF: All ideals involved are accessible by the results in section 21; therefore the quotient formula can be applied and all four formulas become a simple consequence of the lemma; for example, the last one:

$$\mathfrak{L}_{p,q}^{inj\ sur} \sim /\alpha_{p,q}\backslash = (\backslash \alpha_{p,q}^* /)^* \sim (\backslash \mathfrak{D}_{p',q'}/)^* = (\backslash(\mathfrak{P}_{q'}^{dual} \circ \mathfrak{P}_{p'})/)^* =$$
$$= (\mathfrak{P}_{q'}^{dual} / \circ \backslash \mathfrak{P}_{p'})^* = ((\backslash \mathfrak{P}_{q'})^{dual} \circ \backslash \mathfrak{P}_{p'})^* = (\mathfrak{I}_{q'}^{dual} \circ \mathfrak{I}_{p'})^* =$$
$$= \mathfrak{I}_{p'}^* \circ (\mathfrak{I}_{q'}^{dual})^{-1} = \mathfrak{P}_p \circ (\mathfrak{I}_{q'}^{dual})^{-1} =$$
$$= \mathfrak{I}_{p'}^{-1} \circ (\mathfrak{I}_{q'}^{dual})^* = \mathfrak{I}_{p'}^{-1} \circ \mathfrak{P}_q^{dual} \ . \quad \square$$

As a special case, take the ideal $\mathfrak{L}_p = \mathfrak{L}_{p,p'}$ of p-factorable operators

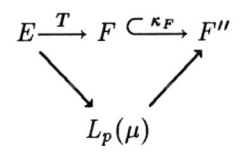

$$E \xrightarrow{T} F \xhookrightarrow{\kappa_F} F''$$
$$\searrow \quad \nearrow$$
$$L_p(\mu)$$

(see section 18). Simple arguments show that
(1) $T \in \mathfrak{L}_p^{inj}$ if and only if T factors through a subspace of some $L_p(\mu)$.
(2) $T \in \mathfrak{L}_p^{sur}$ if and only if $\kappa_F \circ T$ (or, equivalently, T'') factors through a quotient of some $L_p(\mu)$.

(3) $T \in \mathfrak{L}_p^{inj\ sur}$ if and only if T factors through a subspace of a quotient (or, equivalently: a quotient of a subspace) of some $L_p(\mu)$.

(See also Ex 20.8.). Thus, $\mathfrak{L}_p, \mathfrak{L}_p^{inj}, \mathfrak{L}_p^{sur}$ and $\mathfrak{L}_p^{inj\ sur}$ are operator ideals $\mathfrak{A}$ with the property that every $T \in \mathfrak{A}$ factors through some $G \in \text{space}(\mathfrak{A})$ (or at least $\kappa_F \circ T$ or T''). So one may concentrate on $\text{space}(\mathfrak{A})$ when investigating $\mathfrak{A}$. The following result is due to Kwapień [160]. Recall that $\text{space}(\mathfrak{L}_p)$ constists exactly of all $\mathfrak{L}_p^g$-spaces by 23.2..

COROLLARY: *Let E be a Banach space and $1 \le p \le \infty$. Then in each of the statements (1) – (4) the statements (a) – (d) are equivalent:*

(1) (a) $E''(= E$ if $1 < p < \infty)$ *is isomorphic to a complemented subspace of some $L_p(\mu)$.*
 (b) $\mathfrak{P}_p^{dual}(\cdot, E) \subset \mathfrak{I}_p(\cdot, E)$.
 (c) $\mathfrak{P}_{p'}(E, \cdot) \subset \mathfrak{I}_{p'}^{dual}(E, \cdot)$.
 (d) $G \otimes_{d_{p'}^*} E \longrightarrow G \otimes_{g_p} E$ *is continuous for all Banach spaces G.*
 (e) $E \in \text{space}(\mathfrak{L}_p)$.

(2) (a) *E is isomorphic to a subspace of some $L_p(\mu)$.*
 (b) $\mathfrak{P}_p^{dual}(\cdot, E) \subset \mathfrak{P}_p(\cdot, E)$.
 (c) $\mathfrak{I}_{p'}(E, \cdot) \subset \mathfrak{I}_{p'}^{dual}(E, \cdot)$.
 (d) $G \otimes_{d_{p'}^*} E \longrightarrow G \otimes_{g_{p'}^*} E$ *is continuous for all Banach spaces G.*
 (e) $E \in \text{space}(\mathfrak{L}_p^{inj})$.

(3) (a) $E''(= E$ if $1 < p < \infty)$ *is isomorphic to a quotient of some $L_p(\mu)$.*
 (b) $\mathfrak{I}_p^{dual}(\cdot, E) \subset \mathfrak{I}_p(\cdot, E)$.
 (c) $\mathfrak{P}_{p'}(E, \cdot) \subset \mathfrak{P}_{p'}^{dual}(E, \cdot)$.
 (d) $G \otimes_{d_p} E \longrightarrow G \otimes_{g_p} E$ *is continuous for all Banach spaces G.*
 (e) $E \in \text{space}(\mathfrak{L}_p^{sur})$.

(4) (a) *E is isomorphic to a subspace of a quotient of some $L_p(\mu)$ (or equivalently: to a quotient of a subspace).*
 (b) $\mathfrak{I}_p^{dual}(\cdot, E) \subset \mathfrak{P}_p(\cdot, E)$.
 (c) $\mathfrak{I}_{p'}(E, \cdot) \subset \mathfrak{P}_{p'}^{dual}(E, \cdot)$.
 (d) $G \otimes_{d_p} E \longrightarrow G \otimes_{g_{p'}^*} E$ *is continuous for all Banach spaces G.*
 (e) $E \in \text{space}(\mathfrak{L}_p^{inj\ sur})$.

PROOF: $id_E \in \mathfrak{A} \circ \mathfrak{B}^{-1}$ is equivalent to $\mathfrak{B}(\cdot, E) \subset \mathfrak{A}(\cdot, E)$ and $id_E \in \mathfrak{A}^{-1} \circ \mathfrak{B}$ is equivalent to $\mathfrak{A}(E, \cdot) \subset \mathfrak{B}(E, \cdot)$. Hence, the equivalences (a) – (c) follow immediately from the proposition. The equivalence of (c) and (d) is, in each case, a simple consequence of the representation theorem for maximal operator ideals. □

Clearly, the equivalence with (d) could also have been deduced from theorem 25.8..

25.10. The Chevet–Persson–Saphar inequalities (15.10.) say that

$$d_{p'}^* \leq d_p \leq \Delta_p \leq g_{p'}^* \leq g_p \qquad \text{on } L_p(\mu) \otimes E$$

— and there is even equality if $E = L_p(\nu)$. The following result describes conditions on E which imply that equality holds in some of these inequalities; it is due to Kwapień [160] and Gordon–Saphar [87].

PROPOSITION: *Let E be a Banach space, $1 \leq p \leq \infty$ and μ_o a measure such that $L_p(\mu_o)$ is infinite dimensional. Then in each of the statements (1) – (3) all statements are equivalent:*

(1) (a) E *is isomorphic to a subspace of some $L_p(\nu)$.*
 (b) $d_{p'}^* \leq g_{p'}^* \leq c d_{p'}^*$ *on $L_p(\mu_o) \otimes E$ for some $c \geq 1$.*
 (c) $d_{p'}^* \leq \Delta_p \leq c d_{p'}^*$ *on $L_p(\mu_o) \otimes E$ for some $c \geq 1$.*

(2) (a) $E''(= E$ *if $1 < p < \infty$) is isomorphic to a quotient of some $L_p(\nu)$.*
 (b) $d_p \leq g_p \leq c d_p$ *on $L_p(\mu_o) \otimes E$ for some $c \geq 1$.*
 (c) $\Delta_p \leq g_p \leq c \Delta_p$ *on $L_p(\mu_o) \otimes E$ for some $c \geq 1$.*

(3) (a) E *is isomorphic to a subspace of a quotient of some $L_p(\nu)$ (equivalently: quotient of a subspace).*
 (b) $d_p \leq g_{p'}^* \leq c d_p$ *on $L_p(\mu_o) \otimes E$ for some $c \geq 1$.*
 (c) $\Delta_p \leq g_{p'}^* \leq c \Delta_p$ *on $L_p(\mu_o) \otimes E$ for some $c \geq 1$.*
 (d) $d_p \leq \Delta_p \leq c d_p$ *on $L_p(\mu_o) \otimes E$ for some $c \geq 1$.*

PROOF: The statements (d) in the last corollary show that (a) always implies (b). The Chevet–Persson–Saphar inequalities show that (b) implies (c) (and (d) in case (3)). For the converse the fact that d_p and $g_{p'}^*$ are the tensor norms closest to Δ_p will be used, see 15.11.. Assume, for example, (3)(d): then (by the version of 15.11. with fixed E) $g_{p'}^* \leq c d_p$ on $G \otimes E$ for all G — this is statement (4)(d) of the last corollary. The remaining implications follow in the same way. □

A short look at the constants shows that for $c = 1$ one has to replace "isomorphic" by "isometric". Operator forms of this result will be given in 29.9..

If E is isomorphic to a quotient of a subspace of some $L_p(\mu)$, then the right–tensor norm Δ_p on $L_p(\nu) \otimes E$ is equivalent to a tensor norm (namely d_p and $g_{p'}^*$ by (3)(a) of the proposition). It follows that

$$A \otimes id_E : L_p(\eta) \otimes_{\Delta_p} E \longrightarrow L_p(\nu) \otimes_{\Delta_p} E$$

is continuous for all $A \in \mathcal{L}(L_p(\eta), L_p(\nu))$.

COROLLARY: *Let $L_p(\mu_o)$ be infinite dimensional and E a Banach space. If there is a tensor norm α such that*

$$L_p(\mu_o) \otimes_{\Delta_p} E = L_p(\mu_o) \otimes_\alpha E$$

holds isometrically, then E is isometric to a subspace of a quotient of some $L_p(\nu)$.

PROOF: By the Chevet–Persson–Saphar inequality and the remark after proposition 15.11. it follows that

$$g_{p'}^* \leq \alpha \leq d_p \leq \Delta_p \leq g_{p'}^* \qquad \text{on } L_p \otimes E,$$

hence, $\alpha = d_p$ on $L_p \otimes E$. Now (3) of the last proposition gives the result. $\square$

Obviously, one can replace "isometric" by "isomorphic". An operator form of this result will be given in 28.4..

25.11. The description of $\mathfrak{A}^{min}$, as given at the beginning of this section for accessible $\mathfrak{A}$, was crucial for the results which were obtained. It is obvious that it might also be advantageous for the investigation of the minimal kernels of $\mathfrak{A}^{sur}$ and $\mathfrak{A}^{inj}$:

PROPOSITION: *Let $(\mathfrak{A}, \mathbf{A})$ be a p–Banach operator ideal and E, F Banach spaces.*
(1) *If F has the approximation property or: if E' has the approximation property and $(\mathfrak{A}, \mathbf{A})$ is right–accessible, then*

$$(\mathfrak{A}^{sur})^{min}(E, F) = (\mathfrak{A}^{min})^{sur}(E, F) \qquad\qquad \text{isometrically.}$$

(2) *If E' has the approximation property or: if F has the approximation property and $(\mathfrak{A}, \mathbf{A})$ is left–accessible, then*

$$(\mathfrak{A}^{inj})^{min}(E, F) = (\mathfrak{A}^{min})^{inj}(E, F) \qquad\qquad \text{isometrically.}$$

The inclusions $\subset$ are true without any additional assumption — but both are strict in general: For $2 < p < \infty$ Reinov [228] constructed an operator $T_o \in \mathfrak{N}_p^{inj} = (\mathfrak{J}_p^{min})^{inj}$ which is not even approximable; dualizing the factorization of $I \circ T_o : E \to F \hookrightarrow \ell_\infty(\Gamma)$, (which is in $\overline{\mathfrak{F}} \circ \mathfrak{J}_p \circ \overline{\mathfrak{F}}$), it is obvious that the dual operator T_o' is in $((\mathfrak{J}_p^{dual})^{min})^{sur}$ — but $T_o' \notin \overline{\mathfrak{F}} = \overline{\mathfrak{F}}^{dual}$, so T_o' is an example showing that (1) also does not hold in general.

25. Composition of Accessible Operator Ideals   341

PROOF: The details concerning the norms will be omitted — they will be obvious from the constructions.

(1) Take $T \in (\mathfrak{A}^{sur})^{min}(E, F)$. Then there is a factorization

$$\begin{array}{ccccc}
\ell_1(B_E) & \xrightarrow{Q_E} & E & \xrightarrow{T} & F \\
\downarrow R_o & & R \downarrow & & S \uparrow \\
\ell_1(B_{G_1}) & \xrightarrow{Q_{G_1}} & G_1 & \xrightarrow{T_o} & G_2
\end{array} \qquad \begin{array}{l} R, S \in \overline{\mathfrak{F}} \\ T_o \in \mathfrak{A}^{sur} \end{array}$$

and the lifting R_o can be chosen to be compact=approximable by corollary 2(2) in 3.12.. It follows that

$$T \circ Q_E = S \circ (T_o \circ Q_{G_1}) \circ R_o \in \overline{\mathfrak{F}} \circ \mathfrak{A} \circ \overline{\mathfrak{F}}(\ell_1(B_E), F) = \mathfrak{A}^{min}(\ell_1(B_E), F),$$

hence $T \in (\mathfrak{A}^{min})^{sur}$.

Conversely, take $T \in (\mathfrak{A}^{min})^{sur}(E, F)$ and factor $T \circ Q_E = S \circ T_o \circ R$ as follows:

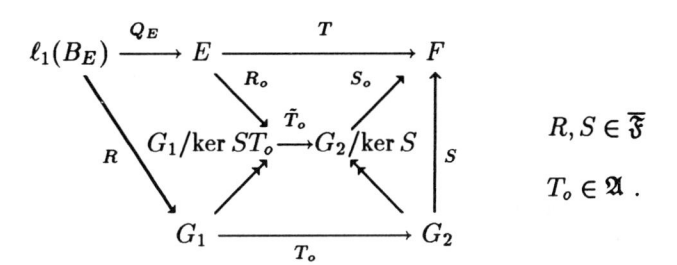

$$\begin{array}{ccc}
\ell_1(B_E) \xrightarrow{Q_E} E & \xrightarrow{T} & F \\
 & & \\
\end{array}$$

with intermediate factorization through $G_1/\ker ST_o \xrightarrow{\tilde{T}_o} G_2/\ker S$, via R_o, S_o, and overall $R: \ell_1(B_E) \to G_1$, $T_o: G_1 \to G_2$, $S: G_2 \to F$, with $R, S \in \overline{\mathfrak{F}}$, $T_o \in \mathfrak{A}$.

Now $\tilde{T}_o \in \mathfrak{A}^{sur}$ and R_o, S_o are compact. If E' has the approximation property, then $R_o \in \overline{\mathfrak{F}}$ and $T \in \mathfrak{A}^{sur} \circ \overline{\mathfrak{F}}$; if, additionally, $\mathfrak{A}$ is right-accessible, then the ideal $\mathfrak{A}^{sur}$ is totally accessible (see Ex 21.1.(e)) and hence $\mathfrak{A}^{sur} \circ \overline{\mathfrak{F}} = (\mathfrak{A}^{sur})^{min}$ by proposition 25.2.. If F has the approximation property, then $S_o \in \overline{\mathfrak{F}}$ and therefore $T \in \overline{\mathfrak{F}} \circ \mathfrak{A}^{sur}$. Since $\mathfrak{A}^{sur}$ is left-accessible, proposition 25.2. again gives $T \in (\mathfrak{A}^{sur})^{min}$.

(2) Any $T \in (\mathfrak{A}^{inj})^{min}(E, F)$ factors

$$\begin{array}{ccccc}
E & \xrightarrow{T} & F & \xhookrightarrow{I} & \ell_\infty(B_{F'}) \\
 & \searrow_{\overline{\mathfrak{F}} \ni R} & & \nearrow_{S \in \mathfrak{A}^{inj}} & \\
 & & G & &
\end{array}$$

Clearly, $I \circ S \in \mathfrak{A}$ and hence $I \circ T \in \mathfrak{A} \circ \overline{\mathfrak{F}}(E, \ell_\infty(B_{F'})) = \mathfrak{A}^{min}(E, \ell_\infty(B_{F'}))$ by corollary 25.2. (an argument like this could have been used in (1) as well).

On the other hand, if $T \in (\mathfrak{A}^{min})^{inj}(E, F)$, then it can be factored as follows:

$$
\begin{array}{ccccc}
E & \xrightarrow{T} & F & \xhookrightarrow{I} & \ell_\infty(B_{F'}) \\
& R_o \searrow & \uparrow S_o & & \\
R \downarrow & \mathrm{im}\, R \xrightarrow{\tilde{T}_o} & S^{-1}(I(F)) & \nearrow S & \\
& \swarrow & \downarrow & & \\
G_1 & \xrightarrow{T_o} & G_2 & &
\end{array}
$$

$R, S \in \overline{\mathfrak{F}}$

$T_o \in \mathfrak{A}$.

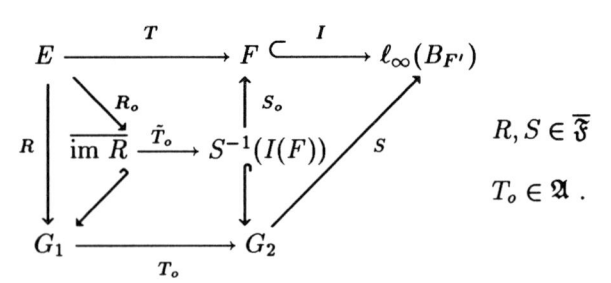

Clearly, $\tilde{T}_o \in \mathfrak{A}^{inj}$. Now the same type of arguments as in part (1) (using 25.2. and Ex 21.1.(d)) give $T \in (\mathfrak{A}^{inj})^{min}$. □

Exercises:

Ex 25.1. (a) A p–Banach operator ideal $\mathfrak{A}$ is accessible if and only if $\mathfrak{A} \circ \overline{\mathfrak{F}} = \overline{\mathfrak{F}} \circ \mathfrak{A}$ holds isometrically.

(b) Let $\mathfrak{A}$ be a maximal normed operator ideal. Show that $\mathfrak{A} \circ \overline{\mathfrak{F}} = \overline{\mathfrak{F}} \circ \mathfrak{A}$ if and only if $\mathfrak{A}^* \circ \overline{\mathfrak{F}} = \overline{\mathfrak{F}} \circ \mathfrak{A}^*$.

Ex 25.2. Every compact operator with values in an $\mathcal{L}_p^g$–space is p–compact (see 22.3.). Hint: $\mathcal{L}_p \circ \overline{\mathfrak{F}}$. This result was already shown in Ex 23.9..

Ex 25.3. Each p–Banach operator ideal $\mathfrak{A}$ of the form $\mathfrak{A} = \overline{\mathfrak{F}} \circ \mathfrak{B}$ is right–accessible, each $\mathfrak{A} = \mathfrak{B} \circ \overline{\mathfrak{F}}$ is left–accessible.

Ex 25.4. Let α be a finitely generated tensor norm and $\mathfrak{A}$ its associated maximal operator ideal. The canonical metric surjection $E' \tilde{\otimes}_\alpha F \to \mathfrak{A}^{min}(E, F)$ is injective for all Banach spaces E and F if and only if $\mathfrak{A}^{min}$ is totally accessible. Hint: Recall $M' \otimes_\alpha N \stackrel{1}{=} \mathfrak{A}^{min}(M, N)$. In particular, $\overline{\mathfrak{F}}$ is totally accessible, but $\mathfrak{N}$ is not.

Ex 25.5. Grothendieck's inequality in operator form 23.10. says that $\mathcal{L}_1 \circ \mathcal{L}_\infty \subset \mathfrak{D}_2$ and $\mathcal{L}_2 \circ \mathcal{L}_1 \subset \mathfrak{P}_1$.

(a) Use the cyclic composition theorem to show that these two results are equivalent.

(b) Show that $\mathcal{L}_2 \circ \mathcal{L}_\infty \subset \mathfrak{D}_2$ is false. Hint: Rotate.

Ex 25.6. Let $1 \leq p \leq 2$. Use Grothendieck's result $\mathcal{L}_p \circ \mathcal{L}_\infty \subset \mathfrak{P}_2$ (see 23.10.) and the cyclic composition theorem to show that:

(a) $\mathfrak{P}_2 \circ \mathcal{L}_p \subset \mathfrak{P}_1$; in particular, every abolutely 2–summing operator on ℓ_p is absolutely 1–summing.

(b) $\mathcal{L}_\infty \circ \mathfrak{P}_2 \subset \mathfrak{D}_{p'}$.

Ex 25.7. Let $2 \leq p < q < \infty$. Show that:

(a) The dual of an absolutely q'–summing operator $T \in \mathcal{L}(\ell_p, F)$ is absolutely 1–summing.

(b) Every absolutely q'–summing operator $T \in \mathcal{L}(E, \ell_1)$ is p'–dominated.

Hint: Rotate Saphar's theorem 24.6..

Ex 25.8. For $1 \leq q < p \leq 2$ every absolutely q-summing operator $E \to \ell_\infty$ is p-dominated. Hint: Rotate Maurey's theorem 24.7., which implies that $\mathfrak{L}_{p'} \circ \mathfrak{L}_\infty \subset \mathfrak{J}_{q'}$ by 8.6..

Ex 25.9. Prove that $\mathfrak{A}^{-1} \circ \mathfrak{B}$ consists of all operators T such that $ST \in \mathfrak{B}$ whenever $S \in \mathfrak{A}$, and the analogous statement for $\mathfrak{B} \circ \mathfrak{A}^{-1}$. Hint: $\ell_2(G_n)$.

Ex 25.10. If $\mathfrak{A}$ is accessible, then $\mathfrak{A}^* = \mathfrak{J} \circ \mathfrak{A}^{-1}$. This is a sort of converse of 17.19.. Hint: $\mathfrak{A} = \mathfrak{A} \circ \mathfrak{L}$. Compare this proof with the one in Ex 17.15..

Ex 25.11. If $\mathfrak{A}$ and $\mathfrak{A}^*$ are accessible, then

$$\mathfrak{A}^* \circ \mathfrak{A} \subset \mathfrak{J}, \quad \mathfrak{A}^* \circ \mathfrak{A}^{min} \subset \mathfrak{N}, \quad \mathfrak{A}^{min} \circ \mathfrak{A}^* \subset \mathfrak{N}.$$

Ex 25.12. Let $\mathfrak{A}$ be a regular quasi-Banach ideal and E, F Banach spaces. If $\mathfrak{A}$ is right-accessible and E' has the approximation property, then: $T \in \mathfrak{A}^{min}(E, F)$ if and only if $\kappa_F \circ T \in \mathfrak{A}^{min}(E, F'')$; moreover, $\mathbf{A}^{min}(\kappa_F \circ T) = \mathbf{A}^{min}(T)$. Hint: $\mathfrak{A}^{min}(E, F) = \mathfrak{A} \circ \overline{\mathfrak{F}}(E, F)$ and $\mathrm{im}(R) \subset S^{-1}(F)$ for a factorization $\kappa_F \circ T = S \circ R \in \mathfrak{A} \circ \overline{\mathfrak{F}}$. For Banach operator ideals $\mathfrak{A}$ there were results of this type in 22.8..

Ex 25.13. Let $1 \leq p \leq \infty$ and E be a Banach space. Prove the following results (Takahashi-Ohazaki [265]):

(a) E is isomorphic to a subspace of some $L_p(\mu)$ if and only if $\mathfrak{N}_{p'}(E, \cdot) \subset \mathfrak{K}_{1,p'}(E, \cdot)$.

(b) E'' is isomorphic to a quotient of some $L_p(\mu)$ if and only if $\mathfrak{K}_{1,p}(\cdot, E) \subset \mathfrak{N}_p(\cdot, E)$.

Hint: Corollary 25.9.(3), (a) equivalent to (d).

Ex 25.14. Reprove Kwapień's test 11.8. for an operator to be absolutely p-summing by using the quotient formula. Hint: $\mathfrak{L}_{p'} \circ \mathfrak{J}_{p'}$.

Ex 25.15. Let $(\mathfrak{A}, \mathbf{A})$ and $(\mathfrak{B}, \mathbf{B})$ be p- resp. q-normed operator ideals and G a Banach space. Use proposition 25.2. to show that:

(a) If $(\mathfrak{A}, \mathbf{A})$ is accessible and $\mathfrak{B}^{min}(E, G) \stackrel{1}{=} \mathfrak{B}(E, G)$ for all Banach spaces E, then

$$(\mathfrak{B} \circ \mathfrak{A})^{min}(E, G) = \mathfrak{B} \circ \mathfrak{A}(E, G)$$

holds isometrically for all Banach spaces E.

(b) If $(\mathfrak{B}, \mathbf{B})$ is accessible and $\mathfrak{A}^{min}(G, F) \stackrel{1}{=} \mathfrak{A}(G, F)$ for all Banach spaces F, then

$$(\mathfrak{B} \circ \mathfrak{A})^{min}(G, F) = \mathfrak{B} \circ \mathfrak{A}(G, F)$$

holds isometrically for all Banach spaces F.

Ex 25.16. If $\mathfrak{A}, \mathfrak{B}, \mathfrak{C}$ and $\mathfrak{D}$ are accessible quasi-Banach operator ideals such that $\mathfrak{A} \circ \mathfrak{B} \circ \mathfrak{C} \subset \mathfrak{D}$, then $\mathfrak{D}^* \circ \mathfrak{A} \circ \mathfrak{B} \subset \mathfrak{C}^*$, $\mathfrak{C} \circ \mathfrak{D}^* \circ \mathfrak{A} \subset \mathfrak{B}^*$ and $\mathfrak{B} \circ \mathfrak{C} \circ \mathfrak{D}^* \subset \mathfrak{A}^*$. Hint: Ex 21.2..

26. More About L_p and Hilbert Spaces

This section provides some additional information about the behaviour of $\alpha_{p,q}$, its associated operator ideals and Δ_p on L_p and Hilbert spaces. In particular, it will turn out that for Hilbert spaces the situation is quite simple.

26.1. Most of the results in this section will be stated for spaces $E = \ell_p$ only. The $\mathfrak{L}_p$-local techniques easily show that these statements also hold for arbitrary $\mathfrak{L}_p^g$-spaces E — in particular, for $L_p(\mu)$-spaces: in this latter case the constants are the same as for ℓ_p. Recall first from 12.8. that $\alpha_{p,q} \le b_{q'} b_{p'} w_2$ (if $p, q \in]1, \infty[$), where b_r is the constant from the Khintchine inequality 8.5..

PROPOSITION: *Let $p, q \in]1, \infty[$ with $1/p + 1/q \ge 1$ and $r, s \in [1, 2]$. Then*

$$\varepsilon \le \alpha_{p,q} \le K_G b_{q'} b_{p'} \varepsilon \qquad \text{on } \ell_r \otimes \ell_s$$
$$\alpha'_{p,q} \le \pi \le K_G b_{q'} b_{p'} \alpha'_{p,q} \qquad \text{on } \ell_{r'} \otimes \ell_{s'}.$$

PROOF: Grothendieck's inequality (in tensor form) gives, by duality, that $w_2^* \le K_G \varepsilon$ on $\ell_1^n \otimes \ell_1^m$, and hence on $L_1(\mu) \otimes L_1(\nu)$. Since $w_2 \le w_2^*$ holds (see Ex 19.5.), the tensor norms w_2 and ε are injective and $\ell_r \hookrightarrow L_1(\mu_r)$ by the Lévy embedding 24.4., one gets

$$\ell_r \otimes_\alpha \ell_s \xrightarrow{1} L_1(\mu_r) \otimes_\alpha L_1(\mu_s)$$

for $\alpha = w_2$ and $\alpha = \varepsilon$. It follows that $w_2 \le K_G \varepsilon$ on $\ell_r \otimes \ell_s$. Since $\alpha_{p,q} \le b_{q'} b_{p'} w_2$ by 12.8., the first result is proven. The second follows by dualization. □

A more general result due to Pisier will be given in section 31: $\varepsilon \sim \alpha_{p,q}$ on tensor products of cotype 2 spaces ($p, q \ne 1$). In terms of operator ideals this proposition is a result of Lindenstrauss–Pełczyński [173].

COROLLARY: *Let $1 \le s \le 2 \le r \le \infty$ and $p, q \in]1, \infty[$ with $1/p + 1/q \ge 1$. Then*

$$\mathfrak{L}(\ell_r, \ell_s) = \mathfrak{L}_p(\ell_r, \ell_s) = \mathfrak{L}_{p,q}(\ell_r, \ell_s)$$

and

$$\mathbf{L}_{p,q}(T : \ell_r \longrightarrow \ell_s) \le K_G b_{q'} b_{p'} \|T\| .$$

In particular, every $T \in \mathfrak{L}(\ell_r, \ell_s)$ factors through a Hilbert space.

This is immediate by the representation theorem for maximal operator ideals. In Ex 26.2. an "adjoint" form of this result is given; see also Ex 26.1.. For factorization through L_p if $1 \le s \le p \le r \le \infty$ see 26.5..

26.2. Operators $\ell_r \to \ell_s$ do *not* automatically factor through a Hilbert space if the condition $1 \leq s \leq 2 \leq r \leq \infty$ is violated: There are three cases to consider. First, let $r, s \in [1, 2[$ (the case $r, s \in]2, \infty]$ follows by duality); if $\mathfrak{L}(\ell_r, \ell_s) = \mathfrak{L}_2(\ell_r, \ell_s)$ were to hold, then there would be a constant c such that

$$\mathbf{L}_2(id : \ell_r^n \longrightarrow \ell_s^n) \leq c \, \|id : \ell_r^n \longrightarrow \ell_s^n\| = c \begin{cases} n^{1/s - 1/r} & \text{if } s < r \\ 1 & \text{if } s \geq r \end{cases}.$$

But this contradicts the following result — which is itself of interest (see Ex 26.4.):

PROPOSITION: $\mathbf{L}_2(id : \ell_r^n \to \ell_s^n) = n^{1/s - 1/2}$ for $s, r \in [1, 2]$.

PROOF: One inequality is trivial:

$$\mathbf{L}_2(\ell_r^n \hookrightarrow \ell_s^n) \leq \|\ell_r^n \hookrightarrow \ell_2^n\| \, \|\ell_2^n \hookrightarrow \ell_s^n\| \leq 1 \cdot n^{1/s - 1/2} .$$

Since $\mathfrak{P}_2^{dual} \circ \mathfrak{P}_2 \circ \mathfrak{L}_2 = \mathfrak{D}_2 \circ \mathfrak{L}_2 = \mathfrak{D}_2 \circ \mathfrak{D}_2^* \subset \mathfrak{J}$ (see 19.3., 19.4.), it follows that

$$n = \mathbf{N}(id_{\ell_r^n}) \leq \mathbf{L}_2(\ell_r^n \hookrightarrow \ell_s^n) \mathbf{P}_2(\ell_s^n \hookrightarrow \ell_2^n) \mathbf{P}_2(\ell_{r'}^n \hookrightarrow \ell_2^n) .$$

Now Ex 11.5. and 11.2. show that

$$\mathbf{P}_2(\ell_s^n \hookrightarrow \ell_2^n) \leq \|\ell_s^n \hookrightarrow \ell_1^n\| \mathbf{P}_2(\ell_1^n \hookrightarrow \ell_2^n) = n^{1-1/s}$$
$$\mathbf{P}_2(\ell_{r'}^n \hookrightarrow \ell_2^n) \leq \|\ell_{r'}^n \hookrightarrow \ell_\infty^n\| \mathbf{P}_2(\ell_\infty^n \hookrightarrow \ell_2^n) = n^{1/2} ,$$

which gives the reverse inequality. □

To see that $\mathfrak{L}(\ell_r, \ell_s) \neq \mathfrak{L}_2(\ell_r, \ell_s)$ if $1 \leq r < 2 < s \leq \infty$, consider the Fourier matrices (see Ex 4.3.)

$$A_n := n^{-1/2} \sum_{k, \ell = 1}^n \exp\left(2\pi i \frac{k \cdot \ell}{n}\right) e_k \otimes e_\ell : \mathbb{C}^n \longrightarrow \mathbb{C}^n .$$

The A_n are unitary and hence $\|A_n : \ell_2^n \to \ell_2^n\| = 1$; moreover, $\|A_n : \ell_1^n \to \ell_\infty^n\| = n^{-1/2}$ obviously holds. The interpolation theorem of Riesz–Thorin (see Bergh–Löfström [11]) gives

$$\|A_n : \ell_p^n \longrightarrow \ell_{p'}^n\| \leq n^{-1/2} n^{1/p'}$$

if $1 \leq p \leq 2$, which easily implies

$$\|A_n : \ell_r^n \longrightarrow \ell_s^n\| \leq n^{-1/2} n^{max\{1/r', 1/s\}} \xrightarrow{n \to \infty} 0$$

for $1 \leq r < 2 < s \leq \infty$. Clearly, the adjoint matrix A_n^* has the same properties as A_n; moreover, $A_n \circ A_n^* = id$ since A_n is unitary. Grothendieck's inequality (see 17.14.) gives

$$\mathbf{P}_2(A_n^* : \ell_\infty^n \longrightarrow \ell_1^n) \leq K_G \|A_n^* : \ell_\infty^n \longrightarrow \ell_1^n\| \leq K_G \cdot n$$

and the little Grothendieck theorem (11.11.) implies that

$$\mathbf{P}_2(A_n : \ell_1^n \longrightarrow \ell_\infty^n) \leq K_{LG}\mathbf{L}_2(A_n : \ell_1^n \longrightarrow \ell_\infty^n) .$$

Since $\mathfrak{P}_2 \circ \mathfrak{P}_2 \subset \mathfrak{N}$ (see 19.4.),

$$n = \mathbf{N}(id : \ell_\infty^n \longrightarrow \ell_\infty^n) \leq \mathbf{P}_2(A_n : \ell_1^n \longrightarrow \ell_\infty^n)\mathbf{P}_2(A_n^* : \ell_\infty^n \longrightarrow \ell_1^n) \leq$$
$$\leq K_G \cdot n \cdot K_{LG}\mathbf{L}_2(A_n : \ell_1^n \longrightarrow \ell_\infty^n) ,$$

and hence $(K_G \cdot K_{LG})^{-1} \leq \mathbf{L}_2(A_n : \ell_1^n \longrightarrow \ell_\infty^n) \leq \mathbf{L}_2(A_n : \ell_r^n \longrightarrow \ell_s^n)$. It follows that there is no constant c such that for all n

$$\mathbf{L}_2(A_n : \ell_r^n \longrightarrow \ell_s^n) \leq c\|A_n : \ell_r^n \longrightarrow \ell_s^n\|$$

and hence, by the closed graph theorem, $\mathfrak{L}(\ell_r, \ell_s) \neq \mathfrak{L}_2(\ell_r, \ell_s)$ if $1 \leq r < 2 < s \leq \infty$.

26.3. Let $T_i \in \mathfrak{L}(L_q(\mu_i), L_p(\nu_i))$; when is

$$T_1 \otimes T_2 : L_q(\mu_1) \otimes_{\Delta_q} L_q(\mu_2) \longrightarrow L_p(\nu_1) \otimes_{\Delta_p} L_p(\nu_2)$$

continuous? If one of the operators is positive, then $\|T_1 \otimes T_2\| = \|T_1\| \|T_2\|$ by theorem 7.3.; Beckner's result (7.9. and 15.12.) gives the same norm estimate if the operators T_i are arbitrary and $1 \leq q \leq p \leq \infty$. Maurey's factorization theorem treated the case $1 \leq p \leq q \leq \infty$. Other results follow from the factorization 26.1. and the

PROPOSITION 1: *For arbitrary $p, q \in [1, \infty]$ and operators $T_1 \in \mathfrak{L}(L_q(\mu_1), L_p(\nu_1))$ and $T_2 \in \mathfrak{L}(E, F)$*

$$\|T_1 \otimes T_2 : L_q(\mu_1) \otimes_{\Delta_q} E \longrightarrow L_p(\nu_1) \otimes_{\Delta_p} F\| \leq K_G \|T_1\| \, \mathbf{L}_2(T_2) .$$

Note that $\mathbf{L}_2(T_2 : L_r(\mu_2) \longrightarrow L_s(\nu_2)) \leq K_G \|T_2\|$ if $1 \leq s \leq 2 \leq r \leq \infty$ by corollary 26.1.(together with local techniques), hence $\|T_1 \otimes T_2 : ...\| \leq K_G^2 \|T_1\| \|T_2\|$ in this case.

PROOF: Assume $\|T_1\| = 1$ and consider the maximal Banach operator ideal $(\mathfrak{M}, \mathbf{M})$ of all $T \in \mathfrak{L}(E, F)$ such that

$$\mathbf{M}(T) := \|T_1 \otimes T : L_q(\mu_1) \otimes_{\Delta_q} E \longrightarrow L_p(\nu_1) \otimes_{\Delta_p} F\| < \infty$$

(see 17.4.). $\mathfrak{M}$ is surjective and injective by 7.4.; since the ideal $\mathfrak{J}$ of integral operators is contained in $\mathfrak{M}$ (it is contained in every maximal normed ideal because of $\varepsilon \leq \alpha$) and $\mathbf{M} \leq \mathbf{I}$, it follows that

$$\mathbf{M} \leq (\mathbf{I}^{sur})^{inj} \leq K_G \mathbf{L}_2$$

by the operator form of Grothendieck's inequality in 20.17. □

(For other results in this direction see 28.4., Ex 28.10., 29.12., 32.3., 32.6., 32.8. and 32.9..) It follows, in particular, that for Hilbert spaces H the operator

$$T \otimes id_H : L_q \otimes_{\Delta_q} H \longrightarrow L_p \otimes_{\Delta_p} H$$

is always continuous and has norm $\leq K_G \|T\|$ — for $p, q < \infty$ a result of Marcinkiewicz–Zygmund [182] (clearly, with another norm estimate). For $p = q$ this was already shown in Ex 23.14.. For $p = 1$ and $q = \infty$ it gives Grothendieck's inequality (see Ex 28.12.), which indicates that the use of Grothendieck's inequality in the proof was natural. In the case $q \leq p$ the constant can be taken to be 1 — but even a little bit more is true:

REMARK 1: Take $p, q, r \in [1, \infty]$. Then, for any $T \in \mathcal{L}(L_q(\mu), L_p(\nu))$, the operator

$$T \otimes id_{\ell_r} : L_q(\mu) \otimes_{\Delta_q} \ell_r \longrightarrow L_p(\nu) \otimes_{\Delta_p} \ell_r$$

has norm $\|T\|$ whenever $q \leq p$ and r satisfies one of the following five conditions:

(1) $r = p$
(2) $r = 2$
(3) $q < r < 2$
(4) $r = q$
(5) $2 < r < p$.

PROOF: The first three cases follow immediately from the version 7.9. of Beckner's result and an isometric embedding of ℓ_r into some L_q (see corollary 3 in 24.5.) by taking $E_1 = F = \ell_r$. The last two cases follow by dualization of (1) and (3). □

This result has a nice application observed by Figiel–Iwaniec–Pełczyński [65]: For any operator

$$T : L_q^{\mathbb{R}}(\mu) \longrightarrow L_p^{\mathbb{R}}(\nu) , \text{ real spaces}$$

denote its natural *complexification* by

$$T^{\mathbb{C}} : L_q^{\mathbb{C}}(\mu) \longrightarrow L_p^{\mathbb{C}}(\nu) , \quad T^{\mathbb{C}}(f) := T(\text{Re } f) + i\, T(\text{Im } f) .$$

Since $L_p^{\mathbb{C}}(\mu) = L_p^{\mathbb{R}}(\mu) \otimes_{\Delta_p} \mathbb{R}_2^2$ holds isometrically and $T^{\mathbb{C}} = T \otimes id_{\mathbb{R}_2^2}$, it follows from theorem 7.3. that *for positive operators T and all $p, q \in [1, \infty]$*

$$\|T^{\mathbb{C}} : L_q^{\mathbb{C}}(\mu) \longrightarrow L_p^{\mathbb{C}}(\nu)\| = \|T : L_q^{\mathbb{R}}(\mu) \longrightarrow L_p^{\mathbb{R}}(\nu)\|$$

holds. Remark 1, case (2) gives:

REMARK 2: $\|T^{\mathbb{C}}\| = \|T\|$ *for all* $T \in \mathcal{L}(L_q^{\mathbb{R}}(\mu), L_p^{\mathbb{R}}(\nu))$ *if* $1 \leq q \leq p \leq \infty$.

PROOF: Since $L_q^{\mathbb{C}}(\mu) = L_q^{\mathbb{R}}(\mu) \otimes_{\Delta_q} \mathbb{R}_2^2$ and $T^{\mathbb{C}} = T \otimes id$, remark 1, case (2), implies

$$\|T^{\mathbb{C}}\| = \|T \otimes id : L_q^{\mathbb{R}}(\mu) \otimes_{\Delta_q} \mathbb{R}_2^2 \longrightarrow L_p^{\mathbb{R}}(\nu) \otimes_{\Delta_p} \mathbb{R}_2^2\| = \|T\|. \square$$

Recall Ex 14.4. showing that this result is false for $q = \infty$ and $p = 1$. For more results about the norm of the complexification see Ex 26.9. and Ex 28.14..

In the cases $1 \leq p = s < q = r < 2$ and $2 < p = s < q = r \leq \infty$ the operator $T_1 \otimes T_2$ in the proposition need not be continuous. The following counterexample (as well as the last proposition for $p = s$ and $q = r$) is due to Bennett [10], see also Rosenthal–Szarek [233]. For $1 \leq q < p < 2$ let μ_q be the Lévy measure on $\mathbb{K}^\infty$ and $I : \ell_q \hookrightarrow L_p(\mu_q)$ the Lévy embedding 24.4.:

PROPOSITION 2: $I \otimes I : \ell_q \otimes_{\Delta_q} \ell_q \to L_p(\mu_q) \otimes_{\Delta_p} L_p(\mu_q)$ is not continuous ($1 \leq p < q < 2$).

PROOF: As in section 24 the k-th projection $\mathbb{K}^\infty \to \mathbb{K}$ will be denoted by h_k and therefore $I(\sum_{k=1}^n \alpha_k e_k) = \sum_{k=1}^n \alpha_k h_k$. Take $z_n := n^{-1/q} \sum_{k=1}^n e_k \otimes e_k$; then $\Delta_q(z_n; \ell_q, \ell_q) = 1$. On the other hand,

$$\Delta_p(I \otimes I(z_n); L_p, L_p) = \left(\int \int \left| \sum_{k=1}^n n^{-1/q} h_k(w) h_k(v) \right|^p \mu_q(dv) \mu_q(dw) \right)^{1/p} =$$

$$= c_{q,p} \left(\int \left(\frac{1}{n} \sum_{k=1}^n |h_k(w)|^q \right)^{p/q} \mu_q(dw) \right)^{1/p}$$

holds. The functions $|h_k|^q$ are identically distributed, stochastically independent functions on the probability space $(\mathbb{K}^\infty, \mu_q)$. Since they are not integrable (see 24.3.), the converse of Kolmogorov's law of large numbers (see e.g. Krickeberg [153], III 3.7.) implies that

$$\frac{1}{n} \sum_{k=1}^n |h_k(w)|^q$$

converges to ∞ on a subset $A \subset \mathbb{K}^\infty$ of positive measure. Fatou's lemma implies that the sequence $(\Delta_p(I \otimes I(z_n); L_p, L_p))$ is unbounded, which shows that $I \otimes I$ is not continuous. $\square$

Simple duality (see 15.10.) shows that

$$I' \otimes I' : L_{p'}(\mu_q) \otimes_{\Delta_{p'}} L_{p'}(\mu_q) \longrightarrow \ell_{q'} \otimes_{\Delta_{q'}} \ell_{q'}$$

is also not continuous ($2 < q' < p' \leq \infty$). For estimates of non–continuity of $I \otimes I$ on $\ell_q^n \otimes \ell_q^n$ see Rosenthal–Szarek [233] and Carl–D. [22].

26.4. To investigate the tensor norm $g_p = \alpha_{p,1}$ it is advisable to first study the ideals of absolutely p-summing operators. Let us recall some results which were already proven: In 23.10. it was shown that for all Banach spaces F

(1) $\quad \mathfrak{P}_p(\ell_s, F) = \mathfrak{P}_1(\ell_s, F)$ and $\mathbf{P}_1 \leq K_G \mathbf{P}_p \quad$ for $1 \leq p \leq 2$ and $1 \leq s \leq 2$

and in 24.7.

(2) $\quad \mathfrak{P}_p(\ell_r, F) = \mathfrak{P}_1(\ell_r, F)$ and $\mathbf{P}_1 \leq \bar{c}_{p',r} \mathbf{P}_p \quad$ for $1 \leq p < 2$ and $2 \leq r < p'$

(the constant can be easily deduced from the proof of theorem 24.7.).

PROPOSITION:

(3) $\quad \mathfrak{P}_q(F, E) = \mathfrak{P}_2(F, E)$ and $\mathbf{P}_2 \leq C_2(E) b_q \mathbf{P}_q$ for $2 \leq q < \infty$ and E of cotype 2.

Recall that $C_2(\ell_s) \leq a_s$ if $1 \leq s \leq 2$ (see 8.6.) — where a_s and b_q are the constants from the Khintchine inequality.

PROOF: Take $T \in \mathfrak{P}_q(F, E)$ and $x_1, ..., x_n \in F$. The definition of cotype 2 and the Grothendieck–Pietsch domination theorem give

$$\left(\sum_{k=1}^n \|Tx_k\|^2\right)^{1/2} \leq C_2(E) \left(\int_D \left\|\sum_{k=1}^n \varepsilon_k(w) Tx_k\right\|_E^2 \mu(dw)\right)^{1/2} \leq$$

$$\leq C_2(E) \left(\int_D \left\|T\left(\sum_{k=1}^n \varepsilon_k(w) x_k\right)\right\|_E^q \mu(dw)\right)^{1/q} \leq$$

$$\leq C_2(E) \mathbf{P}_q(T) \left(\int_D \int_{B_{F'}} \left|\langle x', \sum_{k=1}^n \varepsilon_k(w) x_k\rangle\right|^q \nu(dx') \mu(dw)\right)^{1/q} \leq$$

$$\leq C_2(E) \mathbf{P}_q(T) \sup_{x' \in B_{F'}} \left(\int_D \left|\sum_{k=1}^n \varepsilon_k(w) \langle x', x_k\rangle\right|^q \mu(dw)\right)^{1/q} \leq$$

$$\leq C_2(E) \mathbf{P}_q(T) b_q w_2(x_k; F)$$

by Khintchine's inequality 8.5.. □

An obvious modification of the proof gives

(3') $\quad\quad\quad\quad \mathfrak{C}_2 \circ \mathfrak{P}_q \subset \mathfrak{P}_2 \quad$ for $2 \leq q < \infty$.

The cyclic composition theorem 25.4. gives

(3'') $\quad\quad\quad\quad \mathfrak{J}_2 \circ \mathfrak{C}_2 \subset \mathfrak{J}_q \quad$ for $1 < q \leq 2$.

Since all tensor norms involved are accessible, the transfer argument 17.7. and the embedding lemma allow us to translate these results into inequalities about tensor norms.

COROLLARY 1: *Let F be a Banach space and $s, p \in [1, 2]$ and $r, q \in [2, \infty]$.*

(1) $\quad d_\infty \leq d_q \leq K_G d_\infty \quad$ on $\ell_s \otimes F$
$\qquad g_q^* \leq g_\infty^* \leq K_G g_q^* \quad$ on $\ell_r \otimes F$.

(2) $\quad d_\infty \leq d_q \leq \overline{c}_{q,r} d_\infty \quad$ on $\ell_r \otimes F$ *if additionally* $r < q$
$\qquad g_q^* \leq g_\infty^* \leq \overline{c}_{q,s'} g_q^* \quad$ on $\ell_s \otimes F$ *if additionally* $s > q'$.

(3) $\quad g_2 \leq g_p \leq a_{r'} b_{p'} g_2 \quad$ on $\ell_r \otimes F$ *if additionally* $p \neq 1$
$\qquad d_p^* \leq d_2^* \leq a_s b_{p'} d_p^* \quad$ on $\ell_s \otimes F$ *if additionally* $p \neq 1$.

26.5. It was shown in 26.1. that every operator $\ell_r \to \ell_s$ factors through a Hilbert space if $1 \leq s \leq 2 \leq r \leq \infty$. More generally,

COROLLARY 2 (Kwapień [160]): *Let $1 \leq s \leq p \leq r \leq \infty$. Then*

$$\mathcal{L}(\ell_r, \ell_s) = \mathcal{L}_p(\ell_r, \ell_s) .$$

PROOF: One may assume $s < p < r$. Take $p \leq 2$ (the case $p > 2$ follows by duality) and an operator $T \in \mathcal{L}(\ell_r, \ell_s)$. According to proposition 25.9.(1) one has to show that $TS \in \mathfrak{I}_p(E, \ell_s)$ for each $S \in \mathfrak{P}_p^{dual}(E, \ell_r)$. The statements 26.4.(1) and (2) imply

$$S' \in \mathfrak{P}_p(\ell_{r'}, E') \subset \mathfrak{P}_1(\ell_{r'}, E') \subset \mathfrak{P}_s(\ell_{r'}, E') .$$

Ex 17.7. yields $TS \in \mathfrak{P}_s^{dual}(E, \ell_s) \subset \mathfrak{I}_p(E, \ell_s)$, which is exactly what is needed. □

This time we find it worthwhile to explicitly note that the $\mathcal{L}_p$-local techniques imply that for $1 \leq s \leq p \leq r \leq \infty$ every operator $L_r(\mu) \to L_s(\nu)$ factors

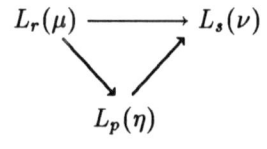

$$\begin{array}{ccc} L_r(\mu) & \longrightarrow & L_s(\nu) \\ & \searrow \quad \nearrow & \\ & L_p(\eta) & \end{array}$$

(recall that $L_1(\nu)$ is complemented in its bidual — see Appendix B10.).

26.6. For Hilbert spaces many of the tensor norms/operator ideals that have been considered coincide. First, recall from 11.6. that

$$\mathfrak{P}_2(H_1, H_2) = \mathfrak{HS}(H_1, H_2) ,$$

the Hilbert-Schmidt operators, holds isometrically for Hilbert spaces H_1 and H_2. Hence,

$$H_1 \otimes_{g_2^*} H_2 \xrightarrow{\ 1\ } \mathfrak{P}_2(H_1', H_2) = \mathfrak{H}\mathfrak{S}(H_1', H_2)$$

and therefore — for finite orthonormal systems (e_i) and (f_j) —

$$g_2^*\left(\sum_{i,j} \alpha_{ij} e_i \otimes f_j; H_1, H_2\right) = \left(\sum_{i,j} |\alpha_{ij}|^2\right)^{1/2},$$

which implies that $g_2 = g_2^* \overset{1}{=} d_2^* = d_2$ on $H_1 \otimes H_2$ and g_2^* is the Hilbert-Schmidt norm. These relations follow also from the Chevet-Persson-Saphar "equality" (15.10., corollary 2). For $\ell_2 \otimes \ell_2$, and hence for all Hilbert spaces, 26.1. and corollary 1 now imply:

COROLLARY 3: *On the tensor product $H_1 \otimes H_2$ of two Hilbert spaces the following inequalities hold:*

$$\begin{aligned}
\varepsilon &\leq \alpha_{p,q} \leq K_G b_{q'} b_{p'} \varepsilon & &\text{for } p, q \in]1, \infty[\\
\alpha_{p,q}^* &\leq \pi \leq K_G b_{q'} b_{p'} \alpha_{p,q}^* & &\text{for } p, q \in]1, \infty[\\
g_q &\leq g_2 \leq K_G g_q & &\text{for } q \in [2, \infty] \\
g_2 &\leq g_p \leq b_{p'} g_2 & &\text{for } p \in]1, 2] \\
g_2 &\leq g_q^* \leq K_G g_2 & &\text{for } q \in [2, \infty] \\
g_p^* &\leq g_2 \leq b_{p'} g_p^* & &\text{for } p \in]1, 2] .
\end{aligned}$$

— and, by transposition, the same for d_p and d_p^*. So there are, up to equivalences, only three tensor norms among the $\alpha_{p,q}$ and $\alpha_{p,q}^*$ on Hilbert spaces: ε, π and the Hilbert-Schmidt norm g_2. In terms of *operators between Hilbert spaces* this means that:

$$\begin{aligned}
p, q \in]1, \infty[&: & \mathfrak{L}_{p,q} &= \mathfrak{L}_p = \mathfrak{L} & &\text{all operators} \\
p, q \in]1, \infty[&: & \mathfrak{D}_{p,q} &= \mathfrak{D}_p = \mathfrak{J} = \mathfrak{N} & &\text{nuclear operators} \\
p \in [1, \infty[&: & \mathfrak{P}_p &= \mathfrak{P}_p^{dual} = \mathfrak{L}_1 = \mathfrak{L}_\infty = \\
q \in]1, \infty[&: & &= \mathfrak{J}_q = \mathfrak{J}_q^{dual} = \mathfrak{H}\mathfrak{S} & &\text{Hilbert–Schmidt operators.}
\end{aligned}$$

(That $\mathfrak{J} = \mathfrak{N}$ follows from the Radon-Nikodým property of Hilbert spaces). Pełczyński's result [198]

$$\mathfrak{P}_p(H_1, H_2) = \mathfrak{H}\mathfrak{S}(H_1, H_2) \qquad \text{for } 1 \leq p < \infty$$

is a special case. For the associated minimal operator ideals see Ex 26.7..

26.7. By what was said in section 12, the definition of a tensor norm on the class HILB of all Hilbert spaces is clear. Since every closed subspace of a Hilbert space is a Hilbert space and 1–complemented, it is clear that tensor norms on HILB are projective, injective and totally accessible. The most prominent tensor norm on HILB is the *Hilbert-Schmidt norm* σ defined by the embedding

$$H_1 \otimes H_2 \hookrightarrow \mathfrak{H}\mathfrak{S}(H_1', H_2) .$$

Note that the dual H' of a Hilbert space is a Hilbert space with the scalar product

$$(x'|y')_{H'} := (R^{-1}y'|R^{-1}x')_H \quad \text{for } x', y' \in H',$$

where $R: H \to H'$ is the "Riesz–mapping" — the (antilinear) mapping from the Riesz' representation theorem. As has already been said

$$\sigma\Big(\sum_{k=1}^{n}\sum_{\ell=1}^{m} \alpha_{k\ell} e_k \otimes f_\ell; H_1, H_2\Big) = \Big(\sum_{k=1}^{n}\sum_{\ell=1}^{m} |\alpha_{k\ell}|^2\Big)^{1/2}$$

if $(e_k)_{k=1}^n$ and $(f_\ell)_{\ell=1}^m$ are orthonormal systems.

If α is a tensor norm on HILB, then its dual α' is defined as in section 15: first on the class HILB $\cap$ FIN and then take the finite (= cofinite) hull. Clearly,

$$H_1 \otimes_\alpha H_2 \xrightarrow{\;1\;} (H_1' \otimes_{\alpha'} H_2')'$$

always holds — hence, $|\langle z', z\rangle| \leq \alpha(z; H_1, H_2)\alpha'(z'; H_1', H_2')$. Recall that $\sigma = g_2 = g_2^* = d_2^* = g_2' = \sigma'$.

PROPOSITION: *Let α be a tensor norm on* HILB. *Then*

$$\sigma(z; H_1, H_2) \leq (\alpha(z; H_1, H_2))^{1/2} (\alpha'(R_1 \otimes R_2(z); H_1', H_2'))^{1/2}$$

for all $z \in H_1 \otimes H_2$ (where $R_k: H_k \to H_k'$ is the Riesz–mapping).

See Ex. 2.10.(b) concerning the antilinear map $R_1 \otimes R_2$.

PROOF: Since $\langle e_k \otimes f_\ell, R_1 e_i \otimes R_2 f_j \rangle = \delta_{k,i} \cdot \delta_{\ell,j}$ for orthonormal systems, it follows for $z = \sum_{k,\ell} \alpha_{k\ell} e_k \otimes f_\ell$ that

$$\sigma(z; H_1, H_2)^2 = \sum_{k,\ell} |\alpha_{k\ell}|^2 = \Big\langle \sum_{k,\ell} \alpha_{k\ell} e_k \otimes f_\ell, \sum_{i,j} \overline{\alpha_{ij}} R_1 e_i \otimes R_2 f_j \Big\rangle =$$

$$= \langle z, R_1 \otimes R_2(z)\rangle \leq \alpha(z; H_1, H_2)\alpha'(R_1 \otimes R_2(z); H_1', H_2') . \quad \square$$

COROLLARY 1: $\sigma \leq (\alpha \cdot \alpha')^{1/2}$ *holds for all tensor norms α on* HILB *in the real case — and in the complex case for all α which satisfy the Riesz–condition*

$$\alpha(z; H_1, H_2) = \alpha(R_1 \otimes R_2(z); H_1', H_2') .$$

It is easy to see that α', α^t and α^* satisfy the Riesz–condition if α does. Since $w_p(x_k; H) = w_p(Rx_k; H')$, it follows that all $\alpha_{p,q}$ (and hence all $\alpha_{p,q}^*$) satisfy this condition. Actually, we do not know of any tensor norm which does not have this property.

COROLLARY 2: *Let α be a tensor norm on HILB (which satisfies the Riesz-condition in the complex case). If $\alpha = \alpha'$, then α is the Hilbert–Schmidt norm σ.*

PROOF: Corollary 1 shows that $\sigma \leq \alpha$, which implies that $\alpha = \alpha' \leq \sigma'$. Now recall that $\sigma = \sigma'$. □

26.8. The Hilbert–Schmidt norm σ plays a central role in the theory of Hilbert spaces; in particular, $\sigma = \sigma' = \sigma^* = \sigma^t$. Is there an extension α of σ from HILB to the class NORM of all normed spaces which is *symmetric* ($\alpha = \alpha^t$) and *self-adjoint* ($\alpha = \alpha^*$)? Since $g_2 = g_2^* \neq g_2^{*t} = d_2^* = d_2$ on NORM (see 11.12. for this statement in terms of operators), none of the self-adjoint extensions g_2 and d_2 is symmetric. Does a self-adjoint and symmetric tensor norm α on NORM exist? Since $\alpha = (\alpha^t)^* = \alpha'$, such a tensor norm necessarily coincides with σ on HILB by corollary 2. In terms of operators: Is there a (maximal) Banach operator ideal $(\mathfrak{A}, \mathbf{A})$ which is self-adjoint ($\mathfrak{A} = \mathfrak{A}^*$) and symmetric ($\mathfrak{A} = \mathfrak{A}^{dual}$)?

If there were a self-adjoint, symmetric tensor norm/operator ideal which were unique — or at least with a natural description — this would undoubtedly enrich the theory of Banach spaces. From this point of view it is "unfortunate" that there are many self-adjoint, symmetric tensor norms, none of which seem to have a nice description.

Self-adjoint and symmetric tensor norms on NORM were constructed in 1943 by Schatten [245].

THEOREM (Schatten): *Let α be a finitely generated, symmetric tensor norm on NORM which satisfies $\alpha \leq \alpha^*$. Then there is a self-adjoint, symmetric and finitely generated tensor norm γ on NORM with*

$$\alpha \leq \gamma \leq \alpha^* .$$

γ is totally accessible if α and α^* are. If, in the complex case, α satisfies the Riesz-condition on HILB, then γ does as well.

It follows from corollary 2 that all of these tensor norms γ coincide on Hilbert spaces with the Hilbert–Schmidt norm σ. The injective norm $\alpha = \varepsilon$ satisfies the assumptions; the tensor norm $\alpha = w_2$ even gives a totally accessible tensor norm γ. For accessibility see Ex 26.8..

PROOF: For the convenience of the construction, assume $\alpha \geq \alpha^*$ instead of $\alpha \leq \alpha^*$.

1. If α and β are tensor norms on FIN, then, obviously, $\frac{1}{2}(\alpha + \beta)$ is a tensor norm.

Moreover, using $E \ni z \rightsquigarrow (z,z) \in (E,\alpha/2) \oplus_1 (E,\beta/2)$ for $E = M' \otimes N'$, gives that

$$\left(\frac{\alpha+\beta}{2}\right)'(z;M,N) = 2\inf\{\max\{\alpha'(z_1;M,N),\beta'(z_2;M,N)\} \mid z = z_1+z_2\} \le$$
$$\le 2\inf\{\max\{\lambda\alpha'(z),(1-\lambda)\beta'(z)\} \mid 0 \le \lambda \le 1\} \le$$
$$\le \frac{1}{2}(\alpha'(z;M,N)+\beta'(z;M,N))$$

by taking an appropriate λ. It follows that

$$\left(\frac{\alpha+\beta}{2}\right)' \le \frac{\alpha'+\beta'}{2} \quad \text{and} \quad \left(\frac{\alpha+\beta}{2}\right)^* \le \frac{\alpha^*+\beta^*}{2}$$

on FIN and hence on NORM. Note that if α and β are finitely (or cofinitely) generated, then $\frac{1}{2}(\alpha+\beta)$ is also.

2. Now take a finitely generated tensor norm α with $\alpha = \alpha^t \ge \alpha^*$ and define inductively

$$\alpha_1 := \alpha, \quad \alpha_n := \frac{\alpha_{n-1}+\alpha^*_{n-1}}{2}.$$

Then by part 1 of the proof
(a) $\alpha_n^* \le (\alpha^*_{n-1}+\alpha^{**}_{n-1})/2 = \alpha_n$
(b) $\alpha_n - \alpha_{n+1} = \alpha_n - (\alpha_n+\alpha_n^*)/2 = (\alpha_n-\alpha_n^*)/2 \ge 0$ by (a), in particular,
(c) $\alpha_n \ge \alpha_{n+1}$ and $\alpha_n^* \le \alpha^*_{n+1}$
(d) $0 \le \alpha_n - \alpha_n^* \le (\alpha_{n-1}+\alpha^*_{n-1})/2 - \alpha^*_{n-1} = (\alpha_{n-1}-\alpha^*_{n-1})/2 \le \ldots \le (\alpha-\alpha^*)/2^{n-1}$
by induction. Define

$$\gamma(z;E,F) := \inf \alpha_n(z;E,F) = \sup \alpha_n^*(z;E,F).$$

It is easy to see that γ is a tensor norm. Moreover,

$$\overrightarrow{\gamma}(z;E,F) = \inf\{\gamma(z;M,N) \mid z \in M \otimes N, (M,N) \in \text{FIN}(E) \times \text{FIN}(F)\} =$$
$$= \inf_{M,N} \inf_n \alpha_n(z;M,N) = \inf_n \inf_{M,N} \alpha_n(z;M,N) = \inf_n \alpha_n(z;E,F) =$$
$$= \gamma(z;E,F)$$

and hence γ is finitely generated. A calculation of the same type, using $\gamma = \sup \alpha_n^*$, shows that γ is cofinitely generated if α and α^* (and hence all α_n^*) are. Since $\alpha_n = \alpha_n^t$, it follows that $\gamma = \gamma^t$. Moreover,

$$\gamma^*(z;M,N) = \sup\{|\langle u,z\rangle| \mid \gamma^t(u;M',N') < 1\} =$$
$$= \sup_n \sup\{|\langle u,z\rangle| \mid \alpha_n^t(u;M',N') < 1\} =$$
$$= \sup_n \alpha_n^*(z;M,N) = \gamma(z;M,N),$$

which shows that γ is self-adjoint. Finally, if α satisfies the Riesz–condition, then α^* does and so do all α_n: this also shows that γ does. $\square$

26.9. It is not a priori clear whether or not the Schatten construction may produce different γ for different α. It was Puhl [227] who noticed that one actually obtains infinitely many operator ideals extending the Hilbert–Schmidt ideal. In order to distinguish the various operator ideals he used limit orders. We shall give the main steps of his construction — in terms of tensor norms.

PROPOSITION: *For every finitely generated tensor norm α on NORM there is a symmetric, finitely generated tensor norm β such that*

$$\beta \leq \beta^* \leq (\alpha \cdot \alpha' \cdot \alpha^* \cdot \alpha^t)^{1/4} \ .$$

β can be chosen to satisfy the Riesz–condition on HILB if α does.

PROOF: It is enough to show this on FIN. Define for $M, N \in$ FIN

$$f(u; M, N) := [\alpha(u)\alpha'(u)\alpha^*(u)\alpha^t(u)]^{1/4} \qquad \text{for } u \in M \otimes N.$$

Then $\varepsilon \leq f \leq \pi$, and

$$\beta(z; M, N) := \sup\{|\langle u, z\rangle| \mid f(u; M', N') \leq 1\}$$

is easily seen to be a symmetric tensor norm on FIN. Since f is positive homogeneous, it follows that

$$|\langle u, z\rangle| \leq f(u)\beta(z) \ ,$$

and hence $\beta' \leq f$. The fact that $|\langle u, z\rangle| \leq \alpha(u)\alpha'(z)$ and $\leq \alpha'(u)\alpha(z)$ (and so on) implies that

$$|\langle u, z\rangle| \leq f(u)f(z) \ ,$$

which shows that $\beta(u) \leq f(u)$; therefore,

$$\beta(z) = \sup\{|\langle u, z\rangle| \mid f(u) \leq 1\} \leq \sup\{|\langle u, z\rangle| \mid \beta(u) \leq 1\} = \beta'(z) = \beta^*(z) \ .$$

The statement concerning the Riesz–condition follows from the fact that f satisfies it whenever α (and hence α', α^* and α^t) does. $\square$

26.10. For $u^{(n)} := \sum_{k=1}^n e_k \otimes e_k \in \mathbb{K}^n \otimes \mathbb{K}^n$ and $p, q \in [1, \infty]$ the inequality

$$\alpha(u^{(n)}; \ell_p^n, \ell_q^n) \leq \pi(u^{(n)}; \ell_p^n, \ell_q^n) \leq n$$

holds. The *limit order* $\lambda(\alpha; p, q)$ of a tensor norm α on FIN is the infimum of all $\lambda \in [0, 1]$ such that there is a $c_\lambda \geq 0$ with

$$\alpha(u^{(n)}; \ell_p^n, \ell_q^n) \leq c_\lambda n^\lambda \qquad \text{for all } n \in \mathbb{N}.$$

The *limit order* $\lambda(\mathfrak{A}; p, q)$ of a Banach operator ideal $(\mathfrak{A}, \mathbf{A})$ is defined to be the infimum of all $\lambda \in [0, 1]$ with a $c_\lambda \geq 0$ such that

$$\mathbf{A}(id : \ell_p^n \longrightarrow \ell_q^n) \leq c_\lambda n^\lambda \qquad \text{for all } n \in \mathbb{N}.$$

This notion is due to Pietsch; see [214], sections 14 and 22, for the explicit calculation of various limit orders. Obviously,

$$\lambda(\alpha; p, q) = \lambda(\mathfrak{A}; p', q)$$

if α and $\mathfrak{A}$ are associated and clearly, $\lambda(\alpha; p, q) = \lambda(\alpha^t; q, p)$. Since

$$n = \langle u^{(n)}, u^{(n)} \rangle \leq \alpha(u^{(n)}; \ell_p^n, \ell_q^n) \alpha'(u^{(n)}; \ell_{p'}^n, \ell_{q'}^n),$$

it follows that

$$1 \leq \lambda(\alpha; p, q) + \lambda(\alpha'; p', q') = \lambda(\alpha; p, q) + \lambda(\alpha^*; q', p') \leq 2.$$

It may happen that the sum is even 2 (see König [144]), but in many cases it is 1. For example (Pietsch [214], 22.4.12.),

(*) $$\lambda(g_p^*; r, s) + \lambda(g_p; s', r') = \lambda(\mathfrak{P}_{p'}; r', s) + \lambda(\mathfrak{I}_p; s, r') = 1$$

if $2 \leq p \leq \infty$ and $(r, s) \in [1, 2] \times [1, \infty]$.

If α is a tensor norm and β_α the one constructed in proposition 26.9.

$$\beta_\alpha \leq \beta_\alpha^* \leq (\alpha \cdot \alpha' \cdot \alpha^* \cdot \alpha^t)^{1/4},$$

then, clearly,

(**) $$\lambda(\beta_\alpha; p, q) \leq \lambda(\beta_\alpha^*; p, q) \leq \frac{1}{4}[\lambda(\alpha; p, q) + \lambda(\alpha^*; q, p) + \lambda(\alpha^*; p, q) + \lambda(\alpha; q, p)] =$$
$$=: \Lambda(\alpha; p, q).$$

LEMMA: *If* $\Lambda(\alpha; p, q) + \Lambda(\alpha; q', p') = 1$, *then*

$$\lambda(\gamma; p, q) = \Lambda(\alpha; p, q)$$

for all tensor norms γ *with* $\beta_\alpha \leq \gamma \leq \beta_\alpha^*$.

PROOF: Inequality (**) gives

$$1 \leq \lambda(\beta_\alpha; p, q) + \lambda(\beta_\alpha^*; q', p') \leq \Lambda(\alpha; p, q) + \Lambda(\alpha, q', p') = 1,$$

so the first terms in the two sums coincide. Again by (**) it follows that

$$\Lambda(\alpha;p,q) = \lambda(\beta_\alpha;p,q) \leq \lambda(\beta_\alpha^*;p,q) \leq \Lambda(\alpha;p,q) .$$

This implies the statement. □

Take $\alpha = g_p^*$ for $2 \leq p \leq \infty$. Then

$$1 = \lambda(g_p^*;2,1) + \lambda(g_p;\infty,2) = \lambda(g_p^*;1,2) + \lambda(g_p;2,\infty) = \lambda(g_p^*;2,\infty) + \lambda(g_p;1,2)$$

by the result (*) mentioned above. Moreover,

$$1 \leq \lambda(g_p^*;\infty,2) + \lambda(g_p;2,1) = \lambda(\mathfrak{P}_{p'};1,2) + \lambda(g_p;2,1) \leq 0 + 1$$

by Grothendieck's theorem. It follows that

$$\Lambda(g_p^*;1,2) + \Lambda(g_p^*;2,\infty) = 1$$

and

$$\Lambda(g_p^*;1,2) = \frac{1}{4} \left[2 + \lambda(g_p^*;1,2) + \lambda(g_p^*;2,1) - \lambda(g_p^*;\infty,2) - \lambda(g_p^*;2,\infty)\right] =$$
$$= \frac{1}{4}\left(2 + \frac{1}{p'} + 1 - 0 - \frac{1}{2}\right) = \frac{1}{4}\left(\frac{7}{2} - \frac{1}{p}\right)$$

— the latter by Pietsch [214], 22.4.12 and Grothendieck's theorem.

Denote by γ_p the self-adjoint, symmetric, finitely generated tensor norm which comes from the Schatten–construction 26.8. when starting with $\alpha = \beta_{g_p^*}$ from Puhl's proposition 26.9.. Then the lemma implies that

$$\lambda(\gamma_p;1,2) = \Lambda(g_p^*;1,2) = \frac{1}{4}\left(\frac{7}{2} - \frac{1}{p}\right)$$

for $2 \leq p \leq \infty$, which yields that the tensor norms γ_p are pairwise non–equivalent.

COROLLARY (Schatten–Puhl): *There are infinitely many non–equivalent self–adjoint, symmetric, finitely generated tensor norms which coincide on Hilbert spaces with the Hilbert–Schmidt norm.*

26.11. In Ex 12.9. it was shown that the unit vectors $e_k \otimes e_\ell \in \ell_2 \tilde{\otimes}_\alpha \ell_2$ form a basis for every tensor norm α, provided the ordering of $(e_k \otimes e_\ell)_{k,\ell}$ is the rectangular ordering: $e_1 \otimes e_1, e_2 \otimes e_1, e_2 \otimes e_2, e_1 \otimes e_2, \ldots$ Since

$$\sigma\left(\sum_{k,\ell=1}^\infty \alpha_{k,\ell} e_k \otimes e_\ell; \ell_2, \ell_2\right) = \left(\sum_{k,\ell=1}^\infty |\alpha_{k,\ell}|^2\right)^{1/2},$$

it is obvious that this natural basis is even unconditional for the Hilbert–Schmidt norm, i.e. the series expansion along the basis converges unconditionally. Gelbaum–Gil de Lamadrid [80] observed that for $\alpha = \varepsilon$ the basis is not unconditional. But much more is true; Kwapień–Pełczyński [163] showed that

THEOREM: *If α is a tensor norm on HILB such that $(e_k \otimes e_\ell)_{k,\ell \in \mathbb{N}}$ is an unconditional basis in $\ell_2 \tilde{\otimes}_\alpha \ell_2$, then α is equivalent to the Hilbert–Schmidt norm σ.*

The key to the proof is the following well-known

LEMMA: *Let (x_n) be an unconditional basis of a Banach space E. Then there is a constant $c > 0$ such that*

$$c^{-1} \left\| \sum_{n=1}^{\infty} |\lambda_n| x_n \right\| \leq \left\| \sum_{n=1}^{\infty} \lambda_n x_n \right\| \leq c \left\| \sum_{n=1}^{\infty} |\lambda_n| x_n \right\|.$$

PROOF: $\ell_1 \tilde{\otimes}_\varepsilon E$ is the space of unconditionally convergent series in E and

$$\varepsilon\left(\sum_{n=1}^{\infty} e_n \otimes \lambda_n x_n; \ell_1, E \right) = w_1(\lambda_n x_n; E) = \sup\left\{ \left\| \sum_{n=1}^{\infty} \alpha_n \lambda_n x_n \right\| \, \bigg| \, |\alpha_n| \leq 1 \right\}$$

by the results of section 8. Since

$$E \longrightarrow \ell_1 \tilde{\otimes}_\varepsilon E$$
$$\sum_{n=1}^{\infty} \lambda_n x_n \rightsquigarrow \sum_{n=1}^{\infty} e_n \otimes \lambda_n x_n$$

is continuous (closed graph theorem), it follows that there is a $c > 0$ such that

$$\left\| \sum_{n=1}^{\infty} |\lambda_n| x_n \right\| \leq w_1(\lambda_n x_n) \leq c \left\| \sum_{n=1}^{\infty} \lambda_n x_n \right\|$$
$$\left\| \sum_{n=1}^{\infty} \lambda_n x_n \right\| \leq w_1(|\lambda_n| x_n) \leq c \left\| \sum_{n=1}^{\infty} |\lambda_n| x_n \right\|,$$

which is the result. □

PROOF (of the theorem): It is clear that it is enough to show that α and σ are equivalent on $\ell_2^n \otimes \ell_2^n$ with constants independent of n. Define for $z := \sum_{k,\ell=1}^{n} \alpha_{k,\ell} e_k \otimes e_\ell \in \ell_2^n \otimes \ell_2^n$

$$|z| := \sum_{k,\ell=1}^{n} |\alpha_{k,\ell}| e_k \otimes e_\ell \in \ell_2^n \otimes \ell_2^n \subset \ell_2 \otimes \ell_2$$

(canonical embedding). Then, by the lemma, there is a constant $c > 0$ such that

$$c^{-1}\alpha(|z|; \ell_2, \ell_2) \leq \alpha(z; \ell_2, \ell_2) \leq c\alpha(|z|; \ell_2, \ell_2) \ .$$

Now fix $z \in \ell_2^n \otimes \ell_2^n$ and choose two isometries $U_1, U_2 : \ell_2^n \to \ell_2^n$ such that

$$U_1 \otimes U_2(z) = \sum_{k=1}^{n} \lambda_k e_k \otimes e_k$$

with $\lambda_k \geq 0$ — this follows, e.g., from the polar decomposition UA of the operator $T_z : \ell_2^n \to \ell_2^n$ (where U is an isometry and A positive) and the diagonalization $A = DV$, hence $U^{-1} T_z V^{-1} = D$.

Isometries (onto) preserve tensor norms, therefore,

$$\sigma(z) = \left(\sum_{k=1}^{n} \lambda_k^2\right)^{1/2} \ .$$

Now take the Fourier matrix (see Ex 4.3.)

$$A_n := \sum_{k,\ell=1}^{n} n^{-1/2} \exp(2\pi i \, k\ell/n) e_k \otimes e_\ell$$

and

$$w := (A_n \otimes id_{\ell_2^n}) \circ (U_1 \otimes U_2)(z) = \sum_{k,\ell=1}^{n} \lambda_k n^{-1/2} \exp(2\pi i \, k\ell/n) e_\ell \otimes e_k \ .$$

Since all operators involved are isometries, it follows that $\alpha(w) = \alpha(z)$. But

$$\alpha(|w|) = \alpha\left(\sum_{k,\ell=1}^{n} \lambda_k n^{-1/2} e_\ell \otimes e_k\right) = \alpha\left(\sum_{\ell=1}^{n} n^{-1/2} e_\ell \otimes \sum_{k=1}^{n} \lambda_k e_k\right) =$$

$$= \left\|\sum_{\ell=1}^{n} n^{-1/2} e_\ell\right\|_{\ell_2^n} \left\|\sum_{k=1}^{n} \lambda_k e_k\right\|_{\ell_2^n} = \left(\sum_{k=1}^{n} \lambda_k^2\right)^{1/2} = \sigma(z) \ ,$$

which implies

$$\sigma(z) = \alpha(|w|) \leq c\alpha(w) = c\alpha(z)$$
$$\sigma(z) = \alpha(|w|) \geq c^{-1}\alpha(w) = c^{-1}\alpha(z)$$

— and this was to be shown. □

Exercices:

Ex 26.1. Let $p, q \in]1, \infty[$ with $1/p + 1/q \geq 1$ and $1 \leq s \leq 2 \leq r \leq \infty$.

(a) $\mathcal{L}_s \circ \mathcal{L}_r \subset \mathcal{L}_{p,q}$.

(b) $\mathfrak{D}_{p',q'} \circ \mathcal{L}_s \subset \mathfrak{D}_{r'}$.

(c) $\mathcal{L}_r \circ \mathfrak{D}_{p',q'} \subset \mathfrak{D}_{s'}$.

Hint: 26.1., 23.1., cyclic composition theorem.

Ex 26.2. Let $p, q \in]1, \infty[$ and $1/p + 1/q \leq 1$ and $1 \leq s \leq 2 \leq r \leq \infty$. Then

$$\mathfrak{D}_{p,q}(\ell_s, \ell_r) = \mathfrak{D}_p(\ell_s, \ell_r) = \mathfrak{I}(\ell_s, \ell_r).$$

Hint: 26.1..

Ex 26.3. For $p, q \in]1, \infty[$ and $1/p + 1/q \geq 1$ the projective associate $\backslash\alpha_{p,q}/$ is equivalent to w_2^* and $/\alpha_{p,q}^*\backslash$ is equivalent to w_2. Hint: Grothendieck's inequality, 20.9. and 26.1..

Ex 26.4. Show that $\mathbf{L}_2(id_{\ell_p^n}) = d(\ell_p^n, \ell_2^n)$. Deduce from this and 26.2. that

$$d(\ell_p^n, \ell_2^n) = n^{|1/p - 1/2|}$$

for all $p \in [1, \infty]$.

Ex 26.5. Take $g_1^1, ..., g_n^1 \in L_p(\mu_1)$ and $g_1^2, ..., g_m^2 \in L_p(\mu_2)$. Then

$$w_q(g_k^1 \otimes g_\ell^2; L_p(\mu_1 \otimes \mu_2)) \leq c w_q(g_k^1) w_q(g_\ell^2)$$

if $1 \leq q' \leq p \leq \infty$ (with constant $c = 1$) or if $1 \leq p \leq 2 \leq q' \leq \infty$ (with constant $c = K_G^2$). Hint: Use $\|\sum e_k \otimes g_k : \ell_{q'}^n \to L_p\| = w_q(g_k; L_p)$ from 8.1. and 26.3..

Ex 26.6. Use the foregoing exercise to show: If $(g_k^i)_{k \in \mathbb{N}}$ are weakly q-summable sequences in $L_p(\mu_i)$ and $1 \leq q' \leq p \leq \infty$ or $1 \leq p \leq 2 \leq q' \leq \infty$, then $(g_k^1 \otimes g_\ell^2)_{k,\ell \in \mathbb{N}}$ is weakly q-summable in $L_p(\mu_1 \otimes \mu_2)$. In particular (see 8.3.), if $p \leq 2$ and (g_k^i) are unconditionally summable in $L_p(\mu_i)$, then $(g_k^1 \otimes g_\ell^2)_{k,\ell \in \mathbb{N}}$ is unconditionally summable in $L_p(\mu_1 \otimes \mu_2)$. Using the counterexamples in 26.3., check that the result is false if $1 \leq p < q' < 2$ or $2 < p < q' \leq \infty$. These results are due to Rosenthal-Szarek [233].

Ex 26.7. Deduce from 26.6. that on Hilbert spaces

$$p, q \in]1, \infty[\quad : \quad \mathfrak{K}_{p,q} = \mathfrak{K} \qquad \text{compact operators}$$
$$p \in]1, \infty[\quad : \quad \mathfrak{K}_1 = \mathfrak{K}_\infty = \mathfrak{N}_p = \mathfrak{H}\mathfrak{S} \qquad \text{Hilbert–Schmidt operators.}$$

Hint: $H_1' \otimes_\alpha H_2 \overset{1}{\hookrightarrow} \mathfrak{A}(H_1, H_2)$ for maximal normed operator ideals.

Ex 26.8. Show that the Schatten–construction in 26.8. gives an accessible tensor norm γ whenever α is accessible.

Ex 26.9. (a) Take $1 \leq q \leq p \leq \infty$. If $E \subset L_q^R(\mu)$ and $T \in \mathfrak{L}(E, L_p^R(\nu))$, then $\|T^C\| = \|T\|$. Hint: Use 7.9. as in 26.3..

(b) The best constants in the Khintchine inequality are the same in the real and complex case.

27. Grothendieck's Fourteen Natural Tensor Norms

The second chapter ends with a reference to Grothendieck's Résumé: It will be shown that if one starts with the injective norm ε and takes the adjoint, transposed, right or left injective or projective associate finitely many times, then one obtains, up to equivalence, only 14 different tensor norms. This Grothendieck result is, in a certain sense, a capsule summary of the results in the Résumé.

27.1. Let Φ be the smallest set of (finitely generated) tensor norms which contains the injective norm ε and is stable under the operations $\alpha \leadsto \alpha', \alpha^t, \alpha\backslash$, and hence stable under $\alpha \leadsto \alpha^*, \alpha/, \backslash\alpha, \backslash\alpha/, /\alpha$ and $/\alpha\backslash$. A finitely generated tensor norm α is called *natural* if it is equivalent to a tensor norm in Φ. Grothendieck's inequality states that

$$/\pi\backslash \sim w_2 \quad \text{and} \quad \backslash\varepsilon/ \sim w_2^*,$$

hence the set Φ_o, consisting of the tensor norms

$$\varepsilon, \ \varepsilon/ = w_1, \ w_1\backslash, \ \backslash\varepsilon = w_\infty, \ /w_\infty$$
$$\pi, \ \pi\backslash = w_\infty^*, \ w_\infty^*/, \ /\pi = w_1^*, \ \backslash w_1^*$$
$$w_2, \ w_2^*, \ w_2/[\sim d_2 = d_2^*], \ \backslash w_2 [\sim g_2 = g_2^*],$$

(see 20.14. and the table in 20.15.) contains only natural norms.

THEOREM (Grothendieck): *Every natural tensor norm is equivalent to one of the fourteen tensor norms in Φ_o.*

27.2. Actually a more precise statement can be made, which (due to an idea of Grothendieck's) may be summarized in the following table of natural norms (and their associated maximal Banach operator ideals):

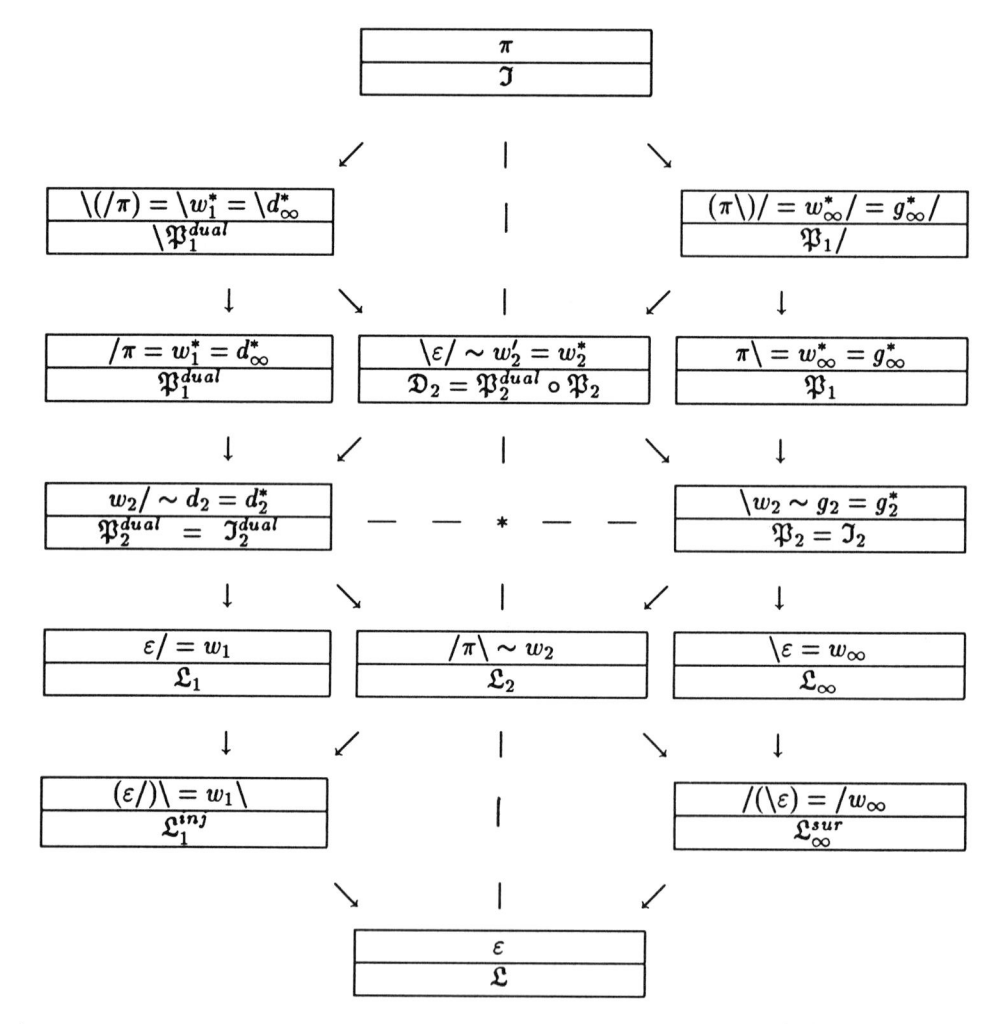

$$\begin{array}{c} \pi \\ \mathfrak{J} \end{array}$$

$$\begin{array}{c} \backslash(/\pi) = \backslash w_1^* = \backslash d_\infty^* \\ \backslash \mathfrak{P}_1^{dual} \end{array}$$

$$\begin{array}{c} (\pi\backslash)/ = w_\infty^*/ = g_\infty^*/ \\ \mathfrak{P}_1/ \end{array}$$

$$\begin{array}{c} /\pi = w_1^* = d_\infty^* \\ \mathfrak{P}_1^{dual} \end{array}$$

$$\begin{array}{c} \backslash\varepsilon/ \sim w_2' = w_2^* \\ \mathfrak{D}_2 = \mathfrak{P}_2^{dual} \circ \mathfrak{P}_2 \end{array}$$

$$\begin{array}{c} \pi\backslash = w_\infty^* = g_\infty^* \\ \mathfrak{P}_1 \end{array}$$

$$\begin{array}{c} w_2/ \sim d_2 = d_2^* \\ \mathfrak{P}_2^{dual} = \mathfrak{J}_2^{dual} \end{array}$$

$$\begin{array}{c} \backslash w_2 \sim g_2 = g_2^* \\ \mathfrak{P}_2 = \mathfrak{J}_2 \end{array}$$

$$\begin{array}{c} \varepsilon/ = w_1 \\ \mathfrak{L}_1 \end{array}$$

$$\begin{array}{c} /\pi\backslash \sim w_2 \\ \mathfrak{L}_2 \end{array}$$

$$\begin{array}{c} \backslash\varepsilon = w_\infty \\ \mathfrak{L}_\infty \end{array}$$

$$\begin{array}{c} (\varepsilon/)\backslash = w_1\backslash \\ \mathfrak{L}_1^{inj} \end{array}$$

$$\begin{array}{c} /(\backslash\varepsilon) = /w_\infty \\ \mathfrak{L}_\infty^{sur} \end{array}$$

$$\begin{array}{c} \varepsilon \\ \mathfrak{L} \end{array}$$

(See 20.12. for the factorization properties of the ideals $\mathfrak{P}_1/$ and $\backslash\mathfrak{P}_1^{dual}$.) This table of the fourteen norms in Φ_o has the following properties, the first of which proves theorem 27.1. and the third of which shows (in particular) that all the fourteen norms in Φ_o are mutually non-equivalent:

(1) If $\alpha \in \Phi_o$, then α', α^t and $\alpha\backslash$ are equivalent to a norm in Φ_o.

(2) *An arrow $\alpha \to \beta$ means that α dominates β, i.e. there is a constant c with $\beta \leq c\alpha$. In other words, $\mathfrak{A} \subset \mathfrak{B}$ for the associated operator ideals.*

(3) *There are no other dominations in the table than those indicated.*

(4) *If the locations of α and β are symmetric with respect to the point $*$ in the middle, then α' and β are equivalent.*

(5) *If the locations of α and β are symmetric with respect to the horizontal dotted line, then α^* and β are equivalent.*

(6) *If the locations of α and β are symmetric with respect to the vertical dotted line, then $\alpha^t = \beta$.*

PROOF: The symmetries (4) – (6) are obvious — the only necessary work was already done in section 20 when showing that $\backslash w_2 \sim g_2 = g_2^*$.

To see the stability property (1) — by the symmetries — it is only necessary to show that $\alpha \backslash$ is equivalent to a norm in Φ_o for each $\alpha \in \Phi_o$: For $\alpha = \backslash(/\pi)$, observe first that $\backslash w_2 = (\backslash w_2)\backslash$ (by Ex 20.11.); now use the injectivity of w_2, Ex 20.5.(b) and Grothendieck's inequality to see that

$$\backslash w_2 = (\backslash w_2)\backslash \leq (\backslash(/\pi))\backslash \leq \backslash(/\pi\backslash) \leq K_G \backslash w_2 \ .$$

The remaining non–obvious cases are simpler:

$(/\pi)\backslash \sim w_2$ by Grothendieck's inequality

$(w_2/)\backslash \sim w_2$ by 20.15.

$(\backslash\varepsilon/)\backslash \sim w_2^*\backslash \sim g_2$ by 20.15.

$((\pi\backslash)/)\backslash = \pi\backslash$ by Ex 20.5.

$(\backslash w_2)\backslash \sim \backslash w_2$ by 20.15.

$(\backslash\varepsilon)\backslash = (\backslash d_1')\backslash = ((/d_1)/)' = d_1' = \varepsilon$ by 20.14.

$(/(\backslash\varepsilon))\backslash = /w_\infty\backslash = /g_\infty\backslash = /\varepsilon = \varepsilon$ by 20.14..

The domination arrows on the right hand side are either trivial or can be deduced from the following (the others are a consequence of transposition; recall $\alpha_{p_2,q_2} \leq \alpha_{p_1,q_1}$ for $p_1 \leq p_2$ and $q_1 \leq q_2$ from 12.5.):

$w_2^* \leq K_{LG} g_2/ = K_{LG} g_2^*/ \leq K_{LG} g_\infty^*/$ by Ex 20.10.

$\backslash w_2 \sim g_2^* \leq g_\infty^*$

$\backslash w_2 \sim g_2^* \leq w_2^*$

$w_2 \leq g_2$

$w_\infty = g_\infty \leq g_2$

$/w_\infty = (g_\infty^*/)^* \leq K_{LG} w_2^{**} = K_{LG} w_2$ by the first case.

It remains to show that there are no other dominations in the table than those indicated; the arguments will mainly use operator ideals: It will be shown that each ideal on the right side of the diagram is only contained in those ideals indicated by the arrows. The rest then follows by transposing.

Since the space ℓ_1 is not in $\mathfrak{L}_\infty$ (see e.g. 23.3.), it is (by the definition of the surjective hull) also not in $\mathfrak{L}_\infty^{sur}$, hence $\mathfrak{L} \not\subset \mathfrak{L}_\infty^{sur}$: Using the symmetry with respect to adjoints, this implies $\mathfrak{P}_1/ \not\subset \mathfrak{I}$.

By corollary 3 in 23.3. the ideals $\mathfrak{L}_1, \mathfrak{L}_2$, and $\mathfrak{L}_\infty$ are mutually uncomparable: It follows that

$$\mathfrak{L}_1^* = \mathfrak{P}_1^{dual}, \quad \mathfrak{L}_2^* = \mathfrak{D}_2, \quad \text{and } \mathfrak{L}_\infty^* = \mathfrak{P}_1$$

are incomparable as well. In particular, $\mathfrak{P}_1 \not\subset \mathfrak{P}_1/ \subset \mathfrak{D}_2$. By again using the symmetry with respect to adjoints, one gets $\mathfrak{L}_\infty^{sur} \not\subset \mathfrak{L}_\infty$. Clearly, $\mathfrak{L}_\infty \not\subset \mathfrak{P}_2 \not\subset \mathfrak{P}_1$ — and hence all the vertical arrows on the right side are not reversable.

As for the righthand side it remains to show that

$$\mathfrak{P}_1/ \not\subset \mathfrak{P}_1^{dual}, \quad \mathfrak{P}_1 \not\subset \mathfrak{L}_1, \quad \mathfrak{P}_2 \not\subset \mathfrak{L}_1, \quad \mathfrak{L}_\infty \not\subset \mathfrak{L}_1^{inj}, \quad \mathfrak{L}_\infty^{sur} \not\subset \mathfrak{L}_1^{inj}.$$

It is a result of Gordon–Lewis [84] that $\mathfrak{P}_1 \not\subset \mathfrak{L}_1$, and hence $\mathfrak{P}_2 \not\subset \mathfrak{L}_1$ as well. Moreover, the fact that $\ell_1 \not\in \text{space}(\mathfrak{L}_\infty^{sur})$ implies that $\mathfrak{L}_1 \not\subset \mathfrak{L}_\infty^{sur}$; this gives $\mathfrak{L}_\infty \not\subset \mathfrak{L}_1^{inj}$ and also $\mathfrak{L}_\infty^{sur} \not\subset \mathfrak{L}_1^{inj}$. Using the symmetry with respect to the point $*$, one obtains $\mathfrak{P}_1/ \not\subset \mathfrak{P}_1^{dual}$ from $\mathfrak{L}_\infty \not\subset \mathfrak{L}_1^{inj}$. □

The deep result of Gordon and Lewis which was used solved one of the open problems in the Résumé. But note that $\mathfrak{P}_2 \not\subset \mathfrak{L}_1$ is much easier to show: If the non–compact embedding $C[0,1] \hookrightarrow L_2[0,1]$ actually factored through some L_1

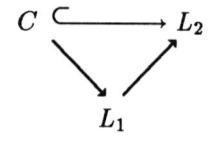

$$\begin{array}{ccc} C & \hookrightarrow & L_2 \\ & \searrow \nearrow & \\ & L_1 & \end{array},$$

then it would be in $\mathfrak{P}_1 \circ \mathfrak{D}_2 \subset \mathfrak{P}_2 \circ \mathfrak{P}_2 \subset \mathfrak{N}$ by Grothendieck's theorem.

27.3. For the convenience of the reader — here is a list of Grothendieck's notation:

$$\begin{aligned}
\wedge &:= \pi \sim \mathfrak{I} & \vee &:= \varepsilon \sim \mathfrak{L} \\
\underline{C} &:= \backslash\varepsilon = w_\infty \sim \mathfrak{L}_\infty & \gamma &:= /\underline{C} = /(\backslash\varepsilon) \sim \mathfrak{L}_\infty^{sur} \\
\underline{L} &:= \varepsilon/ = w_1 \sim \mathfrak{L}_1 & \lambda &:= \underline{L}\backslash = (\varepsilon/)\backslash \sim \mathfrak{L}_1^{inj} \\
\underline{H} &:= w_2 \sim \mathfrak{L}_2 & \underline{H}' &:= w_2' = w_2^* \sim \mathfrak{D}_2.
\end{aligned}$$

Chapter III

Special Topics

In the first two chapters we presented the theory of tensor norms, the theory of operator ideals and their intimate relationship with each other. With this knowledge at hand, the final chapter touches some questions of a more specific character.

Section 28 discusses examples of more tensor norms which are related to Lapresté's $\alpha_{p,q}$. Section 29 presents the powerful technique of traced tensor norms, which mimics for tensor norms the composition of operator ideals. Though at first glance a little bit formal, it also provides a very general setting for understanding the various tensor product characterizations of operator ideals given in Chapter II.

Sections 30 and 31 deal with the factorization of operators through Hilbert spaces in terms of type and cotype. In particular, this leads to the construction of a non-accessible tensor norm/operator ideal due to Pisier. Maurey's mixing operators are studied in some detail in section 32.

An important motivation for the sections 28 to 32 is to answer the question: Under what conditions are operators of the form $S \otimes T : L_p \otimes_{\Delta_p} E \longrightarrow L_q \otimes_{\Delta_q} F$ continuous? The results given include various characterizations and the study of two special operators S : the Fourier transform and the projection onto the Rademacher functions.

Sections 33 and 34 treat two interesting topics from the theory of operator ideals: Lewis' Radon–Nikodým property for tensor norms, which serves to decide when $\mathfrak{A}^{min}(E,F) = \mathfrak{A}^{max}(E,F)$ holds, and the tensor stability of operator ideals.

The last section introduces the so-called tensor norm topologies on tensor products of locally convex spaces. It demonstrates that tensor norm techniques can be successfully applied to questions from the theory of locally convex spaces; in other words, there is also some type of a "local theory" (i.e. in terms of finite dimensional spaces) for locally convex spaces.

28. More Tensor Norms

In this section some additional tensor norms will be studied. They are closely related to Lapresté's tensor norms $\alpha_{p,q}$ which were so important. The associated operator ideals have interesting descriptions in terms of factorization, inequalities and the continuity of operators $A \otimes T : \ell_{p'} \otimes E \to \ell_q \otimes F$.

28.1. For $p, q \in [1, \infty]$ with $1/p + 1/q \geq 1$ (equivalently: $q' \geq p$) and $z \in E \otimes F$ define

$$\beta_{p,q}(z; E, F) := \inf \|(a_{k,\ell}) : \ell_{q'}^n \to \ell_p^n\| w_{q'}(x_\ell; E) w_{p'}(y_k; F)$$
$$\gamma_{p,q}(z; E, F) := \inf \|(a_{k,\ell}) : \ell_{q'}^n \to \ell_p^n\| \ell_{q'}(x_\ell; E) \ell_{p'}(y_k; F)$$
$$\delta_{p,q}(z; E, F) := \inf \|(a_{k,\ell}) : \ell_{q'}^n \to \ell_p^n\| \ell_{q'}(x_\ell; E) w_{p'}(y_k; F) ,$$

where the infima are taken over all finite representations

$$z = \sum_{k,\ell=1}^n a_{k,\ell} x_\ell \otimes y_k \in E \otimes F ;$$

note that $Ae_\ell := \sum_{k=1}^n a_{k,\ell} e_k$ for the mapping $A := (a_{k,\ell}) : \ell_{q'}^n \to \ell_p^n$. With exactly the same proof as for $\alpha_{p,q}$ (see 12.5. — or Ex 12.8.) one can show that these are finitely generated tensor norms satisfying

$$\beta_{p_2, q_2} \leq \beta_{p_1, q_1} \qquad \text{if } p_1 \leq p_2 \text{ and } q_1 \leq q_2$$

(analoguously for $\gamma_{p,q}$ and $\delta_{p,q}$) and $\beta_{p,q}^t = \beta_{q,p}$ (and this also for $\gamma_{p,q}$). Moreover,

$$\beta_{p,q} \leq \alpha_{p,q} \quad \text{and} \quad \beta_{p,q} \leq \delta_{p,q} \leq \gamma_{p,q} .$$

Note that for p or $q = 1$ some of these norms coincide.

Recall (Ex 20.2.) that $\gamma_{p,q}$ is projective and hence $\gamma'_{p,q}$ is injective. It follows from 21.1. that $\gamma_{p,q}$ is accessible and $\gamma'_{p,q}$ totally accessible. It will be seen (28.6. and Ex 28.7.) that the tensor norms $\beta_{p,q}$ and $\delta_{p,q}$ are accessible as well.

28.2. For the study of the associated maximal (and minimal) operator ideals note that $z = \sum_{k,\ell} a_{k,\ell} x'_\ell \otimes y_k \in E' \otimes F = \mathfrak{F}(E, F)$ means that the associated operator T_z factors

$$\begin{array}{ccc}
E & \xrightarrow{T_z} & F \\
{\scriptstyle \sum_{\ell=1}^n x'_\ell \otimes e_\ell} \downarrow & & \uparrow {\scriptstyle \sum_{k=1}^n e_k \otimes y_k} \\
\ell_{q'}^n & \xrightarrow{A} & \ell_p^n
\end{array} \qquad A = (a_{k,\ell}) .$$

Recall from section 8 (in particular Ex 8.19.) that

$$w_s(x'_k; E') = \left\| \sum_{k=1}^n x'_k \otimes e_k : E \to \ell_s^n \right\|$$
$$\ell_s(x'_k; E') = \min \|S : E \to \ell_\infty^n\| \, \|D_\lambda : \ell_\infty^n \to \ell_s^n\| ,$$

where $D_\lambda \circ S = \sum_{k=1}^n x'_k \otimes e_k : E \to \ell_s^n$ is a factorization with a diagonal operator D_λ with non–negative entries. Since D_λ is a positive operator, the s-integral norm satisfies

$$\mathbf{I}_s(D_\lambda : \ell_\infty^n \to \ell_s^n) = \|D_\lambda : \ell_\infty^n \to \ell_s^n\|$$

by 17.17.. In particular,

$$\ell_{p'}(y_k; F) = \min \|S : F' \longrightarrow \ell_\infty^n\| \, \|D_\mu : \ell_\infty^n \longrightarrow \ell_{p'}^n\| =$$
$$= \min \|R : \ell_1^n \longrightarrow F''\| \, \|D_\mu : \ell_p^n \longrightarrow \ell_1^n\| =$$
$$= \min \|R : \ell_1^n \longrightarrow F''\| \, \mathbf{I}_{p'}^{dual}(D_\mu : \ell_p^n \longrightarrow \ell_1^n) \, ,$$

with D_μ a positive diagonal operator such that $R \circ D_\mu = \sum_{k=1}^n e_k \otimes \kappa_F(y_k) : \ell_p^n \to F''$. It is clear now that the maximal operator ideals associated with $\beta_{p,q}, \gamma_{p,q}$ and $\delta_{p,q}$ are related with composition ideals of $\mathfrak{L}_p, \mathfrak{I}_p$ and $\mathfrak{I}_p^{dual}$. We shall treat $\beta_{p,q}$ and $\gamma_{p,q}$ in order to demonstrate the methods and leave $\delta_{p,q}$ as an exercise.

28.3. First $\gamma_{p,q}$: Take for finite dimensional spaces M, N

$$\sum_{k,\ell=1}^n a_{k,\ell} x_\ell' \otimes y_k \in M' \otimes_{\gamma_{p,q}} N \, .$$

Then for the associated operator $T_z : M \to N$ and any factorization

$$\begin{array}{ccc}
M & \xrightarrow{T_z} & N \\
S \downarrow & & \uparrow R \\
\ell_\infty^n \xrightarrow{D_\lambda} \ell_{q'}^n & \xrightarrow{A} \ell_p^n & \xrightarrow{D_\mu} \ell_1^n
\end{array}$$

the inequality $(\mathbf{I}_{p'}^{dual} \circ \mathbf{I}_{q'})(T_z) \leq \|S\| \mathbf{I}_{q'}(D_\lambda) \|A\| \mathbf{I}_{p'}^{dual}(D_\mu) \|R\|$ holds. Passing to the infimum over all representations, it follows that

$$(\mathbf{I}_{p'}^{dual} \circ \mathbf{I}_{q'})(T_z : M \longrightarrow N) \leq \gamma_{p,q}(z; M', N)$$

by what was said before. To see that this is actually an equality, take for $z \in M' \otimes N$ a factorization $T_z = R \circ S$ through some Banach space G. Since $\mathfrak{I}_{q'} \sim \alpha_{q',1}$ and $\mathfrak{I}_{p'}^{dual} \sim \alpha_{1,p'}$ are accessible (section 21), the operators can be factored (see 18.1.) as follows:

$$\begin{array}{ccccc}
M & \xrightarrow{S} & G & \quad G & \xrightarrow{R} N \\
S_1 \downarrow & & \uparrow S_2 & R_1 \downarrow & \uparrow R_2 \\
\ell_\infty^n & \xrightarrow{D_\lambda} & \ell_{q'}^n & \ell_p^n & \xrightarrow{D_\mu} \ell_1^n
\end{array}$$

with $\|S_1\| \|D_\lambda\| \|S_2\| \leq (1+\varepsilon)\mathbf{I}_{q'}(S)$ and $\|R_1\| \|D_\mu\| \|R_2\| \leq (1+\varepsilon)\mathbf{I}_{p'}^{dual}(R)$. Therefore,

$$\gamma_{p,q}(z; M', N) \leq \|S_1\| \|D_\lambda\| \cdot \|R_1 \circ S_2\| \cdot \|D_\mu\| \|R_2\| \leq$$
$$\leq (1+\varepsilon)^2 \mathbf{I}_{q'}(S) \mathbf{I}_{p'}^{dual}(R) \, .$$

It follows that $M' \otimes_{\gamma_{p,q}} N = (\mathfrak{J}_{p'}^{dual} \circ \mathfrak{J}_{q'})(M, N)$ holds isometrically. This is in some sense the core of the following

PROPOSITION: *Let* $1/p + 1/q \geq 1$. *Then*

$$\gamma_{p,q} = \backslash \alpha'_{p,q} /$$

and the associated maximal normed operator ideal is

$$\backslash \mathfrak{D}_{q',p'} / = (\mathfrak{J}_{p'}^{dual} \circ \mathfrak{J}_{q'})^{reg} .$$

Recall from 25.9. that the regular hull $\mathfrak{A}^{reg}$ of a quasi–Banach operator ideal $(\mathfrak{A}, \mathbf{A})$ is the class of all $T \in \mathfrak{L}(E, F)$ such that $\kappa_F \circ T \in \mathfrak{A}$, and $\mathbf{A}^{reg}(T) := \mathbf{A}(\kappa_F \circ T)$. The ideal $\backslash \mathfrak{D}_{q',p'} /$ is, by 20.12., the ideal of all operators T which factor

$$\begin{array}{ccc} E & \xrightarrow{T} & F \hookrightarrow F'' \\ R \downarrow & \nearrow S & \\ C(K) & \xrightarrow{U} L_1(\mu) & \end{array} \quad U \in \mathfrak{D}_{q',p'}$$

for some compact K and measure μ.

PROOF: $\mathfrak{D}_{q',p'} = \mathfrak{P}_{p'}^{dual} \circ \mathfrak{P}_{q'}$ holds isometrically by Kwapień's factorization theorem 19.3.. Moreover, the formula $\backslash g_q^* = g_{q'}$ from 20.14. implies that

$$\backslash \mathfrak{P}_{q'} \sim \backslash g_q^* = g_{q'} \sim \mathfrak{J}_{q'}$$
$$\mathfrak{P}_{p'}^{dual} / \sim g_{p'}^{*t} / = (\backslash g_p^*)^t = g_{p'}^t \sim \mathfrak{J}_{p'}^{dual}$$

by the relationship between tensor norms and operator ideals. 25.9. now gives that

$$\backslash \mathfrak{D}_{q',p'} / = \backslash (\mathfrak{P}_{p'}^{dual} \circ \mathfrak{P}_{q'}) / = ((\mathfrak{P}_{p'}^{dual} /) \circ (\backslash \mathfrak{P}_{q'}))^{reg} = (\mathfrak{J}_{p'}^{dual} \circ \mathfrak{J}_{q'})^{reg}$$

holds isometrically. Since $\alpha'_{p,q} \sim \mathfrak{D}_{q',p'}$ and $(\mathfrak{J}_{p'}^{dual} \circ \mathfrak{J}_{q'})^{reg} = \backslash \mathfrak{D}_{q',p'} /$ is maximal, it remains to verify that

$$M' \otimes_{\gamma_{p,q}} N = (\mathfrak{J}_{p'}^{dual} \circ \mathfrak{J}_{q'})(M, N)$$

— and this was done before. □

In particular, it follows that $(\mathfrak{J}_{p'}^{dual} \circ \mathfrak{J}_{q'})^{reg}$ is a *normed* operator ideal — a fact which is not a priori clear.

28.4. The "adjoint" version of this result describes the operator ideal

$$[(\mathfrak{J}_{p'}^{dual} \circ \mathfrak{J}_{q'})^{reg}]^* = (\mathfrak{J}_{p'}^{dual} \circ \mathfrak{J}_{q'})^* \sim \gamma_{p,q}^* = \gamma'_{q,p} = (\backslash \alpha'_{q,p} /)' = /\alpha_{q,p} \backslash \sim \mathfrak{L}_{q,p}^{inj\ sur}$$

— and this implies a description of the operators in
$$\mathfrak{L}_{q,p}^{inj\ sur} = (\mathfrak{J}_{p'}^{dual} \circ \mathfrak{J}_{q'})^* \sim \gamma_{p,q}^* ,$$
which for $p = q'$ is an operator form of a result of Kwapień's [160] and completes the characterization of subspaces of quotients of L_q–spaces given in 25.9. and 25.10..

THEOREM: *Let $1/p+1/q \geq 1$. For each operator $T \in \mathfrak{L}(E,F)$ the following statements are equivalent:*
(a) $T \in \mathfrak{L}_{q,p}^{inj\ sur}(E,F)$.
(b) *For all $A \in \mathfrak{L}(\ell_{p'}, \ell_q)$ the operator*
$$A \otimes T : \ell_{p'} \otimes_{\Delta_{p'}} E \longrightarrow \ell_q \otimes_{\Delta_q} F$$
is continuous.
(c) *There is a constant $c \geq 0$ such that for all natural numbers $n \in \mathbb{N}$, all matrices $(a_{k,\ell})$, all $x_1, ..., x_n \in E$ and all $y'_1, ..., y'_n \in F'$*
$$\left| \sum_{k,\ell=1}^n a_{k,\ell} \langle Tx_\ell, y'_k \rangle \right| \leq c \ \|(a_{k,\ell}) : \ell_{p'}^n \longrightarrow \ell_q^n \| \|\ell_{p'}(x_\ell)\ell_{q'}(y'_k) .$$

In this case,
$$\mathbf{L}_{q,p}^{inj\ sur}(T) = \sup_{\|A\| \leq 1} \|\ell_{p'} \otimes_{\Delta_{p'}} E \xrightarrow{A \otimes T} \ell_q \otimes_{\Delta_q} F\| =$$
$$= \sup_{\substack{\|A\| \leq 1 \\ n \in \mathbb{N}}} \|\ell_{p'}^n \otimes_{\Delta_{p'}} E \xrightarrow{A \otimes T} \ell_q^n \otimes_{\Delta_q} F\| = \inf c .$$

PROOF: An obvious application of the closed graph theorem shows that (b) implies the existence of a constant c with $\|A \otimes T\| \leq c\|A\|$. Since
$$\ell_q(y_k) = \sup \left\{ \left| \sum_{k=1}^n \langle y_k, y'_k \rangle \right| \ \middle| \ \ell_{q'}(y'_k) \leq 1 \right\}$$
and, for $A = (a_{k,\ell})$,
$$A \otimes T \left(\sum_{\ell=1}^n e_\ell \otimes x_\ell \right) = \sum_{k=1}^n \left(e_k \otimes \left(\sum_{\ell=1}^n a_{k,\ell} T x_\ell \right) \right) ,$$
it is easy to see that (b) implies (c). The converse is true by the density lemma 7.3. for $p > 1$; for $p = 1$ one has $\Delta_\infty = \varepsilon$, and hence this implication follows by $\mathcal{L}_\infty$-local technique.
By what was said before, the operator ideal $\mathfrak{L}_{q,p}^{inj\ sur}$ is associated with $\gamma'_{q,p}$. The representation theorem for maximal operator ideals shows that $T \in \mathfrak{L}_{q,p}^{inj\ sur}$ if and only if its associated bilinear form β_T satisfies $\beta_T \in (E \otimes_{\gamma_{q,p}} F')'$. Therefore the definition of $\gamma_{q,p}$ gives the equivalence of (a) and (c). □

In statement (b) the sequence spaces ℓ_p appear: they can be replaced by arbitrary infinite dimensional $L_p(\mu)$; this could be proved directly using $\mathfrak{L}_{q,p}^{inj\ sur} = \mathfrak{J}_{q'}^{-1} \circ \mathfrak{P}_p^{dual}$ (see 25.9.) — a fact which is related to the statement anyway. But there will be a general result in 29.12. showing this.

28.5. Since $w_2 \leq \alpha_{p,q} \leq \pi$ for $p,q \in [1,2]$, it follows that $\mathfrak{J} = \mathfrak{L}_{1,1} \subset \mathfrak{L}_{p,q} \subset \mathfrak{L}_2$. Moreover, $\mathfrak{L}_2$ is surjective and injective and $\mathfrak{J}^{inj\ sur} = \mathfrak{L}_2$ by Grothendieck's inequality in the operator form 20.17., hence

(1) *For $p,q \in [1,2]$ the relation $\mathfrak{L}_{q,p}^{inj\ sur} = \mathfrak{L}_2$ holds isomorphically.*

In particular, w_2 and $\gamma'_{p,q}$ (for $1 \leq p,q \leq 2$) are equivalent tensor norms. For $p \in [1,\infty]$ the relation $g_{p'}^* = g_p\backslash$ (see 20.14.) implies that

$$\mathfrak{L}_{1,p}^{inj\ sur} \sim /\alpha_{1,p}\backslash = (g_p\backslash)^t\backslash = g_{p'}^{*t}\backslash \sim \mathfrak{P}_p^{dual\ inj} = (\mathfrak{P}_p^{sur})^{dual}$$

and therefore,

(2) $\mathfrak{L}_{1,p}^{inj\ sur} = \mathfrak{P}_p^{dual\ inj}$ and $\mathfrak{L}_{p,1}^{inj\ sur} = \mathfrak{P}_p^{sur}$ hold isometrically $(1 \leq p \leq \infty)$.

Clearly, the theorem is related with proposition 1 in 26.3.: Since $\mathfrak{L}(L_r, L_s) = \mathfrak{L}_2(L_r, L_s)$ if $s \leq 2 \leq r$ (by 26.1.), it follows from (1) that every operator $T: L_r \to L_s$ is in $\mathfrak{L}_{q,p}^{inj\ sur}$ and hence, by (b) of the theorem (and 29.12.),

$$A \otimes T : L_{p'}(\mu_1) \otimes_{\Delta_{p'}} L_r(\mu_2) \longrightarrow L_q(\nu_1) \otimes_{\Delta_q} L_s(\nu_2)$$

is continuous. Note that in 26.3. this was shown for all p,q.

Certainly, one of the most interesting cases is $w_2 = \gamma'_{2,2} \sim \mathfrak{L}_2$, the operators factoring through a Hilbert space.

COROLLARY: *$T \in \mathfrak{L}(E,F)$ factors through a Hilbert space if and only if there is a constant $c \geq 0$ such that for all isometric $n \times n$ matrices $A = (a_{k,\ell})$ and $x_1, ..., x_n \in E$*

$$\sum_{k=1}^{n} \left\| \sum_{\ell=1}^{n} a_{k,\ell} T x_\ell \right\|^2 \leq c^2 \sum_{\ell=1}^{n} \|x_\ell\|^2 .$$

PROOF: By condition (b) or (c) of the theorem the condition is necessary. Conversely, fix x_ℓ; then the convex function $g(A)$ on the left side attains its maximum on the extreme points of the unit ball of $\mathfrak{L}(\ell_2^n, \ell_2^n)$ — which are the isometries (see Halmos [99], p.263). It follows that the inequality holds for all matrices $A = (a_{k,\ell})$ of norm ≤ 1 and this easily implies (b) or (c) of the theorem . □

This result, due to Lindenstrauss–Pełczyński [173], is in some sense central to the theory of operators factoring through a Hilbert space — e.g. in the proof of Kwapień's important type/cotype theorem 30.5.. Note also that w_2^* is projective, hence $w_2^* = \gamma_{2,2}$ (by 28.3.) is an explicit description of the tensor norm associated with the adjoint ideal $\mathfrak{L}_2^* = \mathfrak{D}_2$ of 2–dominated operators. See Ex 28.1.(d) for another description of $\mathfrak{L}_2$.

28.6. Next, the tensor norm

$$\beta_{p,q}(z; E, F) = \inf\left\{\|(a_{k,\ell}) : \ell_{q'}^n \to \ell_p^n\| w_{q'}(x_\ell) w_{p'}(y_k) \,\Big|\, z = \sum_{k,\ell=1}^n a_{k,\ell} x_\ell \otimes y_k\right\}$$

will be investigated. Since $\varepsilon \leq \beta_{p,q}$ and $z \in \ell_q^n \otimes_\varepsilon \ell_p^n = \mathfrak{L}(\ell_{q'}^n, \ell_p^n)$ has the canonical representation

$$z = \sum_{k,\ell=1}^n a_{k,\ell} e_\ell \otimes e_k,$$

the relation $w_{s'}((e_k)_{k=1}^n; \ell_s^n) = 1$ implies that

$$(\ell_{q'}^n \otimes_{\beta'_{p,q}} \ell_{p'}^n)' = \ell_q^n \otimes_{\beta_{p,q}} \ell_p^n = \ell_q^n \otimes_\varepsilon \ell_p^n = \mathfrak{L}(\ell_{q'}^n, \ell_p^n) = (\ell_{q'}^n \otimes_\pi \ell_{p'}^n)'$$

holds isometrically. Recalling that

$$w_{q'}(x'_k; M') = \left\|\sum_{k=1}^n x'_k \otimes e_k : M \longrightarrow \ell_{q'}^n\right\|$$

and $M' \otimes_{\beta_{p,q}} N' = (M \otimes_{\beta'_{p,q}} N)'$, gives that for $\varphi \in (M \otimes N)^*$

$$\|\varphi\|_{(M \otimes_{\beta'_{p,q}} N)'} = \inf \|\delta\| \, \|R\| \, \|S\|,$$

where the infimum is taken over all factorizations $\varphi = \delta \circ (R \otimes S)$ with an operator $R : M \to \ell_{q'}^n$, an operator $S : N \to \ell_{p'}^n$ and $\delta \in (\ell_{q'}^n \otimes_\pi \ell_{p'}^n)'$ for some $n \in \mathbb{N}$. Following the ultraproduct–technique–proof of theorem 18.5. word by word (with the exception that the forms δ are positive there) gives the implication (a) $\curvearrowright$ (b) of the

THEOREM: *Let $1/p + 1/q \geq 1$ and $\varphi \in (E \otimes F)^*$. Then the following statements are equivalent:*

(a) *$\varphi \in (E \otimes_{\beta'_{p,q}} F)'$.*
(b) *There are measures μ and ν, operators $R \in \mathfrak{L}(E, L_{q'}(\mu)), S \in \mathfrak{L}(F, L_{p'}(\nu))$ and $\delta \in (L_{q'}(\mu) \otimes_\pi L_{p'}(\nu))'$ with*

$$\langle \varphi, x \otimes y \rangle = \langle \delta, Rx \otimes Sy \rangle \qquad \text{for all } (x, y) \in E \times F.$$

The measures can be taken to be Borel–Radon measures on locally compact spaces and

$$\|\varphi\|_{(E \otimes_{\beta'_{p,q}} F)'} = \min \|\delta\| \, \|R\| \, \|S\|.$$

PROOF: It remains to show that (b) implies (a): since $(\ell_{q'}^n \otimes_\pi \ell_{p'}^n)' \stackrel{1}{=} (\ell_{q'}^n \otimes_{\beta'_{p,q}} \ell_{p'}^n)'$, the $\mathfrak{L}_p$-local technique easily implies that

$$(L_{q'}(\mu) \otimes_\pi L_{p'}(\nu))' = (L_{q'}(\mu) \otimes_{\beta'_{p,q}} L_{p'}(\nu))'$$

holds isometrically, which gives the result. □

Simple factorization arguments show the

COROLLARY 1: *For $1/p + 1/q \geq 1$ the maximal normed operator ideal associated with the tensor norm $\beta_{p,q}$ is isometrically equal to the ideal*

$$(\mathfrak{L}_p \circ \mathfrak{L}_{q'})^{reg} .$$

Again: $(\mathfrak{L}_p \circ \mathfrak{L}_{q'})^{reg}$ is normed — a fact which is not obvious. If $\mathfrak{A}$ and $\mathfrak{B}$ are accessible operator ideals, straightforward factorization and the strong principle of local reflexivity give that $(\mathfrak{B} \circ \mathfrak{A})^{reg}$ is accessible as well. The ideals $\mathfrak{L}_s \sim w_s$ are accessible, hence:

COROLLARY 2: *The tensor norms $\beta_{p,q}$ are accessible, the tensor norms $\beta_{2,q}$ are even totally accessible.*

The latter follows from the fact that $\mathfrak{L}_2 \circ \mathfrak{L}_{q'} = (\mathfrak{L}_2 \circ \mathfrak{L}_{q'})^{reg}$ is an injective operator ideal, hence $\beta_{2,q}$ is accessible and right–injective, hence totally accessible by 21.1.(2). Note that $\mathfrak{L}_p \circ \mathfrak{L}_2 = (\mathfrak{L}_p \circ \mathfrak{L}_2)^{reg}$ (by Ex 28.5.) and $\mathfrak{L}_2 \circ \mathfrak{L}_{p'} = (\mathfrak{L}_2 \circ \mathfrak{L}_{p'})^{reg}$ are maximal normed operator ideals ($1 \leq p \leq 2$).

28.7. This explicit description of the operator ideal associated with $\beta_{p,q}$ gives further information about $\beta_{p,q}$ for special (p,q) :

COROLLARY 3: (1) $\beta_{p,1} = \backslash w_p$ *for* $1 \leq p \leq \infty$.
(2) $\beta_{p,1} \sim g_2$ *for* $1 < p \leq 2$.
(3) $\beta_{1,1} = \backslash w_1 = \backslash \varepsilon / \sim w_2^*$.
(4) $\beta_{p,q} \sim w_2$ *for* $1 < p, q \leq 2$.
(5) $\beta_{p,p'} = w_p$ *for* $1 \leq p \leq \infty$.

PROOF: The description of $\backslash \mathfrak{A}$ in proposition 20.12. shows that

$$\beta_{p,1} \sim (\mathfrak{L}_p \circ \mathfrak{L}_\infty)^{reg} = (\backslash \mathfrak{L}_p)^{reg} = \backslash \mathfrak{L}_p \sim \backslash w_p .$$

This gives (1) — and (3) follows from this using 20.14. and Grothendieck's inequality (tensor form 20.17.). $(\mathfrak{L}_p \circ \mathfrak{L}_\infty)^{reg} \subset \mathfrak{P}_2 \sim g_2$ for $1 \leq p \leq 2$ is corollary 1 in 23.10. (again Grothendieck's inequality), whereas

$$\mathfrak{P}_2 \subset \mathfrak{L}_2 \circ \mathfrak{L}_\infty \subset (\mathfrak{L}_p \circ \mathfrak{L}_\infty)^{reg}$$

for $1 < p < \infty$ follows from the Grothendieck–Pietsch factorization theorem and the inclusion $\mathfrak{L}_2 \subset \mathfrak{L}_p$.

Statement (4) is a consequence of $(\mathfrak{L}_p \circ \mathfrak{L}_{q'})^{reg} \subset \mathfrak{L}_2$ for $p \leq 2 \leq q'$ (see 26.1.) and $\mathfrak{L}_2 \subset \mathfrak{L}_p \circ \mathfrak{L}_{q'}$ for $1 < p, q' < \infty$. The last claim is obvious by corollary 1. □

28.8. The adjoint operator ideal $(\mathfrak{L}_p \circ \mathfrak{L}_{q'})^* = [(\mathfrak{L}_p \circ \mathfrak{L}_{q'})^{reg}]^* \sim \beta_{p,q}^*$ has also a nice description in terms of inequalities:

COROLLARY 4: *For $1/p + 1/q \geq 1$ and $T \in \mathfrak{L}(E,F)$ the following statements are equivalent:*
(a) $T \in (\mathfrak{L}_p \circ \mathfrak{L}_{q'})^*(E,F)$.
(b) *For all $A \in \mathfrak{L}(\ell_{p'}, \ell_q)$ the operator*
$$A \otimes T : \ell_{p'} \otimes_\varepsilon E \longrightarrow \ell_q \otimes_\pi F$$
is continuous.
(c) *There is a constant $c \geq 0$ such that for all $n \times n$ matrices $(a_{k,\ell})$, all $x_1, \dots, x_n \in E$ and all $y'_1, \dots, y'_n \in F'$*
$$\left| \sum_{k,\ell=1}^n a_{k,\ell} \langle Tx_\ell, y'_k \rangle \right| \leq c \, \|(a_{k,\ell}) : \ell_{p'}^n \longrightarrow \ell_q^n\| w_{p'}(x_\ell) w_{q'}(y'_k) \, .$$

In this case,
$$(\mathbf{L}_p \circ \mathbf{L}_{q'})^*(T) = \sup_{\|A\| \leq 1} \|A \otimes T : \otimes_\varepsilon \longrightarrow \otimes_\pi\| = \inf c \, .$$

PROOF: Since $\ell_q \otimes_\pi F \overset{1}{\hookrightarrow} (\ell_{q'} \otimes_\varepsilon F')'$ and $\varepsilon(\Sigma e_n \otimes u_n; \ell_r, G) = w_r(u_n; G)$ by 6.4. and 12.9., it is clear that (b) and (c) are equivalent by the closed graph theorem and $\mathfrak{L}_p$-local techniques.

The representation theorem for maximal operator ideals and
$$\beta'_{q,p} = \beta^*_{p,q} \sim (\mathfrak{L}_p \circ \mathfrak{L}_{q'})^*$$
show that $T \in (\mathfrak{L}_p \circ \mathfrak{L}_{q'})^*(E,F) \overset{1}{\hookrightarrow} (E \otimes_{\beta_{q,p}} F')'$ if and only if
$$\left| \langle \beta_T, \sum_{k,\ell=1}^n a_{k,\ell} x_\ell \otimes y'_k \rangle \right| \leq c \, \|(a_{k,\ell}) : \ell_{p'}^n \longrightarrow \ell_q^n\| w_{p'}(x_\ell) w_{q'}(y'_k)$$
— and this is the equivalence of (a) and (c). □

The relations in corollary 3 (and 17.12.) give:

$$\begin{aligned}
(\mathfrak{L}_p \circ \mathfrak{L}_\infty)^* &\overset{1}{=} \mathfrak{D}_{p'}^{inj} && \text{for } 1 \leq p \leq \infty \\
(\mathfrak{L}_p \circ \mathfrak{L}_\infty)^* &= \mathfrak{P}_2 && \text{for } 1 < p \leq 2 \\
(\mathfrak{L}_1 \circ \mathfrak{L}_\infty)^* &= \mathfrak{L}_2 \\
(\mathfrak{L}_p \circ \mathfrak{L}_{q'})^* &= \mathfrak{D}_2 && \text{for } 1 < p, q \leq 2 \, .
\end{aligned}$$

28.9. One of the main objectives of this section was to establish conditions under which for a given $T \in \mathfrak{L}(E, F)$ the operator

$$A \otimes T : \ell_{p'} \otimes E \longrightarrow \ell_q \otimes F$$

is continuous (for certain norms) for all $A \in \mathfrak{L}(\ell_{p'}, \ell_q)$. The results of 28.4., 28.8., Ex 28.8. and Ex 28.11. can be summarized in the following table:

tensor norm	operator ideal	$A \otimes T : \ell_{p'} \otimes E \longrightarrow \ell_q \otimes F$	
$\gamma^*_{p,q}$	$(\mathfrak{J}^{dual}_{p'} \circ \mathfrak{J}_{q'})^*$	$\Delta_{p'}$	Δ_q
$\beta^*_{p,q}$	$(\mathfrak{L}_p \circ \mathfrak{L}_{q'})^*$	ε	π
$\delta^*_{p,q}$	$(\mathfrak{L}_p \circ \mathfrak{J}_{q'})^*$	ε	Δ_q
$(\delta^t_{q,p})^*$	$(\mathfrak{J}^{dual}_{p'} \circ \mathfrak{L}_{q'})^*$	$\Delta_{p'}$	π

Recall that $(\mathfrak{J}^{dual}_{p'} \circ \mathfrak{J}_{q'})^* = \mathfrak{L}^{inj\ sur}_{q,p}$. It will be shown in 29.12. that for these four characterizations $A \in \mathfrak{L}(\ell_{p'}, \ell_q)$ can be replaced by $A \in \mathfrak{L}(L_{p'}(\mu), L_q(\nu))$ — with fixed measures μ and ν such that $L_{p'}(\mu)$ and $L_q(\nu)$ are infinite dimensional. For related characterizations of this type for $\mathfrak{L}^{inj}_{p,q}$ and $\mathfrak{L}^{sur}_{p,q}$ see Ex 29.4. and Ex 29.5..

Exercises:

Ex 28.1. Let $1 \leq p < \infty$. For $x_1, ..., x_m, y_1, ..., y_m \in E$ write $(x_k) <_p (y_k)$ if

$$\sum_{k=1}^{m} |\langle x', x_k \rangle|^p \leq \sum_{k=1}^{m} |\langle x', y_k \rangle|^p$$

for all $x' \in E$. An operator $T \in \mathfrak{L}(E, F)$ is called (for the purpose of this exercise) *p–strong* if there is a $c \geq 0$ such that $(x_k) <_p (y_k)$ implies $\ell_p(Tx_k) \leq c\,\ell_p(y_k)$.
(a) The identity of ℓ_p^n is p–strong with constant 1.
(b) Every $T \in \mathfrak{L}^{inj}_p$ is p–strong with constant $\mathbf{L}^{inj}_p(T)$.
(c) If T is p–strong with constant c, then $\mathbf{L}^{inj\ sur}_p(T) \leq c$. Hint: $y_k := \sum a_{k,\ell} x_\ell$ and 28.4.(c).
(d) T is 2–strong if and only if it factors through a Hilbert space.
The converse of (b) is also true for arbitrary p: This will be shown in 33.7..

Ex 28.2. Let μ be a measure and $f, g \in L_2(\mu, E)$ such that

$$\int |\langle x', f(t) \rangle|^2 \mu(dt) \leq \int |\langle x', g(t) \rangle|^2 \mu(dt)$$

for all $x' \in E'$.
(a) If E is a Hilbert space, then $\|f\|_{L_2(E)} \leq \|g\|_{L_2(E)}$. Hint: f and g are separably valued, Parseval's identity.
(b) If $T \in \mathfrak{L}_2(E,F)$, then $\int \|T(f(t))\|^2 dt \leq \mathbf{L}_2(T)^2 \int \|g(t)\|^2 dt$.

Ex 28.3. Let $1 \leq p \leq q' < s < 2$ or: $1 \leq p \leq q' < \infty$ and $s = 2$.
(a) Use the Lévy embedding to show that $\mathfrak{L}_s \subset \mathfrak{L}_{p,q}^{inj}$ and $\alpha_{p,q} \backslash \leq c_{s,q'} c_{s,p}^{-1} w_s$.
(b) Take $T \in \mathfrak{L}(E,F)$ and $c \geq 0$ such that

$$\left(\sum_{k=1}^{\infty}\left\|\sum_{\ell=1}^{\infty} a_{k,\ell} T x_{\ell}\right\|^{s}\right)^{1/s} \leq c \, \|(a_{k,\ell}) : \ell_s \longrightarrow \ell_s\| \|\ell_s(x_k)$$

for all $(a_{k,\ell}) \in \mathfrak{L}(\ell_s, \ell_s)$ and sequences (x_k) in $\ell_{s'}(E)$. Verify that

$$\left(\sum_{k=1}^{\infty}\left\|\sum_{\ell=1}^{\infty} a_{k,\ell} T x_{\ell}\right\|^{p}\right)^{1/p} \leq c_{s,q'} c_{s,p}^{-1} c \, \|(a_{k,\ell}) : \ell_{q'} \longrightarrow \ell_p\| \|\ell_{q'}(x_k)$$

for all $(a_{k,\ell}) \in \mathfrak{L}(\ell_{q'}, \ell_p)$ and sequences (x_k) in $\ell_{q'}(E)$. Hint: (a) and 28.4..

Ex 28.4. Show that $\varepsilon \leq \beta_{p,q} \leq \lambda\mu\varepsilon$ on $E \otimes F$ if E is an $\mathfrak{L}_{q,\lambda}^g$-space and F an $\mathfrak{L}_{p,\mu}^g$-space.

Ex 28.5. If $(\mathfrak{A}, \mathbf{A})$ and $(\mathfrak{B}, \mathbf{B})$ are quasi–Banach operator ideals, $(\mathfrak{A}, \mathbf{A})$ is regular and $(\mathfrak{B}, \mathbf{B})$ injective, then $(\mathfrak{A} \circ \mathfrak{B}, \mathbf{A} \circ \mathbf{B})$ is regular.

Ex 28.6. Show that the finite dimensional Grothendieck constants $K_G(n)$ from 20.17. satisfy:

$$K_G(n) = n \left[\beta_{1,1}\left(\sum_{k=1}^{n} e_k \otimes e_k; \ell_2^n, \ell_2^n\right)\right]^{-1}.$$

Hint: 17.19. and $(\mathfrak{P}_1^{sur})^* \sim \beta_{1,1}$.

The following series of exercises will treat the tensor norm

$$\delta_{p,q}(z; E, F) := \inf\left\{\|(a_{k,\ell}) : \ell_{q'}^n \longrightarrow \ell_p^n\| \|\ell_{q'}(x_{\ell}) w_{p'}(y_k) \,\Big|\, z = \sum_{k,\ell=1}^{n} a_{k,\ell} x_{\ell} \otimes y_k\right\}.$$

The methods to be used are exactly those which were used for $\beta_{p,q}$ and $\gamma_{p,q}$ — in particular, the ultraproduct technique of section 18. The maximal normed operator ideal associated with $\delta_{p,q}$ will be denoted by $(\mathfrak{C}_{p,q}, \mathbf{C}_{p,q})$.

Ex 28.7. (a) Show for $T \in \mathfrak{L}(M,N)$ (finite dimensional spaces) that $\mathbf{C}_{p,q}(T) = \inf \|S\| \|A\| \|D_\lambda\| \|R\|$, where the infimum is taken over all $n \in \mathbb{N}$ and factorizations

$$\begin{array}{ccc}
M & \xrightarrow{T} & N \\
R \downarrow & & \uparrow S \\
\ell_\infty^m & \xrightarrow{D_\lambda} \ell_{q'}^n \xrightarrow{A} & \ell_p^n
\end{array}.$$

Hint: 28.2..

Recall that $\mathbf{I}_{q'}(D_\lambda) = \|D_\lambda\|$.

(b) Show that for arbitrary $T \in \mathcal{L}(E, F)$ the estimate $\mathbf{C}_{p,q}(T) \leq \mathbf{L}_p \circ \mathbf{I}_{q'}(T)$ holds. Hint: Use the maximality of $\mathfrak{C}_{p,q}$ and accessibility.

(c) Use the ultraproduct decomposition of an operator given in Ex 18.8. and 17.17. to show that $(\mathfrak{L}_p \circ \mathfrak{I}_{q'})^{reg} = \mathfrak{C}_{p,q} \sim \delta_{p,q}$ holds isometrically.

(d) $\mathfrak{C}_{p,q}$ and $\delta_{p,q}$ are accessible.

Ex 28.8. Prove that for any $T \in \mathcal{L}(E, F)$ the following statements are equivalent:

(a) $T \in (\mathfrak{L}_p \circ \mathfrak{I}_{q'})^*(E, F)$.

(b) For all $A \in \mathcal{L}(\ell_{p'}, \ell_q)$

$$A \otimes T : \ell_{p'} \otimes_\varepsilon E \longrightarrow \ell_q \otimes_{\Delta_q} F$$

is continuous.

(c) There is a constant $c \geq 0$ such that for all $n \times n$ matrices $(a_{k,\ell})$ and all $x_1, ..., x_n \in E$ and $y'_1, ..., y'_n \in F'$

$$\left|\sum_{k,\ell=1}^{n} a_{k,\ell} \langle Tx_\ell, y'_k \rangle\right| \leq c \|(a_{k,\ell}) : \ell_{p'}^n \longrightarrow \ell_q^n\| w_{p'}(x_\ell) \ell_{q'}(y'_k).$$

In this case, $(\mathbf{L}_p \circ \mathbf{I}_{q'})^*(T) = \sup_{\|A\| \leq 1} \|A \otimes T : \otimes_\varepsilon \to \otimes_{\Delta_q}\| = \inf c$.

Ex 28.9. Verify the following relations with the aid of the associated operator ideals:

(1) $\delta_{p,1} = \backslash w_p$ if $1 \leq p \leq \infty$.   Hint: $\mathfrak{I}_\infty = \mathfrak{L}_\infty$, 20.12. or 28.7.(1).
(2) $\delta_{1,q} = g_{q'}/$ if $1 \leq q \leq \infty$.   Hint: 20.12. or 28.5.(2).
(3) $\delta_{p,1} \sim g_2$ if $1 < p \leq 2$.   Hint: 28.7.(2).
(4) $\delta_{1,q} \sim w_2^*$ if $1 \leq q \leq 2$.   Hint: 28.5.(1).
(5) $\delta_{p,q} \sim g_2$ if $p, q \in]1, 2]$.   Hint: Proposition 26.4. and $\mathfrak{L}_2 \subset \mathfrak{L}_p$.
(6) $\delta_{p,p'} = g_p$ if $1 \leq p \leq \infty$.

Ex 28.10. (a) $T \in \mathcal{L}(E, F)$ is absolutely 2–summing if and only if for some (for all) $1 \leq q \leq 2 \leq s < \infty$ and all $A \in \mathcal{L}(\ell_s, \ell_q)$ the operator $A \otimes T : \ell_s \otimes_\varepsilon E \longrightarrow \ell_q \otimes_{\Delta_q} F$ is continuous. Hint: Ex 28.9.(5), (3) and Ex 28.8..

(b) $T \in \mathcal{L}(E, F)$ factors through a Hilbert space if and only if for some (for all) $1 \leq q \leq 2$ and all $A \in \mathcal{L}(\ell_\infty, \ell_q)$ the operator $A \otimes T : \ell_\infty \otimes_\varepsilon E \longrightarrow \ell_q \otimes_{\Delta_q} F$ is continuous. Hint: Ex 28.9.(4).

Ex 28.11. Show that $\delta^t_{q,p} \sim (\mathfrak{I}^{dual}_{p'} \circ \mathfrak{L}_{q'})^{reg}$ and that $T \in (\mathfrak{I}^{dual}_{p'} \circ \mathfrak{L}_{q'})^*(E, F)$ if and only if $A \otimes T : \ell_{p'} \otimes_{\Delta_{p'}} E \longrightarrow \ell_q \otimes_\pi F$ is continuous for all $A \in \mathcal{L}(\ell_{p'}, \ell_q)$.

Ex 28.12. In 26.3. it was shown (with the help of Grothendieck's inequality) that for every $S \in \mathcal{L}(\ell_\infty, \ell_1)$ the operator $S \otimes id_{\ell_2} : \ell_\infty \otimes_\varepsilon \ell_2 \longrightarrow \ell_1 \otimes_\pi \ell_2$ is continuous. Show, vice–versa, with 28.4. that this statement implies $\mathfrak{L}_2 \subset \mathfrak{L}^{inj\ sur}_{1,1}$, which by 20.17. is equivalent to Grothendieck's inequality.

Ex 28.13. Use 28.4., 29.12. and the definition $K_G(n) := \mathbf{P}_1^{sur}(id_{\ell_2^n})$ of the finite dimensional Grothendieck constants (20.17.) to show that

$$K_G(n) = \sup\{\|S \otimes id : L_\infty(\mu) \otimes_\varepsilon \ell_2^n \longrightarrow L_1(\nu) \otimes_\pi \ell_2^n\| \mid \|S\| \leq 1\}$$

whenever $L_\infty(\mu)$ and $L_1(\nu)$ are infinite dimensional.

Ex 28.14. *Complexification of operators:* For $p, q \in [1, \infty]$ denote by $k_{q,p}$ the best constant satisfying $\|S^\mathbb{C} : L_q^\mathbb{C}(\mu) \to L_p^\mathbb{C}(\nu)\| \leq k_{q,p}\|S : L_q^R(\mu) \to L_p^R(\nu)\|$ for all measures μ and ν and operators $S \in \mathcal{L}(L_q^R(\mu), L_p^R(\nu))$. Recall from 26.3. that $S^\mathbb{C} = S \otimes id_{R_2^2}$ and that $k_{q,p} = 1$ if $1 \leq q \leq p \leq \infty$.

(a) Use 28.4. (as well as 29.12.) and 20.4.(2) to show that

$$k_{q,p} = \mathbf{L}_{p,q'}^{inj\, sur}(id_{R_2^2}) \leq \mathbf{L}_{p,q'}^{inj}(id_{R_2^2}) \text{ for } 1 \leq p \leq q \leq \infty$$

and even equality whenever $q = 2$. Note that Ex 20.16. implies

$$k_{q,p} = \alpha_{p,q'}(s_2; C(S_1), C(S_1)) .$$

(b) Conclude from the foregoing exercise that $k_{\infty,1} = K_G^R(2)$ and note that $K_G^R(2) \leq \pi/2$ by Ex 14.4.; actually $k_{\infty,1} = K_G^R(2) = \sqrt{2}$ by Krivine's result (see 20.17.).

(c) Show that $k_{q,p} = k_{p',q'}$ and $k_{r,s} \leq k_{q,p}$ if $1 \leq p \leq s \leq r \leq q \leq \infty$.

(d) Use the embedding $\ell_2^n \hookrightarrow L_p(S_{n-1}, \eta_{n-1})$ from Ex 11.24. and $\mathcal{L}_{p,q'}^{inj} \circ \mathfrak{P}_q^{dual} \subset \mathfrak{P}_p$ (see 25.9.) to prove that for $1 \leq p \leq q \leq \infty$

$$\mathbf{L}_{p,q'}^{inj}(id_{\ell_2^n}) = \frac{c_{2,q}^1}{c_{2,q}^n} \cdot \frac{c_{2,p}^n}{c_{2,p}^1} = \frac{\mathbf{P}_p(id_{\ell_2^n})}{\mathbf{P}_q(id_{\ell_2^n})}$$

(this result is taken from Pietsch [214], 22.1.5.).

(e) Deduce that the "complexification constants" $k_{q,p}$ satisfy for $1 \leq p \leq q \leq \infty$

$$k_{q,p} \leq \min\left\{\sqrt{2}, (\sqrt{\pi})^{1/p-1/q} \left(\frac{\Gamma((1+q)/2)}{\Gamma((2+q)/2)}\right)^{1/q} \left(\frac{\Gamma((2+p)/2)}{\Gamma((1+p)/2)}\right)^{1/p}\right\}$$

$$k_{2,p} = k_{p',2} = \frac{\sqrt{2}}{2}\left(\sqrt{\pi}\frac{\Gamma((2+p)/2)}{\Gamma((1+p)/2)}\right)^{1/p} \text{ for } 1 \leq p \leq 2$$

$$k_{2,1} = k_{\infty,2} = \frac{\sqrt{2}}{4}\pi \approx 1,11072 .$$

29. The Calculus of Traced Tensor Norms

The basic philosophy behind our investigations is that finitely generated tensor norms and maximal normed operator ideals are in some sense the same — via the representation theorem for maximal operator ideals. In many cases the composition of operator ideals is a powerful tool for obtaining results about tensor norms; the reader is certainly aware of the important role of Kwapień's factorization theorem and the results of section 26. So it is natural to ask whether or not something like the "composition" of tensor norms exists. In fact, it is possible to define such compositions; they are the so-called "traced" tensor norms. They are quite simple and unify many of the results which have already been presented. However, their use may sometimes appear a bit formalistic (a sort of "calculus" is used with them) — and this is why we did not introduce them until now, although it might have shortened some proofs. The first systematic study of traced tensor norms appeared in [41].

29.1. For normed spaces E, F and G (where $G \neq \{0\}$) the *tensor contraction* (or: vector valued trace)

$$C_G : (E \otimes G') \otimes (G \otimes F) \longrightarrow E \otimes F$$
$$(x \otimes u') \otimes (u \otimes y) \rightsquigarrow \langle u', u \rangle_{G', G} x \otimes y$$

is defined and surjective. If F is replaced by its dual, this mapping is just the composition of operators:

$$(E \otimes G') \otimes (G \otimes F') \longrightarrow E \otimes F'$$
$$\mathfrak{F}(G, E) \otimes \mathfrak{F}(F, G) \longrightarrow \mathfrak{F}(F, E)$$
$$T \otimes S \rightsquigarrow T \circ S .$$

See also Ex 12.11.. If α is a tensor norm, then

(*) $$\|C_G : (E \otimes_\varepsilon G') \otimes_\pi (G \otimes_\alpha F) \longrightarrow E \otimes_\alpha F\| \leq 1$$

since

$$\alpha\Big(C_G\Big[\Big(\sum_{m=1}^{k} x_m \otimes u'_m\Big) \otimes \Big(\sum_{n=1}^{\ell} u_n \otimes y_n\Big)\Big]; E, F\Big) =$$

$$= \alpha\Big(\Big[\Big(\sum_{m=1}^{k} u'_m \otimes x_m\Big) \otimes id_F\Big]\Big(\sum_{n=1}^{\ell} u_n \otimes y_n\Big); E, F\Big) \leq$$

$$\leq \Big\|\sum_{m=1}^{k} u'_m \otimes x_m\Big\| \alpha\Big(\sum_{n=1}^{\ell} u_n \otimes y_n; G, F\Big) = \varepsilon\Big(\sum_{m=1}^{k} x_m \otimes u'_m; E, G'\Big) \alpha\Big(\sum_{n=1}^{\ell} u_n \otimes y_n; G, F\Big)$$

(this was Ex 12.11.(c)). Now let α be a right–tensor norm with respect to G (see 17.4., e.g. α a tensor norm or $\alpha = \Delta_p$ if $G = L_p(\mu)$) and β a left–tensor norm with respect to G' (i.e. β^t is a right–tensor norm for $G' \otimes \cdot$; example: $\beta = \Delta_{p'}^t$). Then

$$C_G : (E \otimes_\beta G') \otimes_\pi (G \otimes_\alpha F) \longrightarrow (E \otimes_\varepsilon G') \otimes_\pi (G \otimes_\varepsilon F) \longrightarrow E \otimes_\varepsilon F$$

is continuous, and hence $\ker C_G$ is closed in the tensor product $(E \otimes_\beta G') \otimes_\pi (G \otimes_\alpha F)$. The projective norm

$$\pi(\,\cdot\,; E \otimes_\beta G', G \otimes_\alpha F)$$

now defines a quotient norm $\delta(\cdot; E, F)$ on $E \otimes F$ via the tensor contraction C_G. Straightforward arguments (e.g. by use of the criterion 12.2.) show that the assignment

$$\alpha \otimes_G \beta : (E, F) \rightsquigarrow \delta(\,\cdot\,; E, F)$$

is a tensor norm on NORM which is called the *traced tensor norm of α and β along G*. It is obvious from $(*)$ (and $\|\otimes_\beta \otimes_\pi \otimes_\varepsilon \longrightarrow \otimes_\beta \| \leq 1$ by a similar argument) that

$$\alpha \leq \alpha \otimes_G \beta \qquad \text{if } \alpha \text{ is a tensor norm}$$
$$\beta \leq \alpha \otimes_G \beta \qquad \text{if } \beta \text{ is a tensor norm.}$$

The reader might feel uncomfortable with the notation: the order of α and β in the four-fold tensor product and in $\alpha \otimes_G \beta$ is different. This was done for purpose: it will turn out that the tensor norm $\alpha \otimes_{C_2} \beta$ is related to the operator ideal $\mathfrak{A} \circ \mathfrak{B}$, if $\alpha \sim \mathfrak{A}$ and $\beta \sim \mathfrak{B}$ are associated.

By definition of the projective tensor norm π,

$$\alpha \otimes_G \beta(z; E, F) = \inf\{\pi(w; E \otimes_\beta G', G \otimes_\alpha F) \mid C_G(w) = z\} =$$
$$= \inf\Big\{\sum_{k=1}^n \beta(u_k; E, G')\alpha(v_k; G, F) \mid C_G\Big(\sum_{k=1}^n u_k \otimes v_k\Big) = z\Big\}.$$

This implies the

REMARK: *If α and β^t are finitely generated from the right (see 17.4.), then the tensor norm $\alpha \otimes_G \beta$ is finitely generated.*

Recall that Δ_p is finitely generated from the right (see Ex 7.16.).

29.2. The following lemma will be helpful for some calculations.

LEMMA: *Let α be a right–tensor norm with respect to the dual G' of a normed space G and β a finitely generated tensor norm. Then $\alpha \otimes_{G'} \beta$ on $E \otimes F$ is the quotient norm of the canonical map*

$$(E \otimes_\beta G) \otimes_\pi (G' \otimes_\alpha F) \longrightarrow E \otimes F.$$

PROOF: Denote by δ this quotient seminorm. Since $E \otimes_\beta G$ is a subspace of $E \otimes_\beta G''$ (by the embedding lemma), it is obvious that $\alpha \otimes_{G'} \beta \leq \delta$. Conversely, take $z \in E \otimes F$ and
$$w = \sum_{k=1}^{n} v_k^1 \otimes v_k^2 \in (E \otimes_\beta G'') \otimes_\pi (G' \otimes_\alpha F),$$
with $C_{G'}(w) = z$ and $\sum_{k=1}^{n} \beta(v_k^1; E, G'')\alpha(v_k^2; G', F) \leq (1+\varepsilon)\alpha \otimes_{G'} \beta(z)$. Since β is finitely generated, there is an $M \in \text{FIN}(G'')$ with
$$\beta(v_k^1; E, M) \leq (1+\varepsilon)\beta(v_k^1; E, G'')$$
for all $k = 1, ..., n$. Take representations
$$v_k^1 = \sum_{j=1}^{m} x_{k,j} \otimes u''_{k,j} \in E \otimes M$$
$$v_k^2 = \sum_{j=1}^{m} u'_{k,j} \otimes y_{k,j} \in G' \otimes F$$
and, by the weak principle of local reflexivity 6.5., an operator $R \in \mathcal{L}(M, G)$ with $\|R\| \leq 1+\varepsilon$ and
$$\langle R u''_{k,j}, u'_{\ell,i}\rangle_{G,G'} = \langle u''_{k,j}, u'_{\ell,i}\rangle_{G'',G'}$$
for all i, j, k, ℓ. It follows that z is the image of
$$\sum_{k=1}^{n}[(id \otimes R)(v_k^1)] \otimes v_k^2 \in (E \otimes_\beta G) \otimes_\pi (G' \otimes_\alpha F)$$
and
$$\delta(z) \leq \sum_{k=1}^{n} \beta(id \otimes R(v_k^1); E, G)\alpha(v_k^2) \leq \|R\| \sum_{k=1}^{n} \beta(v_k^1; E, M)\alpha(v_k^2) \leq$$
$$\leq (1+\varepsilon)^2 \sum_{k=1}^{n} \beta(v_k^1; E, G'')\alpha(v_k^2; G', E) \leq (1+\varepsilon)^3 \alpha \otimes_{G'} \beta(z) . \quad \square$$

29.3. Some examples:

PROPOSITION 1: Let $1 \leq p \leq \infty$.
(1) $w_p = \varepsilon \otimes_{\ell_{p'}} \varepsilon$ and $w_1 = \varepsilon \otimes_{c_o} \varepsilon$.
(2) $d_p = \Delta_p \otimes_{\ell_p} \varepsilon$.
(3) $g_p = \varepsilon \otimes_{\ell_{p'}} \Delta_p^t$ and $\pi = \varepsilon \otimes_{c_o} \pi$.

PROOF: Statement (1) is just a reformulation of corollary 12.9. (with the help of lemma 29.2. for $p = 1$). Since
$$d_\infty = w_1 = \varepsilon \otimes_{\ell_\infty} \varepsilon = \Delta_\infty \otimes_{\ell_\infty} \varepsilon ,$$

(2) holds for $p = \infty$. For $1 \le p < \infty$ recall that

$$d_p(z; E, F) = \inf\left\{w_{p'}(x_k; E)\ell_p(y_k; F) \mid z = \sum_{k=1}^n x_k \otimes y_k\right\}$$

and, again from remark 12.9. and section 8,

$$w_{p'}(x_k; E) = \varepsilon\left(\sum x_k \otimes e_k; E, \ell_{p'}\right)$$
$$\ell_p(y_k; F) = \Delta_p\left(\sum e_k \otimes y_k; \ell_p, F\right).$$

This implies that $\Delta_p \otimes_{\ell_p} \varepsilon \le d_p$. The converse follows from $d_p \le \Delta_p$:

$$d_p \le d_p \otimes_{\ell_p} \varepsilon \le \Delta_p \otimes_{\ell_p} \varepsilon.$$

(This was Ex 12.11.(b).) The last statement follows in the same way (for $p = 1$ use the previous lemma or Ex 29.1.). □

29.4. The main result on traced tensor norms is the following

THEOREM: *Let α and β be two finitely generated tensor norms, G a normed space and denote by $(\mathfrak{D}, \mathbf{D})$ the maximal normed operator ideal associated with the traced tensor norm $\alpha \otimes_G \beta$. Then for each $T \in \mathcal{L}(E, F)$ the following statements are equivalent:*
(a) $T \in \mathfrak{D}^*(E, F)$.
(b) $id_G \otimes T : G \otimes_\alpha E \longrightarrow G \otimes_{\overleftarrow{\beta^*}} F$ *is continuous.*

In this case, $\mathbf{D}^*(T) = \|id_G \otimes T : G \otimes_\alpha E \longrightarrow G \otimes_{\overleftarrow{\beta^*}} F\|$.

If $G = L_p(\mu)$ for some measure μ, then the equivalence also holds in the following three cases:
(1) $\alpha = \Delta_p$ *and β a finitely generated tensor norm,*
(2) $\beta = \Delta_{p'}^t$ *and α a finitely generated tensor norm,*
(3) $\alpha = \Delta_p$ *and $\beta = \Delta_{p'}^t$, where $\overleftarrow{(\Delta_{p'}^t)^*} := \Delta_p$.*

PROOF: It was proved in 17.4. that the class of all T satisfying (b) forms a maximal operator ideal $\mathcal{L}_G$ with the norm

$$\mathbf{L}_G(T) := \|id_G \otimes T : G \otimes_\alpha E \longrightarrow G \otimes_{\overleftarrow{\beta^*}} F\|.$$

Thus, it suffices to prove that $\mathbf{D}^*(T : M \to N) = \mathbf{L}_G(T : M \to N)$ for $M, N \in \text{FIN}$. The definition of the traced tensor norm $\delta := \alpha \otimes_G \beta$ and

$$G \otimes_{\overleftarrow{\beta^*}} N' \xrightarrow{\quad 1 \quad} (N \otimes_\beta G')'$$

(by the duality theorem 15.5. for tensor norms and 15.10. for $\beta = \Delta_{p'}^t$,) give the commutative diagram

$$\begin{array}{ccc} (N \otimes_\delta M)' & \xrightarrow{C'_G} & ((N \otimes_\beta G') \otimes_\pi (G \otimes_\alpha M))' \\ \| & & \uparrow 1 \\ \mathfrak{D}^*(M, N') & \longrightarrow & \mathcal{L}(G \otimes_\alpha M, G \otimes_{\overset{\sim}{\beta^*}} N') \\ T & \rightsquigarrow & id_G \otimes T \end{array}$$

and the fact that C'_G is a metric injection. This implies that the lower horizontal map is a metric injection, which is the statement. □

This result unifies many of the tensor product characterizations of operator ideals which have been presented so far. Recall from 15.7. that for each Banach space G

$$G \otimes_{\overset{\sim}{\beta^*}} F = G \otimes_{\beta^*} F$$

if β^* is totally accessible or: if β is left–accessible and G has the metric approximation property or: β is right–accessible and F has the metric approximation property.

29.5. Using $w_{p'} = \varepsilon \otimes_{\ell_p} \varepsilon$ from 29.3. and $\mathfrak{D}_p \sim w_{p'}^*$ (see 17.12.), a first application concerns p–dominated operators:

PROPOSITION: *For $1 \leq p \leq \infty$ and $T \in \mathcal{L}(E, F)$ the following are equivalent:*
(a) $T \in \mathfrak{D}_p(E, F)$.
(b) $id_{\ell_p} \otimes T : \ell_p \otimes_\varepsilon E \longrightarrow \ell_p \otimes_\pi F$ *is continuous.*
In this case, $\mathbf{D}_p(T) = \|id_{\ell_p} \otimes T : \ell_p \otimes_\varepsilon E \longrightarrow \ell_p \otimes_\pi F\|$.

For $p = 2$ this is a result of Grothendieck's ([93], p.42).
It is easy to see from this that a Banach space G which contains the ℓ_p^n's uniformly complemented has the property that $T \in \mathcal{L}(E, F)$ is p–dominated if

$$id_G \otimes T : G \otimes_\varepsilon E \longrightarrow G \otimes_\pi F$$

is continuous. This property of G implies (by the theorem for $E, F \in \text{FIN}$ only) that $\varepsilon \otimes_G \varepsilon \leq c w_{p'}$ which gives (see Ex 29.9.(b)) that G contains nearly the ℓ_p^n's uniformly complemented: There exist a $\lambda \geq 0$ and, for every n, operators $I_k : \ell_p^n \to G$ and $P_k : G \to \ell_p^n$ such that

$$\sum_{k=1}^m P_k \circ I_k = id_{\ell_p^n} \quad \text{and} \quad \sum_{k=1}^m \|P_k\| \|I_k\| \leq \lambda$$

(m depending on n as well) — these spaces are called T_p-spaces (see also Ex 29.9.). It was shown in [42] that T_p-spaces for $p = 1, 2$ and ∞ contain the ℓ_p^n's uniformly complemented.

29.6. The following observations establish a sort of "calculus" for the traced tensor norms:

REMARK 1: *For appropriate right- and left-tensor norms*
$$(\alpha \otimes_G \beta) \otimes_H \gamma = \alpha \otimes_G (\beta \otimes_H \gamma) =: \alpha \otimes_G \beta \otimes_H \gamma .$$

This follows immediately from the facts that the projective norm π is associative (Ex 4.4.) and projective, and from the transitivity of forming quotient norms.

REMARK 2: *Let α and β be tensor norms. Then*
(1) $(\alpha \otimes_G \beta)/ = \alpha/ \otimes_G \beta$.
(2) $\backslash(\alpha \otimes_G \beta) = \alpha \otimes_G \backslash\beta$.
(3) $\backslash(\alpha \otimes_G \beta)/ = \alpha/ \otimes_G \backslash\beta$.

PROOF: Only (1) will be proved; the proof of the other two statements is the same. It suffices to check that for $\delta := \alpha \otimes_G \beta$ and $\gamma := \alpha/ \otimes_G \beta$
$$E \otimes_\delta \ell_1(B_F) \longrightarrow E \otimes_\gamma F$$

is a metric surjection: This follows from

$$\begin{array}{ccc}
(E \otimes_\beta G') \otimes_\pi (G \otimes_\alpha \ell_1(B_F)) & \xrightarrow{1} & E \otimes_\delta \ell_1(B_F) \\
\downarrow & & \downarrow \\
(E \otimes_\beta G') \otimes_\pi (G \otimes_{\alpha/} F) & \xrightarrow{1} & E \otimes_\gamma F
\end{array}$$

since π respects quotients. □

Finally, some useful formulas:

REMARK 3: *For every tensor norm α the following holds:*
(1) $\alpha \otimes_G \pi = \pi \otimes_G \alpha = \pi$.
(2) $\alpha \otimes_G \alpha^* = \alpha^* \otimes_G \alpha = \pi$ *if α is finitely generated.*

PROOF: (1) follows from $\pi \leq \alpha \otimes_G \pi$ (see 29.1.). (2) follows from theorem 29.4.: If $\mathfrak{D}^*$ is the maximal normed operator ideal associated with $(\alpha \otimes_G \alpha^*)^*$, then
$$\|T\| \leq \mathbf{D}^*(T) = \|id_G \otimes T : \otimes_\alpha \longrightarrow \otimes_{\widetilde{\alpha}}\| \leq$$
$$\leq \|id_G \otimes T : \otimes_\alpha \longrightarrow \otimes_\alpha\| = \|T\| .$$

This shows that $(\alpha \otimes_G \alpha^*)^* \sim \mathfrak{L} \sim \varepsilon$ and hence

$$\pi = (\alpha \otimes_G \alpha^*)^{**} = \alpha \otimes_G \alpha^*$$

since $\alpha \otimes_G \alpha^*$ is finitely generated (see remark 29.1.). □

Though it may appear quite formal, the calculus of traced tensor norms will turn out to be very useful: it allows one, for example, to determine without any effort those $T \in \mathfrak{L}(E, F)$ such that

$$A \otimes T : L_p[0, 1] \otimes_{\Delta_p} E \longrightarrow G \otimes_\alpha F$$

is continuous for all Banach spaces G and all $A \in \mathfrak{L}(L_p[0,1], G)$ — see Ex 29.7. (the answer is: $T \in (\mathfrak{I}_p^{dual} \circ \mathfrak{A}^*)^*(E, F)$ if α is right–accessible). It is for results like this that traced tensor norms along a "general" space G have to be introduced.

29.7. For two finitely generated tensor norms α and β and a Johnson space C_2 (see 6.3.) define the tensor norm

$$\alpha \otimes \beta := \alpha \otimes_{C_2} \beta \ ;$$

it is called the *tensor product* of α and β. By 29.1. it is finitely generated. The maximal normed operator ideal associated with $\alpha \otimes \beta$ is denoted by $(\mathfrak{A} \otimes \mathfrak{B}, \mathbf{A} \otimes \mathbf{B})$, the *tensor product of* $(\mathfrak{A}, \mathbf{A}) \sim \alpha$ and $(\mathfrak{B}, \mathbf{B}) \sim \beta$. It follows from theorem 29.4. and lemma 17.16. that $T \in (\mathfrak{A} \otimes \mathfrak{B})^*(E, F)$ if and only if for all Banach spaces G

$$id_G \otimes T : G \otimes_\alpha E \longrightarrow G \otimes_{\beta^*} F$$

is continuous; in this case $(\mathbf{A} \otimes \mathbf{B})^*(T) = \sup\{\|id_G \otimes T : ...\| \mid G \in \text{BAN}\}$.

In particular, the definitions of $\alpha \otimes \beta$ and $\mathfrak{A} \otimes \mathfrak{B}$ are independent of the particular choice of Johnson space C_2. Recall that C_2 is reflexive, has the metric approximation property and is isometric to its dual C_2'. It follows that one may take C_2' in the definition of $\alpha \otimes \beta$.

The name "tensor product" for $\mathfrak{A} \otimes \mathfrak{B}$ comes from

$$\begin{array}{ccc}
(M' \otimes_\beta C_2) \otimes_\pi (C_2' \otimes_\alpha N) & \longrightarrow & M' \otimes_{\alpha \otimes \beta} N \\
\|1 & & \|1 \\
\mathfrak{B}(M, C_2) \otimes_\pi \mathfrak{A}(C_2, N) & \longrightarrow & (\mathfrak{A} \otimes \mathfrak{B})(M, N) \\
R \otimes S & \rightsquigarrow & S \circ R
\end{array}$$

for $M, N \in \text{FIN}$. Since every $L \in \text{FIN}$ is $(1 + \varepsilon)$–isomorphic to some 1–complemented subspace of C_2 and $R \in \mathfrak{B}(M, C_2)$ factors

$$M \xrightarrow{R} C_2$$
$$R_o \searrow \quad \uparrow I$$
$$L$$

$$\mathbf{B}(R_o) \leq (1+\varepsilon)\mathbf{B}(R) ,$$

it follows that

$$\mathbf{A} \otimes \mathbf{B}(T : M \to N) = \inf \sum_{k=1}^{n} \mathbf{A}(S_k)\mathbf{B}(R_k) ,$$

where the infimum is taken over all factorizations $T = \sum_{k=1}^{n} S_k \circ R_k$ with operators $R_k \in \mathfrak{L}(M, L_k)$ and $S_k \in \mathfrak{L}(L_k, N)$ for some $L_k \in \text{FIN}$.

29.8. If β is accessible, then $C_2 \otimes_{\beta^*} F = C_2 \otimes_{\beta^*} F$ and theorem 25.8. and 29.4. imply that $(\mathfrak{A} \otimes \mathfrak{B})^* = (\mathfrak{A} \circ \mathfrak{B})^*$ holds isometrically, the operators $T \in (\mathfrak{A} \otimes \mathfrak{B})^*$ being those for which for all G (or only some C_2) the map

$$id_G \otimes T : G \otimes_\alpha E \longrightarrow G \otimes_{\beta^*} F$$

is continuous; $(\mathbf{A} \otimes \mathbf{B})^*(T) = \|id_{C_2} \otimes T : ...\|$. This result also follows from the following proposition, which is independent of 25.8.:

PROPOSITION: *Let $(\mathfrak{A}, \mathbf{A}) \sim \alpha$ and $(\mathfrak{B}, \mathbf{B}) \sim \beta$ be such that $\mathfrak{A}$ is left-accessible or $\mathfrak{B}$ right-accessible. Then*

(1) $\mathfrak{A} \otimes \mathfrak{B}$ *is the smallest maximal normed operator ideal containing the quasi-Banach ideal $\mathfrak{A} \circ \mathfrak{B}$.*

(2) $(\mathfrak{A} \otimes \mathfrak{B})^* = (\mathfrak{A} \circ \mathfrak{B})^*$ *holds isometrically.*

PROOF: (1) If $(\mathfrak{C}, \mathbf{C})$ is a maximal normed ideal containing $\mathfrak{A} \circ \mathfrak{B}$, then there is a $\rho \geq 0$ with $\mathbf{C} \leq \rho \mathbf{A} \circ \mathbf{B}$. For $T \in \mathfrak{L}(M, N)$ with a representation

$$T = \sum_{k=1}^{n} S_k \circ R_k$$

as at the end of 29.7. it follows that

$$\mathbf{C}(T) \leq \sum_{k=1}^{n} \mathbf{C}(S_k \circ R_k) \leq \rho \sum_{k=1}^{n} \mathbf{A}(S_k)\mathbf{B}(R_k)$$

and hence $\mathbf{C} \leq \rho \mathbf{A} \otimes \mathbf{B}$. Thus, it remains to prove that $\mathfrak{A} \circ \mathfrak{B} \subset \mathfrak{A} \otimes \mathfrak{B}$: again it is enough to check this for finite dimensional spaces. Take $T \in \mathfrak{A} \circ \mathfrak{B}(M, N)$ and a factorization

$$\begin{array}{ccc} M & \xrightarrow{T} & N \\ {}_R\searrow & \nearrow{}_S & \\ & G & \end{array} \qquad \mathbf{A}(S)\mathbf{B}(R) \leq (1+\varepsilon)\mathbf{A}\circ\mathbf{B}(T);$$

then G can be taken to be finite dimensional as well since $\mathfrak{A}$ is left- or $\mathfrak{B}$ right-accessible. It follows that

$$\mathbf{A}\otimes\mathbf{B}(T) \leq \mathbf{A}\circ\mathbf{B}(T)$$

for arbitrary $T \in \mathfrak{L}(E,F)$. This completes the proof of (1). To see (2), first observe that the last statement implies that

$$(\mathbf{A}\circ\mathbf{B})^* \leq (\mathbf{A}\otimes\mathbf{B})^* .$$

Since $(\mathfrak{A}\circ\mathfrak{B})^{**} \supset \mathfrak{A}\circ\mathfrak{B}$ and $(\mathbf{A}\circ\mathbf{B})^{**} \leq \mathbf{A}\circ\mathbf{B}$, the first part of the proof of (1) implies that $(\mathbf{A}\circ\mathbf{B})^{**} \leq \mathbf{A}\otimes\mathbf{B}$, which gives

$$(\mathbf{A}\otimes\mathbf{B})^* \leq (\mathbf{A}\circ\mathbf{B})^{***} = (\mathbf{A}\circ\mathbf{B})^* . \qquad \square$$

Clearly, this result (together with theorem 29.4.) gives another proof of 25.8..

COROLLARY: *Let α and $\mathfrak{A}$ be associated. The following statements hold isometrically:*
(1) $\mathfrak{A}\otimes\mathfrak{L} = \mathfrak{L}\otimes\mathfrak{A} = \mathfrak{A}$ and $\alpha\otimes\varepsilon = \varepsilon\otimes\alpha = \alpha$.
(2) $\mathfrak{A}\otimes\mathfrak{J} = \mathfrak{J}\otimes\mathfrak{A} = \mathfrak{J}$ and $\alpha\otimes\pi = \pi\otimes\alpha = \pi$.
(3) $\mathfrak{A}\otimes\mathfrak{A}^* = \mathfrak{A}^*\otimes\mathfrak{A} = \mathfrak{J}$ and $\alpha\otimes\alpha^* = \alpha^*\otimes\alpha = \pi$.

PROOF: The first statement follows from $\varepsilon \sim \mathfrak{L}$ and

$$(\mathfrak{A}\otimes\mathfrak{L})^* = (\mathfrak{A}\circ\mathfrak{L})^* = \mathfrak{A}^*$$

and the others were already shown in 29.6. since $\pi \sim \mathfrak{J}$. $\square$

Note that by 29.3. $\varepsilon\otimes\varepsilon = \varepsilon \neq w_p = \varepsilon\otimes_{\ell_{p'}} \varepsilon$. If $\mathfrak{A}\circ\mathfrak{B}$ is maximal normed, then the proposition implies that

$$\mathfrak{A}\circ\mathfrak{B} = \mathfrak{A}\otimes\mathfrak{B} .$$

In general, $\mathfrak{A}\circ\mathfrak{B}$ may be "much smaller" than $\mathfrak{A}\otimes\mathfrak{B}$:

$$\mathfrak{J}\circ\mathfrak{J} \subset \mathfrak{P}_2\circ\mathfrak{P}_2 \subset \mathfrak{N} \underset{\neq}{\subset} \mathfrak{J} = \mathfrak{J}\otimes\mathfrak{J} \subset \mathfrak{P}_2\otimes\mathfrak{P}_2 = \mathfrak{J}_2\otimes\mathfrak{P}_2 = \mathfrak{J} .$$

Using ultrastability one can show (see Ex 29.8.): If $\mathfrak{A}$ and $\mathfrak{B}$ are two maximal normed operator ideals such that $\mathfrak{A}$ is left–accessible or $\mathfrak{B}$ right–accessible and $\mathfrak{A}\circ\mathfrak{B}$ is normed, then $(\mathfrak{A}\circ\mathfrak{B})^{reg} \stackrel{1}{=} \mathfrak{A}\otimes\mathfrak{B}$. In particular, $\mathfrak{A}\circ\mathfrak{B} \stackrel{1}{=} \mathfrak{A}\otimes\mathfrak{B}$ if and only if $\mathfrak{A}\circ\mathfrak{B}$ is normed and regular.

The result about $(\mathfrak{A} \circ \mathfrak{B})^{reg}$, by the way, also shows that $\mathfrak{J} \circ \mathfrak{J}$ is not normed, since

$$(\mathfrak{J} \circ \mathfrak{J})^{reg} \subset \mathfrak{N}^{reg} \subset \mathfrak{N}^{dual} \underset{\neq}{\subset} \mathfrak{J} = \mathfrak{J} \otimes \mathfrak{J}.$$

The same type of argument shows that $\mathfrak{P}_2 \circ \mathfrak{P}_2$ is not normed.

29.9. Kwapień's factorization theorem 19.3. now gives

$$\alpha^*_{p,q} \sim \mathfrak{D}_{p',q'} = \mathfrak{P}^{dual}_{q'} \circ \mathfrak{P}_{p'} = \mathfrak{P}^{dual}_{q'} \otimes \mathfrak{P}_{p'} \sim g^{*t}_q \otimes g^*_p$$

— and this is the first of the following formulas:

PROPOSITION: *Let $p, q \in [1, \infty]$ and $1/p + 1/q \geq 1$. Then*
(1) $\alpha^*_{p,q} = d^*_q \otimes g^*_p$.
(2) $\backslash \alpha^*_{p,q} = d^*_q \otimes g_{p'} = d^*_q \otimes_{\ell_p} \Delta^t_{p'}$.
(3) $\alpha^*_{p,q}/ = d_{q'} \otimes g^*_p = \Delta_{q'} \otimes_{\ell_{q'}} g^*_p$.
(4) $\backslash \alpha^*_{p,q}/ = d_{q'} \otimes g_{p'} = d_{q'} \otimes_{\ell_p} \Delta^t_{p'} = \Delta_{q'} \otimes_{\ell_{q'}} g_{p'}$.

PROOF: Using the formulas from 29.3. and 29.6. and $\backslash g^*_p = g_{p'}$ (see 20.14.), one obtains

$$\backslash \alpha^*_{p,q} = \backslash (d^*_q \otimes g^*_p) = d^*_q \otimes \backslash g^*_p = d^*_q \otimes g_{p'}$$
$$= d^*_q \otimes (\varepsilon \otimes_{\ell_p} \Delta^t_{p'}) = (d^*_q \otimes \varepsilon) \otimes_{\ell_p} \Delta^t_{p'} = d^*_q \otimes_{\ell_p} \Delta^t_{p'}.$$

(3) and (4) follow in the same way. □

Via the main theorem 29.4. for traced tensor norms (and 29.7.) each of these formulas can be interpreted in terms of the adjoints of the associated operator ideals:

$$\mathcal{L}^{inj}_{p,q} \sim \alpha_{p,q}\backslash = (\backslash \alpha^*_{p,q})^* = (d^*_q \otimes g_{p'})^* = (d^*_q \otimes_{\ell_p} \Delta^t_{p'})^*$$
$$\mathcal{L}^{sur}_{p,q} \sim /\alpha_{p,q} = (\alpha^*_{p,q}/)^* = ...$$
$$\mathcal{L}^{inj\ sur}_{p,q} \sim /\alpha_{p,q}\backslash = (\backslash \alpha^*_{p,q}/)^* = ...\ .$$

COROLLARY: *Take $T \in \mathcal{L}(E, F)$. Then, in (1) – (3), all the statements (a) – (c) (resp. (a) – (d)) are equivalent:*
(1) (a) $T \in \mathcal{L}^{inj}_{p,q}(E, F)$.
  (b) $id_{\ell_p} \otimes T : \ell_p \otimes_{d^*_q} E \longrightarrow \ell_p \otimes_{\Delta_p} F$ *is continuous.*
  (c) $id_G \otimes T : G \otimes_{d^*_q} E \longrightarrow G \otimes_{g^*_{p'}} F$ *is continuous for all $G \in$ BAN (or only $G = \ell_p$).*
(2) (a) $T \in \mathcal{L}^{sur}_{p,q}(E, F)$.
  (b) $id_{\ell_{q'}} \otimes T : \ell_{q'} \otimes_{\Delta_{q'}} E \longrightarrow \ell_{q'} \otimes_{g_p} F$ *is continuous.*

(c) $id_G \otimes T : G \otimes_{d_{q'}} E \longrightarrow G \otimes_{g_p} F$ is continuous for all $G \in$ BAN (or only $G = \ell_{q'}$).

(3) (a) $T \in \mathfrak{L}^{inj\ sur}_{p,q}(E, F)$.
(b) $id_{\ell_p} \otimes T : \ell_p \otimes_{d_{q'}} E \longrightarrow \ell_p \otimes_{\Delta_p} F$ is continuous.
(c) $id_{\ell_{q'}} \otimes T : \ell_{q'} \otimes_{\Delta_{q'}} E \longrightarrow \ell_{q'} \otimes_{g^*_{p'}} F$ is continuous.
(d) $id_G \otimes T : G \otimes_{d_{q'}} E \longrightarrow G \otimes_{g^*_{p'}} F$ is continuous for all $G \in$ BAN (or only $G = \ell_{q'}$ or only $G = \ell_p$).

Note that for $q = p'$ one obtains characterizations of $\mathfrak{L}^{inj}_p$, $\mathfrak{L}^{sur}_p$ and $\mathfrak{L}^{inj\ sur}_p$, in particular, Kwapień's results in 25.10.. Moreover, it will shown in a moment (29.11.) that the sequence space ℓ_s in the statements above can be replaced by any infinite dimensional $L_s(\mu)$. We omitted the obvious results about the norms.

29.10. Another application of the "calculus" of traced tensor norms is the characterization of $\mathfrak{A}^{inj}$ and $\mathfrak{A}^{sur}$ in terms of tensor product formulas, which was already announced in 20.13.: The formulas give

$$\backslash \alpha^* = \backslash (\alpha^* \otimes \varepsilon) = \alpha^* \otimes \backslash \varepsilon = \alpha^* \otimes w_\infty = \alpha^* \otimes (\varepsilon \otimes_{\ell_1} \varepsilon) = (\alpha^* \otimes \varepsilon) \otimes_{\ell_1} \varepsilon = \alpha^* \otimes_{\ell_1} \varepsilon$$

$$\alpha^*/ = (\varepsilon \otimes \alpha^*)/ = \varepsilon/ \otimes \alpha^* = w_1 \otimes \alpha^* = (\varepsilon \otimes_{\ell_\infty} \varepsilon) \otimes \alpha^* = \varepsilon \otimes_{\ell_\infty} (\varepsilon \otimes \alpha^*) = \varepsilon \otimes_{\ell_\infty} \alpha^*$$

and hence, for the maximal normed operator ideal $\mathfrak{A} \sim \alpha$:

$$\mathfrak{A}^{inj} \sim \alpha \backslash = (\backslash \alpha^*)^* = (\alpha^* \otimes_{\ell_1} \varepsilon)^* \text{ and } \mathfrak{A}^{sur} \sim /\alpha = (\alpha^*/)^* = (\varepsilon \otimes_{\ell_\infty} \alpha^*)^* \ .$$

The main theorem 29.4. shows:

PROPOSITION: *Let α be a finitely generated tensor norm, $(\mathfrak{A}, \mathbf{A})$ its associated maximal normed operator ideal and $T \in \mathfrak{L}(E, F)$.*
(1) $T \in \mathfrak{A}^{inj}$ *if and only if*

$$id_{\ell_1} \otimes T : \ell_1 \otimes_{\alpha^\bullet} E \longrightarrow \ell_1 \otimes_\pi F$$

is continuous. In this case, $\mathbf{A}^{inj}(T) = \|id_{\ell_1} \otimes T : ...\|$.
(2) *If α is right-accessible, then $T \in \mathfrak{A}^{sur}$ if and only if*

$$id_{\ell_\infty} \otimes T : \ell_\infty \otimes_\varepsilon E \longrightarrow \ell_\infty \otimes_\alpha F$$

is continuous. In this case, $\mathbf{A}^{sur}(T) = \|id_{\ell_\infty} \otimes T : ...\|$.

29.11. In many of the formulas and results above the sequence space ℓ_p can be replaced by any infinite dimensional $L_p(\mu)$. The key to this is the following

29. The Calculus of Traced Tensor Norms

PROPOSITION: *If α is a finitely generated tensor norm and $1 \leq p \leq \infty$, then*

$$\Delta_p \otimes_{\ell_p} \alpha = \Delta_p \otimes_{L_p(\mu)} \alpha = d_p \otimes \alpha$$

$$\alpha \otimes_{\ell_p} \Delta_{p'}^t = \alpha \otimes_{L_p(\mu)} \Delta_{p'}^t = \alpha \otimes g_{p'}$$

for every infinite dimensional $L_p(\mu)$. One may replace ℓ_∞ by c_0.

PROOF: If (A_n) is a sequence of pairwise disjoint sets with $\mu(A_n) > 0$, then the mapping $I_s : \ell_s \to L_s(\mu)$ defined by

$$I_s(\xi_n) := \sum_{n=1}^{\infty} \xi_n \, [\mu(A_n)]^{-1/s} \chi_{A_n}$$

is an isometry and its range is 1–complemented with the averaging operator

$$Pf := \sum_{n=1}^{\infty} [\mu(A_n)]^{-1} \int_{A_n} f d\mu \; \chi_{A_n}$$

as a projector. It follows that $I_s \otimes id_E : \ell_s \otimes_\alpha E \hookrightarrow L_s(\mu) \otimes_\alpha E$ is an isometry onto a 1–complemented subspace — and this is also true for $\alpha = \Delta_s$ (by Ex 7.1.). Since

$$\langle I_{p'}x, I_p y\rangle_{L_{p'},L_p} = \langle x, y\rangle_{\ell_{p'},\ell_p}$$

by construction, it follows that

$$(E \otimes_\alpha L_{p'}(\mu)) \otimes_\pi (L_p(\mu) \otimes_{\Delta_p} F) \xrightarrow{C_{L_p}}$$
$$\uparrow 1 \qquad\qquad E \otimes F$$
$$(E \otimes_\alpha \ell_{p'}) \otimes_\pi (\ell_p \otimes_{\Delta_p} F) \xrightarrow{C_{\ell_p}}$$

commutes and hence that $\Delta_p \otimes_{L_p(\mu)} \alpha \leq \Delta_p \otimes_{\ell_p} \alpha$. (Note that the projection does not commute with the tensor contractions!)

To see the other inequality the calculus for traced tensor norms helps:

$$\Delta_p \otimes_{\ell_p} \alpha = \Delta_p \otimes_{\ell_p} (\varepsilon \otimes \alpha) = (\Delta_p \otimes_{\ell_p} \varepsilon) \otimes \alpha = d_p \otimes \alpha \leq$$
$$\leq (d_p \otimes_{L_p(\mu)} \varepsilon) \otimes \alpha = d_p \otimes_{L_p(\mu)} (\varepsilon \otimes \alpha) = d_p \otimes_{L_p(\mu)} \alpha \leq$$
$$\leq \Delta_p \otimes_{L_p(\mu)} \alpha$$

— which also gives the equation with $d_p \otimes \alpha$. The second relations follow in the same way. That ℓ_∞ can be replaced by c_0 follows from Ex 29.1.. □

This shows again how "close" d_p is to Δ_p: Recall that the proof of 15.11. concerning the closest tensor norms to Δ_p used the tensor contraction as well.

29.12. This result has an interesting consequence for the continuity of all

$$A \otimes T : L_q(\mu) \otimes_{\Delta_q} E \longrightarrow L_p(\nu) \otimes_{\Delta_p} F$$

for a given $T \in \mathfrak{L}(E, F)$. For this a little lemma concerning $\alpha \otimes_G \varepsilon \otimes_H \beta$ will be presented — a lemma which is also useful in other situations (see Ex 29.4. – Ex 29.7.).

LEMMA: *Let G and H be normed spaces and α and β finitely generated tensor norms (or $\alpha = \Delta_q$ if $G = L_q(\mu)$, or $\beta = \Delta_{p'}^t$ if $H = L_p(\nu)$). If $\mathfrak{D}$ is the maximal normed operator ideal associated with $\alpha \otimes_G \varepsilon \otimes_H \beta$, then $T \in \mathfrak{D}^*(E, F)$ if and only if there is a constant $c \geq 0$ such that*

$$\|A \otimes T : G \otimes_\alpha E \longrightarrow H \otimes_{\beta^\bullet} F\| \leq c \|A\|$$

for all $A \in \mathfrak{F}(G, H)$. In this case,

$$\mathbf{D}^*(T) = \sup\{\|A \otimes T : \ldots\| \mid A \in \mathfrak{F}(G, H), \|A\| \leq 1\}.$$

PROOF: Define $\gamma := \alpha \otimes_G \varepsilon$. The main theorem says that $T \in \mathfrak{D}^*(E, F)$ if and only if $id_H \otimes T : H \otimes_\gamma E \to H \otimes_{\beta^\bullet} F$ is continuous. By the definition of traced tensor norms this operator is continuous if and only if the map $\Phi := (id_H \otimes T) \circ C_G$ is continuous (same norm):

$$\begin{array}{ccc}
H \otimes_\gamma E & \xrightarrow{id_H \otimes T} & H \otimes_{\beta^\bullet} F \\
C_G \uparrow & \nearrow \Phi & \\
(H \otimes_\varepsilon G') \otimes_\pi (G \otimes_\alpha E) & &
\end{array}$$

Identifying $H \otimes_\varepsilon G'$ and $\mathfrak{F}(G, H)$ this leads to an operator

$$\hat{\Phi} : \mathfrak{F}(G, H) \longrightarrow \mathfrak{L}(G \otimes_\alpha E, H \otimes_{\beta^\bullet} F)$$

which is $A \rightsquigarrow A \otimes T$. In other words, $id_H \otimes T$ is continuous if and only if $\hat{\Phi}$ is continuous (and then with same norm). This is the claim. □

COROLLARY: *For $p, q \in [1, \infty]$ let α be a finitely generated tensor norm or Δ_q and: β a finitely generated right-accessible tensor norm or Δ_p. Take measures μ and ν such that $L_q(\mu)$ and $L_p(\nu)$ are infinite dimensional. Then for $T \in \mathfrak{L}(E, F)$ the following statements are equivalent:*

(a) *For all $A \in \mathfrak{L}(L_q(\mu), L_p(\nu))$ the operator*

$$A \otimes T : L_q(\mu) \otimes_\alpha E \longrightarrow L_p(\nu) \otimes_\beta F$$

is continuous.

(b) *For all* $A \in \mathcal{L}(\ell_q, \ell_p)$ *the operator*

$$A \otimes T : \ell_q \otimes_\alpha E \longrightarrow \ell_p \otimes_\beta F$$

is continuous. One may replace ℓ_∞ *by* c_0.

In this case,

$$\sup_{\|A\| \leq 1} \|A \otimes T : L_q(\mu) \otimes_\alpha E \to L_p(\nu) \otimes_\beta F\| = \sup_{\|A\| \leq 1} \|A \otimes T : \ell_q \otimes_\alpha E \to \ell_p \otimes_\beta F\| \, .$$

PROOF: Since ℓ_s is "nicely" complemented in $L_s(\eta)$ (as was pointed out in the proof of 29.11.), (a) always implies (b). If α and β are finitely generated tensor norms, then simple $\mathcal{L}_p$-local technique gives the converse; note $\Delta_1 = \pi$ and $\Delta_\infty = \varepsilon$.

For the remaining cases, note first that $L_q(\mu)$ has the metric approximation property. It follows that for a net $R_\iota \in \mathfrak{F}(L_q(\mu), L_q(\mu))$ with $\|R_\iota\| \leq 1$ converging pointwise to the identity, $R_\iota \otimes id_E$ also converges on $L_q(\mu) \otimes E$ to the identity pointwise with respect to π and hence with respect to α. Therefore, $AR_\iota \otimes T(z)$ converges to $A \otimes T(z)$ with respect to π and hence for β. Consequently,

$$(*) \quad \begin{aligned} &\sup\{\|A \otimes T : ...\| \mid A \in \mathcal{L}(L_q(\mu), L_p(\nu)), \|A\| \leq 1\} = \\ &= \sup\{\|A \otimes T : ...\| \mid A \in \mathfrak{F}(L_q(\mu), L_p(\nu)), \|A\| \leq 1\} \in [0, \infty] \, . \end{aligned}$$

To apply the previous lemma take $\alpha = \Delta_q$ and $\beta = \Delta_p$ and $p, q \in]1, \infty[$. Then proposition 29.11. implies that

$$\Delta_q \otimes_{L_q(\mu)} \varepsilon \otimes_{L_p(\nu)} \Delta_{p'}^t = (d_q \otimes \varepsilon) \otimes_{L_p(\nu)} \Delta_{p'}^t = d_q \otimes (\varepsilon \otimes_{L_p(\nu)} \Delta_{p'}^t) =$$
$$= d_q \otimes (\varepsilon \otimes g_{p'}) = d_q \otimes g_{p'} \, ,$$

which is independent of μ and ν: the lemma and $(*)$ give the equivalence of (a) and (b) in this case.

The ideal of operators satisfying (b) is associated with $(\alpha \otimes_{\ell_q} \varepsilon \otimes_{\ell_p} \beta^*)^*$ by the lemma. Moreover,

$$\alpha \otimes_{\ell_q} \varepsilon \otimes_{\ell_p} \Delta_{p'}^t = \alpha \otimes_{\ell_q} (\varepsilon \otimes_{L_p(\nu)} \Delta_{p'}^t)$$
$$\Delta_q \otimes_{\ell_q} \varepsilon \otimes_{\ell_p} \beta = (\Delta_q \otimes_{L_q(\mu)} \varepsilon) \otimes_{\ell_p} \beta \quad .$$

In the first case the lemma shows that the operators satisfying (b) are those for which

$$\|A \otimes T : \ell_q \otimes_\alpha E \longrightarrow L_p(\nu) \otimes_{\Delta_p} F\| \leq c\|A\|$$

for $A \in \mathfrak{F}(\ell_q, L_p(\nu))$. If α is a tensor norm, $\mathcal{L}_p$-local technique and $(*)$ give (a). The same type of argument applies in the last case: $\alpha = \Delta_q$ and β a right-accessible finitely generated tensor norm. □

Note that the lemma together with

$$\backslash\alpha^*_{p,q}/ = d_{q'} \otimes g_{p'} = (\Delta_{q'} \otimes L_{q'}(\mu)\, \varepsilon) \otimes (\varepsilon \otimes L_p(\nu)\, \Delta^t_{p'}) =$$
$$= \Delta_{q'} \otimes L_{q'}(\mu) (\varepsilon \otimes \varepsilon) \otimes L_p(\nu)\, \Delta^t_{p'} = \Delta_{q'} \otimes L_{q'}(\mu)\, \varepsilon \otimes L_p(\nu)\, \Delta^t_{p'}$$

gives an alternative proof of the characterization theorem 28.4. of the operators in $\mathfrak{L}^{inj\ sur}_{p,q}$, the ideal associated with $(\backslash\alpha^*_{p,q}/)^*$. Recall from 28.3. that $\backslash\alpha^*_{p,q}/ = \gamma_{q,p}$. The tensor norm $\beta_{p,q}$ and $\delta_{p,q}$ will be calculated in Ex 29.6. as traced tensor norms.

Exercises:

Ex 29.1. Take α and β finitely generated tensor norms and G a normed space. Show that

(a) $\qquad\qquad (\alpha \otimes_G \beta)^t = \beta^t \otimes_{G'} \alpha^t$. Hint : 29.2..

(b) $\qquad\qquad \alpha \otimes_G \beta = \alpha \otimes_{G''} \beta$.

Ex 29.2. Let $\mathfrak{A} \sim \alpha$ be a maximal normed operator ideal and $1 \leq p \leq \infty$. For $T \in \mathfrak{L}(E,F)$ the following three statements are equivalent:

(a) $T \in (\mathfrak{A} \circ \mathfrak{J}_p)^*(E,F)$.

(b) $id_G \otimes T: G \otimes_\alpha E \to G \otimes_{g_p^*} F$ is continuous for all $G \in \mathrm{BAN}$.

(c) $id_{\ell_{p'}} \otimes T: \ell_{p'} \otimes_\alpha E \to \ell_{p'} \otimes_{\Delta_{p'}} F$ is continuous.

Hint: 29.11., the main theorem and 29.7..

Ex 29.3. Deduce theorem 17.15. and corollary 17.16. from the main theorem.

Ex 29.4. Show that $T \in \mathfrak{L}^{inj}_{p,q}(E,F)$ if and only if for all Banach spaces G and all $A \in \mathfrak{L}(G, \ell_p)$ the operator

$$A \otimes T: G \otimes_{d_q^*} E \longrightarrow \ell_p \otimes_{\Delta_p} F$$

is continuous. Hint: Reduce to $A \in \mathfrak{F}(C_2, \ell_p)$, prove $(\alpha_{p,q}\backslash)^* = \ldots = d_q^* \otimes \varepsilon \otimes_{\ell_p} \Delta^t_{p'}$ and use lemma 29.12..

Ex 29.5. Show that $T \in \mathfrak{L}^{sur}_{p,q}(E,F)$ if and only if for all Banach spaces G and all $A \in \mathfrak{L}(\ell_{q'}, G)$ the operator

$$A \otimes T: \ell_{q'} \otimes_{\Delta_{q'}} E \longrightarrow G \otimes_{g_p} F$$

is continuous. Hint: As in Ex 29.4..

Note that corollary 29.12. allows one to replace ℓ_s by infinite dimensional $L_s(\mu)$ in the previous exercises.

Ex 29.6. Use 28.8. to show that

$$\beta_{p,q} = \varepsilon \otimes_{\ell_{p'}} \varepsilon \otimes_{\ell_q} \varepsilon$$

and Ex 28.8. for

$$\delta_{p,q} = \varepsilon \otimes_{\ell_{p'}} \varepsilon \otimes_{\ell_q} \Delta^t_{q'} .$$

Hint: Lemma 29.12..

Ex 29.7. Let α be an accessible finitely generated tensor norm. Describe the maximal normed ideal of all T such that for all Banach spaces G and $A \in \mathcal{L}(L_p[0,1], G)$

$$A \otimes T : L_p[0,1] \otimes_{\Delta_p} E \longrightarrow G \otimes_\alpha F$$

is continuous. Hint: $\Delta_p \otimes_{L_p} \varepsilon \otimes \alpha^* = ...$, metric approximation property of L_p.

Ex 29.8. An operator ideal $(\mathfrak{A}, \mathbf{A})$ is called *ultrastable* if for every ultrafilter $\mathfrak{U}$ on I and $T_\iota \in \mathfrak{A}(E_\iota, F_\iota)$ for $\iota \in I$ with $\sup_I \mathbf{A}(T_\iota) < \infty$

$$\mathbf{A}((T_\iota)_\mathfrak{U}) \leq \lim_\mathfrak{U} \mathbf{A}(T_\iota) .$$

(a) Use the canonical decomposition of an operator $T \in \mathcal{L}(E, F)$ into its finite parts to show that for every ultrastable quasinormed operator ideal $(\mathfrak{A}, \mathbf{A})$

$$(\mathfrak{A}, \mathbf{A})^{max} = (\mathfrak{A}, \mathbf{A})^{reg} .$$

In particular, every ultrastable, regular quasinormed ideal is maximal. Kürsten [156] (see also Heinrich [104]) showed the converse: p–normed maximal operator ideals are ultrastable (and regular).

(b) The composition $\mathfrak{A} \circ \mathfrak{B}$ of two ultrastable operator ideals $\mathfrak{A}$ and $\mathfrak{B}$ (p–normed resp. q–normed) is ultrastable.

(c) Deduce from Kürsten's result, (a), (b) and proposition 29.8.(2) the following: If $\mathfrak{A}$ and $\mathfrak{B}$ are maximal normed operator ideals such that $\mathfrak{A} \circ \mathfrak{B}$ is normed and: $\mathfrak{A}$ is left–accessible or $\mathfrak{B}$ is right–accessible, then $(\mathfrak{A} \circ \mathfrak{B})^{reg} \stackrel{1}{=} \mathfrak{A} \otimes \mathfrak{B}$. This is due to [41], 3.4.5..

Ex 29.9. (a) Let G be a Banach space which contains the ℓ_p^n uniformly complemented. Then there is a $\lambda \geq 1$ such that

$$w_{p'} \leq w_{p'} \otimes_G w_{p'} \leq \lambda w_{p'} .$$

Hint: Definition of $w_{p'}$ and $id : \ell_p^n \xrightarrow{R_n} G \xrightarrow{S_n} \ell_p^n$ with $\|R_n\| \|S_n\| \leq \lambda$.

(b) If a Banach space G satisfies $\varepsilon \otimes_G \varepsilon \leq \lambda w_{p'}$, then it is a T_p–space (in the sense of 29.5.). Hint: Consider

$$(\ell_p^n \otimes_\varepsilon G') \otimes_\pi (G \otimes_\varepsilon \ell_{p'}^n) \longrightarrow \ell_p^n \otimes_\varepsilon \ell_{p'}^n .$$

By the result from [42] mentioned in 29.5. and $\varepsilon \otimes_G \varepsilon \leq w_{p'} \otimes_G w_{p'}$, the converse of (a) is true if $p = 1, 2$ or ∞.

30. The Vector Valued Fourier Transform

The Fourier transform $\mathfrak{F} : L_2(\mathbb{R}) \to L_2(\mathbb{R})$ is a powerful tool in Analysis — so it is worthwhile to determine the conditions under which the vector valued Fourier transform is continuous: Unfortunately, the Fourier transform for E-valued functions is continuous only if E is a Hilbert space. This is the main result of this section — a result which can also be formulated in terms of vector valued Fourier series (i.e. in terms of the Fourier transform $L_2[-\pi, \pi] \to \ell_2(\mathbb{Z})$). For the proof one first uses Gaussian averaging to get Kwapień's important type/cotype factorization theorem $\mathfrak{C}_2 \circ \mathfrak{T}_2 \subset \mathfrak{L}_2$: The composition of a type 2 with a cotype 2 operator factors through a Hilbert space. All the main results of this section are due to Kwapień [161].

30.1. The Fourier transform $\mathfrak{F} : L_2(\mathbb{R}) \to L_2(\mathbb{R})$, defined by

$$[\mathfrak{F}(f)](t) := (2\pi)^{-1/2} \int_{\mathbb{R}} f(s) \exp(-ist) ds$$

if $f \in L_2(\mathbb{R})$ vanishes outside of a compact set, is an isometry onto $L_2(\mathbb{R})$. Its inverse is, up to a sign, of the same form:

$$[\mathfrak{F}^{-1}(f)](t) = (2\pi)^{-1/2} \int_{\mathbb{R}} f(s) \exp(ist) ds = [\mathfrak{F} \circ \mathfrak{J}(f)](t) \,,$$

where $\mathfrak{J}$ is the isometry of $L_2(\mathbb{R})$ defined by

$$[\mathfrak{J}(f)](t) = f(-t) \,.$$

The maximal normed operator ideal $(\mathfrak{L}_\mathfrak{F}, \mathbf{L}_\mathfrak{F})$ of all operators $T \in \mathcal{L}(E, F)$ (between *complex* Banach spaces) such that

$$\mathbf{L}_\mathfrak{F}(T) := \|\mathfrak{F} \otimes T : L_2(\mathbb{R}) \otimes_{\Delta_2} E \longrightarrow L_2(\mathbb{R}) \otimes_{\Delta_2} F\| < \infty$$

(see 17.4.) is injective and surjective: this is a direct consequence of proposition 7.4.. The operators in $\mathfrak{L}_\mathfrak{F}$ are sometimes called *Fourier operators*.

REMARK: $(\mathfrak{L}_\mathfrak{F}, \mathbf{L}_\mathfrak{F}) = (\mathfrak{L}_{\mathfrak{F}^{-1}}, \mathbf{L}_{\mathfrak{F}^{-1}})$.

PROOF: This follows immediately from the fact that for $f \in L_2(\mathbb{R}) \otimes G$

$$\int_{\mathbb{R}} \|f(-t)\|_G^2 dt = \int_{\mathbb{R}} \|f(t)\|_G^2 dt$$

for all Banach spaces G and $\mathfrak{F}^{-1} \otimes T = (\mathfrak{F} \otimes T) \circ (\mathfrak{J} \otimes id_E)$ — and the analoguous formula for $\mathfrak{F} \otimes T$. □

Since
$$(\mathfrak{F}f|g)_{L_2} = (f|\mathfrak{F}^{-1}g)_{L_2}$$
for $f, g \in L_2(\mathbb{R})$, it follows that $\mathfrak{F}' = \mathfrak{F} : L_2(\mathbb{R}) \to L_2(\mathbb{R})$ (watch the complex conjugate in the scalar product!). It is easy to see (use 15.10.) that
$$\mathcal{L}_{\mathfrak{F}}^{dual} = \mathcal{L}_{\mathfrak{F}'}$$
holds isometrically and hence:

COROLLARY: $\mathcal{L}_{\mathfrak{F}} = \mathcal{L}_{\mathfrak{F}}^{dual}$ holds isometrically, i.e. the ideal $\mathcal{L}_{\mathfrak{F}}$ of Fourier operators is completely symmetric.

It was shown in 7.5. that the identity map on ℓ_1 is not a Fourier operator — and in Ex 7.8. that $\mathbf{L}_{\mathfrak{F}}(id_H) = 1$ for Hilbert spaces H. The ideal property implies for the ideal $\mathcal{L}_2$ of operators factoring through a Hilbert space that $\mathcal{L}_2 \subset \mathcal{L}_{\mathfrak{F}}$ and
$$\mathbf{L}_{\mathfrak{F}}(T) \leq \mathbf{L}_2(T)$$
for all $T \in \mathcal{L}(E, F)$.

30.2. Obviously,
$$[\mathfrak{F} \otimes T(h)](t) = (2\pi)^{-1/2} \int_{\mathbb{R}} \exp(-ist) T(h(s)) ds \in F$$
holds for all $h \in L_2(\mathbb{R}) \otimes E$. The following simple result is basic:

LEMMA: For $a > 0$, a complex Banach space E and $x_{-n}, x_{-n+1}, ..., x_n \in E$ define
$$h := \sum_{k=-n}^{n} \chi_{[ka,(k+1)a[} \otimes \frac{x_k}{\sqrt{a}} \in L_2(\mathbb{R}) \otimes E \ .$$

Then
$$\Delta_2(h; L_2(\mathbb{R}), E) = \Big(\sum_{k=-n}^{n} \|x_k\|^2 \Big)^{1/2}$$
$$\Delta_2(\mathfrak{F} \otimes id_E(h); L_2(\mathbb{R}), E) = \Big(\frac{1}{2\pi} \int_{-\pi}^{\pi} \Big\| \sum_{k=-n}^{n} \exp(ikt) x_k \Big\|^2 dt \Big)^{1/2} \ .$$

PROOF: The first formula is obvious. For the second one, observe first that
$$2 \exp(-i\frac{a}{2}) \sin(\frac{a}{2}) = i[\exp(-ia) - 1]$$

and hence, for all $k \in \mathbb{Z}$ and $t \in \mathbb{R}$,

$$\int_{ka}^{(k+1)a} \exp(-ist)ds = \frac{i}{t}[\exp(-i(k+1)at) - \exp(-ikat)] =$$

$$= \frac{i}{t}\exp(-ikat)[\exp(-iat) - 1] = \frac{2}{t}\exp(-ikat)\exp(-i\frac{at}{2})\sin(\frac{at}{2}).$$

With the substitution $t = 2\pi u/a$ this gives

$$\Delta_2(\mathfrak{F} \otimes id_E(h))^2 = \int_{\mathbb{R}} \left\| (2\pi)^{-1/2} \int_{\mathbb{R}} \exp(-ist)h(s)ds \right\|_E^2 dt =$$

$$= \int_{\mathbb{R}} \frac{1}{2\pi} \left\| \sum_{k=-n}^{n} \int_{ka}^{(k+1)a} \exp(-ist) ds \frac{x_k}{\sqrt{a}} \right\|_E^2 dt =$$

$$= \int_{\mathbb{R}} \frac{2}{\pi t^2 a} \left\| \sum_{k=-n}^{n} \exp(-ikat) \sin(\frac{at}{2}) x_k \right\|_E^2 dt =$$

$$= \int_{\mathbb{R}} \frac{\sin^2(\pi u)}{(\pi u)^2} \left\| \sum_{k=-n}^{n} \exp(-2\pi iku) x_k \right\|_E^2 du =$$

$$= \sum_{m \in \mathbb{Z}} \int_0^1 \frac{\sin^2(\pi(u+m))}{\pi^2(u+m)^2} \left\| \sum_{k=-n}^{n} \exp(-2\pi ik(u+m)) x_k \right\|_E^2 du.$$

But for all $u \in \mathbb{R} \setminus \mathbb{Z}$

$$\sum_{m \in \mathbb{Z}} \frac{\sin^2(\pi u)}{\pi^2(u+m)^2} = 1$$

(consider Parseval's identity for the Fourier coefficients of the function $\exp(iu\cdot)$) and therefore,

$$\Delta_2(\mathfrak{F} \otimes id_E(h))^2 = \int_0^1 \left\| \sum_{k=-n}^{n} \exp(-2\pi iku) x_k \right\|_E^2 du.$$

Another obvious subsitution gives the result. □

30.3. For $\mathbf{T} := [-\pi, \pi[$ and $f \in L_2(\mathbf{T}) \tilde{\otimes}_{\Delta_2} E = L_2(\mathbf{T}, E) \subset L_1(\mathbf{T}, E)$ the *Fourier coefficients* are defined by

$$\hat{f}(n) := (2\pi)^{-1/2} \int_{-\pi}^{\pi} f(s) \exp(-ins) ds \in E$$

for $n \in \mathbb{Z}$; here E is again a complex Banach space. Recall that $((2\pi)^{-1/2} \exp(in\cdot))_{n \in \mathbb{Z}}$ is an orthonormal basis in $L_2(\mathbf{T})$ and that for $f \in L_2(\mathbf{T})$ Parseval's identity

$$\sum_{n \in \mathbb{Z}} |\hat{f}(n)|^2 = \int_{-\pi}^{\pi} |f(t)|^2 dt$$

holds. It is now quite interesting to see that T being a Fourier operator means that T satisfies certain Parseval/Bessel-like inequalities.

THEOREM: *Let E and F be complex Banach spaces, $T \in \mathcal{L}(E,F)$ and $c \geq 0$. Then the following five statements are equivalent:*

(a) $T \in \mathcal{L}_{\mathfrak{F}}(E,F)$ and $\mathbf{L}_{\mathfrak{F}}(T) \leq c$.

(b) *For all $n \in \mathbb{N}$ and $x_{-n}, ..., x_n \in E$*

$$\left(\sum_{k=-n}^{n} \|Tx_k\|^2\right)^{1/2} \leq c\left(\int_{-\pi}^{\pi} \left\|\sum_{k=-n}^{n} (2\pi)^{-1/2} \exp(ikt) x_k\right\|^2 dt\right)^{1/2}.$$

(b') *For all $f \in L_2(\mathbf{T}, E)$*

$$\left(\sum_{k \in \mathbb{Z}} \|T(\hat{f}(k))\|^2\right)^{1/2} \leq c\left(\int_{-\pi}^{\pi} \|f(t)\|^2 dt\right)^{1/2}.$$

(c) *For all $n \in \mathbb{N}$ and $x_{-n}, ..., x_n \in E$*

$$\left(\int_{-\pi}^{\pi} \left\|\sum_{k=-n}^{n} (2\pi)^{-1/2} \exp(ikt) Tx_k\right\|^2 dt\right)^{1/2} \leq c\left(\sum_{k=-n}^{n} \|x_k\|^2\right)^{1/2}.$$

(c') *For all $f \in L_2(\mathbf{T}, E)$*

$$\left(\int_{-\pi}^{\pi} \|T(f(t))\|^2 dt\right)^{1/2} \leq c\left(\sum_{k \in \mathbb{Z}} \|\hat{f}(k)\|^2\right)^{1/2}.$$

PROOF: T is a Fourier operator if and only if

$$\Delta_2(\mathfrak{F} \otimes T(h)) \leq c \Delta_2(h)$$

for all $h \in L_2(\mathbb{R}) \otimes E$. Since the functions h of the form treated in lemma 30.2. are dense in $L_2(\mathbb{R}) \otimes_{\Delta_2} E$, the equivalence of (a) and (c) is obvious. Moreover, for h as in the lemma

$$\left(\sum_{h=-n}^{n} \|Tx_k\|^2\right)^{1/2} = \Delta_2((\mathfrak{F}^{-1} \otimes T) \circ (\mathfrak{F} \otimes id_E)(h); L_2(\mathbb{R}), F)$$

holds. Since $\mathcal{L}_{\mathfrak{F}} \doteq \mathcal{L}_{\mathfrak{F}^{-1}}$, this shows that (a) implies (b) — and the converse is also true because the functions of the form $(\mathfrak{F} \otimes id_E)(h)$ are dense in $L_2(\mathbb{R}) \otimes_{\Delta_2} E$ (recall that $\mathfrak{F}$ is an onto isometry).

The Fourier coefficients of the function

$$f(\cdot) = \sum_{k=-n}^{n} (2\pi)^{-1/2} \exp(ik \cdot) x_k \in L_2(\mathbf{T}, E)$$

are $\hat{f}(k) = x_k$ for $|k| \leq n$ and zero otherwise; moreover, these vector valued exponential polynomials are dense in $L_2(\mathbf{T}) \tilde{\otimes}_\pi E$ and hence in $L_2(\mathbf{T}) \tilde{\otimes}_{\Delta_2} E$: these two observations prove the equivalences (b) ↔ (b') and (c) ↔ (c'). □

30.4. For $T = id_E$ the theorem implies that if in Parseval's identity

$$\sum_{k \in \mathbf{Z}} \|\hat{f}(k)\|^2 \stackrel{?}{=} \int_{-\pi}^{\pi} \|f(t)\|_E^2 dt$$

one inequality holds for all $f \in L_2(\mathbf{T}, E)$, then there is equality — and this is equivalent to $\mathbf{L}_{\mathfrak{F}}(id_E) = 1$. A neat trick due to Kandil [141] (see Ex 30.2.) shows that then E satisfies the parallelogram identity and hence is (isometrically) a Hilbert space:

COROLLARY 1: *A complex Banach space E satisfies $\mathbf{L}_{\mathfrak{F}}(id_E) = 1$ if and only if it is isometric to a Hilbert space:*

The isomorphic version of this result is more complicated and will follow in 30.6. from Kwapień's type/cotype theorem 30.5.. Before entering this topic, an example:

COROLLARY 2: *For $1 \leq p \leq \infty$ and $n \in \mathbf{N}$*

$$\mathbf{L}_{\mathfrak{F}}(id_{\ell_p^n}) = n^{|1/p - 1/2|} .$$

PROOF: Since $\mathcal{L}_{\mathfrak{F}}$ is completely symmetric, it is enough to check this for $1 \leq p \leq 2$. By 26.2.

$$\mathbf{L}_{\mathfrak{F}}(id_{\ell_p^n}) \leq \mathbf{L}_2(id_{\ell_p^n}) = n^{1/p - 1/2}$$

holds. For the converse note first that

$$\int_{-\pi}^{\pi} \Big\| \sum_{k=1}^{n} (2\pi)^{-1/2} \exp(-ikt) e_k \Big\|_{\ell_1^n}^2 dt = n^2$$

and $\sum_{k=1}^{n} \|e_k\|_{\ell_1^n}^2 = n$. Theorem 30.3.(c) gives

$$n^{1/2} \leq \mathbf{L}_{\mathfrak{F}}(id_{\ell_1^n}) \leq \|id : \ell_p^n \to \ell_1^n\| \mathbf{L}_{\mathfrak{F}}(id_{\ell_p^n}) \|id : \ell_1^n \to \ell_p^n\| \leq n^{1-1/p} \mathbf{L}_{\mathfrak{F}}(id_{\ell_p^n}) ,$$

which is the other estimate. □

In particular, $\ell_p \notin \mathrm{space}(\mathcal{L}_{\mathfrak{F}})$ for $1 \leq p \leq \infty$ and $p \neq 2$ — but as announced, a more general result is true.

30.5. A rather general theorem about factorization of operators through Hilbert spaces will be proved next. It is based on Gauss averaging. The following lemma shows that sometimes Rademacher averages can be replaced by Gauss averages.

LEMMA: *Let E, F be Banach spaces (real or complex), $T \in \mathcal{L}(E, F)$ and define $d_{I\!R} := 1$ and $d_{\mathcal{C}} := \sqrt{2}$.*

(1) *If T is of type 2, then for all $x_1, ..., x_n \in E$*

$$\left(\int_{I\!K^n} \Big\|\sum_{k=1}^n g_k(w) T x_k\Big\|^2 \gamma_n(dw)\right)^{1/2} \leq d_{I\!K} \mathbf{T}_2(T) \left(\sum_{k=1}^n \|x_k\|^2\right)^{1/2}.$$

(2) *If T is of cotype 2, then for all $x_1, ..., x_n \in E$*

$$\left(\sum_{k=1}^n \|T x_k\|^2\right)^{1/2} \leq d_{I\!K}^{-1} \mathbf{C}_2(T) \left(\int_{I\!K^n} \Big\|\sum_{k=1}^n g_k(w) x_k\Big\|^2 \gamma_n(dw)\right)^{1/2}.$$

This result can be expressed as follows: Rademacher type 2 (resp. cotype 2) implies Gauss type 2 (resp. cotype 2). The additional constant $\sqrt{2}$ in the complex case is natural since the *complex* Gauss functions g_k have norm $\sqrt{2}$.

PROOF: The proof is — as in Ex 8.9. — a simple application of Fubini's theorem and the invariance of the Gauss measure γ_n under isometries: Assume T to be of type 2. Then

$$\int_{D_n} \Big\|\sum_{k=1}^n \varepsilon_k(w) g_k(v) T x_k\Big\|^2 \mu_n(dw) \leq \mathbf{T}_2(T)^2 \sum_{k=1}^n \|x_k\|^2 |g_k(v)|^2$$

for all $v \in I\!K^n$. Integration with respect to γ_n gives

$$\int_{I\!K^n}\int_{D_n} \Big\|\sum_{k=1}^n \varepsilon_k(w) g_k(v) T x_k\Big\|^2 \mu_n(dw)\gamma_n(dv) = \int_{I\!K^n} \Big\|\sum_{k=1}^n g_k(v) T x_k\Big\|^2 \gamma_n(dv) \leq$$

$$\leq \mathbf{T}_2(T)^2 \sum_{k=1}^n \|x_k\|^2 \|g_k\|_{2,I\!K}^2,$$

where $\|g_k\|_{2,I\!K} = d_{I\!K}$ by 8.7.. Statement (2) follows in the same way. $\square$

For $T = id_E$ Gauss cotype 2 implies Rademacher cotype 2, but for arbitrary operators this is false (see Pisier [225], p.36/37). Ex 8.9. implies that Gauss type 2 operators are of Rademacher type 2. For Gauss type/cotype p see Ex 30.3..
Note that with $G_n := \mathrm{span}\{g_k \mid 1 \leq k \leq n\} \subset L_2(\gamma_n)$ and

$$I : \ell_2^n \longrightarrow G_n \quad ; \quad Ie_k := g_k$$

Gauss type 2 means that

$$\sup_n \|I \otimes T : \ell_2^n \otimes_{\Delta_2} E \longrightarrow G_n \otimes_{\Delta_2} F\| < \infty$$

and Gauss cotype 2

$$\sup_n \|I^{-1} \otimes T : G_n \otimes_{\Delta_2} E \longrightarrow \ell_2^n \otimes_{\Delta_2} F \| < \infty,$$

where, as in section 7, the norm Δ_2 on $G_n \otimes E$ is the one induced from $L_2(\gamma_n) \otimes_{\Delta_2} E$. The following *type/cotype theorem* is quite important.

THEOREM: *The composition $S \circ R$ of a type 2 operator R and a cotype 2 operator S factors through a Hilbert space, i.e. $\mathfrak{C}_2 \circ \mathfrak{T}_2 \subset \mathfrak{L}_2$. Moreover,*

$$\mathbf{L}_2 \leq \mathbf{C}_2 \circ \mathbf{T}_2 .$$

In particular, a Banach space is isomorphic to a Hilbert space if and only if it has type 2 and cotype 2.

This characterization of Hilbert spaces is due to Kwapień [161] — and it was observed by Maurey (see Kwapień [162]) that the result also holds for operators.

PROOF: Let $T = S \circ R \in \mathfrak{L}(E, F)$ with $R \in \mathfrak{T}_2(E, X)$ and $S \in \mathfrak{C}_2(X, F)$. Using the characterization 28.5. of 2-factorable operators, it suffices to check that for all $n \in \mathbb{N}$ and all isometries $A \in \mathfrak{L}(\ell_2^n, \ell_2^n)$

$$\|A \otimes (S \circ R) : \ell_2^n \otimes_{\Delta_2} E \longrightarrow \ell_2^n \otimes_{\Delta_2} F\| \leq \mathbf{C}_2(S) \mathbf{T}_2(R) .$$

Fix such an operator A and let $\hat{A} : L_2(\mathbb{K}^n, \gamma_n) \to L_2(\mathbb{K}^n, \gamma_n)$ be the map $f \rightsquigarrow f \circ A'$. It is obvious that $\hat{A}(G_n) \subset G_n$ and

$$A = I^{-1} \circ \hat{A}|_{G_n} \circ I : \ell_2^n \longrightarrow \ell_2^n .$$

Moreover, $\hat{A} \otimes id_X$ is an isometry on $L_2(\gamma_n) \otimes_{\Delta_2} X$ (by the invariance of γ_n under isometries). It follows that the diagram

$$\begin{array}{ccc}
\ell_2^n \otimes_{\Delta_2} E & \xrightarrow{A \otimes T} & \ell_2^n \otimes_{\Delta_2} F \\
{\scriptstyle I \otimes R} \downarrow & & \uparrow {\scriptstyle I^{-1} \otimes S} \\
G_n \otimes_{\Delta_2} X & \xrightarrow{\hat{A} \otimes id_X} & G_n \otimes_{\Delta_2} X
\end{array}$$

commutes — and the remarks made before the theorem show that

$$\|A \otimes T\| \leq \mathbf{T}_2(R) \, \|\hat{A} \otimes id_X\| \, \mathbf{C}_2(S) ,$$

which ends the proof. □

The natural quotient map $Q : \ell_1 \twoheadrightarrow \ell_p$ (for $2 < p < \infty$) does not factor through a Hilbert space. Since ℓ_1 has cotype 2 and ℓ_p type 2, it follows that in general neither the operators in $\mathfrak{T}_2 \circ \mathfrak{C}_2$ nor those in $\mathfrak{T}_2 \cap \mathfrak{C}_2$ factor through a Hilbert space.

Note also that Kwapień's type/cotype theorem improves one of the results in 26.1. stating that every $T : \ell_r \to \ell_s$ factors through ℓ_2 if $1 \leq s \leq 2 \leq r < \infty$. On the other hand, Pisier's factorization theorem (to be proven in the next section) extends Kwapień's result in certain situations since E' has cotype 2 whenever E has type 2 (by Ex 8.14.).

30.6. Fourier operators are of type 2 and cotype 2 — this is the content of the next

PROPOSITION: $\max\{\mathbf{T}_2(T), \mathbf{C}_2(T)\} \leq 2\, \mathbf{L}_{\mathfrak{F}}(T)$ for all $T \in \mathcal{L}(E, F)$.

PROOF: In order to show that $T \in \mathcal{L}_{\mathfrak{F}}(E, F)$ has type 2 fix $x_1, ..., x_n \in E$. Theorem 30.3.(c) shows that for all $w \in D_n$

$$\int_{-\pi}^{\pi} \Big\| \sum_{k=1}^{n} (2\pi)^{-1/2} \exp(ikt) \varepsilon_k(w) T x_k \Big\|^2 dt \leq \mathbf{L}_{\mathfrak{F}}(T)^2 \sum_{k=1}^{n} \|x_k\|^2 .$$

Integrating with respect to μ_n gives

$$\int_{D_n} \int_{-\pi}^{\pi} \Big\| \sum_{k=1}^{n} (2\pi)^{-1/2} \exp(ikt) \varepsilon_k(w) T x_k \Big\|^2 dt\, \mu_n(dw) \leq \mathbf{L}_{\mathfrak{F}}(T)^2 \sum_{k=1}^{n} \|x_k\|^2 .$$

Kahane's contraction principle (see Ex 8.16.) implies that

$$\int_{D_n} \Big\| \sum_{k=1}^{n} \varepsilon_k(w) T x_k \Big\|^2 \mu_n(dw) \leq 4 \int_{D_n} \Big\| \sum_{k=1}^{n} \exp(ikt) \varepsilon_k(w) T x_k \Big\|^2 \mu_n(dw) ,$$

so that $\mathbf{T}_2(T) \leq 2\, \mathbf{L}_{\mathfrak{F}}(T)$. The proof that the operator T has cotype 2 follows along the same lines. □

The constant 2 is actually superfluous: at the end of this section a more general result (with constant 1) will be proven — but we preferred for the sake of simplicity and elegance also to give the proof above.

Now the main result of this section follows easily:

THEOREM (Kwapień): *A complex Banach space E is isomorphic to a Hilbert space if and only if the E-valued Fourier transform*

$$\mathfrak{F} \otimes id_E : L_2(\mathbb{R}) \otimes_{\Delta_2} E \longrightarrow L_2(\mathbb{R}) \otimes_{\Delta_2} E$$

is continuous.

PROOF: If this is so, $id_E \in \mathfrak{L}_{\mathfrak{F}}$ and hence E has type 2 and cotype 2 by the previous proposition. Therefore, $id_E = id_E \circ id_E$ is 2-factorable by Kwapień's type/cotype theorem, which concludes the proof. □

Recall that the isometric version of this result was already given in 30.4.: If $\mathbf{L}_{\mathfrak{F}}(id_E) = 1$, then E is isometric to a Hilbert space.

It is not known whether the operator version of this result holds: Does every Fourier operator factor through a Hilbert space? In other words, is $\mathfrak{L}_{\mathfrak{F}} \subset \mathfrak{L}_2$? A partial answer is the

REMARK: $\mathfrak{L}_2 = \mathfrak{L}_{\mathfrak{F}} \circ \mathfrak{L}_{\mathfrak{F}}$.

PROOF: Clearly $\mathfrak{L}_2 \subset \mathfrak{L}_2 \circ \mathfrak{L}_2 \subset \mathfrak{L}_{\mathfrak{F}} \circ \mathfrak{L}_{\mathfrak{F}}$ and conversely $\mathfrak{L}_{\mathfrak{F}} \circ \mathfrak{L}_{\mathfrak{F}} \subset \mathfrak{C}_2 \circ \mathfrak{T}_2 \subset \mathfrak{L}_2$ by Kwapień's type/cotype theorem. □

30.7. It is worthwhile to have a look at the structure of the proof of the results above: If one says that an operator $T \in \mathcal{L}(E, F)$ has *Fourier type 2* if it satisfies for some $c \geq 0$ and all $x_k \in E$

$$\left(\int_{-\pi}^{\pi} \left\|\sum_{k=-n}^{n} (2\pi)^{-1/2} \exp(ikt) T x_k \right\|^2 dt\right)^{1/2} \leq c \left(\sum_{k=-n}^{n} \|x_k\|^2\right)^{1/2}$$

and — only for the moment (since this notion is not the usual one) — *Fourier cotype 2* if

$$\left(\sum_{k=-n}^{n} \|T x_k\|^2\right)^{1/2} \leq c \left(\int_{-\pi}^{\pi} \left\|\sum_{k=-n}^{n} (2\pi)^{-1/2} \exp(ikt) x_k \right\|^2 dt\right)^{1/2},$$

then the following has been shown in this section:

(a) An operator is a Fourier operator if and only if it is of Fourier type 2 and if and only if it is of Fourier cotype 2 (theorem 30.3.).

(b) Fourier type 2 implies Rademacher type 2, the same for cotype (proof of proposition 30.6.).

(c) Rademacher type 2 implies Gauss type 2, the same for cotype (lemma 30.5.).

(d) The composition $S \circ R$ of a Gauss type 2 operator R and a Gauss cotype 2 operator S factors through a Hilbert space (proof of Kwapień's type/cotype theorem).

30.8. In light of this structure the following type/cotype result with respect to an arbitrary orthonormal basis of an $L_2(\mu)$ is of some interest (the result is also due to Kwapień [161]). For this, take an orthonormal basis $(f_n)_{n \in \mathbb{N}}$ of some separable $L_2(\Omega, \mu)$ and consider the isometry

$$I : L_2(\mu) \longrightarrow \ell_2$$

defined by $g \rightsquigarrow ((g|f_n))_{n \in \mathbb{N}}$. For $T \in \mathcal{L}(E,F)$ continuity of $I \otimes T : L_2 \otimes_{\Delta_2} E \to \ell_2 \otimes_{\Delta_2} F$ means a sort of cotype 2 inequality:

$$\sum_{k=1}^{n} \|Tx_k\|^2 \leq c^2 \int_\Omega \Big\| \sum_{k=1}^{n} f_k(t)x_k \Big\|^2 \mu(dt)$$

— and the continuity of $I^{-1} \otimes T$ means a type 2 inequality (with respect to (f_n)). It is obvious that every isometry I from $L_2(\Omega, \mu)$ onto ℓ_2 fixes (via the unit vector basis of ℓ_2) an orthonormal basis of $L_2(\Omega, \mu)$.

PROPOSITION: *Let (Ω, μ) be a non-atomic, separable measure space, $I : L_2(\Omega, \mu) \to \ell_2$ a surjective isometry and $T \in \mathcal{L}(E, F)$.*
(1) *If $I \otimes T : L_2(\Omega, \mu) \otimes_{\Delta_2} E \to \ell_2 \otimes_{\Delta_2} F$ is continuous, then T is of (Rademacher) cotype 2 and $\mathbf{C}_2(T) \leq \|I \otimes T : ...\|$.*
(2) *If $I^{-1} \otimes T : \ell_2 \otimes_{\Delta_2} E \to L_2(\Omega, \mu) \otimes_{\Delta_2} F$ is continuous, then T is of (Rademacher) type 2 and $\mathbf{T}_2(T) \leq \|I^{-1} \otimes T : ...\|$.*

PROOF: Clearly, one may assume that μ is normalized. Then, by the Halmos–von Neumann theorem (see Brown–Pearcy [18], p.181) the measure spaces (Ω, μ) and $([0,1], \lambda)$ (Lebesgue measure) are isomorphic, i.e. there is a measure preserving, monotone, additive and bijective map Φ from the *classes* of μ-measurable sets onto the classes of λ-measurable sets. Clearly, Φ induces a bijective isometry

$$S(\mu) \otimes_{\Delta_2} E \longrightarrow S(\lambda) \otimes_{\Delta_2} E$$

between the classes of E-valued step–functions, hence a surjective isometry $L_2(\mu, E) \to L_2(\lambda, E)$. This means one may restrict one's attention to the Lebesgue measure on $[0,1]$. Thus, in case (1) there is an orthonormal basis (f_n) of $L_2[0,1]$ such that

$$\sum_{k=1}^{n} \|Tx_k\|^2 \leq c^2 \int_0^1 \Big\| \sum_{k=1}^{n} f_k(t)x_k \Big\|^2 dt$$

for all $x_1, ..., x_n \in E$. One has to show that the f_k can be replaced by the Rademacher functions r_k; this will be done with a gliding hump approximation of the functions r_k by certain f_m.

Fix $x_1, ..., x_n \in E$. Since for the discrete Rademacher functions ε_k

$$\int_D \Big\| \sum_{k=1}^{n} \varepsilon_k(w)x_k \Big\|^2 \mu(dw) = \int_D \Big\| \sum_{k=1}^{n} \varepsilon_{\varphi(k)}(w)x_k \Big\|^2 \mu(dw)$$

holds for every strictly monotone $\varphi : \mathbb{N} \to \mathbb{N}$, the same is true for the sequence (r_k) and so it is enough to find a subsequence of the (r_k).

Take $\varepsilon > 0$ and $\psi(1) \in \mathbb{N}$ such that for

$$g := \sum_{k=1}^{\psi(1)} (r_1|f_k) f_k$$

the function $h_1 := \|g\|_2^{-1} g$ satisfies $\|r_1 - h_1\|_2 \leq 2^{-1}\varepsilon$. Assume, by induction, that there are

$$\varphi(1) := 1 < \varphi(2) < ... < \varphi(n) \text{ and } \psi(0) := 0 < \psi(1) < ... < \psi(n)$$
$$h_k \in \text{span}\{f_m \mid \psi(k-1) < m \leq \psi(k)\}, \quad \|h_k\|_2 = 1$$

such that $\|r_{\varphi(k)} - h_k\|_2 \leq 2^{-k}\varepsilon$: Since $(r_j|f_m) \to 0$ for $j \to \infty$, there is an index $\varphi(n+1) > \varphi(n)$ such that

$$\sum_{k=1}^{\psi(n)} |(r_{\varphi(n+1)}|f_k)|^2 \text{ is small };$$

this easily leads to some $\psi(n+1) > \psi(n)$ and some normalized

$$h_{n+1} \in \text{span}\{f_m \mid \psi(n) < m \leq \psi(n+1)\}$$

such that $\|r_{\varphi(n+1)} - h_{n+1}\|_2 \leq 2^{-(n+1)}\varepsilon$.

The construction showed that $h_n \perp h_m$ whenever $n \neq m$. Now

$$\left(\int_0^1 \left\|\sum_{k=1}^n h_k(t) x_k\right\|^2 dt\right)^{1/2} \leq$$

$$\leq \left(\int_0^1 \left\|\sum_{k=1}^n r_{\varphi(k)}(t) x_k\right\|^2 dt\right)^{1/2} + \underbrace{\sum_{k=1}^n \|r_{\varphi(k)} - h_k\|_2}_{\leq \varepsilon} \max_{k=1,...,n} \|x_k\|$$

and

$$\sum_{k=1}^n \sum_{m=\psi(k-1)+1}^{\psi(k)} f_m (h_k|f_m) x_k = \sum_{k=1}^n h_k x_k .$$

$\|h_k\|_2 = 1$ yields that

$$\sum_{k=1}^n \|T x_k\|^2 = \sum_{k=1}^n \sum_{m=\psi(k-1)+1}^{\psi(k)} \|T((h_k|f_m) x_k)\|^2 \leq$$

$$\leq c^2 \int_0^1 \left\|\sum_{k=1}^n h_k(t) x_k\right\|^2 dt \leq c^2 \int_0^1 \left\|\sum_{k=1}^n r_{\varphi(k)}(t) x_k\right\|^2 dt + c^2 \varepsilon \max \|x_k\|$$

— for all $\varepsilon > 0$. The second claim follows in the same way. $\square$

The result does not hold for arbitrary orthogonal sequences: see Ex 30.7..

One application: If G is an infinite abelian compact group with normalized Haar measure, then the dual group $\hat{G}$ is a discrete topological group and the Fourier transform $\mathfrak{F} : L_2(G) \to \ell_2(\hat{G})$ is an isometry. If $L_2(G)$ is separable, then $\hat{G}$ is countable.

COROLLARY: *Let G be a compact, abelian infinite group such that $L_2(G)$ is separable. If for a Banach space E the E-valued Fourier transform*

$$\mathfrak{F} \otimes id_E : L_2(G) \otimes_{\Delta_2} E \longrightarrow \ell_2(\hat{G}) \otimes_{\Delta_2} E$$

is continuous, then E is isomorphic to a Hilbert space.

PROOF: Since G has no discrete points and each atom of a Borel–Radon measure contains a one–point atom, the Haar measure has no atoms in this case. Therefore the proposition shows that id_E is of (Rademacher) type 2 and cotype 2, hence the claim follows from Kwapień's characterization 30.5. of Hilbert spaces. □

The assumption that $L_2(G)$ is separable is actually superfluous: this was shown by Rubio de Francia and Torrea [234].

30.9. Peetre defined a Banach space E to be of Fourier type p (for $1 \leq p \leq 2$) if the Fourier transform satisfies the Hausdorff–Young inequality

$$\left(\int_{\mathbb{R}} \|(\mathfrak{F} \otimes id_E) f(t)\|_E^{p'} dt \right)^{1/p'} \leq c \left(\int_{\mathbb{R}} \|f(t)\|_E^p dt \right)^{1/p}$$

for all $f \in L_p(\mathbb{R}) \otimes E$. It is obvious that for $p = 2$ this means $id_E \in \mathfrak{L}_{\mathfrak{F}}$, which is the same as having Fourier type 2 in the sense of 30.7.; it follows that these are the spaces isomorphic to a Hilbert space. A deep result of Bourgain states that each Banach space of type $p > 1$ has Fourier type r for r with $r' = 18 T_p (id_E)^{p'}$. See König [149] for references and the behaviour of Fourier coefficients of certain differentiable E–valued functions.

Exercises:

Ex 30.1. Recall that a norm (on a real or complex vector space E) comes from a scalar product if and only if for all $x, y \in E$ it satisfies the parallelogram identity

$$\|x + y\|^2 + \|x - y\|^2 = 2(\|x\|^2 + \|y\|^2) .$$

(a) If $\|x+y\|^2 + \|x-y\|^2 \leq 2(\|x\|^2 + \|y\|^2)$ for all $x, y \in E$ (or $\geq$ for all $x, y \in E$), then the norm $\|\ \|$ satisfies the parallelogram identity.

(b) If for a Banach space E the type 2 constant $\mathbf{T}_2(id_E)$ or the cotype 2 constant $\mathbf{C}_2(id_E)$ is 1, then E is isometric to a Hilbert space.

Ex 30.2. Let E be a complex Banach space.
(a) For $x, y \in E$ define $f(t) := x$ for $-\pi \leq t < 0$ and $f(t) = y$ for $0 \leq t < \pi$. Show that $f \in L_2(\mathbf{T}, E)$ and

$$\sum_{n \in \mathbb{Z}} \|\hat{f}(n)\|^2 = (\|x+y\|^2 + \|x-y\|^2)\pi/2 \; .$$

(b) Deduce from (a) and Ex 30.1.: If

$$\sum_{n \in \mathbb{Z}} \|\hat{f}(n)\|^2 \leq \int_{-\pi}^{\pi} \|f(t)\|^2 dt$$

for all $f \in L_2(\mathbf{T}, E)$ (or $\geq$ for all such f), then E is isometric to a Hilbert space. This argument is due to Kandil [141].

Ex 30.3. Define Gauss type p and Gauss cotype q for operators. Use Kahane's inequality for Gauss averaging (8.7.) to show that for $T \in \mathcal{L}(E, F)$ and $1 \leq p \leq 2 \leq q < \infty$:
(a) T is of Gauss type p if and only if it is of Rademacher type p.
(b) If T is of Rademacher cotype q, then it is of Gauss cotype q.
Hint: 30.5. and Ex 8.9..

Ex 30.4. Use Kwapień's type/cotype theorem to reprove the result from 26.1. that for $1 \leq q \leq 2 \leq p < \infty$ every operator from an $\mathcal{L}_p^g$-space into an $\mathcal{L}_q^g$-space factors through a Hilbert space.

Ex 30.5. If $1 \leq q \leq p \leq \infty$, then every Fourier operator $\ell_p \to \ell_q$ factors through a Hilbert space. Hint: First $q \leq 2$ and 30.5..

Ex 30.6. Show that for $1 \leq q \leq 2 \leq p \leq \infty$, every operator $T \in \mathcal{L}(\ell_p, \ell_q)$ is a Fourier operator. Hint: 26.1..

Ex 30.7. If $f_n := 2^{-n/2}\chi_{[2^{-n}, 2^{-n+1}]}$, then (f_n) is an orthonormal system in $L_2([0,1])$ and

$$\int_0^1 \left\|\sum_{k=1}^n f_k(t) x_k\right\|^2 dt = \sum_{k=1}^n \|x_k\|^2$$

holds for all Banach spaces E and $x_1, ... x_n \in E$. This example (showing that proposition 30.8. does not hold for arbitrary orthonormal systems) is also due to Kwapień.

Ex 30.8. Let E be a complex Banach space.
(a) Show that $\|\hat{f}(n)\|_E \leq \|f\|_2$ for all $f \in L_2(\mathbf{T}, E)$ and $n \in \mathbb{N}$.
(b) The E-valued exponential polynomials are norm–dense in $L_2(\mathbf{T}, E)$. Hint: Use $C(\mathbf{T}, E) = C(\mathbf{T}) \tilde{\otimes}_\varepsilon E$.
(c) *Riemann–Lebesgue lemma:* If $f \in L_2(\mathbf{T}, E)$, then the sequences $(\hat{f}(n))_{n \in \mathbb{N}}$ and $(\hat{f}(-n))_{n \in \mathbb{N}}$ converge (in E) to zero.

Ex 30.9. Show that for each Banach space the following statements are equivalent:
(a) E is isomorphic to a Hilbert space.
(b) If $\sum_{n=0}^{\infty} \|x_n\|^2 < \infty$, then $\sum_{n=0}^{\infty} \exp(in \cdot) x_n \in L_2(\mathbf{T}, E)$.
(c) The sequence of Fourier coefficients of every $f \in L_2(\mathbf{T}, E)$ is in $\ell_2(E)$.

Ex 30.10. If H is a Hilbert space and $f \in L_2(\mathbf{T}, H)$, then the Fourier series

$$\sum (\hat{f}(n)(2\pi)^{-1/2} \exp(in \cdot))_{n \in \mathbb{Z}}$$

converges in $L_2(\mathbf{T}, H)$ to f.

31. Pisier's Factorization Theorem

Kwapień's type/cotype theorem states that every operator T from a space of type 2 into a space of cotype 2 factors through a Hilbert space. Pisier's factorization theorem — in some sense an "abstract" version of Grothendieck's theorem — extends this to compactly approximable operators $T \in \mathfrak{L}(E, F)$ if E' and F have cotype 2; note that E having type 2 is more than E' having cotype 2. The proof of the theorem is taken from Pisier's book [225]. Moreover, a non–accessible tensor norm, also due to Pisier, will be presented.

31.1. Let $R : L_2(D, \mu) \to L_2(D, \mu)$ be the orthogonal projection onto the closed space Rad of the discrete Rademacher functions (see 8.5. and Ex 9.16.). As an orthogonal projection R is self–adjoint in the Hilbert space $L_2(D, \mu)$. Identifying $L_2(D, \mu)$ with its dual through the usual duality bracket $\int fg d\mu$, it can easily be seen that

$$R'f = \overline{R^* \overline{f}} = \overline{R \overline{f}} .$$

Since the Rademacher functions ε_n are real–valued, it is obvious that $R\overline{f} = \overline{Rf}$ and hence $R' = R$.

An operator $T \in \mathfrak{L}(E, F)$ is called K-convex if

$$\mathbf{K}(T) := \|R \otimes T : L_2(D, \mu) \otimes_{\Delta_2} E \longrightarrow L_2(D, \mu) \otimes_{\Delta_2} F\| < \infty .$$

It follows (see 17.4. and 7.4.) that the class $\mathfrak{KC}$ of K-convex operators together with $\mathbf{K}$ forms a maximal, injective and surjective Banach operator ideal. Since

$$L_2 \otimes_{\Delta_2} F' \xrightarrow{R' \otimes T' = R \otimes T'} L_2 \otimes_{\Delta_2} E'$$
$$\downarrow 1 \qquad\qquad\qquad\qquad \downarrow 1$$
$$(L_2 \otimes_{\Delta_2} F)' \xrightarrow{(R \otimes T)'} (L_2 \otimes_{\Delta_2} E)'$$

commutes, $\mathfrak{KC} \subset \mathfrak{KC}^{dual}$ holds and hence, $\mathfrak{KC} = \mathfrak{KC}^{dual}$ isometrically (17.8.): The operator ideal $(\mathfrak{KC}, \mathbf{K})$ of K-convex operators is completely symmetric.

It is easy to see (using Δ_2^t) that the identity operator on a Hilbert space and hence all 2-factorable are K-convex; moreover, $\mathbf{K} \leq \mathbf{L}_2$. Recall from Ex 9.16. that id_{ℓ_1} is not K-convex (proposition 7.8. stated that if there is a non-K-convex operator, then id_{ℓ_1} is not K-convex) and Pisier's result that E is not K-convex (if and) only if E contains the ℓ_1^n uniformly. By another result of Pisier's (mentioned in Ex 7.13.) it follows that E is K-convex if and only if it has proper type.

31.2. The dual of a type p operator has cotype p' (see Ex 8.14.) — but the converse is false: c_o does not have type 2 (Ex 7.14.), but ℓ_1 has cotype 2 (see 8.6.). It is clear from the definitions of type p and cotype q operators that this gap in the duality relation may be filled by K-convexity; more precisely,

PROPOSITION: *For $1 < p \leq 2$ the following two relations hold:*
(1) $\mathfrak{KC} \circ \mathfrak{C}_{p'}^{dual} \subset \mathfrak{T}_p$ *and* $\mathbf{T}_p \leq \mathbf{K} \circ \mathbf{C}_{p'}^{dual}$.
(2) $\mathfrak{C}_{p'} \circ \mathfrak{KC} \subset \mathfrak{T}_p^{dual}$ *and* $\mathbf{T}_p^{dual} \leq \mathbf{C}_{p'} \circ \mathbf{K}$.

PROOF: Since $\mathfrak{KC}^{dual} = \mathfrak{KC}$, it is clear that (2) implies (1). For (2) use the operators

$$L_2 \xrightarrow{\hat{R}} \text{Rad} \xrightarrow{J_{p'}} \ell_{p'}$$
$$L_2 \xleftarrow{J} \text{Rad} \xleftarrow{I_p} \ell_p ,$$

which satisfy $(J_{p'} \circ \hat{R})' = J \circ I_p$. If $S \in \mathfrak{KC}(E, F)$ and $T \in \mathfrak{C}_{p'}(F, G)$, the definitions show that the operator

$$(J_{p'} \otimes T) \circ (\hat{R} \otimes S) : L_2 \otimes_{\Delta_2} E \longrightarrow \ell_{p'} \otimes_{\Delta_{p'}} G$$

has norm $\leq \mathbf{C}_{p'}(T)\mathbf{K}(S)$. The diagram

$$\ell_p \otimes_{\Delta_p} G' \xrightarrow{(J \circ I_p) \otimes (T \circ S)'} L_2 \otimes_{\Delta_2} E'$$
$$\downarrow \qquad\qquad [(J_{p'} \circ \hat{R}) \otimes (T \circ S)]' \qquad \downarrow$$
$$(\ell_{p'} \otimes_{\Delta_{p'}} G)' \xrightarrow{\qquad} (L_2 \otimes_{\Delta_2} E)'$$

yields that $\mathbf{T}_p((T \circ S)')$ satisfies the same norm estimate — which is the claim. $\square$

In particular, if E is K-convex (i.e. — by the result of Pisier — does not contain the ℓ_1^n uniformly), then E has type p if (and only if) E' has cotype p'.

31.3. The following two lemmas prepare the way for the proof of Pisier's factorization theorem. Recall from Ex 8.15. that the Walsh functions ε_A (where A runs through the finite subsets of $\mathbb{N}$) form an orthonormal basis of $L_2(D,\mu)$.

LEMMA 1: *For every $\delta \in]0,1]$ there is a signed Borel measure ν_δ on the compact group $D = \{-1,1\}^{\mathbb{N}}$ such that*

(1) $\qquad \|\nu_\delta\| \leq \sqrt{1/\delta}$

(2) $\qquad \int_D \varepsilon_n \, d\nu_\delta = 1 \text{ for all } n \in \mathbb{N}$

(3) $\qquad \left|\int_D \varepsilon_A \, d\nu_\delta\right| \leq \delta \text{ if } A \subset \mathbb{N} \text{ and } |A| > 1$.

PROOF: For $\Theta \in [-1,1]$ and $n \in \mathbb{N}$ define

$$f_\Theta^n := \prod_{k=1}^n (1 + \Theta \varepsilon_k) \geq 0 \ .$$

It follows that

$$\int_D |f_\Theta^n| d\mu = \prod_{k=1}^n (1 + \Theta \int_{D_1} \varepsilon_k d\mu_1) = 1$$

and, for $A \subset \{1, ..., n\}$,

$$\langle \varepsilon_A, f_\Theta^n d\mu \rangle = \int_{D_n} \varepsilon_A \prod_{k=1}^n (1+\Theta\varepsilon_k) d\mu_n = \prod_{m \in A} \int_{D_1}(w_m + \Theta w_m^2)\mu_1(dw_m) = \Theta^{|A|} \ .$$

For each $n \in \mathbb{N}$ put

$$F_\delta^n := \frac{1}{2\sqrt{\delta}} \left(f_{\sqrt{\delta}}^n - f_{-\sqrt{\delta}}^n\right) \ .$$

Then the sequence $F_\delta^n d\mu$ in $C(D)'$ is norm bounded (by $\sqrt{1/\delta}$). Since $C(D)$ is separable, it has a weak-∗-convergent subsequence with limit ν_δ : It is obvious that this measure satisfies all the requirements. □

$\mathbf{K} \leq \mathbf{L}_2$ — but this estimate can be improved in a certain sense: This will be a crucial point in the proof of Pisier's factorization theorem.

LEMMA 2: *Take $T \in \mathcal{L}_2(E, F)$ and $\delta \in]0,1]$. Then*

$$\mathbf{K}(T) \leq \delta \mathbf{L}_2(T) + \frac{1}{\sqrt{\delta}} \|T\| \ .$$

PROOF: Fix
$$f = \sum_{k=1}^{n} \varepsilon_{A_k} \otimes x_k \in L_2(D,\mu) \otimes E$$
with distinct $A_1,...,A_n$. The idea of the proof is to construct a decomposition
$$R \otimes id_E(f) = f_1 - f_2 \in L_2(D,\mu) \otimes E$$
such that
$$\Delta_2(id \otimes T(f_1); L_2, F) \leq \frac{1}{\sqrt{\delta}} \|T\| \Delta_2(f; L_2, E)$$
$$\Delta_2(id \otimes T(f_2); L_2, F) \leq \delta \mathbf{L}_2(T) \Delta_2(f; L_2, E) ;$$
these estimates imply that
$$\Delta_2(R \otimes T(f); L_2, F) = \Delta_2(id \otimes T(R \otimes id_E(f)); L_2, F) =$$
$$= \Delta_2(id \otimes T(f_1 - f_2); L_2, F) \leq \left[\frac{1}{\sqrt{\delta}}\|T\| + \delta \mathbf{L}_2(T)\right] \Delta_2(f; L_2, E)$$
as desired.

Let ν_δ be the measure constructed in lemma 1 and define
$$f_1 := \sum_{k=1}^{n} \int_D \varepsilon_{A_k} d\nu_\delta \cdot \varepsilon_{A_k} \otimes x_k$$
$$f_2 := f_1 - R \otimes id_E(f) = f_1 - \sum_{|A_k|=1} \varepsilon_{A_k} \otimes x_k = \sum_{|A_k|>1} \int_D \varepsilon_{A_k} d\nu_\delta \cdot \varepsilon_{A_k} \otimes x_k ,$$

since $\int_D \varepsilon_m d\nu_\delta = 1$. Note that the Rademacher functions ε_m (and hence all Walsh functions) are multiplicative on the group D — and μ is actually the normalized Haar measure on D, i.e. invariant under multiplication. It follows that

$$\Delta_2(f_1; L_2, E) = \left(\int_D \left\|\int_D \sum_{k=1}^n \varepsilon_{A_k}(s)\varepsilon_{A_k}(t)x_k \nu_\delta(ds)\right\|_E^2 \mu(dt)\right)^{1/2} \leq$$
$$\leq \left(\int_D \left(\int_D \left\|\sum_{k=1}^n \varepsilon_{A_k}(st)x_k\right\|_E |\nu_\delta|(ds)\right)^2 \mu(dt)\right)^{1/2} \leq$$
$$\leq \int_D \left(\int_D \left\|\sum_{k=1}^n \varepsilon_{A_k}(st)x_k\right\|_E^2 \mu(dt)\right)^{1/2} |\nu_\delta|(ds) =$$
$$= \int_D \left(\int_D \left\|\sum_{k=1}^n \varepsilon_{A_k}(t)x_k\right\|_E^2 \mu(dt)\right)^{1/2} |\nu_\delta|(ds) \leq \frac{1}{\sqrt{\delta}} \Delta_2(f; L_2, E) ,$$

since $\|\nu_\delta\| \leq \sqrt{1/\delta}$, and hence $\Delta_2(id \otimes T(f_1)) \leq \|T\|\sqrt{1/\delta}\Delta_2(f)$. The operator T is 2–factorable — so Ex 28.2. can be used to obtain the estimate for f_2 : Indeed, take $x' \in E'$; then

$$\int_D |\langle x', f_2(t)\rangle|^2 \mu(dt) = \sum_{|A_k|>1} \left|\int_D \varepsilon_{A_k} d\nu_\delta\right|^2 |\langle x', x_k\rangle|^2 \leq$$

$$\leq \delta^2 \sum_{k=1}^n |\langle x', x_k\rangle|^2 = \delta^2 \int_D |\langle x', f(t)\rangle|^2 \mu(dt)$$

by the biorthonormality of the Walsh functions and lemma 1. It follows from Ex 28.2. that

$$\Delta_2(id \otimes T(f_2); L_2, F) \leq \delta \mathbf{L}_2(T)\Delta_2(f; L_2, E) ,$$

which is the remaining estimate. □

31.4. Now the main result of this section can be proven. It was obtained by Pisier [222] in 1980.

PISIER'S FACTORIZATION THEOREM: *Let E and F be Banach spaces such that E' and F have cotype 2. Then every operator $T \in \mathfrak{L}_2(E,F)$ (in particular, every finite rank operator) satisfies*

$$\mathbf{L}_2(T) \leq (2\mathbf{C}_2(E')\mathbf{C}_2(F))^{3/2}\|T\| .$$

In terms of tensor norms:

$$\varepsilon \leq w_2 \leq (2\mathbf{C}_2(E)\mathbf{C}_2(F))^{3/2}\varepsilon \qquad\qquad \text{on } E \otimes F$$

whenever E and F have cotype 2.

PROOF: Since w_2 and ε are totally accessible, $\mathbf{C}_2(E) = \mathbf{C}_2(E'')$ and $\mathfrak{L}_2$ and w_2 are associated, it is clear that the claim about the tensor norms follows from the one about the operators.

Given $T \in \mathfrak{L}_2(E,F)$ Kwapień's type/cotype theorem 30.5. yields

$$\mathbf{L}_2(T) \leq \mathbf{T}_2(T)\mathbf{C}_2(F) .$$

Proposition 31.2. and lemma 2 imply that

$$\mathbf{T}_2(T) \leq \mathbf{C}_2(E')\mathbf{K}(T) \leq \mathbf{C}_2(E')\left[\delta\mathbf{L}_2(T) + \frac{1}{\sqrt{\delta}}\|T\|\right]$$

for every $\delta \in]0,1]$, and hence

$$\mathbf{L}_2(T)[1 - \delta\mathbf{C}_2(E')\mathbf{C}_2(F)] \leq \frac{1}{\sqrt{\delta}}\mathbf{C}_2(E')\mathbf{C}_2(F)\|T\| .$$

Choosing $\delta := (2\mathbf{C}_2(E')\mathbf{C}_2(F))^{-1}$, gives the conclusion. $\square$

31.5. An operator $T \in \mathfrak{L}(E, F)$ is called *compactly approximable* whenever there is a net (T_ι) of finite rank operators converging to T uniformly on all compact subsets of E (see Ex 5.17. for a characterization of these operators). It is called λ-*compactly approximable* if the T_ι can be chosen of norm $\leq \lambda \|T\|$.

COROLLARY 1: *Let E' and F have cotype 2.*
(1) *Then every λ-compactly approximable operator $T \in \mathfrak{L}(E, F)$ is 2-factorable and*
$$\mathbf{L}_2(T) \leq (2\mathbf{C}_2(E')\mathbf{C}_2(F))^{3/2} \lambda \|T\| \ .$$
(2) *In particular, if one of the spaces E or F has the bounded approximation property, then $\mathfrak{L}(E, F) = \mathfrak{L}_2(E, F)$.*

Pisier [225] actually proved that this result holds for all compactly approximable operators — and therefore (2) is also valid if E or F only has the approximation property.

PROOF: The fact that the unit ball of $\mathfrak{L}_2(E, F)$ is closed in $\mathfrak{L}(E, F)$ with respect to the weak operator topology (see 17.21.) together with the theorem shows immediately part (1) of the claim. $\square$

Using the fact that $\mathfrak{L}(E, E) = \mathfrak{L}_2(E, E)$ implies that E is isomorphic to a Hilbert space, gives the

COROLLARY 2: *A Banach space E with the (bounded) approximation property such that E and E' have cotype 2, is isomorphic to a Hilbert space.*

As has already been mentioned at various times (see 4.8.) Pisier has constructed infinite dimensional Banach spaces P such that $P \otimes_\pi P = P \otimes_\varepsilon P$; his construction even gave that P and P' have cotype 2. Obviously, such a space cannot be isomorphic to a Hilbert space and hence id_P shows that in corollary 1 (and 2) an approximability hypothesis is indispensable.

31.6. Pisier's space P and his factorization theorem serve also to find a non-accessible tensor norm, or equivalently, a non-accessible maximal normed operator ideal. The following construction was given to us by Pisier. For $M, N \in $ FIN define
$$\mathbf{A}_P(T : M \to N) := \inf\{\, \|R\| \|S\| \mid T : M \xrightarrow{R} \ell_2(P) \xrightarrow{S} N \,\} \ ;$$
it is easy to see that this defines a normed operator ideal on FIN. Denote by $(\mathfrak{A}_P, \mathbf{A}_P)$ its maximal hull and by α_P the associated (finitely generated) tensor norm.

THEOREM (Pisier): $(\mathfrak{A}_P, \mathbf{A}_P)$ and α_P are neither left- nor right-accessible.

PROOF: By 21.3. it is enough to treat the operator ideal. Assume that $\mathfrak{A}_P$ is right-accessible; it will be shown that this implies that P is isomorphic to a Hilbert space. By the maximality of the ideal of 2-factorable operators it is enough to check that $\mathbf{L}_2(I_M^P : M \hookrightarrow P)$ is uniformly bounded for $M \in \mathrm{FIN}(P)$. Obviously, $\mathbf{A}_P(I_M^P) = 1$ and so accessibility gives

$$\begin{array}{ccc} M & \xrightarrow{I_M^P} & P \\ R \downarrow & T \searrow & \uparrow I_N^P \\ \ell_2(P) & \xrightarrow{S} & N \end{array} \qquad \|R\| = 1 \\ \|S\| \leq 2\mathbf{A}_P(I_M^P) = 2 .$$

It is straightforward from the defining inequality of cotype 2 that with P' also $\ell_2(P') = (\ell_2(P))'$ has cotype 2. Since $I_N^P \circ S$ has finite rank, Pisier's factorization theorem gives

$$\mathbf{L}_2(I_M^P) \leq \|R\|\mathbf{L}_2(I_N^P \circ S) \leq 2(2\mathbf{C}_2(\ell_2(P'))\mathbf{C}_2(P))^{3/2} .$$

To see that $\mathfrak{A}_P$ is not left-accessible one considers in the same way $\mathbf{L}_2(Q_L^P : P \longrightarrow P/L)$ for $L \in \mathrm{COFIN}(P)$. □

The injective hull $\mathfrak{A}_P^{inj}$ of $\mathfrak{A}_P$ (which is associated with the tensor norm $\alpha_P\backslash$) is right-accessible (by 21.1.); however, the same argument as before, the metric approximation property of ℓ_∞ and the fact that the ideal $\mathfrak{L}_2$ of 2-factorable operators is injective imply the

COROLLARY: $(\mathfrak{A}_P^{inj}, \mathbf{A}_P^{inj})$ and $\alpha_P\backslash$ are right- but not left-accessible.

These were non-accessible maximal ideals; see Ex 31.4. for non-accessible ideals consisting of approximable operators.

31.7. Since ℓ'_∞ and ℓ_1 are $\mathfrak{L}_1^g$-spaces and hence have cotype 2 (see 8.6.), Pisier's factorization theorem (more precisely, its corollary 1) implies that

(∗) $$\mathfrak{L}(\ell_\infty, \ell_1) = \mathfrak{L}_2(\ell_\infty, \ell_1) .$$

This is nothing other than Grothendieck's inequality, which can be seen as follows: The little Grothendieck theorem ($\mathfrak{L}(\ell_\infty, \ell_2) = \mathfrak{P}_2(\ell_\infty, \ell_2)$) and Kwapień's factorization theorem give

$$\mathfrak{L}(\ell_\infty, \ell_1) = \mathfrak{L}_2(\ell_\infty, \ell_1) = \mathfrak{P}_2^{dual} \circ \mathfrak{P}_2(\ell_\infty, \ell_1) = \mathfrak{D}_2(\ell_\infty, \ell_1) .$$

Since $w'_2 = w_2^* \sim \mathfrak{D}_2$, the representation theorem for maximal operator ideals yields $(\ell_\infty \otimes_\pi c_o)' = (\ell_\infty \otimes_{w_2} c_o)'$ which is $\pi \leq cw_2$ on $\ell_\infty \otimes c_o$ for some c: This is Grothendieck's inequality in tensor form. See Ex 31.3. for the constant c.

Thus, Pisier's factorization theorem provides a new proof of Grothendieck's inequality — and this justifies labelling this factorization theorem as an "abstract" form of Grothendieck's inequality (in the form (*)). Note that ℓ_∞ does not have type 2 — and hence Kwapień's type/cotype theorem does not give (*).

The use of the little Grothendieck theorem also implies part (1) of

COROLLARY 3 (Maurey): *Let E have cotype 2. Then*
(1) $\mathfrak{L}(\ell_\infty, E) = \mathfrak{P}_2(\ell_\infty, E)$.
(2) $\mathfrak{P}_2(E, F) = \mathfrak{P}_1(E, F)$ *for each Banach space F.*

The second part follows from the first through Saphar's result 20.19.. The content of the corollary was obtained in 1974 by Maurey [186]. Other proofs of Maurey's result will be given in 32.9. and Ex 32.7..

31.8. The last result and proposition 26.4. read:
$$\mathfrak{P}_p(E, F) = \mathfrak{P}_1(E, F) \text{ if } p \in [1, 2] \text{ and } E \text{ has cotype 2}$$
$$\mathfrak{P}_q(E, F) = \mathfrak{P}_2(E, F) \text{ if } q \in [2, \infty[\text{ and } F \text{ has cotype 2}.$$

Since $\mathfrak{P}_p \sim g_{p'}^*$ is totally accessible, the embedding lemma implies the following result concerning the equivalence of tensor norms:

COROLLARY 4: *Let $p \in]1, 2]$ and $q \in [2, \infty]$ and E, F Banach spaces. Then on $E \otimes F$*

$$g_\infty^* \sim g_q^* \text{ if } E' \text{ has cotype 2}$$
$$d_\infty \sim d_q \text{ if } E \text{ has cotype 2}$$
$$g_2^* \sim g_p^* \text{ if } F \text{ has cotype 2}$$
$$d_2 \sim d_p \text{ if } F' \text{ has cotype 2}.$$

Exercises:

Ex 31.1. If E' and F have cotype 2, then $\overline{\mathfrak{F}}(E, F) \subset \mathcal{L}_2(E, F)$.

Ex 31.2. (a) If E and F have cotype 2 and $p, q \in]1, \infty[$ satisfy $1/p + 1/q \geq 1$, then $E \otimes_{\alpha_{p,q}} F = E \otimes_\varepsilon F$ holds isomorphically.

(b) For the dual result let E' and F' have cotype 2: Then $\overleftarrow{\pi}$ and w_2^* are equivalent on $E \otimes F$, but π and w_2^* are in general not; a sufficient condition for π and w_2^* being equivalent on $E \otimes F$ is that E or F has the (bounded) approximation property. If E and F' have cotype 2, then it follows in the same way that π and w_2^* are, in general, not equivalent on $E' \otimes F$, but Pisier ([225], p.47) showed that the nuclear norm $\mathbf{N}$ and the 2-dominated norm $\mathbf{D}_2$ (associated with w_2^*) are equivalent on $\mathfrak{F}(E, F)$.

Ex 31.3. Show that the "abstract" proof 31.7. of Grothendieck's inequality gives the estimate $K_G \leq 8K_{LG}^2$ for the Grothendieck constant. Hint: 8.6..

Ex 31.4. (a) Show for the factorization ideal $\mathfrak{A}_P$ from 31.6. that $\mathfrak{A}_P \circ \overline{\mathfrak{F}}$ is left- but not right-accessible. Hint: Ex 25.3. and proof of 31.6.. Recall that the minimal hull $\mathfrak{A}_P^{min} = \overline{\mathfrak{F}} \circ \mathfrak{A}_P \circ \overline{\mathfrak{F}}$ is accessible by 25.3..

(b) Define the quasi-Banach operator ideal $(\mathfrak{C}_P, \mathbf{C}_P)$ by

$$\mathfrak{C}_P(E, F) := \{T \in \mathfrak{L}(E, F) \,|\, T = T_1 + T_2, T_1 \in \mathfrak{A}_P \circ \overline{\mathfrak{F}}, T_2 \in \overline{\mathfrak{F}} \circ \mathfrak{A}_P\}$$
$$\mathbf{C}_P(T) := \inf\{(\mathbf{A}_P \circ \|\ \|)(T_1) + (\|\ \| \circ \mathbf{A}_P)(T_2)\} \ .$$

Show with Ex 21.2. (b) that $(\mathfrak{C}_P, \mathbf{C}_P)$ is an ideal of approximable operators which is neither right- nor left-accessible.

32. Mixing Operators

The following situation has appeared already at various times: An operator T (or id_E of a Banach space) has the property that $S \circ T$ is absolutely p-summing whenever S is absolutely q-summing. This section is devoted to a systematic study of these "mixing" operators/spaces. Mixing operators are characterized by interesting integral inequalities and by a certain splitting property (which explains the name "mixing"); moreover, they are almost equal to absolutely (r, s)-summing operators for appropriate r and s. Many of the results to be presented are at least implicitly due to Maurey [186]; the ideal of mixing operators was studied systematically by Pietsch in his monograph on operator ideals and by Puhl [226].

32.1. Let $p, q \in [1, \infty]$. An operator $T \in \mathfrak{L}(E, F)$ is said to be (q, p)-*mixing* if for all Banach spaces G and $S \in \mathfrak{P}_q(F, G)$ the composition $S \circ T$ is absolutely p-summing. With

$$\mathbf{M}_{q,p}(T) := \sup\{\mathbf{P}_p(S \circ T) \,|\, \mathbf{P}_q(S) \leq 1\}$$

it is clear that, in the terminology of quotient ideals introduced in section 25, the ideal $\mathfrak{M}_{q,p}$ of (q, p)-mixing operators is isometrically $\mathfrak{P}_q^{-1} \circ \mathfrak{P}_p$:

$$(\mathfrak{M}_{q,p}, \mathbf{M}_{q,p}) = (\mathfrak{P}_q^{-1} \circ \mathfrak{P}_p, \mathbf{P}_q^{-1} \circ \mathbf{P}_p) \ .$$

The quotient formula 25.7. implies that $(\mathfrak{M}_{q,p}, \mathbf{M}_{q,p})$ is a maximal Banach operator ideal — and later on it will be easy to see that it is injective but not surjective (Ex 32.1.). The name "mixing", due to Pietsch, will become clear in 32.7..

Since $\mathfrak{P}_p$ is injective and $\mathfrak{I}_q^{inj} = \mathfrak{P}_q$, it is also easy to see that

$$\mathfrak{M}_{q,p} = \mathfrak{I}_q^{-1} \circ \mathfrak{P}_p$$

holds isometrically. The ideals $\mathfrak{P}_p$ and $\mathfrak{I}_p$ are accessible and $\mathfrak{N}_p = \mathfrak{I}_p^{min}$ is the ideal of p–nuclear operators, hence the quotient formulas 25.7. give a result due to Puhl [226]:

PROPOSITION: *The following relations hold isometrically:*

$$\mathfrak{M}_{q,p} := \mathfrak{P}_q^{-1} \circ \mathfrak{P}_p = \mathfrak{I}_q^{-1} \circ \mathfrak{P}_p = \mathfrak{P}_{q'} \circ \mathfrak{I}_{p'}^{-1} = \mathfrak{I}_{q'} \circ \mathfrak{I}_{p'}^{-1} = \mathfrak{N}_{q'} \circ \mathfrak{N}_{p'}^{-1} =$$
$$= (\mathfrak{I}_{p'} \circ \mathfrak{P}_q)^* = (\mathfrak{I}_{p'} \circ \mathfrak{I}_q)^* \ .$$

Note that this result implies various composition formulas, as for example

$$\mathfrak{M}_{q,p} \circ \mathfrak{I}_{p'} \subset \mathfrak{I}_{q'} \quad , \quad \mathfrak{M}_{q,p} \circ \mathfrak{N}_{p'} \subset \mathfrak{N}_{q'} \ .$$

32.2. Some observations — and translations of former results into the language of mixing operators deserve to be stated explicitly:
(1) It is obvious that $\mathfrak{M}_{q,p} = \mathfrak{L}$ whenever $q \leq p$ and $\mathfrak{M}_{\infty,p} = \mathfrak{P}_p$. So only the case $1 \leq p < q < \infty$ gives something new.
(2) Pietsch's composition formula $\mathfrak{P}_q \circ \mathfrak{P}_r \subset \mathfrak{P}_p$ from 11.5. gives:

$$\mathfrak{P}_r \subset \mathfrak{M}_{q,p} \text{ and } \mathbf{M}_{q,p} \leq \mathbf{P}_r \quad \text{if } p, q, r \in [1, \infty] \text{ with } 1/q + 1/r = 1/p \ ;$$

this will be extended to absolutely (r, s)–summing operators at the end of this section. Since in Pietsch's composition formula the index p is best possible (see Ex 11.23.), it follows that

$$\mathfrak{M}_{q,p} \not\subset \mathfrak{M}_{q,p-\varepsilon} \quad \text{if } 1 \leq p - \varepsilon < p \leq q \ ;$$

see also Ex 32.14..
(3) Saphar's result 20.19. means that:

$$E \in \text{space}(\mathfrak{M}_{q,1}) \quad \text{if and only if } \mathfrak{L}(\ell_\infty, E) = \mathfrak{P}_{q'}(\ell_\infty, E) \ ;$$

moreover, $\mathbf{M}_{q,1}(id_E) = \sup\{\mathbf{P}_{q'}(T) \mid \|T : \ell_\infty \to E\| \leq 1\}$. This also follows from the quotient formula $\mathfrak{M}_{q,1} = \mathfrak{I}_{q'} \circ \mathfrak{I}_\infty^{-1} = \mathfrak{P}_{q'} \circ \mathfrak{L}_\infty^{-1}$. Furthermore, this formula also implies that

$$\mathfrak{M}_{q,1}(\ell_\infty, F) = \mathfrak{P}_{q'}(\ell_\infty, F)$$

holds isomorphically for all F.
(4) The little Grothendieck theorem says that $\mathfrak{L}(\ell_\infty, H) = \mathfrak{P}_2(\ell_\infty, H)$ for every Hilbert space H, hence $id_H \in \text{space}(\mathfrak{M}_{2,1})$ by (3). Looking at the constants in Saphar's result, one gets

$$\mathbf{M}_{2,1}(id_H) \leq K_{LG}$$

— even $= K_{LG}$, see Ex 32.4.. Much more generally:

$$E \in \text{space}(\mathfrak{M}_{2,1}) \quad \text{for all cotype 2 spaces } E$$

— this is just a reformulation of Maurey's result in 31.7.. Conversely, every Banach lattice in space($\mathfrak{M}_{2,1}$) has cotype 2 (see Pisier [225], p.108) — but it seems to be unknown whether or not this holds for arbitrary Banach spaces. In 32.9. and Ex 32.7. alternative proofs of Maurey's result are indicated.

(5) For cotype $q > 2$ less is true. Theorem 24.7. gives:

$$E \in \text{space}(\mathfrak{M}_{q'-\epsilon,1}) \quad \text{if } E \text{ is of cotype } q > 2 \, .$$

Since L_q is of cotype $\max\{q,2\}$, it follows for the r–factorable operators that

$$\mathfrak{L}_r \subset \mathfrak{M}_{p,1} \quad \text{if } 2 < r < p' < \infty$$

and

$$\mathfrak{L}_r \subset \mathfrak{M}_{2,1} \quad \text{if } 1 \leq r \leq 2 \, .$$

(All this was already stated in 26.4.(1) and (2) — including the constants involved.)

(6) In particular, the sequence space ℓ_1 is in space($\mathfrak{M}_{2,1}$). One can show, however, that ℓ_1 is not in space($\mathfrak{M}_{q,1}$) for $q > 2$; this will follow – as will some other counterexamples — from considerations about the ideal of absolutely (r,s)–summing operators: $\mathfrak{M}_{q,1} \subset \mathfrak{P}_{q',1}$ by 32.10. and $\ell_1 \not\in \text{space}(\mathfrak{P}_{q',1})$ by proposition 34.11.. Since $C[0,1] \hookrightarrow L_q(\lambda)$ (Lebesgue measure) is in $\mathfrak{P}_q$ but not in $\mathfrak{P}_p$ for $p < q$ (this was mentioned in 11.3.),

$$C([0,1]) \not\in \text{space}(\mathfrak{M}_{q,p}) \quad \text{if } 1 \leq p < q < \infty \, ,$$

and (by local techniques) also $\ell_\infty \not\in \text{space}(\mathfrak{M}_{q,p})$. Later on (32.8.) it will be shown that

$$\ell_{q'} \not\in \text{space}(\mathfrak{M}_{q,p}) \quad \text{if } 1 \leq p < q < 2$$

— this reveals that the result in (5) (about cotype $q > 2$) cannot be improved.

32.3. The calculus of traced tensor norms and tensor products of operator ideals developed in section 29 (and the quotient formula 25.7.) shows that (see 29.8.)

$$\mathfrak{M}_{q,p} = \mathfrak{P}_q^{-1} \circ \mathfrak{P}_p = (\mathfrak{P}_p^* \circ \mathfrak{P}_q)^* = (\mathfrak{I}_{p'} \circ \mathfrak{P}_q)^* = (\mathfrak{I}_{p'} \otimes \mathfrak{P}_q)^*$$
$$\mathfrak{M}_{q,p} = \mathfrak{I}_q^{-1} \circ \mathfrak{P}_p = (\mathfrak{P}_p^* \circ \mathfrak{I}_q)^* = (\mathfrak{I}_{p'} \circ \mathfrak{I}_q)^* = (\mathfrak{I}_{p'} \otimes \mathfrak{I}_q)^* \, ,$$

and hence the associated tensor norm is, by definition of $\mathbf{A} \otimes \mathbf{B}$ (see 29.7.):

$$\mathfrak{M}_{q,p} \sim (g_{p'} \otimes g_{q'}^*)^* = (g_{p'} \otimes g_q)^* \, .$$

Moreover, by 29.11.

$$g_{p'} \otimes g_{q'}^* = g_{p'} \otimes g_q = g_{p'} \otimes_{L_{q'}(\mu)} \Delta_q^t$$

for every infinite dimensional $L_{q'}(\mu)$. The tensor product characterization 29.4./25.8. of operators in $(\mathfrak{A} \otimes \mathfrak{B})^*$ and the accessibility of g_q and $g_{q'}^*$ give the

PROPOSITION: *Let $p, q \in [1, \infty]$ and $\beta = g_q$ or $g_{q'}^*$. For $T \in \mathfrak{L}(E, F)$ the following statements are equivalent:*
(a) $T \in \mathfrak{M}_{q,p}(E, F)$.
(b) $id_G \otimes T : G \otimes_{g_{p'}} E \longrightarrow G \otimes_\beta F$ *is continuous for all Banach spaces G.*
(c) *For one infinite dimensional $L_{q'}(\mu)$ (and then for all)*

$$id_{L_{q'}(\mu)} \otimes T : L_{q'}(\mu) \otimes_{g_{p'}} E \longrightarrow L_{q'}(\mu) \otimes_{\Delta_{q'}} F$$

is continuous.
In this case, $\mathbf{M}_{q,p}(T) = \|id_{L_{q'}(\mu)} \otimes T\| = \|id_{C_2} \otimes T\| \geq \|id_G \otimes T\|$.

This result has a nice application to the continuity of tensor product operators between spaces of Bochner integrable functions — in some sense an extension of Beckner's result 7.9. (see also 26.3.), which is due to Carl–D. [21] :

COROLLARY: *For $S \in \mathfrak{L}(E, L_p(\mu))$ and $T \in \mathfrak{L}(L_q(\nu), F)$*

$$\|S \otimes T : E \otimes_{\Delta_q^t} L_q(\nu) \longrightarrow L_p(\mu) \otimes_{\Delta_p} F\| \leq \mathbf{M}_{s,p}(S') \mathbf{M}_{s',q'}(T)$$

whenever $p, q, s \in [1, \infty]$.

PROOF: Using the Chevet–Persson–Saphar inequalities 15.10., a look at the diagram

$$\begin{array}{ccc}
E \otimes_{\Delta_q^t} L_q(\nu) & \xrightarrow{S \otimes T} & L_p(\mu) \otimes_{\Delta_p} F \\
\text{id} \downarrow & & \uparrow \text{id} \\
E \otimes_{g_q} L_q(\nu) & \xrightarrow{id \otimes T} E \otimes_{g_{p'}^*} F \xrightarrow{S \otimes id} & L_p(\mu) \otimes_{g_{p'}^*} F \\
& \downarrow{}^1 & \downarrow{}^1 \\
& (E' \otimes_{g_{p'}^t} F')' \xrightarrow{(S' \otimes id)'} & (L_{p'}(\mu) \otimes_{g_{p'}^t} F')'
\end{array}$$

gives the result. □

Various special cases of this result are important enough to be stated explicitly:

(1) *If $q \leq p$, then $\|S \otimes T\| = \|S\| \, \|T\|$.*

(Take $s = p$.) Certainly the most important case is $s = 2$. Since cotype 2 spaces are in space($\mathfrak{M}_{2,1}$) $\subset$ space($\mathfrak{M}_{2,p}$) for all p (by 32.2.(4) and Ex 32.2.(a)), this gives

(2) $S \otimes T$ is continuous in each of the following three cases:
  — F of cotype 2 and $p \geq 2$.
  — E' of cotype 2 and $q \leq 2$.
  — E' and F of cotype 2 .

Note the special case: $E = L_r$ and $F = L_s$ for $s \leq 2 \leq r$. This gives a variant of the result in 26.3..

The fact that
$$\ell_{q'-\varepsilon} \in \text{space}(\mathfrak{M}_{q,1}) \subset \text{space}(\mathfrak{M}_{q,p})$$
whenever $q \leq 2$ (see 32.2.(5)) has another nice consequence:

(3) *Let* $1 \leq p \leq q \leq 2$ *and* $\varepsilon > 0$. *For all operators* $S \in \mathfrak{L}(L_{q+\varepsilon}(\mu_1), L_p(\nu))$ *and* $T \in \mathfrak{L}(L_q(\mu_2), F)$ *the operator*
$$S \otimes T : L_{q+\varepsilon}(\mu_1) \otimes_{\Delta_q^i} L_q(\mu_2) \longrightarrow L_p(\nu) \otimes_{\Delta_p} F$$
is continuous.

PROOF: Take $s = q$. □

32.4. There are two characterizations of mixing operators — one in terms of integral inequalities and one in terms of a splitting property. These descriptions of the quotient ideal $\mathfrak{M}_{q,p} = \mathfrak{P}_q^{-1} \circ \mathfrak{P}_p$ of (q,p)–mixing operators make them important in many situations. First the integral description, which has the flavour of the Grothendieck–Pietsch domination theorem:

PROPOSTION: *Let* $1 \leq p \leq q < \infty$ *and* $T \in \mathfrak{L}(E,F)$. *The following statements are equivalent:*

(a) T *is* (q,p)-*mixing.*

(b) *There is a constant* $c \geq 0$ *such that for each probability measure* μ *on* $B_{F'}$ *there is a probability measure* ν *on* $B_{E'}$ *such that*
$$\left(\int_{B_{F'}} |\langle y', Tx \rangle|^q \mu(dy') \right)^{1/q} \leq c \left(\int_{B_{E'}} |\langle x', x \rangle|^p \nu(dx') \right)^{1/p}$$
for all $x \in E$.

(c) *There is a constant* $c \geq 0$ *such that*
$$\left(\sum_{j=1}^m \left(\sum_{k=1}^n |\langle y'_k, Tx_j \rangle|^q \right)^{p/q} \right)^{1/p} \leq c w_p(x_j) \ell_q(y'_k)$$

for all $x_1, ..., x_m \in E$ and $y'_1, ..., y'_n \in F'$.
In this case, $\mathbf{M}_{q,p}(T) = \min c$, where the minimum is taken over all c satisfying (b) or all satisfying (c).

PROOF: If $T \in \mathfrak{M}_{q,p}(E, F) = \mathfrak{P}_q^{-1} \circ \mathfrak{P}_p(E, F)$ and μ is a probability measure on $B_{F'}$, then
$$I : F \longrightarrow L_q(\mu)$$
$$y \rightsquigarrow \langle \cdot, y \rangle$$
is absolutely q-summing and hence $I \circ T \in \mathfrak{P}_p(E, L_q(\mu))$: The Grothendieck–Pietsch domination theorem gives a probability measure ν on $B_{E'}$ such that for all $x \in E$

$$\left(\int_{B_{F'}} |\langle y', Tx \rangle|^q \mu(dy')\right)^{1/q} = \|I \circ T(x)\| \leq \mathbf{P}_p(I \circ T)\left(\int_{B_{E'}} |\langle x', x \rangle|^p \nu(dx')\right)^{1/p}.$$

Since $\mathbf{P}_p(I \circ T) \leq \mathbf{M}_{q,p}(T)\mathbf{P}_q(I) = \mathbf{M}_{q,p}(T)$, statement (b) follows.
In order to show that (b) implies (c) assume $\ell_q(y'_k) = 1$. Then

$$\mu := \sum_{k=1}^{n} \|y'_k\|^q \delta_{\|y'_k\|^{-1} y'_k}$$

is a probability measure on $B_{F'}$, hence there is a ν as in (b). Therefore,

$$\sum_{j=1}^{m}\left(\sum_{k=1}^{n} |\langle y'_k, Tx_j \rangle|^q\right)^{p/q} = \sum_{j=1}^{m}\left(\int_{B_{F'}} |\langle y', Tx_j \rangle|^q \mu(dy')\right)^{p/q}$$
$$\leq c^p \sum_{j=1}^{m} \int_{B_{E'}} |\langle x', x_j \rangle|^p \nu(dx') \leq c^p (w_p(x_j))^p.$$

Finally, it remains to prove that (c) implies that $T \in \mathfrak{P}_q^{-1} \circ \mathfrak{P}_p$: By what was just said, condition (c) means that all *discrete* probability measures μ on $B_{F'}$ satisfy

$$\left(\sum_{j=1}^{m}\left(\int_{B_{F'}} |\langle y', Tx_j \rangle|^q \mu(dy')\right)^{p/q}\right)^{1/p} \leq cw_p(x_j)$$

for all $x_1, ..., x_m \in E$. Since the set of all discrete probability measures on $B_{F'}$ is $\sigma(M(B_{F'}), C(B_{F'}))$–dense in all probability measures μ, it follows that this inequality holds for all such μ. Hence, if $S \in \mathfrak{P}_q(F, G)$, then another appeal to the domination theorem gives $\mathbf{P}_p(S \circ T) \leq c\mathbf{P}_q(S)$. □

Note that for $q = \infty$ condition (b) can be interpreted as the domination theorem for $\mathfrak{P}_p = \mathfrak{M}_{\infty,p}$.

32.5. Using this characterization, one can "improve" the index p for E in space($\mathfrak{M}_{q,p}$) : this is the content of the following result which one might call the *Maurey-Pisier extrapolation theorem* (see Pisier [225], p.60 and p.69).

THEOREM: *For $1 \leq p < q < \infty$ let E be a Banach space such that*
$$\mathfrak{P}_q(E,F) \subset \mathfrak{P}_p(E,F)$$
for all Banach spaces F. Then
$$\mathfrak{P}_q(E,F) \subset \mathfrak{P}_1(E,F)$$
for all Banach spaces F. In other words, space($\mathfrak{M}_{q,p}$) = space($\mathfrak{M}_{q,1}$).

Recall from 32.2.(2) that $\mathfrak{M}_{q,p} \not\subset \mathfrak{M}_{q,1}$ whenever $p > 1$, hence the result does not hold for operators. Moreover only the case $q < 2$ is interesting by 32.10..

PROOF: Fix a probability measure μ on $B_{E'}$. By the previous proposition one has to find a probability measure ν such that for all $x \in E$
$$\left(\int_{B_{E'}} |\langle \cdot, x\rangle|^q d\mu\right)^{1/q} \leq c \int_{B_{E'}} |\langle \cdot, x\rangle| d\nu$$
— with a constant c independent of μ. Let $\nu_o := \mu$. By assumption and induction there are probability measures ν_n such that
$$\left(\int_{B_{E'}} |\langle \cdot, x\rangle|^q d\nu_n\right)^{1/q} \leq c_o \left(\int_{B_{E'}} |\langle \cdot, x\rangle|^p d\nu_{n+1}\right)^{1/p}$$
where $c_o := \mathfrak{M}_{q,p}(id_E)$. Define $\nu := \sum_{n=1}^{\infty} 2^{-n} \nu_n$ and $\Theta \in]0,1[$ by $1/p = \Theta + (1-\Theta)/q$. Then Hölder's inequality gives
$$\sum_{n=0}^{\infty} \frac{1}{2^n} \|\langle \cdot, x\rangle\|_{L_q(\nu_n)} \leq c_o \sum_{n=0}^{\infty} \frac{1}{2^n} \|\langle \cdot, x\rangle\|_{L_p(\nu_{n+1})} \leq$$
$$\leq c_o \sum_{n=0}^{\infty} \frac{1}{2^n} \|\langle \cdot, x\rangle\|_{L_1(\nu_{n+1})}^{\Theta} \|\langle \cdot, x\rangle\|_{L_q(\nu_{n+1})}^{1-\Theta} \leq$$
$$\leq c_o \Big(\sum_{n=0}^{\infty} \frac{1}{2^n} \|\langle \cdot, x\rangle\|_{L_1(\nu_{n+1})}\Big)^{\Theta} \Big(\sum_{n=0}^{\infty} \frac{1}{2^n} \|\langle \cdot, x\rangle\|_{L_q(\nu_{n+1})}\Big)^{1-\Theta} \leq$$
$$\leq c_o \big(2\|\langle \cdot, x\rangle\|_{L_1(\nu)}\big)^{\Theta} \Big(2 \sum_{n=0}^{\infty} \frac{1}{2^n} \|\langle \cdot, x\rangle\|_{L_q(\nu_n)}\Big)^{1-\Theta}.$$

Since $\mu = \nu_o$, this implies that
$$\big(\|\langle \cdot, x\rangle\|_{L_q(\mu)}\big)^{\Theta} \leq \Big(\sum_{n=0}^{\infty} \frac{1}{2^n} \|\langle \cdot, x\rangle\|_{L_q(\nu_n)}\Big)^{\Theta} \leq c_o 2 \big(\|\langle \cdot, x\rangle\|_{L_1(\nu)}\big)^{\Theta},$$

which gives $\mathbf{M}_{q,1}(id_E) \leq (2\mathbf{M}_{q,p}(id_E))^{1/\Theta}$. □

For an application to spaces which satisfy Grothendieck's theorem (the G.T. spaces from 20.18.) see Ex 32.9..

32.6. Maurey's factorization theorem 18.9. gives a very useful characterization of operators in $\mathfrak{M}_{q,p}^{dual}$, which — for spaces — is also due to Maurey [186], théorème 23.

THEOREM: *Let $1 \leq p \leq q \leq \infty$ and $1/p = 1/r + 1/q$. For $T \in \mathcal{L}(E, F)$ the following five statements are equivalent:*
(a) *For every $S \in \mathcal{L}(L_q(\nu), G)$ and $U \in \mathcal{L}(F, L_p(\mu))$ (for some measures μ and ν) the operator*
$$S \otimes UT : L_q(\nu) \otimes_{\Delta_q} E \longrightarrow G \otimes_{\Delta_p^t} L_p(\mu)$$
is continuous.
(b) *For every $U \in \mathcal{L}(F, \ell_p)$ the operator*
$$id_{\ell_q} \otimes UT : \ell_q \otimes_{\Delta_q} E \longrightarrow \ell_q \otimes_{\Delta_p^t} \ell_p$$
is continuous.
(c) *For every $U \in \mathcal{L}(F, L_p(\mu))$ there are a $g \in L_r(\mu)$ and $R \in \mathcal{L}(E, L_q(\mu))$ such that $U \circ T = M_g \circ R$.*
(d) *For every $U \in \mathcal{L}(F, \ell_p)$ there are a $\lambda \in \ell_r$ and $R \in \mathcal{L}(E, \ell_q)$ such that $U \circ T = D_\lambda \circ R$.*
(e) *T' is (q, p)-mixing.*
In this case,
$$\mathbf{M}_{q,p}(T') = \sup\{\|id_{\ell_q} \otimes UT\| \mid \|U : F \to \ell_p\| \leq 1\} =$$
$$= \sup\{\min\{\|\lambda\|_r \|R\| \mid U \circ T = D_\lambda \circ R\} \mid \|U : F \to \ell_p\| \leq 1\}.$$

PROOF: Maurey's factorization theorem shows the equivalences of (a) and (c), as well as of (b) and (d). Since obviously (a) ↷ (b) and (c) ↷ (d) the implications (b) ↷ (e) and (e) ↷ (a) remain to be shown.

Assume (b): A closed graph argument garanties the existence of a constant $c \geq 0$ such that for all $m, n \in \mathbb{N}$
$$\|id_{\ell_q^n} \otimes UT : \ell_q^n \otimes_{\Delta_q} E \longrightarrow \ell_q^n \otimes_{\Delta_p^t} \ell_p^m\| \leq c\|U\|$$
for all $U \in \mathcal{L}(F, \ell_p^m)$. For $U = \sum_{j=1}^m y'_j \otimes e_j$ one has $\|U\| = w_p(y'_j)$, and hence for all $x_1, \ldots, x_n \in E$
$$\Big(\sum_{j=1}^m \Big(\sum_{k=1}^n |\langle x_k, T'y'_j\rangle|^q\Big)^{p/q}\Big)^{1/p} = \Delta_p^t\Big(\big(id_{\ell_q^n} \otimes UT\big)\Big(\sum_{k=1}^n e_k \otimes x_k\Big)\Big) \leq$$
$$\leq c w_p(y'_j) \ell_q(x_k).$$

Since B_E is weak-$*$-dense in $B_{E''}$, this inequality also holds for $x_k'' \in E''$ instead of $x_k \in E$ — and proposition 32.4. gives that T' is (q,p)-mixing.

Now assume $T' \in \mathfrak{M}_{q,p}(E,F)$ and take U and S as in (a). It follows from the tensor product characterization 32.3.(c) that

$$id_{L_{q'}(\nu)} \otimes T' : L_{q'}(\nu) \otimes_{g_{p'}} F' \longrightarrow L_{q'}(\nu) \otimes_{\Delta_{q'}} E'$$

has norm $\leq \mathbf{M}_{q,p}(T')$. Since $g'_{p'}$ is accessible, this implies that

$$\|id_{L_q(\nu)} \otimes T : L_q(\nu) \otimes_{\Delta_q} E \longrightarrow L_q(\nu) \otimes_{g'_{p'}} F\| \leq \mathbf{M}_{q,p}(T'),$$

hence the diagram

$$
\begin{array}{ccc}
L_q(\nu) \otimes_{\Delta_q} E & \xrightarrow{S \otimes UT} & G \otimes_{\Delta_p^t} L_p(\mu) \\
{\scriptstyle id \otimes T}\downarrow & & \uparrow{\scriptstyle id} \\
L_q(\nu) \otimes_{g'_{p'}} F & \xrightarrow{S \otimes U} & G \otimes_{g'_{p'}} L_p(\mu)
\end{array}
$$

and the Chevet–Persson–Saphar inequality $\Delta_p^t \leq g'_{p'}$ give (a). □

32.7. A consequence of this is *Maurey's splitting theorem:*

COROLLARY 1: *Let $1 \leq p \leq q \leq \infty$ and $1/r + 1/q = 1/p$. An operator $T \in \mathcal{L}(E,F)$ is (q,p)-mixing if and only if for each weakly p-summable sequence (x_n) in E there is a sequence $(\lambda_n) \in \ell_r$ and a weakly q-summable sequence (y_n) in F such that*

$$T x_n = \lambda_n y_n$$

for all $n \in \mathbb{N}$. In this case,

$$\mathbf{M}_{q,p}(T) = \sup_{w_p(x_n) \leq 1} \inf\{\|\lambda_n\|_r \, w_q(y_n) \mid T x_n = \lambda_n y_n\}.$$

This splitting property of mixing operators is sometimes taken as a definition of (p,q)-mixing operators. Pietsch calls weakly p-summable sequences (z_n), which split into $z_n = \lambda_n y_n$ as above, (q,p)-*mixed sequences* — and this explains why he gave operators in $\mathfrak{M}_{q,p}$ a related name: (q,p)-mixing.

PROOF: Let T be (q,p)-mixing and $w_p(x_n; E) \leq 1$. Since T'' is (q,p)-mixing, (d) of the preceding theorem gives that for $V : \ell_{p'} \to E$ defined by $V e_n := x_n$ there is a factorization (read c_o instead of ℓ_∞ if $p = 1$)

$$\begin{array}{ccc}
F' \xrightarrow{T'} E' \xrightarrow{V'} \ell_p \\
R \downarrow \quad \nearrow D_\lambda \\
\ell_q
\end{array} \qquad \|R\|\,\|\lambda\|_r \le \mathbf{M}_{q,p}(T),$$

since $\|V'\| = w_p(x_n) \le 1$. It follows that $\lambda_n R' e_n = Tx_n \in F$, which gives the splitting because of $w_q(R'e_n) \le \|R\| w_q(e_n;\ell_{q'}) = \|R\|$. Moreover, $c := \sup\inf\{...\} \le \mathbf{M}_{q,p}(T)$. Conversely, take $S \in \mathfrak{P}_q(F,G)$ and $w_p(x_n;E) \le 1$. Then

$$\ell_p(STx_n) = \ell_p(\lambda_n Sy_n) \le \ell_r(\lambda_n)\ell_q(Sy_n) \le \ell_r(\lambda_n)\mathbf{P}_q(S)w_q(y_n),$$

which shows that $S \circ T$ is absolutely p-summing, and hence $T \in \mathfrak{M}_{q,p}$. The calculation also gives $\mathbf{M}_{q,p}(T) \le c$. □

32.8. The theorem has another quite interesting consequence:

COROLLARY 2: *Let E be a Banach space and $1 \le p \le q \le 2$. Then $E' \in \mathrm{space}(\mathfrak{M}_{q,p})$ if and only if for all $L_q(\nu_1)$ and $L_p(\nu_2)$ (or only ℓ_q and ℓ_p) and all $L_p(\mu)$ (or only ℓ_p)*

$$S \otimes T : L_q(\nu_1) \otimes_{\Delta_q} E \longrightarrow L_p(\nu_2) \otimes_{\Delta_p} L_p(\mu)$$

is continuous whenever $S \in \mathfrak{L}(L_q(\nu_1), L_p(\nu_2))$ and $T \in \mathfrak{L}(E, L_p(\mu))$.

Note that for $p < q$, the Maurey–Pisier extrapolation theorem says that $\mathrm{space}(\mathfrak{M}_{q,p}) = \mathrm{space}(\mathfrak{M}_{q,1})$.

PROOF: By (a) of the theorem the condition is clearly necessary. To see the converse for $p < q$ (for $p = q$ there is nothing to prove since $\mathfrak{M}_{p,p} = \mathfrak{L}$) observe first that the condition (with ℓ_q and ℓ_p) implies the continuity of all

$$A \otimes T : \ell_q \otimes_{\Delta_q} E \longrightarrow L_p(\nu_2) \otimes_{\Delta_p} \ell_p = L_p(\nu_2) \otimes_{\Delta_p^t} \ell_p$$

by corollary 29.12.. Taking for A a Lévy embedding

$$I : \ell_q \xhookrightarrow{1} L_p(\mu)$$

(see 24.5.) gives that

$$id_{\ell_q} \otimes T : \ell_q \otimes_{\Delta_q} E \longrightarrow \ell_q \otimes_{\Delta_p^t} \ell_p \left[\xhookrightarrow{1} L_p(\mu) \otimes_{\Delta_p^t} \ell_p\right]$$

is continuous. Again, an appeal to part (b) of the theorem gives the claim. □

The corollary has also a negative flavour: Since for $1 \le p < q < 2$ and $I : \ell_q \hookrightarrow L_p(\mu_q)$ (Lévy embedding) the operator

$$I \otimes id_{\ell_q} : \ell_q \otimes_{\Delta_q} \ell_q \longrightarrow L_p(\mu_q) \otimes_{\Delta_p} \ell_q$$

is not continuous by proposition 2 of 26.3., one obtains that

$$\ell_{q'} \not\subseteq \text{space}(\mathfrak{M}_{q,p}) \qquad \text{if } 1 \le p < q < 2 .$$

32.9. The Banach spaces in space($\mathfrak{M}_{2,1}$) have an interesting characterization in terms of the coincidence of tensor norms.

COROLLARY 3: *A Banach space E is in* space($\mathfrak{M}_{2,1}$) *if and only if for one (and then for all) $p \in [1, 2[$*

$$E \otimes_\varepsilon \ell_p = E \otimes_{w_2} \ell_p$$

holds isomorphically.

In Ex 32.8. this result will be interpreted in terms of operator ideals.

PROOF: $E \in$ space($\mathfrak{M}_{2,1}$) implies, by Saphar's characterization (20.19. or 32.2.(3)), that

$$\mathcal{L}(\ell_\infty, E) = \mathfrak{P}_2(\ell_\infty, E) = \mathcal{L}_2(\ell_\infty, E) .$$

Since $\mathcal{L}_2 \sim w_2$ is accessible, this yields that

$$\ell_1 \otimes_\varepsilon E = \ell_1 \otimes_{w_2} E$$

holds isomorphically. It follows that this is true for every $L_1(\mu) \otimes E$; using a Lévy embedding $\ell_p \hookrightarrow L_1(\mu)$ (see 24.5.) and the injectivity of ε and w_2, gives that $\varepsilon \sim w_2$ on $\ell_p \otimes E$ for all $1 \le p \le 2$.

Conversely: If $E \otimes_\varepsilon \ell_p = E \otimes_{w_2} \ell_p$ for one $p \in [1, 2[$, then

$$E' \otimes_\pi \ell'_p = E' \otimes_{w'_2} \ell'_p \hookrightarrow (E \otimes_\varepsilon \ell_p)' = (E \otimes_{w_2} \ell_p)'$$

and hence

$$\mathcal{L}(E', \ell_p) = \mathcal{L}_2(E', \ell_p) \subset \mathcal{J}^{inj \ sur}(E', \ell_p) \subset \mathcal{L}^{inj \ sur}_{p,2}(E', \ell_p)$$

by Grothendieck's inequality, operator form. Theorem 28.4. implies that

$$S \otimes T : \ell_2 \otimes_{\Delta_2} E' \longrightarrow \ell_p \otimes_{\Delta_p} \ell_p$$

is continuous for all S and T: But now corollary 2 implies that $E'' \in$ space($\mathfrak{M}_{2,p}$), and hence $E \in$ space($\mathfrak{M}_{2,p}$). Since $p < 2$, the Maurey–Pisier extrapolation theorem gives the result. □

Since ℓ_p has cotype 2 for $1 \le p \le 2$, Pisier's factorization theorem 31.4. gives that $\varepsilon \sim w_2$ on $\ell_p \otimes E$ if E has cotype 2 — and hence, by the corollary, E is in space($\mathfrak{M}_{2,1}$); this is Maurey's result 31.7. (see also 32.2.(4)).

This corollary can be used to give, as in corollary 2, a characterization of those E such that for all S and T
$$S \otimes T : L_q \otimes_{\Delta_q} E \longrightarrow L_p \otimes_{\Delta_p} L_p$$
is continuous — this time for $p < 2 \leq q$.

COROLLARY 4: *Let $1 \leq p < 2 \leq q \leq \infty$. Then the dual E' of a Banach space is in* space$(\mathfrak{M}_{2,1})$ *if and only if for all $S \in \mathfrak{L}(L_q(\nu_1), L_p(\nu_2))$ and $T \in \mathfrak{L}(E, L_p(\mu))$ the operator*
$$S \otimes T : L_q(\nu_1) \otimes_{\Delta_q} E \longrightarrow L_p(\nu_2) \otimes_{\Delta_p} L_p(\mu)$$
is continuous.

PROOF: By 28.4. and 28.5. the condition is equivalent to the fact that
$$\mathfrak{L}(E, \ell_p) \subset \mathfrak{L}_{p,q'}^{inj,\,sur}(E, \ell_p) = \mathfrak{L}_2(E, \ell_p) .$$
Simple transfer arguments show that this is equivalent to $\varepsilon \sim w_2$ on $E' \otimes \ell_p$, hence corollary 3 gives the result. □

32.10. Recall from Ex 8.12. and Ex 11.20. – Ex 11.22. the ideal of absolutely (r,s)–summing operators $(1 \leq s \leq r \leq \infty)$ — in particular, the inequality
$$\ell_r(Tx_k; F) \leq \mathbf{P}_{r,s}(T) w_s(x_k; E) .$$
The splitting theorem immediately implies that:

REMARK 1: $\mathfrak{M}_{q,p} \subset \mathfrak{P}_{r,p}$ *and* $\mathbf{P}_{r,p} \leq \mathbf{M}_{q,p}$ *if* $1/q + 1/r = 1/p$.

Note the special case $\mathfrak{M}_{2,1} \subset \mathfrak{P}_{2,1}$ — and recall that $E \in$ space$(\mathfrak{P}_{2,1})$ means that E has the Orlicz property (see 8.9.).

REMARK 2: *For $1 \leq r < \infty$ and $1/p - 1/r < 1/2$ the space ideal of $\mathfrak{P}_{r,p}$ (and hence for $q > \max\{2,p\}$ the space ideal of $\mathfrak{M}_{q,p}$) consists of finite dimensional spaces only.*

PROOF: If $E \in$ space$(\mathfrak{P}_{r,p})$ is infinite dimensional, it contains all ℓ_2^n uniformly by Dvoretzky's spherical sections theorem (see [61]). Since $\mathfrak{P}_{r,p}$ is injective, this implies that
$$\mathbf{P}_{r,p}(id_{\ell_2^n}) \leq c$$
for some c independent of n. But
$$n^{1/r} = \left(\sum_{k=1}^n \|e_k\|_{\ell_2^n}^r \right)^{1/r} \leq c w_p(e_k; \ell_2^n) =$$
$$= c \|id : \ell_2^n \to \ell_p^n\| = c \cdot \begin{cases} 1 & \text{if } p \geq 2 \\ n^{1/p - 1/2} & \text{if } p < 2 , \end{cases}$$

which is a contradiction. □

32.11. The inclusion $\mathfrak{M}_{q,p} \subset \mathfrak{P}_{r,p}$ (if $1/q+1/r = 1/p$) is in general strict: For $1 < q < 2$ it is known from 32.8. that $\ell_{q'} \notin \text{space}(\mathfrak{M}_{q,1})$, but $\ell_{q'} \in \text{space}(\mathfrak{P}_{q',1})$ according to Ex 11.22. (or remark 24.7.). The inclusion $\mathfrak{M}_{q,1} \subset \mathfrak{P}_{q',1}$ is even strict for all $1 < q < \infty$; see 32.12..

Talagrand's Banach lattice (mentioned in 8.9.) which has the Orlicz property (hence is in space($\mathfrak{P}_{2,1}$)) but does not have cotype 2, shows that space($\mathfrak{M}_{2,1}$) $\neq$ space($\mathfrak{P}_{2,1}$) since a Banach lattice is of cotype 2 if and only if it is in space($\mathfrak{M}_{2,1}$) (see 32.2.(4)).

Nevertheless, the ideals $\mathfrak{M}_{q,p}$ and $\mathfrak{P}_{r,p}$ almost coincide: This result is contained implicitly in Maurey's thesis [186] and was stated and proved explicitly by Puhl [226]:

THEOREM: $\mathfrak{P}_{r,p} \subset \mathfrak{M}_{q-\varepsilon,p}$ if $1/q + 1/r = 1/p$ and $\varepsilon > 0$.

PROOF: Recall first from 32.1. that

$$\mathfrak{M}_{s,p} = \mathfrak{I}_{s'} \circ \mathfrak{I}_{p'}^{-1}$$

and from proposition 24.7.

$$\mathfrak{P}_{q',1} \circ \mathfrak{L}_\infty \subset \mathfrak{P}_{q'+\eta}$$

(since $\mathfrak{P}_{q',1} \subset \mathfrak{P}_{r_1,s_1}$ if $1 < s_1 < q' < r_1 < q' + \eta$ with appropriate r_1, s_1 — see Ex 11.21.).

Take $T \in \mathfrak{P}_{r,p}(E,F)$ and $S \in \mathfrak{I}_{p'}(G,E)$. By the factorization theorem 18.7. for p'-integral operators and the fact that $U \in \mathfrak{A}$ if and only if $U'' \in \mathfrak{A}$ for maximal normed ideals one may assume that $G = C(K)$, which is an $\mathfrak{L}_\infty^g$-space. It follows that

$$T \circ S \in \mathfrak{P}_{r,p} \circ \mathfrak{I}_{p'} \circ \mathfrak{L}_\infty \subset \mathfrak{P}_{r,p} \circ \mathfrak{P}_{p'} \circ \mathfrak{L}_\infty \subset \mathfrak{P}_{r,p} \circ \mathfrak{M}_{p,1} \circ \mathfrak{L}_\infty$$

and hence by Ex 32.2.(c)

$$\subset \mathfrak{P}_{q',1} \circ \mathfrak{L}_\infty \subset \mathfrak{P}_{q'+\eta} \circ \mathfrak{L}_\infty \subset \mathfrak{I}_{q'+\eta} \ .$$

For appropriately chosen η this gives the result. □

32.12. The "border case" $\varepsilon = 0$ in the last statement was investigated by Carl–D. [21] : If $T \in \mathfrak{L}(E,F)$ has rank n, then

$$\mathbf{M}_{q,p}(T) \leq 5(1 + \log n)^{1/q} \mathbf{P}_{r,p}(T)$$

where $1/q + 1/r = 1/p$. For $p = 1$ and $1 < q < \infty$ this "growth" is sharp in the following sense: There exist a constant $c_q > 0$ and operators $T_n : \ell_\infty^n \longrightarrow \ell_\infty^n$ with

$$c_q(1 + \log n)^{1/q} \leq \mathbf{M}_{q,1}(T_n) \text{ and } \mathbf{P}_{q',1}(T_n) = 1 \ .$$

(For $(q,p) = (2,1)$ these results were first proven by Jameson [120]). In particular, it follows that
$$\mathfrak{M}_{q,1} \underset{\neq}{\subset} \mathfrak{P}_{q',1}$$
for $1 < q < \infty$. For $q < 2$ this result was already obtained in 32.11. — but with totally different methods (32.8. and 26.3.).

Exercises:

Ex 32.1. Show that the operator ideal $\mathfrak{M}_{q,p}$ is injective but in general not surjective.
Hint: Splitting theorem, ℓ_1.

Ex 32.2. Show that
(a) $\mathfrak{M}_{q_1,p_1} \subset \mathfrak{M}_{q_2,p_2}$ if $p_1 \leq p_2$ and $q_2 \leq q_1$.
(b) $\mathfrak{M}_{q,s} \circ \mathfrak{M}_{s,p} \subset \mathfrak{M}_{q,p}$ if $p,q,s \in [1,\infty]$.
(c) $\mathfrak{P}_{p,s} \circ \mathfrak{M}_{s,r} \subset \mathfrak{P}_{q,r}$ if $r,s,p,q \in [1,\infty]$ such that $1/s + 1/q = 1/p + 1/r$.
Hint: Maurey's splitting theorem for $s > r$ and Ex 11.21. for $s \leq r$. In all three statements there are no additional constants in the norm estimate.

Ex 32.3. If E is an $\mathcal{L}^g_\infty$-space, then $\mathfrak{P}_{q'}(E,F) = \mathfrak{M}_{q,1}(E,F)$ for every Banach space F. Hint: $\mathfrak{I}_\infty = \mathcal{L}_\infty$.

Ex 32.4. Show that
$$\mathbf{M}_{2,1}(id_{\ell_2}) = K_{LG} \quad \text{and} \quad \mathbf{M}_{2,1}(id_{\ell_p}) \leq K_G$$
if $1 \leq p \leq 2$. Hint: 32.2.(3), 11.11. and 23.10..

Ex 32.5. Show for $1 \leq p \leq q \leq \infty$:
$$\mathfrak{M}^{dual}_{q,p} = \mathfrak{P}^{dual}_p \circ (\mathfrak{P}^{dual}_q)^{-1} = (\mathfrak{I}^{dual}_{p'})^{-1} \circ \mathfrak{I}^{dual}_{q'} =$$
$$= \mathfrak{P}^{dual}_p \circ (\mathfrak{I}^{dual}_q)^{-1} = (\mathfrak{I}^{dual}_{p'})^{-1} \circ \mathfrak{P}^{dual}_{q'} \subset$$
$$\subset \mathcal{L}_p^{-1} \circ \mathcal{L}_{p,q'} = \mathfrak{D}_{p'} \circ \mathfrak{D}^{-1}_{p',q} .$$

Hint: Quotient formula, Ex 29.1., 32.1., 32.6. and 18.10..

Ex 32.6. Show that $T \in \mathcal{L}(E, F')$ is (q,p)-mixing (for $1 \leq p \leq q < \infty$) if and only if there is a $c \geq 0$ such that
$$\Big(\sum_{j=1}^m \Big(\sum_{k=1}^n |\langle y_k, Tx_j\rangle|^q\Big)^{p/q}\Big)^{1/p} \leq c w_p(x_j; E) \ell_q(y_k; F) .$$

In this case, $\mathbf{M}_{q,p}(T) = \min c$.

Ex 32.7. Use 26.4.(3') and the Maurey–Pisier extrapolation theorem to prove Maurey's result from 31.7.: $E \in \text{space}(\mathfrak{M}_{2,1})$ if E has cotype 2. Note that this again shows that $\ell_1 \in \text{space}(\mathfrak{M}_{2,1})$ which, together with the little Grothendieck theorem, implies $\mathfrak{L}(\ell_1, \ell_2) = \mathfrak{P}_1(\ell_1, \ell_2)$. This is another proof of Grothendieck's inequality.

Ex 32.8. Show that the following statements are equivalent for a Banach space E:

(a) $E \in \text{space}(\mathfrak{M}_{2,1})$.

(b) For one (and then for all) $q \in \,]2, \infty]$
$$\mathfrak{D}_2(E, \ell_q) = \mathfrak{I}(E, \ell_q).$$

(c) For one (and then for all) $p \in [1, 2[$: Every compact operator $E' \to \ell_p$ is 2-factorable. Hint: 32.9. and representation theorems for ideals.

Ex 32.9. (a) If E is a Banach space such that
$$\mathfrak{L}(E, \ell_2) = \mathfrak{P}_p(E, \ell_2)$$
for some $1 \leq p < 2$, then $\mathfrak{L}(E, \ell_2) = \mathfrak{P}_1(E, \ell_2)$, i.e.: E is a G.T. space (Grothendieck theorem space) in the sense of 20.18.. Hint: $E \in \text{space}(\mathfrak{M}_{2,p})$.

(b) G.T. spaces are in $\text{space}(\mathfrak{M}_{2,1})$, in particular, they have the Orlicz property.

Ex 32.10. Let $1 \leq p < q \leq \infty$ and $q > 2$. If a Banach space E is such that for all $T \in \mathfrak{L}(E, \ell_p)$ the operator
$$T \otimes id_{\ell_q} : E \otimes_{\Delta_q^t} \ell_q \longrightarrow \ell_p \otimes_{\Delta_p} \ell_q$$
is continuous, then it is finite dimensional. Hint: 32.6. and 32.10..

Ex 32.11. $\text{space}(\mathfrak{M}_{q,p}) = \text{FIN}$ if and only if $q > \max\{2, p\}$.

Ex 32.12. Show that $\mathfrak{M}_{q,p}(\ell_1, F) = \mathfrak{M}_{q,1}(\ell_1, F)$ if $p \leq 2$.

Ex 32.13. (a) Show that $T \in \mathfrak{L}(E, F)$ is (q,p)-mixing if and only if
$$\Phi_T : \mathfrak{P}_q(F, \ell_\infty) \longrightarrow \mathfrak{P}_p(E, \ell_\infty) \qquad \Phi_T(S) := S \circ T$$
is defined (and hence continuous). In this case,
$$\|\Phi_T\| = \mathbf{M}_{q,p}(T).$$

(b) Show that in (a) the space ℓ_∞ can be replaced by $L_\infty[0,1]$.

Ex 32.14. Use Ex 11.23. to show that
$$\mathfrak{M}_{q,r} \not\subset \bigcup_{1 \leq s < r} \mathfrak{M}_{q,s}$$
whenever $1 < r \leq q < \infty$.

33. The Radon–Nikodým Property for Tensor Norms and Reflexivity

The Radon–Nikodým property led to an understanding of the full duality of the projective and injective tensor norms, namely, describing the conditions under which $E' \tilde{\otimes}_\pi F' = (E \otimes_\varepsilon F)'$ holds. Noticing the central role of the sequence space ℓ_1 for this question, Lewis [169] succeeded in obtaining many results of the form $E' \tilde{\otimes}_\alpha F' = (E \otimes_{\alpha'} F)'$ or, in other words, results about $\mathfrak{A}^{min}(E, F') = \mathfrak{A}(E, F')$ if $\mathfrak{A}$ is the maximal normed operator ideal associated with α. This section presents Lewis' ideas including applications to the reflexivity of tensor products and operator ideals. The main results are due to Lewis, but the proofs are sometimes taken from Meile [189].

33.1. Let α be a (finitely generated) tensor norm and $(\mathfrak{A}, \mathbf{A})$ its associated maximal normed operator ideal. The objective of this section is to find conditions under which the natural map

$$J : E' \tilde{\otimes}_\alpha F' \xrightarrow{1} \mathfrak{A}^{min}(E, F') \hookrightarrow \mathfrak{A}(E, F') \stackrel{1}{=} (E \otimes_{\alpha'} F)'$$

is a metric surjection (the reader may already be accustomed to recognize the use of the representation theorem for minimal and maximal ideals). Note that J is injective if E' or F' has the approximation property (see 17.20.) or if F' has the α-approximation property (see section 21) or if α is totally accessible.

For $\alpha = \pi$ this question was investigated in section 16: It turned out that in order for

$$J : E' \tilde{\otimes}_\pi F' \xrightarrow{1} \mathfrak{N}(E, F') \hookrightarrow \mathfrak{I}(E, F') \stackrel{1}{=} (E \otimes_\varepsilon F)'$$

to be a metric surjection a sufficient condition is that F' (or E') has the Radon–Nikodým property (16.5.). An alternative proof to the one given in section 16 (or Appendix D7.) uses two ingredients:

(a) The Lewis–Stegall theorem: *If the Banach space G has the Radon–Nikodým property, then every $T \in \mathfrak{L}(L_1(\mu), G)$ factors through $\ell_1(B_G)$ (or through ℓ_1 if μ is finite), and $\|T\| = \inf \|R\| \|S\|$, where the infimum is taken over all these factorizations* (Appendix D2., proposition 2; one may take R the natural quotient map $\ell_1(B_G) \twoheadrightarrow G$).

(b) $\mathfrak{I}(E, \ell_1) = (E \otimes_\varepsilon c_o)' = (c_o(E))' = \ell_1(E') = \ell_1 \tilde{\otimes}_\pi E' = \mathfrak{N}(E, \ell_1)$ isometrically.

Now take $T \in \mathfrak{I}(E, F')$ with F' having the Radon–Nikodým property. The factorization theorem 10.5. for integral operators and (a) give

$$E \xrightarrow{T} F'$$

$$R \downarrow \quad S \uparrow \nearrow^{S_2} \ell_1$$

$$C \xrightarrow{I} L_1(\mu)$$

$\|R\| = \|S\| = 1$
$\|S_1\| \, \|S_2\| \leq 1 + \varepsilon$
$\mathbf{I}(T) = \|I\|$,

and it follows from (b) that T factors through

$$T_o = S_1 I R \in \mathfrak{I}(E, \ell_1) \stackrel{1}{=} \mathfrak{N}(E, \ell_1) ,$$

which implies that $T \in \mathfrak{N}(E, F')$ and $\mathbf{N}(T) \leq \mathbf{I}(T)(1 + \varepsilon)$.

33.2. This simply structured proof leads to the following definition, due to Lewis:

DEFINITION: *A finitely generated tensor norm α has the Radon–Nikodým property if*

$$E' \tilde{\otimes}_\alpha \ell_1 = (E \otimes_{\alpha'} c_o)'$$

holds isometrically for all Banach spaces E.

It was just shown that π has the Radon–Nikodým property — but ε does not: This follows from

$$E' \tilde{\otimes}_\varepsilon \ell_1 \stackrel{1}{=} \mathfrak{K}(E, \ell_1) \hookrightarrow \mathfrak{L}(E, \ell_1) \stackrel{1}{=} (E \otimes_\pi c_o)' .$$

The fact that ℓ_1 has the metric approximation property implies that the mapping $J : E' \tilde{\otimes}_\alpha \ell_1 \to (E \otimes_{\alpha'} c_o)'$ is a metric injection whenever α is left–accessible. Since the usual tensor norms are accessible, the important part of the definition is the surjectivity of J — in other words, it will be important to know when

$$\mathfrak{A}^{min}(E, \ell_1) = \mathfrak{A}(E, \ell_1)$$

holds (where $\mathfrak{A}$ is the associated operator ideal). Since

$$E' \otimes_{\alpha/} \ell_1 \stackrel{1}{=} E' \otimes_\alpha \ell_1 \quad \text{and} \quad E' \otimes_{\alpha'\backslash} c_o \stackrel{1}{=} E' \otimes_{\alpha'} c_o$$

(see 20.8.), it follows that

PROPOSITION: *α has the Radon–Nikodým property if and only if $\alpha/$ has it.*

In particular, $\varepsilon/ = d_\infty = w_1 = \alpha_{1,\infty}$ does not have the Radon–Nikodým property, but it will soon be seen that all the other $\alpha_{p,q}$ do!

33.3. The following is simple:

LEMMA: *If $(\mathfrak{A}, \mathbf{A})$ is a quasinormed operator ideal and E a Banach space such that $\mathfrak{A}^{min}(E, \ell_1) = \mathfrak{A}(E, \ell_1)$ holds isometrically, then*

$$\mathfrak{A}^{min}(E, \ell_1(\Gamma)) = \mathfrak{A}(E, \ell_1(\Gamma))$$

holds isometrically for all sets Γ.

PROOF: Since ℓ_1^n is 1-complemented in ℓ_1, the finite case is obvious. If Γ is infinite and $T \in \mathfrak{A}(E, \ell_1(\Gamma))$, then for each bounded sequence (x_n) in E the sequence (Tx_n) is contained in some 1-complemented subspace $\ell_1(\Gamma_o)$ with countable $\Gamma_o \subset \Gamma$. It follows from the assumption that (Tx_n) has a convergent subsequence, hence T is compact. Therefore, T has separable range, which — again — is in some 1-complemented $\ell_1(\Gamma_o)$, as before. This implies that $T \in \mathfrak{A}^{min}(E, \ell_1(\Gamma))$ (with the same norm). □

This observation and the fact that operators in $\mathfrak{A}/$ factor through some $L_1(\mu)$ (see 20.12.) give a result which can be seen as a justification of Lewis' definition:

THEOREM (Lewis): *Let α be a tensor norm with the Radon–Nikodým property and $(\mathfrak{A}, \mathbf{A})$ its associated maximal normed operator ideal. If F' has the Radon–Nikodým property, then*

$$E' \tilde{\otimes}_{\alpha/} F' \twoheadrightarrow (E \otimes_{(\alpha/)'} F)'$$

is a metric surjection and

$$(\mathfrak{A}/)^{min}(E, F') = \mathfrak{A}/(E, F')$$

holds isometrically for all Banach spaces E.

PROOF: The definition of $T \in \mathfrak{A}/(E, F')$ and the Lewis–Stegall factorization theorem give

$$\begin{array}{ccc} E & \xrightarrow{T} & F' \\ R \downarrow & \nearrow S & \uparrow Q \\ L_1(\mu) & \xrightarrow{S_o} & \ell_1(B_{F'}) \end{array}$$

$\mathbf{A}/(T) = \mathbf{A}(R)\|S\|$

$\|S_o\| \le \|S\|(1+\varepsilon)$.

Since clearly, $\mathbf{A}/(R) = \mathbf{A}(R)$ and since $\alpha/$ has the Radon–Nikodým property (by proposition 33.2.), it follows from the lemma that

$$S_o \circ R \in \mathfrak{A}/(E, \ell_1(B_{F'})) \stackrel{1}{=} (\mathfrak{A}/)^{min}(E, \ell_1(B_{F'}))$$

and hence $T \in (\mathfrak{A}/)^{min}(E, F')$ and $(\mathbf{A}/)^{min}(T) \le \mathbf{A}/(S_o \circ R)\|Q\| \le \mathbf{A}/(T)(1+\varepsilon)$. □

This result does not hold when $\alpha/$ is replaced by α: It will be seen in a moment that $\alpha := w_2 \sim \mathfrak{L}_2$ has the Radon–Nikodým property, but

$$\ell_2 \tilde{\otimes}_{w_2} \ell_2 = \mathfrak{L}_2^{min}(\ell_2, \ell_2) \subset \mathfrak{K}(\ell_2, \ell_2) \ne \mathfrak{L}(\ell_2, \ell_2) = (\ell_2 \otimes_{w_2'} \ell_2)' .$$

33.4. It is interesting to note that whenever $E'\tilde{\otimes}_\beta F' \twoheadrightarrow (E \otimes_{\beta'} F)'$ holds for some tensor norm β, then this holds also for β replaced by some injective or projective associates of β. This will be an easy consequence of the

LEMMA: *Let β be a finitely generated tensor norm, E and F Banach spaces.*
(1) *If $\ell_\infty(B_{E'})'\tilde{\otimes}_\beta F' \twoheadrightarrow (\ell_\infty(B_{E'}) \otimes_{\beta'} F)'$ is a metric surjection, then*

$$E'\tilde{\otimes}_{\backslash\beta} F' \twoheadrightarrow (E \otimes_{(\backslash\beta)'} F)'$$

is a metric surjection as well.
(2) *If $\ell_\infty(B_E)\tilde{\otimes}_\beta F' \twoheadrightarrow (\ell_1(B_E) \otimes_{\beta'} F)'$ is a metric surjection and F' has the approximation property or: β is right-accessible and E' has the approximation property, then*

$$E'\tilde{\otimes}_{/\beta} F' \twoheadrightarrow (E \otimes_{(/\beta)'} F)'$$

is a metric surjection.

PROOF: With $I: E \hookrightarrow \ell_\infty(B_{E'})$, part (1) follows from the diagram

$$\begin{array}{ccc}
\ell_\infty(B_{E'})'\tilde{\otimes}_\beta F' & \xrightarrow{1} & (\ell_\infty(B_{E'}) \otimes_{\beta'} F)' \\
{\scriptstyle I'\tilde{\otimes}id_{F'}}\downarrow\ {\scriptstyle 1} & & {\scriptstyle 1}\uparrow {\scriptstyle (I\otimes id_F)'} \\
E'\tilde{\otimes}_{\backslash\beta} F' & \longrightarrow & (E \otimes_{/\beta'} F)'
\end{array}$$

To see (2) take $\mathfrak{B} \sim \beta$ and observe that $\mathfrak{B}^{sur} \sim /\beta$. The assumption implies immediately that

$$(\mathfrak{B}^{min})^{sur}(E, F') \stackrel{1}{=} \mathfrak{B}^{sur}(E, F') \ .$$

Proposition 25.11.(1) implies that $(\mathfrak{B}^{min})^{sur}(E, F') \stackrel{1}{=} (\mathfrak{B}^{sur})^{min}(E, F')$ under the conditions stated, which means that

$$E'\tilde{\otimes}_{/\beta} F' \stackrel{1}{\twoheadrightarrow} (\mathfrak{B}^{sur})^{min}(E, F') \stackrel{1}{=} \mathfrak{B}^{sur}(E, F') = (E \otimes_{(/\beta)'} F)' \ . \square$$

Taking $F := c_0$, and recalling that $\backslash\alpha$ and $/\alpha$ are always left-accessible, this shows the

COROLLARY 1: *If α has the Radon-Nikodým property, then $\backslash\alpha$ and $/\alpha$ also have it.*

Together with proposition 33.2. this implies that along with α

$$\alpha/, \backslash\alpha, /\alpha \quad \text{and} \quad \backslash\alpha/\ , \ /(\alpha/) \ , \ (\alpha/)/$$

(and so on) also have the Radon-Nikodým property. But note that $\alpha\backslash$ need not have it: This follows from $\varepsilon = w_\infty\backslash$ (see 20.14.) and from 33.5. below.

Combining this with the lemma and theorem 33.3., leads to another result of Lewis':

COROLLARY 2: *If the tensor norm α has the Radon-Nikodým property, then so do $\beta = \alpha/, \backslash\alpha/, /(\alpha/)$ and $(/\alpha)/$. If E and F are Banach spaces (E' or F' with the approximation property if $\beta = /(\alpha/)$) such that F' has the Radon-Nikodým property, then*
$$E' \tilde{\otimes}_\beta F' \twoheadrightarrow (E \otimes_{\beta'} F)'$$
is a metric surjection and $\mathfrak{B}^{min}(E, F') = \mathfrak{B}(E, F')$ holds isometrically for the maximal operator ideal $(\mathfrak{B}, \mathbf{B})$ associated with β.

33.5. Lapresté's tensor norms and their adjoints (this means: the ideals $\mathfrak{L}_{p,q}$ of (p,q)-factorable and $\mathfrak{D}_{p,q}$ of (p,q)-dominated operators) nearly all have the Radon-Nikodým property. The result is due to Lewis:

THEOREM: *Let $p, q \in [1, \infty]$ such that $1/p + 1/q \geq 1$. The following tensor norms have the Radon-Nikodým property:*
(1) $\alpha_{p,q}$ *for* $(p,q) \neq (1, \infty)$.
(2) $\alpha_{p,q}\backslash$ *for* $1 < p < \infty$.
(3) $\alpha_{p,q}^*$ *for* $(p,q) \neq (\infty, 1)$ *and* $(p,q) \neq (1,1)$.

The tensor norms $\alpha_{1,1}^* = \varepsilon$ and $\alpha_{1,\infty} = w_1 = \varepsilon/$ do not have the Radon-Nikodým property. Unfortunately, we do not know whether $\alpha_{\infty,1}^* = g_\infty^*$ has it or not; since $g_\infty^* \sim \mathfrak{P}_1 = \mathfrak{J}^{inj}$, this is the same as asking whether or not

$$\mathfrak{P}_1(E, \ell_1) \stackrel{?}{=} \mathfrak{P}_1^{min}(E, \ell_1) = (\mathfrak{J}^{inj})^{min}(E, \ell_1) = \mathfrak{N}^{inj}(E, \ell_1)$$

(by proposition 25.11.(2)) holds: Is every absolutely summing operator $E \to \ell_1$ quasi-nuclear (see Ex 9.13.)?

PROOF: Since all these tensor norms (and their associated ideals $\mathfrak{A}$) are accessible, it remains only to show that
$$\overline{\mathfrak{F}} \circ \mathfrak{A}(E, \ell_1) = \mathfrak{A}(E, \ell_1)$$
(see the initial remarks in 33.2. and 25.2.).
(1) Each operator $T \in \mathfrak{L}_{p,q}(E, \ell_1)$ factors as follows

$$\begin{array}{ccc} E & \xrightarrow{T} & \ell_1 \\ R \downarrow & & \uparrow S \\ L_{q'}(\mu) & \xrightarrow{I} & L_p(\mu) \end{array} \qquad \|S\| = 1 .$$

Thus, by the very properties of ℓ_1, it is enough to show that S is weakly compact (=compact=approximable): For $1 < p < \infty$ this follows from the reflexivity of L_p, for $p = \infty$ this is a consequence of the fact that every operator $L_\infty \to \ell_1$ factors through a Hilbert space (e.g. by Grothendieck's inequality, see 17.14.). For $p = 1$ this is not true — a more complicated argument is necessary.

So assume that $p = 1$ and $1 < q < \infty$ (the case $\alpha_{1,1} = \pi$ is already clear). It is enough to show that

$$L_{q'}(\mu) \xhookrightarrow{I} L_1(\mu) \xrightarrow{S} \ell_1$$

(μ a finite measure) is in $\mathfrak{L}_{1,q}^{min}$. Since ℓ_1 has the Radon–Nikodým property, the operator S has a Riesz–density $g = (g_n)$ with $g_n \in L_1(\Omega, \mu)$ and

$$h(w) := \sum_{n=1}^{\infty} |g_n(w)| = \|g(w)\|_{\ell_1} \leq \|S\| = 1$$

for all $w \in \Omega$. The sequence $\lambda = (\lambda_n)$ defined by

$$\lambda_n := \left(\int_\Omega |g_n| h^{q/q'} d\mu \right)^{1/q}$$

is in ℓ_q. It suffices to prove that the operator $S \circ I$ factors through the diagonal operator $D_\lambda : \ell_{q'} \to \ell_1$ and that

$$D_\lambda = \sum_{n=1}^{\infty} e_n \otimes \lambda_n e_n \overset{!}{\in} \mathfrak{L}_{1,q}^{min}(\ell_{q'}, \ell_1) \overset{!}{=} \ell_q \tilde{\otimes}_{d_q} \ell_1 \ .$$

The latter follows from

$$d_q\left(\sum_{n=1}^{\infty} e_n \otimes \lambda_n e_n; \ell_q, \ell_1 \right) \leq w_{q'}(e_n; \ell_q) \ell_q(\lambda_n e_n; \ell_1) = 1 \cdot \|\lambda\|_q \ .$$

For the factorization define $U : \ell_q \to L_q(\mu)$ by $U e_n := \lambda_n^{-1} g_n$ — and this definition makes sense since for $(\alpha_n) \in \ell_q$

$$\int \left| \sum_{n=1}^{\infty} \alpha_n \lambda_n^{-1} g_n \right|^q d\mu \leq \int \left(\sum_{n=1}^{\infty} |\alpha_n| \lambda_n^{-1} |g_n|^{1/q} |g_n|^{1/q'} \right)^q d\mu \leq$$

$$\leq \int \sum_{n=1}^{\infty} |\alpha_n|^q \lambda_n^{-q} |g_n| h^{q/q'} d\mu = \sum_{n=1}^{\infty} |\alpha_n|^q \ .$$

It is easy to verify that $S \circ I = D_\lambda \circ U'$ — and this ends the proof of (1).
(2) For $1 < p < \infty$ take $T \in \mathfrak{L}_{p,q}^{inj}(E, \ell_1)$ and factor

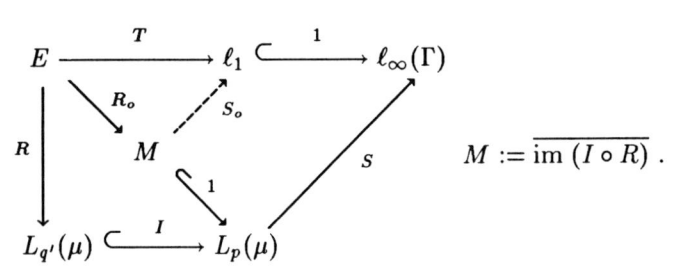

$$
\begin{CD}
E @>T>> \ell_1 @>1>> \ell_\infty(\Gamma) \\
@VRVV \nearrow_{R_o} \nearrow_{S_o} \nearrow_S \\
& M & \\
L_{q'}(\mu) @>I>> L_p(\mu)
\end{CD}
\qquad M := \overline{\operatorname{im}(I \circ R)}.
$$

Then it is obvious that $R_o \in \mathcal{L}_{p,q}^{inj}$ and $S_o \in \overline{\mathfrak{F}}(M, \ell_1)$ since M is reflexive. It follows that

$$T = S_o \circ R_o \in \overline{\mathfrak{F}} \circ \mathcal{L}_{p,q}^{inj}(E, \ell_1) = (\mathcal{L}_{p,q}^{inj})^{min}(E, \ell_1)$$

by the accessibility of $\mathcal{L}_{p,q}^{inj} \sim \alpha_{p,q}\backslash$.

(3) First recall that $\alpha_{p,q}^* \sim \mathfrak{D}_{p',q'} = \mathfrak{P}_{q'}^{dual} \circ \mathfrak{P}_{p'}$ by Kwapień's factorization theorem 19.3.. Since $d_q^*/ = d_{q'}$ (by 20.14.), it follows from part (1) of the proof that for $1 < q \leq \infty$ the tensor norm $d_q^*/$ and hence d_q^* has the Radon–Nikodým property (33.2.). Since $d_q^* \sim \mathfrak{P}_{q'}^{dual}$, this means that

$$\mathfrak{P}_{q'}^{dual}(E, \ell_1) \stackrel{1}{=} (\mathfrak{P}_{q'}^{dual})^{min}(E, \ell_1)$$

— and Ex 25.15.(a) (about $(\mathfrak{B} \circ \mathfrak{A})^{min}(E, G) = \mathfrak{B} \circ \mathfrak{A}(E, G)$) gives the result.

In the remaining case $q = 1$ one has $\alpha_{p,1}^* = g_p^* = g_{p'}\backslash$, and therefore $\alpha_{p,1}^*$ has the Radon–Nikodým property by (2) whenever $1 < p < \infty$. $\square$

33.6. The fact that the (right–projective) tensor norm d_p has the Radon–Nikodým property (for $1 \leq p < \infty$) implies that

$$F' \tilde{\otimes}_{d_p} E' \twoheadrightarrow (F \otimes_{d_p'} E)'$$

is a metric surjection if E' has the Radon–Nikodým property (theorem 33.3.). Since $d_p^t = g_p \sim \mathfrak{I}_p$ (the ideal of p–integral operators) and $\mathfrak{N}_p = \mathfrak{I}_p^{min}$ (the ideal of p–nuclear operators), it follows that

COROLLARY 1: *Let $1 \leq p < \infty$ and E, F Banach spaces. If E' has the Radon–Nikodým property, then*

$$\mathfrak{I}_p(E, F') = \mathfrak{N}_p(E, F')$$

holds isometrically.

For $p = 1$ this was shown in Appendix D. Recall from 21.10. the Pietsch p–integral operators, which factor as follows ($1 \leq p < \infty$) :

$$\begin{array}{ccc} E & \xrightarrow{T} & F \\ R \downarrow & & \uparrow S \\ C(K) & \xrightarrow{I} & L_p(\mu) \end{array} \qquad \mathbf{PI}_p(T) = \inf \|R\| \, \|I\| \, \|S\| \; .$$

Since L_p is 1–complemented in L_p'' for $1 \leq p < \infty$, it follows from corollary 1 that

$$\mathbf{PI}_p(I \circ R) = \mathbf{I}_p(I \circ R) = \mathbf{N}_p(I \circ R)$$

whenever E' has the Radon–Nikodým property, and hence,

COROLLARY 2: *For $1 \leq p < \infty$ and E' with the Radon–Nikodým property*

$$\mathfrak{PI}_p(E,F) = \mathfrak{N}_p(E,F)$$

holds isometrically.

33.7. One application of the coincidence of p–nuclear and p–integral operators: In Ex 28.1. an operator $T \in \mathfrak{L}(E,F)$ was called *p–strong* (for $1 \leq p < \infty$) if there is a constant $c \geq 0$ such that

$$\sum_{k=1}^{m} \|Tx_k\|^p \leq c^p \sum_{k=1}^{m} \|y_k\|^p \; ,$$

whenever $x_1, ..., x_m, y_1, ..., y_m \in E$ are such that

$$\sum_{k=1}^{m} |\langle x', x_k \rangle|^p \leq \sum_{k=1}^{m} |\langle x', y_k \rangle|^p$$

for all $x' \in E'$. It was shown that every operator $T \in \mathfrak{L}_p^{inj}$ is p–strong — and that all p–strong operators are in $\mathfrak{L}_p^{inj \; sur}$. Even more is true:

THEOREM: *An operator is in $\mathfrak{L}_p^{inj}$ if and only if it is p–strong ($1 \leq p < \infty$).*

For $T = id_E$ this is a result of Lindenstrauss–Pełczyński [173] (see Ex 33.5.) — and for arbitrary T the following proof was given to us by Tomczak–Jaegermann.

PROOF: That each $T \in \mathfrak{L}_p^{inj}$ is p–strong with constant $c = \mathbf{L}_p^{inj}(T)$ was shown in Ex 28.1.. For the converse take a p–strong $T \in \mathfrak{L}(E,F)$ with constant c. It will be shown that $\mathbf{L}_p^{inj}(T) \leq c$. Recall from 25.9. that $\mathfrak{L}_p^{inj} = \mathfrak{P}_p \circ (\mathfrak{P}_p^{dual})^{-1}$. Therefore it is sufficient to show that for $S \in \mathfrak{P}_p^{dual}(G,E)$ with $\mathbf{P}_p^{dual}(S) = 1$ the composition TS satisfies $\mathbf{P}_p(TS) \leq c$. By the definition of T being p–strong and the fact that $\mathfrak{L}_p^{inj}$ and $\mathfrak{P}_p^{dual}$ are maximal operator ideals, one may assume that E and G are finite dimensional. Since $\mathfrak{J}_p^{inj} = \mathfrak{P}_p$ holds isometrically, it follows that

$$I \circ S' : E' \xrightarrow{S'} G' \xhookrightarrow{I} \ell_\infty(B_G)$$

is p–integral and $\mathbf{I}_p(I \circ S') = \mathbf{P}_p(S') = 1$, hence

$$I \circ S' \in \mathfrak{I}_p(E', \ell_\infty(B_G)) \stackrel{1}{=} \mathfrak{N}_p(E', \ell_\infty(B_G)) \stackrel{1}{=} E \otimes_{g_p} \ell_\infty(B_G)$$

with $\mathbf{N}_p(I \circ S') = 1$ by corollary 1 before (or by the representation theorems 22.5. since E is finite dimensional and g_p accessible). Therefore it is a consequence of the definition of g_p (see e.g. 17.10.) that $I \circ S'$ factors as follows

$$\begin{array}{ccccc} E' & \xrightarrow{S'} & G' & \xhookrightarrow{I} & \ell_\infty(B_G) \\ {\scriptstyle R_1}\downarrow & & & \nearrow {\scriptstyle R_2} & \\ \ell_\infty^n & \xrightarrow{D_\lambda} & \ell_p^n & & \end{array} \qquad \begin{array}{l} \|R_1\| = \|R_2\| = 1 \\ \sum_{k=1}^n |\lambda_k|^p \leq (1+\varepsilon)^p \end{array}$$

— or dually

$$\begin{array}{ccccc} \ell_\infty(B_G)' & \xtwoheadrightarrow{I'} & G & \xrightarrow{S} & E \\ {\scriptstyle R_2'}\downarrow & & & \nearrow {\scriptstyle R_1'} & \\ \ell_{p'}^n & \xrightarrow{D_\lambda} & \ell_1^n & & \end{array} \; .$$

Now take $z_1, \ldots, z_m \in G$ with $w_p(z_k; G) \leq 1$, assume without loss of generality $m = n$ and define

$$x_k := Sz_k \in E \text{ and } y_k := R_1' D_\lambda e_k = \lambda_k R_1' e_k \in E \; .$$

These two m–tuples of elements in E satisfy for each $x' \in E'$

$$\sum_{k=1}^m |\langle x', x_k\rangle|^p = \sum_{k=1}^m |\langle S'x', z_k\rangle|^p \leq [w_p(z_k; G)]^p \|S'x'\|^p \leq$$

$$\leq \|R_2 D_\lambda R_1 x'\|^p \leq \|D_\lambda R_1 x'\|^p = \sum_{k=1}^m |\langle D_\lambda R_1 x', e_k\rangle|^p =$$

$$= \sum_{k=1}^m |\langle x', R_1' D_\lambda e_k\rangle|^p = \sum_{k=1}^m |\langle x', y_k\rangle|^p \; .$$

Since T is p–strong, it follows that

$$\sum_{k=1}^m \|TSz_k\|^p \leq c^p \sum_{k=1}^m \|y_k\|^p = c^p \sum_{k=1}^m |\lambda_k|^p \|R_1' e_k\|^p \leq$$
$$\leq c^p (1+\varepsilon)^p \|R_1'\|^p = c^p (1+\varepsilon)^p \; ,$$

which shows that $\mathbf{P}_p(TS) \leq c$. $\square$

33.8. At the end of this section conditions under which the tensor product

$$E \tilde{\otimes}_\alpha F$$

and/or the component

$$\mathfrak{A}(E, F)$$

of the associated operator ideal is reflexive will be investigated. If E and F are $\neq \{0\}$ (which will be assumed from now on) it is a necessary condition that the complemented subspaces E (or E') and F of $E\tilde{\otimes}_\alpha F$ (or $\mathfrak{A}(E,F)$) are reflexive. Clearly, this is not sufficient since, for example,

$$\mathfrak{L}(\ell_2, \ell_2) = (\ell_2 \tilde{\otimes}_\pi \ell_2)' = (\mathfrak{N}(\ell_2, \ell_2))' = (\mathfrak{K}(\ell_2, \ell_2))'' = (\ell_2 \tilde{\otimes}_\varepsilon \ell_2)''$$

(see section 16) is not reflexive. Note that

$$\mathfrak{L}(\ell_2, \ell_2) = \mathfrak{L}_{p,q}(\ell_2, \ell_2) \qquad \text{for } p, q \in\,]1, \infty[$$
$$\mathfrak{L}(\ell_2, \ell_2) = \mathfrak{P}_1^{sur}(\ell_2, \ell_2) = \mathfrak{P}_p^{sur}(\ell_2, \ell_2) \qquad \text{for } p \in [1, \infty]$$

(see 26.1. and Grothendieck's inequality, operator form) — so the spaces $\mathfrak{L}_{p,q}(E, F)$ and $\mathfrak{P}_p^{sur}(E, F)$ are also not necessarily reflexive if E and F are. Recall that a reflexive space E has the approximation property if and only if its dual has — and this implies that E and E' both have even the metric approximation property.

THEOREM: *Let α be a finitely generated accessible tensor norm, E and F Banach spaces one of which has the approximation property. The tensor product*

$$E \tilde{\otimes}_\alpha F$$

is reflexive if and only if E and F are reflexive and

$$E \tilde{\otimes}_\alpha F = (E' \otimes_{\alpha'} F')'$$
$$E' \tilde{\otimes}_{\alpha'} F' = (E \otimes_\alpha F)'$$

hold isometrically.

PROOF: Clearly, the condition is sufficient. Assume now that $E\tilde{\otimes}_\alpha F$ is reflexive. Then E and F are reflexive and, e.g., E and E' have the metric approximation property. It follows that the natural injections

$$I_1 : E'' \tilde{\otimes}_\alpha F'' \hookrightarrow (E' \tilde{\otimes}_{\alpha'} F')'$$
$$I_2 : E' \tilde{\otimes}_{\alpha'} F' \hookrightarrow (E \tilde{\otimes}_\alpha F)'$$

are isometries. The commutativity of the diagram

$$\begin{CD} E\tilde{\otimes}_\alpha F @>{\kappa_{E\tilde{\otimes}_\alpha F}}>> (E\tilde{\otimes}_\alpha F)'' \\ @| @VV{1}V{I'_2}V \\ E''\tilde{\otimes}_\alpha F'' @>{I_1}>> (E'\tilde{\otimes}_{\alpha'} F')' \end{CD}$$

shows that I_1 is surjective and I'_2 injective: This is the claim. $\square$

33.9. If $\mathfrak{A}$ is the associated maximal operator ideal, then $(\alpha')^t = \alpha^* \sim \mathfrak{A}^*$ and the representation theorems for maximal and minimal operator ideals give

COROLLARY 1: *Let $(\mathfrak{A}, \mathbf{A})$ be an accessible, maximal normed operator ideal and E, F Banach spaces one of which has the approximation property. Then $\mathfrak{A}(E,F)$ is reflexive if and only if E and F are reflexive and*

$$\mathfrak{A}^{min}(E,F) = \mathfrak{A}(E,F)$$
$$(\mathfrak{A}^*)^{min}(F,E) = \mathfrak{A}^*(F,E)$$

hold isometrically.

Using again $E \otimes_\alpha F = F \otimes_{\alpha^t} E$, another consequence is immediate as well:

COROLLARY 2: *Let $\alpha \sim (\mathfrak{A}, \mathbf{A})$ be accessible and E, F Banach spaces one of which has the approximation property.*
(1) *If $E\tilde{\otimes}_\alpha F$ is reflexive, then*

$$E'\tilde{\otimes}_{\alpha'} F' , \quad F\tilde{\otimes}_{\alpha^t} E \quad \text{and} \quad F'\tilde{\otimes}_{\alpha^*} E'$$

are reflexive as well.
(2) *If $\mathfrak{A}(E,F)$ is reflexive, then*

$$\mathfrak{A}^*(F,E) \quad \text{and} \quad \mathfrak{A}^{dual}(F', E')$$

are reflexive.
(3) *$\mathfrak{A}(E,F)$ is reflexive if and only if $E'\tilde{\otimes}_\alpha F$ is reflexive.*

33.10. Theorem 33.8. shows that Lewis' theorem 33.3. and its corollary 2 in 33.4. provide conditions for reflexivity. Recall again that reflexive spaces have the Radon–Nikodým property.

COROLLARY 3 (Lewis): *Let α be a tensor norm such that α and α^* have the Radon–Nikodým property. For $\beta := /(\alpha/)$ or $\beta := (/\alpha)/$ and E, F reflexive Banach spaces, one with the approximation property, the spaces*

$$E\tilde{\otimes}_\beta F , \quad E\tilde{\otimes}_{\beta^t} F , \quad E\tilde{\otimes}_{\beta'} F , \quad E\tilde{\otimes}_{\beta^*} F$$

are reflexive.

The analogous statements for operator ideals are obvious.

PROOF: By the foregoing corollary and $(/(\alpha/))^* = (/(\alpha^*))/$ it is enough to prove that $E \tilde{\otimes}_\beta F$ is reflexive for $\beta = (/\alpha)/$. The tensor norms $\beta = (/\alpha)/$ and $\beta^* = /(\alpha^*/)$ are tensor norms as in 33.4., corollary 2, hence (using the approximation property),

$$E \tilde{\otimes}_\beta F = (E' \tilde{\otimes}_{\beta'} F')'$$
$$E' \tilde{\otimes}_{\beta'} F' = F' \tilde{\otimes}_{\beta^*} E' = (F \otimes_{\beta^t} E)' = (E \otimes_\beta F)' \ .$$

Since β^* is (totally) accessible (by proposition 21.1.(3)), and therefore β accessible, theorem 33.8. gives the result. □

33.11. Some examples: It has already been seen that

$$\mathfrak{L}_{p,q}(\ell_2, \ell_2)$$

is not reflexive whenever $p, q \in]1, \infty[$ and hence $\ell_2 \tilde{\otimes}_{\alpha_{p,q}} \ell_2$, by corollary 2(3), is not reflexive as well. What about g_p and $d_p = g_p^t$?

PROPOSITION 1: *Let E and F be reflexive Banach spaces, one of which having the approximation property.*
(1) *$E \tilde{\otimes}_{g_p} F$ is reflexive for $1 < p \leq \infty$, but not in general for $p = 1$.*
(2) *$\mathfrak{I}_p(E, F), \mathfrak{I}_p^{dual}(E, F)$ are reflexive for $1 < p \leq \infty$, but not in general for $p = 1$.*
(3) *$\mathfrak{P}_p(E, F), \mathfrak{P}_p^{dual}(E, F)$ are reflexive for $1 \leq p < \infty$, but not in general for $p = \infty$.*

For $p = 1$ the statement (3) is due to Gordon–Lewis–Retherford [85], the other results seem to be due to Lewis [169].

PROOF: Since $g_p \sim \mathfrak{I}_p$ and $g_{p'}^* \sim \mathfrak{P}_p$, it is enough (by corollary 2) to deal with g_p or $g_p^t = d_p = (/d_p)/$ (see 20.14.). For $1 < p < \infty$ the tensor norms d_p and d_p^* have the Radon–Nikodým property by 33.5., therefore the result follows from corollary 3.
The space $\ell_2 \tilde{\otimes}_\pi \ell_2 = \ell_2 \tilde{\otimes}_{g_1} \ell_2$ is certainly not reflexive, so it remains to prove that $E \tilde{\otimes}_{d_\infty} F$ is reflexive: It follows from Ex 33.6. that

$$E' \tilde{\otimes}_{d'_\infty} F' = E' \tilde{\otimes}_{g_\infty^*} F' = \mathfrak{P}_1^{min}(E, F') = \mathfrak{P}_1(E, F') = (E \otimes_{d_\infty} F)'$$

hold isometrically — this is one condition of the reflexivity theorem 33.8.. For the other one it remains to check (recall $d_\infty = w_1 \sim \mathfrak{L}_1$) that

$$E \tilde{\otimes}_{d_\infty} F = \mathfrak{L}_1^{min}(E', F) \xrightarrow{1} \mathfrak{L}_1(E', F) = (E' \tilde{\otimes}_{d'_\infty} F')'$$

is surjective ($d_\infty = w_1$ is accessible, one of the spaces has the metric approximation property). Take $T \in \mathcal{L}_1(E', F)$, use the Radon–Nikodým property of F and the Lewis–Stegall factorization theorem (33.1.) to see that T factors as follows:

$$\begin{array}{ccc} E & \xrightarrow{T} & F \\ {\scriptstyle R}\searrow & {\scriptstyle S}\nearrow\quad & \nwarrow{\scriptstyle Q} \\ & L_1(\mu) \xrightarrow{S_o} \ell_1(B_F) & \end{array}$$

where $Q \in \mathcal{L}_1$ and $S_o \circ R$ is weakly compact=approximable since E is reflexive. It follows that $T \in \mathcal{L}_1 \circ \tilde{\mathfrak{F}} = \mathcal{L}_1^{min}$ by 25.2.. □

More results in this direction follow by considering the tensor norm $\alpha_{p,q}/$ (see also Ex 33.11.).

PROPOSITION 2: *Let $p, q \in [1, \infty]$ such that $1/p + 1/q \geq 1$ and E, F reflexive Banach spaces with E having the approximation property.*
(1) *For $q > 1$ the spaces*

$$E \tilde{\otimes}_{\alpha_{p,q}/} F \text{ and } E \tilde{\otimes}_{\alpha_{q,p}^*\backslash} F$$
$$\mathcal{L}_{p,q}/(E, F) \text{ and } \backslash \mathcal{L}_{q,p}(F, E)$$
$$\mathfrak{D}^{inj}_{q',p'}(E, F) \text{ and } \mathfrak{D}^{sur}_{p',q'}(F, E)$$

are reflexive.
(2) *For $q = 1$ each of these statements fails for $E = F = \ell_2$.*

PROOF: It is clear from 33.9., corollary 2, that it is enough to show the reflexivity of

$$E \tilde{\otimes}_{\alpha_{p,q}/} F \ .$$

The case $\alpha_{1,\infty}/ = d_\infty/ = d_\infty$ was settled in proposition 1. For $1 < q < \infty$ the tensor norm $\alpha_{p,q}$ has the Radon–Nikodým property (33.5.), hence Lewis' theorem 33.3. gives

$$E \tilde{\otimes}_{\alpha_{p,q}/} F = (E' \otimes_{(\alpha_{p,q}/)'} F')' \ .$$

Applying theorem 33.8. again, it remains to show that

$$(\mathfrak{D}^{inj}_{q',p'})^{min}(E, F') = E' \tilde{\otimes}_{\alpha_{q,p}^*\backslash} F' \stackrel{!}{=} (E \otimes_{(\alpha_{q,p}^*\backslash)'} F)' = \mathfrak{D}^{inj}_{q',p'}(E, F') \ ;$$

(note that $\alpha_{p,q}^*\backslash$ is totally accessible by 21.1.). Since

$$(\mathfrak{D}^{inj}_{q',p'})^{min}(E, F') = (\mathfrak{D}^{min}_{q',p'})^{inj}(E, F')$$

by 25.11.(2), it is enough to show that

$$\mathfrak{D}_{q',p'}^{min}(E,\ell_\infty(\Gamma)) = \mathfrak{D}_{q',p'}(E,\ell_\infty(\Gamma)) .$$

Since $\mathfrak{P}_{q'}(E,G) = \mathfrak{P}_{q'}^{min}(E,G)$ for all G (see Ex 33.6., here the approximation property of E — and not of E or F — is needed), this follows from Kwapień's factorization theorem

$$\mathfrak{D}_{q',p'} = \mathfrak{P}_{p'}^{dual} \circ \mathfrak{P}_{q'}$$

and Ex 25.15.(b) (concerning $(\mathfrak{B} \circ \mathfrak{A})^{min} = \mathfrak{B} \circ \mathfrak{A}$).
For $q = 1$ it is a consequence of Grothendieck's theorem (this was already mentioned in 33.8.) that

$$\mathfrak{L}(\ell_2,\ell_2) = \mathfrak{P}_1^{sur}(\ell_2,\ell_2) = \mathfrak{P}_{p'}^{sur}(\ell_2,\ell_2) = \mathfrak{D}_{p',\infty}^{sur}(\ell_2,\ell_2)$$

is not reflexive. □

Exercises:

Ex 33.1. An operator $T \in \mathfrak{L}(E,F)$ is called a *Radon–Nikodým operator* if for every $S \in \mathfrak{L}(L_1[0,1],E)$ the operator $T \circ S$ is representable. Denote by $\mathfrak{Y}$ the ideal of all Radon–Nikodým operators and give it the operator norm $\|\ \|$. Show that $(\mathfrak{Y}, \|\ \|)$ is an injective Banach operator ideal which is not surjective and not maximal. Hint: Identity of ℓ_1 and $L_1[0,1]$.

Ex 33.2. If μ is a finite measure, $S \in \mathfrak{L}(L_1(\mu),E)$ and $T \in \mathfrak{Y}(E,F)$, then $T \circ S$ is representable. Hint: Proof of theorem D1..

Ex 33.3. The ideal of operators $T \in \mathfrak{L}(E,F)$ which factor

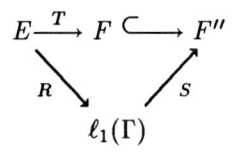

$$E \xrightarrow{T} F \hookrightarrow F''$$
$$R \searrow \quad \nearrow S$$
$$\ell_1(\Gamma)$$

through some $\ell_1(\Gamma)$ with the norm $\inf\|R\|\,\|S\|$ is not maximal.

Ex 33.4. Use corollary 2 from 33.4. in order to give conditions under which

$$\mathfrak{A}(E,F) = \mathfrak{A}^{min\ reg}(E,F) .$$

Give an example for which $\mathfrak{A}(E,F) = \mathfrak{A}^{min\ reg}(E,F) \neq \mathfrak{A}^{min}(E,F)$. Hint: $\mathfrak{A} = \mathfrak{J}$ and 22.9. for the counterexample.

Ex 33.5. For $1 \leq p < \infty$ a Banach space E is isomorphic (resp. isometric) to a subspace of some $L_p(\mu)$ if and only if there is a $c \geq 1$ (resp. $c = 1$) such that for all $m \in \mathbb{N}$ and $x_k, y_k \in E$ the condition

$$\sum_{k=1}^{m} |\langle x', x_k \rangle|^p \leq \sum_{k=1}^{m} |\langle x', y_k \rangle|^p \qquad \text{for all } x' \in E'$$

implies that $\sum_{k=1}^{m} \|x_k\|^p \leq c^p \sum_{k=1}^{m} \|y_k\|^p$. Hint: 33.7..

Ex 33.6. If E' has the Radon–Nikodým property and: E' or F has the approximation property, then for all $1 \leq p < \infty$

$$\mathfrak{P}_p(E, F) = \mathfrak{N}_p^{inj}(E, F) = \mathfrak{P}_p^{min}(E, F)$$

holds isometrically. Hint: 33.6. and 25.11.(2).

Ex 33.7. Let α be a finitely generated left–projective tensor norm with associated maximal operator ideal $\mathfrak{A}$. If α^t and E' have the Radon–Nikodým property, then

$$\mathfrak{A}(E, F') = \mathfrak{A}^{min}(E, F')$$

holds isometrically for all Banach spaces F.

Ex 33.8. Let $\mathfrak{A}$ be an accessible maximal normed operator ideal. Show that $\mathfrak{A}(E, F)$ is reflexive for all reflexive E, F such that E has the approximation property if and only if $\mathfrak{A}^*(E, F)$ is reflexive for all reflexive E, F such that F has the approximation property.

Ex 33.9. Prove again the result from 16.7. that for reflexive Banach spaces E and F (one with the approximation property) $\mathfrak{L}(E, F)$ is reflexive if and only if $\mathfrak{L}(E, F) = \mathfrak{K}(E, F)$. Hint: 33.9., corollary 1.

Ex 33.10. Let E, F be separable Banach spaces, one of which has the approximation property, and $\mathfrak{A}$ an accessible maximal normed operator ideal. If $\mathfrak{A}(E, F)$ is reflexive, then it is separable.

Ex 33.11. Use the fact that the tensor norm $\alpha_{p,2}$ is left–injective (for $1 \leq p < \infty$) to show that

$$E \tilde{\otimes}_{\alpha_{p,2}} F$$

is reflexive if E and F are reflexive Banach spaces and E or F has the approximation property. Hint: $(/\alpha)/$.

Note that proposition 2 in 33.11. did not cover the case that F has the approximation property.

Ex 33.12. Use corollary 26.1. to show that for $p, q \in]1, \infty[$ the spaces

$$\mathfrak{L}_{p,q}^{inj}(\ell_2, \ell_2), \mathfrak{L}_{p,q}^{sur}(\ell_2, \ell_2), \mathfrak{D}_{p,q}/(\ell_2, \ell_2) \text{ and } \backslash \mathfrak{D}_{p,q}(\ell_2, \ell_2)$$

are not reflexive.

34. Tensorstable Operator Ideals

This section is devoted to the following natural questions: Given an operator ideal $\mathfrak{A}$ and a tensor norm β; is $S \tilde{\otimes}_\beta T \in \mathfrak{A}$ for all $S, T \in \mathfrak{A}$? Is $E \tilde{\otimes}_\beta F \in \text{space}(\mathfrak{A})$ if E and F are? Although this happens rarely for general $\mathfrak{A}$ and β, there are some positive situations which have interesting applications. On the other hand, the number of results which handle these questions in the special case where $\beta = \varepsilon$ or $\beta = \pi$ is too large to cover all of them here. We only present a systematic approach for the more general setting. At the end of this section a method will be presented which improves inequalities with the help of tensor stability.

34.1. Let β be a tensor norm. A quasi–Banach operator ideal $(\mathfrak{A}, \mathbf{A})$ is called β-*tensorstable* if for all $T_i \in \mathfrak{A}(E_i, F_i)$ the continuous tensor product operator

$$T_1 \tilde{\otimes}_\beta T_2 : E_1 \tilde{\otimes}_\beta E_2 \longrightarrow F_1 \tilde{\otimes}_\beta F_2$$

is in the ideal $\mathfrak{A}$. It can easily be seen using an ℓ_1–sum argument that in this case there is a constant $c \geq 1$ satisfying

$$\mathbf{A}(T_1 \tilde{\otimes}_\beta T_2) \leq c \cdot \mathbf{A}(T_1) \cdot \mathbf{A}(T_2)$$

— we say: $\mathfrak{A}$ is β-*tensorstable with constant c*. For $c = 1$ the term *metrically β-tensorstable* will be used. See Ex 34.2.(c) for an example showing that not every β-tensorstable ideal is metrically β-tensorstable.

The metric mapping property of tensor norms says that $(\mathfrak{L}, \|\cdot\|)$ is metrically β-tensorstable for all tensor norms. The relation

$$S_n \tilde{\otimes}_\beta T_n - S \tilde{\otimes}_\beta T = (S_n - S) \tilde{\otimes}_\beta T_n + S \tilde{\otimes}_\beta (T_n - T)$$

also shows that the ideal $\overline{\mathfrak{F}}$ of approximable operators is metrically β-tensorstable for all β. If $T_i \in \mathfrak{L}(E_i, F_i)$ is compact, then $T_i(B_{E_i})$ is relatively compact and hence

$$T_1 \otimes_\pi T_2(\Gamma(B_{E_1} \otimes B_{E_2})) = \Gamma(T_1(B_{E_1}) \otimes T_2(B_{E_2})) \subset F_1 \tilde{\otimes}_\pi F_2$$

is also relatively compact, which shows that the ideal $\mathfrak{K}$ of compact operators is metrically π-tensorstable.

The ideal $(\mathfrak{N}, \mathbf{N})$ of nuclear operators is easily seen to be metrically β-tensorstable for all β : just use expansions.

Since $\ell_2 \tilde{\otimes}_\pi \ell_2$ and $\ell_2 \tilde{\otimes}_\varepsilon \ell_2$ are not reflexive, the ideal $\mathfrak{W}$ of weakly compact operators is neither π– nor ε–tensorstable.

34.2. The metric mapping property implies that even $\|T_1 \tilde{\otimes}_\beta T_2\| = \|T_1\| \|T_2\|$. Under what conditions does a β-tensorstable operator ideal $(\mathfrak{A}, \mathbf{A})$ satisfy an estimate from below: $\mathbf{A}(T_1)\mathbf{A}(T_2) \leq c\mathbf{A}(T_1 \tilde{\otimes}_\beta T_2)$? Using an argument of Pełczyński ([203], lecture 14), the following can be shown for the adjoint operator ideal $\mathfrak{A}^*$:

PROPOSITION: *Let $(\mathfrak{A}, \mathbf{A})$ be a quasi-Banach operator ideal which is β-tensorstable with constant $c \geq 1$. Then*

$$\mathbf{A}^*(T_1)\mathbf{A}^*(T_2) \leq c\mathbf{A}^*(T_1 \tilde{\otimes}_\beta T_2) \in [0, \infty] \ .$$

PROOF: Using the definition 17.9. of $\mathbf{A}^*$, take $N_i \in \mathrm{FIN}(E_i)$ and $L_i \in \mathrm{COFIN}(F_i)$ as well as $R_i : F_i/L_i \longrightarrow N_i$ with $\mathbf{A}(R_i) \leq 1$. Then $\mathbf{A}(R_1 \tilde{\otimes}_\beta R_2) \leq c$, and hence

$$\mathbf{A}^*(T_1 \tilde{\otimes}_\beta T_2) \geq \frac{1}{c} |tr \left[(R_1 \tilde{\otimes}_\beta R_2) \circ (Q^{F_1}_{L_1} \tilde{\otimes}_\beta Q^{F_2}_{L_2}) \circ (T_1 \tilde{\otimes}_\beta T_2) \circ (I^{E_1}_{N_1} \tilde{\otimes}_\beta I^{E_2}_{N_2})\right]| =$$
$$= \frac{1}{c} |tr \left[(R_1 \circ Q^{F_1}_{L_1} \circ T_1 \circ I^{E_1}_{N_1}) \tilde{\otimes}_\beta (R_2 \circ Q^{F_2}_{L_2} \circ T_2 \circ I^{E_2}_{N_2})\right]| =$$
$$= \frac{1}{c} |tr(R_1 \circ Q^{F_1}_{L_1} \circ T_1 \circ I^{E_1}_{N_1})| \cdot |tr(R_2 \circ Q^{F_2}_{L_2} \circ T_2 \circ I^{E_2}_{N_2})|$$

since $tr(S_1 \otimes S_2) = trS_1 \cdot trS_2$ (check with rank–one operators). The definition of $\mathbf{A}^*$ now yields the claim. □

Since $\mathfrak{A}^{**} = \mathfrak{A}$ for maximal normed ideals, this result (applied to $\mathfrak{A}^*$) implies the

COROLLARY: *Let $(\mathfrak{A}, \mathbf{A})$ be a maximal Banach ideal. If $(\mathfrak{A}, \mathbf{A})$ and $(\mathfrak{A}^*, \mathbf{A}^*)$ are metrically β-tensorstable, then*

$$\mathbf{A}(T_1 \tilde{\otimes}_\beta T_2) = \mathbf{A}(T_1)\mathbf{A}(T_2)$$

for all $T_1, T_2 \in \mathfrak{A}$.

Operator ideals satisfying this relation will be called *strongly β-tensorstable*.

REMARK: *The ideal $(\mathfrak{J}^{inj\ sur}, \mathbf{I}^{inj\ sur})$ is g_2-tensorstable, but not strongly g_2-tensorstable.*

PROOF: By Grothendieck's inequality (in the operator form 20.17.) $\mathfrak{J}^{inj\ sur} = \mathcal{L}_2$ holds with equivalent norms. Since $H \tilde{\otimes}_{g_2} H = H \tilde{\otimes}_\sigma H$ (Hilbert–Schmidt norm; 26.7.) is a Hilbert space, simple factorization shows that $\mathcal{L}_2$ and hence $\mathfrak{J}^{inj\ sur}$ is g_2-tensorstable. Assume now that $\mathbf{I}_o := \mathbf{I}^{inj\ sur}$ satisfies

$$\mathbf{I}_o(T_1)\mathbf{I}_o(T_2) \leq \mathbf{I}_o(T_1 \tilde{\otimes}_{g_2} T_2) \ .$$

Since $\mathbf{I}_o(id_{\ell_2}) = K_G$, the Grothendieck constant, this would imply

$$K_G^2 = \mathbf{I}_o(id_{\ell_2})\mathbf{I}_o(id_{\ell_2}) \leq \mathbf{I}_o(id_{\ell_2 \tilde{\otimes}_{g_2} \ell_2}) = \mathbf{I}_o(id_{\ell_2}) = K_G ,$$

which is not possible because $K_G > 1$. □

We do not know whether every β–tensorstable ideal $\mathfrak{A}$ necessarily admits a lower estimate $\mathbf{A}(T_1)\mathbf{A}(T_2) \leq c\mathbf{A}(T_1 \tilde{\otimes}_\beta T_2)$.

34.3. More examples: Since $\|T_1 \tilde{\otimes}_\beta T_2\| = \|T_1\| \, \|T_2\|$, the ideals $\mathfrak{L}$ and $\overline{\mathfrak{F}}$ are strongly β–tensorstable and $\mathfrak{K}$ is strongly π–tensorstable. Using

$$E_1 \tilde{\otimes}_\beta E_2 \xrightarrow{T_1 \tilde{\otimes}_\beta T_2} F_1 \tilde{\otimes}_\beta F_2 \longrightarrow \ell_\infty(B_{F_1'}) \tilde{\otimes}_\beta \ell_\infty(B_{F_2'})$$

and the tensor stability of $\overline{\mathfrak{F}}$, it follows that the ideal $\mathfrak{K}$ of compact operators is strongly β–tensorstable whenever β is injective, e.g. $\beta = \varepsilon$ or w_2. A similar argument with $\ell_1(B_{E_1}) \tilde{\otimes}_\beta \ell_1(B_{E_2})$ shows that $\mathfrak{K}$ is strongly β–tensorstable if β is projective, for example, $\beta = w_2^*$. It seems to be unknown whether or not $\mathfrak{K}$ is tensorstable with respect to arbitrary tensor norms.

LEMMA: *For $p = 1, 2$ the mapping $e_n \rightsquigarrow e_n \otimes e_n$ extends to an isometry*

$$\ell_1 \xrightarrow{\;\;1\;\;} \ell_p \tilde{\otimes}_\pi \ell_p .$$

PROOF: $\pi(\sum_{n=1}^\infty \lambda_n e_n \otimes e_n; \ell_p, \ell_p) \leq \sum_{n=1}^\infty |\lambda_n|$ gives one inequality. The other one follows from $\ell_1 \tilde{\otimes}_\pi \ell_1 = \ell_1(\mathbb{N} \times \mathbb{N})$ and, for $p = 2$, from (see Ex 3.28.)

$$\mathbf{N}\left(\sum_{n=1}^\infty \lambda_n e_n \underline{\otimes} e_n : \ell_2 \longrightarrow \ell_2\right) = \sum_{n=1}^\infty |\lambda_n| . \quad □$$

If $I : \ell_1 \hookrightarrow \ell_2$, then obviously

$$\begin{array}{ccc} \ell_1 \tilde{\otimes}_\pi \ell_1 & \xrightarrow{I \tilde{\otimes}_\pi I} & \ell_2 \tilde{\otimes}_\pi \ell_2 \\ {\scriptstyle 1}\uparrow & & \uparrow{\scriptstyle 1} \\ \ell_1 & \xrightarrow{\;\;id\;\;} & \ell_1 \end{array}$$

commutes and this implies the

COROLLARY 1: *If the operator ideal $\mathfrak{A}$ is injective and $\ell_1 \notin$ space($\mathfrak{A}$), then the operator $I \tilde{\otimes}_\pi I : \ell_1 \tilde{\otimes}_\pi \ell_1 \longrightarrow \ell_2 \tilde{\otimes}_\pi \ell_2$ is not in $\mathfrak{A}$.*

In particular, for $1 \leq p < \infty$ the ideal $\mathfrak{P}_p$ of absolutely p-summing operators is not π-tensorstable.

The latter follows from $I \in \mathfrak{P}_1 \subset \mathfrak{P}_p$ (see Ex 11.5.).

Since the ideal $\mathfrak{QN}$ of quasinuclear operators is the injective hull $\mathfrak{N}^{inj}$ of the nuclear operators (Ex 9.13.), the same argument which was used for $\mathfrak{K} = \overline{\mathfrak{F}}^{inj}$ gives that $\mathfrak{QN}$ is metrically β-tensorstable whenever β is an injective tensor norm. But Holub [113] observed that:

COROLLARY 2: *The ideal $\mathfrak{QN}$ of quasinuclear operators is not π-tensorstable.*

PROOF: Take $(\lambda_n) \in c_o$ and $T := \sum_{n=1}^{\infty} \lambda_n e_n \otimes e_n : \ell_1 \longrightarrow \ell_2$. Then the operator

$$\ell_1 \xrightarrow{\sum \lambda_n e_n \otimes e_n} \ell_1 \xhookrightarrow{I} \ell_2 \hookrightarrow \ell_\infty(B_{\ell_2})$$

is nuclear by Ex 11.5. and Appendix C8., corollary (recall that $\mathfrak{J}^{inj} = \mathfrak{P}_1$), and therefore T is quasinuclear. Now look at the diagram

$$\begin{array}{ccc}
\ell_1 \tilde{\otimes}_\pi \ell_1 & \xrightarrow{T \tilde{\otimes}_\pi T} & \ell_2 \tilde{\otimes}_\pi \ell_2 \\
{\scriptstyle 1}\Big\downarrow & & {\scriptstyle 1}\Big\downarrow \\
\ell_1 & \xrightarrow{T_o} & \ell_1
\end{array} \qquad T_o e_n := \lambda_n^2 e_n \;.$$

If $T \tilde{\otimes}_\pi T$ were quasinuclear, then T_o would be quasinuclear and hence

$$T_o^2 = \sum_{n=1}^{\infty} \lambda_n^4 e_n \otimes e_n : \ell_1 \longrightarrow \ell_1$$

nuclear ($T_o^2 \in \mathfrak{QN}^2 \subset \mathfrak{P}_1^2 \subset \mathfrak{P}_2^2 \subset \mathfrak{N}$). But this is only the case if the sequence (λ_n) is in ℓ_4 (Ex 3.28.). □

For a positive result in this direction see Ex 34.3..

34.4. Many of the arguments already used are valid in a more general setting. The following statements (taken von Carl–D.–Ramanujan [23]) allow a more systematic treatment of tensor stability.

PROPOSITION: *Let $(\mathfrak{A}, \mathbf{A})$ and $(\mathfrak{B}, \mathbf{B})$ be Banach operator ideals and β a tensor norm.*
(1) *If $(\mathfrak{A}, \mathbf{A})$ and $(\mathfrak{B}, \mathbf{B})$ are β-tensorstable (with constants a and b, respectively), then $(\mathfrak{A} \circ \mathfrak{B}, \mathbf{A} \circ \mathbf{B})$ is β-tensorstable (with constant $a \cdot b$).*

(2) If $(\mathfrak{A}, \mathbf{A})$ is β-tensorstable, then its minimal kernel $(\mathfrak{A}^{min}, \mathbf{A}^{min})$ is β-tensorstable (with the same constant).

(3) If β is injective, $(\mathfrak{A}^{max}, \mathbf{A}^{max})$ right-accessible and

$$\mathbf{A}(T_1 \otimes_\beta T_2) \le c \cdot \mathbf{A}(T_1) \cdot \mathbf{A}(T_2)$$

for operators T_1 and T_2 between finite dimensional Banach spaces, then the maximal hull $(\mathfrak{A}^{max}, \mathbf{A}^{max})$ of $\mathfrak{A}$ is β-tensorstable (with constant c).

(4) Let $(\mathfrak{A}, \mathbf{A})$ be β-tensorstable (with constant c). If β is injective, then $(\mathfrak{A}^{inj}, \mathbf{A}^{inj})$ is β-tensorstable; if β is projective, then $(\mathfrak{A}^{sur}, \mathbf{A}^{sur})$ is β-tensorstable (same constants).

(5) If β is projective, $(\mathfrak{A}, \mathbf{A})$ is β-tensorstable (with constant c) and $(\mathfrak{A}^{max}, \mathbf{A}^{max})$ is left-accessible, then

$$(\mathfrak{A}^{max\ dual}, \mathbf{A}^{max\ dual})$$

is β'-tensorstable (with constant c).

PROOF: (1) and (2) are obvious since $\mathfrak{A}^{min} = \mathfrak{F} \circ \mathfrak{A} \circ \mathfrak{F}$ and $\mathfrak{F}$ is β-tensorstable. Statement (4) follows as in the case of $\mathfrak{K} = \mathfrak{F}^{inj} = \mathfrak{F}^{sur}$ at the beginning of 34.3..
To prove (3) take $T_i \in \mathfrak{A}^{max}(E_i, F_i)$ and put $E := E_1 \otimes_\beta E_2$. Then, by the density lemma for maximal operator ideals (Ex 17.17.), it is enough to show that

$$\mathbf{A}\left((T_1 \tilde\otimes_\beta T_2) \circ I_M^{\tilde E}\right) \le c \cdot \mathbf{A}^{max}(T_1) \cdot \mathbf{A}^{max}(T_2)$$

for cofinally many $M \in \mathrm{FIN}(E)$. Since for each such M there are $M_i \in \mathrm{FIN}(E_i)$ with $M \subset M_1 \otimes M_2$ and β is injective, it follows that

$$M \xhookrightarrow{1} M_1 \otimes_\beta M_2 \xhookrightarrow{1} E_1 \otimes_\beta E_2$$

— and this implies that M of the form $M_1 \otimes_\beta M_2$ suffice. $\mathfrak{A}^{max}$ is supposed to be right-accessible, hence there are $N_i \in \mathrm{FIN}(F_i)$ and $S_i \in \mathcal{L}(M_i, N_i)$ with

$$T_i \circ I_{M_i}^{E_i} = I_{N_i}^{F_i} \circ S_i \quad \text{and} \quad \mathbf{A}(S_i) \le (1+\varepsilon)\mathbf{A}^{max}(T_i) .$$

Now the desired inequality is a consequence of

$$\mathbf{A}\left((T_1\tilde\otimes_\beta T_2)\circ I_{M_1\otimes M_2}^{\tilde E}\right) = \mathbf{A}\left((T_1 \circ I_{M_1}^{E_1})\tilde\otimes_\beta(T_2 \circ I_{M_2}^{E_2})\right) =$$
$$= \mathbf{A}\left((I_{N_1}^{F_1}\circ S_1)\tilde\otimes_\beta(I_{N_2}^{F_2}\circ S_2)\right) = \mathbf{A}\left(I_{N_1\otimes N_2}^{F_1\tilde\otimes_\beta F_2}\circ(S_1\otimes_\beta S_2)\right) \le$$
$$\le c\cdot \mathbf{A}(S_1)\cdot \mathbf{A}(S_2) \le c(1+\varepsilon)^2 \mathbf{A}^{max}(T_1)\cdot \mathbf{A}^{max}(T_2) .$$

To see (5) first observe that for operators T_1 and T_2 on finite dimensional spaces, one has $(T_1 \otimes_{\beta'} T_2)' = T_1' \otimes_\beta T_2'$, and therefore,

$$\mathbf{A}^{dual}(T_1 \otimes_{\beta'} T_2) = \mathbf{A}(T_1' \otimes_\beta T_2') \le c\cdot \mathbf{A}(T_1')\cdot \mathbf{A}(T_2') \le c\cdot \mathbf{A}^{dual}(T_1)\cdot \mathbf{A}^{dual}(T_2) .$$

Now (3) applied to $\mathfrak{A}^{dual}$ and β' gives the result. □

Note that (1), (2) and (4) also hold for quasi–Banach operator ideals.

34.5. The most important positive examples come from factorization techniques (apparently first used by Holub [113] in this context). Factorization of an operator in $\mathcal{L}_{p,q}(E,F)$ usually goes into F''' (rather than F), and therefore the following simple observation is useful: If β is finitely generated, then any of the two maps Φ_j from the extension lemma 13.2. make the diagram

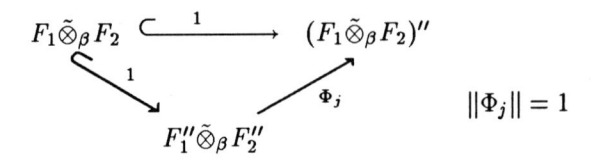

$$\begin{array}{ccc} F_1 \tilde{\otimes}_\beta F_2 & \xrightarrow{1} & (F_1 \tilde{\otimes}_\beta F_2)'' \\ {\scriptstyle 1} \searrow & \nearrow {\scriptstyle \Phi_j} & \\ & F_1'' \tilde{\otimes}_\beta F_2'' & \end{array} \qquad \|\Phi_j\| = 1$$

commutative; recall that we do not know whether or not the Φ_j are isometries.

If $T_i \in \mathfrak{I}(E_i, F_i)$, then the factorization 10.5. for integral operators yields for each finitely generated tensor norm β

$$\begin{array}{ccccc} E_1 \tilde{\otimes}_\beta E_2 & \xrightarrow{T_1 \tilde{\otimes}_\beta T_2} & F_1 \tilde{\otimes}_\beta F_2 & \xrightarrow{1} & F_1'' \tilde{\otimes}_\beta F_2'' \\ \downarrow & & & \nearrow & \\ C(K_1 \times K_2) = C(K_1) \tilde{\otimes}_\varepsilon C(K_2) & \longrightarrow & L_1(\mu_1) \tilde{\otimes}_\pi L_1(\mu_2) = L_1(\mu_1 \otimes \mu_2) & & \end{array}$$

— and the observation above implies the

REMARK: *The ideal $(\mathfrak{I}, \mathbf{I})$ of integral operators is metrically β–tensorstable for all finitely generated tensor norms β.*

Exploring this factorization idea for $T \in \mathcal{L}_{p,q}(E,F)$, the factorization theorem 18.11. for (p,q)–factorable operators leads to the diagram

$$\begin{array}{ccccc} E_1 \tilde{\otimes}_\gamma E_2 & \xrightarrow{T_1 \tilde{\otimes} T_2} & F_1 \tilde{\otimes}_\beta F_2 & \xrightarrow{1} & F_1'' \tilde{\otimes}_\beta F_2'' \\ \downarrow & & & \nearrow & \\ L_{q'}(\mu_1 \otimes \mu_2) = L_{q'}(\mu_1) \tilde{\otimes}_{\Delta_{q'}} L_{q'}(\mu_2) & \longrightarrow & L_p(\mu_1) \tilde{\otimes}_{\Delta_p} L_p(\mu_2) = L_p(\mu_1 \otimes \mu_2) \, , & & \end{array}$$

and hence the

PROPOSITION: *Let β and γ be finitely generated tensor norms such that*

$$\Delta_{q'} \leq \gamma \quad on \quad L_{q'} \otimes L_{q'}$$
$$\beta \leq \Delta_p \quad on \quad L_p \otimes L_p .$$

Then for $T_i \in \mathfrak{L}_{p,q}(E_i, F_i)$ the operator

$$T_1 \tilde{\otimes} T_2 : E_1 \tilde{\otimes}_\gamma E_2 \longrightarrow F_1 \tilde{\otimes}_\beta F_2$$

is defined, (p,q)-factorable and $\mathbf{L}_{p,q}(T_1 \tilde{\otimes} T_2) \leq \mathbf{L}_{p,q}(T_1) \cdot \mathbf{L}_{p,q}(T_2)$.

It follows from 29.11. and 29.4. that in the hypotheses $L_{q'}$ and L_p can be replaced by $\ell_{q'}$ and ℓ_p, respectively. The result is interesting because of its special cases $\mathfrak{I}_p = \mathfrak{L}_{p,1}$ and $\mathfrak{I}_q^{dual} = \mathfrak{L}_{1,q}$: the fact that $d_{p'}^* = d_p = \Delta_p = g_{p'}^* = g_p$ on $L_p \otimes L_p$ (see 15.10.) implies the

COROLLARY 1:

(1) *For $1 \leq p \leq \infty$ the ideal $\mathfrak{I}_p$ of p-integral operators is metrically β-tensorstable for all finitely generated tensor norms satisfying $\beta \leq \Delta_p$ on $L_p \otimes L_p$. Examples: $\varepsilon, d_{p'}^*, d_p$, $g_{p'}^*, g_p, w_p, w_{p'}$ and all smaller tensor norms.*

(2) *For $1 \leq q \leq \infty$ the ideal $\mathfrak{I}_q^{dual}$ is metrically β-tensorstable for all finitely generated tensor norms satisfying $\beta \geq \Delta_{q'}$ on $L_{q'} \otimes L_{q'}$. Examples: $\pi, d_q^*, d_{q'}, g_q^*, g_{q'}, w_{q'}^*, w_q^*$ and all larger tensor norms.*

It follows that for $1 \leq p \leq 2$ the ideal $\mathfrak{I}_p$ is metrically w_r-tensorstable for all $r \in [1, \infty]$. Proposition 34.4. allows one to verify tensor stability for many other operator ideals related to $\mathfrak{I}_p$ and $\mathfrak{I}_p^{dual}$ (see Ex 34.2.). Restricting to ε and π, it gives the

COROLLARY 2: *Take $1 \leq p \leq \infty$.*

(1) *The ideals $\mathfrak{I}_p$ of p-integral, $\mathfrak{P}_p$ of absolutely p-summing, and $\mathfrak{N}_p^{inj}$ of quasi-p-nuclear operators are strongly ε-tensorstable, the ideal $\mathfrak{N}_p$ of p-nuclear operators is metrically ε-tensorstable.*

(2) *The ideals $\mathfrak{I}_p^{dual}, \mathfrak{P}_p^{dual}$ and $\mathfrak{K}_{1,p}^{sur}$ are strongly π-tensorstable, the ideal $\mathfrak{K}_{1,p}$ is metrically π-tensorstable.*

Because of its application to the theory of eigenvalue distribution (see 34.10.) perhaps the most important statement here is that $\mathfrak{P}_p$ is ε-tensorstable; this result is due to Holub [113].

PROOF: Corollary 1 and proposition 34.4. give that all these ideals are metrically tensorstable with respect to ε or π. Since $\mathfrak{I}_p^* = \mathfrak{P}_{p'}$ and $\mathfrak{P}_p^* = \mathfrak{I}_{p'}$, corollary 34.2. gives that the maximal ideals in (1) are even strongly ε-tensorstable. Since $g_p = g_p\backslash = g_{p'}^*$

on $E' \otimes \ell_\infty$ and $g_{p'}^*$ is totally accessible, the representation theorems for maximal and minimal ideals imply that

$$\mathfrak{N}_p^{inj}(E,F) \xhookrightarrow{1} \mathfrak{P}_p(E,F)$$

for all E, F — and this shows that with $\mathfrak{P}_p$ also $\mathfrak{N}_p^{inj}$ is strongly ε-tensorstable. The relations
$$(\mathfrak{J}_p^{dual})^* = \mathfrak{J}_p^{*dual} = \mathfrak{P}_{p'}^{dual}$$
$$(\mathfrak{P}_p^{dual})^* = \mathfrak{J}_{p'}^{dual}$$

give that the maximal ideals in (2) are strongly π-tensorstable and

$$\mathfrak{K}_{1,p}^{sur}(E,F) \xhookrightarrow{1} \mathfrak{P}_p^{dual}(E,F)$$

that this is also true for $\mathfrak{K}_{1,p}^{sur}$. $\square$

34.6. Two applications of these results. First, recall from 23.6. that the projection constant $\lambda(E)$ of a finite dimensional Banach space is $\mathbf{L}_\infty(id_E) = \mathbf{I}_\infty(id_E)$. Hence the relation
$$\mathbf{I}_p(T_1 \tilde\otimes_\varepsilon T_2) = \mathbf{I}_p(T_1)\mathbf{I}_p(T_2) ,$$
which was just proved, implies that

$$\lambda(E \tilde\otimes_\varepsilon F) = \lambda(E)\lambda(F)$$

for all finite dimensional Banach spaces E and F. This is a well-known result. However, there is no such relation for infinite dimensional spaces! Even more:

PROPOSITION: *Let E and F be infinite dimensional spaces. Then the Banach space $E \tilde\otimes_\varepsilon F$ is not injective — in other words: $\lambda(E \tilde\otimes_\varepsilon F) = \infty$.*

This result was given to us by Pełczyński.

PROOF: First take $E = F = \ell_\infty$: By a result of Kalton [140] the space $\ell_\infty \tilde\otimes_\varepsilon \ell_\infty = \mathfrak{K}(\ell_1, \ell_\infty)$ is not complemented in $\mathfrak{L}(\ell_1, \ell_\infty)$ — and is therefore not an injective Banach space.

Now, take arbitrary E and F and assume $E \tilde\otimes_\varepsilon F$ to be injective. Then, as complemented subspaces, E and F are injective as well. A result of Rosenthal's states (see [177], I, p.105) that ℓ_∞ is isomorphic to subspaces $E_o \subset E$ and $F_o \subset F$; they are complemented (since ℓ_∞ is injective), and hence $\ell_\infty \tilde\otimes_\varepsilon \ell_\infty$ is isomorphic to the complemented subspace $E_o \tilde\otimes_\varepsilon F_o$ of the injective space $E \tilde\otimes_\varepsilon F$. It follows that $\ell_\infty \tilde\otimes_\varepsilon \ell_\infty$ is injective as well — which is a contradiction. $\square$

Another application for bases (see Ex 12.9.) is also due to Holub [113]:

REMARK: *The unit vector basis* $(e_n \otimes e_m)$ *in* $\ell_1 \tilde{\otimes}_\varepsilon \ell_1$ *is not unconditional.*

PROOF: The embedding $I : \ell_1 \hookrightarrow \ell_2$ is absolutely 1–summing. By corollary 2 the operator $I \tilde{\otimes}_\varepsilon I : \ell_1 \tilde{\otimes}_\varepsilon \ell_1 \longrightarrow \ell_2 \tilde{\otimes}_\varepsilon \ell_2$ is also in $\mathfrak{P}_1$ and hence maps unconditionally convergent series to absolutely convergent series. If $(e_n \otimes e_m)$ were an unconditional basis of $\ell_1 \tilde{\otimes}_\varepsilon \ell_1$, then every expansion

$$\sum_{n,m=1}^\infty \lambda_{n,m} e_n \otimes e_m \in \ell_1 \tilde{\otimes}_\varepsilon \ell_1$$

would converge absolutely in $\ell_2 \tilde{\otimes}_\varepsilon \ell_2$ which means that $\sum |\lambda_{n,m}| < \infty$ would hold. In other words, every operator in $\mathfrak{K}(c_o, \ell_1) = \ell_1 \tilde{\otimes}_\varepsilon \ell_1$ would be nuclear; but this is not true (see e.g. Ex 4.3.). □

34.7. What about tensor stability of space($\mathfrak{A}$)? Negative results will follow from the

LEMMA: *Let β be an injective tensor norm and E_1, E_2 infinite dimensional Banach spaces. Then $E_1 \tilde{\otimes}_\beta E_2$ contains the ℓ_∞^n uniformly.*

PROOF: By Dvoretzky's spherical section theorem, which was already mentioned in 32.10., the spaces E_i contain subspaces M_i^n with $d(M_i^n, \ell_2^n) \le 2$. Since β is injective, all $M_1^n \otimes_\beta M_2^n$ are isometric subspaces of $E_1 \tilde{\otimes}_\beta E_2$; therefore, it is enough to show that ℓ_∞^n is isomorphic to a subspace of $\ell_2^n \otimes_\beta \ell_2^n$ with control of the norm. For $\beta = \varepsilon$

$$I : \ell_\infty^n \xrightarrow{1} \ell_2^n \otimes_\varepsilon \ell_2^n \ ; \quad I(\lambda_k) := \sum_{k=1}^n \lambda_k e_k \otimes e_k$$

is obviously an isometry. But $\varepsilon = w_2$ on $\ell_2^n \otimes \ell_2^n$, hence Grothendieck's inequality (and again the injectivity of β) give

$$\varepsilon \le \beta \le /\pi\backslash \le K_G w_2 = K_G \varepsilon \qquad \text{on } \ell_2^n \otimes \ell_2^n$$

— this ends the proof. □

Note that the lemma is false if β is not injective: Take the Hilbert space $\ell_2 \tilde{\otimes}_{g_2} \ell_2$ as an example. The following theorem and its corollaries are due to Carl–D.–Ramanujan [23]:

THEOREM: *Let $(\mathfrak{A}, \mathbf{A})$ be a Banach operator ideal and β a tensor norm such that*
(a) *β is injective and $\sup \mathbf{A}(id_{\ell_\infty^n}) = \infty$, or:*
(b) *β is projective and $\sup \mathbf{A}(id_{\ell_1^n}) = \infty$.*
Then $E_1 \tilde{\otimes}_\beta E_2 \notin$ space($\mathfrak{A}$) whenever E_1 and E_2 are infinite dimensional Banach spaces.

PROOF: Assume $E_1 \tilde{\otimes}_\beta E_2 \in \text{space}(\mathfrak{A})$ and the first condition (a): The lemma implies that there is a $c \geq 1$ such that

$$\mathbf{A}(id_{\ell^n_\infty}) = \mathbf{A}^{inj}(id_{\ell^n_\infty}) \leq c\mathbf{A}(id_{E_1 \tilde{\otimes}_\beta E_2}),$$

which is not possible.

Under condition (b) the dual β' of β is injective, and therefore totally accessible. It follows that

$$E'_1 \tilde{\otimes}_{\beta'} E'_2 \xrightarrow{1} (E_1 \tilde{\otimes}_\beta E_2)' \in \text{space}(\mathfrak{A}^{max \ dual})$$

(recall 17.8., corollary 4) and hence,

$$E'_1 \tilde{\otimes}_{\beta'} E'_2 \in \text{space}(\mathfrak{A}^{max \ dual \ inj}).$$

Since

$$\sup \mathbf{A}^{max \ dual \ inj}(id_{\ell^n_\infty}) = \sup \mathbf{A}(id_{\ell^n_1}) = \infty,$$

the tensor norm β' and the Banach ideal $\mathfrak{A}^{max \ dual \ inj}$ satisfy condition (a), therefore $E'_1 \tilde{\otimes}_\beta E'_2 \notin \text{space}(\mathfrak{A}^{max \ dual \ inj})$ — a contradiction. □

Since $\sup \mathbf{A}(id_{\ell^n_p}) < \infty$ implies that $\mathfrak{L}_p \subset \mathfrak{A}$ whenever $\mathfrak{A}$ is maximal, the theorem yields the

COROLLARY 1: *Let $\mathfrak{A}$ be a maximal Banach operator ideal such that* $\text{space}(\mathfrak{A})$ *contains an infinite dimensional space.*
(1) *If $\mathfrak{A}$ is β-tensorstable for some injective tensor norm β, then $\mathfrak{L}_\infty \subset \mathfrak{A}$.*
(2) *If $\mathfrak{A}$ is β-tensorstable for some projective tensor norm β, then $\mathfrak{L}_1 \subset \mathfrak{A}$.*

This has various consequences:

The ideal $\mathfrak{L}_p$ of p–factorable operators $(1 < p < \infty)$ is not ε–tensorstable and not π–tensorstable (since neither $\mathfrak{L}_1$ nor $\mathfrak{L}_\infty$ are in $\mathfrak{L}_p$).

The ideal $\mathfrak{C}_q$ of cotype q operators $(2 \leq q < \infty)$ is not ε–tensorstable (since ℓ_∞ has no proper cotype, Ex 7.14.). It is not known whether or not $\mathfrak{C}_2$ is π–tensorstable: Tomczak–Jaegermann [271] showed that $\ell_2 \tilde{\otimes}_\pi \ell_2$ has cotype 2, but it has not yet been decided whether $\ell_2 \tilde{\otimes}_\pi \ell_2 \tilde{\otimes}_\pi \ell_2$ has cotype 2 or not.

The ideal $\mathfrak{T}_p$ of type p operators $(1 < p \leq 2)$ is neither ε– nor π–tensorstable (since ℓ_1 and ℓ_∞ do not have proper type, Ex 7.13. and Ex 7.14.).

The ideal $\mathfrak{M}_{2,1}$ of $(2,1)$–mixing operators is not ε–tensorstable (since $\ell_\infty \notin \text{space}(\mathfrak{M}_{2,1})$ by 32.2.(6)). See also Ex 34.4., Ex 34.8. and Ex 34.9..

More generally, one has the

COROLLARY 2: *There is no maximal Banach operator ideal* $\mathfrak{A} \neq \mathcal{L}$ *such that* space($\mathfrak{A}$) *contains an infinite dimensional Banach space and which satisfies* (1) *or* (2):

(1) $\mathfrak{A}$ *is injective and* β-*tensorstable for some injective tensor norm.*

(2) $\mathfrak{A}$ *is surjective and* β-*tensorstable for some projective tensor norm.*

PROOF: Corollary 1 would imply

$$\mathcal{L} = \mathcal{L}_\infty^{inj} \subset \mathfrak{A}^{inj} = \mathfrak{A} \quad \text{in case (1)}$$

and

$$\mathcal{L} = \mathcal{L}_1^{sur} \subset \mathfrak{A}^{sur} = \mathfrak{A} \quad \text{in case (2)} \quad \cdot \square$$

For another interesting application of the lemma see Ex 34.6..

34.8. Since $\ell_2 \tilde{\otimes}_\pi \ell_2$ and $\ell_2 \tilde{\otimes}_\varepsilon \ell_2$ are not reflexive, they are not contained in any L_p (for $1 < p < \infty$). It will be shown below that this is also true for $p = 1$. The proof of this result requires some preparation.

DOUBLE KHINTCHINE INEQUALITY:

$$a_p^{-2} \Big(\sum_{i,j=1}^n |\alpha_{i,j}|^2\Big)^{1/2} \le \Big(\int_D \int_D \Big|\sum_{i,j=1}^n \alpha_{i,j}\varepsilon_i(w)\varepsilon_j(w')\Big|^p \mu(dw)\mu(dw')\Big)^{1/p} \le$$

$$\le b_p^2 \Big(\sum_{i,j=1}^n |\alpha_{i,j}|^2\Big)^{1/2}$$

for all $\alpha_{i,j} \in \mathbb{C}$ and $1 \le p < \infty$, the a_p and b_p being the constants from the Khintchine inequality.

PROOF: The left inequality is obvious for $p \ge 2$, the right one for $p \le 2$. Denote by $S_p : \ell_2 \longrightarrow \text{Rad}_p \subset L_p(\mu)$ the identification of ℓ_2 with the closed subspace Rad_p of $L_p(\mu)$ spanned by the Rademacher functions; the Khintchine inequality says that $\|S_p\| \le b_p$ and $\|S_p^{-1}\| \le a_p$. The variant of Beckner's result given in 7.9. shows that for $p \ge 2$

$$\|S_p \otimes S_p : \ell_2 \otimes_{\Delta_2} \ell_2 \longrightarrow \text{Rad}_p \otimes_{\Delta_p} \text{Rad}_p \xhookrightarrow{1} L_p(\mu) \otimes_{\Delta_p} L_p(\mu)\| \le b_p^2$$

and for $p \leq 2$

$$\|S_p^{-1} \otimes S_p^{-1} : \text{Rad}_p \otimes_\Delta, \text{Rad}_p \longrightarrow \ell_2 \otimes_{\Delta_2} \ell_2\| \leq a_p^2$$

— and these are the missing inequalities. □

The next result is due to McCarthy [188].

PROPOSITION: *Let (Ω, ν) be a measure space, $1 \leq p < \infty$ and E a subspace of $L_p(\Omega, \nu)$. For given n-tuples $(P_1, ..., P_n)$ and $(Q_1, ..., Q_n)$ of pairwise orthogonal projections in E which commute $(P_i Q_j = Q_j P_i)$ define*

$$a := \max\left\{ \left\| \sum_{i=1}^n \varepsilon_i P_i \right\| \,\middle|\, \varepsilon_i \in \{-1, 1\} \right\}$$

$$b := \max\left\{ \left\| \sum_{j=1}^n \varepsilon_j Q_j \right\| \,\middle|\, \varepsilon_j \in \{-1, 1\} \right\}.$$

Then

$$\max\left\{ \left\| \sum_{i,j=1}^n \varepsilon_{i,j} P_i Q_j \right\| \,\middle|\, \varepsilon_{i,j} \in \{-1, 1\} \right\} \leq a_p^2 b_p^2 a^3 b^3,$$

where a_p and b_p are the constants from the Khintchine inequality.

PROOF: Note first that $R := \sum_{i,j=1}^n P_i Q_j$ is a projection of norm $\leq a \cdot b$ and that the family $(P_i Q_j)_{i,j=1}^n$ of projections is pairwise orthogonal. It follows that it is enough to show for all $f \in E$ and $f_{i,j} := P_i Q_j f \in P_i Q_j E$ that

$$\left\| \sum_{i,j=1}^n \varepsilon_{i,j} f_{i,j} \right\|_p \leq a_p^2 b_p^2 a^2 b^2 \left\| \sum_{i,j=1}^n f_{i,j} \right\|_p$$

since $Rf = \sum_{i,j=1}^n f_{i,j}$. In order to use Rademacher averaging take $w, w' \in D = \{-1, 1\}^N$. Then for $g_{i,j} \in P_i Q_j E$

$$\left\| \sum_{i,j} g_{i,j} \right\|_p = \left\| \sum_k \varepsilon_k(w) \sum_j \varepsilon_k(w) \varepsilon_j(w')^2 g_{k,j} \right\|_p =$$

(*) $$= \left\| \sum_k \varepsilon_k(w) P_k \left[\sum_{i,j} \varepsilon_i(w) \varepsilon_j(w')^2 g_{i,j} \right] \right\|_p \leq$$

$$\leq a \left\| \sum_{i,j} \varepsilon_i(w) \varepsilon_j(w')^2 g_{i,j} \right\|_p \leq ... \leq ab \left\| \sum_{i,j} \varepsilon_i(w) \varepsilon_j(w') g_{i,j} \right\|_p.$$

Integrating the p-th power of this inequality for $g_{i,j} = \varepsilon_{i,j} f_{i,j}$ and using the double Khintchine inequality, gives

$$\left\|\sum_{i,j} \varepsilon_{i,j} f_{i,j}\right\|_p \leq ab \left(\int_\Omega \int_D \int_D \left|\sum_{i,j} \varepsilon_i(w)\varepsilon_j(w') \varepsilon_{i,j} f_{i,j}(t)\right|^p \mu(dw)\mu(dw')\nu(dt) \right)^{1/p}$$

$$\leq ab b_p^2 \left(\int_\Omega \left(\sum_{i,j} |\varepsilon_{i,j} f_{i,j}(t)|^2\right)^{p/2} \nu(dt) \right)^{1/p} =$$

$$= ab b_p^2 \left(\int_\Omega \left(\sum_{i,j} |f_{i,j}(t)|^2\right)^{p/2} \nu(dt) \right)^{1/p} \leq$$

$$\leq ab b_p^2 a_p^2 \left(\int_\Omega \int_D \int_D \left|\sum_{i,j} \varepsilon_i(w)\varepsilon_j(w') f_{i,j}(t)\right|^p \mu(dw)\mu(dw')\nu(dt) \right)^{1/p} \leq$$

$$\leq ab b_p^2 a_p^2 \left(\int_D \int_D \left\|\sum_{i,j} \varepsilon_i(w)\varepsilon_j(w') f_{i,j}\right\|_p^p \mu(dw)\mu(dw') \right)^{1/p} \leq$$

$$\leq a_p^2 b_p^2 a^2 b^2 \left\|\sum_{i,j} f_{i,j}\right\|_p ,$$

again by inequality (∗). This ends the proof. □

Note that the constant can be taken to be $a_p^2 b_p^2 a^2 b^2$ if

$$id_E = \sum_{i=1}^n P_i = \sum_{j=1}^n Q_j .$$

Everything is ready for proving the announced result:

THEOREM: *If β is a tensor norm on HILB such that $\ell_2 \tilde{\otimes}_\beta \ell_2$ is isomorphic to a subspace of some L_p (for $1 \leq p < \infty$), then β and the Hilbert-Schmidt norm σ are equivalent on $\ell_2 \otimes \ell_2$.*

The result is a special case of a result of Lewis [168] (for $p = 1$) and of Pisier [218]. The argument for the present proof was given to us by Pełczyński.

PROOF: By the Kwapień-Pełczyński result 26.11. it is enough to show that the unit vector basis $(e_i \otimes e_j)$ (rectangular ordering, see Ex 12.9.) is unconditional in the tensor product $E = \ell_2 \tilde{\otimes}_\beta \ell_2$ whenever this space can be embedded into some L_p. A basis (x_n) is unconditional if there is a constant $c \geq 1$ such that

$$\left\|\sum_{k=1}^n \varepsilon_k \lambda_k x_k\right\| \leq c \left\|\sum_{k=1}^n \lambda_k x_k\right\|$$

for all $\varepsilon_k \in \{-1,1\}$, all $n \in \mathbb{N}$ and all $\lambda_k \in \mathbb{K}$; this follows from 8.3., Ex 8.1. and a closed graph argument. The basis (e_k) of ℓ_2 satisfies this inequality with $c = 1$. Denote by $U_k := e_k \otimes e_k$ the k-th projection $\ell_2 \longrightarrow \ell_2$. Then

$$P_i := U_i \tilde{\otimes}_\beta id_{\ell_2} \quad \text{and} \quad Q_j := id_{\ell_2} \tilde{\otimes}_\beta U_j$$

satisfy the assumptions of McCarthy's result with $a = b = 1$, and hence

$$\left\| \sum_{i,j=1}^n \varepsilon_{i,j} U_i \tilde{\otimes}_\beta U_j \right\| \leq a_p^2 b_p^2 .$$

Since, obviously, $U_i \tilde{\otimes}_\beta U_j$ is the (i,j)-th projection of the basis $(e_i \otimes e_j)$, this shows its unconditionality. □

COROLLARY: *Let β be a tensor norm which is not equivalent to the Hilbert–Schmidt norm on $\ell_2 \otimes \ell_2$ (for example: ε, π) and $1 \leq p < \infty$. Then the ideals $\mathfrak{L}_p^{inj}$ and $\mathfrak{L}_{p'}^{sur}$ are not β-tensorstable.*

PROOF: Since ℓ_2 is a subspace of L_p, the first case follows immediately from the theorem. Dually, ℓ_2 is a quotient of $L_{p'}$; assume that $E = \ell_2 \tilde{\otimes}_\beta \ell_2 \in \text{space}(\mathfrak{L}_{p'}^{sur})$. Then $E' \in \text{space}(\mathfrak{L}_{p'}^{sur \ dual}) = \text{space}(\mathfrak{L}_p^{inj})$ and therefore

$$\ell_2 \tilde{\otimes}_{\beta'} \ell_2 \xrightarrow{1} (\ell_2 \otimes_\beta \ell_2)' \hookrightarrow L_p ,$$

hence β' is equivalent to the Hilbert–Schmidt norm σ on $\ell_2 \otimes \ell_2$ by the theorem. Since $\sigma' = \sigma$ (see 26.7.) and $\beta'' = \beta$ on HILB, this is not possible. □

34.9. The tensor stability with respect to ε and π of the ideal $\mathfrak{L}_{p,q}$ of (p,q)-factorable operators, its injective and surjective hull and its minimal kernel $\mathfrak{K}_{p,q} = \mathfrak{L}_{p,q}^{min}$ is completely known:

THEOREM: *The following two tables show which of the ideals in the first column are ε- or π-tensorstable:*

	ε	π
$\mathfrak{L}_{p,q}$	$\mathfrak{J}_p$	$\mathfrak{J}_p^{dual}$
$\mathfrak{L}_{p,q}^{inj}$	$\mathfrak{P}_p$	$\mathfrak{L}$
$\mathfrak{L}_{p,q}^{sur}$	$\mathfrak{L}$	$\mathfrak{P}_p^{dual}$

	ε	π
$\mathfrak{K}_{p,q}$	$\mathfrak{N}_p$	$\mathfrak{K}_{1,p}$
$\mathfrak{K}_{p,q}^{inj}$	$\mathfrak{N}_p^{inj}$	$\mathfrak{K}$
$\mathfrak{K}_{p,q}^{sur}$	$\mathfrak{K}$	$\mathfrak{K}_{1,p}^{sur}$

For example, among the ideals $\mathfrak{L}_{p,q}$ only the $\mathfrak{L}_{p,1} = \mathfrak{J}_p$ are ε-tensorstable, all others are not. The result is taken from Carl–D.–Ramanujan [23].

PROOF: The positive results have already been established in 34.5., corollary 2.
First the negative results for the maximal ideals will be settled. Since, by proposition 34.4.(5), the ideal $\mathfrak{A}^{dual}$ is ε-tensorstable if $\mathfrak{A}$ is π-tensorstable, it is easy to see that it must only be checked that the ideals not mentioned in the first column are not ε-tensorstable. Moreover, by 34.4.(4), the negative statements about $\mathfrak{L}_{p,q}^{inj}$ imply those about $\mathfrak{L}_{p,q}$.

So take $\mathfrak{L}_{p,q}^{inj}$ with $q > 1$ (since $\mathfrak{L}_{p,1}^{inj} = \mathfrak{P}_p$ is the positive case). For the ideal $\mathfrak{L}_{1,\infty}^{inj} = \mathfrak{L}_1^{inj}$ ε-tensorstability would imply (see 34.7., corollary 1) that $\ell_\infty \in \text{space}(\mathfrak{L}_1^{inj})$ — but this is not true since, for example, ℓ_∞ does not have cotype 2.

If $p, q \in]1, \infty[$, then $\mathfrak{L}_{p,q}^{inj} \subset \mathfrak{W}$, the weakly compact operators. Now $\ell_2 \in \text{space}(\mathfrak{L}_{p,q}) \subset \text{space}(\mathfrak{L}_{p,q}^{inj})$ by 17.13., but $\ell_2 \tilde{\otimes}_\varepsilon \ell_2$ is not reflexive.

It remains to check the case $\mathfrak{L}_{1,q}^{inj} = \mathfrak{J}_q^{sur \ dual} \subset \mathfrak{W}$ for $1 < q < \infty$. Since the natural quotient mapping $\ell_1 \twoheadrightarrow \ell_2$ is in

$$\mathfrak{P}_2(\ell_1, \ell_2) = \mathfrak{J}_2(\ell_1, \ell_2) \subset \mathfrak{J}_q(\ell_1, \ell_2)$$

by the little Grothendieck theorem and 26.4.(3''), it follows that $\ell_2 \in \text{space}(\mathfrak{J}_q^{sur})$ and hence $\ell_2 \in \text{space}(\mathfrak{J}_q^{sur \ dual})$ also. The same argument as before gives that $\mathfrak{L}_{1,q}^{inj}$ is not ε-tensorstable.

Now $\mathfrak{L}_{p,q}^{sur}$ will be treated for $(p, q) \neq (1, \infty)$. It will be shown that for all these (p, q) the ideal $\mathfrak{L}_{p,q}^{sur \ inj}$ is not ε-tensorstable (which is enough by 34.4.(4)). Grothendieck's inequality shows that

$$\mathfrak{L}_2 \subset \mathfrak{J}^{sur \ inj} \subset \mathfrak{L}_{p,q}^{sur \ inj}$$

— and the latter ideal is contained in $\mathfrak{W}$ if p or $q \in]1, \infty[$; the space $\ell_2 \tilde{\otimes}_\varepsilon \ell_2$ destroys ε-tensorstability. The same applies to

$$\mathfrak{L}_{1,1}^{sur \ inj} = \mathfrak{J}^{sur \ inj} \subset \mathfrak{L}_2 \subset \mathfrak{W} \ .$$

So it remains to consider $\mathfrak{L}_{\infty,1}^{sur} = \mathfrak{L}_\infty^{sur}$ — but this is not ε-tensorstable as was shown in 34.8., corollary.

The negative results for the minimal ideals follow from the negative results for the maximal ones: Those in the ε-column using 34.4.(3), those in the π-column using 34.4.(5). □

34.10. Tensor stability has interesting consequences for the distribution of eigenvalues of power compact operators (more generally: of Riesz operators, see e.g. Pietsch [216]). These operators $T : E \to E$ (in a complex Banach space) have at most countably many eigenvalues $\lambda \neq 0$, each of which has finite *algebraic* multiplicity $n(T, \lambda)$ defined to be the dimension of

$$N_\infty(\lambda id_E - T) := \bigcup_{n=1}^\infty \ker(\lambda id_E - T)^n \ .$$

With each such operator $T \in \mathcal{L}(E, E)$ one can associate a (not necessarily unique) sequence $(\lambda_n(T))_{n \in \mathbb{N}}$ of eigenvalues such that

(1) Each eigenvalue $\lambda \neq 0$ appears in the sequence exactly $n(T, \lambda)$-times and consecutively.

(2) $|\lambda_1(T)| \geq |\lambda_2(T)| \geq \ldots$.

(3) If T has exactly k eigenvalues $\lambda \neq 0$ (counted according to their multiplicity), then $\lambda_n(T) = 0$ for all $n > k$.

Using Weyl's inequality in Hilbert spaces, one can show that Hilbert–Schmidt operators have eigenvalues $(\lambda_n(T))$ in ℓ_2. Since every absolutely 2–summing operator $T : E \to E$ admits a factorization $T = R \circ S$ through a Hilbert space such that $S \circ R$ is Hilbert–Schmidt, it follows from the principle of related operators (due to Pietsch) that T has eigenvalues in ℓ_2 as well. In particular, nuclear operators have eigenvalues in ℓ_2; in Hilbert spaces they have eigenvalues in ℓ_1 and it can be shown that a Banach space E is isomorphic to a Hilbert space if all $T \in \mathfrak{N}(E, E)$ have summable eigenvalues (Johnson–König–Maurey–Retherford [131]).

In order to prove eigenvalue estimates the following result, due to Pietsch [215], is sometimes helpful:

PROPOSITION: *Let $(\mathfrak{A}, \mathbf{A})$ be a quasi–Banach operator ideal of Riesz operators such that for all Banach spaces E and all $T \in \mathfrak{A}(E, E)$ the eigenvalue sequence $(\lambda_n(T))$ is in ℓ_p (for $0 < p < \infty$). If there is a tensor norm β such that $(\mathfrak{A}, \mathbf{A})$ is β–tensorstable with constant c, then*

$$(*) \qquad \left(\sum_{n=1}^{\infty} |\lambda_n(T)|^p \right)^{1/p} \leq c \mathbf{A}(T)$$

for all $T \in \mathfrak{A}(E, E)$.

PROOF: First observe that there is a $c_1 \geq 1$ such that the inequality $(*)$ holds with c_1: If not there would be $T_m \in \mathfrak{A}(E_m, E_m)$ with $\mathbf{A}(T_m) \leq (2d)^{-m}$ (where d is the constant in the quasi–triangle inequality for $\mathbf{A}$) and $\|(\lambda_n(T_m))\|_p \geq m$. Considering $\sum_m T_m$ in $\ell_2(E_m)$, gives a contradiction.

Now take c_1 to be the best constant in the inequality $(*)$. Since for eigenvalues λ and μ of $T \in \mathfrak{A}(E, E)$ one has

$$N_\infty(\lambda id_E - T) \otimes N_\infty(\mu id_E - T) \subset N_\infty(\lambda \mu id_{E \otimes E} - T \otimes T)$$

(see Pietsch [216], 3.4.7., for a proof), it follows that

$$\left(\sum_{n=1}^{\infty} |\lambda_n(T)|^p \right)^{2/p} = \left(\sum_{n,m=1}^{\infty} |\lambda_n(T) \lambda_m(T)|^p \right)^{1/p} \leq$$

$$\leq \left(\sum_{k=1}^{\infty} |\lambda_k(T \tilde{\otimes}_\beta T)|^p \right)^{1/p} \leq c_1 \mathbf{A}(T \tilde{\otimes}_\beta T) \leq c_1 c \mathbf{A}(T)^2 ,$$

and hence that $c_1 \leq (c_1 c)^{1/2}$ since c_1 was chosen to be the best possible constant. It follows that $c_1 \leq c$. □

Note that for this "trick" only $T \tilde{\otimes}_\beta T \in \mathfrak{A}$ was used and that *the constant c does not depend on p*: This will be important in a moment. By the way, the trick was already used in remark 34.2..

Using König's Weyl inequality for Banach spaces (see [145]), one can show that in the case $2 \leq p < \infty$ there are constants c_p with

$$\sup_n |\lambda_n(T)| n^{1/p} \leq c_p \mathbf{P}_p(T)$$

for all $T \in \mathfrak{P}_p(E, E)$ (see [145],[131]). In particular, this "weak inequality" implies that $(\lambda_n(T)) \in \ell_{p+\varepsilon}$ for all $\varepsilon > 0$. Since $\mathfrak{P}_p$ is metrically ε-tensorstable (see 34.5.), it follows from the proposition that

$$\Big(\sum_{n=1}^\infty |\lambda_n(T)|^{p+\varepsilon}\Big)^{1/p+\varepsilon} \leq \mathbf{P}_p(T)$$

for all $\varepsilon > 0$ and $T \in \mathfrak{P}_p(E, E)$ and hence,

THEOREM: *If $T \in \mathfrak{L}(E, E)$ is absolutely p-summing for $2 \leq p < \infty$, then*

$$\Big(\sum_{n=1}^\infty |\lambda_n(T)|^p\Big)^{1/p} \leq \mathbf{P}_p(T) \, .$$

This important theorem is due to Johnson-König-Maurey-Retherford [131], the idea of proving it with the "tensor product trick" is due to Pietsch [215]. For more details, see the books of König [147] and Pietsch [216].

The proposition about the tensor product trick also has a negative flavour. For example, the inequality

$$\sup_n |\lambda_n(T)| n^{1/p} < \infty$$

actually holds for all $T \in \mathfrak{P}_{p,2}(E, E)$ with $2 \leq p < \infty$, see [131] — but for $2 < p$ the eigenvalues are not in ℓ_p (see König-Retherford-Tomczak-Jaegermann [150]). The above reasoning implies the

REMARK: *For $2 < p < \infty$ the ideal $\mathfrak{P}_{p,2}$ of absolutely $(p,2)$-summing operators is not tensorstable for any tensor norm.*

See also Ex 34.8. and Ex 34.9. for the tensor stability of the ideal $\mathfrak{P}_{r,s}$.

34.11. The ideals $\mathfrak{P}_{p,1}$ and $\mathfrak{M}_{p',1}$ nearly coincide (see 32.10.–32.12.). A variant of the tensor product trick will give a nice improvement of the following theorem, due to Kwapień [157].

THEOREM: For $p \in [1, \infty]$ and $1/r = 1 - |1/p - 1/2|$
$$\mathfrak{L}(\ell_1, \ell_p) = \mathfrak{P}_{r,1}(\ell_1, \ell_p) \text{ and } \mathbf{P}_{r,1}(T) \leq 2^{\Theta/2} K_G^{1-\Theta} \|T\|$$
for $T \in \mathfrak{L}(\ell_1, \ell_p)$ and $\Theta := |1 - 2/p|$.

Note, the special cases

$p = 1$: ℓ_1 has the Orlicz property (see 8.9.).
$p = 2$: Grothendieck's theorem/inequality.
$p = 4/3$: an extension of Littlewood's result mentioned in 10.8. — this will be clarified in a moment.

PROOF (taken from Tomczak–Jaegermann [272]) : Using $\mathfrak{L}_p$-local technique, it suffices to find a constant c_p such that
$$\mathbf{P}_{r,1}(T) \leq c_p \|T\|$$
for all $T \in \mathfrak{L}(\ell_1^n, \ell_p^n)$; equivalently: $\ell_r(Tx_i; \ell_p^n) \leq c_p \|T\|$ for all $x_1, ..., x_m \in \ell_1^n$ with $w_1(x_i; \ell_1^n) = 1$. Fix such $(x_1, ..., x_m)$. Then one has to show that $\Phi_p(T) := (Tx_i)$ satisfies
$$\|\Phi_p : \mathfrak{L}(\ell_1^n, \ell_p^n) \longrightarrow \ell_r^m(\ell_p^n)\| \stackrel{!}{\leq} c_p .$$
Since $\mathfrak{L}(\ell_1^n, \ell_p^n) = \ell_\infty^n \otimes_\varepsilon \ell_p^n = \ell_\infty^n(\ell_p^n)$, this means
$$\|\Phi_p : \ell_\infty^n(\ell_p^n) \longrightarrow \ell_r^m(\ell_p^n)\| \stackrel{!}{\leq} c_p .$$

Φ_p as a map $\mathbb{K}^{n \times n} \longrightarrow \mathbb{K}^{m \times n}$ does not depend on p, so this is a typical situation for using interpolation. By Grothendieck's inequality and $\mathbf{P}_{2,1}(id_{\ell_1}) \leq C_2(\ell_1) \leq a_1 = \sqrt{2}$ (Orlicz's theorem 8.9., 8.6.) one gets
$$\|\Phi_1 : \ell_\infty^n(\ell_1^n) \longrightarrow \ell_2^m(\ell_1^n)\| \leq \sqrt{2}$$
$$\|\Phi_2 : \ell_\infty^n(\ell_2^n) \longrightarrow \ell_1^m(\ell_2^n)\| \leq K_G$$
$$\|\Phi_\infty : \ell_\infty^n(\ell_\infty^n) \longrightarrow \ell_2^m(\ell_\infty^n)\| \leq \sqrt{2} .$$

First take $1 \leq p \leq 2$ and Θ defined by
$$\frac{1}{p} = \frac{1 - \Theta}{2} + \frac{\Theta}{1} .$$

Then $1/r = (1 - \Theta)/1 + \Theta/2$. Complex interpolation of $\ell_\infty^n(\cdot)$ (n-tuples!) and of vector valued L_p (see Bergh–Löfström [11], p.107) gives
$$\|\Phi_p\| \leq 2^{\Theta/2} K_G^{1-\Theta} .$$

For $2 \leq p \leq \infty$ start with $1/p = (1 - \Theta)/2 + \Theta/\infty$ — and do the same. □

Before proving that the index r in this result is best possible the connection with Littlewood's result will be clarified: For $\varphi \in \mathfrak{Bil}(c_o, c_o) = \mathfrak{L}(c_o, \ell_1) = \ell_1^w(\ell_1)$ one has $\|L_\varphi\| = \|\varphi\| = w_1(L_\varphi e_n; \ell_1)$ (see 8.2.(1)) and hence Littlewood's inequality implies

$$\left(\sum_{n=1}^{\infty} \|L_\varphi e_n\|_{\ell_{4/3}}^{4/3}\right)^{3/4} = \left(\sum_{n,m=1}^{\infty} |\varphi(e_n, e_m)|^{4/3}\right)^{3/4} \leq c\|\varphi\|$$

for all $\varphi \in \mathfrak{Bil}(c_o, c_o)$, which means exactly that the embedding $I : \ell_1 \hookrightarrow \ell_{4/3}$ is in $\mathfrak{P}_{4/3,1}$. So for $p = r = 4/3$ the theorem generalizes Littlewood's result from I to all operators $T \in \mathfrak{L}(\ell_1, \ell_{4/3})$. The theorem gives $c \leq 2^{1/4} K_G^{1/2}$ for the best constant in Littlewood's inequality.

PROPOSITION: *Let $p, r \in [1, \infty]$. If $\mathfrak{L}(\ell_1, \ell_p) = \mathfrak{P}_{r,1}(\ell_1, \ell_p)$, then $1/r \leq 1 - |1/p - 1/2|$. For $p \leq 2$ this conclusion holds if only $\mathbf{P}_{r,1}(I : \ell_1 \hookrightarrow \ell_p) < \infty$.*

In particular, this means that the index r in the theorem cannot be improved (a result also due to Kwapień [157]) and that in the Littlewood inequality the exponent $p = 4/3$ is best possible as well.

PROOF: Walsh matrices will be used. They are defined by induction:

$$A_1 := \begin{pmatrix} 1 & 1 \\ 1 & -1 \end{pmatrix} \quad \text{and} \quad A_{n+1} := \begin{pmatrix} A_n & A_n \\ A_n & -A_n \end{pmatrix}.$$

$2^{-n/2} A_n : \ell_2^{2^n} \longrightarrow \ell_2^{2^n}$ is orthogonal and self-adjoint, hence $\|A_n : \ell_2^{2^n} \longrightarrow \ell_2^{2^n}\| = 2^{n/2}$ and $A_n \circ A_n = 2^n \text{id}$. Moreover, $\|A_n : \ell_1^{2^n} \longrightarrow \ell_\infty^{2^n}\| = 1$ and the factorization

$$A_n : \ell_\infty^{2^n} \xrightarrow{\text{id}} \ell_2^{2^n} \xrightarrow{A_n} \ell_2^{2^n} \xrightarrow{\text{id}} \ell_1^{2^n}$$

give $\|A_n : \ell_\infty^{2^n} \longrightarrow \ell_1^{2^n}\| \leq (2^n)^{3/2}$. To prove the result, observe first that there is a $c \geq 1$ satisfying

$$\mathbf{P}_{r,1}(T : \ell_1^{2^n} \longrightarrow \ell_p^{2^n}) \leq c\|T\|$$

for all $T \in \mathfrak{L}(\ell_1^{2^n}, \ell_p^{2^n})$ or only $T = I_n : \ell_1^{2^n} \longrightarrow \ell_p^{2^n}$ if $1 \leq p \leq 2$ (and all $n \in \mathbb{N}$). Take the latter case first: For $A_n : \ell_\infty^{2^n} \longrightarrow \ell_1^{2^n}$ one has

$$\left(\sum_{k=1}^{2^n} \|I_n A_n e_k\|_p^r\right)^{1/r} \leq c \cdot w_1(A_n e_k; \ell_1^{2^n}) = c \cdot \|A_n : \ell_\infty^{2^n} \longrightarrow \ell_1^{2^n}\| \leq c(2^n)^{3/2},$$

hence, $((2^n)^{r/p} 2^n)^{1/r} \leq c(2^n)^{3/2}$, which is only possible if the claim about r is satisfied. If $2 \leq p \leq \infty$, consider

$$2^n (2^n)^{1/r} = \left(\sum_{k=1}^{2^n} \|A_n A_n e_k\|_p^r\right)^{1/r} \leq c \cdot \|A_n : \ell_1^{2^n} \longrightarrow \ell_p^{2^n}\| \, w_1(A_n e_k; \ell_1^{2^n}) =$$
$$= c \cdot \sup_k \|A_n e_k\|_p \|A_n : \ell_\infty^{2^n} \longrightarrow \ell_1^{2^n}\| \leq c \cdot (2^n)^{1/p + 3/2}$$

— and again r satisfies the desired inequality. □

34.12. Next the announced improvement (due to Carl–D. [21]) of Kwapień's theorem 34.11. will be given. Recall that $\mathfrak{M}_{q,1} \subset \mathfrak{P}_{q',1}$.

THEOREM: *Let $p, q \in [1, \infty]$ with $1/q = |1/2 - 1/p|$. Then*
$$\mathfrak{L}(\ell_1, \ell_p) = \mathfrak{M}_{q,1}(\ell_1, \ell_p)$$
and $\mathbf{M}_{q,1}(T) \leq K_G^2 \|T\|$ for all $T \in \mathfrak{L}(\ell_1, \ell_p)$.

In particular, this space coincides with $\mathfrak{P}_{q',1}(\ell_1, \ell_p)$ and therefore the index q cannot be improved by proposition 34.11.. See Ex 34.11. for two interesting reformulations.

As was already mentioned, the proof will use a trick involving tensor stability — this time with respect to the right–tensor norms Δ_p. The idea is condensed into the following lemma which shows how to improve weak inequalities.

LEMMA 1: *Let $\mathfrak{A} \subset \bigcup_{n=1}^\infty \mathfrak{L}(\mathbb{K}^n, \mathbb{K}^n)$ be a set of operators and $\mathbf{A}, \mathbf{B} : \mathfrak{A} \longrightarrow [0, \infty[$ be two functions such that the following three conditions are satisfied:*

(1) *For every $\varepsilon > 0$ there is a $c(\varepsilon) > 0$ such that*
$$\mathbf{A}(T) \leq c(\varepsilon) n^\varepsilon \mathbf{B}(T) \qquad \text{for all } T \in \mathfrak{L}(\mathbb{K}^n, \mathbb{K}^n) \cap \mathfrak{A}.$$

(2) $[T \otimes T : \mathbb{K}^n \otimes \mathbb{K}^n = \mathbb{K}^{n^2} \longrightarrow \mathbb{K}^{n^2}] \in \mathfrak{A}$ *whenever $T \in \mathfrak{A}$.*

(3) *There are constants a and b such that for all $T \in \mathfrak{A}$*
$$\mathbf{A}(T)^2 \leq a\,\mathbf{A}(T \otimes T) \text{ and } \mathbf{B}(T \otimes T) \leq b\,\mathbf{B}(T)^2.$$

Then $\mathbf{A}(T) \leq ab\,\mathbf{B}(T)$ for all $T \in \mathfrak{A}$.

PROOF: Fix $\varepsilon > 0$. Then for each $T \in \mathfrak{L}(\mathbb{K}^n, \mathbb{K}^n)$ in $\mathfrak{A}$
$$\mathbf{A}(T)^2 \leq a\mathbf{A}(T \otimes T) \leq ac(\varepsilon) n^{2\varepsilon} \mathbf{B}(T \otimes T) \leq abc(\varepsilon) n^{2\varepsilon} \mathbf{B}(T)^2,$$
hence $\mathbf{A}(T) \leq \sqrt{abc(\varepsilon)} n^\varepsilon \mathbf{B}(T)$. Replacing $c(\varepsilon)$ by $\sqrt{abc(\varepsilon)}$ and iterating the argument, leads to
$$\mathbf{A}(T) \leq (ab)^{\sum_{i=1}^k 2^{-i}} \cdot (c(\varepsilon))^{2^{-k}} \cdot n^\varepsilon \cdot \mathbf{B}(T).$$

Taking first $k \to \infty$ and then $\varepsilon \to 0$ shows the result. □

The reader may have noticed the similarity between the proofs of lemma 1 and proposition 34.10..

LEMMA 2: For $p, q \in [1, \infty]$ and $T \in \mathcal{L}(\ell_1^n, \ell_p^n)$

$$\mathbf{M}_{q,1}(T)^2 \leq K_G^2 \mathbf{M}_{q,1}(T \otimes T : \ell_1^n \otimes_{\Delta_1} \ell_1^n \longrightarrow \ell_p^n \otimes_{\Delta_p} \ell_p^n) .$$

PROOF: Recall first from Ex 26.5. that for $x_1, ..., x_m \in \ell_1^n$

$$w_1(x_i \otimes x_j; \ell_1^n \otimes_{\Delta_1} \ell_1^n) \leq K_G^2 w_1(x_i; \ell_1^n) w_1(x_j; \ell_1^n) .$$

In order to use the discrete description of the $(q, 1)$–mixing norm from 32.4.(c), take $y_1', ..., y_m' \in \ell_{p'}^n$. Then

$$\left[\sum_{j=1}^m \left(\sum_{k=1}^m |\langle y_k', Tx_j\rangle|^q\right)^{1/q}\right]^2 = \sum_{i,j=1}^m \left(\sum_{k,\ell=1}^m |\langle y_k' \otimes y_\ell', T \otimes T(x_i \otimes x_j)\rangle|^q\right)^{1/q} \leq$$

$$\leq \mathbf{M}_{q,1}(\ell_1^n \otimes_{\Delta_1} \ell_1^n \xrightarrow{T \otimes T} \ell_p^n \otimes_{\Delta_p} \ell_p^n) w_1(x_i \otimes x_j; \ell_1^n \otimes_{\Delta_1} \ell_1^n) \ell_q(y_k' \otimes y_\ell'; \ell_{p'}^n \otimes_{\Delta_{p'}} \ell_{p'}^n) \leq$$
$$\leq \mathbf{M}_{q,1}(T \otimes T) K_G^2 [w_1(x_j; \ell_1^n) \cdot \ell_q(y_k'; \ell_{p'}^n)]^2$$

and this concludes the proof. □

PROOF of the theorem: One may exclude the cases $p = 2$ (Grothendieck's theorem) and $p = \infty$ (since $id_{\ell_1} \in \mathfrak{M}_{2,1}$). For $T \in \mathcal{L}(\mathbb{K}^n, \mathbb{K}^n)$ define

$$\mathbf{A}(T) := \mathbf{M}_{q,1}(T : \ell_1^n \longrightarrow \ell_p^n) \text{ and } \mathbf{B}(T) := \|T : \ell_1^n \longrightarrow \ell_p^n\| .$$

Beckner's result (7.9. or 15.12.) and $\ell_p^n \otimes_{\Delta_p} \ell_p^n = \ell_p^{n^2}$ show that $\mathbf{B}(T \otimes T) \leq \mathbf{B}(T)^2$ and lemma 2 gives that $\mathbf{A}(T)^2 \leq K_G^2 \mathbf{A}(T \otimes T)$. In order to apply the tensor product trick from lemma 1 one needs the weak inequality

$$\mathbf{A}(T) \leq c(\varepsilon) n^\varepsilon \mathbf{B}(T) .$$

For this, Maurey's theorem 32.11.

$$\mathbf{M}_{q,1} \leq c(r) \mathbf{P}_{r,1} \qquad\qquad \text{if } q < r'$$

and Kwapień's theorem 34.11.

$$\mathbf{P}_{r,1}(S : \ell_1^n \longrightarrow \ell_{p+\delta}^n) \leq 4\|S : \ell_1^n \longrightarrow \ell_{p+\delta}^n\| \qquad \text{if } 1/r' = |1/(p+\delta) - 1/2|$$

will be used: For $\varepsilon \in\ ]0, 1[$ choose $\delta \in \mathbb{R}$ with $p + \delta \geq 1$ such that

$$\varepsilon = \left|\frac{1}{p+\delta} - \frac{1}{p}\right| \text{ and } \frac{1}{r'_\varepsilon} := \left|\frac{1}{p+\delta} - \frac{1}{2}\right| < \frac{1}{q} = \left|\frac{1}{p} - \frac{1}{2}\right| .$$

It follows for $T \in \mathfrak{L}(\ell_1^n, \ell_p^n)$ and the identity $I : \ell_p^n \longrightarrow \ell_{p+\delta}^n$ that

$$\mathbf{M}_{q,1}(T) \leq c(r_\varepsilon)\mathbf{P}_{r_\varepsilon,1}(I^{-1} \circ I \circ T) \leq c(r_\varepsilon) \cdot \|I^{-1}\| \cdot \mathbf{P}_{r_\varepsilon,1}(I \circ T : \ell_1^n \longrightarrow \ell_{p+\delta}^n) \leq$$
$$\leq c(r_\varepsilon) \cdot \|I^{-1}\| \cdot 4 \cdot \|I \circ T\| \leq 4 \cdot c(r_\varepsilon) \|I^{-1}\| \, \|I\| \, \|T\| \leq 4c(r_\varepsilon) n^\varepsilon \|T\|$$

— and this is the desired weak estimate. Now apply lemma 1. □

Littlewood's result was that $I \in \mathfrak{P}_{4/3,1}(\ell_1, \ell_{4/3})$ (see 34.11.) — the theorem shows that this operator is even $(4,1)$-mixing which means: Every absolutely 4-summing operator $T : \ell_{4/3} \longrightarrow F$ is absolutely 1-summing when restricted to ℓ_1.
The $\mathfrak{L}_p$-local technique lemma for operator ideals 23.1. shows that the theorem implies $\mathfrak{L}_p \circ \mathfrak{L}_1 \subset \mathfrak{M}_{q,1}$ — or, by the definition of $(q,1)$-mixing operators

$$\mathfrak{P}_q \circ \mathfrak{L}_p \circ \mathfrak{L}_1 \subset \mathfrak{P}_1 \, .$$

Cyclic composition (see Ex 25.16.) gives the

COROLLARY: *Let $p, q \in [1, \infty]$ and $1/q = |1/2 - 1/p|$. Then*

$$\mathfrak{L}_\infty \circ \mathfrak{P}_q \circ \mathfrak{L}_p \subset \mathfrak{P}_1^{dual} \quad , \quad \mathfrak{L}_1 \circ \mathfrak{L}_\infty \circ \mathfrak{P}_q \subset \mathfrak{D}_{p'}$$
$$\mathfrak{L}_p \circ \mathfrak{L}_1 \circ \mathfrak{L}_\infty \subset \mathfrak{I}_{q'} \quad , \quad \mathfrak{L}_p \circ \mathfrak{K}_1 \circ \mathfrak{L}_\infty \subset \mathfrak{N}_{q'}$$
$$\mathfrak{K}_p \circ \mathfrak{L}_1 \circ \mathfrak{L}_\infty \subset \mathfrak{N}_{q'} \quad , \quad \mathfrak{L}_p \circ \mathfrak{L}_1 \circ \mathfrak{K}_\infty \subset \mathfrak{N}_{q'} \, .$$

For an application of this to the question of which absolutely q-summing operators factor through a Hilbert space see Ex 34.12..

Bennett [9] and Carl [20] have investigated the norms $\mathbf{P}_{r,s}(I : \ell_u \longrightarrow \ell_v)$ and recently Carl-D. [21] the norms $\mathbf{M}_{q,p}(I : \ell_u \longrightarrow \ell_v)$ for other indices u, v. The latter paper also gives further applications of the tensor product trick presented in lemma 1.

Exercises:

Ex 34.1. Let $(\mathfrak{A}, \mathbf{A})$ be a quasi-Banach operator ideal. If $T_1 \tilde{\otimes}_\beta T_2 \in \mathfrak{A}$ and $T_2 \neq 0$, then $T_1 \in \mathfrak{A}$ and $\mathbf{A}(T_1)\|T_2\| \leq \mathbf{A}(T_1 \tilde{\otimes}_\beta T_2)$.

Ex 34.2. (a) Show that $\mathfrak{L}_p$ is metrically β-tensorstable for $\beta = g_p, g_{p'}^*, d_p, d_{p'}^*$, and that for these β the β-tensor product of two $\mathcal{L}_p^g$-spaces is an $\mathcal{L}_p^g$-space. See also Ex 23.17.. Hint: 34.5..

(b) $\mathfrak{D}_2$ is strongly β-tensorstable for $\beta = d_2$ and g_2. Hint: 34.5., corollary 1.

(c) $(\mathfrak{L}_1 \circ \mathfrak{L}_\infty)^{reg}$ is β-tensorstable, but not metrically β-tensorstable for the tensor norms $\beta = d_2$ and g_2. Hint: (b), $\mathbf{D}_2 \leq K_G(\mathbf{L}_1 \circ \mathbf{L}_\infty)^{reg}$ and the tensor product trick 34.12..

Ex 34.3. (a) Use factorization and the fact that $L_1 \otimes_\pi \cdot$ respects subspaces to show that $T_1 \tilde{\otimes}_\beta T_2 : E_1 \tilde{\otimes}_\beta E_2 \longrightarrow F_1 \tilde{\otimes}_\beta F_2$ is absolutely 1–summing if $T_1 \in \mathfrak{J}$ and $T_2 \in \mathfrak{P}_1$.

(b) Use the factorization $E_1 \longrightarrow c_0 \xrightarrow{D_\lambda} \ell_1 \longrightarrow F_1$ of nuclear operators to show that $T_1 \tilde{\otimes}_\beta T_2$ is quasinuclear whenever $T_1 \in \mathfrak{N}$ and $T_2 \in \mathfrak{QN}$.

These results are due to Holub [117].

Ex 34.4. (a) For $1 \leq p \leq \infty$ the ideal $\mathfrak{L}_p$ is neither w_2 nor w_2'-tensorstable. Hint: For $1 \leq p < \infty$ use the fact that $\mathfrak{L}_\infty \not\subset \mathfrak{L}_p$ and for $p = \infty$ Grothendieck's inequality, tensor form, and 34.7..

(b) For $1 < p \leq 2$ the ideal $\mathfrak{T}_p$ of type p operators is neither w_2 nor w_2'-tensorstable.

(c) For $2 \leq q < \infty$ the ideal $\mathfrak{C}_q$ of cotype q operators is not w_2-tensorstable. However, it is not known whether or not $\mathfrak{C}_2$ is π-tensorstable; recall from 34.7. the open question of whether or not $\ell_2 \tilde{\otimes}_\pi \ell_2 \tilde{\otimes}_\pi \ell_2$ has cotype 2.

Ex 34.5. (a) Prove the following extension of Schauder's theorem due to Vala [273] : If $T \in \mathfrak{L}(E_1, E_2)$ and $S \in \mathfrak{L}(E_3, E_4)$ are compact, then

$$\text{Hom}(T,S) : \mathfrak{L}(E_2, E_3) \longrightarrow \mathfrak{L}(E_1, E_4)$$
$$R \rightsquigarrow SRT$$

is compact as well. Hint: Use the compactness of $E_1 \tilde{\otimes}_\pi E_4' \xrightarrow{T \otimes S'} E_2 \tilde{\otimes}_\pi E_3'$ to show that $\text{Hom}(T, S'')$ is compact.

(b) Use the same type of argument to show that for $1 \leq p < \infty$

$$\text{Hom}(T, S) \in \mathfrak{P}_p(\mathfrak{L}(E_2, E_3), \mathfrak{L}(E_1, E_4))$$

if $S \in \mathfrak{P}_p$ and $T \in \mathfrak{P}_p^{dual}$.

Ex 34.6. (a) Let $1 \leq p \leq \infty$. If a Banach space E contains all ℓ_p^n uniformly complemented and $E \otimes_\varepsilon G = E \otimes_\pi G$ holds topologically, then G is finite dimensional. Hint: Use 29.5. to show that $id_G \in \mathfrak{D}_p$.

(b) There are no infinite dimensional Banach spaces E, F and G such that $E \otimes_\varepsilon F \otimes_\varepsilon G = E \otimes_\pi F \otimes_\pi G$ holds topologically. Hint: $\ell_\infty^n \hookrightarrow E \otimes_\varepsilon F = E \otimes_\pi F$.

Ex 34.7. Let $(\mathfrak{A}, \mathbf{A})$ and $(\mathfrak{B}, \mathbf{B})$ be two quasi–Banach ideals such that $\mathfrak{A} \subset \mathfrak{B}$ and $\mathbf{B}(\cdot) \leq c\mathbf{A}(\cdot)$. If both ideals are β-tensorstable and for all $T \in \mathfrak{B}$

$$\mathbf{B}(T)^2 \leq \mathbf{B}(T \tilde{\otimes}_\beta T) \text{ and } \mathbf{A}(T \tilde{\otimes}_\beta T) \leq \mathbf{A}(T)^2 ,$$

then c can be chosen to be 1.

Ex 34.8. Show that $\mathfrak{P}_{r,1}$ is not ε-tensorstable for $2 \leq r < \infty$. Hint: $\ell_2 \in \text{space}(\mathfrak{P}_{r,1})$. This result even holds for all $r \in]1, \infty[$ which will be shown in the next exercise:

Ex 34.9. (a) Show that
$$w_p((x_i \otimes y_j)_{i,j}; E \otimes_\varepsilon F) = w_p(x_i; E) w_p(y_j; F)$$
$$\ell_p((x_i \otimes y_j)_{i,j}; E \otimes_\varepsilon F) = \ell_p(x_i; E) \ell_p(y_j; F) .$$

Hint: Use 4.6. or Ex 8.1.(a) and $\ell_{p'}^n \otimes_{\Delta_{p'}} \ell_{p'}^m = \ell_{p'}^{nm}$.

(b) Use this and 32.4.(c) to prove that
$$\mathbf{M}_{q,p}(T_1) \mathbf{M}_{q,p}(T_2) \leq \mathbf{M}_{q,p}(T_1 \tilde\otimes_\varepsilon T_2) \in [1, \infty] .$$

(c) Assume that $1/q + 1/r = 1/p$. If $\mathfrak{P}_{r,p}$ is ε-tensorstable, then $\mathfrak{M}_{q,p} = \mathfrak{P}_{r,p}$. Hint: Use the first estimate in 32.12. and lemma 1 of 34.12..

(d) Use (c) and 32.12. to show that $\mathfrak{P}_{r,1}$ is not ε-tensorstable for all $1 < r < \infty$.

Ex 34.10. (a) Deduce from Kwapień's result 34.11. and Ex 11.21. that
$$\mathfrak{P}_{r,s}(\ell_1, \ell_p) = \mathfrak{L}(\ell_1, \ell_p)$$
for $p, r, s \in [1, \infty]$ such that $1/r \leq 1/s - |1/p - 1/2|$.

(b) Show that this equality also holds for $1 \leq p \leq 2 \leq s \leq \infty$ and $r := p's/2$ by interpolation between the case $p = 1$ (trivial) and $p = 2$ (little Grothendieck theorem). Hint: Proceed as in 34.11.. The second result is taken from Bennett [10].

Ex 34.11. Let $p, q \in [1, \infty], 1/q = |1/2 - 1/p|$ and $T \in \mathfrak{L}(\ell_1, \ell_p)$.

(a) For every probability measure μ on $B_{\ell_p'}$ there is a probability measure ν on B_{ℓ_∞} such that
$$\left(\int_{B_{\ell_p'}} |\langle y', Tx \rangle|^q \mu(dy') \right)^{1/q} \leq K_G^2 \|T\| \int_{B_{\ell_\infty}} |\langle x', x \rangle| \nu(dx')$$
for all $x \in \ell_1$.

(b) For each weakly summable sequence (x_n) in ℓ_1 there is a sequence $(\lambda_n) \in \ell_{q'}$ and a weakly q-summable sequence (y_n) in ℓ_p such that $Tx_n = \lambda_n y_n$ for all $n \in \mathbb{N}$.

In particular, for every unconditionally summable sequence (x_n) in ℓ_1 there is a sequence $(\lambda_n) \in \ell_{4/3}$ and a weakly 4-summable sequence in $\ell_{4/3}$ such that $x_n = \lambda_n y_n$ for all n.

Hint: 34.12. and section 32.

Ex 34.12. Show that for $p, q \in [1, \infty]$
$$\mathfrak{P}_q \circ \mathfrak{L}_p \subset \mathfrak{L}_2$$
(every absolutely q-summing operator on L_p factors through a Hilbert space) if and only if $1/q \geq |1/p - 1/2|$. Hint: $\mathfrak{L}_2 = \mathfrak{D}_2^* = (\mathfrak{L}_1 \circ \mathfrak{L}_\infty)^*$.

35. Tensor Norm Techniques for Locally Convex Spaces

The final section shows that tensor norm techniques can be successfully used for the study of problems in the theory of locally convex spaces. As in the case of Banach spaces, tensor norms (different from ε and π) and local $\mathfrak{L}_p$-techniques help one to better understand various phenomena concerning the projective and injective tensor product of locally convex spaces — such as the investigation of topological–geometric properties, Grothendieck's (DF)-problem and the negative solution of the "problème des topologies" given by Taskinen. Moreover, the so-called tensor norm topologies serve to classify locally convex spaces. Tensor norm topologies were introduced by Harksen in his dissertation 1979 (see also [102]). We assume that the reader is familiar with the basics of the theory of locally convex spaces (see Jarchow [123], Köthe [152]).

35.1. The topology and the uniform structure of a locally convex space are given by the family $cs(E)$ of all continuous seminorms on E — or by the family $\mathfrak{U}(E)$ of all absolutely convex closed (or open) neighbourhoods of zero. For $p \in cs(E)$ denote

$$\kappa_p : E \longrightarrow E/\ker p \; ; \; E_p := (E/\ker p, \hat{p}) \; ; \; \hat{p}(\kappa_p(x)) := p(x) \; .$$

E_p is a normed space which sometimes is also denoted by E_U if $U := \{x \, | \, p(x) \leq 1\}$ and one wants to point at the neighbourhoods instead of the seminorms. It is clear that it is enough to consider bases $\mathfrak{U}_o(E)$ of $\mathfrak{U}(E)$ or bases $cs_o(E)$ of $cs(E)$ (this means: for every $p \in cs(E)$ there are a $q \in cs_o(E)$ and $c \geq 0$ such that $p \leq cq$). If $p \leq cq$, then there is a natural map $\kappa_{q,p} : E_q \longrightarrow E_p$, and the completion of a separated locally convex space E can be represented as the projective limit of the net $(\tilde{E}_p)_{p \in cs_o(E)}$:

$$\tilde{E} = \operatorname*{proj}_{p \in cs_o(E)} \tilde{E}_p \; .$$

The strong topologies on the dual E' and the bidual E'' will be indicated by E'_b and E''_b; the space E'' endowed with its natural topology of uniform convergence on equicontinuous subsets of E' will be denoted by E''_e.

35.2. Given a tensor norm α (on NORM) and locally convex spaces E and F, define for $p \in cs_o(E)$ and $q \in cs_o(F)$ the seminorm $p \otimes_\alpha q$ on $E \otimes F$ by

$$p \otimes_\alpha q(z; E, F) := \alpha(\kappa_p \otimes \kappa_q(z); E_p, F_q)$$

for $z \in E \otimes F$. The family

$$\{p \otimes_\alpha q \, | \, p \in cs_o(E), q \in cs_o(F)\}$$

defines a locally convex topology on $E \otimes F$ (notation: $E \otimes_\alpha F$ and $E \tilde{\otimes}_\alpha F$ for the completion), which is certainly independent of the special bases $cs_o(E)$ and $cs_o(F)$ which were taken. This so-called *tensor norm topology* was introduced by Harksen. For $\alpha = \varepsilon$ or π it coincides with the usual injective or projective topology on the tensor product of locally convex spaces. In many respects the tensor norm topologies behave very similarly to the tensor norms on normed spaces.

PROPOSITION: *Let α be a finitely generated tensor norm.*
(1) *If E and F are separated, then so is $E \otimes_\alpha F$.*
(2) *Mapping property: If $\mathfrak{D}_i \subset \mathfrak{L}(E_i, F_i)$ are equicontinuous, then*

$$\{T_1 \otimes T_2 : E_1 \otimes_\alpha E_2 \longrightarrow F_1 \otimes_\alpha F_2 \mid T_i \in \mathfrak{D}_i\}$$

is equicontinuous.
(3) *Embedding lemma: The natural map*

$$E \otimes_\alpha F \hookrightarrow E \otimes_\alpha F''_e$$

is a topological embedding.
(4) $E \otimes_\alpha F \hookrightarrow \tilde{E} \otimes_\alpha \tilde{F}$ *is a dense topological embedding which in particular means that* $E \tilde{\otimes}_\alpha F = \tilde{E} \tilde{\otimes}_\alpha \tilde{F}$ *holds topologically.*
(5) *Extension lemma: If $D \subset (E \otimes_\alpha F)'$ is equicontinuous, then the set*

$$D^\wedge := \{\varphi^\wedge \mid \varphi \in D\}$$

of right-canonical extensions is equicontinuous in $(E \otimes_\alpha F''_e)'$.

The assumption of α being finitely generated is not needed for (1) and (2); the statements (3) and (4) also hold for cofinitely generated tensor norms.

PROOF: For $0 \neq z \in E \otimes F$ it is easy to see with corollary 2.3. that there are seminorms p, q with $\kappa_p \otimes \kappa_q(z) \neq 0$; this implies (1). (2) is obvious from the definition of the seminorms $p \otimes_\alpha q$.

For (3) observe that for $U \in \mathfrak{U}(F)$ the dual of F_U is isometric to the Banach space $[U^0] \subset F'$ (see 3.1. for this notation) and that

$$(F'')_{U^{00}} \xrightarrow{\ 1\ } [U^0]' \overset{1}{=} (F_U)''$$

is an isometric embedding. The embedding lemma 13.3. for normed spaces shows that

$$E_V \otimes_\alpha F_U \xrightarrow{\ 1\ } E_V \otimes_\alpha (F'')_{U^{00}} ,$$

hence the conclusion.

It follows from 13.4. (with an obvious notation) that the mapping

$$E_V \otimes_\alpha F_U \hookrightarrow (\tilde{E})_{\tilde{V}} \otimes_\alpha (\tilde{F})_{\tilde{U}}$$

is dense and isometric — and this implies (4).

(5) also follows directly from the normed case since every equicontinuous $D \subset (E \otimes_\alpha F)'$ factors through an equicontinuous subset of some $(E_U \otimes_\alpha F_V)'$. □

It is easy to see that the trace $tr_E : E'_b \otimes_\pi E \longrightarrow \mathbb{K}$ is continuous if and only if the locally convex space E is normable. Therefore using the trace requires more precautions than in the Banach space case. Nevertheless, under certain circumstances the trace can be defined on the space $\mathfrak{N}(E, E)$ of all nuclear operators (i.e., for complete E, those operators which factor through a nuclear operator between Banach spaces):

$$\operatorname*{ind}_{p, B \to} E'_p \tilde{\otimes}_\pi [B] \twoheadrightarrow \mathfrak{N}(E, E) ,$$

where the inductive limit runs over all $p \in cs(E)$ and Banach discs $B \subset E$; see [44] for details concerning this topic.

35.3. For the study of right–projective tensor norms (to be done in 35.6.) and traced tensor norms a little lemma is helpful. For a given surjective map $Q : F \twoheadrightarrow G$ and a seminorm p on F, define the quotient seminorm p^Q on G to be

$$p^Q(Q(x)) := \inf\{p(x + y) \mid y \in \ker Q\} .$$

LEMMA: *Let $Q : F \longrightarrow G$ be continuous and surjective. Then Q is a topological surjection if and only if for every $p \in cs_0(F)$ there is a $q \in cs(G)$ such that the canonical map*

$$F_p \longrightarrow G_q$$

is defined, a metric surjection and $\ker \kappa_q \subset \ker p^Q \subset G$.

PROOF: If Q is topological, then $q := p^Q$ is continuous on G — and this gives the conclusion. Conversely, it is enough to show that $p^Q \leq q$. Note that $(p^Q)^{\kappa_q} = \hat{q}$ on G_q. Take $q(x) < 1$. Then

$$(p^Q)^{\kappa_q}(\kappa_q(x)) = \hat{q}(\kappa_q(x)) < 1 ,$$

hence there is $y \in \ker \kappa_q \subset \ker p^Q$ such that $p^Q(x + y) < 1$. It follows that $p^Q(x) \leq p^Q(x + y) + p^Q(y) < 1$. □

Recall the definition of a traced tensor norm $\alpha \otimes_G \beta$ from 29.1.. Its fundamental property also holds for locally convex spaces; the following result is taken from [41]:

THEOREM: Let α and β be two tensor norms (on NORM), G a normed space and $\delta := \alpha \otimes_G \beta$. Then for all locally convex spaces E and F the tensor contraction

$$C_G : (E \otimes_\beta G') \otimes_\pi (G \otimes_\alpha F) \longrightarrow E \otimes_\delta F$$
$$(x \otimes u') \otimes (u \otimes y) \rightsquigarrow \langle u', u \rangle x \otimes y$$

is a topological surjection.

PROOF: For $p_1 \in cs(E)$ and $p_2 \in cs(F)$ consider the following commutative diagram:

$$\begin{array}{ccc}
(E \otimes_\beta G') \otimes_\pi (G \otimes_\alpha F) & \xrightarrow{C_G} & E \otimes_\delta F \\
\downarrow & & \downarrow {\kappa_{p_1} \otimes \kappa_{p_2}} \\
(E_{p_1} \otimes_\beta G') \otimes_\pi (G \otimes_\alpha F_{p_2}) & \xrightarrow{C_G^{p,q}} & E_{p_1} \otimes_\delta F_{p_2}
\end{array}$$

For $p := (p_1 \otimes_\beta \| \cdot \|_{G'}) \otimes_\pi (\| \cdot \|_G \otimes_\alpha p_2)$ and $q := p_1 \otimes_\delta p_2$ the lower tensor contraction $C_G^{p,q}$ is a metric surjection — and, by the lemma, it remains to show that $p^{C_G}(z) = 0$ for all

$$z \in \ker(\kappa_{p_1} \otimes \kappa_{p_2}) = (\ker p_1 \otimes F) + (E \otimes \ker p_2) .$$

Decompose $z = \sum_{i=1}^m x_i \otimes y_i + \sum_{i=m+1}^n x_i \otimes y_i$ according to this equation and choose $(u', u) \in G' \times G$ with $\langle u', u \rangle = 1$. Then

$$C_G\left(w := \sum_{i=1}^n (x_i \otimes u') \otimes (u \otimes y_i) \right) = z$$

and

$$p^{C_G}(z) \leq p(w) \leq \sum_{i=1}^n p_1(x_i) \, \|u'\| \, \|u\| \, p_2(y_i) = 0 . \;\square$$

35.4. Let α and β be finitely generated tensor norms with α accessible. For a given locally convex space E, under which conditions is the operator

$$id_G \otimes id_E : G \otimes_\beta E \longrightarrow G \otimes_\alpha E$$

continuous for all locally convex spaces G? It is enough to check this only for Banach spaces G — or even just $G = C_2$ a Johnson space: Apply lemma 17.16. to

$$id_{C_2} \otimes \kappa_{q,p} : C_2 \otimes_\beta E_q \longrightarrow C_2 \otimes_\alpha E_p .$$

However, using traced tensor norms, continuity of this latter mapping means that $\tilde{\kappa}_{q,p} \in \mathfrak{D}^*$ where

$$\mathfrak{D} \sim \beta \otimes \alpha^*$$

(see theorem 29.4.). Since $\mathfrak{D}^* = (\mathfrak{B} \circ \mathfrak{A}^*)^*$ by 29.8., one obtains

PROPOSITION: *Let α and β be finitely generated tensor norms, α accessible, with associated maximal normed operator ideals $\mathfrak{A}$ and $\mathfrak{B}$, and E a locally convex space. Then*

$$id_G \otimes id_E : G \otimes_\beta E \longrightarrow G \otimes_\alpha E$$

is continuous for all locally convex spaces G (or only C_2) if and only if for every $p \in cs(E)$ there is a $q \in cs(E)$ such that $\tilde{\kappa}_{q,p} : \tilde{E}_q \longrightarrow \tilde{E}_p$ is in $(\mathfrak{B} \circ \mathfrak{A}^*)^*$, the operator ideal associated with $(\beta \otimes \alpha^*)^*$.

If $\mathfrak{A}$ is a quasi–Banach operator ideal, call a locally convex space E to be in space($\mathfrak{A}$) if for all $p \in cs_o(E)$ there is a $q \in cs(E)$ such that $\tilde{\kappa}_{q,p} \in \mathfrak{A}$. For Banach spaces this coincides with the notation which was used so far. For locally convex spaces the books of Jarchow [123], Junek [133] and Pietsch [214] give much information about space($\mathfrak{A}$). Assume $\mathfrak{A}$ to be maximal and normed and take α to be its associated tensor norm. Then

$$\mathfrak{A} \sim \alpha = (\alpha^*)^* = (\varepsilon \otimes \alpha^*)^* = (\alpha^* \otimes \pi^*)^*$$

(see 29.8.), hence the proposition implies the

COROLLARY: *Let $\mathfrak{A}$ be a maximal normed accessible operator ideal with associated tensor norm α. For each locally convex space E the following statements are equivalent:*

(a) *E is in space($\mathfrak{A}$).*

(b) *$G \otimes_{\alpha^*} E = G \otimes_\pi E$ (topologically) for all locally convex spaces G (or only $G = C_2$).*

(c) *$G \otimes_\varepsilon E = G \otimes_\alpha E$ (topologically) for all locally convex spaces G (or only $G = C_2$).*

For the equivalence (a) $\leftrightarrow$ (b) accessibility of $\mathfrak{A}$ is not needed.

35.5. Some examples: The spaces in space($\mathfrak{N}$) are usually called *nuclear spaces*. Since for all $1 \leq p < \infty$ there is an $n \in \mathbb{N}$ with $\mathfrak{P}_p \circ .. \circ \mathfrak{P}_p \subset \mathfrak{N}$ (n–times composition, see 11.7.), it follows that

$$\text{space}(\mathfrak{N}) = \text{space}(\mathfrak{P}_p) = \text{space}(\mathfrak{I}_p) = \text{space}(\mathfrak{I})$$

for all $1 \leq p < \infty$. To be in space($\mathfrak{I}$) means that $\cdot \otimes_\varepsilon E = \cdot \otimes_\pi E$ (topologically) since $\mathfrak{I} \sim \pi$, which is Grothendieck's original definition of nuclear spaces. See also Ex 35.2.– Ex 35.4..

The locally convex spaces E in space($\mathfrak{L}_p$) have the property that for all $p \in cs_o(E)$ there is a $q \in cs(E)$ such that

$$\tilde{E}_q \longrightarrow \tilde{E}_p \hookrightarrow \tilde{E}_p''$$

factors through some $L_p(\mu)$. We do not know whether or not each complete locally convex space $E \in \text{space}(\mathfrak{L}_p)$ is the reduced projective limit of a net of $\mathfrak{L}_p^g$-spaces (the Banach spaces in $\text{space}(\mathfrak{L}_p)$) — except in the case $p = 2$: The spaces in $\text{space}(\mathfrak{L}_2)$ are exactly the *Hilbertizable* locally convex spaces.

The $\mathfrak{L}_p$-local technique reads as follows:

REMARK: *Let α and β be tensor norms, β finitely generated.*
(1) *If for $T \in \mathcal{L}(F_1, F_2)$ (locally convex spaces)*

$$id_{\ell_p} \otimes T : \ell_p \otimes_\beta F_1 \longrightarrow \ell_p \otimes_\alpha F_2$$

is continuous, then

$$id_E \otimes T : E \otimes_\beta F_1 \longrightarrow E \otimes_\alpha F_2$$

is continuous for each locally convex space $E \in \text{space}(\mathfrak{L}_p)$.
(2) *If $id : \ell_p \otimes_\beta \ell_q \longrightarrow \ell_p \otimes_\alpha \ell_q$ is continuous, then*

$$id : E \otimes_\beta F \longrightarrow E \otimes_\alpha F$$

is continuous for all locally convex spaces $E \in \text{space}(\mathfrak{L}_p)$ and $F \in \text{space}(\mathfrak{L}_q)$.

PROOF: (2) is a consequence of (1). For (1) take for $U_2 \in \mathfrak{U}(F_2)$ a $U_1 \in \mathfrak{U}(F_1)$ such that

$$id_{\ell_p} \otimes T_{U_1,U_2} : \ell_p \otimes_\beta (F_1)_{U_1} \longrightarrow \ell_p \otimes_\alpha (F_2)_{U_2}$$

is continuous. Then Ex 13.7. gives that ℓ_p can be replaced by $L_p(\mu)$. Thus, if the operator $\tilde{\kappa}_{V_1,V_2} : \tilde{E}_{V_1} \longrightarrow \tilde{E}_{V_2}$ is p-factorable, factorization into the bidual gives

$$\begin{array}{ccc}
\tilde{E}_{V_1} \otimes_\beta (F_1)_{U_1} & \longrightarrow \tilde{E}_{V_2} \otimes_\alpha (F_2)_{U_2} \hookrightarrow (\tilde{E}_{V_2})'' \otimes_\alpha (F_2)_{U_2} \\
\downarrow & \nearrow \\
L_p(\mu) \otimes_\beta (F_1)_{U_1} & \longrightarrow L_p(\mu) \otimes_\alpha (F_2)_{U_2}
\end{array}$$

and hence the result. □

In [44], 4.4., it was shown that *if α^* is totally accessible, then all locally convex spaces $E \in \text{space}(\mathfrak{A})$ have the approximation property.* Banach spaces in $\text{space}(\mathfrak{A})$ even have the bounded approximation property (see 21.6.) — but for general locally convex spaces (and $\alpha = \pi$) this is already false for Fréchet spaces since there are nuclear Fréchet spaces without the bounded approximation property.

Since w_p^* is totally accessible, locally convex spaces in $\text{space}(\mathfrak{L}_p)$ have the approximation property.

35.6. Corollary 35.4. showed that $E \in \text{space}(\mathfrak{L}_p)$ if and only if

$$G \otimes_\varepsilon E = G \otimes_{w_p} E \quad \text{or} \quad G \otimes_{w_p^*} E = G \otimes_\pi E$$

for all locally convex G or only $G = C_2$ (see also section 23). Since $w_\infty = \backslash \varepsilon$ and $w_1^* = /\pi$ (see 20.14.), it follows, as in the case of Banach spaces, that:

$$E \in \text{space}(\mathfrak{L}_1) \text{ if and only if } E \otimes_\pi \cdot = E \otimes_{\pi \backslash} \cdot$$
$$E \in \text{space}(\mathfrak{L}_\infty) \text{ if and only if } E \otimes_\varepsilon \cdot = E \otimes_{\varepsilon/} \cdot \; .$$

Thus it will be important to know that injective norms respect subspaces topologically and that projective norms respect quotients topologically also for locally convex spaces. The following result is due to Harksen [101], [102] :

THEOREM: *Let α be a tensor norm (on NORM) and E, F, G locally convex spaces.*
(1) *If α is right–injective, then*

$$id_E \otimes I : E \otimes_\alpha G \hookrightarrow E \otimes_\alpha F$$

is a topological embedding whenever $I : G \hookrightarrow F$ is.
(2) *If α is right–projective, then*

$$id_E \otimes Q : E \otimes_\alpha F \longrightarrow E \otimes_\alpha G$$

is a topological surjection whenever $Q : F \twoheadrightarrow G$ is.

PROOF: (1) follows from the fact that the sets $V \cap G$ with $V \in \mathfrak{U}(F)$ are the typical neighbourhoods of G and

$$E_U \otimes_\alpha G_{V \cap G} \xhookrightarrow{1} E_U \otimes_\alpha F_V$$

for all $U \in \mathfrak{U}(E)$.
For (2) lemma 35.3. will be applied. Take a typical seminorm $p = p_1 \otimes_\alpha p_2$ on $E \otimes_\alpha F$. Then, by the lemma, there is a $q \in cs(G)$ such that $\ker \kappa_q \subset \ker p_2^Q$ and $F_{p_2} \xrightarrow{1} G_q$ is a metric surjection. Since

$$(E \otimes_\alpha F)_p = E_{p_1} \otimes_\alpha F_{p_2} \xrightarrow{1} E_{p_1} \otimes_\alpha G_q$$

is a metric surjection, it remains to show (again by the lemma) that each element $z \in \ker(\kappa_{p_1} \otimes \kappa_q)$ satisfies $p^{id \otimes Q}(z) \leq \varepsilon$ for all $\varepsilon > 0$. Write

$$z = \sum_{i=1}^m x_i \otimes y_i + \sum_{i=m+1}^n x_i \otimes y_i \in \ker(\kappa_{p_1} \otimes \kappa_q) = \ker \kappa_{p_1} \otimes G + E \otimes \ker \kappa_q$$

and choose elements $\bar{y}_i \in F$ such that $Q(\bar{y}_i) = y_i$; since $\ker \kappa_q \subset \ker p_2^Q$ one can achieve $p_2(\bar{y}_i) \leq c\varepsilon$ for $i = m+1, ..., n$ with a constant $c \geq 0$ to be chosen in a moment ($\downarrow$). It follows that

$$(p_1 \otimes_\alpha p_2)^{id \otimes Q}(z) \leq (p_1 \otimes_\alpha p_2)\left(\sum_{i=1}^{n} x_i \otimes \bar{y}_i\right) \leq$$

$$\leq \sum_{i=1}^{m} 0 \cdot p_2(\bar{y}_i) + \sum_{i=m+1}^{n} p_1(x_i) c \varepsilon \overset{\downarrow}{\leq} \varepsilon . \quad \square$$

The theorem generalizes the well-known (and simple) facts that ε respects topological embeddings and π topological surjections. The characterizations of space($\mathfrak{L}_1$) and space($\mathfrak{L}_\infty$) given above allow one to extend the characterization 23.5. of $\mathcal{L}_1^g$ and $\mathcal{L}_\infty^g$-spaces from Banach to locally convex spaces; this result is due to Hollstein, [107] and [109]:

COROLLARY: *Let E be a locally convex space.*
(1) $E \in \text{space}(\mathfrak{L}_1)$ *if and only if $E \otimes_\pi \cdot$ respects topological embeddings.*
(2) $E \in \text{space}(\mathfrak{L}_\infty)$ *if and only if $E \otimes_\varepsilon \cdot$ respects topological surjections.*

PROOF: It follows from the theorem that both conditions are necessary. For the sufficiency take $G = C_2$ a Johnson space and look at the diagrams

$$\begin{array}{ccc} E \otimes_\varepsilon \ell_1(B_{C_2}) = E \otimes_{\varepsilon/} \ell_1(B_{C_2}) & \quad & E \otimes_\pi C_2 \hookrightarrow E \otimes_\pi \ell_\infty(B_{C'_2}) \\ \downarrow \qquad \qquad \downarrow ! & & \| ! \qquad \qquad \| \\ E \otimes_\varepsilon C_2 \quad = \quad E \otimes_{\varepsilon/} C_2 & & E \otimes_{\pi\backslash} C_2 \hookrightarrow E \otimes_{\pi\backslash} \ell_\infty(B_{C'_2}) . \quad \square \end{array}$$

These results have many applications to the extension and lifting of operators and to the theory of vector valued functions.

35.7. For the duality theory of $E \otimes_\varepsilon F$ and also for the study of vector valued functions it is interesting to know for an inductive limit $F = \text{ind} F_n$ of locally convex spaces (see F.–Wloka [75] for the definition and basic properties of inductive limits) and a locally convex space E under which conditions the algebraic identity

$$E \otimes_\alpha \underset{n \to}{\text{ind}} F_n = \underset{n \to}{\text{ind}} E \otimes_\alpha F_n$$

holds topologically; the mapping property implies that the identity $\longleftarrow$ is always continuous. A locally convex space is said to have the *countable neighbourhood property* if for each sequence (p_n) in $cs_o(E)$ there are a $p \in cs(E)$ and $c_n \geq 0$ such that $p_n \leq c_n p$. (DF)–spaces E enjoy this property (this is easy: look at $U_n^0 \subset E'_b$ which is a Fréchet space).

PROPOSITION: *Let E and F_n be locally convex spaces and α a tensor norm. If E has the countable neighbourhood property, then*

$$E \otimes_\alpha \left(\bigoplus_{n=1}^\infty F_n\right) = \bigoplus_{n=1}^\infty (E \otimes_\alpha F_n)$$

holds topologically.

PROOF: The typical seminorm on $\bigoplus_n G_n$ is $\sum r_n$ with $r_n \in cs(G_n)$. It has to be shown that for $p_n \in cs(E)$ and $q_n \in cs(F_n)$ there is a $p \in cs(E)$ and $q \in cs(\bigoplus_n F_n)$ such that

$$\sum_{n=1}^\infty (p_n \otimes_\alpha q_n) \leq p \otimes_\alpha q \, .$$

Observe first that for all $r_m \in cs(F_m)$ the space $(F_m)_{r_m}$ is 1–complemented in the space $(\bigoplus_n F_n)_{\sum r_n}$, hence for each $p \in cs(E)$

$$p \otimes_\alpha r_m(z_m) \leq \left[p \otimes_\alpha \left(\sum_{n=1}^\infty r_n\right)\right] \left(\sum_{n=1}^\infty z_n\right)$$

for all $\sum_n z_n \in E \otimes \bigoplus_n F_n$. Choose $p_n \leq c_n p$ according to the countable neighbourhood property and define $q := \sum_{n=1}^\infty 2^n c_n q_n$. Then

$$\sum_{m=1}^\infty p_m \otimes_\alpha q_m(z_m) \leq \sum_{m=1}^\infty [(c_m p) \otimes_\alpha q_m](z_m) = \sum_{m=1}^\infty 2^{-m} [p \otimes_\alpha 2^m c_m q_m](z_m) \leq$$
$$\leq \sum_{m=1}^\infty 2^{-m} p \otimes_\alpha q\left(\sum_{n=1}^\infty z_n\right) = p \otimes_\alpha q\left(\sum_{n=1}^\infty z_n\right) . \square$$

An inductive limit $\mathrm{ind}_{n\to} F_n$ is (topologically) a quotient of the direct sum $\bigoplus_{n=1}^\infty F_n$, so

$$E \otimes_\alpha \mathrm{ind}_{n\to} F_n = \mathrm{ind}_{n\to} E \otimes_\alpha F_n$$

holds topologically if E has the countable neighbourhood property and one of the following three conditions is satisfied:

(1) The kernel of the canonical map $\bigoplus_{n=1}^\infty F_n \longrightarrow \mathrm{ind}_n F_n$ is complemented (following De Wilde such inductive limits are said to have a partition of unity — for example Schwartz' space $\mathcal{D}(\Omega)$ of test functions on $\Omega \subset \mathbb{R}^n$ has such a partition; see Pérez Carreras–Bonet [207] for details).

(2) α is right–projective.

(3) $E \otimes_\alpha \cdot = E \otimes_{\alpha/} \cdot$.

These results are due to Harksen [101], [102] (see also [69]).

For $\alpha = \varepsilon$ this implies, e.g., that each (DF)–space E in space($\mathcal{L}_\infty$) satisfies the relation $E \otimes_\varepsilon \operatorname{ind} F_n = \operatorname{ind} E \otimes_\varepsilon F_n$ — and this is even characteristic for the $\mathcal{L}_\infty^g$-spaces among the Banach spaces (see Ex 35.11.). To obtain $E\tilde\otimes_\varepsilon \operatorname{ind} F_n = \operatorname{ind} E\tilde\otimes_\varepsilon F_n$ one needs additional assumptions (see Hollstein [107] for information in this direction).

35.8. If E is a Banach space and $F = \operatorname{ind} F_n$ an inductive limit of Banach spaces such that $E \otimes_\alpha \operatorname{ind} F_n = \operatorname{ind} E \otimes_\alpha F_n$, then $E \otimes_\alpha F$ is *quasibarrelled* (i.e. bounded sets in the strong dual are equicontinuous). There is a good amount of literature investigating topological–geometric properties (such as quasibarrelled, bornological, barrelled) of $E \otimes_\alpha F$ and $C(K) \otimes_\varepsilon F$ or $C(K, F) = C(K)\tilde\otimes_\varepsilon F$. We want to show how traced tensor norms can successfully be used in this context.

A pair (E, F) of locally convex spaces is said to satisfy *property* (BB) (for bibounded) if for each bounded $A \subset E \otimes_\pi F$ there are bounded sets $B \subset E$ and $C \subset F$ which "lift" A, i.e.
$$A \subset \overline{\Gamma(B \otimes C)}.$$

If E and F are Fréchet spaces, then compact sets in $E\tilde\otimes_\pi F$ can be lifted (this follows as in 3.4.) — but, in general, bounded sets not; this (deep) result will be treated in a moment. The pair (ℓ_1, E) having (BB) is also known as property (B) of Pietsch of E. Endow the space $c_o(E)$ of zero sequences in E and the space $\ell_1(E)$ of absolutely summable sequences in E with their natural topologies coming from the seminorms $\sup p(x_n)$ and $\Sigma p(x_n)$ (for $p \in cs_o(E)$), respectively. Note that $c_o(\tilde E) = c_o\tilde\otimes_\varepsilon \tilde E$ and $\ell_1(\tilde E) = \ell_1\tilde\otimes_\pi E$ hold topologically. It is easy to see that the space $c_o(E)$ has a strong dual which is topologically embedded into $\ell_1(E_b')$:

$$\ell_1 \otimes_\pi E_b' \xrightarrow{top.} (c_o(E))_b' \xrightarrow{top.} \ell_1(E_b'),$$

the duality bracket being $\langle (x_n'), (x_n) \rangle := \sum_{n=1}^\infty \langle x_n', x_n \rangle$. Since for $U \in \mathfrak{U}(E)$
$$(c_o(E_U))' = (c_o\tilde\otimes_\varepsilon E_U)' = \ell_1\tilde\otimes_\pi E_U' = \ell_1(\llbracket U^0 \rrbracket)$$
holds isometrically, the equicontinuous sets in $(c_o(E))'$ are exactly those which are in some unit ball
$$B_{\ell_1(\llbracket U^0 \rrbracket)} \subset \ell_1(E_b').$$

Since every $(x_n') \in \ell_1(E_b')$ is the limit of its finite parts, $(c_o(E))'$ is dense in $\ell_1(E_b')$. It also follows that
$$(c_o(E))_b' = \ell_1(E_b')$$
holds topologically if every countable bounded set in $(c_o(E))_b'$ is equicontinuous. This is the case, for example, if $c_o(E)$ is quasibarrelled.

PROPOSITION: *Let E be a locally convex space. Then the following statements are equivalent:*
(a) $C(K, E)$ *is quasibarrelled for all compact sets K (or only one infinite compact K).*

(b) $c_o(E)$ is quasibarrelled.

(c) $G \otimes_\varepsilon E$ is quasibarrelled for all $\mathcal{L}^g_\infty$-spaces G (or only one infinite dimensional $\mathcal{L}^g_\infty$-space G).

(d) E is quasibarrelled and (ℓ_1, E'_b) has property (BB).

PROOF: Using finite sequences (and the above remarks) it is easily seen that (b) and (d) are equivalent. Moreover, since $c_o \otimes_\varepsilon E$ is a *large* subspace of $c_o(E)$ (each bounded set in $c_o(E)$ is in the closure of a bounded subset of $c_o \otimes_\varepsilon E$, see Ex 35.7.), $c_o(E)$ is quasibarrelled if and only if $c_o \otimes_\varepsilon E$ is quasibarrelled (call this (b')).
The crucial point in the proof is now the statement of Ex 29.9.(a): If F is an infinite dimensional $\mathcal{L}^g_\infty$-space (hence contains the ℓ^n_∞ uniformly complemented by 23.3.), then
$$\varepsilon/ \otimes_F \varepsilon/ = w_1 \otimes_F w_1 \sim w_1 = \varepsilon/ \ .$$
(b') $\curvearrowright$ (c): Take $F = c_o$ and G an $\mathcal{L}^g_\infty$-space. Then
$$(G \otimes_{\varepsilon/} \ell_1) \otimes_\pi (c_o \otimes_{\varepsilon/} E) \longrightarrow G \otimes_{\varepsilon/} E$$
is a topological surjection. Since G and c_o are $\mathcal{L}^g_\infty$-spaces, one can replace $\varepsilon/$ by ε in all three places. Now it is enough to note that $X \otimes_\pi Y$ is quasibarrelled if X is normed and Y is quasibarrelled (take $A \subset (X \otimes_\pi Y)'$ bounded; then A is bounded in $\mathcal{L}_b(X, Y'_b)$ and hence $A(B_X) \subset Y'_b$ is bounded = equicontinuous).

(c) $\curvearrowright$ (b'): Assume the space $G_o \otimes_\varepsilon E$ to be quasibarrelled for an $\mathcal{L}^g_\infty$-space G_o. Since $\varepsilon/ \otimes_{G_o} \varepsilon/ \sim \varepsilon/$, it follows that the surjection
$$(c_o \otimes_{\varepsilon/} G'_o) \otimes_\pi (G_o \otimes_{\varepsilon/} E) \longrightarrow c_o \otimes_{\varepsilon/} E$$
is topological, hence the same argument applies.
Since $C(K) \otimes_\varepsilon E$ is also a large subspace of $C(K, E)$ (this again follows from Ex 35.7.), the equivalence with (a) follows easily. □

These results are due to Marquina–Sanz Serna [184], Mendoza Casas [190] and D.-Govaerts [48]. In his paper Mendoza also proved that for barrelled spaces E the space $C(K, E)$ is barrelled if it is quasibarrelled. Hence, the proposition also gives a characterization of barrelled $C(K, E)$ (see also Ex 35.11.). For more information about topological–geometric properties of $C(\Omega, E)$ (when Ω is completely regular) see Schmets [252].

35.9. Let E and F be Fréchet spaces; then Baire's category theorem implies that every continuous operator $E \longrightarrow F'_b$ factors through some $[\![U^0]\!]$ with $U \in \mathfrak{U}(F)$, hence $\mathfrak{L}(E, F'_b) = (E \otimes_\pi F)'$ holds algebraically. Grothendieck's "problème des topologies" ([92], I, p.43) is the question of whether or not the strong topologies coincide:
$$\mathfrak{L}_b(E, F'_b) \stackrel{?}{=} (E \otimes_\pi F)'_b \ .$$

The strong topology on $\mathfrak{L}(E, F_b')$ comes from pairs (A, B) of bounded sets, the strong topology on $(E \otimes_\pi F)'$ from bounded sets in $E \otimes_\pi F$. It follows that the canonical map $(E \otimes_\pi F)_b' \longrightarrow \mathfrak{L}_b(E, F_b')$ is always continuous and that *the topologies coincide if and only if the pair (E, F) has property (BB)*, defined in 35.8.. For separable E and F it is not difficult to verify that $\mathfrak{L}_b(E, F_b') = (E \otimes_\pi F)_b'$ if and only if $\mathfrak{L}_b(E, F_b')$ is a (DF)-space. Moreover, the "problème des topologies" is closely related to another one of Grothendieck's problems namely, whether or not $E \otimes_\varepsilon F$ is a (DF)-space if E and F are; see 35.10. below.

The problem was solved in the negative by Taskinen [267] in 1986, see also [268]. He constructed a Fréchet–Montel space F such that (ℓ_2, F) and (F, F) do not have property (BB). The space $F \tilde{\otimes}_\pi F$ contains ℓ_1 — hence is not even reflexive. On the other hand it is easy to see that $E \tilde{\otimes}_\pi F$ is Fréchet–Schwartz whenever E and F are: since compact sets are liftable, (E, F) has (BB) in this case.

We shall concentrate on the case in which one space is a Banach space.

PROPOSITION 1: *If E is an $\mathfrak{L}_1^g$-space, then for every Fréchet space F the pair (E, F) has property (BB).*

PROOF: Since local technique will be applied, the case of $E = \ell_1$ will be treated first. Topologically $\ell_1 \tilde{\otimes}_\pi E = \ell_1(E)$, hence, for each bounded set $A \subset \ell_1(E)$ it is enough to find a bounded $C \subset E$ such that $A \subset \ell_1([C])$ is bounded: If (p_k) is a basis of seminorms of F, put

$$c_k := \sup\left\{\sum_{n=1}^\infty p_k(x_n) \mid (x_n) \in A\right\},$$

choose $\lambda_k > 0$ such that $\sum_{k=1}^\infty c_k \lambda_k \leq 1$ and define

$$C := \sup\left\{x \in E \mid \sum_{k=1}^\infty \lambda_k p_k(x) \leq 1\right\}.$$

Obviously, C is absolutely convex and bounded. Then for $(x_n) \in A$

$$\sum_{n=1}^\infty m_C(x_n) = \sum_{n=1}^\infty \sum_{k=1}^\infty \lambda_k p_k(x_n) \leq \sum_{k=1}^\infty \lambda_k \sum_{n=1}^\infty p_k(x_n) \leq \sum_{k=1}^\infty \lambda_k c_k \leq 1$$

— and the first claim is proved.

Now take E to be an arbitrary $\mathfrak{L}_{1,\lambda}^g$-space. It has the λ-bounded approximation property (see 21.6.), hence there is a net (T_η) of operators $T_\eta \in \mathfrak{F}(E, E)$ with $\|T_\eta\| \leq \lambda$ converging pointwise to id_E. The very definition of $\mathfrak{L}_{1,\lambda}^g$-spaces gives operators

$$T_\eta(E) \hookrightarrow E$$
$$R_\eta \searrow \nearrow S_\eta$$
$$\ell_1$$

$\|S_\eta\| = 1$

$\|R_\eta\| \leq 2\lambda$.

It follows that $\|R_\eta \circ T_\eta\| \leq 2\lambda^2$ and the net
$$(R_\eta \circ T_\eta) \otimes id_F : E \otimes_\pi F \longrightarrow \ell_1 \otimes_\pi F$$
is equicontinuous. Now take $A \subset E \otimes_\pi F$ bounded; then
$$A_o := \bigcup_\eta \{(R_\eta \circ T_\eta) \otimes id_F(z) \mid z \in A\}$$
is bounded in $\ell_1 \otimes_\pi F$, hence contained in $\overline{\Gamma(B_{\ell_1} \otimes C)}$ for some bounded $C \subset F$ by the first part of the proof. Since $S_\eta(B_{\ell_1}) \subset B_E$, it follows that for all $z \in A$
$$(T_\eta \otimes id_F)(z) = (S_\eta \otimes id_F) \circ (R_\eta \circ T_\eta \otimes id_F)(z) \in S_\eta \otimes id_F(\overline{\Gamma(B_{\ell_1} \otimes C)}) \subset \overline{\Gamma(B_E \otimes C)}\,.$$
The fact that $T_\eta \otimes id_F(z)$ converges to z concludes the proof. □

This result is due to Taskinen [269], but the present simple proof by local technique is taken from Bonet–D.–Galbis [12]; see Ex 35.6. for more applications of this technique to the problem of lifting bounded sets in $E \otimes_\pi F$.

Astonishingly enough, a converse of the above result is also true. The following is a slight generalization of Taskinen's famous original example (see [269]):

PROPOSITION 2: *For every Banach space E which is not an $\mathcal{L}_1^g$-space there is a Fréchet space F and a countable set $A \subset E \otimes_\pi F$ which is not contained in any set of the form*
$$\overline{\Gamma(B_E \otimes C)}$$
where $C \subset F$ is bounded. The Fréchet space F can be chosen to have the bounded approximation property and to be the reduced projective limit of a sequence of separable reflexive Banach spaces with the approximation property.

PROOF (taken from D.–F.–Taskinen [47]) : Note first that the $\mathcal{L}_1^g$-spaces E are exactly those Banach spaces for which $E \otimes_\pi \cdot$ respects subspaces (with a constant λ, see 23.5., corollary 5). Fix a Banach space E which is not an $\mathcal{L}_1^g$-space. Then this observation together with an argument as in the proof of 20.3.(1) shows that there are $G_n \in \mathrm{FIN}$ with subspaces $M_n \subset G_n$ such that there exist $z_n \in E \otimes M_n$ with
$$\pi(z_n; E, G_n) < 1 \quad \text{and} \quad 2n \leq \pi(z_n; E, M_n) \leq 2^n\,.$$
Fix a continuous projection $P_n : G_n \longrightarrow G_n$ onto M_n and define for all $n, k \in \mathbb{N}$
$$\|y\|_{n,k} := \|P_n y\|_n + k\|y - P_n y\|_n \qquad \text{for } y \in G_n\,.$$
It follows that $\|\cdot\|_n \leq \|\cdot\|_{n,k}$ and $\|y\|_n = \|y\|_{n,k}$ for $y \in M_n$. Notation:
$$G_{n,k}^0 := (G_n, \|\cdot\|_{n,k}) \quad , \quad G_{n,k} := (G_n, \|\cdot\|_n)$$
$$H^0 := \ell_2((G_{n,k}^0)_{n,k \in \mathbb{N}}) \quad , \quad H := \ell_2((G_{n,k})_{n,k \in \mathbb{N}})\,.$$

Clearly, $H^0 \subset H$ and $\|\cdot\|_H \le \|\cdot\|_{H^0}$. If $I_{n,k}: G_{n,k} \hookrightarrow H$ is the canonical injection, then $z_{n,k} := (id_E \otimes I_{n,k})(z_n)$ satisfies

(*) $\qquad\qquad \pi(z_{n,k}; E, H) \le 1 \quad\text{and}\quad \pi(z_{n,k}; E, H^0) \le 2^n$

— the latter since $\pi(z_n; E, G^0_{n,k}) \le \pi(z_n; E, M_n) \le 2^n$.

Now assume that for $k, m, n \in \mathbb{N}$

$$z_{n,k} \in \Gamma(B_E \otimes (mB_{H^0} \cap nB_H));$$

then (by projection on the (n,k)–th component of H) there is a representation

$$z_n = \sum_{i=1}^{N} \lambda_i x_i \otimes y_i = \sum_{i=1}^{N} \lambda_i x_i \otimes P_n y_i \in E \otimes G_n$$

such that

$$\sum_{i=1}^{N} |\lambda_i| \le 1, \quad \|x_i\|_E \le 1, \quad \|y_i\|_{n,k} \le m, \quad \|y_i\|_n \le n.$$

Since $P_n y_i \in M_n$, it follows that

$$2n \le \pi(z_n; E, M_n) \le \sum_{i=1}^{N} |\lambda_i|\, \|x_i\|_E \|P_n y_i\|_n \le$$

$$\le \sum_{i=1}^{N} |\lambda_i|(\|y_i\|_n + \|y_i - P_n y_i\|_n) \le n + m/k.$$

This is impossible for large k (and given m). Thus, it has been shown that for all $m, n \in \mathbb{N}$ there is a $k \in \mathbb{N}$ such that

(**) $\qquad\qquad z_{n,k} \notin \Gamma(B_E \otimes (mB_{H^0} \cap nB_H))$.

Everything is prepared for defining the needed space:

$$H_n^m := \begin{cases} H & \text{for } n \ge m \\ H^0 & \text{for } n < m \end{cases}$$

$$F_m := \ell_2((H_n^m)_{n \in \mathbb{N}}) \subset H^{\mathbb{N}}$$

$$F := \bigcap_{m=1}^{\infty} F_m.$$

F is a Fréchet space of the kind expressed in the statement of the proposition. Denote by J_n the canonical injection of the n–th component into $H^{\mathbb{N}}$ and by Q_n the corresponding projection. The set

$$A := \{id_E \otimes J_n(z_{n,k}) \mid n, k \in \mathbb{N}\} \subset E \otimes_\pi F$$

is bounded — this follows immediately from (∗). Assume now that A could be "lifted", i.e. there were $c_n \geq 1$ such that the bounded set

$$C := \bigcap_{m=1}^{\infty} c_m B_{F_m} \subset F$$

satisfied $A \subset \overline{\Gamma(B_E \otimes C)}$. Since $\Gamma(B_E \otimes U)$ is a zero neighbourhood of $E \otimes_\pi F$ for $U := c_1 B_{F_1} \cap c_{n+1} B_{F_{n+1}}$, it would follow that

$$\overline{\Gamma(B_E \otimes C)} \subset \Gamma\left(B_E \otimes \bigcap_{m=1}^{\infty} c_m B_{F_m}\right) + \Gamma(B_E \otimes (c_1 B_{F_1} \cap c_{n+1} B_{F_{n+1}})) \subset$$
$$\subset \Gamma(B_E \otimes (2c_1 B_{F_1} \cap 2c_{n+1} B_{F_{n+1}})) \ .$$

Consequently, $id_E \otimes J_n(z_{n,k}) \in A \subset \overline{\Gamma(B_E \otimes C)}$ would imply that

$$z_{n,k} = (id_E \otimes Q_n)(id_E \otimes J_n)(z_{n,k}) \in id_E \otimes Q_n \left[\Gamma(B_E \otimes (2c_1 B_{F_1} \cap 2c_{n+1} B_{F_{n+1}}))\right] =$$
$$= \Gamma(B_E \otimes (2c_1 B_H \cap 2c_{n+1} B_{H^\circ})) \ ,$$

which, by (∗∗), is impossible for all k. This contradiction finishes the proof. □

Propositions 1 and 2 give the

THEOREM (Taskinen): *A Banach space E is an $\mathcal{L}_1^g$-space if and only if (E, F) has property (BB) for all Fréchet spaces F.*

35.10. Besides the Fréchet spaces the most important class of locally convex spaces are the (DF)-spaces introduced by Grothendieck; they simulate the structure of the strong dual of a Fréchet space. A locally convex space E is called a *(DF)-space* if it has a countable basis for the bounded sets and every strongly bounded subset $A \subset E'$ is equicontinuous if it is of the form $A = \bigcup_{n=1}^{\infty} A_n$ with equicontinuous A_n. If E is metrizable its strong dual is (DF) and the strong dual of every (DF)-space is a Fréchet space. It was one of the problems which Grothendieck left open in his study of (DF)-spaces whether or not $E \otimes_\epsilon F$ is (DF) whenever E and F are. As was already mentioned, this problem is intimately related with the "problème des topologies" just solved in the negative. It is in some sense dual (the proofs will show this) — so it is not surprising that the $\mathcal{L}_\infty^g$-spaces are the "good" Banach spaces for this problem. As before in 35.9. (proposition 1) the "typical" $\mathcal{L}_\infty^g$-space c_o will be treated first.

PROPOSITION: *Let F be a (DF)-space. Then the space $c_o(F)$ of all zero sequences in F is a (DF)-space as well.*

PROOF: That $c_o(F)$ has a fundamental sequence of bounded sets is straightforward. Moreover, note that $(c_o(F))'_b \subset \ell_1(F'_b)$ (topologically) by 35.8. — and that the equicontinuous subsets in $(c_o(F))'$ are those which are bounded in $\ell_1([U^0])$ for some $U \in \mathfrak{U}(F)$.

So take a bounded $A = \bigcup_{n=1}^{\infty} A_n \subset (c_o(F))'_b$ with equicontinuous A_n. Then there are $U_n \in \mathfrak{U}(F)$ such that

$$\sum_{m=1}^{\infty} m_{U_n^\circ}(x'_m) \leq 1 \qquad \text{for all } (x'_m) \in A_n.$$

Since (ℓ_1, F'_b) has property (BB), it follows that there is a bounded $C \subset F'_b$ such that

$$\sum_{m=1}^{\infty} m_C(x'_m) \leq 1 \qquad \text{for all } (x'_m) \in A.$$

Take $C_n := C \cap U_n^\circ$. Then $\bigcup_{n=1}^{\infty} C_n$ is equicontinuous since F is a (DF)-space. For the equicontinuous set $D = \Gamma \bigcup_{n=1}^{\infty} C_n$ it follows that

$$m_D \leq m_{C_n} \leq m_C + m_{U_n^\circ}$$

which implies that $\sum_{m=1}^{\infty} m_D(x'_m) \leq 2$ for all $x' = (x'_m) \in A_n$, and hence for all $x' \in A$. This shows that A is bounded in $\ell_1[D])$ and hence is equicontinuous. $\square$

This result implies (by what was said in 35.8.) that $(c_o(F))'_b = \ell_1(F'_b)$ if F is a (DF)-space. Moreover, since $c_o \otimes_\varepsilon F$ is a large subspace of $c_o(F)$ (see Ex 35.7.), the space $c_o \otimes_\varepsilon F$ is also a (DF)-space in this case. Now, as in 35.8., the traced tensor norm technique

$$(E \otimes_{\varepsilon/} \ell_1) \otimes_\pi (c_o \otimes_{\varepsilon/} F) \twoheadrightarrow E \otimes_{\varepsilon/} F,$$

the fact that the π-tensor product of two (DF)-spaces is (DF) (see e.g. Jarchow [123], p.335, for a proof), and the fact that $E \otimes_\varepsilon \cdot = E \otimes_{\varepsilon/} \cdot$ whenever E is a $\mathcal{L}^g_\infty$-space give immediately: $E \otimes_\varepsilon F$ is (DF) if F is a (DF)-space and the Banach space E an $\mathcal{L}^g_\infty$-space (see also Ex 35.8. and Ex 35.9.). The converse statement is also true — a result due to D.-F.-Taskinen [47]:

THEOREM: *For each Banach space E the following statements are equivalent:*
(a) *E is an $\mathcal{L}^g_\infty$-space.*
(b) *For every (DF)-space F the space $E \otimes_\varepsilon F$ is a (DF)-space.*
(c) *For every inductive limit F of separable reflexive Banach spaces with the approximation property the space $E \tilde{\otimes}_\varepsilon F$ is a (DF)-space.*

Note that the completion $\tilde{G}$ of a (DF)-space G is a (DF)-space.

PROOF: It remains to show that (c) implies (a). For this it is necessary to show that E' is an $\mathcal{L}^g_1$-space (by 23.2.). Assume not, then proposition 2 in 35.9. gives a Fréchet space G, which is a reduced projective limit

$$G := \underset{\leftarrow n}{\text{proj}}\, G_n$$

of a sequence of reflexive separable Banach spaces with the approximation property and a *countable* $A \subset E' \otimes_\pi G$ which is not liftable. Then

$$F := G'_b = \operatorname*{ind}_{n \to} G'_n \quad ; \quad F_n := G'_n$$

holds topologically since G is distinguished and hence G'_b is bornological. This space F is as in (c), therefore $E \tilde\otimes_\varepsilon F$ is a (DF)-space.

(1) Consider the natural map $J : E \otimes_\varepsilon F \hookrightarrow \mathcal{L}_b(E', F)$. Since for $U \in \mathfrak{U}(F)$ and $z \in E \otimes F$

$$\|\cdot\|_{E \otimes_\varepsilon} m_U(z; E, F) = \sup\{|\langle z, x' \otimes y'\rangle| \mid x' \in B_{E'}, y' \in U^0\} =$$
$$= \sup\{m_U(J(z)(x')) \mid x' \in B_{E'}\},$$

it follows that J is continuous. $\mathcal{L}_b(E', F)$ is complete and therefore also the dual mapping $\tilde{J}' : (\mathcal{L}_b(E', F))'_b \longrightarrow (E \tilde\otimes_\varepsilon F)'_b$ is continuous. The continuity of the canonical bilinear map $E' \times F'_b \longrightarrow (\mathcal{L}_b(E', F))'_b$ gives that the natural map (note that $(E \tilde\otimes_\varepsilon F)'_b$ is a Fréchet space by (c))

$$\Psi : E' \tilde\otimes_\pi F'_b \longrightarrow (E \tilde\otimes_\varepsilon F)'_b \; ; \; \langle \Psi(x' \otimes y'), x \otimes y \rangle := \langle x', x \rangle \langle y', y \rangle$$

is defined and continuous. The diagram

$$\begin{array}{ccc} E' \tilde\otimes_\pi F'_b = E' \tilde\otimes_\pi G & \xrightarrow{\Psi} & (E \tilde\otimes_\varepsilon G'_b)' = (E \tilde\otimes_\varepsilon F)' \\ \downarrow & & \downarrow \text{restriction} \\ E' \tilde\otimes_\pi G_n & \hookrightarrow & (E \tilde\otimes_\varepsilon G'_n)' \end{array}$$

shows that Ψ is injective since the approximation property of G_n implies that the lower map is always injective.

(2) The equicontinuous subsets $D \subset (E \tilde\otimes_\varepsilon F)'$ can be described as follows: There is a $U \in \mathfrak{U}(F)$ such that D is bounded in the Banach space

$$(E \otimes_\varepsilon F_U)' = \mathfrak{I}(E, F'_U) = \mathfrak{I}(E, \llbracket U^0 \rrbracket).$$

Since $G = F'_b$ is reflexive, U^0 is weakly compact, hence there is another bounded = equicontinuous Banach disc V^0 such that $I : \llbracket U^0 \rrbracket \hookrightarrow \llbracket V^0 \rrbracket$ is a weakly compact operator (see e.g. Jarchow [123], p. 215). $\mathfrak{W} \circ \mathfrak{I} \subset \mathfrak{N}$ (see Appendix C8.), therefore D is bounded in

$$\mathfrak{I}(E, \llbracket U^0 \rrbracket) \hookrightarrow \mathfrak{N}(E, \llbracket V^0 \rrbracket) \subset \mathfrak{I}(E, F'_V) \subset (E \tilde\otimes_\varepsilon F)'.$$

It is clear (elementary tensors) that

$$E' \tilde{\otimes}_\pi F'_b \xrightarrow{\Psi} (E \tilde{\otimes}_\varepsilon F)'_b$$
$$I_V \uparrow \qquad \qquad \uparrow$$
$$E' \tilde{\otimes}_\pi [V^0] \xrightarrow{1} \mathfrak{N}(E, [V^0])$$

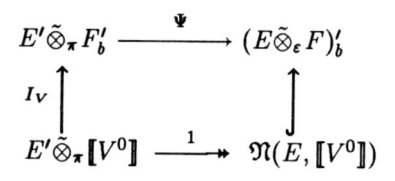

commutes.

Now take the countable bounded $A \subset E' \otimes_\pi F'_b$ which cannot be lifted. This means, in particular, that A is not contained in any $\overline{\Gamma B}$ with $B \subset E' \tilde{\otimes}_\pi [V^0]$ being bounded. However, $\Psi(A) \subset (E \tilde{\otimes}_\varepsilon F)'_b$ is countable and bounded, hence equicontinuous since $E \tilde{\otimes}_\varepsilon F$ is a (DF)-space. The description of the equicontinuous sets in (2) and the injectivity of Ψ give a contradiction: E' must be an $\mathcal{L}^g_1$-space and E an $\mathcal{L}^g_\infty$-space. □

The main work in the proof was showing that

$$E' \tilde{\otimes}_\pi F'_b = (E \otimes_\varepsilon F)'$$

holds — with a good description of the equicontinuous sets in $(E \otimes_\varepsilon F)'$. In essence this type of duality result is due to Grothendieck's investigation of what he called Phillip's property ([92], I, p.108) — nowadays referred to as the Radon–Nikodým property or the local Radon–Nikodým property; for more information about the duality of ε and π in the locally convex setting, see Collins–Ruess [34], D.–F.–Taskinen [47], and [40], [44].

35.11. The theory of tensor norms represents a good part of the so-called "local" theory of Banach space (i.e. the study in terms of finite dimensional subspaces). The tensor norm techniques just presented for locally convex spaces may mark the beginning of a "local" theory of locally convex spaces.

Exercises:

Ex 35.1. Let $\mathfrak{A}$ and $\mathfrak{B}$ be two quasi–Banach operator ideals.
(a) If $\mathfrak{A} \subset \mathfrak{B}$, then space($\mathfrak{A}$) $\subset$ space($\mathfrak{B}$).
(b) space($\mathfrak{A}$) $\cap$ space($\mathfrak{B}$) = space($\mathfrak{A} \circ \mathfrak{B}$) = space($\mathfrak{B} \circ \mathfrak{A}$).
Clearly, space($\mathfrak{A}$) is used here in its locally convex meaning.

Ex 35.2. Prove:
(a) For $(p,q) \neq (\infty, \infty)$ space($\mathfrak{D}_{p,q}$) coincides with the class of all nuclear spaces. The same holds for space($\mathfrak{J}_p$) = space($\mathcal{L}_{p,1}$) and space($\mathcal{L}_{1,p}$) if $1 \leq p < \infty$.
Hint: For the first statement use Kwapień's factorization theorem.
(b) space($\mathcal{L}_{p,q}$) = space($\mathcal{L}_2$), all Hilbertizable locally convex spaces, if $1 < p, q \leq 2$.
Hint: 12.8..

Ex 35.3. Use the previous exercise to show that for a locally convex space E and $(p,q) \neq (1,1)$ the following are equivalent:
(a) E is nuclear.
(b) $G \otimes_{\alpha_{p,q}} E = G \otimes_\pi E$ for all locally convex spaces G (or only $G = C_2$).
(c) $G \otimes_\varepsilon E = G \otimes_{\alpha^*_{p,q}} E$ for all locally convex spaces G (or only $G = C_2$).

Ex 35.4. Show that
$$\text{space}(\mathfrak{L}_p) \cap \text{space}(\mathfrak{L}_\infty) = \text{nuclear spaces if } 1 \leq p < \infty$$
and
$$\text{space}(\mathfrak{L}_p) \cap \text{space}(\mathfrak{L}_1) = \text{nuclear spaces if } 1 < p \leq \infty.$$
Hint: For the first equality use that $\mathfrak{L}_p \circ \mathfrak{L}_\infty$ is contained in $\mathfrak{P}_2$ if $1 \leq p \leq 2$, and in $\mathfrak{P}_{p+\varepsilon}$ if $2 < p < \infty$ (see e.g. 23.10. and 24.7.). For the second look at $\mathfrak{L}_1 \circ \mathfrak{L}_p \subset (\mathfrak{L}_{p'} \circ \mathfrak{L}_\infty)^{dual}$. This result is taken from Pietsch [214], 29.7.6..

Ex 35.5. (a) Let $\mathfrak{A} \sim \alpha$ be accessible and E a locally convex space. Then $E \otimes_\alpha \cdot$ respects quotients isomorphically if and only if $E \otimes_\alpha \cdot = E \otimes_{\alpha/} \cdot$ holds and if and only if $E \in \text{space}(\mathfrak{A}^{dual} \circ (\mathfrak{A}^{dual} \circ \mathfrak{L}_\infty)^*)^*$. Hint: For the first equivalence use the ideas of 35.6. and for the second 35.4. and 20.12..

(b) Let $p = 2$ or ∞. Show that a locally convex space E is nuclear if and only if $E \otimes_{w_p} \cdot$ respects quotients topologically. Hint: Check the equalities $(\mathfrak{L}_2 \circ (\mathfrak{L}_2 \circ \mathfrak{L}_\infty)^*)^* = (\mathfrak{L}_2 \circ \mathfrak{P}_2)^* = \mathfrak{P}_2$ (little Grothendieck) and $(\mathfrak{L}_1 \circ (\mathfrak{L}_1 \circ \mathfrak{L}_\infty)^*)^* = (\mathfrak{L}_1 \circ \mathfrak{D}_2^*)^* = (\mathfrak{L}_1 \circ \mathfrak{L}_2)^* = (\mathfrak{L}_2/)^* = /\mathfrak{D}_2 = \mathfrak{P}_2^{dual}$ (Grothendieck's inequality, 20.12. and 20.15.).

Note that the tensor norm w_1 is right-projective.

Ex 35.6. (a) Let E be an $\mathfrak{L}_p^g$-space and F a Fréchet space. Show that if (ℓ_p, F) has property (BB), then (E, F) has it as well. Hint: Use the technique from 35.9., proposition 1.

(b) Prove that if (ℓ_2, F) does not have (BB), then (ℓ_p, F) does not have (BB) for any $p \in\,]1, \infty[$. Hint: 23.2., corollary 2.

These results are taken from Bonet–D.–Galbis [12].

Ex 35.7. (a) Let E and F be two locally convex spaces, F complete, E with the bounded approximation property and α a tensor norm. Then $E \otimes_\alpha F$ is a *large* subspace of $E \tilde{\otimes}_\alpha F$, i.e. every bounded subset B of $E \tilde{\otimes}_\alpha F$ is contained in the closure of a bounded subset C of $E \otimes_\alpha F$. Hint: Take an equicontinuous net (T_η) of finite rank operators on E which converges pointwise to id_E, and look at $C := \bigcup T_\eta \tilde{\otimes}_\alpha id_F(B) \subset E \otimes_\alpha F$.

Note that, in general, $\mathbb{K} \otimes_\alpha F = F$ is not a large subspace of $\tilde{F} = \mathbb{K} \tilde{\otimes}_\alpha F$.

(b) Show that $c_o \otimes_\varepsilon F$ is a large subspace of $c_o(F)$ for any locally convex space F. Hint: Finite sequences.

(c) Let K be compact and F a locally convex space. Using partitions of unity, show that $C(K) \otimes_\varepsilon F$ is a large subspace of $C(K, F)$.

Ex 35.8. Take a normed space E, a (DF)–space F and a finitely generated tensor norm α. Then $E \otimes_{\alpha/} F$ is a (DF)–space. Hint: Use the formula $\alpha/ = \varepsilon \otimes_{\ell_\infty} \alpha$ from 29.10..

Ex 35.9. Let E_1 and E_2 be (DF)-spaces.

(a) If $E_k \otimes_\varepsilon G$ is (DF) for all Banach spaces G (or only C_2), then $E_1 \otimes_\varepsilon E_2$ is (DF). Hint: $\varepsilon \otimes \varepsilon = \varepsilon$.

(b) If $E_1, E_2 \in \text{space}(\mathfrak{L}_1)$, then $E_1 \otimes_\varepsilon E_2$ is a (DF)-space. Hint: $G \otimes_\varepsilon E_k = G \otimes_{\varepsilon/} E_k$ and the previous exercise.

Statement (a) can be applied to Hilbertizable bornological (DF)-spaces, which by a result of Hollstein [111] have the property that their ε-tensor product with each normed space is (DF). In [12] this was used to prove that pairs of Hilbertizable Fréchet spaces have property (BB) — a result stated but not proved by Grothendieck [92].

Ex 35.10. (a) Use the ideas of 12.9. to show that for normed spaces E and F the natural map

$$(E \otimes_\varepsilon \ell_{q'}) \otimes_\pi \ell_r \otimes_\pi (\ell_{p'} \otimes_\varepsilon F) \longrightarrow E \otimes_{\alpha_{p,q}} F$$

$$(x_n) \otimes (\lambda_n) \otimes (y_n) \rightsquigarrow \sum_{n=1}^{\infty} \lambda_n x_n \otimes y_n$$

(with $1/p' + 1/q' + 1/r = 1$) is a metric surjection.

(b) Extend (a) to locally convex spaces.

This result simplifies the study of $\otimes_{\alpha_{p,q}}$-stability of classes of locally convex spaces.

Ex 35.11. Use 35.7., 35.8. and 35.10. to show that for a Banach space E the following are equivalent:

(a) E is an $\mathfrak{L}_\infty^g$-space.

(b) For every inductive limit $\text{ind } F_n$ of Banach spaces $E \otimes_\varepsilon \text{ind } F_n = \text{ind } E \otimes_\varepsilon F_n$ holds topologically.

(c) $E \otimes_\varepsilon F$ is quasibarrelled for each quasibarrelled F such that (ℓ_1, F'_b) has (BB).

Hint: $E \otimes_\varepsilon F$ has a fundamental sequence of bounded sets if E and F are (DF)-spaces.

Ex 35.12. Let $T : E \longrightarrow F$ be a homomorphism between locally convex spaces, i.e. a topological surjection onto its image. Then the following operators are homomorphisms as well:

(a) $id_G \otimes_\pi T$ whenever G is a locally convex space in $\text{space}(\mathfrak{L}_1)$.

(b) $id_G \otimes_\varepsilon T$ whenever G is a locally convex space in $\text{space}(\mathfrak{L}_\infty)$.

Appendix A: Some Structural Properties of Banach Spaces

This appendix collects some results from Banach space theory which may be unfamiliar to the reader. These facts are used in various places throughout the book without always calling this to the attention of the reader. References will only be given for those facts which are not in nearly every introductory book on Functional Analysis.

A1. A *normed space* is a pair $(E, \|\cdot\|)$ of a vector space E over the field $\mathbb{K} = \mathbb{R}$ or $\mathbb{C}$ of real or complex scalars, together with a norm $\|\cdot\| : E \longrightarrow [0, \infty[$. If it is clear what the norm is, then we write simply E for the normed space — and $\|\ \|_E$ for the norm if necessary. If every Cauchy sequence in E converges, E is called a *Banach space*. The notation $\overset{\circ}{B}_E$ (open unit ball), B_E (closed unit ball), $\mathcal{L}(E, F)$ (continuous linear operators) and E' (dual space) is used (see list of symbols). Obviously, the reader should be familiar with basic examples of Banach spaces — such as: $c_o, \ell_p, C(K)$, Hilbert spaces and L_p; for the latter see Appendix B. Moreover, we do not believe it is necessary to explicitly state the open-mapping theorem, the closed graph theorem and the Hahn–Banach extension theorem. However, in the last case the reader might like to be reminded of the real version in which sublinear functionals w are involved: Let E be a real vector space and F a linear subspace, $w : E \longrightarrow \mathbb{R}$ sublinear, i.e. for all $x, y \in E$ and $\lambda \geq 0$:

$$w(x+y) \leq w(x) + w(y) \quad , \quad w(\lambda x) = \lambda w(x) \;,$$

and $\varphi : F \longrightarrow \mathbb{R}$ linear with $\varphi \leq w|_F$. Then there is a linear extension $\tilde\varphi : E \longrightarrow \mathbb{R}$ of φ with $\tilde\varphi \leq w$. It is worthwhile to recall that this version of the Hahn–Banach theorem has powerful applications even in the case where $F = \{0\}$!

A mapping $I \in \mathcal{L}(E, F)$ is called an *isometry* or a *metric injection* if $\|Ix\|_F = \|x\|_E$ for all $x \in E$. A surjective mapping $Q \in \mathcal{L}(E, F)$ is called a *metric surjection* if

$$\|Q(x)\|_F = \inf\{\|y\|_E \mid Q(y) = Q(x)\}$$

for all $x \in E$, i.e. $\|\ \|_F$ is the quotient norm of $\|\ \|_E$ along Q. Notation: $\overset{1}{\hookrightarrow}$ and $\overset{1}{\twoheadrightarrow}$ to indicate a metric injection or surjection, respectively. For every Banach space E there are natural mappings

$$\ell_1(B_E) \overset{1}{\twoheadrightarrow} E \qquad \text{and} \qquad E \overset{1}{\hookrightarrow} \ell_\infty(B_{E'})$$

$$(\lambda_x) \rightsquigarrow \sum_{x \in B_E} \lambda_x x \qquad\qquad x \rightsquigarrow \langle \cdot, x\rangle \quad .$$

Obviously, B_E can be replaced by any dense subset of the unit ball and $B_{E'}$ by any norming subset of $B_{E'}$. In particular, for every separable Banach space E there are mappings

$$\ell_1 \xrightarrow{1} E \xhookrightarrow{1} \ell_\infty .$$

The natural metric injection of a normed space into its bidual E''

$$E \xhookrightarrow{1} E'' , \quad x \rightsquigarrow [x' \rightsquigarrow \langle x', x \rangle]$$

will be denoted by κ_E. If κ_E is surjective, E is called reflexive. Hilbert spaces H are reflexive — and in the complex case the reader should be careful not to mix up the bilinear duality bracket $\langle \cdot, \cdot \rangle$ with the scalar product $(\cdot | \cdot)$ (linear in the first variable, antilinear in the second variable) when identifying H with H'.

A2. A *separating dual system* $\langle E, F \rangle$ is a pair of vector spaces E and F (over $\mathbb{K}$) with a bilinear map $\langle \cdot, \cdot \rangle : E \times F \longrightarrow \mathbb{K}$ such that for each $0 \neq x \in E$ there is a $y \in F$ such that $\langle x, y \rangle \neq 0$ and for each $0 \neq y \in F$ there is an $x \in E$ with $\langle x, y \rangle \neq 0$. The topology on E of pointwise convergence with respect to F is denoted by $\sigma(E, F)$ — and, by symmetry, $\sigma(F, E)$ on F : the *weak topologies of the dual system*. $(E, \sigma(E, F))' = F$ holds with the obvious identification. For $A \subset E$ define the *absolute polar* A^0 of A by

$$A^0 := \{ y \in F \mid |\langle a, y \rangle| \leq 1 \text{ for all } a \in A \}$$

and analoguously for $B \subset F$. Polars A^0 are absolutely convex and $\sigma(F, E)$–closed; the Hahn–Banach theorem implies that

$$A^{00} = \overline{\Gamma A}^{\sigma(E,F)}$$

holds for the bipolar A^{00} of A *(bipolar theorem)*. The most important example of a separating dual system is $\langle E, E' \rangle$ if E is a normed space — the duality bracket being $\langle x, x' \rangle := x'(x)$ (this notation was already used in A1.). $\sigma(E, E')$ is called the *weak topology* on E and $\sigma(E', E)$ is sometimes called the *weak-∗-topology* on E'; note that on E' there is also the weak topology $\sigma(E', E'')$. Though they are quite different (if $\dim E = \infty$) the norm and the weak topology on a normed space E have two things in common: the *convex* closed sets and the bounded sets:

(1) A *convex* subset $A \subset E$ is norm–closed if and only if it is $\sigma(E, E')$–closed.

(2) $A \subset E$ is norm–bounded (i.e. $\sup \|A\| < \infty$) if and only if it is $\sigma(E, E')$–bounded (i.e. $\sup |\langle x', A \rangle| < \infty$ for all $x' \in E'$).

(3) Let E be a Banach space. $A \subset E'$ is norm–bounded if and only if it is $\sigma(E', E)$–bounded.

(2) and (3) are refered to as the *Mackey theorem* or the *uniform boundedness principle*. It is important to know that the unit ball $B_{E'} = B^0_E$ of the dual space is always $\sigma(E', E)$-compact (Banach–Alaoglu–Bourbaki theorem).

A3. Compact convex sets with respect to the norm or a weak topology — more generally: in a locally convex space — enjoy many interesting properties.

LEMMA OF KY FAN: *Let $K \neq \emptyset$ be a convex, compact subset of a topological vector space and*

$$\mathfrak{K} \subset \{f : K \longrightarrow [-\infty, \infty[\;|\; f \text{ is concave, upper semicontinuous}\}$$

a convex set such that for each $f \in \mathfrak{K}$ there is an $x \in K$ with $f(x) \geq 0$. Then there is an $x_o \in K$ such that $f(x_o) \geq 0$ for all $f \in \mathfrak{K}$.

See Pietsch [214], p.40, for a proof.

A4. A *basis* (x_n) of an infinite dimensional Banach space E is a sequence of elements of E such that for each $x \in E$ there exists a *unique* sequence (α_n) of scalars such that $x = \sum_{n=1}^{\infty} \alpha_n x_n$. The coefficient functionals $x'_n : x \rightsquigarrow \alpha_n$ are well-defined and linear and one can show that they are continuous. Clearly, $\langle x'_n, x_m \rangle = \delta_{n,m}$. For an orthonormal basis (e_n) in a Hilbert space the coefficient functionals are $e'_n := (\cdot | e_n)$ and hence $\|e_n\| = \|e'_n\| = 1$. Results concerning the existence of bases and related questions are treated in section 5. Clearly, finite dimensional Banach spaces E always have a finite algebraic basis $(x_n)_{n=1}^{N}$. Such a basis can even be chosen with

$$\|x_n\|_E = \|x'_n\|_{E'} = 1$$

— in this case, $(x_n)_{n=1}^{N}$ is called an *Auerbach basis* (see [177], I, 1.c.3., for a proof).

A5. If the complex Banach space A also has a multiplication $A \times A \longrightarrow A$ which makes A an algebra and satisfies $\|x \cdot y\| \leq \|x\| \, \|y\|$, then A is called a *Banach algebra*. Examples: $\mathcal{L}(E, E), \ell_\infty, L_\infty(\mu), C(K)$ with their natural multiplications. If, in addition, there is an "involution" $* : A \longrightarrow A$ which is antilinear and satisfies for all $x, y \in A$

(a) $x^{**} = x$, (b) $(xy)^* = y^* x^*$, (c) $\|xx^*\| = \|x\|^2$,

then A is called a *C^*-algebra*. Commutative C^*-algebras with a unit for the multiplication (such as ℓ_∞ and $L_\infty(\mu)$) are isomorphic (with norm and algebraic structure) to a space $C(K)$ where K is compact (*Gelfand representation theorem*). K is the weak-$*$-compact subset of $B_{A'}$ consisting of all non-trivial, multiplicative, linear functionals on A.

A6. A *real Banach lattice* E is a real Banach space E with an order $\leq$ (transitive, reflexive and antisymmetric) which satisfies

(1) $x \leq y$ implies $x + z \leq y + z$ for all $z \in E$, and $\lambda x \leq \lambda y$ for all $\lambda \geq 0$.

(2) The minimum $x \wedge y$ exists for all $x, y \in E$.

(3) $|x| \leq |y|$ implies that $\|x\| \leq \|y\|$ (where $|x| := x \vee (-x)$).

A *complex* Banach lattice E is a complex Banach space which has a real linear subspace E_o such that

(1) $E_o \oplus iE_o = E$; the unique decomposition of x is written as $x = \text{Re}(x) + i\,\text{Im}(x)$ and E_o is called the real part of E.

(2) E_o has an order such that (with the induced norm) E_o is a real Banach lattice.

(3) $|z| := \sup\{|\text{Re}(e^{it}z)| \mid 0 \le t \le 2\pi\}$ exists in E_o for all $z \in E$.

(4) $|z_1| \le |z_2|$ implies $\|z_1\| \le \|z_2\|$ for all $z_1, z_2 \in E$.

Basic references for Banach lattices: Schaefer [244], Schwarz [256], and Lacey [164].

A closed, linear subspace F of a (real or complex) Banach lattice E is called *solid* if $x \in F, y \in E$ and $|y| \le |x|$ implies $y \in F$. In this case, the quotient E/F is a Banach lattice under the natural ordering: $\hat{x} \le \hat{y}$ if $x \le y$ (see Schaefer [244], p.85 and p.137).

The dual space E' of a Banach lattice has a natural order given by $x' \ge 0$ whenever $\langle x', x\rangle \ge 0$ for all $x \ge 0$. With this order E' is a Banach lattice. The canonical embedding $\kappa_E : E \hookrightarrow E''$ preserves the order.

A7. The most common examples of Banach lattices are L_p and $C(K)$. A (real or complex) Banach lattice is called an *abstract L_p-space* if for all $x, y \in E$ with $x, y \ge 0$ and $x \wedge y = 0$ the relation

$$\|x+y\|^p = \|x\|^p + \|y\|^p \qquad \text{(if } 1 \le p < \infty\text{)}$$
$$\|x+y\| = \max\{\|x\|, \|y\|\} \qquad \text{(if } p = \infty\text{)}$$

holds. For $p = \infty$ these Banach lattices are also called *abstract M-spaces*, for $p = 1$ *abstract L-spaces*. In this latter case the definition is equivalent to

$$\|x+y\| = \|x\| + \|y\|$$

for all $x, y \in E$ with $x, y \ge 0$ (see Schwarz [256], p.98). It is obvious that a complex Banach lattice is an abstract L_p if and only if its real part is. The following representations hold:

THEOREM: (1) *For $1 \le p < \infty$ every abstract L_p-space is linearly isomorphic (with norm and order) to some $L_p(\mu)$, where μ is a Borel–Radon measure on a locally compact space.*

(2) *Every abstract M-space with a unit (i.e. an element $e \in B_E$ such that $|x| \le e$ for all $x \in B_E$) is linearly isomorphic (with norm and order) to some $C(K)$, where K is a compact space.*

For $p = 1$ and $p = \infty$ these results are usually attributed to Kakutani, for $1 < p < \infty$ to Bohnenblust and Nakano. Proofs can be found in Schaefer [244] (p.104, 114, 138 and 149) or Lacey [164] (p.59 and 135).

In particular, these results show that a *Banach lattice E is an abstract L_p-space if and only if its dual E' is an abstract $L_{p'}$-space*. For $p = 1$ even more is true:

THEOREM: *Let E be a Banach space such that E' is isometric to an abstract M–space with unit. Then E is isometric to an abstract L–space, hence to an $L_1(\mu)$ for some Borel–Radon measure.*

This result is due to Grothendieck [91]; more recent references for a proof are Schaefer [244], p.150 and Lacey [164], p.95–96 and 99.

Appendix B: Integration Theory

In this appendix a summary of the measure and integration theory which is used throughout the book is given. An outline of the Daniell–Stone integration theory will be presented — the details can be found in many textbooks; we follow [71], where the reader will find all proofs of the following statements on real–valued integration. The necessary modifications for the complex case will be indicated. At the end of this appendix the basics on Bochner integration will be given.

Extension Procedure and the Basic Theorems

B1. Let Ω be a non–empty set and $\mathcal{L}$ a vector lattice of functions $\Omega \to \mathbb{R}$ which is *Stonean*, i.e.

$$\min\{1, f\} \in \mathcal{L} \qquad \text{whenever } f \in \mathcal{L}.$$

A *complex (or signed) Daniell functional* on $\mathcal{L}$ is a linear functional $\mu : \mathcal{L} \to \mathbb{C}$ which is σ–*continuous*, i.e.: If $f_n \in \mathcal{L}$ and $f_n \downarrow 0$ pointwise, then

$$\lim_n \mu(f_n) = 0 .$$

If μ is real–valued, it is called a *real Daniell functional* — and if it is positive (i.e. $0 \leq f \in \mathcal{L}$ implies $\mu(f) \geq 0$), it is simply called a *Daniell functional*. If μ is a complex Daniell functional, then

$$|\mu|(f) := \sup\{|\mu(h)| \mid 0 \leq |h| \leq f, h \in \mathcal{L}\} \qquad \text{if } 0 \leq f \in \mathcal{L}$$
$$|\mu|(f) := |\mu|(f^+) - |\mu|(f^-) \qquad \text{if } f \in \mathcal{L}$$

defines a Daniell functional: The *total variation* of μ. Clearly,

$$|\mu(f)| \leq |\mu|(|f|) \qquad \text{for all } f \in \mathcal{L}.$$

If μ is a real Daniell functional, then its positive and negative parts
$$\mu^+ := \frac{1}{2}(|\mu| + \mu) \qquad \mu^- := \frac{1}{2}(|\mu| - \mu)$$
are (positive) Daniell functionals and $\mu = \mu^+ - \mu^-$.

Examples: (1) If Ω is a locally compact topological space and $\mathcal{L} := C_{oo}(\Omega, \mathbb{R})$, the space of all real–valued continuous functions on Ω with compact support, then, by Dini's theorem, every positive linear $\mu : C_{oo}(\Omega, \mathbb{R}) \to \mathbb{R}$ is σ–continuous. Moreover, if $\mu : C_{oo}(\Omega, \mathbb{R}) \to \mathbb{C}$ is linear and of bounded variation, i.e. $|\mu|(f) < \infty$ (defined as before) for all $f \geq 0$, then $\operatorname{Re}(\mu)^+$, $\operatorname{Re}(\mu)^-$, $\operatorname{Im}(\mu)^+$ and $\operatorname{Im}(\mu)^-$ are well–defined positive linear functionals and hence σ–continuous; it follows that μ is σ–continuous as well. These μ are sometimes called (signed) *Radon measures* (see the fundamental theorem B3. below).

(2) Let Σ be a (non–empty) *ring* of subsets of Ω (i.e.: $A \cup B$, $A \cap B$, $A \setminus B \in \Sigma$ whenever $A, B \in \Sigma$). Then $\mathfrak{S}(\Sigma)$ is the Stonean vector lattice of all *step functions*
$$f := \sum_{m=1}^{n} \alpha_m \chi_{A_m},$$
where $m \in \mathbb{N}$, $\alpha_m \in \mathbb{R}$, and χ_{A_m} is the characteristic function of $A_m \in \Sigma$. An additive function $\mu : \Sigma \to \mathbb{C}$ is called a *complex* (or *signed*) *measure* if it is σ–additive, i.e.
$$\mu\left(\bigcup_{m=1}^{\infty} A_m\right) = \sum_{m=1}^{\infty} \mu(A_m)$$
whenever $A_n, \bigcup_{m=1}^{\infty} A_m \in \Sigma$ and the A_n are pairwise disjoint. If μ is *positive* (i.e. $\mu(\Sigma) \subset [0, \infty[$), then μ is σ–additive if and only if its "linearization" $\mu^L : \mathfrak{S}(\Sigma) \to \mathbb{C}$
$$\mu^L(f) := \sum_{m=1}^{n} \alpha_m \mu(A_m)$$
is well–defined and σ–continuous; this equivalence holds for arbitrary additive μ if Σ is a *local σ–ring* (i.e. a ring closed under countable intersections). The terms *real* measure and *measure* ($:=$ positive measure) will be used.

B2. It is the aim of measure and integration theory to extend μ to a larger class of functions which satisfies good convergence theorems. Let $\mu : \mathcal{L} \to \mathbb{R}$ be a (positive) Daniell functional. Then
$$\mathfrak{U} := \{u : \Omega \to]-\infty, \infty] \mid \text{ there exist } f_n \in \mathcal{L} : f_n \uparrow u \text{ pointwise}\}$$
$$\mu'(u) := \lim_{n \to \infty} \mu(f_n)$$
and μ' is well–defined. A function $f : \Omega \to [-\infty, \infty]$ is called μ–*integrable* (in symbols: $f \in \mathcal{L}_1(\Omega, \mathcal{L}, \mu) =: \mathcal{L}_1(\Omega, \mu) =: \mathcal{L}_1(\mu)$) if for every $\varepsilon > 0$ there are $u, -v \in \mathfrak{U}$ with $v \leq f \leq u$ and
$$\mu'(u - v) \leq \varepsilon.$$

In this case,
$$\mu(f) := \int f d\mu := \int_\Omega f(x)\mu(dx) := \inf\{\mu'(u) \mid u \in \mathfrak{U}, \, u \geq f\} \in \mathbb{R}$$
is the μ-*integral* of f. A complex-valued function is μ-integrable if $\text{Re}(f)$ and $\text{Im}(f)$ are μ-integrable,
$$\int f d\mu := \int \text{Re}(f) d\mu + i \int \text{Im}(f) d\mu \, .$$
This integral is linear, monotone and σ-continuous on $\mathfrak{L}_1(\mu)$ (more precisely, on the Stonean vector lattice $\mathfrak{L}_1^f(\mu)$ of real-valued integrable functions), hence a Daniell functional. The extension procedure applied to this functional $\mu : \mathfrak{L}_1^f \to \mathbb{R}$ gives $\mathfrak{L}_1(\mu)$ again, nothing new. The convergence theorems:

(1) BEPPO LEVI THEOREM (MONOTONE CONVERGENCE): *If $f_n \in \mathfrak{L}_1(\mu)$ is an increasing sequence such that $\sup \int f_n d\mu < \infty$, then $\lim_{n \to \infty} f_n$ is μ-integrable and*
$$\lim_{n \to \infty} \int f_n d\mu = \int \lim_{n \to \infty} f_n d\mu \, .$$

(2) FATOU LEMMA: *Let $f_n \in \mathfrak{L}_1(\mu)$ be non-negative functions with $\sup \int f_n d\mu < \infty$. Then the function $\liminf_{n \to \infty} f_n$ is μ-integrable and the inequality*
$$\int \liminf_{n \to \infty} f_n d\mu \leq \liminf_{n \to \infty} \int f_n d\mu$$
holds.

(3) LEBESGUE DOMINATED CONVERGENCE THEOREM: *If $f_n, g \in \mathfrak{L}_1(\mu)$ are such that $|f_n| \leq g$ and $f := \lim_{n \to \infty} f_n$ exists pointwise, then f is μ-integrable and*
$$\int \lim_{n \to \infty} f_n d\mu = \lim_{n \to \infty} \int f_n d\mu \, .$$

B3. A function $f : \Omega \to [-\infty, \infty]$ is μ-*measurable* (in symbols: $f \in \mathfrak{M}(\mu)$ or simply $\mathfrak{M}$) if for all $0 \leq h \in \mathfrak{L}_1(\mu)$ the function
$$\text{md}(f, h, -h) := \max\{-h, \min\{f, -h\}, \min\{f, h\}\}$$
(the "middle" of f, h, and $-h$) is μ-integrable. This definition (due to Stone) says that a function is measurable if its graph looks "locally" like the graph of an integrable function. Complex-valued functions f are μ-measurable if $\text{Re}(f), \text{Im}(f) \in \mathfrak{M}$. The trivial criterion
$$f \,\, \mu\text{-measurable}, g \in \mathfrak{L}_1(\mu), |f| \leq g \,\curvearrowright\, f \,\, \mu\text{-integrable}$$

gains its importance from the fact that $\mathfrak{M}$ is very large: It is stable under the formation of pointwise limits and suprema of sequences and for $f_1, ..., f_n \in \mathfrak{M}$ the function

$$\varphi(f_1, ..., f_n)$$

is also μ-measurable whenever $\varphi : [-\infty, \infty]^n \to [-\infty, \infty]$ is continuous (more generally: Borel measurable). $\mathfrak{M}$ is so large that for the Lebesgue measure it is not possible — without using the axiom of choice — to construct a non-measurable function.

If Σ is a σ-algebra on Ω (i.e. a local σ-ring with $\Omega \in \Sigma$, hence closed under countable unions and intersections), then a function $f : \Omega \to [-\infty, \infty]$ is called Σ-measurable whenever

$$[f \leq \alpha] := \{x \in \Omega \mid f(x) \leq \alpha\}$$

is in Σ for all $\alpha \in \mathbb{R}$. For a topological space Ω the smallest σ-algebra which contains all open sets, is called the *Borel σ-algebra* $\mathfrak{B}(\Omega)$; Borel measurable means $\mathfrak{B}(\Omega)$-measurable.

If μ is a Daniell functional, then the *μ-measurable sets*

$$\Sigma_\mu := \{A \subset \Omega \mid \chi_A \in \mathfrak{M}(\mu)\}$$

form a σ-algebra. It follows that $f : \Omega \to [-\infty, \infty]$ is μ-measurable if and only if it is Σ_μ-measurable.

A function $f : \Omega \to [-\infty, \infty]$ is called *μ-null* if $f \in \mathcal{L}_1(\mu)$ and $\int |f| d\mu = 0$; a set A is μ-null if χ_A is μ-null, and it follows that f is μ-null if and only if the set $[f \neq 0]$ is μ-null. One says that two functions f and g are *μ-almost everywhere equal* ($= \mu$-a.e.) if the set $[f \neq g]$ is μ-null. Note that Σ_μ and the local σ-ring

$$I_\mu := \{A \subset \Omega \mid \chi_A \in \mathcal{L}_1(\mu)\} \subset \Sigma_\mu$$

of μ-integrable sets are *μ-complete* in the following sense: If $\mu(\chi_A) = 0$ and $B \subset A$, then $B \in I_\mu$ and $\mu(\chi_B) = 0$. A subring Σ^0 of I_μ is called *μ-dense* if for every μ-integrable set B there is an $A \in \Sigma^0$ with $\mu(\chi_{A \Delta B}) = 0$.

μ is called *finite* if Ω is μ-integrable and *σ-finite* if $\Omega = \cup_{n=1}^\infty A_n$ with $A_n \in I_\mu$.

If μ is a Daniell functional on a Stonean vector lattice $\mathcal{L}$, then

$$\hat{\mu}(A) := \int_A \chi_A d\mu$$

defines a measure on the local σ-ring I_μ. Since a non-negative function $f : \Omega \to [0, \infty]$ is the pointwise limit of the increasing sequence (f_n) of functions

$$f_n := 2^n \chi_{[f \geq 2^n]} + \sum_{m=0}^{2^{2n}-1} \frac{m}{2^n} \chi_{[m 2^{-n} \leq f < (m+1) 2^{-n}]},$$

the convergence theorems imply the

FUNDAMENTAL THEOREM OF THE DANIELL–STONE INTEGRATION THEORY: *If the ring $\Sigma^0 \subset I_\mu$ is μ-dense, then*

$$\mathcal{L}_1(\Omega, \mathcal{L}, \mu) = \mathcal{L}_1(\Omega, \mathfrak{S}(\Sigma^0), \hat{\mu})$$

with equal integrals.

This means that actually every Daniell functional comes from a measure on a certain ring of sets, e.g., the local σ-ring of integrable sets. This is why we always simply speak of a *measure space* (Ω, μ) or $(\Omega, \Sigma_\mu, \mu)$ if we want to draw attention to the μ-measurable sets. Note that Σ_μ is always μ-complete and that μ is only defined on I_μ (and has finite values) and not on all μ-measurable sets.

For Radon measures (i.e. positive linear functionals on $C_{oo}(\Omega)$) this theorem will have a particularly interesting consequence (see B9.).

For real Daniell functionals/real measures μ define

$$\mathcal{L}_1(\mu) := \mathcal{L}_1(|\mu|) = \mathcal{L}_1(\mu^+) \cap \mathcal{L}_1(\mu^-),$$

$$\int f d\mu := \int f d\mu^+ - \int f d\mu^-$$

and in the complex case

$$\mathcal{L}_1(\mu) := \mathcal{L}_1(\operatorname{Re}(\mu)) \cap \mathcal{L}_1(\operatorname{Im}(\mu)),$$

$$\int f d\mu := \int f d(\operatorname{Re}(\mu)) + i \int f d(\operatorname{Im}(\mu)).$$

(Obvious definitions for $f : \Omega \to \mathcal{C}$). It follows that

$$\left| \int f d\mu \right| \le \int |f| d|\mu|$$

whenever f is μ-integrable.

B4. If $(\Omega_i, \Sigma_{\mu_i}, \mu_i)$ are two measure spaces and $\mathcal{L}$ the Stonean vector lattice of all step functions on $\Omega_1 \times \Omega_2$,

$$\sum_{m=1}^n \alpha_m \chi_{A_m^1 \times A_m^2},$$

where $A_m^i \subset \Omega_i$ are μ_i-integrable, then

$$\mu_1 \otimes \mu_2(f) := \int_{\Omega_1} \int_{\Omega_2} f(x_1, x_2) \mu_2(dx_2) \mu_1(dx_1) =$$

$$= \int_{\Omega_2} \int_{\Omega_1} f(x_1, x_2) \mu_1(dx_1) \mu_2(dx_2) = \sum_{m=1}^n \alpha_m \mu_1(A_m^1) \mu_2(A_m^2)$$

defines a Daniell functional/measure on $\mathfrak{L}$. It is called the *tensor product* of μ_1 and μ_2.
If $f : \Omega_1 \times \Omega_2 \to \mathbb{C}$ is such that the function

$$x_1 \rightsquigarrow \int_{\Omega_2} f(x_1, x_2)\mu_2(dx_2)$$

is μ_1–almost everywhere defined and μ_1–integrable, one says that the *iterated integral* $\int \int f d\mu_2 d\mu_1$ exists.

FUBINI THEOREM: *If $f \in \mathfrak{L}_1(\mu_1 \otimes \mu_2)$, then both iterated integrals exist and*

$$\int_{\Omega_1 \times \Omega_2} f d\mu_1 \otimes \mu_2 = \int_{\Omega_1} \int_{\Omega_2} f d\mu_2 d\mu_1 = \int_{\Omega_2} \int_{\Omega_1} f d\mu_1 d\mu_2 \; .$$

Conversely,

FUBINI–TONELLI THEOREM: *If $f : \Omega_1 \times \Omega_2 \to \mathbb{C}$ (or $\to [-\infty, \infty]$) is $\mu_1 \otimes \mu_2$–measurable, $[f \neq 0] \subset \Omega_1 \times \Omega_2$ is σ–finite (with respect to $\mu_1 \otimes \mu_2$) and one of the iterated integrals of $|f|$ exists, then f is $\mu_1 \otimes \mu_2$–integrable.*

None of the assumptions in this last result is indispensable. (A set $A \subset \Omega$ is called σ–finite with respect to μ if there are $A_n \in I_\mu$ with $A \subset \bigcup_{n=1}^{\infty} A_n$.)

The L_p–Spaces

B5. Let (Ω, μ) be a measure space and $0 < p < \infty$. A function $f : \Omega \to \mathbb{K}(= \mathbb{R}$ or $\mathbb{C})$ is called *p–integrable* if it is μ–measurable and $|f|^p$ is μ–integrable; in symbols: $f \in \mathfrak{L}_p(\Omega, \mu) =: \mathfrak{L}_p(\mu)$. Since every integrable function $\Omega \to [-\infty, \infty]$ is almost everywhere equal to a real–valued function, for $p = 1$ this slight difference in the definition will cause no confusion. $\mathfrak{L}_\infty(\mu)$ is the space of μ–measurable functions $f : \Omega \to \mathbb{K}$ which are uniformly bounded outside a *local μ–null set* A_f (i.e. a set A such that $\mu(A \cap B) = 0$ for all μ–integrable B; in general there exist local μ–null sets which are not μ–null). Define, for $f \in \mathfrak{L}_p(\mu)$,

$$\|f\|_p := \left(\int |f|^p d\mu\right)^{1/p} \quad \text{if } 0 < p < \infty$$

$$\|f\|_\infty := \inf\{ r \geq 0 \mid [|f| > r] \text{ local } \mu - \text{null}\}$$

— the latter is called the *essential supremum* of the function f and is sometimes denoted by ess-sup $|f|$. The functions in $\mathfrak{L}_\infty(\mu)$ are called *μ–essentially bounded*.

HÖLDER INEQUALITY: *If $p, q \in [1, \infty]$ satisfy $1/p + 1/q = 1$ and $f \in \mathfrak{L}_p(\mu)$ and $g \in \mathfrak{L}_q(\mu)$, then $f \cdot g \in \mathfrak{L}_1(\mu)$ and*

$$\|f \cdot g\|_1 \leq \|f\|_p \|g\|_q \; .$$

This allows one to show that $\|\cdot\|_p$ is a seminorm on $\mathcal{L}_p(\mu)$ for $1 \leq p \leq \infty$, and therefore

$$L_p(\Omega,\mu) := L_p(\mu) := \mathcal{L}_p(\mu) \,/\, \{f \mid \|f\|_p = 0\}$$

(or briefly: L_p) with $\|\cdot\|_p$ is a normed space. The convergence theorems show that L_p is a Banach space — even a Banach lattice with its natural ordering. Note that $\|f\|_p = 0$ is the same as f being a μ-null function (for $p < \infty$) or a local μ-null function (if $p = \infty$). The fact that the elements of L_p are classes of functions should not be forgotten: If necessary the notation $\tilde{f}$ or $\tilde{f}^{loc}$ for the class containing f will be used. For $1 \leq p < \infty$ the space L_p is an order complete Banach lattice (but for $p = \infty$ in general not: see B6. below), even more: every norm-bounded increasing net in L_p has a supremum.

For $p < \infty$ the triangle inequality reads as follows

$$\left(\int_\Omega \left|\sum_{m=1}^n f(x,m)\right|^p \mu(dx)\right)^{1/p} \leq \sum_{m=1}^n \left(\int_\Omega |f(x,m)|^p \mu(dx)\right)^{1/p}.$$

The sum can be replaced by an integral — this means:

CONTINUOUS TRIANGLE (OR MINKOWSKI) INEQUALITY: *Let (Ω_i, μ_i) be measure spaces, $1 \leq p < \infty$ and $f : \Omega_1 \times \Omega_2 \to [-\infty, \infty]$ (or $\mathbb{C}$) $\mu_1 \otimes \mu_2$-measurable such that $[f \neq 0]$ is σ-finite. If in the following inequality the right iterated integral exists, then the left one does also exist and*

$$\left(\int_{\Omega_1} \left|\int_{\Omega_2} f(x_1, x_2)\mu_2(dx_2)\right|^p \mu_1(dx_1)\right)^{1/p} \leq \int_{\Omega_2} \left(\int_{\Omega_1} |f(x_1,x_2)|^p \mu_1(dx_1)\right)^{1/p} \mu_2(dx_2)$$

holds.

B6. For $1/p + 1/q = 1$ Hölder's inequality gives a natural separating duality bracket between $L_p(\Omega, \mu)$ and $L_q(\Omega, \mu)$:

$$\langle \tilde{f}, \tilde{g} \rangle := \int_\Omega fg\,d\mu \qquad \text{whenever } \tilde{f} \in L_p, \tilde{g} \in L_q$$

(replace $\sim$ by $\sim_{loc}$ if p or $q = \infty$), hence an embedding

$$\Phi : L_q \hookrightarrow L_p'.$$

Since

$$\|f\|_p = \sup\{|\int fg\,d\mu| \mid g \in \mathcal{L}_q, \|g\|_q \leq 1\}$$

holds, the unit ball of L_q is norming for the norm $\|\cdot\|_p$ of the Banach space L_p; in particular, Φ is an isometry. For $1 < p < \infty$ this isometry is even onto

$$L_p(\Omega,\mu)' = L_q(\Omega,\mu).$$

This follows from the Radon–Nikodým theorem below. In particular, all L_p are reflexive ($1 < p < \infty$); this is not true for $p = 1$ or ∞ if L_1 is infinite dimensional. Moreover, the dual of L_∞ is strictly larger than L_1 if L_1 is infinite dimensional.

The Radon–Nikodým theorem is the appropriate tool for determining the conditions under which $L'_1 = L_\infty$ holds: For this, take $\mathcal{L}$ a Stonean vector lattice of *bounded* functions (this is not really a restriction by the fundamental theorem B3.) and μ and ν (positive) Daniell functionals = measures on $\mathcal{L}$. The measure ν is called μ-*absolutely continuous* if every ν-measurable μ-null set is a ν-null set. ν is called μ-*singular* if there exists a ν-measurable μ-null set N such that

$$\int f d\nu = \int_N f d\nu$$

for all $f \in \mathcal{L}$ (and then for all $f \in \mathcal{L}_1(\nu)$). If $g_o \geq 0$ is a *locally μ-integrable function* (i.e. $f \cdot g_o \in \mathcal{L}_1(\mu)$ for all $f \in \mathcal{L}$), then

$$\nu(f) := \int f g_o d\mu \qquad \text{for all } f \in \mathcal{L}$$

is μ-absolutely continuous; such a g_o is called a μ-*density* of ν.

SEGAL'S LOCALIZATION THEOREM: *Let $\mathcal{L}$ be a Stonean vector lattice of bounded functions on a set Ω and μ a Daniell functional on $\mathcal{L}$. The following statements are equivalent:*

(a) *$L_\infty(\mu)$ is order complete.*
(b) *$L_1(\mu)' = L_\infty(\mu)$ isometrically.*
(c) *Every Daniell functional $\nu \leq \mu$ on $\mathcal{L}$ (i.e. $\nu(f) \leq \mu(f)$ if $0 \leq f \in \mathcal{L}$) has a μ-density.*
(d) *The RADON–NIKODÝM THEOREM holds for μ : Every μ-absolutely continuous Daniell functional ν on $\mathcal{L}$ has a μ-density.*

The main part of the proof is to show that the Radon–Nikodým theorem holds for finite measures — the rest is to pass in L_∞ from the increasing net $(\tilde{\chi}_A^{loc})$ coming from the μ-integrable sets A to all of Ω; this is why measures with $L_\infty(\mu)$ being order complete are called *localizable*. σ-finite and Borel–Radon measures (see B9. below) are localizable, but there are non-localizable measures.

In proving this theorem one also gets

LEBESGUE'S DECOMPOSITION THEOREM: *Let μ and ν be Daniell functionals on a Stonean vector lattice of bounded functions and let $|\nu(f)| \leq c$ if $f \in \mathcal{L}$ and $|f| \leq 1$. Then there is a unique decomposition (ν_a, ν_s) of Daniell functionals on $\mathcal{L}$ such that*

$$\nu = \nu_a + \nu_s$$

where ν_a is μ-absolutely continuous and ν_s is μ-singular.

B7. When is a measure localizable? A measure space $(\Omega, \Sigma_\mu, \mu)$ is called *strictly localizable* if it admits a μ-decomposition $\mathfrak{D}$, i.e. a set of pairwise disjoint μ-integrable sets K such that for every μ-integrable set A of positive measure there is a $K \in \mathfrak{D}$ with $\mu(A \cap K) > 0$. It follows that $\Omega \setminus \cup \mathfrak{D}$ is a local μ-null set and a function f is μ-measurable if (and only if) all functions $f\chi_K$ with $K \in \mathfrak{D}$ are μ-measurable: (Ω, μ) is decomposed into the family $(K, \mu|_K)$ of finite measure spaces. It is easy to see that every strictly localizable measure is localizable.

Maharam's theorem says that every finite measure μ admits a linear, multiplicative, positive mapping
$$\lambda_\infty : L_\infty(\Omega, \mu) \longrightarrow \mathcal{L}_\infty(\Omega, \mu)$$
which is a right inverse of the quotient mapping $\mathcal{L}_\infty \to L_\infty$ (i.e. $\lambda_\infty(\tilde{f}^{loc}) \subset \tilde{f}^{loc}$) and such that $\lambda_\infty(\tilde{1}^{loc}) = 1$. Such a λ_∞ is called a (multiplicative) *lifting*. For $p < \infty$ there are no liftings (even without multiplicativity) for the Lebesgue measure. It is obvious that the existence of liftings facilitates the analysis of a measure enormously. It can be shown that a measure space (Ω, μ) is *strictly localizable if and only if it admits a lifting*.

Localizability in some sense gives the possibility of decomposing L_∞ into finite measures, strict localizability gives the possibility of decomposing $\mathcal{L}_\infty$. Since integrable functions vanish μ-a.e. outside a σ finite set, it is always possible to decompose the Banach space $L_1(\mu)$:

REMARK: *For every measure space (Ω, μ) there is a family $(\Omega_\iota, \mu_\iota)$ of finite measure spaces $(\iota \in I)$ such that*
$$L_1(\mu) = \ell_1(I; L_1(\mu_\iota))$$
holds isometrically.

For the proof take a maximal family $\mathfrak{D} = \{\Omega_\iota | \iota \in I\}$ of μ-integrable sets such that $\mu(\Omega_{\iota_1} \cap \Omega_{\iota_2}) = 0$ if $\iota_1 \neq \iota_2$ (Zorn's lemma), observe that for every μ-integrable A (and then also for every σ-finite A) the set of indices
$$\{\iota \in I \mid \mu(A \cap \Omega_\iota) > 0\}$$
is countable and that
$$A = \bigcup \{A \cap \Omega_\iota \mid \mu(A \cap \Omega_\iota) > 0\} \quad \mu\text{-a.e.}\,.$$
This clearly shows that the linear mapping
$$L_1(\Omega, \mu) \longrightarrow \ell_1(I; L_1(\Omega_\iota, \mu|_{\Omega_\iota}))$$
$$f \rightsquigarrow (f\chi_{\Omega_\iota})_{\iota \in I}$$

is isometric — and onto anyhow.

Borel–Radon Measures and the Riesz Representation Theorem

B8. Assume now that the (positive) Daniell functional μ on the Stonean vector lattice $\mathfrak{L}$ is not only σ-continuous but even τ-*continuous*: This means that for every pointwise decreasing net $f_\gamma \downarrow 0$

$$\lim_\gamma \mu(f_\gamma) = 0 \,.$$

One can now follow the same extension procedure as in the σ-case — just replace increasing sequences in the definition of $\mathfrak{U}$ by increasing nets. This way one obtains $\mathfrak{U}^\tau$ and (with the same definitions as before) the spaces $\mathcal{L}_1(\mu^\tau)$ and $\mathfrak{M}(\mu^\tau)$ of μ^τ-integrable and μ^τ-measurable functions ("Bourbaki extension procedure"). The integral μ^τ is σ-continuous on $\mathcal{L}_1(\mu^\tau)$ but not in general τ-continuous. The Daniell extension procedure applied to μ^τ on $\mathcal{L}_1^f(\mu^\tau)$ (finite valued functions) gives nothing new. Clearly, $\mathcal{L}_1(\mu) \subset \mathcal{L}_1(\mu^\tau)$, with

$$\int f d\mu = \int f d\mu^\tau \,.$$

Moreover, for every $f \in \mathcal{L}_1(\mu^\tau)$ there is an $f_o \in \mathcal{L}_1(\mu)$ with $f = f_o$ μ^τ-almost everywhere; in particular,

$$L_p(\mu) = L_p(\mu^\tau)$$

holds isometrically for $1 \leq p < \infty$. Since $L_\infty(\mu^\tau)$ is always order complete (see the following theorem), one can show that $L_\infty(\mu) \neq L_\infty(\mu^\tau)$ may occur.

KÖLZOW'S THEOREM: *μ^τ is strictly localizable. It follows that $L_\infty(\mu^\tau)$ admits a lifting, $L_1(\mu^\tau)' = L_\infty(\mu^\tau)$, the space $L_\infty(\mu^\tau)$ is order complete and the Radon–Nikodým theorem holds for μ^τ (and μ^τ-absolutely continuous measures ν on I_{μ^τ}).*

B9. Certainly the most important examples of τ-continuous Daniell functionals are Radon measures, i.e. positive linear functionals

$$\mu : C_{oo}(\Omega, \mathbb{R}) \longrightarrow \mathbb{R}$$

(Ω locally compact) — a consequence of Dini's theorem. In this case,

$$\mathfrak{U}^\tau = \left\{ u : \Omega \to \,]-\infty, \infty] \,\middle|\, \begin{array}{l} \text{lower semicontinuous, bounded from} \\ \text{below, } [u < 0] \text{ relatively compact} \end{array} \right\} \,.$$

It follows that all Borel sets are μ^τ-measurable and the measure

$$\nu(A) := \int \chi_A d\mu^\tau$$

on the local σ-ring $\mathfrak{B}_c(\Omega)$ of all relatively compact Borel sets is *regular*, i.e.
$$\nu(A) = \sup\{\nu(K) \mid K \subset A \text{ compact}\} = \inf\{\nu(V) \mid V \supset A \text{ open}, \nu\text{-integrable}\} .$$

Measures on $\mathfrak{B}_c(\Omega)$ are called *Borel measures* and hence ν is a regular Borel measure. Regularity and the fundamental theorem (B 3.) imply that $\mathcal{L}_1(\mu^\tau) = \mathcal{L}_1(\nu)$ (with equal integrals): There is a one-to-one correspondence between positive linear functionals on $C_{oo}(\Omega, \mathbb{R})$ and regular Borel measures on $\mathfrak{B}_c(\Omega)$. This is why these functionals/regular Borel measures are labelled *Borel-Radon measures*.

If $\varphi : C_{oo}(\Omega, \mathbb{K}) \to \mathbb{K}$ is $\mathbb{K}$-linear and of bounded variation, i.e.
$$|\varphi|(f) := \sup\{|\varphi(h)| \mid 0 \leq |h| \leq f, h \in C_{oo}(\Omega, \mathbb{K})\} < \infty$$
for every $0 \leq f \in C_{oo}(\Omega, \mathbb{R})$, one obtains a Borel-Radon measure $|\varphi| : C_{oo}(\Omega, \mathbb{R}) \to \mathbb{R}$ from the definition
$$|\varphi|(f) := |\varphi|(f^+) - |\varphi|(f^-)$$
(and obvious extension to $C_{oo}(\Omega, \mathbb{C})$ if $\mathbb{K} = \mathbb{C}$). Clearly,
$$|\varphi(f)| \leq |\varphi|(f) \qquad \text{if } 0 \leq f \in C_{oo}(\Omega, \mathbb{R}).$$

Considering $\varphi = (\text{Re}\varphi)^+ - (\text{Re}\varphi)^- + i(\text{Im}\varphi)^+ - i(\text{Im}\varphi)^-$, the Radon-Nikodým theorem (applied to $(\text{Re}\varphi)^+ \leq |\varphi|$ and so on) gives a locally $|\varphi|^\tau$-integrable function g_o such that
$$\varphi(f) = \int f g_o d|\varphi|^\tau$$
for all $f \in C_{oo}(\Omega, \mathbb{K})$ (the *polar decomposition* of φ). Clearly, μ defined by $d\mu := g_o d|\varphi|^\tau$ is a signed Borel measure which is regular (i.e. $|\mu|$ is regular).

RIESZ' REPRESENTATION THEOREM: *If Ω is locally compact and $\varphi : C_{oo}(\Omega, \mathbb{K}) \to \mathbb{K}$ is $\mathbb{K}$-linear and of bounded variation, then there is a unique signed regular Borel measure μ on Ω such that*
$$\varphi(f) = \int_\Omega f d\mu \qquad \qquad \text{for all } f \in C_{oo}(\Omega, \mathbb{K}) .$$

Conversely, every signed Borel measure μ defines such a φ — but note that there are non-regular Borel measures.

For Banach space theory the case of compact Ω is the most important: If μ is a regular signed Borel measure and φ its associated linear functional
$$\langle \varphi, f \rangle = \int_\Omega f d\mu$$
on the Banach space $C(\Omega, \mathbb{K})$, then
$$\|\varphi\| \leq |\mu|(\Omega) = |\varphi|(1) = \|\varphi\|$$

(by the definition of $|\varphi|$). Denoting by $M(\Omega, \mathbb{K})$ the space of all $\mathbb{K}$-valued regular Borel measures μ with norm $\|\mu\| := |\mu|(\Omega)$, one obtains the

COROLLARY: *If Ω is compact, then $C(\Omega, \mathbb{K})' = M(\Omega, \mathbb{K})$ holds isometrically.*

B10. The following result is quite important for the theory of Banach spaces:

PROPOSITION: *If (Ω, μ) is a measure space, then $L_1(\mu)$ is 1-complemented in its bidual.*

PROOF: Since $L_1(\mu)$ is an abstract L-space, Kakutani's representation theorem (Appendix A7.) shows that μ may be assumed to be a Borel–Radon measure on a locally compact Ω, hence $L_1(\mu)' = L_\infty(\mu)$. Take $\varphi \in L_1(\mu)'' = L_\infty(\mu)'$ of norm 1; the functional

$$\varphi_o(f) := \langle \varphi, \tilde{f} \rangle \qquad\qquad f \in \mathfrak{S}(I_\mu)$$

has, by definition, a μ–absolutely continuous total variation $|\varphi_o|$: The Radon–Nikodým theorem (for μ applied to $(\operatorname{Re}\varphi_o)^+$ etc.) gives a locally μ–integrable function g_o such that

$$\varphi_o(f) = \int f g_o d\mu$$

for all $f \in \mathfrak{S}(I_\mu)$. Since

$$\left|\int f g_o d\mu\right| \leq \|\tilde{f}\|_{L_\infty(\mu)},$$

it follows that g_o is μ–integrable and $\|\tilde{g}_o\|_{L_1(\mu)} \leq 1$. Clearly, the class $\tilde{g}_o$ representing φ_o is unique, and hence $P\varphi := \tilde{g}_o$ defines a projection

$$P : L_1(\mu)'' \longrightarrow L_1(\mu)$$

of norm ≤ 1. □

Bochner Integration

B11. If (Ω, μ) is a measure space and E a (real or complex) normed space, then a function $f : \Omega \to E$ is called a μ–*step function* if it has a representation

$$f(\cdot) = \sum_{k=1}^{n} \chi_{A_k}(\cdot) x_k$$

with $x_k \in E$ and μ–integrable $A_k \subset \Omega$; in this case, the *integral*

$$\int_\Omega f d\mu := \sum_{k=1}^{n} \mu(A_k) x_k \in E$$

is well-defined. $f : \Omega \to E$ is called μ-measurable (or: strongly μ-measurable) if for every μ-integrable $A \subset \Omega$ there is a sequence (f_n) of μ-step functions such that

$$f(w) = \lim_{n \to \infty} f_n(w)$$

for μ-almost all $w \in A$. It is easy to see that on every σ-finite set $f : \Omega \to E$ is the a.e.-limit of a sequence of step functions. The following criterion is of fundamental importance:

THEOREM: Let (Ω, μ) be a measure space, E a Banach space and $F \subset E'$ a subspace such that $F \cap B_{E'}$ is norming for E. For each function $f : \Omega \to E$ which vanishes outside a σ-finite set the following statements are equivalent:

(a) f is μ-measurable.

(b) f is μ-essentially separably valued (i.e. there is a separable subspace $E_o \subset E$ with $f(w) \in E_o$ μ-a.e.) and Borel measurable (i.e. $f^{-1}(B)$ is μ-measurable for each Borel set $B \subset E$).

(c) f is μ-essentially separably valued and $\sigma(E, F)$-measurable (with respect to μ), i.e. $\langle x', f(\cdot) \rangle$ is μ-measurable for all $x' \in F$.

(d) For every $\varepsilon > 0$ there exists a μ-measurable, countably valued function $f_\varepsilon : \Omega \to E$ with $\|f_\varepsilon(\cdot) - f(\cdot)\| \leq \varepsilon$ μ-a.e.

The equivalence of (a) and (c) is called the *Pettis' measurability criterion*. $\sigma(E, E')$-measurable functions are labelled *weakly measurable* — and $\sigma(E', E)$-measurable functions $\Omega \to E'$ *weak-$*$-measurable*. There are many references for this theorem: First observe that it is obviously enough to prove it for finite μ; Dunford–Schwartz [59], III 6.9. and 6.11. show the first three equivalences (for $F = E'$, but the proof is easily adapted to $F \subset E'$ as stated). For Pettis' criterion see also Yosida [277], V.4., and Diestel-Uhl [55], p.42; the proof in this last reference even yields (d).

B12. For $1 \leq p < \infty$ the μ-measurable functions $f : \Omega \to E$ such that the function $\|f(\cdot)\|_E^p$ is μ-integrable, form a vector space $\mathcal{L}_p(\mu, E)$. Its quotient with respect to the μ-null-functions (i.e. $\|f(\cdot)\|_E = 0$ μ-a.e.)

$$L_p(\mu, E) := \mathcal{L}_p(\mu, E)/\sim$$

(or briefly: $L_p(E)$) equipped with

$$\|\tilde{f}\|_{L_p(E)} := \left(\int_\Omega \|f(w)\|_E^p \mu(dw) \right)^{1/p}$$

is a normed space — even complete if E is a Banach space. It can be seen (the proof of Yosida [277], V.5, for $p = 1$ can easily be modified) that $f : \Omega \to E$ is in $\mathcal{L}_p(\mu, E)$

if and only if there is a sequence (f_n) of μ–step functions such that $\|f_n(\cdot) - f(\cdot)\|_E^p$ is μ–integrable and
$$\int_\Omega \|f_n(w) - f(w)\|_E^p \mu(dw) \longrightarrow 0 .$$
In particular, the (equivalence classes of) step functions are norm–dense in $L_p(E)$. For $p = 1$ the *integral*
$$\int_\Omega f d\mu := \lim_{n \to \infty} \int_\Omega f_n d\mu$$
is well–defined. The functions in $\mathcal{L}_1(\mu, E)$ are called *Bochner integrable*, those in $\mathcal{L}_p(\mu, E)$ *Bochner p–integrable*.

As an immediate consequence of the scalar case one obtains *Lebesgue's dominated convergence theorem*: If (f_n) is a sequence of μ–measurable functions $\Omega \to E$ which converges pointwise to a function $f : \Omega \to E$ and if there is a $g \in \mathcal{L}_1(\mu)$ such that $\|f_n(w)\| \le g(w)$ for all $w \in \Omega$, then f is Bochner integrable and
$$\lim_{n \to \infty} \int_\Omega f_n d\mu = \int_\Omega f d\mu .$$

For $p = \infty$ define $\mathcal{L}_\infty(\mu, E)$ to be the space of all μ–essentially bounded, μ–measurable functions $\Omega \to E$ and the normed space
$$L_\infty(\mu, E) := \mathcal{L}_\infty(\mu, E)/\sim_{loc}$$
with $\|\tilde{f}^{loc}\|_\infty := \text{ess-sup}\|f(\cdot)\|_E$; this space is complete if E is.

Take $p, q \in [1, \infty]$ such that $1/p + 1/q = 1$. If $f \in \mathcal{L}_p(\mu, E)$ and $g \in \mathcal{L}_q(\mu, E')$, then the function
$$\langle f(\cdot), g(\cdot) \rangle_{E, E'}$$
is μ–integrable, $\|\langle f, g \rangle\|_1 \le \|f\|_p \|g\|_q$ by Hölder's inequality and the mapping
$$\Phi : L_q(\mu, E') \hookrightarrow (L_p(\mu, E))'$$
$$\tilde{g} \rightsquigarrow [\tilde{f} \rightsquigarrow \int \langle f(w), g(w) \rangle \mu(dw)]$$
is even an isometry (into); to prove this, first use step functions. The question of under what conditions Φ is surjective will be treated in D4..

B13. It is obvious that for $T \in \mathcal{L}(E, F)$ and $f \in \mathcal{L}_1(\mu, E)$ the function $T \circ f : \Omega \to F$ is also Bochner integrable and
$$T\left(\int_\Omega f(w)\mu(dw)\right) = \int_\Omega T(f(w))\mu(dw) .$$
In particular, this implies that
$$\left\|\int_\Omega f(w)\mu(dw)\right\|_E \le \int_\Omega \|f(w)\|_E \mu(dw)$$

for all $f \in \mathcal{L}_1(\mu, E)$.

B14. If $f : \Omega \to E$ is such that $\langle x', f(\cdot)\rangle$ is μ-integrable for all $x' \in E'$, then for each μ-measurable $A \subset \Omega$ the functional

$$E' \ni x' \rightsquigarrow \int_A \langle x', f(w)\rangle \mu(dw)$$

is linear, hence in E'^*. If this functional (for all such A) can be represented by an element $x_A \in E \hookrightarrow E'^*$, then f is called $\sigma(E, E')$-integrable or *Pettis integrable* and

$$x_\Omega =: \int_\Omega f(w)\mu(dw)$$

is called the *Pettis integral* (or *weak-* or $\sigma(E, E')$-*integral*) of f. Bochner integrable functions are Pettis integrable and the Bochner integral of $f\chi_A$ is x_A.
For a separating duality system $\langle E, F\rangle$ it is obvious how to define the notion of $\sigma(E, F)$-integrability. In particular, it is clear what weak-$*$-integrable means.

B15. The unit ball of $L_\infty(\mu)$ is norming for $L_1(\mu)$. Even though $L_1(\mu, E)'$ is, in general, much larger than $L_\infty(\mu, E')$, this is also true in the vector valued case: If $\Sigma^0 \subset I_\mu$ is a μ-dense ring, then the E'-valued Σ^0-step functions are dense in $L_p(\mu, E')$ (for $1 \leq p < \infty$). Since

$$L_q(\mu, E) \xrightarrow{1} L_q(\mu, E'') \xrightarrow{1} (L_p(\mu, E'))'$$

($1/p + 1/q = 1$), this implies that the set of all E'-valued Σ^0-step functions in the unit ball of $L_p(\mu, E')$ is norming for $L_q(\mu, E)$ — and so is the whole unit ball. For finite measures μ this argument applies also for $q = 1$. The following result shows that this is true for arbitrary measures and gives another, more convenient formulation of this fact:

VARIATION LEMMA: *Let (Ω, μ) be a measure space and $\Sigma^0 \subset I_\mu$ a μ-dense ring. If E is normed and $f \in \mathcal{L}_1(\mu, E)$, then*

$$\int_\Omega \|f(w)\|_E \mu(dw) = \sup\left\{\sum_{m=1}^n \left\|\int_{A_m} f(w)\mu(dw)\right\|_E \;\middle|\; \begin{array}{l} A_1, ..., A_n \in \Sigma^0 \\ \text{pairwise disjoint} \end{array}\right\}.$$

The right side might be denoted by $V(fd\mu)$ — the variation of the vector measure $A \rightsquigarrow \int_A fd\mu$.

PROOF: Clearly, $V(fd\mu) \leq \|f\|_{L_1(E)}$. Conversely, first take a Σ^0-step function

$$f_o := \sum_{m=1}^n \chi_{A_m} x_m$$

with pairwise disjoint $A_m \in \Sigma^0$. Then

$$V(f_o d\mu) \leq \|f_o\|_{L_1(E)} = \sum_{m=1}^{n} \mu(A_m)\|x_m\| \leq V(f_o d\mu) .$$

If $f \in \mathcal{L}_1(\mu, E)$ is arbitrary and $\varepsilon > 0$, then there is a Σ^0-step function f_o with $\|f - f_o\|_{L_1(E)} \leq \varepsilon$. It follows that

$$\left| \int \|f\| d\mu - \int \|f_o\| d\mu \right| \leq \int \|f - f_o\| d\mu \leq \varepsilon$$

and, for a representation of f_o as before,

$$\int \|f_o\| d\mu = \sum_{m=1}^{n} \left\| \int_{A_m} f_o d\mu \right\| \leq \sum_{m=1}^{n} \left\| \int_{A_m} (f - f_o) d\mu \right\| + \sum_{m=1}^{n} \left\| \int_{A_m} f d\mu \right\| \leq$$
$$\leq \varepsilon + V(f d\mu) .$$

This implies that $\int \|f\| d\mu \leq 2\varepsilon + V(f d\mu)$. □

Appendix C: Representable Operators

This appendix deals with operators $L_1(\mu) \to E$ (for finite measures μ) which have a representation $Tf = \int fg d\mu$ for some bounded Bochner integrable function $g \in L_1(\mu, E)$. The strong version of the Dunford–Pettis theorem will be proved. Since we use some results about the projective tensor norm π this is in some sense an appendix to Chapter I. In contrast to Appendices A and B in this appendix complete proofs will be given.

Grothendieck's Characterization

C1. Throughout this appendix the measure μ will be finite. If E is a (real or complex) Banach space, an operator $T \in \mathcal{L}(L_1(\Omega, \mu), E)$ is called *Riesz representable* (briefly: representable) if there is a μ-essentially bounded, μ-measurable function $g : \Omega \to E$ such that

$$Tf = \int_\Omega fg d\mu$$

for all $f \in L_1(\Omega, \mu)$. That g is μ–essentially bounded means (see B12.) that

$$\|g\|_\infty := \| \|g(\cdot)\|_{\widetilde{E}}\|_{L_\infty(\Omega,\mu)} < \infty$$

— measurability implies that g is Bochner integrable. The function g is called a *Riesz–density* of T — and it is clear that g is necessarily a weak density, i.e. weakly measurable and

$$\langle Tf, y'\rangle = \int f(w)\langle g(w), y'\rangle \mu(dw)$$

for all $f \in L_1(\Omega, \mu)$ and $y' \in E'$. For $E = \mathbb{K}$ every operator = functional is representable by the Radon–Nikodým theorem and the density is the Radon–Nikodým density; the study of representable operators will serve to understand vector valued versions of the Radon–Nikodým theorem.

PROPOSITION: *Let $T \in \mathcal{L}(L_1(\Omega, \mu), E)$ for a finite measure μ and a Banach space E.*
(1) *If T is representable, then the range of T is separable.*
(2) *If g is a Riesz–density of T, then*

$$\|g\|_\infty = \|T\| .$$

In particular, the Riesz–density of a representable operator is unique μ–a.e..
(3) *If $g \in L_1(\mu, E)$ and $\Sigma^0 \subset I_\mu$ a μ–dense algebra (see Appendix B3.) such that*

$$T\chi_A = \int_A g\, d\mu$$

for all $A \in \Sigma^0$, then g is μ–essentially bounded and a Riesz–density for T.
(4) *If μ is purely atomic, then T is always representable.*
(5) *If T is representable with density g and $S \in \mathcal{L}(E, F)$, then $S \circ T$ is representable with density $S \circ g$.*
(6) *If g represents T, then $g(w) \in \overline{\mathrm{im}(T)}$ μ–a.e..*

PROOF: Since the measure μ is finite, every μ–measurable function g has μ–a.e. values in a separable closed subspace E_o of E, hence (1) follows. To see (2) and (3) choose $x'_n \in B_{E'}$ such that $\{x'_n\}$ is norming for E_o. The hypothesis of (3) gives

$$\langle x'_n, T\chi_A\rangle = \int_A \langle x'_n, g\rangle d\mu$$

for all $A \in \Sigma^0$, therefore

$$\left|\int_A \langle x'_n, g\rangle d\mu\right| \leq \|T\|\mu(A)$$

and this implies that $|\langle x'_n, g(w)\rangle| \leq \|T\|$ for μ-almost all $w \in \Omega$. Since $\{x'_n\}$ is norming, $\|g\|_\infty \leq \|T\|$. It follows that the operator

$$T_o f := \int fg d\mu$$

is well–defined, continuous and coincides with T on the dense subspace of Σ^0-step functions: $T = T_o$ which proves (3). Clearly, $\|T\| \leq \|g\|_\infty$ — and (2) follows as well. That g is unique a.e. is a consequence of $\|g_1 - g_2\|_\infty = \|T - T\| = 0$.

If μ is purely atomic, then there are pairwise disjoint atoms $A_n \subset \Omega$ with

$$\sum_{n=1}^\infty \mu(A_n) = \mu(\Omega) < \infty.$$

It is easily seen that

$$g := \sum_{n=1}^\infty \mu(A_n)^{-1} \chi_{A_n} \otimes T\chi_{A_n}$$

is a density.

(5) is obvious, and (6) follows from the observation that

$$L_1(\mu) \xrightarrow{T} E \xrightarrow{Q} E/\overline{\mathrm{im}T}$$

is representable and has the μ-a.e. unique density $Q \circ g = 0$ by (2) and (5). □

The interpretation of T being representable as a sort of Radon–Nikodým theorem (and the density as a sort of derivative) is supported by the following observation: If $\mathfrak{C}$ is a finite partition $(A_1, ..., A_n)$ of Ω by μ-integrable sets and

$$g_\mathfrak{C}^T := \sum_{k=1}^n \mu(A_k)^{-1} \chi_{A_k} T\chi_{A_k} \in L_1(\mu, E),$$

then the following holds:

COROLLARY: $T : L_1(\mu) \to E$ is representable if and only if the net $(g_\mathfrak{C}^T)_\mathfrak{C}$ is a Cauchy net in $L_1(\mu, E)$. It is enough to take the partitions in a μ-dense subalgebra Σ^0 of I_μ.

PROOF: If $(g_\mathfrak{C}^T)$ converges in $L_1(\mu, E)$ to g, then

$$\int_A g d\mu = \lim_\mathfrak{C} \int_A g_\mathfrak{C}^T d\mu = T\chi_A$$

for all $A \in \Sigma^0$ and property (3) of the proposition shows that g is a density for T. Conversely, if g is a density, then

$$g_{\mathcal{C}}^T = \sum_{k=1}^n \frac{1}{\mu(A_k)} \chi_{A_k} \int_{A_k} g d\mu = \mathbf{E}(g|\mathcal{C}) ,$$

the conditional expectation of g with respect to $\mathcal{C}$. An easy calculation shows that

$$\|\mathbf{E}(h|\mathcal{C})\|_{L_1} \leq \|h\|_{L_1} \qquad \text{for all } h \in L_1(\mu, E) ,$$

and it is immediate that $\mathbf{E}(h|\mathcal{C}) \to h$ in $L_1(\mu, E)$ if h is a Σ^0-step function: These two facts show that the net $(\mathbf{E}(h|\mathcal{C}))_\mathcal{C}$ converges to h in $L_1(\mu, E)$ for all $h \in L_1(\mu, E)$. This ends the proof. □

C2. It is clear that every nuclear operator $T : L_1(\mu) \to E$ is representable: If

$$T = \sum_{n=1}^\infty g_n \otimes y_n \qquad \text{with} \qquad \sum_{n=1}^\infty \|g_n\| \|y_n\| \leq \mathbf{N}(T)(1+\varepsilon)$$

and $g_n \in L_1(\mu)' = L_\infty(\mu)$, then $g(w) := \sum_{n=1}^\infty g_n(w) y_n$ defines a density.
On the other hand, consider the operator $T : L_1[0,1] \to c_o$ of Fourier coefficients:

$$Tf := \left(\int_0^1 f(t) \exp(2\pi i n t) dt \right)_{n \in \mathbb{N}} .$$

If g were a density of T or of $\kappa_{c_o} \circ T : L_1 \to \ell_\infty$, then (looking at components) it would necessarily follow that

$$g(t) = (\exp(2\pi i n t))_n \in \ell_\infty \qquad \text{a.e.} .$$

But this function is not a.e. separably valued. So neither $T : L_1[0,1] \to c_o$ nor $\kappa_{c_o} \circ T : L_1[0,1] \to \ell_\infty$ is a representable operator; another argument is: $g(t) \notin c_o$ as it should be almost everywhere as a density by C1.(6). Note that $c_o^{\mathbb{R}}$ is isomorphic to $c_o^{\mathbb{C}}$.

C3. The operator $T \in \mathcal{L}(L_1(\mu), E)$ having a density

$$g \in L_1(\mu, E) = L_1(\mu) \tilde{\otimes}_\pi E \xrightarrow{1} \mathfrak{N}(L_\infty(\mu), E)$$

means that T corresponds to a certain nuclear operator $L_\infty(\mu) \to E$. More precisely, the following result holds:

THEOREM (Grothendieck): *Let (Ω, μ) be a finite measure space, E a Banach space and $T \in \mathcal{L}(L_1(\mu), E)$. Then the following statements are equivalent:*

(a) T is Riesz representable.

(b) The operator $T \circ I : L_\infty(\mu) \xrightarrow{I} L_1(\mu) \xrightarrow{T} E$ is nuclear.

If μ is a Borel–Radon measure on a compact space Ω, then these statements are equivalent to:

(c) The operator $T \circ I : C(\Omega) \xrightarrow{I} L_1(\mu) \xrightarrow{T} E$ is nuclear.

In both cases: If g is a Riesz-density of T, then

$$\mathbf{N}(T \circ I) = \int_\Omega \|g\|_E d\mu \leq \|T\| \, \|I\| ,$$

where I is the canonical mapping $L_\infty \hookrightarrow L_1$ or $C \longrightarrow L_1$.

PROOF: Clearly, (b) implies (c). For the converse denote by $I : L_\infty \to L_1$ and $I_o : C(\Omega) \to L_\infty$ the canonical mappings and by

$$D^L : C(\Omega)' \longrightarrow L_1(\Omega, \mu) ; \qquad D^L(\nu) := \nu_a$$

the mapping which assigns to each signed measure $\nu \in C(\Omega)'$ its μ-absolutely continuous part ν_a by Lebesgue's decomposition theorem (see B6.); clearly $\|D^L\| \leq 1$. Moreover, observe that $I : L_\infty \hookrightarrow L_1$ is $\sigma(L_\infty, L_1) - \sigma(L_1, L_\infty)$-continuous and $I' = \kappa_{L_1} \circ I$. It follows that

$$(T \circ I)' : E' \xrightarrow{T'} L_\infty \xrightarrow{I} L_1 \xrightarrow{\kappa_{L_1}} L'_\infty \xrightarrow{I'_o} C(\Omega)' \xrightarrow{D^L} L_1 \xrightarrow{\kappa_{L_1}} L'_\infty$$

since $D^L \circ I'_o \circ \kappa_{L_1} = id_{L_1}$. If $T \circ I \circ I_o = \sum_{n=1}^\infty \nu_n \otimes x_n$ is a nuclear expansion with $\nu_n \in C(\Omega)'$ and $x_n \in E$, then the above decomposition of $(T \circ I)'$ gives

$$(T \circ I)' = \sum_{n=1}^\infty x_n \otimes D^L(\nu_n) ,$$

which implies that $T \circ I = \sum_{n=1}^\infty D^L(\nu_n) \otimes x_n$ and

$$\mathbf{N}(T \circ I) \leq \sum_{n=1}^\infty \|D^L(\nu_n)\| \, \|x_n\| \leq \sum_{n=1}^\infty \|\nu_n\| \, \|x_n\| ,$$

hence $\mathbf{N}(T \circ I) \leq \mathbf{N}(T \circ I \circ I_o)$ — and even equality since the reverse inequality is obvious.

It remains to show that (a) is equivalent to (b) (and the statement about the norm). If $g = \sum_{n=1}^\infty g_n \otimes x_n \in L_1 \tilde{\otimes}_\pi E = L_1(\mu, E)$ represents μ, then

$$(T \circ I)(f) = \sum_{n=1}^\infty \langle g_n, f \rangle_{L_1, L_\infty} x_n ,$$

and hence $\mathbf{N}(T \circ I) \leq \inf \sum_{n=1}^{\infty} \|g_n\| \|x_n\| = \int \|g\|_E d\mu$ by 3.3. and 3.5.. Conversely, the operator $T \circ I : L_\infty \to E$ is $\sigma(L_\infty, L_1) - \sigma(E, E')$-continuous: so if it is nuclear, then it can be represented as

$$T \circ I = \sum_{n=1}^{\infty} g_n \underline{\otimes} x_n ,$$

with $g_n \in L_1$ such that $\sum_{n=1}^{\infty} \|g_n\|_{L_1} \|x_n\| \leq \mathbf{N}(T \circ I)(1+\varepsilon)$ by Ex 3.34. since L_1 is 1-complemented in its bidual (see B10.). The function $g = \sum_{n=1}^{\infty} g_n \underline{\otimes} x_n \in L_1(\mu, E)$ satisfies

$$T\chi_A = \int_A g d\mu \qquad \text{for all } A \in I_\mu ,$$

hence it is a density for T by proposition C1.(3). Since

$$\int \|g\|_E d\mu \leq \sum_{n=1}^{\infty} \|g_n\|_{L_1} \|x_n\| \leq \mathbf{N}(T \circ I)(1+\varepsilon)$$

and g is μ–a.e. unique, it follows that $\int \|g\|_E d\mu = \mathbf{N}(T \circ I)$. Finally, observe that $\int \|g\|_E d\mu \leq \|g\|_\infty \mu(\Omega) = \|T\| \|I\|$. □

As an application, take a finite measure μ which is *not* purely atomic. Then the map $L_\infty(\mu) \hookrightarrow L_1(\mu)$ is not compact (simulate Rademacher functions as on $[0,1]$), hence, by (b) of the theorem, id_{L_1} is not representable.

C4. Another example of representable operators:

PROPOSITION: *Every operator $T : L_1(\mu) \to \ell_1$ is representable (μ a finite measure).*

PROOF: Since $L_1(\mu)' = L_\infty(\mu)$, it follows that there are $g_n \in L_\infty(\mu)$ such that

$$Tf = \sum_{n=1}^{\infty} \int fg_n d\mu \cdot e_n \in \ell_1 \qquad \text{for all } f \in L_1 ,$$

where $e_n = (\delta_{n,m})_m$ as usual. It remains to show that $g := \sum_{n=1}^{\infty} g_n \underline{\otimes} e_n$ is a density. If P_N is the N-th expansion operator of the unit vector basis and $g^N := P_N g = (g_1, ..., g_N, 0, ...)$, then $\|g^N(w)\| \leq \|g^{N+1}(w)\|$ for all w. Since, by the variation lemma (see B15.),

$$\int \|g^N(w)\| \mu(dw) = \sup\left\{ \sum_{k=1}^{n} \left\| \int_{A_k} g^N d\mu \right\| \, \middle| \, A_1, ..., A_n \in I_\mu \text{ pairwise disjoint} \right\} =$$

$$= \sup\left\{ \sum_{k=1}^{n} \|P_N T\chi_{A_k}\| \, | \, ... \right\} \leq \|T\| \mu(\Omega) ,$$

Beppo Levi's monotone convergence theorem implies that $\|g(\cdot)\| = \lim_{N\to\infty} \|g^N(\cdot)\|$ is integrable, and therefore g is Bochner integrable. Since

$$T\chi_A = \sum_{n=1}^{\infty} \int_A g_n d\mu \cdot e_n = \int_A g d\mu$$

by Lebesgue's dominated convergence theorem, proposition C1.(3) implies that g is a density for T. $\square$

The following result, due to Lewis and Stegall [170], says that the operators $L_1 \to \ell_1$ are typical representable operators.

THEOREM: $T : L_1(\mu) \to E$ is representable if and only if it admits a factorization $T = S \circ R$ through ℓ_1. In this case, $\|T\| = \inf \|R\| \|S\|$ over all such factorizations.

PROOF: One direction is obvious by the proposition above. Take T with a density g and recall that $\|g\|_\infty = \|T\|$. Then, by the characterization B11. of measurability, there are $g_n : \Omega \to E$ countably valued and measurable such that $g = \sum_{n=1}^{\infty} g_n$ pointwise,

$$\|g_1\|_\infty \leq \|T\| + \frac{\varepsilon}{2} \quad \text{and} \quad \|g_n\|_\infty \leq \frac{\varepsilon}{2^n} \quad \text{for } n \geq 2 \;.$$

If $g_n = \sum_{k=1}^{\infty} \chi_{A_{k,n}} \otimes x_{k,n}$ with pairwise disjoint $A_{k,n}$ (for fixed n), then

$$S : L_1(\mu) \longrightarrow \ell_1(\mathbb{N}^2) \;;\; Sf := \left(\int_{A_{k,n}} f d\mu \, \|x_{k,n}\|\right)_{k,n}$$

$$R : \ell_1(\mathbb{N}^2) \longrightarrow E \;;\; R(\alpha_{k,n}) := \sum_{k,n} \alpha_{k,n} \frac{x_{k,n}}{\|x_{k,n}\|}$$

define a factorization as desired. $\square$

The Dunford–Pettis Theorem

C5. It was shown in 3.3. that every operator $T : L_1(\Omega, \mu) \to F'$ has a weak-$*$-density $g : \Omega \longrightarrow F'$, i.e.
(1) $\langle g(\cdot), y \rangle$ is measurable for every $y \in F$ and
(2) $\langle Tf, y \rangle = \int f(w) \langle g(w), y \rangle \mu(dw)$ for all $f \in L_1(\mu)$ and $y \in F$.
(weak version of the Dunford–Pettis theorem); moreover, $g(w) \in (TB_{L_1})^{00}$ and hence, in particular, $\|g(w)\|_{F'} \leq \|T\|$ for all $w \in \Omega$.
If μ is finite and g a.e. separably valued, then g is μ-measurable (by the Pettis criterion, B11.) and hence Bochner integrable. This easily implies (see corollary 3.3.) the following result:

PROPOSITION: *Let μ be a finite measure.*
(1) *If F' is separable, then every operator $L_1(\mu) \to F'$ is representable.*
(2) *Every weakly compact operator $L_1(\mu) \to E$ with separable range is representable.*

This result has some nice structural consequences: There is an operator $L_1[0,1] \to c_o$ which is not representable (see C2.), hence the separable space c_o cannot be isomorphic to a dual space. In the same way, Gelfand's famous result that $L_1[0,1]$ is not isomorphic to a dual space follows.

C6. To get the full Dunford-Pettis theorem, two more results are needed which are both of interest in their own right.

LEMMA: *If $T \in \mathfrak{L}(L_1(\mu), E)$ is representable and $D \subset L_1(\mu)$ relatively weakly compact, then TD is relatively compact.*

PROOF: (a) If g is a density for T, there are step functions g_n which converge almost everywhere to g and satisfy $\|g_n\|_\infty \leq \|g\|_\infty = \|T\|$. For each measurable $A \subset \Omega$ define

$$T_A f := \int_A f g d\mu \quad \text{and} \quad T_{n,A} f := \int_A f g_n d\mu.$$

Then the $T_{n,A}$ are finite rank operators and hence compact. For each $\delta > 0$ there is, by Egoroff's theorem, a measurable $A \subset \Omega$ with $\mu(\Omega \setminus A) \leq \delta$ such that $g_n \to g$ uniformly on A. It follows that

$$\|T_A - T_{n,A}\| \leq \|(g_n - g)\chi_A\|_\infty \longrightarrow 0$$

and hence T_A is compact for this $A \subset \Omega$.

(b) Every weakly compact subset $D \subset L_1(\mu)$ is *uniformly integrable* (see for example [70], p.108), which means: For every $\varepsilon > 0$ there is a $\delta > 0$ such that for all $f \in D$

$$\int_B |f| d\mu \leq \varepsilon$$

whenever $\mu(B) \leq \delta$. Choose, for a given $\varepsilon > 0$, such a number $\delta \leq \varepsilon$ and A as in (a). Then

$$\|Tf - T_A f\| = \left\| \int_{\Omega \setminus A} f g d\mu \right\| \leq \varepsilon \|g\|_\infty$$

for all $f \in D$, which easily implies that the set TD is precompact, hence relatively compact. □

Note that (for finite measures μ) the relatively weakly compact subsets $D \subset L_1(\mu)$ are exactly those which are norm-bounded and uniformly integrable (e.g.[70], p.108).

PROPOSITION: *Let $\mathfrak{A}$ be an operator ideal, μ a finite measure and F a Banach space. If every $T \in \mathfrak{A}(L_1(\mu), F)$ with separable range is representable, then every operator $T \in \mathfrak{A}(L_1(\mu), F)$ has separable range and therefore is representable.*

PROOF: Take $T \in \mathfrak{A}(L_1(\mu), F)$. To show that $\operatorname{im} T$ is separable it is enough to prove that $\{T\chi_B | B \in I_\mu\}$ is relatively sequentially compact. Take $B_n \in I_\mu$ and let Σ^0 be the σ-algebra generated by $\{B_n | n \in \mathbb{N}\}$. The Σ^0-measurable functions in $L_1(\mu)$ form a separable subspace and the operator $T_o : L_1(\mu) \to F$ defined by

$$T_o(f) := T(\mathbf{E}(f|\Sigma^0))$$

(the conditional expectation) is in $\mathfrak{A}$, has separable range and is hence representable. Since the unit ball of L_∞ is weakly compact in L_1, it follows from the preceding lemma that the sequence $(T_o\chi_{B_n} = T\chi_{B_n})_{n \in \mathbb{N}}$ has a convergent subsequence. $\square$

C7. This proposition implies the

THEOREM OF DUNFORD–PETTIS (strong version): *Let μ be a finite measure and F a Banach space. Then every weakly compact operator $T : L_1(\mu) \to F$ is Riesz representable.*

PROOF: Apply proposition C5.(2) to $\mathfrak{A} := \mathfrak{W}$ in proposition C6.. $\square$

Note that the weak version implies that the Riesz–density g of T takes its values in $\overline{TB_{L_1}}$ μ-almost everywhere. Note also that representable operators need not be weakly compact: Take a quotient map $L_1[0,1] \to \ell_1$. The lemma in C6. implies

COROLLARY 1: *If $T : L_1(\mu) \to F$ is a weakly compact operator and $D \subset L_1(\mu)$ is a weakly compact subset, then TD is compact.*

This property of $L_1(\mu)$ is called the *Dunford–Pettis property*. If K is compact, then $C(K)$ has the Dunford–Pettis property as well; for more information about this property see Diestel's survey [53].

COROLLARY 2: *(1) If $T : L_1(\mu) \to L_1(\mu)$ is weakly compact, then T^2 is compact.*

(2) $L_1(\mu)$ has no reflexive, infinite dimensional, complemented subspaces.

It follows that the closed span Rad of the Rademacher functions (see 8.6.) is not complemented in $L_1[0,1]$; see also 1.5..

C8. Finally, as an application, a result (also due to Grothendieck) about integral and nuclear operators:

THEOREM: *If* $T \in \mathfrak{L}(E,F)$ *is integral and* $U \in \mathfrak{L}(F,G)$ *weakly compact, then* $U \circ T$ *is nuclear and* $\mathbf{N}(U \circ T) \leq \|U\| \mathbf{I}(T)$. *In short:* $\mathfrak{W} \circ \mathfrak{I} \subset \mathfrak{N}$ *and* $\mathbf{N} \leq \| \| \circ \mathbf{I}$.

PROOF: $U \circ T$ admits a factorization

$$E \xrightarrow{T} F \xhookrightarrow{\kappa_F} F'' \xrightarrow{U^\pi} G$$
$$R \downarrow \qquad \nearrow S$$
$$L_\infty \xhookrightarrow{I} L_1$$

$\|R\| = \|S\| = 1$, $\|I\| = \mathbf{I}(T)$

(see 10.5.). Since $U^\pi \circ S$ is weakly compact, it is representable by the Dunford–Pettis theorem, and therefore $U^\pi \circ S \circ I$ is nuclear by Grothendieck's characterization C3.. It follows that $U \circ T = U^\pi \circ S \circ I \circ R$ is nuclear and

$$\mathbf{N}(U \circ T) \leq \mathbf{N}(U^\pi \circ S \circ I)\|R\| \leq \|U^\pi \circ S\| \, \|I\| \, \|R\| = \mathbf{I}(T)\|U\|$$

by the estimate given in C3. □

COROLLARY: *If* $U \in \mathfrak{L}(E,F)$ *is weakly compact and* $T \in \mathfrak{L}(F,G)$ *integral, then* $U' \circ T'$ *is nuclear and* $\mathbf{N}(U' \circ T') \leq \mathbf{I}(T)\|U\|$. *In particular,* $\kappa_G \circ T \circ U$ *is nuclear; if* E' *has the approximation property, then* $T \circ U$ *is nuclear and* $\mathbf{N}(T \circ U) \leq \mathbf{I}(T)\|U\|$.

The last statement follows from 5.9..

In the following appendix, which deals with the Radon–Nikodým property, a more systematic approach to this method of argumentation will be presented in order to obtain more information concerning the relationship between nuclear and integral operators.

Appendix D: The Radon–Nikodým Property

The study of those Banach spaces which allow a certain vector valued Radon–Nikodým theorem for vector measures to hold goes back to the work of Dunford, Pettis and Phillips in the late thirties and to Grothendieck's "propriété de Phillips" in his thesis [92], p.104. Nowadays, the standard reference is the book of Diestel–Uhl [55] (and its appendix [56]). Their point of view is quite measure theoretic. Since for our purposes the operator theoretic aspects are more important, we orient this short introduction in this direction. This appendix is based on the results of Appendix C and contains proofs of all the results which are given.

Basic Properties and Examples

D1. A (real or complex) Banach space E is said to have the *Radon–Nikodým property* if for each finite measure μ every operator $T \in \mathfrak{L}(L_1(\mu), E)$ is representable.

Since every operator $T : L_1(\mu) \to E$ defines a (σ-additive) vector measure M on the set I_μ of μ-integrable sets by
$$M(A) := T(\chi_A),$$
it is not difficult to see that E has the Radon–Nikodým property if and only if (for every finite μ) E-valued measures of a certain type (μ-absolutely continuous and of bounded variation) have a density with respect to μ; this is shown in Diestel-Uhl [55], p.63 — we do not need this result. In other words, the above definition is equivalent to the one given in Diestel-Uhl's book. The situation is the same as in the scalar case (see Appendix B6.) where the scalar Radon–Nikodým theorem (for μ) is equivalent to the statement that $L_1(\mu)' = L_\infty(\mu)$: this operator theoretic aspect of the Radon–Nikodým theorem will be important for us in the vector valued case; see D4. for more information about this point.

It is not necessary to check for all finite measures μ in order to know that a given Banach space has the Radon–Nikodým property.

THEOREM: *A Banach space E has the Radon–Nikodým property if and only if every operator $T : L_1[0, 1] \to E$ (Lebesgue measure) is representable.*

The following proof of this well-known result is essentially taken from Botelho [13].

PROOF: Clearly, the condition is necessary. Conversely, take an arbitrary finite measure μ on Ω and decompose it into its continuous and purely atomic parts (Ω_c, μ_c) and (Ω_a, μ_a):
$$L_1(\Omega, \mu) = L_1(\Omega_c, \mu_c) \oplus L_1(\Omega_a, \mu_a).$$
By proposition C1.(4) each operator $T \in \mathfrak{L}(L_1(\mu), E)$ has a density on the atomic part — hence one may assume μ to have no atoms (and to be normalized).

If $L_1(\mu)$ is separable, then $L_1(\mu)$ is isomorphic to $L_1[0, 1]$ due to a theorem of Carathéodory (and Halmos–von Neumann, see Brown–Pearcy [18], p.181) and therefore each T is representable by the ℓ_1-factorization theorem of Lewis and Stegall C4..

If $L_1(\mu)$ is arbitrary, then corollary C1. will be applied. One has to show that for $T : L_1(\mu) \to E$ the net $(g_{\mathfrak{C}}^T)$ is Cauchy: this is the case if $(g_{\mathfrak{C}_n}^T)$ converges for every increasing sequence $(\mathfrak{C}_n)$ of partitions. Given such a $(\mathfrak{C}_n)$, take the μ-complete σ-algebra Σ generated by $\bigcup \mathfrak{C}_n$; then $L_1(\Sigma, \mu)$ is separable and, by what was just proven, the restriction of T to $L_1(\Sigma, \mu)$ is representable. Since the algebra $\bigcup \mathfrak{C}_n$ is μ–dense in Σ, corollary C1. gives that $(g_{\mathfrak{C}_n}^T)_n$ is Cauchy. □

D2. The following results will be used frequently when dealing with the Radon–Nikodým property:

PROPOSITION 1: *Let μ be a finite measure and E a Banach space.*
(1) If E has the Radon–Nikodým property, then every operator $T : L_1(\mu) \to E$ has a separable range.
(2) Every $T : L_1(\mu) \to E$ which factors through a space with the Radon–Nikodým property is representable.
(3) Let $I : L_\infty(\mu) \hookrightarrow L_1(\mu)$ be the canonical embedding (or also $I : C(K) \to L_1(\mu)$ if μ is a Borel–Radon measure on a compact K) and let E have the Radon–Nikodým property. Then $T \circ I$ is nuclear for all $T \in \mathcal{L}(L_1(\mu), E)$ and $\mathbf{N}(T \circ I) \leq \|T\| \, \|I\|$.

This is immediate from what was shown in Appendix C about representable operators — in particular Grothendieck's characterization C3.. The *Lewis–Stegall theorem* C4. reads (with a slight extension) as follows:

PROPOSITION 2: *A Banach space E has the Radon–Nikodým property if and only if for each measure μ (finite or not — or: only $\mu = \lambda$ the Lebesgue measure on $[0, 1]$) each operator $T : L_1(\mu) \to E$ factors through some $\ell_1(\Gamma)$. In this case,*

$$\|T\| = \inf \|R\| \, \|S\|$$

— the infimum taken over all such factorizations. One may take $\Gamma = B_E$ and R the natural metric surjection $\ell_1(B_E) \twoheadrightarrow E$.

PROOF: For finite measures this is just C4. (and the result in D1. about the Lebesgue measure; note that $L_1[0, 1]$ is separable). For arbitrary μ recall from B7. that

$$L_1(\mu) = \ell_1(I; L_1(\mu_\iota))$$

for some index set I and finite measures μ_ι. This easily gives the desired factorization through $\ell_1(I; \ell_1) = \ell_1(I \times \mathbb{N})$. The fact that every operator $\ell_1(\Gamma) \to E$ is "liftable" (see 3.12.) to some operator $\ell_1(\Gamma) \to \ell_1(B_E) \twoheadrightarrow E$ implies the result. □

Another characterization is also quite useful:

PROPOSITION 3: *For each Banach space E the following statements are equivalent:*
(a) E has the Radon–Nikodým property.
(b) All closed subspaces of E have the Radon–Nikodým property.

(c) *All separable closed subspaces of E have the Radon–Nikodým property.*

PROOF: That (a) implies (b) follows from C1.(6) and that (c) implies (a) is a consequence of proposition C6. for $\mathfrak{A} := \mathfrak{L}$. □

D3. The following Banach spaces E have the Radon–Nikodým property:

(1) $E = \ell_1$ (see C4.) or more generally: Every Banach space with a boundedly complete basis (i.e. $\{\sum_1^n \alpha_m e_m \mid n \in \mathbb{N}\}$ bounded implies the convergence of $\sum_1^\infty \alpha_n e_n$; inspect the proof of proposition C4. with the equivalent norm $\sup_n \|\sum_{m=1}^n \alpha_m e_m\|$)..

(2) $E = \ell_1(\Gamma)$, where Γ is infinite (proposition 2 in D2.).

(3) E is a separable dual space (this follows from the weak version of the Dunford–Pettis theorem C5.).

(4) E reflexive (this is the Dunford–Pettis theorem, strong version, C7.).

The following spaces E do *not* have the Radon–Nikodým property:

(1) $E = c_o$ (see C2.).

(2) $E = \ell_\infty(\Gamma)$, where Γ is infinite (by D2. since $c_o \subset \ell_\infty(\Gamma)$).

(3) $C(K)$ whenever K is an infinite compact space (same argument).

(4) $E = L_1(\mu)$ whenever μ is not purely atomic (see C3. if μ is finite — but the general case is easily reduced to this).

(5) $M(K) := C(K)'$, the space of signed measures on a compact space K for which there exists a measure μ which is not purely atomic (since $L_1(\mu) \subset M(K)$ isometrically).

It follows, in particular, that neither the Radon–Nikodým property nor its negation is inherited by quotients (take $\ell_1(\Gamma) \twoheadrightarrow c_o$ in the first case and $L_1[0,1] \twoheadrightarrow \ell_1$ in the latter) and that E having the property is not related to E' having the property (consider c_o, ℓ_1, ℓ_∞).

In Diestel's and Uhl's book [55], p.217–219, the reader can find a list of almost 30 equivalent formulations of the Radon–Nikodým property and a list of spaces with or without this property.

D4. It was observed in B6. that the Radon–Nikodým theorem for a (finite) measure μ simply means $L_1(\mu)' = L_\infty(\mu)$. For the vector valued case recall from B12. that

$$L_\infty(\mu, F) := \{g : \Omega \to F \mid \mu\text{-essentially bounded, } \mu\text{-measurable }\}/\sim$$

(equivalence classes with respect to $\|g\|_\infty = 0$). Now 3.3, the universal property of the projective tensor norm, and Appendix B12. imply that

$$L_\infty(\mu, E') \overset{1}{\hookrightarrow} (L_1(\mu, E))' = (L_1(\mu) \otimes_\pi E)' = \mathfrak{L}(L_1(\mu), E') .$$

Obviously, equality holds if and only if every operator $L_1(\mu) \to E$ is representable.

PROPOSITION: *Let E be a Banach space, $p \in [1, \infty[$ and $1/p + 1/q = 1$. Then*

$$L_p(\mu, E)' = L_q(\mu, E')$$

holds isometrically for all finite measures μ (or only the Lebesgue measure on $[0, 1]$) if and only if E' has the Radon-Nikodým property.

PROOF: If this equality holds, take $T : L_1(\mu) \to E'$ and let $\beta_T \in (L_1(\mu) \otimes_\pi E)'$ be its associated functional. Since

$$I : L_p(\mu, E) \hookrightarrow L_1(\mu, E)$$

is continuous, it follows that $\beta_T \circ I \in L_q(\mu, E')$: There is a $g \in L_q(\mu, E') \subset L_1(\mu, E')$ such that, in particular,

$$\langle T\chi_A, x \rangle = \langle \beta_T \circ I, \chi_A \otimes x \rangle = \int_A \langle g(w), x \rangle \mu(dw) = \langle \int_A g d\mu, x \rangle ,$$

hence $T\chi_A = \int_A g d\mu$ and it follows, by C1.(3), that g is a Riesz-density for T. Conversely, take $\varphi \in L_p(\mu, E)'$ and define (as in the scalar case) an E'-valued measure M by

$$\langle M(A), x \rangle := \langle \varphi, \chi_A \otimes x \rangle .$$

This measure (we omit the details since the result is not needed; see Diestel-Uhl [55], p.61) is of the type which has a density $g \in L_1(\mu, E')$ (see D1.). For the sets $B_n := \{w \in \Omega \mid \|g(w)\|_{E'} \leq n\}$ it follows that φ_n defined by

$$\langle \varphi_n, f \rangle = \int_{B_n} \langle f(w), g(w) \rangle \mu(dw)$$

coincides with φ for step functions f which are zero outside B_n. This implies that $\|\varphi_n\| \leq \|\varphi\|$ (in $L_p(\mu, E)'$) and hence $(\int_{B_n} \|g\|^q d\mu)^{1/q} = \|\varphi_n\| \leq \|\varphi\|$; Beppo Levi's monotone convergence theorem and the theorem of Banach-Steinhaus show that the function g belongs to $L_q(\mu, E')$ and that g represents φ. □

Pietsch Integral Operators

D5. Integral operators $T : E \to F$ are characterized by their factorization

$$\begin{array}{ccccc} E & \xrightarrow{T} & F & \xhookrightarrow{\kappa_F} & F'' \\ {\scriptstyle R}\downarrow & & {\scriptstyle S}\nearrow & & \\ L_\infty & \xhookrightarrow{I} & L_1 & & \end{array} \qquad \mathbf{I}(T) = \min \|R\| \, \|I\| \, \|S\|$$

(see 10.5.). In general, there is no such factorization with S taking values in F (see D9. below for an example). If this can be achieved, then T is called *Pietsch integral* (or *strongly integral*). The class $\mathfrak{PI}$ of all Pietsch integral operators together with

$$\mathbf{PI}(T) := \inf\{\|R\|\,\|I\|\,\|S\| \mid T : E \xrightarrow{R} L_\infty(\mu) \xhookrightarrow{I} L_1(\mu) \xrightarrow{S} F, \mu \text{ finite}\,\}$$

is easily seen to be a Banach operator ideal.

PROPOSITION:
(1) $\mathfrak{N} \subset \mathfrak{PI} \subset \mathfrak{I} \subset \mathfrak{P}_1$ *and all these inclusions are strict; moreover,*

$$\mathbf{P}_1 \leq \mathbf{I} \leq \mathbf{PI} \leq \mathbf{N}\;.$$

(2) *If F is λ-complemented in F'', then*

$$\mathfrak{PI}(E, F) = \mathfrak{I}(E, F) \text{ and } \mathbf{PI}(T) \leq \lambda\mathbf{I}(T)\;.$$

In particular, $\mathfrak{PI}(E, G') = \mathfrak{I}(E, G')$ holds isometrically.
(3) *If F has the λ-extension property, then*

$$\mathfrak{PI}(E, F) = \mathfrak{I}(E, F) = \mathfrak{P}_1(E, F) \text{ and } \mathbf{PI}(T) \leq \lambda\mathbf{P}_1(T)\;.$$

In particular, $\mathfrak{PI}(E, F) = \mathfrak{I}(E, F) = \mathfrak{P}_1(E, F)$ hold isometrically if $F = \ell_\infty$ or $L_\infty(\mu)$.
(4) *If $T \in \mathfrak{I}$, then the dual $T' \in \mathfrak{PI}$ and $\mathbf{PI}(T') \leq \mathbf{I}(T)$.*
(5) *If $T \in \mathfrak{I}(E, F)$, then $\kappa_F \circ T \in \mathfrak{PI}$ and $\mathbf{PI}(\kappa_F \circ T) = \mathbf{I}(T)$.*

PROOF: The only non–obvious statement in (1), (2) and (3) is $\mathfrak{PI} \neq \mathfrak{I} \neq \mathfrak{P}_1$: this follows from Ex 11.5. and an example which will be given in D9.. If $T \in \mathfrak{I}$, then $T' \in \mathfrak{I}$ (by 10.2.), hence (4) follows from (2). Since $\mathbf{I}(\kappa_F \circ T) = \mathbf{I}(T)$ (the regularity of the operator ideal $\mathfrak{I}$, see 10.2.), (5) is also a consequence of (2). $\square$

In general, $T \notin \mathfrak{PI}$ if $T' \in \mathfrak{PI}$: this would imply that $\mathfrak{PI} = \mathfrak{I}$. For the same reason, the ideal $\mathfrak{PI}$ is not regular.

D6. The Grothendieck–Pietsch factorization theorem (see 11.3., corollary 2) shows that:

PROPOSITION: *If K is compact and F a Banach space, then*

$$\mathfrak{P}_1(C(K), F) = \mathfrak{I}(C(K), F) = \mathfrak{PI}(C(K), F)$$

holds isometrically.

In particular, the same relation holds for $E = L_\infty(\mu)$ (where μ is an arbitrary measure).

The Radon–Nikodým Property and Operator Ideals

D7. The key to understanding the role of the Radon–Nikodým property in the theory of Banach operator ideals is Grothendieck's theorem C3., which states that an operator $T : L_1(\mu) \to F$ is representable if and only if

$$T \circ I : \begin{array}{c} C(K) \\ \text{or} \\ L_\infty(\mu) \end{array} \xrightarrow{I} L_1(\mu) \xrightarrow{T} F$$

is nuclear; moreover, $\mathbf{N}(T \circ I) = \int \|g\| d\mu \le \|g\|_\infty \mu(\Omega) = \|T\| \, \|I\|$ whenever g is a Riesz–density of T.

THEOREM: F has the Radon–Nikodým property if and only if $\mathfrak{PI}(E, F) = \mathfrak{N}(E, F)$ for all Banach spaces E.

This is a consequence of the following proposition, which contains some interesting details concerning this result.

PROPOSITION: (1) If F has the Radon–Nikodým property, then the relations

$$\mathfrak{P}_1(C(K), F) = \mathfrak{I}(C(K), F) = \mathfrak{PI}(C(K), F) = \mathfrak{N}(C(K), F)$$

(for all compact K) and

$$\mathfrak{PI}(E, F) = \mathfrak{N}(E, F)$$

(for all Banach spaces E) hold isometrically.
(2) *If $\mathfrak{PI}(E, F) = \mathfrak{N}(E, F)$ for $E = C([0, 1])$ or $E = L_\infty[0, 1]$, then F has the Radon–Nikodým property.*

PROOF: To see (2) take $T \in \mathfrak{L}(L_1[0, 1], F)$ (by theorem D1.). Then

$$T \circ I : L_\infty[0, 1] \hookrightarrow L_1[0, 1] \longrightarrow F$$

(or $T \circ I : C[0, 1] \longrightarrow L_1[0, 1] \longrightarrow F$) is Pietsch integral, hence nuclear by assumption. Grothendieck's theorem C3. gives that T is representable.
For (1) one only has to show that $\mathfrak{P}_1(C(K), F) \stackrel{1}{=} \mathfrak{N}(C(K), F)$ by D6. (and the fact that $L_\infty(\mu) \stackrel{1}{=} C(K)$ for some K): Take $T \in \mathfrak{P}_1(C(K), F)$. Then, by the Grothendieck–Pietsch factorization theorem,

$$C(K) \xrightarrow{T} F, \quad I \searrow \; \nearrow S \qquad \mathbf{P}_1(T) = \|I\| \, \|S\|.$$
$$L_1(\mu)$$

Since S is representable, $T = S \circ I$ is nuclear and

$$\mathbf{N}(T) = \mathbf{N}(S \circ I) \leq \|S\|\, \|I\| = \mathbf{P}_1(T) ,$$

and hence equality holds. $\square$

The result D5.(2) implies

COROLLARY 1: *F' has the Radon–Nikodým property if and only if $\mathfrak{I}(E, F') = \mathfrak{N}(E, F')$ for all Banach spaces E. In this case, the equality holds isometrically.*

A nice application is

COROLLARY 2: *If $F \subset G$ and G' has the Radon–Nikodým property, then F' has it as well.*

PROOF: Take $T \in \mathfrak{I}(E, F') = (E \otimes_\varepsilon F)'$. Since $E \otimes_\varepsilon F \subset E \otimes_\varepsilon G$ holds isometrically, by the Hahn–Banach theorem there exists

$$\tilde{T} \in (E \otimes_\varepsilon G)' = \mathfrak{I}(E, G')$$

satisfying $Q \circ \tilde{T} = T$ (for $Q : G' \twoheadrightarrow F'$). It follows that the operator $\tilde{T}$ is nuclear and hence T is also. $\square$

D8. This was the characterization of the Radon–Nikodým property in terms of Pietsch integral = nuclear for the range space of an operator. For the domain space one has the

THEOREM: *For each Banach space E the following statements are equivalent:*
(a) *E' has the Radon–Nikodým property.*
(b) *$\mathfrak{PI}(E, F) = \mathfrak{N}(E, F)$ for all Banach spaces F (or only $F = L_1[0, 1]$).*
(c) *$\mathfrak{I}(E, F') = \mathfrak{N}(E, F')$ for all Banach spaces F.*
In this case, all the relations hold isometrically.

This result, in essence due to Grothendieck, was formulated more or less in this way by various authors and got its final form (removing an assumption about the approximation property) from Alencar [1] only in 1985.

PROOF: Assume E' to have the Radon–Nikodým property, take $T \in \mathfrak{PI}(E, F)$ and factorize as follows:

$$\begin{array}{ccc}
E & \xrightarrow{T} & F \\
\downarrow{\scriptstyle T_o} & \searrow & \uparrow{\scriptstyle S} \\
L_\infty(\mu) & \longrightarrow & L_1(\mu)
\end{array} \qquad \mathbf{PI}(T_o)\|S\| \leq \mathbf{PI}(T)(1+\varepsilon) \; .$$

Then D7. implies that $T'_o \in \mathfrak{I}(L_\infty(\mu), E') = \mathfrak{N}(L_\infty(\mu), E')$ and

$$\mathbf{N}(T''_o) \leq \mathbf{N}(T'_o) = \mathbf{I}(T'_o) = \mathbf{I}(T_o) \leq \mathbf{PI}(T_o) \; .$$

Since L_1 is 1-complemented in its bidual, it follows that

$$\mathbf{N}(T) \leq \mathbf{N}(T_o)\|S\| \leq \mathbf{N}(T''_o)\|S\| \leq \mathbf{PI}(T_o)\|S\| \leq \mathbf{PI}(T)(1+\varepsilon) \; ,$$

which implies (b) and also (c) since $\mathfrak{I}(E, F') \stackrel{1}{=} \mathfrak{PI}(E, F')$. The fact that $L_1[0,1]$ is 1-complemented in its bidual also yields that (c) implies $\mathfrak{PI}(E, L_1[0,1]) \stackrel{1}{=} \mathfrak{N}(E, L_1[0,1])$ — and it remains to verify that this implies (a). In order to apply theorem D1. take $T \in \mathfrak{L}(L_1[0,1], E')$. Since

$$T \circ I : L_\infty[0,1] \xhookrightarrow{I} L_1[0,1] \xrightarrow{T} E'$$

is the dual of

$$S : E \xhookrightarrow{\kappa_E} E'' \xrightarrow{T'} L_\infty[0,1] \xhookrightarrow{I} L_1[0,1]$$

and I is Pietsch integral, it follows that S is nuclear and hence also $T \circ I$: Grothendieck's characterization C3. gives that T is representable. □

COROLLARY: *If the dual E' of a Banach space E has the approximation property, then E' has the Radon–Nikodým property if and only if $\mathfrak{I}(E, F) = \mathfrak{N}(E, F)$ for all Banach spaces F (or only $F = L_1[0,1]$). In this case, the equality holds isometrically.*

PROOF: By the theorem the condition is clearly sufficient. For the converse take $T \in \mathfrak{I}(E, F)$, then

$$T' \in \mathfrak{I}(F', E') = \mathfrak{N}(F', E') \qquad \text{and} \quad \mathbf{N}(T') = \mathbf{I}(T') = \mathbf{I}(T)$$

(by D7., corollary 1). Since $\mathbf{N}(T) = \mathbf{N}(T')$ thanks to the approximation property of E' (see 5.9.), the result follows. □

The example in D9. will show that the assumption that E' has the approximation property is indispensable in this corollary.

For absolutely 1-summing operators there is no such characterization since Lewis and Stegall [170] proved that $\mathfrak{P}_1(E, F) = \mathfrak{N}(E, F)$ for all Banach spaces F if and only if E' is isomorphic to some $\ell_1(\Gamma)$, see 23.9..

D9. Take the Figiel–Johnson example (see 5.2.(5)) of a Banach space E with the approximation property but without the bounded one — and with E' separable: this implies that E' has the Radon–Nikodým property. It is shown in 16.8. (with the help of D7.) that there is a non–nuclear $T \in \mathfrak{L}(E, E)$ with nuclear dual T'. Theorem D8. shows that T cannot be Pietsch integral as well; since $T \in \mathfrak{J}$ if and only if $T' \in \mathfrak{J}$, it follows that:

PROPOSITION: *There is a Banach space E with an integral operator $T \in \mathfrak{L}(E, E)$ which is not Pietsch integral, but $T' \in \mathfrak{N} \subset \mathfrak{PJ} \subset \mathfrak{J}$ and E' has the Radon–Nikodým property.*

This observation is due to Alencar [1]. Note that by corollary D8. the dual E' of any Banach space E which admits such an operator does not have the approximation property.

Bibliography

[1] Alencar, R.: *Multilinear mappings of nuclear and integral type*; Proc. Amer. Math. Soc. 44(1985)33-38

[2] Alencar, R. – M. C. Matos: *Some classes of multilinear mappings between Banach spaces*; Publ. Dpto. Analisis Mat., Univ. Complut. Madrid 12(1989)

[3] Amemiya, I. – K. Shiga: *On tensor products of Banach spaces*; Kodai Math. Sem. Rep. 9(1957)161-178

[4] Banach, S.: *Théorie des opérations linéaires*; Monogr. Mat. 1, Warszawa, 1932

[5] Bartle, R. G. – R. M. Graves: *Mappings between function spaces*; Trans. Amer. Math. Soc. 72(1952)400-413

[6] Barton, T. – Y. Friedman: *Grothendieck's inequality for JB^*-triples and applications*; J. London Math. Soc. 36(1987)513-523

[7] Beauzamy, B.: *Introduction to Banach spaces and their geometry*; Math. Studies/Notas de Matemática 68, North Holland, 1985 (2nd ed.)

[8] Beckner, W.: *Inequalities in Fourier Analysis*; Annals of Math. 102(1975)159-182

[9] Bennett, G.: *Inclusion mappings between ℓ^p-spaces*; J. Funct. Analysis 13(1973)20-27

[10] Bennett, G.: *Schur multipliers*; Duke J. Math. 44(1977)603-639

[11] Bergh, J. – J. Löfström: *Interpolation spaces*; Grundl. math. Wiss. 223, Springer, 1976

Bessaga, C.: See Pełczyński–Bessaga

Bonet, J.: See also Pérez Carreras–Bonet

[12] Bonet, J. – A. Defant – A. Galbis: *Tensor products of Fréchet or (DF)-spaces with a Banach space*; J. Math. Anal. Appl. 166(1992)305-318

[13] Botelho, G.: *Aspectos analíticos e geométricos da propriedade de Radon-Nikodým em espaços de Banach*; Tese de mestrado, Unicamp, 1989

[14] Bourgain, J.: *New classes of L^p-spaces*; Lecture Notes Math. 889, 1981

[15] Bourgain, J.: *Some remarks on Banach spaces in which martingale difference sequences are unconditional*; Ark. Mat. 22(1983)163-168

[16] Bourgain, J.: *New Banach space properties of the disc algebra and H^∞*; Acta Math. 152(1984)1-48

[17] Bourgain, J. – O. Reinov: *On the approximation properties for the space H^∞*; Math. Nachr. 122(1985)19-27

[18] Brown, A. – Pearcy, C.: *Introduction to operator theory I*; Grad. Texts Math. 55, Springer, 1977

[19] Burkholder, D. L.: *A geometrical condition that implies the existence of certain singular integrals of Banach space valued functions*; Proc. Conf. Harmonic Analysis in Honour of Antony Zygmund, Chicago (1981)270-286

[20] Carl, B.: *Absolut $(p,1)$-summierende identische Operatoren von ℓ_u nach ℓ_v*; Math. Nachr. 63(1974)353-360

[21] Carl, B. – A. Defant: *Tensor products and Grothendieck type inequalities of operators in L_p-spaces*; Trans. Amer. Math. Soc. 331(1992)55-76

[22] Carl, B. – A. Defant: *An inequality between the p- and $(p,1)$-summing norm of finite rank operators from $C(K)$-spaces*; Israel J. Math. 74(1991)323-335

[23] Carl, B. – A. Defant – M. S. Ramanujan: *On tensor stable operator ideals*; Michigan Math. J. 36(1989)63-75

[24] Cartier, P.: *Classes de formes bilinéaires sur les espaces de Banach*; Sem. Bourbaki 13e année (1960/61) exp. 211

[25] Castillo, J.: *Weakly p-summable sequences in Banach spaces*; Publ. Dpto. Analysis Mat., Univ. Complut. Madrid 21(1990/91)

[26] Chatterji, S. D.: *Continuous functions representable as sums of independent random variables*; Z. Wahrscheinlichkeitstheorie verw. Gebiete 13(1969)338-341

[27] Chevet, S.: *Sur certains produits tensoriels topologiques d'espaces de Banach*; Z. Wahrscheinlichkeitstheorie verw. Gebiete 11(1969)120-138

[28] Choquet, G.: *Lectures on Analysis*, Vol. I-III; Benjamin, 1969

[29] Chu, C. H. – B. Iochum – G. Loupias: *Grothendieck's theorem and factorization of operators in Jordan triples*; Math. Ann. 284(1989)41-53

[30] Cohen, J. S.: *A characterization of inner product spaces using absolutely 2-summing operators*; Studia Math. 38(1970)271-276

[31] Cohen, J. S.: *Absolutely p-summing, p-nuclear operators and their conjugates*; Math. Ann. 201(1973)177-200

[32] Cohen, J. S. – A. B. Stephens: *Tensor norms on $\ell_1^n \otimes \ell_1^n$*; Math. Ann. 209(1974)277-290

[33] Cohen, P. J.: *A counterexample to the closed graph theorem for bilinear maps*; J. Funct. Analysis 16(1974)235-240

[34] Collins, H. S. – W. Ruess: *Duals of spaces of compact operators*; Studia Math. 74(1982)213-245

[35] Corbacho Rosas, E. – A. Martínez Martínez: *Tensor norms and operator ideals*; Publicationes Universidad de Zaragoza (1989)41-59

[36] Dacunha-Castelle, D. – J. L. Krivine: *Applications des ultraproduits à l'étude des espaces et des algèbres de Banach*; Studia Math. 41(1972)315-334

[37] Davis, W. J. – T. Figiel – W. B. Johnson – A. Pełczyński: *Factoring weakly compact operators*; J. Funct. Analysis 17(1974)311-327

[38] Day, M. M.: *Normed linear spaces*; Ergebn. Math. Grenzgeb. 21, Springer, (3rd ed., 1973)

[39] Dean, D. W.: *The equation $L(E, X^{**}) = L(E, X)^{**}$ and the principle of local reflexivity*; Proc. Amer. Math. Soc. 40(1973)146-148

Defant, A.: See also Bonet–Defant–Galbis, Carl–Defant, Carl–Defant–Ramanujan

[40] Defant, A.: *A duality theorem for locally convex tensor products*; Math. Zeitschr. 190(1985)45-53

[41] Defant, A.: *Produkte von Tensornormen*; Habilitationsschrift, Oldenburg, 1986

[42] Defant, A.: *Absolutely p-summing operators and Banach spaces containing all ℓ_p^n uniformly complemented*; Studia Math. 95(1990)43-53

[43] Defant, A. – K. Floret: *Aspects of the metric theory of tensor products and operator ideals*; Note di Matematica 8(1988)181-281

[44] Defant, A. – K. Floret: *Topological tensor products and the approximation property of locally convex spaces*; Bull. Soc. Roy. Sci. Liège 58(1989)29-51

[45] Defant, A. – K. Floret: *Tensornorm techniques for the (DF)-space problem*; Note di Matematica 10(1990)217-222

[46] Defant, A. – K. Floret: *Continuity of tensor product operators between spaces of Bochner integrable functions*; in: Progress in Functional Analysis, Proc. Peñiscola Meeting 1990 (eds.: Bierstedt, Bonet, Horvath, Maestre), North–Holland Math. Studies 170(1992)367-381

[47] Defant, A. – K. Floret – J. Taskinen: *On the injective tensor product of (DF)-spaces*; Archiv Math. 57(1991)149-154

[48] Defant, A. – W. Govaerts: *Tensor products and spaces of vector valued continuous functions*; manuscripta math. 55(1986)433-449

[49] Defant, A. – V. Mascioni: *Limit orders of ε-π-continuous operators*; Math. Nachr. 145(1990)337-344

[50] Defant, M.: *Zur vektorwertigen Hilbert transformation*; Dissertation, Kiel, 1986

[51] Defant, M.: *On the vector valued Hilbert transform*; Math. Nachr. 141(1989)251-265

[52] Diaz, J. C. – J. A. López Molina – M. J. Rivera Ortun: *The approximation property of order (p,q) in Banach spaces*; Collect. Mat. 41(1990)217-232

[53] Diestel, J.: *A survey of results related to the Dunford-Pettis property*; Contemporary Math. 2(1980)15-60

[54] Diestel, J.: *Sequences and series in Banach spaces*; Grad. Texts Math. 92, Springer, 1984

[55] Diestel, J. – J. J. Uhl: *Vector measures*; Math. Surveys 15, Amer. Math. Soc., 1977

[56] Diestel, J. – J. J. Uhl: *Progress in vector measures 1977-1983*; in: Lecture Notes Math. 1033(1983)144-192

[57] Dineen, S. – R. M. Timoney: *Absolute bases, tensor products and a theorem of Bohr*; Studia Math. 94(1989)227-234

[58] Dunford, N. – R. Schatten: *On the associate and conjugate space for the direct product*; Trans. Amer. Math. Soc. 59(1946)430-436

[59] Dunford, N. – J. Schwartz: *Linear operators, part I*; Interscience Publ., 1958

[60] Duren, P. L.: *Theory of H^p-spaces*; Academic Press, 1970

[61] Dvoretzky, A.: *Some results on convex bodies and Banach spaces*; Proc. Symp. on Linear Spaces, Jerusalem (1961)123-160

[62] Edwards, R. E. – G. I. Gaudry: *Littlewood-Paley and multiplier theory*; Ergebn. Math. Grenzgeb. 90, Springer, 1977

[63] Enflo, P.: *A counterexample to the approximation problem*; Acta Math. 130(1973) 309-317

Figiel, T.: See also Davis–Figiel–Johnson–Pełczyński

[64] Figiel, T.: *Local theory of Banach spaces and some operator ideals*; Proc. IMC Warsaw (1983)961-986

[65] Figiel, T. – T. Iwaniec – A. Pełczyński: *Computing norms of some operators in L^p-spaces*; Studia Math. 79(1984)227-274

[66] Figiel, T. – W. B. Johnson: *The approximation property does not imply the bounded approximation property*; Proc. Amer. Math. Soc. 41(1973)197-200

Floret, K.: See also Defant–Floret, Defant–Floret–Taskinen

[67] Floret, K.: *$\mathcal{L}_1$-Räume und Liftings von Operatoren nach Quotienten lokalkonvexer Räume*; Math. Zeitschr. 134(1973)107-117

[68] Floret, K.: *Der Satz von Dunford-Pettis und die Darstellung von Massen mit Werten in lokalkonvexen Räumen*; Math. Ann. 208(1974)203-212

[69] Floret, K.: *Some aspects of the theory of locally convex inductive limits*; in: Functional Analysis: Surveys and Recent Results II (eds.: Bierstedt, Fuchssteiner), North-Holland Math. Studies 38(1980)205-237

[70] Floret, K.: *Weakly compact sets*; Lecture Notes Math. 801, 1980

[71] Floret, K.: *Maß- und Integrationstheorie*; Teubner, 1981

[72] Floret, K.: *Elementos de posto finito em produtos tensoriais topológicos*; Sem. Brasil. Análise 24(1986)189-195

[73] Floret, K.: *Normas tensoriais e a propriedade de aproximação limitada*; Seminar-Reports UFF Niteroi, 1987

[74] Floret, K.: *Tensornorms and adjoint operator ideals*; Seminar on Functional Analysis, Murcia Ser. Notas de Matematica 5(1988-89)1-8

[75] Floret, K. - J. Wloka: *Einführung in die Theorie der lokalkonvexen Räume*; Lecture Notes Math. 56, 1968

[76] Fourie, J. - J. Swart: *Tensor products and Banach ideals of p-compact operators*; manuscripta math. 35(1981)343-351

Friedman, Y.: See Barton-Friedman

Galbis, A.: See Bonet-Defant-Galbis

[77] Garcia-Cuerva, J. - J. L. Rubio de Francia: *Weighted norm inequalities and related topics*; Math. Studies/Notas de Matemática 104, North Holland, 1985

[78] Garling, D. J. H.: *Operators with large trace and a characterization of ℓ_n^∞*; Proc. Cambridge Phil. Soc. 76(1974)413-414

[79] Garling, D. J. H. - Y. Gordon: *Relations between some constants associated with finite dimensional Banach spaces*; Israel J. Math. 9(1971)346-361

Gaudry, G. I.: See Edwards-Gaudry

[80] Gelbaum, B. - J. Gil de Lamadrid: *Bases of tensor products of Banach spaces*; Pacific J. Math. 11(1961)1281-1286

Gil de Lamadrid, J.: See Gelbaum-Gil de Lamadrid

[81] Gilbert, J. E. - T. Leih: *Factorizaction, tensor products and bilinear forms in Banach space theory*; Notes in Banach spaces, Univ. Texas Press (1980)182-305

Gordon, Y.: See also Garling-Gordon

[82] Gordon, Y.: *On the projection and MacPhail constants of ℓ_n^p spaces*; Israel J. Math. 6(1968)295-302

[83] Gordon, Y.: *On p-absolutely summing constants of Banach spaces*; Israel J. Math. 7(1969)151-163

[84] Gordon, Y. - D. R. Lewis: *Absolutely summing operators and local unconditional structure*; Acta Math. 133(1974)27-48

[85] Gordon, Y. - D. R. Lewis - J. R. Retherford: *Banach ideals of operators with applications*; J. Funct. Analysis 14(1973)85-129

[86] Gordon, Y. – S. Reisner: *Some geometrical properties of Banach spaces of polynomials*; Israel J. Math. 42(1982)99-116

[87] Gordon, Y. – P. Saphar: *Ideal norms on $E \otimes L_p$*; Illinois J. Math. 21(1977)266-285

Govaerts, W.: See Defant–Govaerts

Graves, R. M.: See Bartle–Graves

[88] Graves, W. H. – W. Ruess: *Representing compact sets of compact operators and of compact range vector measures*; Arch. Math. 49(1987)316-325

[89] Grothendieck, A.: *Quelques points de la théorie des produits tensoriels topologiques*; 2° Sympos. Probl. mat. Latino América, Montevideo (1954)173-177

[90] Grothendieck, A.: *Résultats nouveaux dans la théorie des opérations linéaires I, II*; C. R. A. S. Paris 239(1954)577-579 and 607-609

[91] Grothendieck, A.: *Une charactérisation vectorielle métrique des espaces L_1*; Canadian J. Math. 7(1955)552-561

[92] Grothendieck, A.: *Produits tensoriels topologiques et espaces nucléaires*; Memoirs Amer. Math. Soc. 16, 1955

[93] Grothendieck, A.: *Résumé de la théorie métrique des produits tensoriels topologiques*; Bol. Soc. Mat. São Paulo 8(1956)1-79

[94] Grothendieck, A.: *Sur certains classes de suites dans les espaces de Banach et le théorème de Dvoretzky-Rogers*; Bol. Soc. Mat. São Paulo 8(1956)83-110

[95] Grothendieck, A.: *The trace of certain operators*; Studia Math. 20(1961)141-143

[96] Haagerup, U.: *The best constants in the Khintchine inequality*; Studia Math. 70(1982)231-283

[97] Haagerup, U.: *The Grothendieck inequality for bilinear forms on C^*-algebras*; Advances in Math. 56(1985)93-116

[98] Haagerup, U.: *A new upper bound for the complex Grothendieck constant*; Israel J. Math. 60(1987)199-214

[99] Halmos, P. R.: *A Hilbert space problem book*; Grad. Texts Math. 19, Springer, (2nd ed., 1982)

[100] Hardy, G. H. – J. E. Littlewood: *Bilinear forms bounded in space $[p, q]$*; Quart. J. Math. 5(1934)241-254

[101] Harksen, J.: *Tensornormtopologien*; Dissertation, Kiel, 1979

[102] Harksen, J.: *Charakterisierung lokalkonvexer Räume mit Hilfe von Tensorprodukttopologien*; Math. Nachr. 106(1982)347-374

[103] Heinrich, S.: *Weak sequential completeness of Banach operator ideals*; Siberian Math. J. 17(1976)857-862

[104] Heinrich, S.: *Ultraproducts in Banach space theory*; J. reine angew. Math. 313 (1980)72-104

[105] Herz, C. − J. Wick Pelletier: *Dual functors and integral operator in the category of Banach spaces*; J. Pure Appl. Algebra 8(1976)5-22

[106] Hoffmann-Jørgensen, J.: *Sums of independent Banach space valued random variables*; Studia Math. 52(1974)159-186

[107] Hollstein, R.: *Inductive limits and ε-tensor products*; J. reine angew. Math. 319 (1980)38-62

[108] Hollstein, R.: *A sequence characterization of subspaces of $L_1(\mu)$ and quotient spaces of $C(K)$*; Bull. Soc. Roy. Liège 51(1982)403-416

[109] Hollstein, R.: *Extension and lifting of continuous linear mappings in locally convex spaces*; Math. Nachr. 108(1982)273-297

[110] Hollstein, R.: *Locally convex α-tensor products and α-spaces*; Math. Nachr. 120 (1985)73-90

[111] Hollstein, R.: *Tensor sequences and inductive limits with local partition of unity*; manuscripta math. 52(1985)227-249

[112] Holub, J. R.: *Tensor product bases and tensor diagonals*; Trans. Amer. Math. Soc. 151(1970)563-579

[113] Holub, J. R.: *Tensor product mappings*; Math. Ann. 188(1970)1-12

[114] Holub, J. R.: *Hilbertian operators and reflexive tensor products*; Pacific J. Math. 36(1971)185-194

[115] Holub, J. R.: *Compactness of topological tensor products and operator spaces*; Proc. Amer. Math. Soc. 36(1972)398-406

[116] Holub, J. R.: *A characterization of L_p-spaces*; Studia Math. 42(1972)265-270

[117] Holub, J. R.: *Tensor products mappings, II*; Proc. Amer. Math. Soc. 42(1974)437-441

[118] Horowitz, C.: *An elementary counterexample to the open mapping principle for bilinear maps*; Proc. Amer. Math. Soc. 53(1975)293-294

Iochum, B.: See Chu–Iochum–Loupias

Iwaniec, T.: See Figiel–Iwaniec–Pełczyński

[119] Jameson, G. J. O.: *Summing and nuclear norms in Banach space theory*; London Math. Soc. Student Text 8, 1987

[120] Jameson, G. J. O.: *Relations between summing norms of mappings on ℓ_∞^n*; Math. Zeitschr. 194(1987)89-94

[121] Jarchow, H.: *Factorization through inclusion mappings between ℓ_p-spaces*; Math. Ann. 220(1976)123-125

[122] Jarchow, H.: *On a theorem of John and Zizler*; in: Special Topics of Appl. Math. (eds.: Frehse, Pallaschke, Trottenberg), North-Holland (1980)3-13

[123] Jarchow, H.: *Locally convex spaces*; Teubner, 1981

[124] Jarchow, H.: *Factoring absolutely summing operators through Hilbert-Schmidt operators*; Glasgow Math. J. 31(1989)131-135

[125] Jarchow, H. – K. John: *On the equality of injective and projective tensor products*; Czech. Math. J. 38(1988)464-472

[126] Jarchow, H. – R. Ott: *On trace ideals*; Math. Nachr. 108(1982)23-37

John, K.: See also Jarchow–John

[127] John, K.: *Tensor product of several spaces and nuclearity*; Math. Ann. 269(1984) 333-356

[128] John, K.: *On the compact non-nuclear operator problem*; Math. Ann. 287(1990) 509-514

Johnson, W. B.: See also Davis–Figiel–Johnson–Pełczyński, Figiel–Johnson

[129] Johnson, W. B.: *Factoring compact operators*; Israel J. Math. 9(1971)337-345

[130] Johnson, W. B.: *A complementary universal conjugate Banach space and its relation to the approximation problem*; Israel J. Math. 18(1972)301-310

[131] Johnson, W. B. – Her. König – B. Maurey – J. R. Retherford: *Eigenvalues of p-summing and ℓ_p-type operators in Banach spaces*; J. Funct. Analysis 32(1979)353-380

[132] Johnson, W. B. – H. P. Rosenthal – M. Zippin: *On bases, finite dimensional decomposition and weaker structures in Banach spaces*; Israel J. Math. 9(1971)488-506

[133] Junek, H.: *Locally convex spaces and operator ideals*; Teubner-Texte 56, 1983

[134] Kaballo, W.: *Lifting theorems for vector-valued functions and the ε-tensor product*; in: Funct. Analysis: Surveys and Recent Results (eds.: Bierstedt, Fuchssteiner), North-Holland Math. Studies 27(1977)149-166

[135] Kaballo, W.: *Liftingsätze für Vektorfunktionen und (εL)-Räume*; J. reine angew. Math. 309(1979)55-85

[136] Kadec, M. J.: *On linear dimension of spaces L_p*; Uspechi Mat. Nauk 13.6(1958)95-98

[137] Kadec, M. J. – A. Pełczyński: *Bases, lacunary sequences and complemented subspaces in the spaces L_p*; Studia Math. 31(1962)161-176

[138] Kadec, M. J. – M. G. Snobar: *Certain functionals on the Minkowski compactum*; Math. Notes 10(1971)694-696 (russian)

[139] Kadison, R. V. – J. R. Ringrose: *Fundamentals of the theory of operator algebras*; Academic Press, 1983

[140] Kalton, N. J.: *Spaces of compact operators*; Math. Ann. 208(1974)267-278

[141] Kandil, M. A.: *Parseval's identity and Hilbertization of Banach spaces*; Proc. Math. Phys. Soc. Egypt 53(1982)45-46

[142] Kisliakov, S. V.: *On spaces with "small" annihilators*; Sem. Leningrad Otdel Mat. Inst. Steklov 65(1976)192-195 (russian)

[143] Kisliakov, S. V.: *Two remarks on the equality* $\Pi_p(X,\cdot) = I_p(X,\cdot)$; Sem. Leningrad. Otdel. Mat. Inst. Steklov 113(1981)135-148 (russian)

König, Her.: See also Johnson–König–Maurey–Retherford

[144] König, Her.: *Diagonal and convolution operators as elements of operator ideals*; Math. Ann. 218(1975)97-106

[145] König, Her.: *Interpolation of operators ideals with an application to eigenvalue distribution problems*; Math. Ann. 233(1978)35-48

[146] König, Her.: *On the tensor stability of s-number ideals*; Math. Ann. 269(1984)77-93

[147] König, Her.: *Eigenvalue distribution of compact operators*; Birkhäuser, 1986

[148] König, Her.: *On the complex Grothendieck constant in the n-dimensional case*; Proc. Strobl–conference on "Geometry of Banach–spaces" 1989 (eds.: Müller, Schachermayer), Lect. Notes London Math. Soc. 158(1990)181-199

[149] König, Her.: *On the Fourier coefficients of vector valued functions*; Math. Nachr. 152(1991) 215-227

[150] König, Her. – J. R. Retherford – N. Tomczak-Jaegermann: *On the eigenvalues of $(p,2)$-summing operators and constants associated with normed spaces*; J. Funct. Analysis 37(1980)88-126

[151] Köthe, G.: *Hebbare lokalkonvexe Räume*; Math. Ann 165(1960)181-195

[152] Köthe, G.: *Topological vector spaces*, Vol. I and II; Grundl. math. Wiss. 159 and 237, Springer, 1969 and 1979

[153] Krickeberg, K.: *Wahrscheinlichkeitstheorie*; Teubner, 1963

Krivine, J. L.: See also Dacunha-Castelle–Krivine

[154] Krivine, J. L.: *Sur la complexification des opérateurs de L^∞ dans L^1*; C. R. A. S. Paris 284(1977)377-379

[155] Krivine, J. L.: *Constantes de Grothendieck et fonctions de type positif sur les sphères*; Advances in Math. 31(1979)16-30

[156] Kürsten, K. D.: *s-Zahlen und Ultraprodukte von Operatoren in Banachräumen*; Dissertation, Leipzig, 1976

[157] Kwapień, S.: *Some remarks an (p, q)-absolutely summing operators in ℓ_p-spaces*; Studia Math. 29(1968)327-337

[158] Kwapień, S.: *A remark on p-absolutely summing operators in ℓ_r-spaces*; Studia Math. 34(1970)109-111

[159] Kwapień, S.: *On a theorem of L. Schwartz and its applications to absolutely summing operators*; Studia Math. 38(1970)193-201

[160] Kwapień, S.: *On operators factoring through L_p-space*; Bull. Soc. Math. France Mémoire 31-32(1972)215-225

[161] Kwapień, S.: *Isomorphic characterization of inner product spaces by orthogonal series with vector valued coefficients*; Studia Math. 44(1972)583-595

[162] Kwapień, S.: *Isomorphic characterizations of Hilbert spaces by orthogonal series with vector valued coefficients*; Sém. Maurey-Schwartz 1972-73, Ecole Polytéchnique, Exp. 8

[163] Kwapień, S. – A. Pełczyński: *The main triangle projection in matrix spaces and its applications*; Studia Math. 34(1970)43-68

[164] Lacey, H. E.: *The isometric theory of classical Banach spaces*; Grundl. math. Wiss. 208, Springer, 1974

[165] Lapresté, J. T.: *Sur une généralisation de la notion d'opérateurs nucléaires e sommants dans les espaces de Banach*; C. R. A. S. Paris 275(1972)45-48

[166] Lapresté, J. T.: *Opérateurs sommants et factorisations à travers les espaces L^p*; Studia Math. 57(1976)47-83

Leih, T.: See Gilbert–Leih

Lewis, D. R.: See also Gordon–Lewis, Gordon–Lewis–Retherford

[167] Lewis, D. R.: *Conditional weak compactness in certain inductive tensor products*; Math. Ann. 201(1973)201-209

[168] Lewis, D. R.: *An isomorphic characterization of the Schmidt class*; Compositio Math. 30(1975)293-297

[169] Lewis, D. R.: *Duals of tensor products*; in: Lecture Notes Math. 604(1977)57-66

[170] Lewis, D. R. – C. Stegall: *Banach spaces whose duals are isomorphic to $\ell_1(\Gamma)$*; J. Funct. Analysis 12(1973)177-187

[171] Lindenstrauss, J.: *Extension of compact operators*; Memoirs Amer. Math. Soc. 48, 1964

[172] Lindenstrauss, J.: *On James' paper "Separable conjugate spaces"*; Israel J. Math. 9(1971)279-284

[173] Lindenstrauss, J. – A. Pełczyński: *Absolutely summing operators in $\mathcal{L}_p$-spaces and applications*; Studia Math. 29(1968)275-326

[174] Lindenstrauss, J. – H. P. Rosenthal: *The $\mathcal{L}_p$-spaces*; Israel J. Math. 7(1969)325-349

[175] Lindenstrauss, J. – L. Tzafriri: *On the complemented subspace problem*; Israel J. Math 9(1971)263-269

[176] Lindenstrauss, J. – L. Tzafriri: *Classical Banach spaces*; Lecture Notes Math. 338, 1973

[177] Lindenstrauss, J. – L. Tzafriri: *Classical Banach spaces*, Vol. I and II; Ergebn. Math. Grenzgeb. 92 and 97, Springer, 1977 and 1979

Littlewood, J. E.: See also Hardy–Littlewood

[178] Littlewood, J. E.: *On bounded bilinear forms in an infinite number of variables*; Quart. J. Math. 1(1930)164-174

Löfström, J.: See Bergh–Löfström

López Molina, J. A.: See Diaz–López Molina–Rivera Ortun

[179] Losert, V. – P. Michor: *Ausarbeitung der "Résumé de la théorie métrique des produits tensoriels topologiques"*; unpublished seminar notes, University of Wien, 1976

[180] Lotz, H. P.: *Grothendieck ideals of operators in Banach spaces*; unpublished lecture notes, Univ. Illinois, Urbana, 1973

Loupias, G.: See Chu–Iochum–Loupias

[181] Mankiewicz, P. – N. J. Nielsen: *A superreflexive Banach space with a finite dimensional decomposition so that no large subspace has a basis*; Israel J. Math. 70(1990)188–204

[182] Marcinkiewicz, J. – A. Zygmund: *Quelques inégalités pour les opérations linéaires*; Fund. Math. 32(1939)115-121

[183] Marcus, M. B. – G. Pisier: *Random Fourier series with applications to Harmonic Analysis*; Ann. of Math. Studies 101, Princeton, 1981

[184] Marquina, A. – J. M. Sanz Serna: *Barrelledness conditions on $c_0(E)$*; Archiv Math. 31(1978)589-596

Martínez-Martínez, A.: See Corbacho Rosas–Martínez-Martínez

Mascioni, V.: See Defant–Mascioni

Matos, M. C.: See Alencar–Matos

Maurey, B.: See also Johnson–König–Maurey–Retherford

[185] Maurey, B.: *Espaces de cotype p*; Sém. Maurey-Schwartz 1972-1973, exp. 7

[186] Maurey, B.: *Théorèmes de factorisation pour les opérateurs linéaires à valeurs dans les espaces L^p*; Astérisque 11, 1974

[187] Maurey, B. – G. Pisier: *Séries de variables aléatoires vectorielles indépendantes et propriétés géométriques des espaces de Banach*; Studia Math. 58(1976)45-90

[188] McCarthy, C. A.: *Commuting Boolean algebras of projections, II. Boundedness in L_p*; Proc. Amer. Math. Soc. 15(1964)781-787

[189] Meile, M.: *Duale Tensornormen*; Diplomarbeit, Oldenburg, 1988

[190] Mendoza Casas, J.: *Necessary and sufficient conditions for $C(X, E)$ to be barrelled or infrabarrelled*; Simon Stevin 57(1983)103-123

Michor, P.: See also Losert–Michor

[191] Michor, P.: *Functors and categories in Banach-spaces*; Lecture Notes Math. 651, 1978

[192] Milman, V. D. – G. Schechtman: *Asymptotic theory of finite dimensional normed spaces*; Lecture Notes Math. 1200, 1986

[193] Murray, F. J. – J. von Neumann: *On rings of operators*; Annals of Math. 37(1936) 116-229

Neumann, J. von: See Murray–von Neumann, Schatten–von Neumann

Nielsen, N.: See Mankiewicz–Nielsen

Okazaki, Y.: See Takahashi–Okazaki

[194] Oertel, F.: *Konjugierte Operatorenideale und das $\mathfrak{A}$-lokale Reflexivitätsprinzip*; Dissertation, Kaiserslautern, 1990

[195] Orlicz, W.: *Über unbedingte Konvergenz in Funktionenräumen I*; Studia Math. 4(1933)33-37

Ott, R.: See Jarchow–Ott

[196] Oya, E. – O. Reinov: *A counterexample to A. Grothendieck*; Eesti N. SV Tead Akad Toimetised Füüs-Mat. 37(1938)14-17, 121 (russian); M.R. 89g46016

Pearcy, C.: See Brown–Pearcy

Pełczyński, A.: See also Davis–Figiel–Johnson–Pełczyński, Figiel–Iwaniec–Pełczyński, Kadec–Pełczyński, Kwapień–Pełczyński, Lindenstrauss–Pełczyński

[197] Pełczyński, A.: *Projections in certain Banach spaces*; Studia Math. 19(1960)209-228

[198] Pełczyński, A.: *A characterization of Hilbert-Schmidt operators*; Studia Math. 28(1967)355-360

[199] Pełczyński, A.: *On Banach spaces containing $L_1(\mu)$*; Studia Math. 30(1968)231-246

[200] Pełczyński, A.: *Any separable Banach space with the bounded approximation property is a complemented subspace of a Banach space with a basis*; Studia Math. 40(1971)239-242

[201] Pełczyński, A.: *Sur certains propriétés isomorphes nouvelles des espaces de Banach de fonction holomorphes A e H^∞*; C. R. A. S. Paris 279(1974)9-12

[202] Pełczyński, A.: *Banach spaces of analytic functions and absolutely summing operators*; Amer. Math. Soc. Regional Conf. Ser. Math. 30, 1977

[203] Pełczyński, A.: *Geometry of finite dimensional Banach spaces and operator ideals*; Notes in Banach Spaces, Univ. Texas Press (1980)81-181

[204] Pełczyński, A.: *Structural theory of Banach spaces and its interplay with Analysis and Probability*; Proc. IMC Warsaw (1983)237-269

[205] Pełczyński, A. – C. Bessaga: *Some aspects of the present theory of Banach spaces*; in: S. Banach Oeuvres, Vol. II; Pol. Scient. Publ., Warsaw (1979)221-302

Pelletier, J. W.: See also Herz–Wick Pelletier

[206] Pelletier, J. W.: *Tensornorms and operators in the category of Banach spaces*; Int. Eqns. Op. Theory 5(1982)85-113

[207] Pérez Carreras, P. – J. Bonet: *Barrelled locally convex spaces*; Math. Studies/Notas de Matemática 131, North-Holland, 1987

[208] Persson, A.: *On some properties of p-nuclear and p-integral operators*; Studia Math. 23(1969)213-222

[209] Persson, A. – A. Pietsch: *p-nukleare und p-integrale Abbildungen in Banachräumen*; Studia Math. 33(1969)19-62

Pietsch, A.: See also Persson–Pietsch

[210] Pietsch, A.: *Absolut p-summierende Abbildungen in normierten Räumen*; Studia Math. 27(1967)333-353

[211] Pietsch, A.: *Ideale von Operatoren in Banachräumen*; Mitteilungen Math. Ges. DDR 1(1968)1-13

[212] Pietsch, A.: *Adjungierte normierte Operatorenideale*; Math. Nachr. 48(1971)189-211

[213] Pietsch, A.: *Ultraprodukte von Operatoren in Banachräumen*; Math. Nachr. 61 (1974)123-132

[214] Pietsch, A.: *Operator ideals*; Deutscher Verlag der Wiss., 1978 and North-Holland, 1980

[215] Pietsch, A.: *Eigenvalues of absolutely r-summing operators*; in: Aspects of Mathematics and Its Applications (ed.: Barroso), Elsevier (1986)607-617

[216] Pietsch, A.: *Eigenvalues and s-numbers*; Akad. Verlagsges., 1987 and Cambridge Studies in Advanced Mathematics 13, 1987

Pisier, G.: See also Marcus–Pisier, Maurey–Pisier

[217] Pisier, G.: *Sur les espaces qui ne contiennent pas ℓ_1^n uniformément*; in: Sém. Maurey-Schwartz 1973-74, exp.7 and C. R. A. S. Paris 277(1973)991-994

[218] Pisier, G.: *Some results on Banach spaces without local unconditional structure*; Compositio Math. 37(1978)3-19

[219] Pisier, G.: *Une nouvelle classes d'espaces vérifiant le théorème de Grothendieck*; Ann. Inst. Fourier 28(1978)69-90

[220] Pisier, G.: *Grothendieck's theorem for non-commutative C^*-algebras with an appendix to Grothendieck's constants*; J. Funct. Analysis 29(1978)397-415

[221] Pisier, G.: *De nouveaux espaces de Banach sans le propriété d'approximation (d'après A. Szankowski)*; in: Lecture Notes Math. 770 (Sém. Bourbaki 1978-79), 312-327

[222] Pisier, G.: *Un théorème sur les opérateurs entre espaces de Banach qui se factorisent par un espace de Hilbert*; Ann. Ecole Norm. Sup. 13(1980)23-43

[223] Pisier, G.: *Holomorphic semi-groups and the geometry of Banach spaces*; Ann. of Math. 115(1982)379-392

[224] Pisier, G.: *Counterexamples to a conjecture of Grothendieck*; Acta Math. 151(1983) 181-208

[225] Pisier, G.: *Factorization of linear operators and geometry of Banach spaces*; CBMS Regional Conf. Series 60, Amer. Math. Soc., 1986

[226] Puhl, J.: *Quotienten von Operatorenidealen*; Math. Nachr. 79(1977)131-144

[227] Puhl, J.: *Selfadjoint and completely symmetric operator ideals*; Math. Nachr. 93(1979)331-335

Ramanujan, M. S.: See Carl-Defant-Ramanujan

Reinov, O.: See also Bourgain-Reinov, Oya-Reinov

[228] Reinov, O.: *Approximation properties of order p and the existence of non-p-nuclear operators with p-nuclear second adjoints*; Math. Nachr. 109(1982)125-134

[229] Reinov, O.: *A survey of some results in connection with Grothendieck appoximation property*; Math. Nachr. 119(1984)257-264

[230] Reisner, S.: *On Banach spaces having the property G.L.*; Pacific J. Math. 83(1979) 505-521

Reisner, S.: See Gordon-Reisner

Retherford, J. R.: See also Gordon-Lewis-Retherford, Stegall-Retherford, Johnson-König-Maurey-Retherford, König-Retherford-Tomczak-Jaegermann

[231] Retherford, J. R.: *Applications of Banach ideals of operators*; Bull. Amer. Math. Soc. 81(1975)978-1012

Ringrose, J. R.: See Kadison–Ringrose

Rivera Ortun, M. J.: See also Diaz–López Molina–Rivera Ortun

[232] Rivera Ortun, M. J.: *The weak topology in certain right-injective tensor products of Banach spaces*; Note di Matematica 9(1989)263–266

Rosenthal, H. P.: See also Johnson–Rosenthal–Zippin, Lindenstrauss–Rosenthal

[233] Rosenthal, H. P. – S. J. Szarek: *On tensor products of operators from L^p to L^q*; in: Lecture Notes Math. 1470(1991)108–132

Rubio de Francia, J. L.: See also Garcia-Cuerva–Rubio de Francia

[234] Rubio de Francia, J. L. – J. L. Torrea: *Some Banach space techniques in vector valued Fourier Analysis*; Coll. Math. 54(1987)273-284

[235] Ruckle, W. H.: *The tensor product of complete biorthogonal sequences*; Duke Math. J. 38(1971)681-696

Ruess, W.: See Collins–Ruess, Graves–Ruess

Sanz Serna, J. M.: See Marquina–Sanz Serna

Saphar, P.: See also Gordon–Saphar

[236] Saphar, P.: *Applications à puissance nucléaire et applications de Hilbert-Schmidt*; C. R. A. S. Paris 261(1965)867-870

[237] Saphar, P.: *Applications à puissance nucléaire et applications de Hilbert-Schmidt dans les espaces de Banach*; Ann. Scient. Ec. Norm. Sup. 83(1966)113-151

[238] Saphar, P.: *Produits tensoriels topologiques et classes d'applications linéaires*; C. R. A. S. Paris 266(1968)526-528

[239] Saphar, P.: *Comparaison de normes sur des produits tensoriels d'espaces de Banach. Applications*; C. R. A. S. Paris 266(1968)809-811

[240] Saphar, P.: *Quelques propriétés des normes tensorielles g_k et d_k*; C. R. A. S. Paris 268(1969)528-531

[241] Saphar, P.: *Produits tensoriels d'espaces de Banach et classes d'applications linéaires*; Studia Math. 38(1970)71-100

[242] Saphar, P.: *Applications p-décomposantes et p-absolument sommantes*; Israel J. Math. 11(1972)164-179

[243] Saphar, P.: *Hypothèse d'approximation à l'ordre p dans les espaces de Banach et approximation d'applications p-absolument sommantes*; Israel J. Math. 13(1972) 379-399

[244] Schaefer, H. H.; *Banach lattices and positive operators*; Grundl. math. Wiss. 215, Springer, 1974

Schatten, R.: See also Dunford–Schatten

[245] Schatten, R.: *On the direct product of Banach spaces*; Trans. Amer. Math. Soc. 53(1943)195-217

[246] Schatten, R.: *On reflexive norms for the direct product*; Trans. Amer. Math. Soc. 54(1943)498-506

[247] Schatten, R.: *The cross-space of linear transformations*; Annals of Math. 47(1946) 73-84

[248] Schatten, R.: *On projections of bound 1*; Annal. of Math. 48(1947)321-325

[249] Schatten, R.: *A theory of cross-spaces*; Ann. of Math. Studies 26, 1950

[250] Schatten, R.: *Norm ideals of completely continuous operators*; Ergebn. Math. Grenzgeb. 27, Springer, 1960

[251] Schatten, R. – J. von Neumann: *The cross-space of linear transformations II, III*; Annals of Math. 47(1946)608-630 and 49(1948)557-582

Schechtman, G.: See Milman–Schechtman

[252] Schmets, J.: *Spaces of vector valued continuous functions*; Lecture Notes Math. 1003, 1983

[253] Schütt, C.: *Unconditionality in tensor products*; Israel J. Math. 31(1978)209-216

Schwartz, J.: See Dunford–Schwartz

[254] Schwarz, H. U.: *Adjungierte Operatorendiale*; Math. Nachr. 55(1973)293-308

[255] Schwarz, H. U.: *Dualität und Approximation von Normidealen*; Math. Nachr. 66 (1975)305-317

[256] Schwarz, H. U.: *Banach lattices and operators*; Teubner-Texte 71, 1984

Shiga, K.: see Amemiya–Shiga

[257] Sims, B.: *Ultratechniques in Banach space theory*; Queen's papers in Pure and Applied Math. 60, 1982

Snobar, M. G.: See Kadec–Snobar

Stegall, C. P.: See also Lewis–Stegall

[258] Stegall, C. P.: *A proof of the principle of local reflexivity*; Proc. Amer. Math. Soc. 78(1980)154-156

[259] Stegall, C. P. – J. R. Retherford: *Fully nuclear and completely nuclear operators with applications to $\mathcal{L}_1$-and $\mathcal{L}_\infty$-spaces*; Trans. Amer. Math. Soc. 163(1972)457-492

[260] Stein, E. M.: *Singular integrals and differentiability properties of functions*; Princeton, 1970

Stephens, A. B.: See Cohen–Stephens

Swart, J.: See Fourie–Swart

[261] Szankowski, A.: *A Banach lattice without the approximation property*; Israel J. Math 24(1976)329-337

[262] Szankowski, A.: *Subspaces without the approximation property*; Israel J. Math. 30(1978)123-129

[263] Szankowski, A.: *B(H) does not have the approximation property*; Acta Math. 147(1981)89-108

Szarek, S. J.: See also Rosenthal–Szarek

[264] Szarek, S. J.: *A Banach space without a basis which has the bounded approximation property*; Acta Math. 159(1987)81-98

[265] Takahashi, Y. – Y. Okazaki: *On Persson's theorem concerning p-nuclear operators*; Proc. Japan Acad. 62(1986)94-96

[266] Talagrand, M.: *Cotype of operators from C(K)*; Invent. Math. 107(1992)1-40

Taskinen, J.: See also Defant–Floret–Taskinen

[267] Taskinen, J.: *Counterexamples to "Problème des topologies" of Grothendieck*; Ann. Acad. Sci. Fenn. Sci. A I Math. Diss. 63(1986)

[268] Taskinen, J.: *The projective tensor product of Fréchet-Montel spaces*; Studia Math. 91(1988)17-30

[269] Taskinen, J.: *(FBa) and (FBB)-spaces*; Math. Zeitschr. 198(1988)339-365

[270] Terzioglu, T.: *A characterization of compact linear mappings*; Archiv Math. 22 (1971)76-78

Timoney, R. M.: See Dineen–Timoney

Tomczak-Jaegermann, N.: See also König–Retherford–Tomczak-Jaegermann

[271] Tomczak-Jaegermann, N.: *The moduli of smoothness and convexity and the Rademacher averages of trace classes S_p ($1 \leq p < \infty$)*; Studia Math. 50(1974)163-182

[272] Tomczak-Jaegermann, N.: *Banach-Mazur distances and finite dimensional operator ideals*; Pitman Monogr. and Surv. Pure Appl. Math. 38, 1989

Tzafriri, L.: See Lindenstrauss–Tzafriri

Uhl, J. J.: See Diestel–Uhl

[273] Vala, K.: *On compact sets of compact operators*; Ann. Acad. Sci. Fenn. Ser. A I Math. 351(1964)

[274] Valdivia, M.: *A class of locally convex spaces without C-webs*; Ann. Inst. Fourier 32(1982)261-269

[275] Virot, B.: *Extensions vectorielles d'operatéurs linéaires bornés sur L^p*; C. R. A. S. Paris 293(1981)413-415

[276] Vladimirskii, Y. N.: *Compact perturbations of Φ-operators in locally convex spaces*; Sibirian Math. J. 14(1973)511-524

Wloka, J.: See Floret–Wloka

[277] Yosida, K.: *Functional analysis*; Grundl. math. Wiss. 123, Springer, 1968 (2nd ed.)

[278] Zalduendo, I.: *An estimate for multilinear forms on ℓ_p-spaces*; preprint, 1989

Zippin, M.: See Johnson–Rosenthal–Zippin

Zygmund, A.: See Marcinkiewicz–Zygmund

List of Symbols

1. General Symbols

$\mathbb{N}$	set of natural numbers
$\mathbb{Z}$	set of integers
$\mathbb{R}$	set of real numbers
$\mathbb{C}$	set of complex numbers
$\mathbb{K}$	$= \mathbb{R}$ or $\mathbb{C}$
$\mathbb{K}^\infty$	$:= \mathbb{K}^\mathbb{N}$
$[a,b]$	$:= \{x \in \mathbb{R} \cup \{-\infty, \infty\} \mid a \leq x \leq b\}$ for $-\infty \leq a \leq b \leq \infty$
$]a,b[$	$:= \{x \in \mathbb{R} \mid a < x < b\}$ for $-\infty \leq a \leq b \leq \infty$
$[a,b[,]a,b]$	
p'	conjugate exponent of $p \in [1, \infty]$ satisfying $1/p + 1/p' = 1$
$\operatorname{Re}(z), \operatorname{Im}(z)$	real and imaginary part of $z \in \mathbb{C}$
NORM	class of all normed spaces
BAN	class of all Banach spaces
FIN	class of all finite dimensional Banach spaces
FIN(E)	$:= \{F \subset E \mid \dim F < \infty\}$
COFIN(E)	$:= \{F \subset E \mid F$ closed subspace, $\dim E/F < \infty\}$
$\langle E, F \rangle$	separating dual system, 490
$\langle x, y \rangle$	duality bracket, 490
$\sigma(E, F)$	weak topology on E with respect to $\langle E, F \rangle$, 490
$(\,\cdot\,\mid\,\cdot\,)_H$	scalar product of a (pre-)Hilbert space
$R: H \to H'$	Riesz mapping defined by $\langle Rx, \,\cdot\, \rangle_{H', H} := (\,\cdot\,\mid x\,)_H$
$\|x\|, x \wedge y, x \vee y$	if x, y are elements of a lattice, 491

Let A be a subset of a vector space E:

$\operatorname{conv}(A)$	convex hull of A
$\Gamma(A)$	absolutely convex hull of A
$\operatorname{span} A$	linear hull of A
m_A	Minkowski gauge functional of A, 26

$[A]$	$:= (\operatorname{span} A, m_A)$ if A is absolutely convex, 26
A^0	$:= \{y \in F \mid |\langle a, y \rangle| \leq 1$ for all $a \in A\}$ absolute polar of $A \subset E$ with respect to the dual system $\langle E, F \rangle$, 490

Let E and F be normed spaces:

B_E	closed unit ball $\{x \in E \mid \|x\|_E \leq 1\}$
$\overset{\circ}{B}_E$	open unit ball $\{x \in E \mid \|x\|_E < 1\}$
$L(E, F)$	$:= \{T : E \to F \mid \text{linear}\}$
$\mathcal{F}(E, F)$	$:= \{T \in L(E, F) \mid \operatorname{rank} T < \infty\}$
$\mathfrak{L}(E, F)$	$:= \{T \in L(E, F) \mid \text{continuous}\}$
$\mathfrak{F}(E, F)$	$:= \{T \in \mathfrak{L}(E, F) \mid \operatorname{rank} T < \infty\}$
$\tau_{co}, \tau_s, \tau_w$	topologies on $\mathfrak{L}(E, F)$, 58, 190
E^*	$:= L(E, \mathbb{K})$ algebraical dual of E
E'	$:= \mathfrak{L}(E, \mathbb{K})$ (topological) dual of E
$\tilde{E}$	completion of E
$d(E, F)$	Banach–Mazur distance, 70
$\lambda(E)$	projection constant of E, 307
$E \overset{1}{=} F$	E and F are isometric
id_E	identical mapping $E \to E$
$\kappa_E : E \overset{1}{\hookrightarrow} E''$	canonical metric injection defined by $\langle \kappa_E(x), x' \rangle_{E'', E'} := \langle x, x' \rangle_{E, E'}$
$P_E : E''' \overset{1}{\twoheadrightarrow} E'$	canonical restriction defined by $\langle P_E(x'''), x \rangle_{E', E} := \langle x''', \kappa_E(x) \rangle_{E''', E''}$
$I_G^E : G \overset{1}{\hookrightarrow} E$	canonical metric injection of a subspace G of E into E
$Q_G^E : E \overset{1}{\twoheadrightarrow} E/G$	canonical metric surjection for a closed subspace G of E
$Q_E : \ell_1(B_E) \overset{1}{\twoheadrightarrow} E$	canonical metric surjection (E a Banach space), 489
$I_E : E \overset{1}{\hookrightarrow} \ell_\infty(B_{E'})$	canonical metric injection, 489

Let $T : E \longrightarrow F$ be linear and continuous:

$\|T\|$	$:= \sup\{\|Tx\|_F \mid x \in B_E\}$ norm of T
$\lambda(T)$	extension norm of T, 312
$\ker(T)$	$:= \{x \in E \mid Tx = 0\}$ kernel of T
$\operatorname{im}(T)$	$:= \{Tx \in F \mid x \in E\}$ range of T

rank(T)	$:= \dim T(E)$ rank of T
$T: E \hookrightarrow F$	T injective
$T: E \stackrel{1}{\hookrightarrow} F$	T metric injection, i.e. $\|Tx\|_F = \|x\|_E$ for all $x \in E$
$T: E \twoheadrightarrow F$	T surjective
$T: E \stackrel{1}{\twoheadrightarrow} F$	T metric surjection, i.e. $\|y\|_F = \inf\{\|x\|_E \mid Tx = y\}$
$T': F' \longrightarrow E'$	dual operator $T'x' := x' \circ T'$
$T^\times : E'' \longrightarrow F$	astriction of T if $T \in \mathfrak{W}(E,F)$ defined by $T^\times(x'') := T''x''$

2. Sequence spaces and related spaces

E^I	$:= \prod_{i \in I} E$ product
$E^{(I)}$	$:= \bigoplus_{i \in I} E$ direct sum
$\ell_p(I, E_i) := \ell_p(E_i)$	$:= \{(x_i) \in \prod_{i \in I} E_i \mid \|(x_i)\|_p < \infty\}$ where
$\|(x_i)\|_p$	$:= (\sum_{i \in I} \|x_i\|_{E_i}^p)^{1/p}$ if $1 \leq p < \infty$
$\|(x_i)\|_\infty$	$:= \sup_{i \in I} \|x_i\|_{E_i}$
$\ell_p(I)$	$:= \ell_p(I, \mathbb{K})$
ℓ_p	$:= \ell_p(\mathbb{N})$
ℓ_p^n	$:= \ell_p(\{1, ..., n\})$
$\ell_p(E)$	$:= \ell_p(\mathbb{N}, E)$, 90
$\ell_p^w(E)$	91
$\ell_p^{w,o}(E)$	92
$c_o(E)$	$:= \{(x_n) \in \ell_\infty(E) \mid \lim_{n \to \infty} \|x_n\|_E = 0\}$
c_o	$:= c_o(\mathbb{K})$
$\ell_p(x_n; E) := \ell_p(x_n)$	$:= \|(\|x_n\|_E)\|_p$, 90, 107
$w_p(x_n; E) := w_p(x_n)$	$:= \sup\{\|(\langle x', x_n \rangle)\|_p \mid \|x'\|_{E'} \leq 1\}$, 91, 104, 107
$(E_\iota)_\mathfrak{U}$	ultraproduct of Banach spaces E_ι along an ultrafilter $\mathfrak{U}$, 226
$(T_\iota)_\mathfrak{U}$	ultraproduct of operators $T_\iota \in \mathfrak{L}(E_\iota, F_\iota)$, 227
D_λ	diagonal operator for a (finite or infinite) sequence $\lambda = (\lambda_n)$ of scalars, $D_\lambda((x_n)) := (\lambda_n x_n)$

3. Spaces

$C(K, E)$	space of continuous functions $f: K \to E$, where K is compact and E a locally convex space
$C(K)$	$:= C(K, \mathbb{K})$

$C_{00}(\Omega, \mathbb{K})$	space of continuous functions $f : \Omega \longrightarrow \mathbb{K}$ with compact support (Ω locally compact)
$H^p := H^p(D)$	Hardy space on the unit disc $D \subset \mathbb{C}$
C_p	Johnson space, 71
Rad	closed span in $L_2[0,1]$ or $L_2(D, \mu)$ of the Rademacher functions, 10, 97
Rad_p	Rad with norm induced from L_p, 99
$\mathcal{L}^g_{p,\lambda}, \mathcal{L}^g_p$	$\mathcal{L}^g_p$-space, 40

4. Locally convex spaces

$cs(E)$	continuous seminorms on E
$cs_o(t)$	a basis of $cs(E)$
$\mathfrak{U}(E)$	neighbourhoods of zero in E
$\mathfrak{U}_o(E)$	a basis of $\mathfrak{U}(E)$
ker p	$:= \{x \in E \mid p(x) = 0\}$ kernel of a seminorm p
E_p	$:= (E/\ker p, \hat{p})$ if $p \in cs(E)$ where $\hat{p}(\hat{v}) := p(v)$
$\kappa_p : E \to E_p$	canonical quotient mapping
$E_U := E_p$	for $U \in \mathfrak{U}(E)$ absolutely convex if $p = m_U$
$\mathcal{L}_b(E, F)$	$\mathcal{L}(E, F)$ (continuous linear operators) with the topology of uniform convergence on bounded sets
E'_b	$:= \mathcal{L}_b(E, \mathbb{K})$ strong dual, 469
E''_e	the bidual E'' with the topology of uniform convergence on equicontinuous subsets of E', 469
$p \otimes_\alpha q$	469

5. Measure and integration theory

$(\Omega, \mu), (\Omega, \Sigma_\mu, \mu)$	measure space, 497
$|\mu|, \mu^+, \mu^-$	variation, positive and negative part of a measure/Daniell functional μ, 493, 494, 497
$\text{Re}(\mu), \text{Im}(\mu)$	real and imaginary part of μ, 497
$\mu_1 \otimes \mu_2$	product measure, 498
μ^τ	Bourbaki extension of μ, B8.
Σ_μ	set of μ–measurable sets, 496
I_μ	set of μ–integrable sets, 496
$\mathfrak{S}(\Sigma)$	space of Σ–step function, 494
$\mathfrak{S}(\mu)$	$:= \mathfrak{S}(I_\mu)$

$\mathcal{L}_1(\Omega, \mathcal{L}, \mu) = \mathcal{L}_1(\Omega, \mu) =$ $= \mathcal{L}_1(\mu)$	space of μ–integrable functions $f : \Omega \to \mathbb{K}$, 494, 497
$\mathfrak{M}(\mu)$	space of μ–measurable functions, 495
$\mathcal{L}_p(\Omega, \mu), \mathcal{L}_p(\mu)$	space of p–integrable functions, 498
$L_p(\Omega, \mu), L_p(\mu)$	space of classes of p–integrable functions, 499
$S(\mu)$	space of classes of I_μ–step functions
$\|f\|_p$	498
$\|f\|_\infty = \text{ess-sup}|f|$	498
$\tilde{f}, \tilde{f}^{loc}$	class of the function f, 499
$M(\Omega, \mathbb{K})$	space of $\mathbb{K}$-valued regular Borel measures, 504
$L_p(\mu, E), L_p(E)$	space of (classes of) Bochner p–integrable functions $\Omega \to E$, 505
$L_\infty(\mu, E), L_\infty(E)$	57, 88, 506
$V(fd\mu)$	507
μ-a.e.	$= \mu$-almost everywhere, 496
χ_A	characteristic function of a subset $A \subset \Omega$ defined by $\chi_A(x) = 1$ if $x \in A$ and $\chi_A(x) = 0$ if $x \notin A$.
$[f \geq \alpha]$	$:= \{x \in \Omega \mid f(x) \geq \alpha\}$ for $f : \Omega \to [-\infty, +\infty]$
$[f < \alpha], [\alpha < f < \beta], \cdots$	obvious definitions
r_k	Rademacher function on $[0, 1]$, 10
ε_k	(discrete) Rademacher function on D_n or D, 96
ε_A	Walsh function on D_n or D, 106
$D_n := \{-1, 1\}^n$	95
$D := \{-1, 1\}^{\mathbb{N}}$	95
μ_n, μ	Rademacher measure on D_n resp. D, 95
γ_n, γ	Gauss measure on $\mathbb{K}^n$ resp. $\mathbb{K}^{\mathbb{N}}$, 100
g_n	Gauss function, 100
μ_p^n, μ_p	stable measure on $\mathbb{K}^n$ resp. $\mathbb{K}^{\mathbb{N}}$, 318

6. Bilinear mappings and tensor products

$Bil(E, F; G)$	$= L(E \otimes F, G)$ bilinear mappings $E \times F \to G$, 8, 17
$Bil(E, F)$	$:= Bil(E, F; \mathbb{K})$
$\mathfrak{Bil}(E, F; G)$	$= \mathcal{L}(E \otimes_\pi F, G)$ continuous bilinear mappings, 8, 27
$\mathfrak{Bil}(E, F)$	$:= \mathfrak{Bil}(E, F; \mathbb{K})$
$L(E, F^*)$	$= Bil(E, F) = (E \otimes F)^*$, 9, 17
$E' \otimes F$	$= \mathfrak{F}(E, F)$, 20
$\Phi \rightsquigarrow \Phi^L$	linearization of $\Phi \in \mathfrak{Bil}(E, F; G)$, 26

$\|\Phi\|$	$:= \sup\{\|\Phi(x,y)\|_G \mid x \in B_E, y \in B_F\}$ norm of $\Phi \in \mathfrak{Bil}(E,F;G)$, 8
$T \rightsquigarrow \beta_T$	for $T \in L(E, F^*)$, 9
$\varphi \rightsquigarrow L_\varphi$	for $\varphi \in Bil(E, F)$, 9
$\varphi^\wedge, {}^\wedge\varphi$	right- and left-canonical extension of $\varphi \in \mathfrak{Bil}(E, F)$, 12
$E \otimes F$	tensor product of two vector spaces, 17
$E \otimes^K F$	25, 43
$x \otimes y$	elementary tensor, 18
$x \underline{\otimes} y$	20
$\mathrm{rank}(z)$	rank of $z \in E \otimes F$, 158
z^t	transposed of $z \in E \otimes F$, 21
$z \rightsquigarrow T_z$	for $z \in E' \otimes F$, 20
$T \rightsquigarrow z_T$	for $T \in \mathfrak{F}(E, F)$, 201
tr_E	trace on E, 21, 35
$\alpha(\cdot; E, F)$	tensor norm α on $E \otimes F$, 147
$E \otimes_\alpha F$	tensor product $E \otimes F$ equipped with α, 147, 469
$E \tilde{\otimes}_\alpha F$	completion of $E \otimes_\alpha F$, 147, 469
$T_1 \otimes T_2$	tensor product of operators, 22
$A_1 \otimes A_2$	tensor product of sets, 26
C_G	tensor contraction, 378, 471

7. Norms on tensor products

Let α and β be tensor norms:

α^t	transposed, 149
α'	dual, 178
$\alpha^* = (\alpha')^t$	adjoint or contragradient, 178
$\overrightarrow{\alpha}$	finite hull, 149
$\overleftarrow{\alpha}$	cofinite hull, 149, 189
$\alpha^\rightarrow$	right-finite hull, 157
$\alpha^\leftarrow$	right-cofinite hull, 157
$\alpha/, \backslash\alpha, \backslash\alpha/$	right-, left-, injective associate, 259
$\alpha\backslash, /\alpha, /\alpha\backslash$	right-, left-, projective associate, 256
$\alpha \otimes_G \beta$	traced tensor norm along G, 379, 383
$\alpha \otimes \beta$	traced tensor norm, 384
$\alpha \sim \beta$	equivalent norms: there exist $c, d > 0$ with $c\alpha \leq \beta \leq d\alpha$

Projective norm:

π	27
$/\pi\backslash$	267, 362
$\pi\backslash$	265, 306, 362

Injective norm:

ε	46
$\backslash\varepsilon/$	267, 372
$\varepsilon/$	265, 306, 362

other associates of ε and π: Sec.27

Other special norms:

Δ_p	norm on $L_p \otimes E$ coming from $L_p(E)$, see Index
Δ_p^t	transpose of Δ_p, 79
$\alpha_{p,q}$	Lapresté's tensor norms, see Index
$\beta_{p,q}$	158, Sec.28, 393
$\gamma_{p,q}$	158, Sec.28
$\delta_{p,q}$	366, 375, 393
σ	Hilbert–Schmidt norm, 351
σ_1	312
$g_p := d_{p,1}$	see Index
$d_p := \alpha_{1,p} = g_p^t$	see Index
$w_p := \alpha_{p,p'}$	see Index
d_2, g_2, w_2	table, 266

8. Operator ideals

Let $(\mathfrak{A}, \mathbf{A})$ be a (quasi–) Banach operator ideal:

$\mathfrak{A}(E, F)$	108
$\mathfrak{A}^{inj}$	injective hull, 112
$\mathfrak{A}^{sur}$	surjective hull, 113
$\mathfrak{A}/$	264
$\backslash\mathfrak{A}$	264
$\backslash\mathfrak{A}/$	273
$\mathfrak{A}^{dual}$	dual ideal, 114
$\mathfrak{A}^{reg}$	regular hull, 299
$\mathfrak{A}^{max}$	maximal hull, 201
$\mathfrak{A}^{min}$	minimal kernel, 287

$\mathfrak{A}^*$	adjoint ideal, 208
$\mathfrak{A}^\triangle$	conjugate ideal, 221
$\mathfrak{A}^{min\ dual}$	297
$\mathfrak{A}^{min\ reg}$	298
$\mathfrak{A}^{inj\ min}$	299, 340
$\mathfrak{A}^{sur\ min}$	340
$\mathfrak{A}^{min}(E,F) = \mathfrak{A}^{max}(E,F)$	Sec.**33**
$\mathfrak{A} \circ \mathfrak{B}$	composition ideal, 114
$\mathfrak{A} \circ \mathfrak{B} \subset \mathfrak{C}$	cyclic composition theorem, 330
$\mathfrak{A} \otimes \mathfrak{B}$	tensor product, 384
$(\mathfrak{A} \circ \mathfrak{B})^* = (\mathfrak{A} \otimes \mathfrak{B})^*$	385, 333
$\mathfrak{A}^{-1} \circ \mathfrak{B}, \mathfrak{A} \circ \mathfrak{B}^{-1}$	quotient ideals, 332, 333

9. Special operator ideals

$(\mathfrak{C}_q, \mathbf{C}_q)$	cotype q operators, 85
$(\mathfrak{D}_{p,q}, \mathbf{D}_{p,q})$	(p,q)–dominated operators, 209, Sec.**19**
$(\mathfrak{D}_p, \mathbf{D}_p)$	$= (\mathfrak{D}_{p,p'}, \mathbf{D}_{p,p'})$ p–dominated operators, 210
$(\mathfrak{F}, \|\ \|)$	finite rank operators, 20
$(\overline{\mathfrak{F}}, \|\ \|)$	approximable operators, 47, 110
$(\mathfrak{HS}, \mathbf{HS})$	Hilbert–Schmidt operators, 134, 143
$(\mathfrak{I}, \mathbf{I})$	integral operators, Sec.**10**
$(\mathfrak{I}^{sur\ inj}, \mathbf{I}^{sur\ inj})$	267
$(\mathfrak{I}_p, \mathbf{I}_p)$	$= (\mathfrak{L}_{p,1}, \mathbf{I}_{p,1})$ p–integral operators, 209, Sec.**18**
$(\mathfrak{K}, \|\ \|)$	compact operators, 110
$(\mathfrak{KC}, \mathbf{K})$	K–convex operators, 118, Sec.**31**
$(\mathfrak{K}_{p,q}, \mathbf{K}_{p,q})$	$= (\mathfrak{L}_{p,q}, \mathbf{L}_{p,q})^{min}$ (p,q)–compact operators, 291
$(\mathfrak{K}_p, \mathbf{K}_p)$	$= (\mathfrak{L}_p, \mathbf{L}_p)^{min}$ p–compact operators, 291
$(\mathfrak{L}, \|\ \|)$	all continuous linear operators
$(\mathfrak{L}_S, \mathbf{L}_S)$	111, 157, 202, 313
$(\mathfrak{L}_\mathfrak{F}, \mathbf{L}_\mathfrak{F})$	Fourier operators, Sec.**30**
$(\mathfrak{L}_{p,q}, \mathbf{L}_{p,q})$	(p,q)–factorable operators, 208, Sec.**18**
$(\mathfrak{L}_{p,q}^{inj}, \mathbf{L}_{p,q}^{inj})$	337, 377, 387, 392
$(\mathfrak{L}_{p,q}^{sur}, \mathbf{L}_{p,q}^{sur})$	337, 387, 392
$(\mathfrak{L}_{p,q}^{sur\ inj}, \mathbf{L}_{p,q}^{sur\ inj})$	337, 369, 377, 388
$(\mathfrak{L}_p, \mathbf{L}_p)$	$= (\mathfrak{L}_{p,p'}, \mathbf{L}_{p,p'})$ p–factorable operators, 209

$(\mathfrak{L}_2, \mathbf{L}_2)$	2–factorable = Hilbertian operators, 116
$(\mathfrak{L}_p^{inj}, \mathbf{L}_p^{inj})$	238, 437, 458
$(\mathfrak{L}_p^{sur}, \mathbf{L}_p^{sur})$	238, 458
$(\mathfrak{M}_{q,p}, \mathbf{M}_{q,p})$	(q,p)–mixing operators, Sec.32
$(\mathfrak{N}, \mathbf{N})$	nuclear operators, 34
$(\mathfrak{N}_p, \mathbf{N}_p)$	$= (\mathfrak{I}_p, \mathbf{I}_p)^{min}$ p–nuclear operators, 291
$(\mathfrak{P}_{r,s}, \mathbf{P}_{r,s})$	absolutely (r,s)–summing operators, 105, 145
$(\mathfrak{PI}, \mathbf{PI})$	Pietsch integral operators, 522
$(\mathfrak{PI}_p, \mathbf{PI}_p)$	Pietsch p–integral operators, 283
$(\mathfrak{P}_p, \mathbf{P}_p)$	absolutely p–summing operators, Sec.11
$(\mathfrak{P}_1^{sur}, \mathbf{P}_1^{sur})$	267
$(\mathfrak{QN}, \mathbf{QN})$	quasinuclear operators, 116
$(\mathfrak{T}_p, \mathbf{T}_p)$	type p operators, 85
$(\mathfrak{W}, \|\ \|)$	weakly compact operators, 110

10. Space ideals

space$(\mathfrak{A})$	116, 473
space$(\mathfrak{L}_p)$	$= \mathfrak{L}_p^g$–spaces, 40, 338, Sec.23
space$(\mathfrak{L}_p^{inj})$	338, 339, 437
space$(\mathfrak{L}_p^{sur})$	338, 339
space$(\mathfrak{L}_p^{inj\ sur})$	274, 338, 339

11. Relations between maximal normed operator ideals and tensor norms

$\mathfrak{A} \sim \alpha$	202		$\backslash\mathfrak{A} \sim \backslash\alpha$	264
$\mathfrak{A}^{inj} \sim \alpha\backslash$	263		$\mathfrak{A}^{dual} \sim \alpha^t$	206
$\mathfrak{A}^{sur} \sim /\alpha$	263		$\mathfrak{A}^* \sim \alpha^*$	208
$\mathfrak{A}/ \sim \alpha/$	264		$\mathfrak{A} \otimes \mathfrak{B} \sim \alpha \otimes \beta$	384
$\mathfrak{D}_p \sim w_{p'}^*$	210		$\mathfrak{L}_p \sim w_p$	209
$\mathfrak{D}_{p,q} \sim \alpha_{p',q'}^*$	209		$\mathfrak{L}_{p,q} \sim \alpha_{p,q}$	208
$\backslash\mathfrak{D}_{p,q}/ \sim \gamma_{q',p'}$	368, 374		$\mathfrak{L}_{p,q}^{inj\ sur} \sim \gamma_{q,p}^*$	369, 374
$\mathfrak{J} \sim \pi$	204		$(\mathfrak{L}_p \circ \mathfrak{L}_{q'})^{reg} \sim \beta_{p,q}$	372, 374
$\mathfrak{I}_p \sim g_p$	209		$(\mathfrak{L}_p \circ \mathfrak{J}_{q'})^{reg} \sim \delta_{p,q}$	374, 376
$\mathfrak{L} \sim \varepsilon$	211		$\mathfrak{P}_p \sim g_{p'}^*$	210

12. Constants

a_p, b_p	constants in the Khintchine inequality, 96
$c_{p,q}$	constant in the Lévy embedding, 315, 319
$\bar{c}_{p,r}$	323
$c_{2,p}^n$	p-th root of the p-th moment of the Gauss measure γ_n, 145
$k_{q,p}$	complexification constant, 377
K_G	Grothendieck constant, 169
$K_G(n)$	finite dimensional Grothendieck constant, 268
K_{LG}	little Grothendieck constant, 139

Index

$\alpha_{p,q}$, 150, 153, 237, 242, 254, 279, 285, 326, 344, 360, 368, 387, 414, 434, 442, 444
absolutely p-summable sequence, 90, 92, 103, 128, 143, 322
absolutely p-summing operator, Sec. 11, 203, 210, 217, 221, 232, 271, 285, 286, 321, 322, 326, 332, 338, 342, 349, 351, 370, 413, Sec.32, 441, 444, 448, 451, 461, 466, 467
absolutely 1-summing operator, 128, 132, 138, 142, 272, 310, 342, 349, 362, 414, Sec.32, 522
absolutely 2-summing operator, 134, 136, 143, 272, 274, 342, 349, 362, 370, 376, 413, Sec.32, 460
absolutely (r,s)-summing operator, 105, 145, 203, 322, 426, 461, 462, 467, 468
abstract L-space, 38, 89, 492
abstract L_p-space, 492
abstract M-space, 492
accessible, right-, left-,
  operator ideal, Sec.21, Sec.25, 413, 415
  tensor norm, Sec.21, 180, 253, 413
adjoint operator ideal, 207, 221, 278, 293, 330, 333, 446
adjoint tensor norm, 178, 207
Alencar, R., 5, 524, 526
Amemiya, I., 3
approximable operator, 47, 60, 110, 116, 165, 240, 287, 313, 412, 415, 445, 447
approximation lemma, 160
approximation property, Sec.5, 59, 64, 67, 219, 298, 474
  α-, Sec.21, 280
  bounded, Sec.16, 59, 67, 279
  bounded α-, Sec.21, 280
  λ-bounded, 59
  metric, 59, 193, 194
  order p, 280

Aron, R., 125
associated bilinear mapping, 18
Auerbach basis, 491
averaging technique, 95

Banach, S., 315
Banach algebra, 491
Banach lattice, 491
Banach–Mazur distance, 70, 144, 360
Banach operator ideal, 109
p-Banach operator ideal, 109
Banach–Steinhaus theorem, 14
Barton, T., 249
basis, 59, 158, 298, 358, 453, 457, 491
  Auerbach, 491
  unconditional, 358, 457
Beauzamy, B., 89
Beckner, W., 87, 89, 187, 346, 418
Behrens, L., 5
Bennett, G., 348, 466, 468
Beppo Levi monotone convergence theorem, 495
Bergh, J., 462
Bessaga, C., 94, 104, 315
Bessaga–Pełczyński selection principle, 104
Bierstedt, K. D., 6
bilinear, 7
  Banach–Steinhaus, 14
  closed graph theorem, 14
  continuous, 8
  extension, 13
  integral, 53, 55, 58
  linearization, 18
  nuclear, 9, 58
  positive, 225, 228
  separately continuous, 8
  symmetric, 25
  transposition, 11
bipolar theorem, 490
Bochner integrable function, 24, 29, 506
  space of, 24, Sec.7, 29, 505, 507, 521

Bochner's theorem, 316
Bohnenblust, H. F., 492
Bonet, J., 477, 481, 487
Botelho, G., 518
Bourgain, J., 59, 86, 270, 285, 405
Brown, A., 315, 403, 518
Burkholder, D. L., 86

C^*-algebra, 247, 491
canonical decomposition of an operator into its finite parts, 239
canonical extension, right-, left-, 11, 14, 75, 161, 470
Carathéodory, C., 315
Carl, B., 5, 6, 348, 418, 427, 448, 453, 458, 464, 466
Castillo, J., 6, 105
Chatterji, S. D., 97
Chevet, S., 150, 186
Choquet, G., 316
Chu, C. H., 249
closed graph theorem, 14, 43
closed operator ideal, 110
cofinite hull of a tensor norm, 149, 160, 189, 191
  one-sided, 157, 165, 202
cofinitely generated tensor norm, 149
Cohen, P. J., 11
Cohen, J. S., 141, 150
Collins, H. S., 486
compact
  operator, 40, 45, 48, 60, 92, 110, 165, 195, 326, 445, 447, 516
  sets, 31, 33, 42, 159, 165
  weakly conditionally, 189
(p,q)-compact operator, 291, 343, 361, 451, 458
p-compact operator, 313, 466
compact extension property, 45, 53
compactly approximable operator, 69, 412
completely continuous operator, 115
completely symmetric operator ideal, 114

complexification, 176, 347, 361, 377
component of an operator ideal, 108
composition operator ideal, Sec.25, 114, 217, 245, 278, 300, 385, 393
  normed, 115, 290, 386
conjugate operator ideal, 221
continuous functions, space of, 19, 38, 48, 50, 53, 89, 132, 156, 167, 171, 199, 312, 479, 520
continuous trace, 300
continuous triangle inequality, 79, 499
contragradient tensor norm, 178
cotype q
  Fourier, 402
  Gauss, 399, 406
  operator, 85, 89, 106, 111, 117, 203, 277, 349, 399, 402, 403, 408, 454, 467
  space, 93, 103, 322, 324, 349, Sec.31, 417, 419
countable neighbourhood property, 476
crossnorm, 147
cyclic composition theorem, 330

d_p, 152, 153, 178, 185, 238, 254, 265, 271, 338, 350, Sec.27, 380, 387, 389, 414
  table for d_2, 266
Δ_p, Sec.7, 82, 147, 184, 187, 233, 314, 339, 346, 369, 374, 380, 389, 418, 422, 424, 426, 429
Daniell functional, 493
Davie, A. M., 174
Davis, W. J., 110
Day, M. M., 38
Defant, M., 5, 6, 86
Defant, U., 6
dense, μ-, 496
density lemma, 80, 163
  maximal normed operator ideals, 223
(DF)-space, 483
diagonal operator, 45, 145
Diaz, J. C., 280, 287
Diestel, J., 94, 105, 189, 505, 516, 517,

518, 520, 521
Dineen, S., 56
disc algebra, 60
discrete Rademacher functions, 96
(p,q)-dominated operator, 210, Sec.19, 279, 351, 360, 368, 434
p-dominated operator, 210, 309, 351, 360, 382, 429, 466
double Khintchine inequality, 455
dual operator ideal, 114, 206, 264, 278, 297, 299
dual system, 490
dual tensor norm, Sec.15
duality bracket, 490
duality theorem, 179
Dunford, N., 505
Dunford–Pettis property, 516
Dunford–Pettis theorem
  weak version, 30
  strong version, 516
Duren, P. L., 57
Dvoretzky–Rogers theorem, 132, 143

Edwards, R. E., 84
eigenvalues, 459
elementary tensor, 18
embedding lemma, 162, 470
embedding theorem, 205
Enflo, P., 59
equivalent tensor norms, 153
exercises, 4
extension, → canonical
extension lemma, 75, 161, 470
extension norm, 312
extension property, 13, 38, 41, 273, 300, 306
  absolutely 2–summing operators, 143
  → compact
  integral operators, 126
extreme point, 14

(p,q)-factorable operator, 208, Sec.18, 245, 279, 337, 344, 351, 360, 369, 434, 442, 451, 458

p-factorable operator, 209, 230, 303, 307, 350, 351, 360, 362, 372, 454, 466
2–factorable operator, 116, 211, 230, 248, 250, 314, 344, 360, 370, 374, 376, 400, 402, 409, 411, 429, 468
factorization theorem Davis–Figiel–Johnson–Pełczyński, 110
factorization theorem
  p-factorable, 230
  (p,q)-factorable, 236
  Grothendieck–Pietsch, 130
  integral operators, 122
  p-integral operators, 231
  Kwapień, 244
  Maurey, 233, 241, 422
  Pisier, 411
  see also Sec.28
Fan, K., 491
Fatou lemma, 495
Figiel, T., 60, 86, 87, 110, 347
finite hull of a tensor norm, 149, 160, 189
  one–sided, 157, 165, 202, 251
finitely generated tensor norm, 149
finite rank operator, 20, 24, 47, 165, 198, 427
Fourier
  cotype, 402
  matrix, 56, 121, 345
  operator, 203, Sec.30
  series, 397, 407
  transform, Sec.30, 44, 83, 90, 316, 511
  type, 402, 405
Friedman, Y., 249
Fubini theorem, 498
Fubini–Tonelli theorem, 498

g_p, 152, 153, 178, 185, 232, 238, 254, 265, 274, 276, 281, 314, 321, 338, 350, Sec.27, 372 376, 380, 387, 389, 414, 417, 441
  table for g_2, 266, 274

Galbis, A., 481, 487
Garling, D. J. H., 137, 184, 313
Gaudrey, G. I., 84
Gauss cotype, 399, 406
Gauss functions, 99, 105
Gauss measure, 100, 145
Gauss–Khintchine equality, 100, 145
Gauss type, 399, 406
Gelbaum, B., 158, 358
Gelfand, I. M., 491, 515
Gelfand representation theorem, 491
Gilbert, J. E., 3, 5, 279
Gil de Lamadrid, J., 158, 358
Goodner, D. A., 38
Gordon, Y., 2, 5, 137, 145, 184, 186, 221, 273, 285, 313, 339, 364, 441
Govaerts, W., 479
Grothendieck, A., 1, 2, 3, 5, 31, 39, 58, 118, 125, 127, 129, 130, 139, 146, 147, 150, 166, 176, 182, 247, 267, 361, 382, 473, 479, 483, 486, 488, 493, 508, 511, 517, 524
Grothendieck constant K_G, 169, 173, 174, 176, 247, 267, 415, 428
  finite dimensional $K_G(n)$, 268, 274, 375, 377
  little K_{LG}, 139, 173
Grothendieck's inequality, Sec.14, 347, 376, 462
  abstract form, 414
  matrix form, 172
  non–commutative, 247
  nuclear form, 292
  operator form, 212, 267, 311, 342
  proofs, 169, 175, 413, 429
  tensor form, 169, 247, 267
Grothendieck's notation, 364
Grothendieck–Pietsch domination theorem, 129
Grothendieck–Pietsch factorization theorem, 130
Grothendieck theorem, 212, 270, 310, 462
  → little

representable operators, 511
  spaces satisfying, → G.T. space
G.T. space, 270, 429

Haagerup, U., 96, 174, 247, 248, 269
Hahn–Banach theorem, 489
  for operators, 10
  → extension property
Halmos, P. R., 315, 370
Hanson, B., 6
Hardy, G. H., 56
Hardy space, 57, 60, 285, 310
Harksen, J., 5, 469, 470, 475, 477
Harmand, P., 5, 6
Hasumi, M., 38
Hausdorff–Young inequality, 405
Heinrich, S., 228, 240, 393
Helly's lemma, 73
Hilbertian operator, → 2–factorable
Hilbertizable space, 474, 486, 488
Hilbert–Schmidt
  norm, 351, 358, 457
  operator, 134, 143, 351, 361, 460
Hilbert space, 25, 88, 105, 134, 137, 144, 172, 238, 250, 304, 313, Sec.26, 457
  characterizations, 141, 221, 311, 370, 398, 400, 401, 405, 406, 407, 412, 460
Hilbert transform, 84, 86
Hölder inequality, 498
Hoffmann–Jørgensen, J., 85
Hollstein, R., 476, 478, 488
Holub, J. R., 199, 448, 450, 451, 452, 467
Horowitz, C., 11, 14

inductive limit, 476
inequality
  continuous triangle, 79, 499
  double Khintchine, 455
  → Grothendieck
  Hausdorff–Young, 405
  Hölder, 498

Kahane, 98
Khintchine, 96, 361, 455
Littlewood, 125, 462, 463, 466
Minkowski, 79, 499
injective
  Banach space, 13, 38, 306, 312, 452
  hull of an operator ideal, 112, 263, 340
  operator ideal, 111, 388
  tensor norm, right-, left-, Sec.20, 275, 475
    associate, 259, 260
  tensor norm ε, Sec.4, 46, 78, Sec.6, 193, 265, 303, 306, 344, Sec.27, 380, 414, 425, 452, 467
    coincidence with π, 467,
    → Pisier space
integral bilinear form, 53, 55
integral operator, Sec.10, 132, 142, 193, 203, 217, 245, 309, 351, 362, 386, 429, 441, 450, 517, 522
α-integral operator, 204
p-integral operator, 209, 216, 231, 271, 321, 332, 338, 349, 351, 368, 392, Sec.32, 436, 441, 451, 466
integrating functional, 54, 121, 224
Iochum, B., 249
isometry, 489
Iwaniec, T., 87, 347

Jameson, G. J. O., 37, 145, 174, 184, 428
Jarchow, H., 5, 6, 469, 473, 484, 485
John, K., 56
Johnson, W. B., 2, 60, 77, 110, 135, 313, 460, 461
Johnson space, 71, 76, 214
Junek, H., 473

Kaballo, W., 306
Kadec, M. J., 144, 314, 315
Kadison, R. V., 250
Kahane, J. P., 98, 107
Kahane contraction principle, 107

Kahane inequality, 98
Kakutani, S., 492
Kakutani (-Bohnenblust-Nakano) representation theorem, 492
Kalton, N. J., 452
Kandil, M. A., 398, 406
K-convex operator, 86, 118, 203, 407, Sec.31
Kelley, J. L., 38
Khintchine inequality, 96, 361, 455
  double, 455
Kisliakov, S. V., 270, 285
Kölzow's theorem, 502
König, Her., 2, 115, 135, 174, 269, 300, 356, 405, 460, 461
Köthe, G., 39, 197, 469
Krickeberg, K., 348
Krivine, J. L., 166, 176, 269, 377
Kürsten, K. D., 393
Kwapień, S., 2, 5, 84, 136, 141, 144, 242, 314, 323, 336, 338, 339, 350, 358, 369, 394, 400, 401, 402, 406, 457, 461, 463
Kwapień's factorization theorem, 244
Kwapień's theorem on the Fourier transform, 401

$\mathcal{L}_p$-space, 304, 338
$\mathcal{L}_p$-local technique, 163, 166, 301, 313, 474
$\mathcal{L}_p^g$-space, 40, 41, 47, 163, 165, 279, Sec.23, 320, 375, 406, 466
$\mathcal{L}_1^g$-space, 41, 260, 305, 306, 309, 314, 476, 483
$\mathcal{L}_\infty^g$-space, 51, 77, 171, 260, 305, 306, 308, 312, 314, 428, 476, 479, 484
Lacey, H. E., 38, 492, 493
Lapresté, J. T., 150, 208
large subspace, 479, 487
Lebesgue decomposition theorem, 500
Lebesgue dominated convergence theorem, 495, 506
left-tensor norm, 202
Leih, T., 3, 279

lemma
  approximation, 160
  density, 80, 163, 223
  embedding, 162, 470
  extension, 75, 161, 470
  Fatou, 495
  Helly, 73
  Ky Fan, 491
  local technique, 163, 301
  Pietsch, 135
  quotient, 82, 88
  Riemann–Lebesgue, 406
  variation, 507
Levi, B., 495
Lévy, P., 314, 319
Lévy embedding, 319, 348, 375
Lévy's theorem, 319
Lewis, D. R., 2, 5, 190, 221, 273, 310, 329, 364, 430, 432, 434, 440, 441, 457, 514, 525
Lewis–Stegall theorem, 430, 514
lifting
  L_∞–, 501
  operators, 13, 39, 41, 44, 45, 273, 300
    integral operators, 126
limit order, 355
Lindenstrauss, J., 1, 5, 10, 41, 53, 60, 98, 144, 166, 270, 298, 300, 304, 306, 344, 370, 437
Lindenstrauss' compactness argument, 76, 239
linear dimension, 315
little Grothendieck theorem, 139, 173, 213, 272, 428, 468
Littlewood, J. E., 56, 125, 462, 463, 466
localizable measure, 500
  strictly, 501
local technique → $\mathfrak{L}_p$–
Löfström, J., 462
López Molina, J. A., 280, 287
Losert, V., 3, 299
Lotz, H. P., 2, 3, 5, 146, 204
Loupias, G., 249

Mackey theorem, 490
Maharam's theorem, 501
Mankiewicz, P., 60
mapping property, 470
  metric, 147
Marcinkiewicz J., 314, 315, 347
Marcus, M. B., 101, 107
Marquina, A., 479
Matos, M., 6, 312
Matziwitzki, I., 6
Maurey, B., 2, 5, 85, 101, 103, 105, 118, 135, 240, 314, 323, 400, 414, 415, 421, 422, 427, 429, 460, 461
Maurey factorization theorem, 233, 240
Maurey splitting theorem, 423
Maurey–Pisier extrapolation theorem, 421
maximal hull of an operator ideal, 201
maximal operator ideal, Sec.17, 288
  density lemma, 223
McCarthy, C. A., 456
measure, 494
  absolutely continuous, 500
  Borel–Radon, 503
  finite, 496
  Gauss, 100
  Lévy, 318
  localizable, 500
  Rademacher, 95
  signed, 494
  singular, 500
  space, 497
  stable, Sec.24, 319
  strictly localizable, 501
Meile, M., 5, 430
Mendoza Casas, J., 479
metric extension property, 13, 38
metric injection, 489
metric mapping property, 147
metric surjection, 489
Michor, P., 3, 299
minimal kernel, 287
minimal operator ideals, Sec.22, 327, 329, 332, 340, 342, 343, 430, 432,

440, 444
Minkowski inequality, 79, 499
Minkowski gauge functional, 26
Mityagin, B. S., 127, 285
mixed sequence, 423
(p,q)–mixing operator, Sec.**32**, 454, 464, 468
multilinear mapping, 25
multiplication operator, 88, 142, 233, 240
multiplicity, 459
Mujica, J., 6
Murray, F. J., 1

Nachbin, L., 6, 38
Nakano, N., 492
Neumann, J. von, 1, 315
Nielsen, N., 60
p–norm, 108
normed operator ideal, 109
p–normed operator ideal, 109
norming set, 46
nuclear bilinear form, 9, 58
nuclear operator, 34, 35, 43, 45, 64, 68, 72, 76, 101, 111, 112, 115, 117, 120, 142, 193, 195, 245, 247, 300, 308, 310, 330, 351, 445, 460, 471, 512, 517, 523
    α–, 291, 297
    dual, 66, 197, 200, 298, 526
    p–, 291, 326, 332, 340, 343, 361, Sec.**32**, 436, 444, 451, 466
nuclear space, 473, 486, 487

Oertel, F., 331
Ohazaki, Y., 343
open mapping theorem, 11
operator
    → absolutely p–summing
    → absolutely 1–summing
    → absolutely 2–summing
    → absolutely (r,s)–summing
    → approximable
    canonical decomposition, 239

→ compact
compactly approximable, 69, 412
→ (p,q)–compact
→ p–compact
completely continuous, 115
→ cotype q
diagonal, 145
→ (p,q)–dominated
→ p–dominated
eigenvalues, 459
→ (p,q)–factorable
→ p–factorable
→ 2–factorable
→ finite rank
→ Fourier
→ Hilbert–Schmidt
→ integral
α–integral, 204
→ p–integral
→ K–convex
→ lifting
→ (p,q)–mixing
multiplication, 88, 142, 233, 240
multiplicity, 459
→ nuclear
→ Pietsch integral
→ Pietsch p–integral
→ positive
→ quasinuclear
Radon–Nikodým, 443
regular, 81, 90
representable, App.C, App.D
→ p–strong
→ type p
→ weakly compact
→ weak∗–weak continuous
operator ideal, Sec.**9**
    → accessible, right–, left–
    → adjoint
    Banach, 109
    p–Banach, 109
    closed, 110
    completely symmetric, 114
    component, 108

→ composition
conjugate, 221
→ dual
→ injective
→ maximal
→ minimal
non–accessible 413, 415
normed, 109
p–normed, 109
→ product
proper, 110
quasi–Banach, 109
quasinormed, 109
quotient, 332
→ regular
regular hull, 299
→ tensor product characterization
→ surjective
→ tensorstable
→ totally accessible
ultrastable, 393
Orlicz, W., 102, 103, 105
Orlicz property, 103, 462
Oya, E., 299

Paley projection, 86
parallelogram identity, 405
Parseval's identity, 398
Pearcy, C., 315, 403, 518
Peetre, J., 405
Pełczyński, A., 1, 5, 41, 67, 84, 86, 87, 104, 110, 127, 144, 166, 230, 270, 285, 300, 304, 310, 315, 344, 347, 351, 358, 370, 437, 446, 452, 457
Pérez Carreras, P., 477
Persson, A., 186, 332
Pettis integrable function, 507
Pettis integral, 507
Pettis' measurability criterion, 505
Phillip's property, 486, 517
Pietsch, A., 2, 109, 110, 111, 127, 129, 130, 132, 135, 137, 145, 300, 312, 323, 326, 345, 356, 357, 377, 415, 459, 460, 461, 473, 478, 491

Pietsch integral operator, 522, 526
Pietsch p–integral operator, 283, 437
Pietsch lemma, 135
Pisier, G., 2, 5, 56, 59, 86, 89, 101, 103, 105, 106, 107, 118, 147, 166, 174, 240, 247, 248, 270, 344, 399, 407, 409, 411, 412, 413, 414, 417, 421, 457
Pisier's factorization theorem, 411
Pisier, Maurey–P. extrapolation theorem, 421
Pisier space, 56, 59, 147, 412
Pitt's theorem, 104, 197
positive bilinear functional, 225, 228
positive definite, 315
positive operator, 80, 88, 123, 129, 216, 233, 235
principle of local reflexivity, 74
  operator ideals, 286
  weak, 73
problème des topologies, 479
product operator ideal
  → composition operator ideal
projection constant, 44, 218, 307, 312, 452
projective tensor norm, right–, left–, Sec.20, 275, 475
  associate, 256, 258
  π, Sec.3, 33, 36, 78, Sec.6 192, 265, 303, 306, 344, Sec.27, 467
  coincidence with ε, 467
  → Pisier space
Prolla, J., 6
proper cotype, 85
proper operator ideal, 110
proper type, 85
property (B) of Pietsch, 478
property (BB), 478, 483
Puhl, J., 355, 357, 415, 427

quasi–Banach operator ideal, 109
quasi–Banach space, 108
quasibarrelled, 478
quasinorm, 108

quasinormed operator ideal, 109
quasinuclear operator, 116, 308, 434, 444, 448, 467
quotient formula, 333
quotient lemma, 82, 88
quotient operator ideal, 332

Rademacher functions, 10, 84, 95, 105, 107, 117, 326
    discrete, 96
Rademacher measure, 95
Radon–Nikodým operator, 443
Radon–Nikodým theorem, 500
    operator valued measures, 44
Radon–Nikodým property
    local, 486
    spaces, 193, 194, 309, 486, App.D
    tensor norm, Sec.33
Ramanujan, M. S., 448, 453, 458
rank of a tensor, 158
reflexivity,
    local 73, 74
    tensor product and operator spaces, 196, 199, Sec.33, 439, 440, 444
reasonable, 147
regular hull of an operator ideal, 299
regular operator, 81, 90
regular operator ideal, 119, 165, 200, 206, 298, 299, 336, 375, 393, 443
Reinov, O., 59, 281, 283, 285, 299, 340
Reisner, S., 277, 285
related operators, 143
representable operator, App.C, App.D
representation theorem
    Gelfand, 491
    Kakutani (–Bohnenblust–Nakano), 492
    maximal operator ideals, 203
    minimal operator ideals, 290
    Riesz, 503
respects subspaces, 36
respects quotients, 49
Résumé, 1, 2, 3, 127, 150, 182, 247, 267, 273, 299, 360, 364

Retherford, J. R., 2, 5, 135, 221, 306, 308, 441, 460, 461
Riemann–Lebesgue lemma, 406
Riesz–condition, 352
Riesz–density, 509
Riesz–identification, 25, 352
Riesz–representable operator,
    → representable
Riesz representation theorem, 503
right–tensor norm, 202
Ringrose, J. R., 250
Rivera Ortun, M. J., 190, 280, 287
Rosenthal, H. P., 304, 348, 360, 452
Rosenthal's theorem, 189
rotating, 331
Rubio de Francia, J. L., 405
Ruess, W., 486

Sanz Serna, J. M., 479
Saphar, P., 2, 5, 150, 186, 271, 275, 280, 314, 320, 339
Schaefer, H. H., 81, 492, 493
Schatten, R., 1, 2, 146, 147, 182, 353, 357
Schatten's theorem, 353
Schmets, J., 479
Schur product, 325
Schwartz, J., 505
Schwarz, H. U., 293, 492
Shiga, K., 3
Segal's localization theorem, 500
separating dual system, 490
sequence
    → absolutely p–summable
    (q,p)–mixed, 423
    → unconditionally summable
    → weakly p–summable
Sims, B., 228
Snobar, M. G., 144
space
    abstract L–, 38, 89, 492
    abstract L_p–, 492
    abstract M–, 492
    → cotype p

G.T., 270, 429
Hardy, 57, 60, 285, 310
→ Hilbert
Hilbertizable, 474, 486, 488
ideal, 116, 222, 279, 308, 453, 455, 473
injective, 13, 306, 312, 452
Johnson, 71, 76, 214
$\mathfrak{L}_p$-, 304, 338
→ $\mathfrak{L}_p^g$-
→ $\mathfrak{L}_1^g$-
→ $\mathfrak{L}_\infty^g$-
measure, 497
nuclear, 473
Pisier, 56, 59, 147
→ Radon–Nikodým property
Stonean, 38
T_p-, 383, 393
→ type p
stable measures, Sec.24, 319
stable type, 324, 326
Stegall, C. P., 306, 308, 310, 514, 525
Stein, E. M., 96
step function, 494
Stollmann, P., 6
Stonean space, 38
strictly localizable measure, 501
p-strong operator, 374, 437, 444
sublinear, 489
surjective
  hull of an operator ideal, 113, 263, 340
  operator ideal, 112, 388
symmetric bilinear form, 25
symmetric finite dimensional spaces, 184, 189, 218
symmetric tensor norm, 149, 353
symmetry constant, 189
Szankowski, A., 59
Szarek, S. J., 60, 67, 348, 360

Takahashi, Y., 343
Talagrand, M., 103, 427
Taskinen, J., 480, 481, 483, 484, 486

tensor
  associated linear map, 22
  elementary, 18
  rank, 158, 188
tensor contraction, Sec.29, 155, 159, 472
tensor norm, 147
  → accessible
  adjoint, 178, 207
  → canonical extension
  → cofinite hull
  cofinitely generated, 149
  contragradient, 178
  dual, Sec.15
  equivalent, 153
  → finite hull
  finitely generated, 149
  Hilbert–Schmidt, 351
  → injective
  left–, 202
  non–accessible, 413
  → projective
  Radon–Nikodým property, Sec.33
  right–, 202
  self–adjoint, 353
  symmetric, 149, 353
  topology, 470
  → totally accessible
  traced, Sec.29
  transposed, 149
tensor product
  algebraic theory, Sec.2
  measures, 498
  operators, 22
    bidual, 222
    injectivity, 66
  operator ideals, 384, 393
  tensor norms, 384
tensor product characterization of operator ideals, 203, 213, 215, 335, 381, 384, 388, 390, 392, 473
tensorstable operator ideal, Sec.34
  metrically, Sec.34
  strongly, Sec.34

Terzioglu, T., 165
theorem
   Banach–Steinhaus, 14
   Beppo Levi monotone convergence, 495
   bipolar, 490
   Bochner, 316
   Carathéodory–Halmos–von Neumann, 315
   closed graph, 14, 43
   cyclic composition, 330
   Davis–Figiel–Johnson–Pełczyński, 110
   duality, 179
   Dunford–Pettis, 30, 516
   Dvoretzky–Rogers, 132, 143
   embedding, 205
   → factorization
   Fubini, 498
   Fubini–Tonelli, 498
   Gelfand, 491
   Grothendieck, 212, 270, 310, 462
   Grothendieck, representable operator, 511
   Grothendieck–Pietsch, 129, 130
   Hahn–Banach, 489
   Kakutani (–Bohnenblust–Nakano), 492
   Kölzow, 502
   Kwapień factorization, 244
   Kwapień, Fourier transform, 401
   Lebesgue decomposition, 500
   Lebesgue dominated convergence, 495, 506
   Lévy, 319
   Lewis–Stegall, 430, 514
   → little Grothendieck
   Mackey, 490
   Maharam, 501
   Maurey factorization, 233, 240
   Maurey splitting, 423
   Maurey–Pisier extrapolation, 421
   open mapping, 11
   Pisier factorization, 411
   Pitt, 104, 197
   Radon–Nikodým, 44, 500
   → representation
   Riesz, 503
   Rosenthal, 189
   Schatten, 353
   Segal localization, 500
   type/cotype, 400
Timoney, R. M., 56
Tomczak–Jaegermann, N., 5, 437, 454, 461, 462
Torrea, J. L., 405
totally accessible, Sec.21
   operator ideal, 330, 342
   tensor norm, 180
T_p-space, 383, 393
trace, 21, 24, 35, 68, 183, 188, 294, 300, 471
   continuous, 300
   duality, 22, 24, 69, 76, 120, 177, 196, 207, 293, 312, 330
   for operators, 64, 294, 300
traced tensor norm, Sec.29
transfer argument, 206, 211
transposition, 11, 23
type p
   Fourier, 402, 405
   Gauss, 399, 406
   operator, 85, 89, 106, 111, 117, 203, 277, 399, 402, 403, 408, 454, 467
   space, 98, 324, 326, 419
   stable, 324, 326
type/cotype theorem, 400
Tzafriri, L., 10, 60, 98, 304

Uhl, J. J., 505, 517, 518, 520, 521
ultrapower, 227
ultraproduct
   Banach spaces, 226
   C^*-algebras, 238
   operators, 227
ultrastable operator ideal, 393
unconditional basis, 358, 457
unconditionally summable, 93, 103,

105, 132, 322, 360
uniform boundedness principle, 490
uniform crossnorm, 147

Vala, K., 467
Valdivia, M., 158
variation lemma, 507
Virot, B., 81
Voigt, J., 5, 6

w_p, 152, 153, 155, 249, 265, 271, 274, 276, 280, Sec.**23**, 320, 326, Sec.**27**, 372, 376, 380, 393, 414, 425, 475
   table for w_2, 266
Walsh functions, 106, 409, 410
Walsh matrix, 176, 463
weakly compact operator, 31, 110, 445, 516, 517
weakly conditionally compact, 189
weakly measurable function, 505
weakly p-summable, 90, 92, 103, 104, 124, 128, 143, 155, 322, 360, 423
weak-∗-measurable function, 505
weak retraction, 44
weak topology, 490
weak-∗-topology, 490
weak∗-weak continuous operator, 24, 45, 48, 295, 299
Wloka, J., 476

Yosida, K., 505

Zalduendo, I., 125
Zygmund, A., 314, 315, 347